Landolt-Börnstein / New Series

Springer

Berlin
Heidelberg
New York
Barcelona
Budapest
Hong Kong
London
Milan
Paris
Singapore
Tokyo

Landolt-Börnstein

Numerical Data and Functional Relationships
in Science and Technology

New Series
Editor in Chief: W. Martienssen

Units and Fundamental Constants in Physics and Chemistry

Elementary Particles, Nuclei and Atoms (Group I)
(Formerly: Nuclear and Particle Physics)

Molecules and Radicals (Group II)
(Formerly: Atomic and Molecular Physics)

Condensed Matter (Group III)
(Formerly: Solid State Physics)

Physical Chemistry (Group IV)
(Formerly: Macroscopic Properties of Matter)

Geophysics (Group V)

Astronomy and Astrophysics (Group VI)

Biophysics (Group VII)

Some of the group names have been changed to provide a better description of their contents.

Landolt-Börnstein
Numerical Data and Functional Relationships in Science and Technology
New Series / Editor in Chief: W. Martienssen

Group IV: Physical Chemistry
Volume 8

Thermodynamic Properties of Organic Compounds and their Mixtures

Subvolume G
Densities of Alcohols

M. Frenkel, X. Hong, R.C. Wilhoit, K.R. Hall

Edited by K.R. Hall and K.N. Marsh

ISSN 0942-7996 (Physical Chemistry)

ISBN 3-540-66233-2 Springer-Verlag Berlin Heidelberg New York

Library of Congress Cataloging in Publication Data
Zahlenwerte und Funktionen aus Naturwissenschaften und Technik, Neue Serie
Editor in Chief: W. Martienssen
Vol. IV/8G: Editors: K.R. Hall, K.N. Marsh
At head of title: Landolt-Börnstein. Added t.p.: Numerical data and functional relationships in science and technology.
Tables chiefly in English.
Intended to supersede the Physikalisch-chemische Tabellen by H. Landolt and R. Börnstein of which the 6th ed. began publication in 1950 under title: Zahlenwerte und Funktionen aus Physik, Chemie, Astronomie, Geophysik und Technik.
Vols. published after v. 1 of group I have imprint: Berlin, New York, Springer-Verlag
Includes bibliographies.
1. Physics--Tables. 2. Chemistry--Tables. 3. Engineering--Tables.
I. Börnstein, R. (Richard), 1852-1913. II. Landolt, H. (Hans), 1831-1910.
III. Physikalisch-chemische Tabellen. IV. Title: Numerical data and functional relationships in science and technology.
QC61.23 502'.12 62-53136

Product Liability: The data and other information in this handbook have been carefully extracted and evaluated by experts from the original literature. Furthermore, they have been checked for correctness by authors and the editorial staff before printing. Nevertheless, the publisher can give no guarantee for the correctness of the data and information provided. In any individual case of application, the respective user must check the correctness by consulting other relevant sources of information.

Cover layout: Erich Kirchner, Heidelberg
Typesetting: Authors and Redaktion Landolt-Börnstein, Darmstadt
Printing: Computer to plate, Mercedes-Druck, Berlin
Binding: Lüderitz & Bauer, Berlin

SPIN: 10735217 63/3020 - 5 4 3 2 1 0 – Printed on acid-free paper

Editors

K.R. Hall
Thermodynamics Research Center
TheTexas A&M University System
College Station, Texas 77843-3111, USA

K.N. Marsh
former
Thermodynamics Research Center
TheTexas A&M University System
College Station, Texas 77843-3111, USA
now
Department of Chemical and Process Engineering
University of Canterbury
Christchurch, New Zealand

Authors

M. Frenkel
X. Hong
R.C. Wilhoit
K.R. Hall
Thermodynamics Research Center
TheTexas A&M University System
College Station, Texas 77843-3111, USA

Landolt-Börnstein

Editorial Office
Gagernstr. 8, D-64283 Darmstadt, Germany
fax: +49 (6151) 171760
e-mail: lb@springer.de

Internet
http://science.springer.de/newmedia/laboe/lbhome.htm

Helpdesk
e-mail: em-helpdesk@springer.de

Preface

Critically evaluated experimental data covering the densities of organic compounds is essential for both scientific and industrial applications. Knowledge of densities is important in many areas, including custody transfer of materials, product specification, development of various predictive methods, and for characterizing compounds and estimating their purity.

Various compilations of densities for organic compounds have been published. The early Landolt-Börnstein compilation [23-ano] contained recommended values at specific temperatures. International Critical Tables [28-ano-1] provided recommended densities at 0 °C and values of constants for either a second or third order polynomial equation to represent densities as a function of temperature. This compilation also gave the range of validity of the equation and the limits of uncertainty, references used in the evaluation and those not considered. This compilation is one of the most comprehensive ever published. Timmermans [50-tim, 65-tim], Dreisbach [55-dre, 59-dre, 61-dre] and Landolt-Börnstein [71-ano] published additional compilations, primarily of experimental data. These compilations contained experimental data along with reference sources but no estimates of uncertainty for the data nor recommended values.

The Thermodynamics Research Center has published recommended values for the densities of organic compounds since 1942 in its two loose leaf publications: TRC Thermodynamic Tables - Hydrocarbons and Non-Hydrocarbons. These compilations are updated with four supplements per year. References to the literature values used in the selection and those not used in the selection appear in the references for each table. The accuracy of the values is apparent from the number of significant figures provided. More recently, the Design Institute of Physical Property Data, Project 801 has assembled a set of recommended equations for the densities of over 1500 compounds [89-dau/dan, 91-dau/dan, 92-dau/dan, 93-dau/dan, 94-dau/dan, 95-dau/dan, 96-daudan, 97-daudan]. Densities are represented by an equation fit to selected values from the freezing temperature to the critical temperature. References to sources of data used in the evaluation and those not used are given along with a quality assessment. In many cases, the equation does not fit density values at intermediate temperatures, especially at 293.15 K and 298.15 K, within the experimental uncertainty. Thus, the equation is not useful for purity comparisons, custody transfer, or product specification when the highest accuracy is required. Smith and Srivastava [86-smi/sri, 86-smi/sri-1] recently have published a compilation (in two volumes) which contains recommended values in tabular form as well as equations with statistical information regarding the fit. However, this compilation contains no indication of data quality or uncertainties.

The present volume contains densities for non-cyclic alcohols including monoalcohols, diols, and triols both fully saturated and with various extents of unsaturation which have been collected from the literature published from 1870 to early 1999. The various compilations listed above also have been consulted for sources of original data. This volume continues our effort in evaluation of the densities of organic compounds [see volume IV/8B [96-wil/mar] for the densities of alkanes, volume IV/8C [96-wil/mar-1] for the densities of alkenes, alkynes and alkadienes, volume IV/8D [97-wil/hon] for the densities of monocyclic non-aromatic hydrocarbons, volume IV/8E [98-wil/hon] for the densities of aromatic hydrocarbons, volume IV/8F [99-wil/hon] for the densities of polycyclic hydrocarbons]. All experimental density values have been evaluated critically and assigned numerical uncertainties individually. These assessments have been used to derive an equation to fit the data and to obtain recommended values with uncertainties. Detailed evaluation procedures appear in Chapter 1. Algorithms for the automatic selection of data used in the fit and for the selection of the type of equation and order of polynomial have been developed. The algorithms depend upon the assigned uncertainties, the distribution of density values over the experimental temperature range, and the magnitude and distribution of differences between observed and smoothed values. These algorithms can fit any kind of data to a function of independent variables. We have collected data for 829 compounds, consisting of data sets drawn from 1119 sources.

The volume contains six chapters; a list of references, and two indexes (Chemical Abstracts Service Registry Number Index and Chemical Name Index). Chapter 1 contains: a short introduction; a description of the tables; a detailed description of the methods used in the evaluation, selection and smoothing process; a glossary of symbols and a description of the order of compounds in the Tables used. Chapter 2 covers the monoalcohols, Chapter 3 contains the data for diols, and Chapter 4 has the triols. The tables contain the original literature data along with their estimated uncertainties and the evaluated data in both numerical form and as coefficients to equations with selected statistical information. When data cover a sufficient temperature range, graphical plots of the deviations of the experimental data from the recommended equation are given. The chemical name index contains the IUPAC names for the compounds, as well as alternate names that often appear in practice.

This volume should be useful to a wide community of researchers, data specialists, and engineers working in the field of physical and organic chemistry, chemical engineering, material science, environmental chemistry, chemical aspects of energy technology, and those engaged in the development of new predictive procedures. The book should also be of use to students and faculty in Chemistry and Chemical Engineering departments at universities as a reference book of evaluated thermophysical properties.

Acknowledgments

The authors wish to express their sincere appreciation to staff members of the Thermodynamics Research Center (TRC), part of the Chemical Engineering Division of the Texas Engineering Experiment Station in The Texas A&M University System. Our special thanks to Munaf Chasmawala and Cheryl Clark for their assistance in data collection and entry, formatting the text, preparing the graphs, and composing the camera-ready copy of the manuscript.

College Station, June 1999 M. Frenkel, X. Hong, R.C. Wilhoit, K.R. Hall

References

23-ano	Landolt-Börnstein Physikalisch-Chemische Tabellen, 5. Auflage, Hauptwerk, Band 1. Roth, W.A., Scheel, K. (eds.), Berlin: Springer-Verlag, 1923.
28-ano-1	International Critical Tables of Numerical Data, Physics, Chemistry and Technology Vol. III. Washburn, E.W. (ed.), New York: McGraw-Hill, 1928.
50-tim	Timmermans, J.: Physico-Chemical Constants of Pure Organic Compounds, Vol. 1. New York: Elsevier, 1950.
55-dre	Dreisbach, R.R.: Physical Properties of Chemical Compounds, Advances in Chemistry Series No. 15. Washington, D.C.: Am. Chem. Soc., 1955.
59-dre	Dreisbach, R.R.: Physical Properties of Chemical Compounds - II, Advances in Chemistry Series No. 22. Washington, D.C: Am. Chem. Soc., 1959.
61-dre	Dreisbach, R.R.: Physical Properties of Chemical Compounds, Advances in Chemistry Series No. 29. Washington, D.C.: Am. Chem. Soc., 1961.
65-tim	Timmermans, J.: Physico-Chemical Constants of Pure Organic Compounds, Vol. II. New York: Elsevier, 1965.
71-ano	Landolt-Börnstein Sechste Auflage, II. Band, Eigenschaften der Materie in ihren Aggregatzuständen, 1. Teil, Mechanisch-thermische Zustandsgrössen. Berlin: Springer-Verlag, 1971.
86-smi/sri	Smith, B.D., Srivastava, R.: Thermodynamic Data for Pure Compounds, Part B. Halogenated Hydrocarbons and Alcohols. New York: Elsevier, 1986.
86-smi/sri-1	Smith, B D., Srivastava, R.: Thermodynamic Data for Pure Compounds, Part A. Hydrocarbons and Ketones. New York: Elsevier, 1986.
89-dau/dan	Daubert, T.E., Danner, R.P.: Physical and Thermodynamic Properties of Pure Chemicals: Data Compilation. London: Taylor & Francis, 1989.
91-dau/dan	Daubert, T.E., Danner, R.P.: Physical and Thermodynamic Properties of Pure Chemicals: Data Compilation, Supplement 1. London: Taylor & Francis, 1991.
92-dau/dan	Daubert, T.E., Danner, R.P.: Physical and Thermodynamic Properties of Pure Chemicals: Data Compilation, Supplement 2. London: Taylor & Francis, 1992.
93-dau/dan	Daubert, T.E., Danner, R.P.: Physical and Thermodynamic Properties of Pure Chemicals: Data Compilation, Supplement 3. London: Taylor & Francis, 1993.
94-dau/dan	Daubert, T.E., Danner, R.P.: Physical and Thermodynamic Properties of Pure Chemicals: Data Compilation, Supplement 4. London: Taylor & Francis, 1994.
95-dau/dan	Daubert, T.E., Danner, R.P.: Physical and Thermodynamic Properties of Pure Chemicals: Data Compilation, Supplement 5. London: Taylor & Francis, 1995.

96-dau/dan	Daubert, T.E., Danner, R.P.: Physical and Thermodynamic Properties of Pure Chemicals: Data Compilation, Supplement 6. London: Taylor & Francis, 1996.
96-wil/mar	Wilhoit, R.C., Marsh, K.N., Hong, X., Gadalla, N., Frenkel, M.: Densities of Aliphatic Hydrocarbons: Alkanes. Landolt-Börnstein, New Series, Vol. IV/8B, Berlin: Springer-Verlag, 1996.
96-wil/mar-1	Wilhoit, R.C., Marsh, K.N., Hong, X., Gadalla, N., Frenkel, M.: Densities of Aliphatic Hydrocarbons: Alkenes, Alkadienes, Alkynes and Miscellaneous Compounds. Landolt-Börnstein, New Series, Vol. IV/8C, Berlin: Springer-Verlag, 1996.
97-dau/dan	Daubert, T.E., Danner, R.P.: Physical and Thermodynamic Properties of Pure Chemicals: Data Compilation, Supplement 7. London: Taylor & Francis, 1997.
97-wil/hon	Wilhoit, R.C., Hong, X., Frenkel, M., Hall, K.R.: Densities of Monocyclic Hydrocarbons. Landolt-Börnstein, New Series, Vol. IV/8D, Berlin: Springer-Verlag, 1997.
98-wil/hon	Wilhoit, R.C., Hong, X., Frenkel, M., Hall, K.R.: Densities of Aromatic Hydrocarbons. Landolt-Börnstein, New Series, Vol. IV/8E, Berlin: Springer-Verlag, 1998.
99-wil/hon	Wilhoit, R.C., Hong, X., Frenkel, M., Hall, K.R.: Densities of Polycyclic Hydrocarbons. Landolt-Börnstein, New Series, Vol. IV/8F, Berlin: Springer-Verlag, 1999.

Contents

1 Introduction

1.1 Basic Concepts

The mass density of a substance is an intensive quantity defined by:

$$\rho = m / v \tag{1.1}$$

where m is the mass and v is the volume of the sample. Both these quantities are extensive quantities. Other densities often referred to in the literature are the relative density and the molar density. Relative density, also called specific gravity, is:

$$\rho_r = \rho / \rho_0 \tag{1.2}$$

where ρ_0 is the density of a standard substance. It is necessary to specify the conditions of temperature and pressure for the standard substance. The most common reference material is water often at the temperature of its maximum density, 4 °C, at atmospheric pressure. The molar density is defined as

$$\rho_m = 1 / V \tag{1.3}$$

where V is the volume occupied by one mole of the substance. The only densities reported in this volume are the mass densities; relative and molar densities have been converted to mass densities, and all densities have been reported in SI units. If the mass, m, in equation (1.1) has not been corrected for air bouancy it gives the apparent density in air. The API specific gravity is: ρ (API) = 141.5/ρ_r – 131.5, in which ρ_r is ρ(288.68 K)/ρ_0(288.68 K), and the standard substance is water at 288.68 K (60 °F).

The density of a material is a function of temperature and pressure but its value at some standard condition (for example, 293.15 K or 298.15 K at either atmospheric pressure or at the vapor pressure of the compound) often is used to characterize a compound and to ascertain its purity. Accurate density measurements as a function of temperature are important for custody transfer of materials when the volume of the material transferred at a specific temperature is known but contracts specify the mass of material transferred. Engineering applications utilize the density of a substance widely, frequently for the efficient design and safe operation of chemical plants and equipment. The density and the vapor pressure are the most often-quoted properties of a substance, and the properties most often required for prediction of other properties of the substance. In this volume, we do not report the density of gases, but rather the densities of solids as a function of temperature at atmospheric pressure and the densities of liquids either at atmospheric pressure or along the saturation line up to the critical temperature.

The purpose of this compilation is to tabulate the densities of compounds, hence only minimal description of experimental methods used to measure the density of liquids or solids appears. Detailed descriptions of methods for density determination of solids, liquids and gases, along with appropriate density reference standards, appear in a chapter by Davis and Koch in Physical Methods of Chemistry, Volume VI, Determination of Thermodynamic Properties [86-ros/bae].

The two principal experimental apparatuses used to determine the density of a liquid are: the pycnometer and the vibrating tube densimeter. The pycnometer method involves measuring the mass of a liquid in a vessel of known volume. The volume of the pycnometer, either at the temperature of measurement or at some reference temperature, is determined using a density standard, usually water or mercury. Using considerable care and a precision analytical balance accurate to $\pm 10^{-5}$ g, it is possible to achieve densities accurate to a few parts in 10^6 with a pycnometer having a volume of 25 cm^3 to 50 cm^3.

It is common to achieve accuracies of 1 part in 10^5 in using equation (1.1) with pycnometers as small as 5 cm^3 and routine measurements can achieve 1 part in 10^4. However the main sources of error in assigning density to a particular compound in a particular state arise from factors other than the measurment of mass and volume. See Section 1.4.1

The vibrating tube densimeter relies upon the fact that the frequency f of vibration for a U or V shaped tube depends upon the mass of material in the tube:

$$\rho = A / f^2 + B \quad (1.4)$$

Calibration of the apparatus is necessary; usually water and air or nitrogen are the reference materials. Vibrating tube densimeters designed to operate close to atmospheric pressure can achieve repeatability of parts in 10^6. If the reciprocal of the frequency is linear in density, accuracies of 1 part in 10^5 are readily achievable.

The principal experimental method used to measure the density of a solid is determination of the mass of liquid displaced by a known mass of solid. It is essential that the solid have no appreciable solubility in the liquid, that all occluded air be removed from the solid and that the density of the displacement fluid be less than that of the solid lest the solid float. Densities of crystalline solids also can be determined from the dimensions of the unit cell. Davis and Koch discuss other methods for measuring the density of liquids and solids such as: hydrostatic weighing of a buoy and flotation methods.

1.2 Scope of the Compilation

Volume IV/8G presents observed values for the densities of non-cyclic alcohols. These values represent a compilation and evaluation of data from the scientific literature covering approximately the past 100 years. The values presented come from the TRC Source Database. The Thermodynamics Research Center has assembled these data over a period of years and has used them to provide the evaluated density values listed in the TRC Thermodynamic Tables - Non-Hydrocarbons. An additional literature search has been performed immediately before producing this compilation to locate new or missing data and to bring the collection up-to-date. This compilation should include at least 90% of the pertinent data reported in the literature. The usual experimental conditions are in contact with air at one atmosphere below the normal boiling point, and in equilibrium with the vapor phase above the normal boiling point. In the summary tables, temperatures reported on the Kelvin scale have been obtained by adding 273.15 to temperatures originally given on the Celsius scale.

Densities have units of kilograms per cubic meter ($kg \cdot m^{-3}$). Values reported in units of grams per milliliter, where the liter is "the volume of one kilogram of water at its temperature of maximum density" convert to $kg \cdot m^{-3}$ when multiplied by 999.972 (as defined by the 12th General Conference of the International Committee on Weights and Measure, 1964). Values of specific gravity relative to water at a stated reference temperature become density upon multiplication by the accepted density of water at the reference temperature. Most reported densities for liquids below the boiling point apply to the air-saturated liquid.

Compounds are identified by an IUPAC approved name [93-ano-1], the empirical molecular formula, and the Chemical Abstracts Service Registry Number. A summary table is available for each compound which includes the reported temperature and density values, an assigned uncertainty for the density, the difference between the observed and smoothed density values and an index key to the source of the data. A complete list of references, identified by the index keys, appears at the end of the volume.

Where appropriate, tables of smoothed, recommended values are given at integral multiples of 10 K over the experimental range of temperatures. Values at 293.15 K and 298.15 K are included when they are in the range of the original data set. The recommended values also have assigned uncertainties.

1.3 Description of Data Tables

Data for a particular compound are selected, evaluated and smoothed in one of four ways, depending upon the number and accuracy of the reported values and upon their distribution over the temperature range.

Case 1. When the data set consists of at least four acceptable, effectively distinct values (see section 1.5.3), the densities in selected subsets are fit to a function of temperature using the least squares criterion. A summary table for the selected set gives the densities, their estimated uncertainties, the deviations between observed and calculated values, an index key to the list of references and a plotting symbol. If sufficient space remains, some data outside the selected set also are included in the summary table along with reference keys to any remaining data. A plot of the deviations between observed and calculated values is shown for the selected subset. Error bars indicate the size of the estimated uncertainties for the data. Distinct plotting symbols identify the five data sources that have the smallest average estimated uncertainties. A single symbol represents all remaining data in the selected set. A table consisting of smoothed, recommended values (calculated from the fitted functions) is also given. Estimated uncertainties are given for the recommended values which also appear as a continuous line on the deviation plot. Densities of crystal phases are in a separate section of the table. In most cases, these densities have not been fit as a function of temperature. Values of parameters, statistical measures of the fit, and references to sources of critical constants appear at the beginning of these sections.

Case 2. For data sets that do not meet the criteria of Case 1, but contain acceptable values over a temperature range of at least two degrees, the results are smoothed using a linear function of temperature with an estimated coefficient of thermal expansion. A table of smoothed recommended values is presented.

Case 3. For data sets that do not meet the criteria of either Case 1 or 2 but contain two or more values at a single temperature, a recommended value is given for this temperature by taking a weighted average of the observed values.

Case 4. For data sets that contain only single values at one or two temperatures, the reported values are given rather than recommended values.

1.4 Evaluation, Selection and Smoothing of Data

1.4.1 Assignment of Uncertainties

The Thermodynamics Research Center staff have assigned an uncertainty value to each observed and recommended density value listed in the tables. The true value of the property has a 95% probability of being in the range covered by + or – the uncertainty about the reported value. Assignment of uncertainty is a subjective evaluation based upon what is known about the measurement when the value is entered into the database, and includes the effects of all sources of experimental error. The errors have been propagated to the listed density at the reported temperature. Uncertainties reported by the investigators are considered but not necessarily adopted. Often, investigators report repeatability, but they usually do not provide uncertainty.

Errors in density result from errors in temperature measurement or control; calibration of instruments; transfer, handling and weighing of samples; and impurities in the samples. At temperatures well below the critical temperature and near room temperature, standard techniques easily achieve accuracies of ±0.05%. For the compounds in this compilation, that level corresponds to about ±0.4 $kg \cdot m^{-3}$. Under these conditions, errors in temperature are not very significant. This level of accuracy only requires temperatures to be known within ± 0.5 K. At temperatures approaching the critical temperature, measurements become more demanding because of the rapid increase in the magnitude of the coefficient of thermal expansion. Greater accuracy, in general, requires careful attention to calibration, mass determination and sample handling techniques. It is assumed that values obtained by pycnometers have been corrected for buoyancy of air, unless the author specifically says otherwise. This correction increases the apparent density by 0.05 - 0.1%. When this correction has not been made, the estimated uncertainty is greater.

Most measurements of densities of liquids below their normal boiling points are made in the presence of air. Densities reported here refer to liquids in equilibrium with a gas phase consisting of a mixture or air and vapor at a total pressure of one atmosphere below the normal boiling point and of vapor at the equilibrium vapor pressure above the boiling point. Thus air is not regarded as an impurity.

A major source of error in most measurements is the presence of impurities in the sample. The effect of an impurity depends upon its amount in the sample and upon the difference between its density and the density of the principal constituent. Even when the sample purity is provided quantitatively, the impurities often are not identified individually. Nevertheless, a report of sample purity reduces the estimated uncertainty because it can be taken as evidence that the investigator has considered sample purity. The most ubiquitous impurity in liquids is water, and, because its density differs significantly from those of hydrocarbons, it is a common source of error. Exclusion of water requires that the sample be protected from the atmosphere during transfer, and that special precautions be taken to remove the sample from containers.

1.4.2 Quantitative Effect of Impurity on Density of Liquids

The molar volume of a mixture of components, V, in terms of the mole fractions x_i and partial molar volumes of the components V_i is:

$$V = \sum_{i=1}^{c} x_i V_i \tag{1.5}$$

For an ideal solution, the partial molal volumes equal the molar volumes of the pure liquid components. Denoting component the main components as 1 and the impurities as > 1, the volume becomes:

$$V = x_1 V_1 + \sum_{i=2}^{c} x_i V_i \tag{1.6}$$

Then using,

$$\rho = M / V \tag{1.7}$$

and the molar mass of the mixture:

$$M = \sum_{i=1}^{c} x_i M_i \tag{1.8}$$

and assuming that the x_i are small for $i > 1$, then

$$\rho = \frac{\rho_1}{w_1}\left(1 - \rho_1 \sum_{i=2}^{c} w_i v_i\right) \tag{1.9}$$

where $v_i = V_i / M_i$ are partial specific volumes of the impurities and w_i is the mass fraction of component i. Finally, the density of the mixture is related to the density of the main component and the impurities i by:

$$\rho = \frac{\rho_1}{w_1}\left(1 - \rho_1 \sum_{i=2}^{c} \frac{w_i}{\rho_i}\right) \tag{1.10}$$

The observed value of the density of a sample is sometimes presented as evidence of its purity. Assuming the sample contains a single impurity, equation (1.10) can be solved for ρ - ρ_1:

$$\rho - \rho_1 = \rho_1 (1 - w_1 - \rho_1 w_2 / \rho_2) / w_1 \tag{1.11}$$

1.4.3 Procedure for Selection and Smoothing of Density Values - Case 1

A selected subset of the reported densities is fit to functions of temperature using the least squares criterion. Up to a boundary temperature T_b (approximately $0.8T_c$), the calculated density ρ_x is represented by a polynomial in temperature with coefficients a_k of order p,

$$\rho_x = \sum_{k=0}^{p} a_k T^k . \tag{1.12}$$

Above T_b the smoothed values are given by a modification of the Guggenheim equation [67-gug]

$$\rho_x = (1 + 1.75\theta + 0.75\theta^3)\left[\rho_c + b_1 (T_c - T) + b_2 (T_c - T)^2 + b_3 (T_c - T)^3 + b_4 (T_c - T)^4\right] \tag{1.13}$$

where T_c is the critical temperature and $\theta = (1\text{-}T/T_c)^{1/3}$. Selected values of critical constants are constant. Continuity with equation (1.12) results from forcing the two functions and their first derivatives with respect to temperature to be equal at the boundary. When no values are available above this temperature, only the polynomial is used.

The following steps, implemented by a computer program written in C, generate the smoothed, recommended values. Input to the program consists of the set of observed density values, temperatures, estimated uncertainties, critical constants and values of certain parameters used by the program.

Step 1. Separate the initial data into two sets, corresponding to temperatures above and below T_b.

Step 2. Make an initial selection from the low temperature set by rejecting all points with zero uncertainty and all points with uncertainties above a limit determined by the data selection algorithm described in section 1.5.2. Zero uncertainties are assigned to values that are not experimental and are included for comparison only (these are most often values recommended in other compilations).

Step 3. Determine the effective number of data values, n_e, as described in 1.5.3. If the effective number of values is less than four, terminate the calculation. If the total number of values is more than eight and the effective number is greater than or equal to four but less than eight, make another initial data selection with relaxed selection criteria.

Step 4. For the j-th value in the set calculate normalized values, $\rho_{n,j}$ and $T_{n,j,}$ and weighting factors, $w_j = 1/u_j^2$ where u_j is the uncertainty assigned to the j-th observed density and $\rho_{n,j} = \rho_j - \overline{\rho}$ where $\overline{\rho}$ is the mean value of the observed density in the set. and $T_{n,j} = T^k - \overline{T^k}$ where $\overline{T^k}$ is the mean value of the T_j^k value in the set..

Step 5. Using $\rho_n = a_1 T_n$, fit the data subject to least squares with points weighted by w_j.

Step 6. Calculate the standard deviation σ for this fit. Eliminate any points from this set for which $|\delta_j| > 3.5\sigma$, where $\delta_j = \rho_j - \rho_{x,j}$.

Step 7. Fit the remaining normalized values to a series of polynomials, $\rho_n = \sum a_k T^k$, starting with order 1 and increasing in order. Use w_j as weighting factors and stop increasing the order when satisfying one of the following conditions:

1. A value of p given as an input parameter to the program is reached, or
2. $\chi_k^2 < 1.1[1 + 1/(n-k)]^2 \chi_{k-1}^2$ (see glossary of symbols) and the deviations pass the random deviation test (see 1.5.4).

Step 8. If any points have $|\delta_j| > 2.2\sigma$ for the final polynomial, eliminate these points and repeat step 7.

Step 9. Calculate parameter a_0.

Step 10. Apply the initial data selection described in step 2 to the high temperature data set.

Step 11. Fit the selected high temperature data with the modified Guggenheim equation using least squares with weighting factors w_j.

Step 12. The following procedure provides continuity at the boundary. Set equation (1.13) and its first derivative at T_b equal to the corresponding values from equation (1.12) at T_b. Eliminate parameters b_3 and b_4 from these two simultaneous equations to obtain a function containing parameters b_1 and b_2 which can be evaluated for the high temperature range using least squares. Do not use densities at temperatures within 2 K of the critical temperature.

Step 13. Generate the output table of temperature, observed densities, estimated uncertainties, and difference between observed and calculated densities and arrange it in order of year of publication with authors. For data from a particular source, arrange in order of temperature.

Step 14. Calculate the table of smoothed and recommended values with their corresponding estimated uncertainties.

Coefficients A, B, C, D and E listed in the heading of Table 1 for each compound correspond to a_0, a_1, a_2, a_3 and a_4 in equation (1.12) for temperatures below T_b. σ_ℓ is the weighted standard deviation for individual points in this region (see the glossary). If the data set includes values above T_b, the coefficients A, B, C and D correspond to b_1, b_2, b_3 and b_4 in equation (1.13) for this range. The weighted standard deviation, $\sigma_{c,w}$, and the unweighted standard deviation for the fit, $\sigma_{c,uw}$, include both ranges. If the data set covers only values below T_b then $\sigma_{c,w}$ and $\sigma_{c,uw}$ represent that range only.

The uncertainty in the smoothed values depends upon the uncertainties in the original observed values and upon the magnitude of deviations between observed and calculated values. To approximate the contribution of these two effects at the temperature T, the uncertainties $u_x(T)$ for the low temperature range are calculated from:

$$u_x(T) = \left[u(T)^2 + \sum_k \sum_l C_{kl} (T^k - \overline{T^k})(T^l - \overline{T^l}) \right]^{1/2} . \tag{1.14}$$

In this equation, $u(T)$ represents the uncertainty of the observed data in the vicinity of T and is approximated by fitting a polynomial of order 1-3 to the estimated uncertainties as a function of temperature (other symbols appear in the glossary). Uncertainties in the smoothed data for the high temperature range are calculated using:

$$u_x(T) = \left[u_x(T_b)^2 + h(T)^{-2} \right]^{1/2} , \tag{1.15}$$

where $u_x(T_b)$ is the uncertainty calculated using equation (1.14) for the low temperature range at the boundary temperature T_b and $h(T)$ is a polynomial in temperature fit to the reciprocals of the estimated uncertainties in the high temperature region.

The uncertainties in extrapolated data should increase as the extent of extrapolation increases. Since equation (1.15) does not always give this result, manual adjustment is sometime required in this range.

1.4.4 Procedure for Selection and Smoothing of Density Values - Case 2

When the data set for a particular compound satisfies the criteria for Case 2, it is smoothed by a linear function of temperature,

$$\rho_x = a_0 + a_1 T \,. \tag{1.16}$$

The coefficient a_1 is either calculated from two densities of sufficient accuracy reported at different temperatures, preferably by the same investigator, or estimated by examination of the coefficient of expansion of similar compounds obtained from a least squares calculation. The constant term then results from equation (1.17) after eliminating values with large uncertainties

$$a_0 = \sum w_j \left(\rho_j - a_1 T_j\right) / \sum w_j \,. \tag{1.17}$$

The uncertainties for the smoothed values are:

$$u_x(T) = [\sigma_0^2 + \sigma_1^2 (T - \overline{T})^2]^2 , \tag{1.18}$$

where $\overline{T}$ is the weighted mean temperature for the accepted set, $\sigma_0^2 = (\Sigma w_j \delta_j^2) / \Sigma w_j$ and σ_1 is the estimated standard deviation of a_1 .

1.4.5 Procedure for Selection and Smoothing of Density Values - Case 3

The recommended density at a particular temperature is the weighted mean observed density for that temperature. The corresponding uncertainty is the standard deviation from the mean for each value.

1.5 Calculation Procedures

1.5.1 Least Squares Calculation

Parameters of all the smoothing functions are adjusted to minimize the function

$$\chi^2 = \Sigma w_j \delta_j^2 \tag{1.19}$$

by the singular value decomposition of the matrix of independent variables of the function. The parameters are calculated by functions **svdcmp** and **svbksb** described in [88-pre/fla] modified to accept weighting factors. The covariance matrix used in equation (1.14) is calculated by the function **covar** from the same book.

1.5.2 Selection of Data Based upon Estimated Uncertainties

The selection procedure is:

Step 1. Obtain ΔT, the range of temperatures covered by the data set.

Step 2. For each density value, ρ_j, in the set, calculate,

$$x_{jl} = \exp\left(q\left|T_j - T_l\right|\right) \tag{1.20}$$

$$z_1 = \sum_{l \neq j} x_{jl} \tag{1.21}$$

$$z_2 = \sum_{l \neq j} u_l x_{jl} \tag{1.22}$$

$$y = u_j z_1^{1.5} z_2^{-1} \tag{1.23}$$

Accept point j if $y \le d$; reject it otherwise.

Step 3. Repeat steps 1 and 2 with points accepted in the first pass.

The accepted points are those that remain from Step 3. The constants q and d are:

$$q = -2.628 g_1 \left[1 + (\Delta T / 30)^2\right] / \Delta T$$

$$d = g_2 / \log_{10}(1+n).$$

The number z_2 / z_1 is a weighted mean of all points in the set other than the j-th point. The weighting factor decreases exponentially with the difference in temperature of the l-th point from the j-th point. The parameter g_1 determines the rate of decrease. This procedure compares the uncertainty of the j-th point to the weighted mean of other points. The parameter g_2 determines the rejection level from this comparison for the j-th point. Larger values of g_2 are less selective. Values for g_1 and g_2 are supplied to the algorithm. For all cases g_1 is in the range of 1 to 2 (usually 1.8). The value of g_2 is in the range of 2 to 3 ($\text{kg} \cdot \text{m}^{-3}$) (usually 2.5).

1.5.3 Count the Effective Number of Density Values in a Set

The number of degrees of freedom in a least squares fit is the number of distinct data values minus the number of adjustable parameters. To obtain a meaningful smoothing of data, the order of the polynomial function is limited to values which gives three or more degrees of freedom. However, if two or more density values in the set are at the same (or nearly the same) temperature, they should count as only one point in calculating the degrees of freedom. In general, the effective number of density values minus the number of fitting parameters is used as the degrees of freedom. Effective data values are those that are separated by at least 1.2 K.

1.5.4 Testing a Set of Deviations between Observed and Calculated Density Values for a Random Distribution

One of the criteria for acceptance of the order of a polynomial least squares fit is that the deviations between calculated and random values be distributed "randomly" over the range of conditions covered by the data. The concept of randomness for this purpose probably cannot be defined rigorously. However, the following test for randomness is used whenever the original data set contains seven or more values.

Step 1. Sort the values in order of increasing temperature

Step 2. Separate the total range of temperature, ΔT, into s subranges each of size $\Delta T/s$. Form s subsets of data corresponding to these temperature subranges.

Step 3. Make the following comparison for each subset j which has at least four members.

$$0.01 < \frac{\Sigma|\delta_j|}{n_s} \quad \text{and} \quad 0.2 < \frac{|\Sigma\delta_j|}{\Sigma|\delta_j|}.$$

If both comparisons are true for any subset, the test for randomness fails.

Step 4. Apply steps 2 and 3 to the data for s from 2 to an upper limit. The upper limit is determined by the number of values in the original set according to the following table.

n, number of values in original set	maximum number of subsets
7 to 10	2
11 to 20	3
21 to 33	4
> 33	5

1.6 Glossary of Symbols

a_k	parameters in the polynomial function for densities at temperatures $\le T_b$
b_k	k-th parameter in modified Guggenheim equation for density at temperatures $>T_b$
g_1, g_2	parameters used in the data selection algorithm
n	number of accepted values of density in a set
n_e	effective number of accepted values of density in a set
n_s	number of density values in subset s
p	order of the polynomial for density values at temperatures $\le T_b$
s	number of subsets in the random deviation algorithm
u_j	uncertainty assigned to the j-th observed density value in a set
w_j	weighting factor for the j-th density value in a set
$C_{k,l}$	Element k,l of the variance-covariance matrix for the polynomial parameters
T	absolute temperature
T_b	boundary temperature
T_c	critical temperature
T_j	temperature for the j-th observed density
$\overline{T^k}$	mean value of the T_j^k values in a set
$T_{n,j}$	$T_j^k - \overline{T^k}$, normalized value of the j-th temperature raised to the k power
δ_j	$\rho_j - \rho_{x,j}$
θ_j	$(1 - T_j/T_c)^{1/3}$
ρ	density
ρ(API)	API specific gravity
$\overline{\rho}$	mean value of observed densities in a set
ρ_o	density of a standard substance
ρ_c	critical density
ρ_j	observed value of j-th density in a data set
ρ_m	molar density
$\rho_{n,j}$	$\rho_j - \overline{\rho}$, normalized density for the j-th value
ρ_r	relative density
$\rho_{x,j}$	calculated value of the j-th density in a data set
σ	$(\chi^2/n)^{1/2}$, standard deviation for density values in a set
χ_k^2	$\Sigma\delta_j^2$ for all values in a set fit to a polynomial of order k
ΔT	$T_n - T_1$, range of temperatures for data in a set

The following symbols refer to components in a mixture at a fixed temperature

c	number of components in the mixture
i	component number
v_i	partial specific volume of component i in the mixture
w_i	mass fraction of component i in the mixture
x_i	mole fraction of component i in the mixture
M_i	molar mass (molecular weight) of component i
V	molar volume of a mixture
V_i	partial molal volume of component i in the mixture
ρ	density of a mixture of components
ρ_i	density of pure component i

Symbols used in the tables:

A, B, C, D, E	coefficients in function for density (see section 1.4.3)
ρ_{calc}	calculated density, ρ_x
ρ_{exp}	observed value of j-th density in a data set, ρ_j
σ_ℓ	$(\Sigma w_j \delta_j^2 / \Sigma w_j)^{1/2}$, for low temperature range only
$\sigma_{\text{c,w}}$	$(\Sigma w_j \delta_j^2 / \Sigma w_j)^{1/2}$, for low and high temperature range combined
$\sigma_{\text{c,uw}}$	$[\Sigma \delta_j^2 / n(n-p-2)]^{1/2}$, for low and high temperature range combined
$2\sigma_{\text{est}}$	estimated uncertainty, u_j

1.7 Order of Compounds in the Tables

The density tables are organized into 3 main classes of compounds as described in the Table of Contents: monoalcohols, diols, and triols. Within each main class (except for the triols) there are several subclasses. They start with fully saturated compounds and proceed with increasing extents of unsaturation. Within each subclass the compounds are arranged in formula order. First with increasing number of carbon atoms in the empirical formula and then with increasing number of hydrogen atoms. Compounds with the same formula are sorted alphabetically by Table Name.

References

67-gug — Guggenheim, E.A.: Thermodynamics, 5th edition, Amsterdam: North-Holland Publishing Company (1957).

86-ros/bae — Rossiter, B.W., Baetzold, R.C.: Physical Methods of Chemistry, 2nd edition, Vol. VI, Determination of Thermodynamic Properties, New York: John Wiley & Sons (1986).

88-pre/fla — Press, W.H., Flannery, B.P., Teukolsky, S.A., Vetterling, W.T.: Numerical Recipes in C, New York: Cambridge Univ. Press (1988).

2 Tabulated Data on Density - Monoalcohols

2.1 Alkanols

2.1.1 Alkanols, C_1 - C_4

Methanol **[67-56-1]** **CH_4O** **MW = 32.04** **1**

T_c = 512.64 K [89-tej/lee]
ρ_c = 272.00 kg·m^{-3} [89-tej/lee]

Table 1. Coefficients for the polynomial expansion equations. Standard deviations (see introduction): $\sigma_\ell = 5.5832 \cdot 10^{-1}$ (low temperature range), $\sigma_{c,w} = (4.8307 \cdot 10^{-1}$ combined temperature ranges, weighted), $\sigma_{c,uw} = 3.4016 \cdot 10^{-1}$ (combined temperature ranges, unweighted).

Coefficient	T = 176.15 to 400.00 K $\rho = A + BT + CT^2 + DT^3 + \ldots$	T = 400.00 to 512.64 K $\rho = [1 + 1.75(1 - T/T_c)^{1/3} + 0.75(1 - T/T_c)]$ $[\rho_c + A(T_c - T) + B(T_c - T)^2 + C(T_c - T)^3$ $+ D(T_c - T)^4]$
A	$1.14445 \cdot 10^3$	2.46064
B	−1.79908	$-5.82461 \cdot 10^{-2}$
C	$3.16450 \cdot 10^{-3}$	$5.61772 \cdot 10^{-4}$
D	$-3.87839 \cdot 10^{-6}$	$-1.89776 \cdot 10^{-6}$

Table 2. Experimental values with uncertainties and deviation from calculated values.

$\frac{T}{K}$	$\frac{\rho_{exp} \pm 2\sigma_{est}}{kg \cdot m^{-3}}$	$\frac{\rho_{exp} - \rho_{calc}}{kg \cdot m^{-3}}$	Ref. (Symbol in Fig. 1)	$\frac{T}{K}$	$\frac{\rho_{exp} \pm 2\sigma_{est}}{kg \cdot m^{-3}}$	$\frac{\rho_{exp} - \rho_{calc}}{kg \cdot m^{-3}}$	Ref. (Symbol in Fig. 1)
	crystal			473.15	552.50 ± 3.00	−3.21	1887-ram/you(X)
78.15	1041 ± 3.00		30-bil/fis-1	483.15	525.40 ± 3.00	−2.04	1887-ram/you(X)
	liquid			493.15	490.00 ± 3.00	2.14	1887-ram/you(X)
296.09	789.07 ± 0.50	0.56	1887-ram/you[1)]	498.15	468.00 ± 3.00	6.30	1887-ram/you(X)
296.09	789.09 ± 0.60	0.58	1887-ram/you[1)]	503.15	441.80 ± 5.00	12.68	1887-ram/you[1)]
353.15	736.80 ± 1.00	3.85	1887-ram/you[1)]	505.15	430.10 ± 5.00	16.61	1887-ram/you[1)]
363.15	723.90 ± 1.00	1.20	1887-ram/you[1)]	507.15	412.70 ± 5.00	17.09	1887-ram/you[1)]
373.15	713.50 ± 1.00	1.27	1887-ram/you(X)	509.15	395.30 ± 5.00	21.13	1887-ram/you[1)]
383.15	702.80 ± 1.50	1.26	1887-ram/you(X)	510.15	386.60 ± 5.00	25.49	1887-ram/you[1)]
393.15	689.90 ± 1.50	−0.68	1887-ram/you(X)	511.15	370.50 ± 5.00	25.55	1887-ram/you[1)]
403.15	677.70 ± 1.50	−1.36	1887-ram/you(X)	511.65	364.20 ± 0.00	29.63	1887-ram/you[1)]
413.15	664.90 ± 1.50	0.95	1887-ram/you(X)	293.15	790.48 ± 0.00	−0.81	1893-ram/shi-3[1)]
423.15	649.40 ± 1.50	2.71	1887-ram/you(X)	343.15	745.98 ± 0.00	2.97	1893-ram/shi-3[1)]
433.15	632.10 ± 2.00	2.92	1887-ram/you(X)	353.15	735.50 ± 0.60	2.55	1893-ram/shi-3[1)]
443.15	616.00 ± 2.00	3.84	1887-ram/you(X)	363.15	725.00 ± 1.50	2.30	1893-ram/shi-3[1)]
453.16	596.70 ± 2.00	1.43	1887-ram/you(X)	373.15	714.00 ± 1.50	1.77	1893-ram/shi-3(X)
463.15	578.00 ± 2.00	0.76	1887-ram/you(X)	383.15	702.00 ± 1.50	0.46	1893-ram/shi-3(X)

[1]) Not included in Fig. 1.

cont.

Methanol (cont.)

Table 2. (cont.)

T / K	$\rho_{exp} \pm 2\sigma_{est}$ / $kg \cdot m^{-3}$	$\rho_{exp} - \rho_{calc}$ / $kg \cdot m^{-3}$	Ref. (Symbol in Fig. 1)	T / K	$\rho_{exp} \pm 2\sigma_{est}$ / $kg \cdot m^{-3}$	$\rho_{exp} - \rho_{calc}$ / $kg \cdot m^{-3}$	Ref. (Symbol in Fig. 1)
393.15	690.00 ± 1.50	−0.58	1893-ram/shi-3(×)	223.15	857.20 ± 0.50	−0.26	67-kom/man[1)]
403.15	677.00 ± 1.50	−2.06	1893-ram/shi-3(×)	233.15	847.80 ± 0.50	−0.06	67-kom/man[1)]
413.15	664.00 ± 1.50	0.05	1893-ram/shi-3(×)	243.15	838.40 ± 0.50	0.06	67-kom/man[1)]
423.15	649.50 ± 1.50	2.81	1893-ram/shi-3(×)	253.15	829.00 ± 0.40	0.11	67-kom/man[1)]
433.15	634.00 ± 2.00	4.82	1893-ram/shi-3(×)	263.15	819.60 ± 0.40	0.12	67-kom/man[1)]
443.15	616.00 ± 2.00	3.84	1893-ram/shi-3(×)	273.15	810.20 ± 0.40	0.11	67-kom/man[1)]
453.15	598.00 ± 2.00	2.71	1893-ram/shi-3(×)	283.15	800.80 ± 0.40	0.10	67-kom/man[1)]
463.15	577.00 ± 2.00	−0.24	1893-ram/shi-3(×)	293.15	791.30 ± 0.40	0.01	67-kom/man[1)]
473.15	553.00 ± 3.00	−2.71	1893-ram/shi-3(×)	298.15	787.00 ± 0.40	0.44	67-kom/man[1)]
483.15	525.50 ± 3.00	−1.94	1893-ram/shi-3(×)	413.15	636.94 ± 0.00	−27.01	69-zub/bag[1)]
493.15	490.00 ± 3.00	2.14	1893-ram/shi-3(×)	423.15	628.93 ± 0.00	−17.76	69-zub/bag[1)]
503.15	441.00 ± 5.00	11.88	1893-ram/shi-3[1)]	433.15	621.12 ± 0.00	−8.06	69-zub/bag[1)]
507.15	414.50 ± 5.00	18.89	1893-ram/shi-3[1)]	443.15	609.76 ± 0.00	−2.40	69-zub/bag[1)]
509.15	395.50 ± 5.00	21.33	1893-ram/shi-3[1)]	453.15	594.11 ± 1.50	−1.18	69-zub/bag(×)
513.15	271.20 ± 0.00	19.30	1893-ram/shi-3[1)]	463.15	573.39 ± 1.50	−3.85	69-zub/bag(×)
293.15	791.31 ± 0.06	0.02	32-ros(∇)	473.15	550.99 ± 1.50	−4.72	69-zub/bag(×)
323.15	763.20 ± 0.40	0.55	50-hou/mas-1[1)]	483.15	524.49 ± 1.50	−2.95	69-zub/bag(×)
333.15	753.60 ± 0.40	0.70	50-hou/mas-1[1)]	493.15	489.00 ± 2.00	1.14	69-zub/bag(×)
343.15	744.20 ± 0.40	1.19	50-hou/mas-1(×)	498.15	466.90 ± 2.00	5.20	69-zub/bag(×)
353.15	734.70 ± 0.50	1.75	50-hou/mas-1(×)	503.15	440.01 ± 2.00	10.89	69-zub/bag(×)
363.15	725.10 ± 0.60	2.40	50-hou/mas-1[1)]	505.15	426.20 ± 2.00	12.71	69-zub/bag(×)
403.15	671.80 ± 1.00	−7.26	55-kay/don(×)	507.15	409.70 ± 3.00	14.09	69-zub/bag(×)
413.15	657.40 ± 1.00	−6.55	55-kay/don(×)	509.15	387.81 ± 3.00	13.64	69-zub/bag(×)
423.15	642.00 ± 1.00	−4.69	55-kay/don(×)	511.15	346.87 ± 3.00	1.92	69-zub/bag(×)
433.15	625.80 ± 1.50	−3.38	55-kay/don(×)	512.65	274.57 ± 0.00	2.05	69-zub/bag[1)]
443.15	608.10 ± 1.50	−4.06	55-kay/don(×)	183.15	896.80 ± 0.20	−0.47	71-yer/swi(×)
453.15	589.10 ± 1.50	−6.19	55-kay/don(×)	193.15	886.70 ± 0.20	−0.36	71-yer/swi(×)
463.15	568.10 ± 2.00	−9.14	55-kay/don(×)	203.15	876.80 ± 0.20	−0.25	71-yer/swi(×)
473.15	544.80 ± 2.00	−10.91	55-kay/don(×)	213.15	867.30 ± 0.20	0.11	71-yer/swi(×)
483.15	517.00 ± 2.00	−10.44	55-kay/don(×)	223.15	857.40 ± 0.20	−0.06	71-yer/swi(×)
493.15	481.10 ± 3.00	−6.76	55-kay/don(×)	233.15	847.90 ± 0.20	0.04	71-yer/swi(×)
503.15	432.10 ± 5.00	2.98	55-kay/don[1)]	243.15	838.30 ± 0.20	−0.04	71-yer/swi(×)
512.58	272.00 ± 5.00	−23.47	55-kay/don[1)]	253.15	828.90 ± 0.20	0.01	71-yer/swi(×)
283.15	800.45 ± 0.10	−0.25	58-yam/kun(×)	263.15	819.70 ± 0.20	0.22	71-yer/swi(×)
288.15	795.81 ± 0.10	−0.19	58-yam/kun(×)	273.15	810.40 ± 0.20	0.31	71-yer/swi[1)]
293.15	791.32 ± 0.10	0.03	58-yam/kun[1)]	283.15	800.80 ± 0.20	0.10	71-yer/swi[1)]
273.15	809.98 ± 0.10	−0.11	59-mck/ski(×)	281.40	802.32 ± 0.20	−0.03	72-rei/eis(□)
176.15	904.60 ± 0.70	0.07	67-kom/man(×)	284.78	798.99 ± 0.20	−0.18	72-rei/eis(□)
183.15	897.00 ± 0.70	−0.27	67-kom/man[1)]	285.69	798.15 ± 0.20	−0.16	72-rei/eis(□)
193.15	886.80 ± 0.60	−0.26	67-kom/man[1)]	288.72	795.35 ± 0.20	−0.11	72-rei/eis(□)
203.15	876.50 ± 0.60	−0.55	67-kom/man[1)]	289.64	794.49 ± 0.20	−0.11	72-rei/eis(□)
213.15	866.60 ± 0.60	−0.59	67-kom/man[1)]	292.96	791.41 ± 0.20	−0.06	72-rei/eis(□)

[1)] Not included in Fig. 1.

cont.

Table 2. (cont.)

$\frac{T}{K}$	$\frac{\rho_{exp} \pm 2\sigma_{est}}{kg \cdot m^{-3}}$	$\frac{\rho_{exp} - \rho_{calc}}{kg \cdot m^{-3}}$	Ref. (Symbol in Fig. 1)
293.82	790.60 ± 0.20	−0.06	72-rei/eis(□)
296.97	787.67 ± 0.20	−0.01	72-rei/eis(□)
297.87	786.82 ± 0.20	−0.01	72-rei/eis(□)
300.90	783.98 ± 0.20	0.02	72-rei/eis(□)
301.62	783.29 ± 0.20	0.02	72-rei/eis(□)
304.83	780.27 ± 0.20	0.04	72-rei/eis(□)
305.70	779.45 ± 0.20	0.05	72-rei/eis(□)
308.32	776.96 ± 0.20	0.06	72-rei/eis(□)
309.17	776.16 ± 0.20	0.07	72-rei/eis(□)
311.32	774.13 ± 0.20	0.09	72-rei/eis(□)
312.17	773.33 ± 0.20	0.11	72-rei/eis(□)
317.16	768.56 ± 0.20	0.12	72-rei/eis(□)
293.15	791.04 ± 0.20	−0.25	76-hal/ell[1)]
298.15	786.37 ± 0.20	−0.19	76-hal/ell[1)]
303.15	781.65 ± 0.20	−0.17	76-hal/ell[1)]
320.00	765.59 ± 0.20	−0.11	76-hal/ell[1)]
340.00	745.88 ± 0.20	−0.26	76-hal/ell(×)
360.00	725.21 ± 0.20	−0.74	76-hal/ell(×)
380.00	703.03 ± 0.25	−1.90	76-hal/ell(×)
400.00	678.69 ± 0.25	−4.23	76-hal/ell[1)]
410.00	665.40 ± 0.25	−3.66	76-hal/ell(×)
420.00	651.27 ± 0.20	−0.95	76-hal/ell(×)
430.00	636.05 ± 0.30	1.40	76-hal/ell(×)
440.00	619.50 ± 0.30	2.03	76-hal/ell(×)
293.15	791.25 ± 0.05	−0.04	84-cer/bou(○)
298.15	786.57 ± 0.03	0.01	86-oga/mur(□)
298.15	786.55 ± 0.06	−0.01	86-tan/toy(◆)
298.15	786.57 ± 0.05	0.01	87-oga/mur(Δ)
273.15	809.99 ± 0.20	−0.10	88-sun/bis[1)]
283.15	800.70 ± 0.20	−0.00	88-sun/bis[1)]
293.15	791.32 ± 0.20	0.03	88-sun/bis[1)]
303.15	781.86 ± 0.20	0.04	88-sun/bis[1)]
313.15	772.32 ± 0.20	0.03	88-sun/bis[1)]
323.15	762.70 ± 0.20	0.05	88-sun/bis(×)
333.15	753.00 ± 0.20	0.10	88-sun/bis(×)
205.16	875.03 ± 0.20	−0.02	88-sun/sch-1(×)
212.69	867.57 ± 0.20	−0.07	88-sun/sch-1(×)
223.11	857.33 ± 0.20	−0.17	88-sun/sch-1(×)
232.45	848.35 ± 0.20	−0.18	88-sun/sch-1(×)
243.17	838.12 ± 0.20	−0.20	88-sun/sch-1(×)
252.11	829.73 ± 0.20	−0.14	88-sun/sch-1(×)
261.62	820.74 ± 0.20	−0.18	88-sun/sch-1(×)
269.38	813.50 ± 0.20	−0.13	88-sun/sch-1(×)
273.27	809.79 ± 0.20	−0.19	88-sun/sch-1[1)]
281.82	801.87 ± 0.20	−0.08	88-sun/sch-1[1)]
292.38	792.04 ± 0.20	0.03	88-sun/sch-1[1)]
294.36	790.05 ± 0.20	−0.10	88-sun/sch-1[1)]
298.16	786.55 ± 0.20	−0.00	88-sun/sch-1[1)]
303.14	781.79 ± 0.20	−0.04	88-sun/sch-1[1)]
313.14	772.23 ± 0.20	−0.06	88-sun/sch-1[1)]
323.17	762.58 ± 0.20	−0.05	88-sun/sch-1(×)
333.15	752.62 ± 0.20	−0.28	88-sun/sch-1(×)
203.15	877.08 ± 0.20	0.03	90-sun/sch-1(×)
213.15	867.09 ± 0.20	−0.10	90-sun/sch-1(×)
223.15	857.29 ± 0.20	−0.17	90-sun/sch-1(×)
233.15	847.65 ± 0.20	−0.21	90-sun/sch-1(×)
243.15	838.13 ± 0.20	−0.21	90-sun/sch-1(×)
253.15	828.70 ± 0.20	−0.19	90-sun/sch-1(×)
263.15	819.33 ± 0.20	−0.15	90-sun/sch-1(×)
298.15	786.80 ± 0.20	0.24	98-ami/ban(○)
303.15	781.80 ± 0.20	−0.02	98-ami/ban(○)
308.15	777.10 ± 0.20	0.04	98-ami/ban(○)
298.15	786.60 ± 0.20	0.04	98-ami/pat-1(Δ)
303.15	781.70 ± 0.20	−0.12	98-ami/pat-1(Δ)
308.15	776.90 ± 0.20	−0.16	98-ami/pat-1(Δ)

[1)] Not included in Fig. 1.

Further references: [1848-kop, 1854-kop, 1863-gla/dal, 1864-lan, 1880-pry, 1882-sch-1, 1883-sch-3, 1884-gla, 1884-per, 1884-sch-6, 1884-zan, 1886-tra, 1890-gar, 1891-jah, 1891-sch/kos, 1892-lan/jah, 1896-zel/kra, 1898-kah, 1898-roh, 00-loo, 02-you/for, 03-car/cop, 04-bru/sch, 04-cri, 06-car/fer, 06-kla/nor, 06-wal-1, 07-che-1, 07-tim, 08-get, 08-gyr, 08-ric/mat, 09-dor, 09-hol/sag, 10-daw, 10-dor/pol, 10-fon, 10-tim, 11-dor, 12-kor, 12-mal, 12-tim, 12-tim-1, 12-tyr, 13-atk/wal, 13-bri-1, 13-rom, 13-ste, 14-kre/mei, 14-low, 14-mer/tur, 14-tyr, 15-ric/coo, 16-har-2, 16-ric/shi, 16-sei/alt, 16-wro/rei, 17-jae, 18-her-2, 19-eyk, 20-ric/dav, 21-bar/bir, 21-rei/hic, 22-her/sch, 22-mck/sim, 23-wil/smi, 24-bus-1, 24-dan, 24-mar-1, 24-mil, 25-har/rai, 25-lew, 25-nor/ash, 25-par, 25-per, 25-rak, 26-ewa/rai, 26-gol/aar, 26-mat, 26-mun, 26-ris/hic, 26-sch, 27-arb-2, 28-llo/bro, 29-kel-1, 30-bil/fis-1, 30-rak/fro, 30-tim/hen, 31-bea/mcv, 31-fio/gin, 31-lun/bje, 33-but/tho, 33-koz/koz, 33-nat/bac, 33-vos/con, 34-cor/arc, 34-smi, 34-was/spe, 35-gib, 35-hen, 35-kef/mcl, 36-tom]

cont.

Methanol (cont.)

Further references: (cont.)
[37-bet/ham, 37-gib/kin, 37-sta/gil, 38-jon/for, 39-lar/hun, 40-pes-1, 40-was/gra, 42-bri/rin, 42-mor/mun, 42-mul, 43-eck/luc, 44-pes/lag, 45-add, 45-dul, 46-kre/now, 46-sca/woo, 46-sca/woo-1, 47-woo, 48-jon/bow, 48-vog-2, 48-wei, 48-wil/ros, 49-gor/gor, 49-gri/buf, 49-hat, 49-sta/gup, 49-tsc/ric, 49-udo/kal, 50-jac, 50-joe/nik, 50-lar/ver, 50-pic/zie, 50-sac/sau, 50-sad/fuo, 50-tei/gor, 50-wol/sau, 51-car/rid, 51-cli/cam, 51-lyu/ter, 51-sie/cru, 51-tei/gor, 52-hug/mal, 52-sca/tic, 52-sta/spi, 53-ame/pax, 53-ani-1, 53-mck/tar, 53-par/cha, 54-col, 54-cro/spi, 54-gri, 54-kre/wie, 54-pur/bow, 54-sad/fuo, 55-den/col, 55-gre/ven, 55-ham/sto, 56-ame/pax, 56-fai/win, 57-gol, 57-mil, 57-tul/chr, 58-ano-5, 58-cos/bow, 58-lin/van, 58-mur/van, 59-ale, 59-yen/ree, 60-cop/fin, 61-mik/kim, 62-bro/smi, 62-chu/tho, 62-mik/kim, 62-nag-4, 62-par/mis, 63-aki/yos, 63-fis, 63-hov/sea, 63-mcc/lai, 63-raj/ran, 64-ma /koh, 65-for/moo, 66-kat/pra-1, 66-kat/shi, 67-cun/vid, 67-fre, 67-han/hac, 67-nak/nak, 67-nak/shi, 67-sum/tho, 68-ano, 68-bek/hal, 68-nak/shi, 68-pfl/pop, 68-sin/ben, 69-bru/gub, 70-iin/sud, 70-kat/kon, 70-kat/kon-1, 70-kon/lya, 70-nak/shi, 70-str/svo, 71-des/bha, 71-des/bha-1, 71-nag/oht, 72-bou/aim, 72-nie/nov, 72-pol/lu, 72-pol/lu -1, 73-khi/ale, 73-svo/ves, 74-dut/mat, 75-mat/fer, 75-mus/ver, 75-nak/wad, 75-tok, 76-kat/nit, 76-mcg/wil, 76-nag/oht, 76-nak/ash, 76-sri/kul, 76-wes, 77-hwa/rob, 77-sch/pla, 77-tre/ben, 78-och/lu, 79-cha/ses-1, 79-cib/hyn, 79-coc/pis, 79-dia/tar, 79-jim/paz, 79-kiy/ben, 80-aim/cip, 80-arc/bla, 80-ben/kiy, 80-edu/boy, 80-kub/tsu, 80-yos/tak, 81-fre/cri, 81-kum/pra, 81-sjo/dyh, 81-tas/ara, 81-won/chu, 82-diz/mar, 82-dom/rat, 82-ort, 83-fuk/ogi, 83-pik-2, 83-rau/ste, 83-tri, 84-sak/nak, 85-dri/ras, 85-kov/svo, 85-mat/ben, 85-mat/ben-1, 85-nag, 85-ort/paz-1, 85-pat/san, 85-sch/pla, 85-tri/rod, 86-bot/bre, 86-cra, 86-hne/cib, 86-kud/str, 86-lep/mat, 86-lin/ber, 86-miy/hay, 86-san/sha, 86-yer/wor, 86-zha/ben-1, 87-ham/als, 87-kub/tan, 87-lin/ber, 87-man/ami, 87-pik, 88-aww/all, 88-bag/gur, 88-ben/van, 88-man/ami, 88-nag-2, 88-ohg/tak, 88-oka/oga, 89-ala/sal, 89-ale/fer, 89-dou/kha, 89-kat/tan, 89-kat/tan-1, 89-nao/sur, 89-raj/ren, 90-arc/dom, 90-cha/kat, 90-jos/ami-1, 90-lee/hon, 90-let/sch-1, 90-rie/sch, 91-cab/bel, 91-gar/her, 91-kat/tan, 91-liu/pus, 91-ram/muk, 92-hia/yam, 92-lee/wei, 92-pap/zio, 92-pla/ste, 93-ami/ara, 94-arc/bla, 94-kim/lee, 94-pap/pan-1, 95-arc/bla, 95-che/kna, 95-red/ram-1, 96-nik/jad, 96-nik/mah, 97-arc/bla, 97-com/fra, 98-arc/mar, 98-arc/nar, 98-bla/ort, 98-igl/org, 98-nik/shi, 98-zie].

Table 3. Recommended values (fit to the reliable experimental values according to the equations $\rho = A + BT + CT^2 + DT^3 + \ldots$ or $\rho = [1 + 1.75(1 - T/T_c)^{1/3} + 0.75(1 - T/T_c)][\rho_c + A(T_c - T) + B(T_c - T)^2 + C(T_c - T)^3 + D(T_c - T)^4]$).

$\frac{T}{\text{K}}$	$\frac{\rho \pm \sigma_{\text{fit}}}{\text{kg} \cdot \text{m}^{-3}}$	$\frac{T}{\text{K}}$	$\frac{\rho \pm \sigma_{\text{fit}}}{\text{kg} \cdot \text{m}^{-3}}$	$\frac{T}{\text{K}}$	$\frac{\rho \pm \sigma_{\text{fit}}}{\text{kg} \cdot \text{m}^{-3}}$
170.00	911.00 ± 0.33	293.15	791.29 ± 0.06	410.00	669.06 ± 1.88
180.00	900.52 ± 0.33	298.15	786.56 ± 0.07	420.00	652.22 ± 1.91
190.00	890.26 ± 0.32	300.00	784.81 ± 0.07	430.00	634.65 ± 1.94
200.00	880.18 ± 0.31	310.00	775.30 ± 0.10	440.00	617.47 ± 1.99
210.00	870.28 ± 0.29	320.00	765.70 ± 0.15	450.00	600.65 ± 2.05
220.00	860.51 ± 0.26	330.00	755.99 ± 0.23	460.00	583.17 ± 2.14
230.00	850.87 ± 0.23	340.00	746.14 ± 0.34	470.00	563.04 ± 2.29
240.00	841.33 ± 0.19	350.00	736.13 ± 0.47	480.00	537.31 ± 2.83
250.00	831.86 ± 0.15	360.00	725.95 ± 0.64	490.00	501.91 ± 3.95
260.00	822.44 ± 0.12	370.00	715.55 ± 0.85	500.00	450.53 ± 4.84
270.00	813.05 ± 0.09	380.00	704.93 ± 1.11	510.00	363.22 ± 6.34
280.00	803.66 ± 0.07	390.00	694.06 ± 1.40		
290.00	794.26 ± 0.06	400.00	682.92 ± 1.75		

cont.

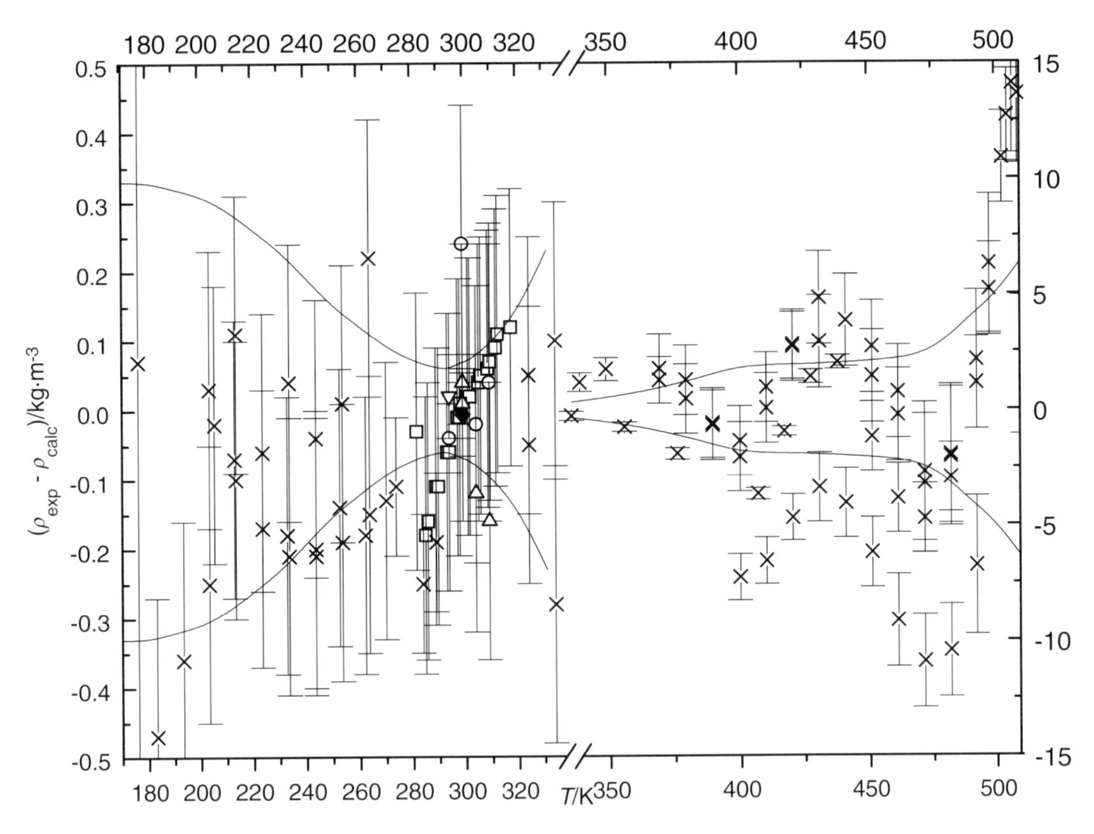

Fig. 1. The symbols show the deviation of the calculated from the experimental values from Table 2. The curves above and below the zero line indicate the calculated error region of the recommended values given in Table 3. The error bars represent the experimental errors. (Error bars smaller than the symbols are omitted for clarity of the figure.)

Ethanol **[64-17-5]** **C_2H_6O** **MW = 46.07** **2**

T_c = 514.10 K [89-tej/lee]
ρ_c = 276.00 kg·m^{-3} [89-tej/lee]

Table 1. Coefficients for the polynomial expansion equations. Standard deviations (see introduction): $\sigma_\ell = 2.7027 \cdot 10^{-1}$ (low temperature range), $\sigma_{c,w} = (1.5405 \cdot 10^{-1}$ combined temperature ranges, weighted), $\sigma_{c,uw} = 1.5778 \cdot 10^{-1}$ (combined temperature ranges, unweighted).

Coefficient	T = 191.15 to 400.00 K $\rho = A + BT + CT^2 + DT^3 + \ldots$	T = 400.00 to 514.10 K $\rho = [1 + 1.75(1 - T/T_c)^{1/3} + 0.75(1 - T/T_c)]$ $[\rho_c + A(T_c - T) + B(T_c - T)^2 + C(T_c - T)^3$ $+ D(T_c - T)^4]$
A	$1.16239 \cdot 10^3$	1.73318
B	−2.25788	$-3.58319 \cdot 10^{-2}$
C	$5.30621 \cdot 10^{-3}$	$3.19777 \cdot 10^{-4}$
D	$-6.63070 \cdot 10^{-6}$	$-1.03287 \cdot 10^{-6}$

cont.

Ethanol (cont.)

Table 2. Experimental values with uncertainties and deviation from calculated values.

$\frac{T}{K}$	$\frac{\rho_{exp} \pm 2\sigma_{est}}{kg \cdot m^{-3}}$	$\frac{\rho_{exp} - \rho_{calc}}{kg \cdot m^{-3}}$	Ref. (Symbol in Fig. 1)
	crystal		
78.15	1021 ± 3.0		30-bilfis-1
	liquid		
288.15	793.75 ± 0.08	0.03	13-osb/mck(×)
298.15	785.03 ± 0.08	−0.12	13-osb/mck[1)]
273.15	806.28 ± 0.08	−0.14	48-kre/now(◆)
298.15	785.04 ± 0.08	−0.11	48-kre/now[1)]
353.15	763.12 ± 0.08	28.38	48-kre/now[1)]
273.15	806.30 ± 0.40	−0.12	58-cos/bow[1)]
293.15	789.40 ± 0.40	−0.05	58-cos/bow[1)]
313.15	772.20 ± 0.40	0.14	58-cos/bow[1)]
333.15	754.10 ± 0.40	0.17	58-cos/bow[1)]
353.15	734.80 ± 0.50	0.06	58-cos/bow[1)]
373.15	715.70 ± 0.50	1.51	58-cos/bow[1)]
393.15	692.50 ± 0.50	0.57	58-cos/bow(×)
413.15	663.10 ± 0.60	−2.49	58-cos/bow(×)
433.15	632.90 ± 0.60	−0.00	58-cos/bow(×)
453.15	598.40 ± 0.60	0.81	58-cos/bow(×)
473.15	556.80 ± 0.70	2.22	58-cos/bow(×)
483.15	518.60 ± 2.00	−7.37	64-ska/kay(×)
488.15	503.40 ± 2.00	−5.29	64-ska/kay(×)
493.15	486.30 ± 2.00	−2.43	64-ska/kay(×)
498.15	466.80 ± 2.00	1.52	64-ska/kay(×)
503.15	443.20 ± 3.00	6.26	64-ska/kay(×)
508.15	411.30 ± 3.00	10.86	64-ska/kay[1)]
513.15	339.50 ± 5.00	1.88	64-ska/kay(×)
293.15	789.20 ± 0.20	−0.25	76-hal/ell[1)]
298.15	784.93 ± 0.20	−0.22	76-hal/ell[1)]
303.15	780.64 ± 0.20	−0.18	76-hal/ell[1)]
320.00	765.83 ± 0.20	−0.12	76-hal/ell(×)
340.00	747.47 ± 0.20	−0.02	76-hal/ell(×)
360.00	727.80 ± 0.20	−0.07	76-hal/ell(×)
380.00	706.17 ± 0.20	−0.60	76-hal/ell(×)
400.00	681.96 ± 0.30	−1.90	76-hal/ell[1)]
420.00	654.27 ± 0.30	−0.49	76-hal/ell(×)
430.00	638.82 ± 0.50	0.60	76-hal/ell(×)
440.00	621.92 ± 0.50	0.76	76-hal/ell(×)
450.00	603.52 ± 0.50	0.09	76-hal/ell(×)
455.00	593.63 ± 0.50	−0.46	76-hal/ell(×)
298.15	785.13 ± 0.02	−0.02	79-kiy/ben(Δ)
288.15	793.50 ± 0.05	−0.22	80-ben/kiy(∇)
293.15	789.24 ± 0.05	−0.21	80-ben/kiy(∇)
298.15	784.96 ± 0.05	−0.19	80-ben/kiy[1)]
303.15	780.67 ± 0.05	−0.15	80-ben/kiy(∇)
308.15	776.38 ± 0.05	−0.08	80-ben/kiy(∇)
223.15	848.91 ± 0.10	−0.18	82-sch/pol(×)
223.17	848.99 ± 0.10	−0.08	82-sch/pol(×)
233.15	840.29 ± 0.10	−0.08	82-sch/pol[1)]
233.15	840.31 ± 0.10	−0.06	82-sch/pol[1)]
233.15	840.39 ± 0.10	0.02	82-sch/pol[1)]
233.15	840.28 ± 0.10	−0.09	82-sch/pol[1)]
243.13	831.74 ± 0.10	−0.06	82-sch/pol[1)]
243.15	831.79 ± 0.10	0.01	82-sch/pol[1)]
243.15	831.71 ± 0.10	−0.07	82-sch/pol[1)]
243.15	831.73 ± 0.10	−0.05	82-sch/pol[1)]
253.14	823.28 ± 0.10	−0.01	82-sch/pol(×)
253.14	823.26 ± 0.10	−0.03	82-sch/pol(×)
253.18	823.25 ± 0.10	−0.01	82-sch/pol(×)
258.15	819.03 ± 0.10	−0.03	82-sch/pol(×)
258.15	819.04 ± 0.10	−0.02	82-sch/pol(×)
263.15	814.83 ± 0.10	−0.01	82-sch/pol(×)
263.15	814.82 ± 0.10	0.02	82-sch/pol(×)
268.15	810.62 ± 0.10	−0.01	82-sch/pol(×)
268.15	810.60 ± 0.10	−0.03	82-sch/pol(×)
273.14	806.30 ± 0.10	−0.12	82-sch/pol[1)]
273.15	806.32 ± 0.10	−0.10	82-sch/pol[1)]
283.15	797.88 ± 0.10	−0.08	82-sch/pol(×)
283.15	797.89 ± 0.10	−0.07	82-sch/pol(×)
293.17	789.28 ± 0.10	−0.15	82-sch/pol[1)]
293.18	789.26 ± 0.10	−0.16	82-sch/pol[1)]
298.15	785.27 ± 0.02	0.12	84-eas/woo(□)
298.15	785.10 ± 0.02	−0.05	86-oga/mur(○)
303.15	780.74 ± 0.15	−0.08	87-pik[1)]
333.15	754.38 ± 0.15	0.45	87-pik(×)
298.15	784.50 ± 0.08	−0.65	87-rat/sin-3[1)]
308.15	776.30 ± 0.08	−0.16	87-rat/sin-3(×)
191.15	878.08 ± 0.20	−0.28	88-sun/sch-1(×)
202.34	867.78 ± 0.20	−0.06	88-sun/sch-1(×)
212.98	858.24 ± 0.20	0.10	88-sun/sch-1(×)
222.18	850.13 ± 0.20	0.19	88-sun/sch-1[1)]
229.95	843.37 ± 0.20	0.23	88-sun/sch-1[1)]
239.35	835.28 ± 0.20	0.25	88-sun/sch-1[1)]
250.85	825.25 ± 0.20	0.02	88-sun/sch-1[1)]
260.49	817.35 ± 0.20	0.27	88-sun/sch-1[1)]

[1)] Not included in Fig. 1.

cont.

Table 2. (cont.)

$\frac{T}{K}$	$\frac{\rho_{exp} \pm 2\sigma_{est}}{kg \cdot m^{-3}}$	$\frac{\rho_{exp} - \rho_{calc}}{kg \cdot m^{-3}}$	Ref. (Symbol in Fig. 1)	$\frac{T}{K}$	$\frac{\rho_{exp} \pm 2\sigma_{est}}{kg \cdot m^{-3}}$	$\frac{\rho_{exp} - \rho_{calc}}{kg \cdot m^{-3}}$	Ref. (Symbol in Fig. 1)
268.58	810.46 ± 0.20	0.19	88-sun/sch-1[1)]	203.15	867.05 ± 0.10	–0.04	91-sun/sch(×)
280.78	800.14 ± 0.20	0.17	88-sun/sch-1[1)]	213.15	858.06 ± 0.10	0.07	91-sun/sch(×)
292.39	790.25 ± 0.20	0.15	88-sun/sch-1[1)]	223.15	849.24 ± 0.10	0.15	91-sun/sch(×)
298.15	785.31 ± 0.20	0.16	88-sun/sch-1[1)]	233.15	840.57 ± 0.10	0.20	91-sun/sch[1)]
303.15	780.93 ± 0.20	0.11	88-sun/sch-1[1)]	243.15	832.00 ± 0.10	0.22	91-sun/sch[1)]
313.17	772.17 ± 0.20	0.13	88-sun/sch-1[1)]	253.15	823.50 ± 0.10	0.22	91-sun/sch(×)
323.21	763.20 ± 0.20	0.15	88-sun/sch-1(×)	263.15	815.04 ± 0.10	0.20	91-sun/sch(×)
333.15	754.15 ± 0.20	0.22	88-sun/sch-1(×)	320.00	766.04 ± 0.20	0.09	96-tak/uem(×)
273.15	806.27 ± 0.20	–0.15	88-sun/ten[1)]	380.00	705.92 ± 0.20	–0.85	96-tak/uem(×)
283.15	797.83 ± 0.20	–0.13	88-sun/ten[1)]	400.00	681.80 ± 0.20	–2.06	96-tak/uem[1)]
293.15	789.34 ± 0.20	–0.11	88-sun/ten[1)]	420.00	654.35 ± 0.30	–0.41	96-tak/uem(×)
303.15	780.81 ± 0.20	–0.01	88-sun/ten[1)]	420.00	654.24 ± 0.20	–0.52	96-tak/uem(×)
313.15	772.23 ± 0.20	0.17	88-sun/ten[1)]	440.00	622.02 ± 0.30	0.86	96-tak/uem(×)
323.15	763.61 ± 0.20	0.50	88-sun/ten(×)	460.00	582.79 ± 0.50	–1.47	96-tak/uem(×)
333.15	754.94 ± 0.20	1.01	88-sun/ten(×)	480.00	535.76 ± 1.00	0.02	96-tak/uem(×)
193.15	876.24 ± 0.10	–0.22	91-sun/sch(×)				

[1)] Not included in Fig. 1.

Further references: [94-gil, 1811-bra, 1864-lan, 1865-men, 1869-dup/pag, 1882-sch-1, 1883-sch-3, 1884-per, 1884-sch-6, 1888-ket, 1890-gar, 1891-jah, 1891-sch/kos, 1892-lan/jah, 1893-eyk, 1897-zec, 1898-kah, 1898-roh, 00-loo, 01-rud, 02-you, 03-car/cop, 04-bru/sch, 04-cri, 04-dun, 05-win, 06-kla/nor, 06-wal-1, 07-che, 07-dun/tho, 07-tim, 08-and, 08-dor/rak, 08-dun/stu, 08-get, 08-ric/mat, 09-hol/sag, 10-daw, 10-dor/pol, 10-pol, 10-tho, 11-del, 11-dor, 11-kai, 12-kor, 12-mal, 12-osb/mck, 12-sch-1, 12-tyr, 12-wad/mer, 12-wre, 13-bri-1, 13-mer, 13-muc, 13-rob/acr, 13-rom, 13-ste, 14-low, 14-mer/tur, 14-tyr, 15-pea, 15-pri, 15-ric/coo, 16-ric/shi, 16-sei/alt, 16-wro/rei, 17-jae, 18-her-2, 19-eyk, 21-bar/bir, 21-bru/cre, 21-rei/hic, 22-her/sch, 22-mck/sim, 22-sch/reg, 22-tro, 23-moe-1, 23-rii, 23-wil/smi, 24-par/sch, 25-gro/kel, 25-nor/ash, 25-pal/con, 25-par/kel-1, 25-per, 25-rak, 25-ric/cha, 26-bar, 26-gol/aar, 26-mat, 26-mun, 26-sch, 27-arb-2, 27-del, 27-krc/wil, 28-llo/bro, 28-mon, 28-par/nel, 29-ber/reu, 29-ham/and, 29-kel-3, 29-smy/sto, 30-bil/fis-1, 30-fro, 30-mon, 31-bea/mcv, 31-fio/gin, 31-lun/bje, 32-ess/cla, 32-ros, 32-sol/mol, 32-swi/zma, 33-har, 33-vos/con, 34-lal, 35-hen, 36-tom, 37-dol/bri, 37-zep, 38-sca/ray, 39-lar/hun, 40-gos-1, 40-mon/qui, 40-tri/ric, 40-was/gra, 42-mul, 43-bru/bog, 45-add, 45-dul, 46-kre/now, 47-rei/dem, 47-sho/pri, 48-jon/bow, 48-vog-2, 48-wei, 48-zas, 49-dre/mar, 49-gri/chu, 49-hat, 49-hat-1, 49-hat-2, 49-kre/wie-1, 49-tsc/ric, 49-tsv/mar, 49-vve/iva, 50-jac, 50-pic/zie, 50-sac/sau, 50-tei/gor, 50-vie, 50-wol/sau, 51-tei/gor, 52-gri, 52-hug/mal, 52-kip/tes, 52-sta/spi, 53-ame/pax, 53-ani, 53-ani-1, 53-bar-5, 53-bar/bro, 53-cha/mou, 53-kha/kud, 53-mck/tar, 53-ots/wil, 54-gri, 54-pur/bow, 54-smi/otv, 54-tha/row, 55-fle/sau, 55-ham/sto, 56-ame/pax, 56-fai/win, 56-kat/new, 58-ano-5, 58-lin/van, 58-muk/gru, 58-mur/van, 58-oga/tak, 58-wag/web, 59-bar/dod, 59-mck/ski, 59-nie/web, 60-cop/fin, 60-fro/shr, 60-oak/web, 61-bel/web, 62-bro/smi, 62-kae/web, 63-aga/men, 63-hov/sea, 63-kud/sus-1, 63-mcc/lai, 64-sca/sat, 67-kom/man, 67-mis/sub, 68-ano, 68-kaw/min, 68-pfl/pop, 68-sin/ben, 69-fin/cop, 70-kat/kon, 70-kon/lya, 70-min/kaw, 70-nak/shi, 70-sus/hol, 71-bra/joh, 71-nag/oht, 72-nie/nov, 72-pol/lu, 72-pol/lu -1, 73-khi/ale, 74-dut/mat, 74-mye/cle, 75-mat/fer, 75-mus/ver, 75-tok, 76-for/ben, 76-kat/nit, 76-nag/oht, 76-wes, 77-gup/han, 77-hwa/rob, 77-tan/yam, 77-tre/ben, 78-nit/fuj, 79-cha/ses-1, 79-dia/tar, 79-jim/paz, 79-tho/nag, 80-arc/bla, 80-edu/boy, 80-fuk/ogi, 80-mar/ric, 80-oza/ooy, 80-pik, 80-yos/tak, 80-yu /ish, 81-kum/pra, 81-oht/koy, 81-sjo/dyh, 81-tas/ara, 81-won/chu, 82-diz/mar, 82-dom/rat, 82-ort, 83-alb/edg, 83-fuk/ogi, 83-pik-1, 83-pik-2, 83-rau/ste, 83-tri, 84-sak/nak, 85-dri/ras, 85-kov/svo, 85-mat/ben, 85-mat/ben-1, 85-nag, 85-nag-3, 85-ogi/ara, 85-ort/paz-1, 86-lep/mat, 86-miy/hay, 86-ort/pen, 86-san/sha, 86-tan/toy, 86-zha/ben-1, 87-ber-1, 87-ber-8, 87-ber-9, 87-kub/tan, 87-mou, 87-oga/mur, 87-ogi/ara, 87-pap/pap, 88-nag-4, 89-ala/sal, 89-kac/rad]

cont.

Ethanol (cont.)

Further references: (cont.)

[89-kat/tan, 89-kat/tan-1, 89-kou/pan, 89-nao/sur, 89-ort/sus, 89-sol/mar, 89-sus/ort, 90-cha/kat, 90-let/sch-1, 90-siv/rao, 91-cab/bel, 91-gar/her, 91-kat/tan, 91-pap/eva, 91-pap/zia, 91-pap/zio, 91-ram/muk, 91-vro/noo, 92-hia/yam, 92-lee/wei, 93-ami/ara, 94-hia/tak-1, 94-kim/lee, 94-pap/pan, 94-pap/pan-1, 94-sol/bar, 95-arc/bla, 95-hia/tak-1, 96-elb, 96-nik/jad, 96-nik/mah, 97-com/fra, 98-ami/ban, 98-ami/pat-1, 98-arc/mar, 98-arc/nar, 98-igl/org, 98-nik/shi, 98-zie].

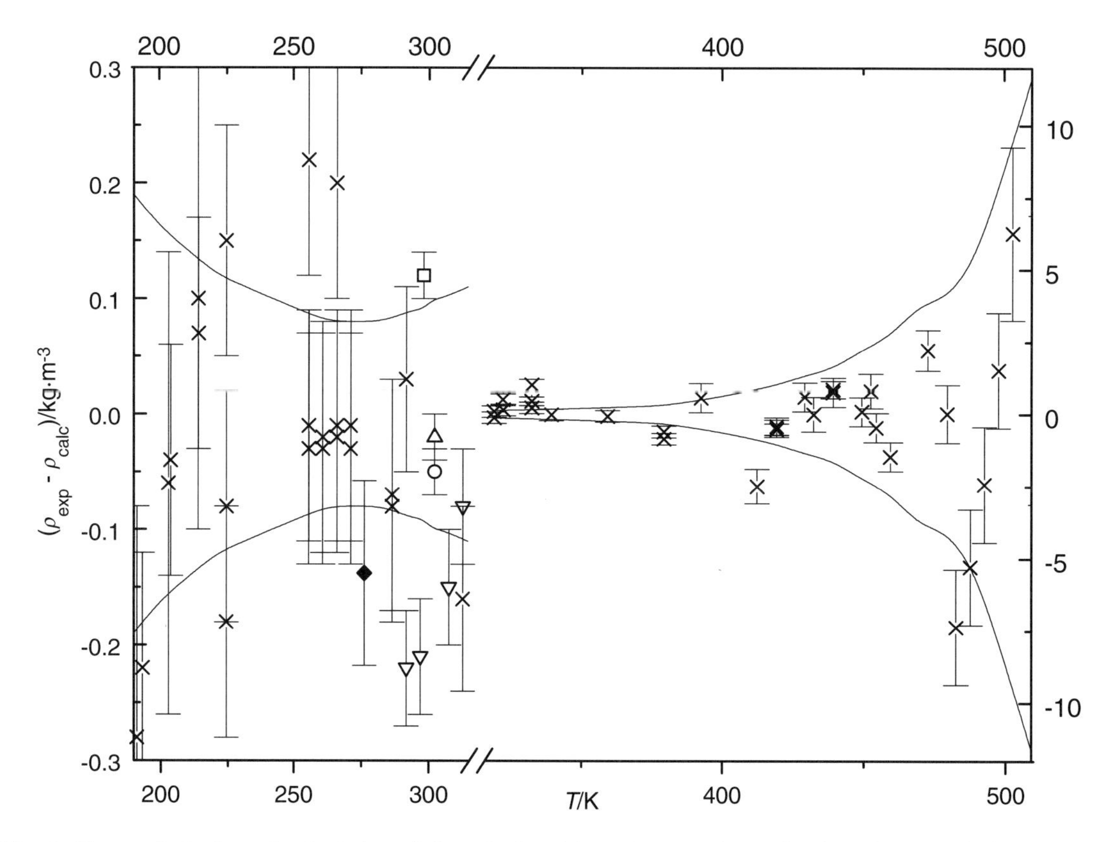

Fig. 1. The symbols show the deviation of the calculated from the experimental values from Table 2. The curves above and below the zero line indicate the calculated error region of the recommended values given in Table 3. The error bars represent the experimental errors. (Error bars smaller than the symbols are omitted for clarity of the figure.)

cont.

Table 3. Recommended values (fit to the reliable experimental values according to the equations $\rho = A + BT + CT^2 + DT^3 + \ldots$ or $\rho = [1 + 1.75(1 - T/T_c)^{1/3} + 0.75(1 - T/T_c)][\rho_c + A(T_c - T) + B(T_c - T)^2 + C(T_c - T)^3 + D(T_c - T)^4]$).

$\frac{T}{\text{K}}$	$\frac{\rho \pm \sigma_{fit}}{\text{kg} \cdot \text{m}^{-3}}$ Z	$\frac{T}{\text{K}}$	$\frac{\rho \pm \sigma_{fit}}{\text{kg} \cdot \text{m}^{-3}}$ Z	$\frac{T}{\text{K}}$	$\frac{\rho \pm \sigma_{fit}}{\text{kg} \cdot \text{m}^{-3}}$ Z
190.00	879.47 ± 0.19	298.15	785.15 ± 0.10	410.00	670.34 ± 0.81
200.00	870.01 ± 0.16	300.00	783.55 ± 0.10	420.00	654.76 ± 1.03
210.00	860.83 ± 0.14	310.00	774.84 ± 0.11	430.00	638.22 ± 1.35
220.00	851.87 ± 0.12	320.00	765.95 ± 0.13	440.00	621.16 ± 1.59
230.00	843.10 ± 0.11	330.00	756.85 ± 0.15	450.00	603.43 ± 2.24
240.00	834.47 ± 0.10	340.00	747.49 ± 0.18	460.00	584.26 ± 2.72
250.00	825.95 ± 0.09	350.00	737.85 ± 0.20	470.00	562.34 ± 3.83
260.00	817.50 ± 0.08	360.00	727.87 ± 0.24	480.00	535.74 ± 4.03
270.00	809.07 ± 0.08	370.00	717.53 ± 0.27	490.00	501.66 ± 5.42
280.00	800.63 ± 0.08	380.00	706.77 ± 0.31	500.00	455.47 ± 8.47
290.00	792.14 ± 0.09	390.00	695.56 ± 0.46	510.00	383.00 ± 11.63
293.15	789.45 ± 0.09	400.00	683.86 ± 0.61		

1-Propanol **[71-23-8]** **C_3H_8O** **MW = 60.1** **3**

T_c = 536.70 K [89-tej/lee]
ρ_c = 275.00 kg·m^{-3} [89-tej/lee]

Table 1. Coefficients for the polynomial expansion equations. Standard deviations (see introduction): $\sigma_\ell = 5.7712 \cdot 10^{-1}$ (low temperature range), $\sigma_{c,w} = (3.4844 \cdot 10^{-1}$ combined temperature ranges, weighted), $\sigma_{c,uw} = 2.0066 \cdot 10^{-1}$ (combined temperature ranges, unweighted).

Coefficient	T = 150.00 to 410.00 K $\rho = A + BT + CT^2 + DT^3 + \ldots$	T = 410.00 to 536.70 K $\rho = [1 + 1.75(1 - T/T_c)^{1/3} + 0.75(1 - T/T_c)]$ $[\rho_c + A(T_c - T) + B(T_c - T)^2 + C(T_c - T)^3 + D(T_c - T)^4]$
A	$1.01077 \cdot 10^3$	1.19032
B	$-3.99649 \cdot 10^{-5}$	$-1.74429 \cdot 10^{-2}$
C	$-6.64923 \cdot 10^{-3}$	$1.13720 \cdot 10^{-4}$
D	$2.16751 \cdot 10^{-5}$	$-2.75663 \cdot 10^{-7}$
E	$-2.46167 \cdot 10^{-8}$	

Table 2. Experimental values with uncertainties and deviation from calculated values.

$\frac{T}{\text{K}}$	$\frac{\rho_{exp} \pm 2\sigma_{est}}{\text{kg} \cdot \text{m}^{-3}}$	$\frac{\rho_{exp} - \rho_{calc}}{\text{kg} \cdot \text{m}^{-3}}$	Ref. (Symbol in Fig. 1)	$\frac{T}{\text{K}}$	$\frac{\rho_{exp} \pm 2\sigma_{est}}{\text{kg} \cdot \text{m}^{-3}}$	$\frac{\rho_{exp} - \rho_{calc}}{\text{kg} \cdot \text{m}^{-3}}$	Ref. (Symbol in Fig. 1)
273.15	819.28 ± 0.10	−0.08	51-kre(×)	328.20	774.64 ± 1.50	−0.54	51-kre[1)]
288.02	807.55 ± 0.10	−0.09	51-kre[1)]	330.62	772.55 ± 1.50	−0.58	51-kre[1)]
298.12	799.50 ± 0.10	−0.15	51-kre[1)]	333.31	770.22 ± 1.50	−0.62	51-kre[1)]
308.05	791.47 ± 0.10	−0.25	51-kre[1)]	335.56	768.25 ± 1.50	−0.66	51-kre[1)]
318.18	783.12 ± 0.10	−0.38	51-kre(×)	338.23	765.89 ± 1.50	−0.71	51-kre[1)]

[1)] Not included in Fig. 1.

cont.

1-Propanol (cont.)

Table 2. (cont.)

T / K	$\rho_{exp} \pm 2\sigma_{est}$ / $kg \cdot m^{-3}$	$\rho_{exp} - \rho_{calc}$ / $kg \cdot m^{-3}$	Ref. (Symbol in Fig. 1)
348.15	756.89 ± 1.50	−0.93	51-kre[1)]
273.15	819.30 ± 0.50	−0.06	58-cos/bow[1)]
293.15	803.50 ± 0.50	−0.09	58-cos/bow[1)]
313.15	787.50 ± 0.50	−0.10	58-cos/bow[1)]
333.15	770.00 ± 0.50	−0.98	58-cos/bow[1)]
353.15	752.00 ± 0.50	−1.25	58-cos/bow[1)]
373.15	732.50 ± 0.50	−1.33	58-cos/bow[1)]
393.15	711.00 ± 0.60	−1.04	58-cos/bow[1)]
413.15	687.50 ± 0.60	0.38	58-cos/bow(×)
433.15	660.00 ± 0.80	0.73	58-cos/bow(×)
453.15	628.50 ± 1.00	−0.43	58-cos/bow(×)
473.15	592.00 ± 1.00	−1.97	58-cos/bow(×)
493.15	548.50 ± 1.00	−0.97	58-cos/bow(×)
273.15	819.47 ± 0.10	0.11	59-mck/ski(×)
293.15	803.50 ± 0.30	−0.09	63-amb/tow[1)]
414.80	687.50 ± 0.50	2.59	63-amb/tow(×)
427.33	670.00 ± 0.50	2.39	63-amb/tow(×)
439.00	653.30 ± 0.50	2.60	63-amb/tow(×)
450.08	635.00 ± 0.50	1.19	63-amb/tow(×)
459.44	620.40 ± 0.50	1.79	63-amb/tow(×)
465.92	607.70 ± 0.50	0.31	63-amb/tow(×)
474.16	592.30 ± 0.50	0.29	63-amb/tow(×)
479.55	581.60 ± 0.50	0.51	63-amb/tow(×)
498.27	537.60 ± 1.00	1.99	63-amb/tow(×)
505.69	516.60 ± 2.00	3.60	63-amb/tow(×)
513.26	499.20 ± 2.00	13.23	63-amb/tow[1)]
518.48	472.40 ± 3.00	8.21	63-amb/tow(×)
523.37	449.90 ± 5.00	9.44	63-amb/tow(×)
527.76	426.70 ± 5.00	11.75	63-amb/tow[1)]
529.84	412.70 ± 5.00	12.07	63-amb/tow[1)]
323.15	779.50 ± 0.20	0.10	63-brz/har[1)]
333.15	771.30 ± 0.20	0.32	63-brz/har[1)]
343.15	763.70 ± 0.20	1.41	63-brz/har(×)
153.15	917.00 ± 0.70	−2.12	67-kom/man[1)]
163.15	908.00 ± 0.70	−2.46	67-kom/man[1)]
173.15	899.60 ± 0.60	−2.21	67-kom/man[1)]
183.15	890.80 ± 0.60	−2.38	67-kom/man[1)]
193.15	882.50 ± 0.60	−2.12	67-kom/man[1)]
203.15	874.50 ± 0.50	−1.64	67-kom/man(×)
213.15	866.80 ± 0.50	−0.96	67-kom/man[1)]
223.15	858.70 ± 0.50	−0.77	67-kom/man(×)
233.15	850.50 ± 0.40	−0.78	67-kom/man(×)
243.15	842.70 ± 0.40	−0.49	67-kom/man(×)
253.15	834.90 ± 0.40	−0.28	67-kom/man(×)
263.15	826.80 ± 0.40	−0.45	67-kom/man[1)]
273.15	819.00 ± 0.30	−0.36	67-kom/man[1)]
283.15	811.10 ± 0.30	−0.38	67-kom/man[1)]
293.15	803.20 ± 0.30	−0.39	67-kom/man[1)]
298.15	799.50 ± 0.30	−0.13	67-kom/man[1)]
303.15	795.96 ± 0.04	0.31	74-rao/nai-1(Δ)
293.15	803.61 ± 0.20	0.02	76-hal/ell[1)]
298.15	799.60 ± 0.20	−0.03	76-hal/ell[1)]
303.15	795.61 ± 0.20	−0.04	76-hal/ell[1)]
320.00	781.71 ± 0.20	−0.29	76-hal/ell[1)]
340.00	764.40 ± 0.20	−0.66	76-hal/ell(×)
360.00	745.71 ± 0.20	−1.11	76-hal/ell(×)
380.00	725.26 ± 0.20	−1.41	76-hal/ell[1)]
400.00	702.59 ± 0.20	−1.31	76-hal/ell[1)]
420.00	677.31 ± 0.30	−0.53	76-hal/ell(×)
440.00	648.88 ± 0.30	−0.33	76-hal/ell(×)
460.00	616.11 ± 0.40	−1.56	76-hal/ell(×)
470.00	597.88 ± 0.50	−2.07	76-hal/ell(×)
483.83	571.43 ± 0.50	−0.40	76-hal/ell(×)
303.15	795.93 ± 0.10	0.28	78-red/nai-1[1)]
313.15	787.62 ± 0.10	0.02	78-red/nai-1(×)
298.15	799.91 ± 0.15	0.28	79-dia/tar[1)]
308.15	791.85 ± 0.15	0.21	79-dia/tar[1)]
318.15	783.59 ± 0.15	0.07	79-dia/tar[1)]
333.15	770.80 ± 0.15	−0.18	79-dia/tar(×)
298.15	799.57 ± 0.02	−0.06	79-kiy/ben(□)
288.15	807.40 ± 0.05	−0.14	80-ben/kiy(◆)
293.15	803.49 ± 0.05	−0.10	80-ben/kiy(◆)
298.15	799.35 ± 0.05	−0.28	80-ben/kiy[1)]
303.15	795.47 ± 0.05	−0.18	80-ben/kiy(◆)
308.15	791.41 ± 0.05	−0.23	80-ben/kiy(◆)
303.15	796.00 ± 0.05	0.35	81-nar/dha(∇)
150.00	922.70 ± 0.18	0.85	82-zak(×)
150.00	922.70 ± 0.18	0.85	82-zak(×)
170.00	903.50 ± 0.18	−1.03	82-zak(×)
170.00	903.60 ± 0.18	−0.93	82-zak(×)
190.00	886.50 ± 0.18	−0.81	82-zak(×)
190.00	886.60 ± 0.18	−0.71	82-zak(×)

[1)] Not included in Fig. 1.

cont.

Table 2. (cont.)

$\frac{T}{K}$	$\frac{\rho_{exp} \pm 2\sigma_{est}}{kg \cdot m^{-3}}$	$\frac{\rho_{exp} - \rho_{calc}}{kg \cdot m^{-3}}$	Ref. (Symbol in Fig. 1)	$\frac{T}{K}$	$\frac{\rho_{exp} \pm 2\sigma_{est}}{kg \cdot m^{-3}}$	$\frac{\rho_{exp} - \rho_{calc}}{kg \cdot m^{-3}}$	Ref. (Symbol in Fig. 1)
210.00	870.60 ± 0.17	0.21	82-zak(×)	350.00	757.60 ± 0.15	1.46	82-zak(×)
210.00	870.10 ± 0.17	–0.29	82-zak(×)	350.00	756.40 ± 0.15	0.26	82-zak(×)
230.00	854.40 ± 0.17	0.55	82-zak(×)	370.00	739.60 ± 0.15	2.57	82-zak[1)]
230.00	854.10 ± 0.17	0.25	82-zak(×)	370.00	736.60 ± 0.15	–0.43	82-zak(×)
250.00	838.50 ± 0.17	0.80	82-zak(×)	283.15	811.45 ± 0.10	–0.03	86-hei/sch(×)
250.00	838.60 ± 0.17	0.90	82-zak(×)	298.15	799.63 ± 0.10	–0.00	86-hei/sch[1)]
270.00	823.00 ± 0.16	1.16	82-zak(×)	313.15	787.46 ± 0.10	–0.14	86-hei/sch(×)
270.00	822.80 ± 0.16	0.96	82-zak(×)	298.15	799.57 ± 0.03	–0.06	86-oga/mur(○)
290.00	807.20 ± 0.16	1.12	82-zak[1)]	293.15	803.78 ± 0.10	0.19	86-wag/hei[1)]
290.00	807.30 ± 0.16	1.22	82-zak[1)]	298.15	799.81 ± 0.10	0.18	86-wag/hei[1)]
310.00	791.40 ± 0.16	1.25	82-zak[1)]	333.15	770.58 ± 0.10	–0.40	86-wag/hei(×)
310.00	791.10 ± 0.16	0.95	82-zak[1)]	303.15	795.61 ± 0.10	–0.04	87-pik[1)]
330.00	774.30 ± 0.15	0.64	82-zak(×)	333.15	770.64 ± 0.10	–0.34	87-pik(×)
330.00	774.80 ± 0.15	1.14	82-zak(×)				

[1)] Not included in Fig. 1.

Further references: [1864-lan, 1871-ros, 1872-lin/von, 1879-bru, 1880-bru-3, 1881-nac/pag, 1882-sch-1, 1882-zan, 1883-sch-3, 1884-per, 1884-sch-6, 1884-zan, 1886-tra, 1889-ram/you, 1890-gar, 1891-jah, 1891-sch/kos, 1892-lan/jah, 1892-sch-1, 1893-eyk-1, 1894-sch, 1898-kah, 1898-lou, 1898-roh, 00-loo, 02-you/for, 05-dun, 08-dor/dvo, 08-ric/mat, 09-dor, 09-dor/roz, 09-hol/sag, 10-dor/pol, 10-dor/pol-1, 10-tim, 11-dor, 12-sch-2, 12-wre, 13-atk/wal, 13-bri-1, 14-eng/tur, 14-kre/mei, 14-low, 16-wro/rei, 17-jae, 18-her-2, 19-eyk, 21-bru/cre, 21-rei/hic, 23-bru, 23-tim, 23-wil/smi, 24-par/sch, 25-nor/ash, 25-pal/con, 25-per, 26-han, 26-mun, 26-par/huf, 26-sch, 27-mou/duf, 28-llo/bro, 29-ber, 30-bil/fis-1, 33-azi/bha, 33-but/tho, 33-kil, 33-tre/wat, 33-vos/con, 34-tim/del, 35-bra/fel, 35-cou/hop, 35-hen, 36-spe, 36-tom, 39-lar/hun, 42-mul, 42-was/bro, 45-add, 48-vog-2, 48-wei, 49-hat, 49-tsc/ric, 49-tsv/mar, 50-jac, 50-mum/phi, 50-pic/zie, 50-sac/sau, 50-tei/gor, 51-dim/lan, 51-tei/gor, 52-coo, 53-ani, 53-ani-1, 53-mcc/jon, 54-pur/bow, 55-bak, 57-mal/mal, 57-rom, 58-lin/van, 58-mur/van, 59-ale, 60-cop/fin, 61-ogi/cor, 62-bro/smi, 62-chu/tho, 62-par/mis, 63-gol/bag, 63-hov/sea, 63-mcc/lai, 63-mik/kim, 63-pra/van, 64-ska/kay, 65-fin/kid, 66-gur/raj, 66-kil/che, 67-gol/per, 68-ano, 68-joh, 68-pfl/pop, 69-bro/foc, 69-fin/cop, 69-smi/kur, 70-gur/raj, 70-kri/kom, 70-str/svo, 72-nie/nov, 72-udo/maz, 73-daw/new, 73-svo/ves, 74-dut/mat, 74-rao/nai, 75-mat/fer, 75-tok, 76-for/ben, 76-kat/nit, 76-kow/kas, 76-nag/oht, 76-red/nai, 76-wes-1, 77-hwa/rob, 77-tre/ben, 78-dap/don, 79-cha/ses-1, 79-ern/gli, 79-jim/paz, 80-arc/bla, 80-fuk/ogi, 80-yos/tak, 81-kum/pra, 81-nai/nai, 81-sjo/dyh, 81-tas/ara, 81-won/chu, 82-diz/mar, 82-kar/red, 82-nai/nai, 82-ort, 82-ven/dha, 83-fuk/ogi, 83-pik-1, 83-pik-2, 83-rau/ste, 83-tri, 84-sak/nak, 85-fer/ber, 85-fer/pin, 85-mat/ben, 85-mat/ben-1, 85-nag-2, 85-ort/paz-1, 85-ped/dav, 85-zhu/dur, 86-ash/sri, 86-ber/wec-1, 86-lep/mat, 86-mah/daw, 86-miy/hay, 86-san/sha, 86-tan/toy, 87-ber-2, 87-ber-3, 87-kub/tan, 87-oga/mur, 87-rat/sin-3, 88-jad/fra, 88-nag-2, 88-ort/mat, 89-ala/sal, 89-nao/sur, 89-ort/sus, 89-pae/con, 89-sus/ort, 90-cha/kat, 90-sri/nai, 91-cab/bel, 91-gar/her, 91-kat/tan, 91-ram/muk, 92-kum/sre, 93-ami/ara, 94-hia/tak, 94-hia/tak-1, 94-kim/lee, 94-pap/pan, 94-pap/pan-1, 94-rom/pel, 94-sin/kal, 94-ven/ven, 95-hia/tak, 95-hia/tak-1, 95-red/ram-1, 96-elb, 96-nik/jad, 96-nik/mah, 97-com/fra, 98-ami/ban, 98-ami/pat-1, 98-nik/shi, 98-pal/sha].

cont.

1-Propanol (cont.)

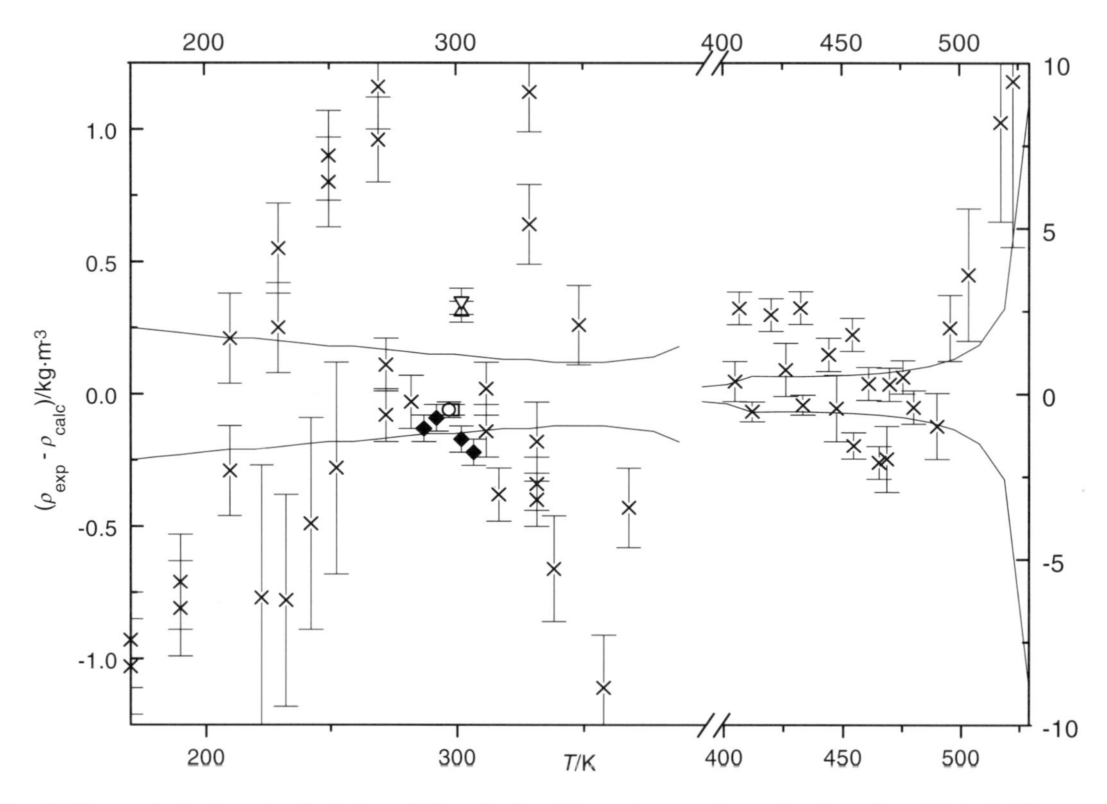

Fig. 1. The symbols show the deviation of the calculated from the experimental values from Table 2. The curves above and below the zero line indicate the calculated error region of the recommended values given in Table 3. The error bars represent the experimental errors. (Error bars smaller than the symbols are omitted for clarity of the figure.)

Table 3. Recommended values (fit to the reliable experimental values according to the equations $\rho = A + BT + CT^2 + DT^3 + \ldots$ or $\rho = [1 + 1.75(1 - T/T_c)^{1/3} + 0.75(1 - T/T_c)][\rho_c + A(T_c - T) + B(T_c - T)^2 + C(T_c - T)^3 + D(T_c - T)^4]$).

$\frac{T}{K}$	$\frac{\rho \pm \sigma_{fit}}{kg \cdot m^{-3}}$	$\frac{T}{K}$	$\frac{\rho \pm \sigma_{fit}}{kg \cdot m^{-3}}$	$\frac{T}{K}$	$\frac{\rho \pm \sigma_{fit}}{kg \cdot m^{-3}}$
150.00	921.85 ± 0.27	290.00	806.08 ± 0.15	410.00	691.28 ± 0.29
160.00	913.19 ± 0.26	293.15	803.59 ± 0.15	420.00	677.84 ± 0.54
170.00	904.53 ± 0.25	298.15	799.63 ± 0.15	430.00	663.81 ± 0.53
180.00	895.89 ± 0.24	300.00	798.16 ± 0.15	440.00	649.21 ± 0.53
190.00	887.31 ± 0.23	310.00	790.15 ± 0.14	450.00	633.93 ± 0.55
200.00	878.81 ± 0.22	320.00	782.00 ± 0.13	460.00	617.67 ± 0.58
210.00	870.39 ± 0.21	330.00	773.66 ± 0.13	470.00	599.95 ± 0.63
220.00	862.07 ± 0.21	340.00	765.06 ± 0.12	480.00	580.14 ± 0.71
230.00	853.85 ± 0.20	350.00	756.14 ± 0.12	490.00	557.41 ± 0.84
240.00	845.73 ± 0.19	360.00	746.82 ± 0.12	500.00	530.62 ± 1.06
250.00	837.70 ± 0.18	370.00	737.03 ± 0.13	510.00	498.18 ± 1.49
260.00	829.74 ± 0.18	380.00	726.67 ± 0.14	520.00	457.21 ± 2.58
270.00	821.84 ± 0.17	390.00	715.66 ± 0.18	530.00	399.44 ± 8.86
280.00	813.96 ± 0.16	400.00	703.90 ± 0.22		

2- Propanol **[67-63-0]** **C_3H_8O** **MW = 60.1** **4**

T_c = 508.00 K [89-tej/lee]
ρ_c = 273.00 kg·m^{-3} [89-tej/lee]

Table 1. Coefficients for the polynomial expansion equations. Standard deviations (see introduction): $\sigma_\ell = 9.7228 \cdot 10^{-1}$ (low temperature range), $\sigma_{c,w} = (9.9994 \cdot 10^{-1}$ combined temperature ranges, weighted), $\sigma_{c,uw} = 5.4125 \cdot 10^{-1}$ (combined temperature ranges, unweighted).

Coefficient	T = 213.15 to 400.00 K $\rho = A + BT + CT^2 + DT^3 + \ldots$	T = 400.00 to 508.00 K $\rho = [1 + 1.75(1 - T/T_c)^{1/3} + 0.75(1 - T/T_c)]$ $[\rho_c + A(T_c - T) + B(T_c - T)^2 + C(T_c - T)^3$ $+ D(T_c - T)^4]$
A	$9.93868 \cdot 10^{2}$	1.42642
B	$-3.40464 \cdot 10^{-5}$	$-2.55851 \cdot 10^{-2}$
C	$-6.95191 \cdot 10^{-3}$	$2.00711 \cdot 10^{-4}$
D	$2.40158 \cdot 10^{-5}$	$-5.71341 \cdot 10^{-7}$
E	$-2.92637 \cdot 10^{-8}$	

Table 2. Experimental values with uncertainties and deviation from calculated values.

$\frac{T}{K}$	$\frac{\rho_{exp} \pm 2\sigma_{est}}{kg \cdot m^{-3}}$	$\frac{\rho_{exp} - \rho_{calc}}{kg \cdot m^{-3}}$	Ref. (Symbol in Fig. 1)	$\frac{T}{K}$	$\frac{\rho_{exp} \pm 2\sigma_{est}}{kg \cdot m^{-3}}$	$\frac{\rho_{exp} - \rho_{calc}}{kg \cdot m^{-3}}$	Ref. (Symbol in Fig. 1)
303.15	778.60 ± 0.45	1.70	33-azi/bha[1)]	443.72	595.70 ± 0.50	−1.33	63-amb/tow(×)
313.15	769.80 ± 0.45	1.59	33-azi/bha[1)]	450.70	581.50 ± 0.50	−2.04	63-amb/tow(×)
323.15	761.70 ± 0.45	2.50	33-azi/bha[1)]	458.45	564.00 ± 0.50	−3.06	63-amb/tow(×)
333.15	750.90 ± 0.45	1.11	33-azi/bha(×)	465.13	548.00 ± 0.50	−3.12	63-amb/tow(×)
297.97	781.04 ± 0.10	−0.25	45-kol/bur(Δ)	469.84	535.70 ± 0.50	−2.93	63-amb/tow(×)
298.15	781.48 ± 0.10	0.34	54-tha/row(∇)	474.89	522.20 ± 1.00	−1.60	63-amb/tow(×)
293.15	785.50 ± 0.20	0.17	56-tor-1[1)]	480.40	505.50 ± 1.00	−0.06	63-amb/tow(×)
313.15	768.40 ± 0.20	0.19	56-tor-1(×)	486.34	485.20 ± 1.00	2.37	63-amb/tow(×)
333.15	749.70 ± 0.20	−0.09	56-tor-1(×)	491.20	465.80 ± 1.00	4.68	63-amb/tow(×)
213.15	849.10 ± 0.50	−1.08	58-cos/bow(×)	495.79	441.80 ± 1.00	4.75	63-amb/tow(×)
233.15	834.10 ± 0.50	0.24	58-cos/bow(×)	500.43	416.50 ± 1.00	9.31	63-amb/tow(×)
253.15	818.30 ± 0.50	0.53	58-cos/bow(×)	504.22	384.90 ± 1.00	10.32	63-amb/tow(×)
273.15	802.70 ± 0.50	0.99	58-cos/bow(×)	506.68	349.30 ± 1.00	7.80	63-amb/tow(×)
293.15	786.20 ± 0.40	0.87	58-cos/bow[1)]	300.00	779.00 ± 1.00	−0.58	64-dan/bah[1)]
313.15	768.80 ± 0.40	0.59	58-cos/bow[1)]	320.00	763.00 ± 1.00	0.92	64-dan/bah[1)]
333.15	750.20 ± 0.50	0.41	58-cos/bow[1)]	340.00	745.00 ± 1.00	1.93	64-dan/bah[1)]
353.15	730.30 ± 0.50	0.88	58-cos/bow(×)	360.00	724.00 ± 1.00	2.15	64-dan/bah[1)]
373.15	708.00 ± 0.60	1.69	58-cos/bow(×)	380.00	702.00 ± 1.00	4.39	64-dan/bah[1)]
393.15	683.00 ± 0.60	3.43	58-cos/bow(×)	400.00	676.00 ± 1.50	6.59	64-dan/bah[1)]
413.15	656.50 ± 0.60	7.69	58-cos/bow(×)	420.00	648.00 ± 1.50	10.24	64-dan/bah[1)]
433.15	626.40 ± 0.60	10.54	58-cos/bow(×)	440.00	619.00 ± 1.50	15.16	64-dan/bah(×)
293.15	785.11 ± 0.10	−0.22	59-bar/dod(○)	460.00	585.00 ± 1.50	21.47	64-dan/bah(×)
407.35	660.50 ± 0.50	2.49	63-amb/tow(×)	480.00	541.00 ± 2.00	34.03	64-dan/bah[1)]
422.33	635.30 ± 0.50	1.34	63-amb/tow(×)	500.00	470.00 ± 2.00	59.70	64-dan/bah[1)]

[1)] Not included in Fig. 1.

cont.

2- Propanol (cont.)

Table 2. (cont.)

$\frac{T}{K}$	$\frac{\rho_{exp} \pm 2\sigma_{est}}{kg \cdot m^{-3}}$	$\frac{\rho_{exp} - \rho_{calc}}{kg \cdot m^{-3}}$	Ref. (Symbol in Fig. 1)	$\frac{T}{K}$	$\frac{\rho_{exp} \pm 2\sigma_{est}}{kg \cdot m^{-3}}$	$\frac{\rho_{exp} - \rho_{calc}}{kg \cdot m^{-3}}$	Ref. (Symbol in Fig. 1)
436.38	610.40 ± 0.50	0.14	63-amb/tow(×)	313.15	768.98 ± 0.20	0.77	66-kat/shi(×)
298.15	780.93 ± 0.20	−0.21	76-hal/ell[1)]	303.15	776.56 ± 0.15	−0.34	83-pik-2(◆)
303.15	776.75 ± 0.20	−0.15	76-hal/ell[1)]	298.15	781.02 ± 0.15	−0.12	83-wec/byl(□)
320.00	761.72 ± 0.20	−0.36	76-hal/ell(×)	298.15	781.01 ± 0.15	−0.13	83-wec/byl(□)
340.00	742.37 ± 0.20	−0.70	76-hal/ell(×)	298.15	781.70 ± 0.30	0.56	87-isl/qua[1)]
360.00	720.76 ± 0.20	−1.09	76-hal/ell(×)	303.15	777.60 ± 0.30	0.70	87-isl/qua[1)]
380.00	696.49 ± 0.20	−1.12	76-hal/ell(×)	308.15	773.50 ± 0.30	0.91	87-isl/qua[1)]
400.00	669.17 ± 0.20	−0.24	76-hal/ell(×)	313.15	769.30 ± 0.30	1.09	87-isl/qua[1)]
420.00	638.28 ± 0.20	0.52	76-hal/ell(×)	318.15	765.20 ± 0.30	1.45	87-isl/qua(×)
430.00	621.15 ± 0.25	−0.07	76-hal/ell(×)	323.15	761.30 ± 0.30	2.10	87-isl/qua(×)
213.15	847.60 ± 0.50	−2.58	76-kat/nit(×)	303.15	776.59 ± 0.40	−0.31	87-pik[1)]
233.15	834.00 ± 0.50	0.14	76-kat/nit(×)	333.15	749.33 ± 0.40	−0.46	87-pik(×)
253.15	817.40 ± 0.50	−0.37	76-kat/nit(×)	278.15	797.26 ± 0.30	−0.40	88-sak(×)
273.15	801.30 ± 0.50	−0.41	76-kat/nit(×)	288.15	798.13 ± 0.30	8.65	88-sak[1)]
275.00	798.80 ± 0.40	−1.41	78-amb/cou-1(×)	298.15	780.80 ± 0.30	−0.34	88-sak[1)]
300.00	779.55 ± 0.40	−0.03	78-amb/cou-1[1)]	308.15	772.20 ± 0.30	−0.39	88-sak[1)]
325.00	757.02 ± 0.40	−0.47	78-amb/cou-1(×)	318.15	763.29 ± 0.30	−0.46	88-sak(×)
350.00	731.84 ± 0.40	−0.95	78-amb/cou-1(×)	293.15	785.10 ± 0.30	−0.23	89-pae/con[1)]
375.00	702.89 ± 0.40	−1.11	78-amb/cou-1(×)	298.15	780.90 ± 0.30	−0.21	89-pae/con[1)]
400.00	669.36 ± 0.40	−0.05	78-amb/cou-1(×)	303.15	776.70 ± 0.30	−0.20	89-pae/con[1)]
425.00	630.00 ± 1.00	0.43	78-amb/cou-1(×)	313.15	767.80 ± 0.30	−0.41	89-pae/con[1)]
450.00	581.99 ± 2.00	−2.96	78-amb/cou-1(×)	323.15	759.10 ± 0.20	−0.10	89-pae/con(×)
475.00	519.76 ± 2.00	−3.70	78-amb/cou-1(×)	298.15	781.23 ± 0.20	0.09	98-nik/shi(×)
500.00	425.22 ± 5.00	14.92	78-amb/cou-1[1)]	303.15	776.95 ± 0.20	0.05	98-nik/shi(×)
508.30	273.17 ± 0.00	5.23	78-amb/cou-1[1)]	308.15	773.46 ± 0.20	0.87	98-nik/shi(×)

[1)] Not included in Fig. 1.

Further references: [1880-bru-1, 1883-sch-3, 21-bru/cre, 21-leb, 23-bru, 25-nor/ash, 25-par/kel, 26-mat, 26-mun, 28-par/kel, 29-kel, 35-but/ram, 38-ols/was, 39-lar/hun, 39-par/moo, 44-ira, 46-kre/now, 50-par/gol, 52-cap/mug, 53-ani-1, 53-par/cha, 56-goe/mcc, 58-ano-5, 58-mur/van, 59-bur/can, 61-ogi/cor, 62-chu/tho, 63-pra/van-1, 65-fin/kid, 66-kat/pra-1, 67-fre, 67-gol/per, 68-ano, 68-ver, 69-kom/kri, 69-ver/lau, 70-ver, 71-nag/oht, 73-nag/oht, 74-pur/pol, 75-tok, 76-nag/oht, 76-sri/kul, 80-edu/boy, 80-kas/izy, 80-yos/tak, 82-kar/red, 82-ven/dha, 83-fuk/ogi, 83-tri, 83-wec, 84-ped/sal, 85-dri/ras, 85-mat/ben, 85-mat/ben-1, 85-nag-1, 85-rao/red, 86-ber/wec, 86-hne/cib, 86-mou/nai, 87-ber-1, 87-ber-4, 88-nag-1, 89-kat/tan, 89-nao/sur, 91-ace/ped-1, 91-kat/tan, 92-ard/say-1, 92-kum/sre, 94-hia/tak, 94-sin/kal, 95-hia/tak, 96-nik/jad, 96-nik/mah].

cont.

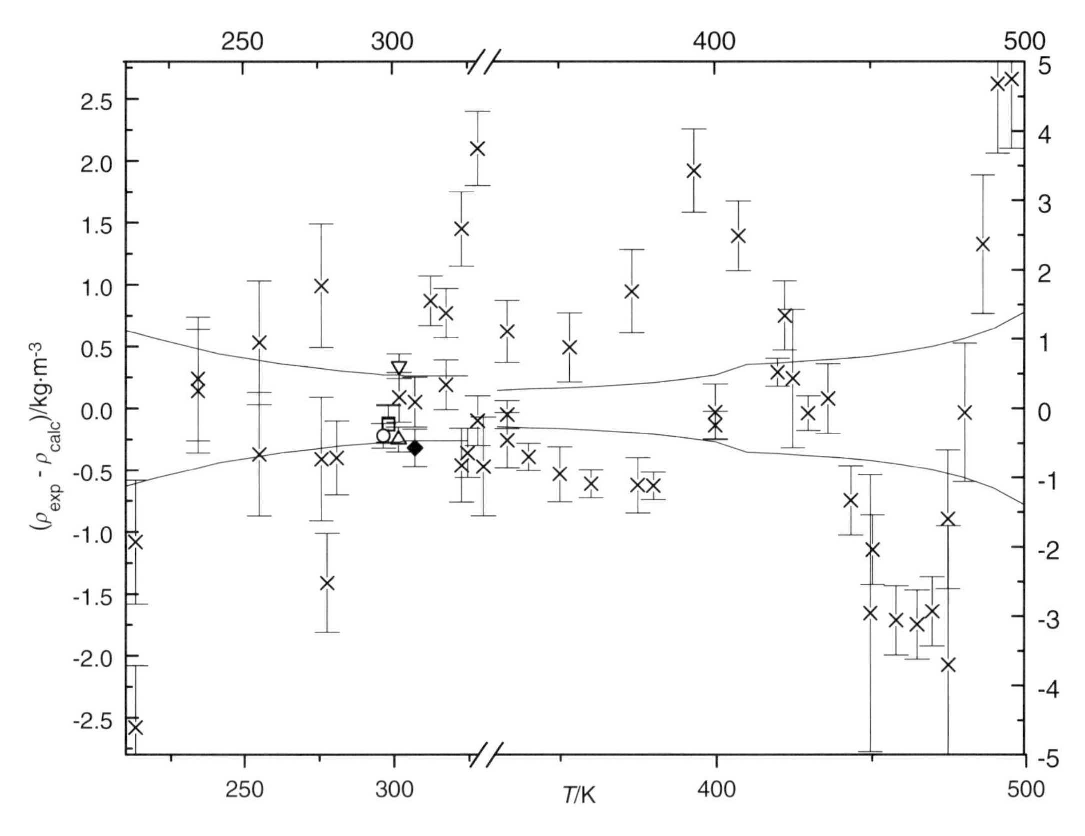

Fig. 1. The symbols show the deviation of the calculated from the experimental values from Table 2. The curves above and below the zero line indicate the calculated error region of the recommended values given in Table 3. The error bars represent the experimental errors. (Error bars smaller than the symbols are omitted for clarity of the figure.)

Table 3. Recommended values (fit to the reliable experimental values according to the equations $\rho = A + BT + CT^2 + DT^3 + \ldots$ or $\rho = [1 + 1.75(1 - T/T_c)^{1/3} + 0.75(1 - T/T_c)][\rho_c + A(T_c - T) + B(T_c - T)^2 + C(T_c - T)^3 + D(T_c - T)^4]$).

$\frac{T}{K}$	$\frac{\rho \pm \sigma_{fit}}{kg \cdot m^{-3}}$	$\frac{T}{K}$	$\frac{\rho \pm \sigma_{fit}}{kg \cdot m^{-3}}$	$\frac{T}{K}$	$\frac{\rho \pm \sigma_{fit}}{kg \cdot m^{-3}}$
210.00	852.78 ± 0.63	300.00	779.58 ± 0.26	410.00	653.83 ± 0.63
220.00	844.56 ± 0.56	310.00	770.98 ± 0.26	420.00	637.76 ± 0.65
230.00	836.41 ± 0.50	320.00	762.08 ± 0.26	430.00	621.22 ± 0.68
240.00	828.33 ± 0.44	330.00	752.80 ± 0.26	440.00	603.84 ± 0.71
250.00	820.30 ± 0.40	340.00	743.07 ± 0.28	450.00	584.95 ± 0.75
260.00	812.28 ± 0.36	350.00	732.79 ± 0.29	460.00	563.53 ± 0.81
270.00	804.25 ± 0.33	360.00	721.85 ± 0.31	470.00	538.18 ± 0.89
280.00	796.15 ± 0.30	370.00	710.16 ± 0.34	480.00	506.97 ± 1.00
290.00	787.95 ± 0.28	380.00	697.61 ± 0.37	490.00	466.79 ± 1.15
293.15	785.33 ± 0.27	390.00	684.06 ± 0.42	500.00	410.30 ± 1.39
298.15	781.14 ± 0.27	400.00	669.41 ± 0.48		

1-Butanol **[71-36-3]** **$C_4H_{10}O$** **MW = 74.12** **5**

T_c = 562.40 K [89-tej/lee]
ρ_c = 270.00 kg·m^{-3} [89-tej/lee]

Table 1. Coefficients for the polynomial expansion equations. Standard deviations (see introduction): σ_ℓ = 3.7574 · 10^{-1} (low temperature range), $\sigma_{c,w}$ = (5.9442 · 10^{-1} combined temperature ranges, weighted), $\sigma_{c,uw}$ = 4.3091 · 10^{-1} (combined temperature ranges, unweighted).

Coefficient	T = 192.35 to 420.00 K $\rho = A + BT + CT^2 + DT^3 + \ldots$	T = 420.00 to 562.40 K $\rho = [1 + 1.75(1 - T/T_c)^{1/3} + 0.75(1 - T/T_c)]$ $[\rho_c + A(T_c - T) + B(T_c - T)^2 + C(T_c - T)^3$ $+ D(T_c - T)^4]$
A	1.15309 · 10^{3}	1.16613
B	−2.13475	−1.80981 · 10^{-2}
C	5.15573 · 10^{-3}	1.25770 · 10^{-4}
D	−6.38112 · 10^{-6}	−3.17257 · 10^{-7}

Table 2. Experimental values with uncertainties and deviation from calculated values.

T/K	$\rho_{exp} \pm 2\sigma_{est}$ / kg·m^{-3}	$\rho_{exp} - \rho_{calc}$ / kg·m^{-3}	Ref. (Symbol in Fig. 1)	T/K	$\rho_{exp} \pm 2\sigma_{est}$ / kg·m^{-3}	$\rho_{exp} - \rho_{calc}$ / kg·m^{-3}	Ref. (Symbol in Fig. 1)
	crystal			197.57	882.41 ± 0.30	−0.95	31-ton/ueh(X)
78.15	1030.0 ± 5.0		30-bilfis-1	204.09	877.29 ± 0.30	−0.62	31-ton/ueh(X)
	liquid			217.71	866.76 ± 0.30	−0.09	31-ton/uch(X)
273.15	824.70 ± 0.20	0.09	28-tim/mar(X)	217.82	866.54 ± 0.30	−0.23	31-ton/ueh(X)
288.15	813.35 ± 0.20	−0.02	28-tim/mar[1)]	227.45	859.32 ± 0.30	0.14	31-ton/ueh(X)
303.15	802.04 ± 0.20	0.07	28-tim/mar[1)]	233.70	854.48 ± 0.20	0.15	31-ton/ueh(X)
193.15	887.20 ± 0.50	0.08	29-smy/sto(X)	234.79	853.49 ± 0.20	−0.00	31-ton/ueh(X)
203.15	879.30 ± 0.50	0.61	29-smy/sto(X)	235.45	853.05 ± 0.20	0.06	31-ton/ueh(X)
213.15	871.30 ± 0.50	0.79	29-smy/sto(X)	240.28	849.36 ± 0.20	0.07	31-ton/ueh(X)
223.15	863.40 ± 0.50	0.85	29-smy/sto[1)]	249.85	841.97 ± 0.20	−0.07	31-ton/ueh(X)
233.15	855.60 ± 0.50	0.84	29-smy/sto[1)]	255.01	838.52 ± 0.20	0.36	31-ton/ueh(X)
243.15	847.90 ± 0.50	0.79	29-smy/sto[1)]	261.87	832.64 ± 0.20	−0.39	31-ton/ueh(X)
253.15	840.20 ± 0.40	0.64	29-smy/sto(X)	273.15	823.80 ± 0.20	−0.81	31-ton/ueh(X)
263.15	832.80 ± 0.40	0.73	29-smy/sto(X)	273.15	824.29 ± 0.20	−0.32	32-ell/rei(X)
273.15	825.00 ± 0.40	0.39	29-smy/sto[1)]	298.15	805.61 ± 0.20	−0.19	32-ell/rei[1)]
283.15	817.30 ± 0.40	0.17	29-smy/sto[1)]	273.15	824.60 ± 0.20	−0.01	39-jon/chr(X)
293.15	809.80 ± 0.40	0.20	29-smy/sto[1)]	298.15	805.70 ± 0.20	−0.10	39-jon/chr[1)]
303.15	802.20 ± 0.40	0.23	29-smy/sto[1)]	463.15	625.10 ± 0.60	−8.32	55-kay/don(X)
313.15	794.50 ± 0.40	0.28	29-smy/sto[1)]	473.15	601.90 ± 0.60	−16.39	55-kay/don(X)
323.15	786.70 ± 0.40	0.40	29-smy/sto[1)]	483.15	595.60 ± 0.60	−6.95	55-kay/don(X)
333.15	778.70 ± 0.40	0.52	29-smy/sto[1)]	493.15	575.90 ± 0.60	−9.92	55-kay/don(X)
343.15	770.30 ± 0.50	0.49	29-smy/sto[1)]	503.15	561.80 ± 0.60	−5.68	55-kay/don(X)
353.15	761.60 ± 0.50	0.45	29-smy/sto[1)]	513.15	542.20 ± 0.70	−4.54	55-kay/don(X)
363.15	752.70 ± 0.50	0.52	29-smy/sto(X)	523.15	519.70 ± 0.70	−2.83	55-kay/don(X)
192.35	886.07 ± 0.30	−1.74	31-ton/ueh(X)	533.15	491.70 ± 0.70	−1.65	55-kay/don(X)

[1)] Not included in Fig. 1.

cont.

Table 2. (cont.)

$\frac{T}{K}$	$\frac{\rho_{exp} \pm 2\sigma_{est}}{kg \cdot m^{-3}}$	$\frac{\rho_{exp} - \rho_{calc}}{kg \cdot m^{-3}}$	Ref. (Symbol in Fig. 1)	$\frac{T}{K}$	$\frac{\rho_{exp} \pm 2\sigma_{est}}{kg \cdot m^{-3}}$	$\frac{\rho_{exp} - \rho_{calc}}{kg \cdot m^{-3}}$	Ref. (Symbol in Fig. 1)
543.15	458.50 ± 0.80	1.70	55-kay/don(X)	548.85	436.50 ± 1.00	5.68	63-amb/tow(X)
553.15	414.60 ± 0.90	7.50	55-kay/don(X)	200.00	876.00 ± 1.50	–5.32	64-dan/bah[1)]
562.89	267.00 ± 0.00	?.00	55-kay/don[1)]	220.00	865.00 ± 1.50	–0.03	64-dan/bah[1)]
293.15	809.90 ± 0.30	0.30	55-kus[1)]	240.00	846.00 ± 1.50	–3.51	64-dan/bah[1)]
298.15	806.10 ± 0.30	0.30	55-kus[1)]	260.00	832.00 ± 1.50	–2.43	64-dan/bah[1)]
303.15	802.20 ± 0.30	0.23	55-kus[1)]	280.00	817.00 ± 1.50	–2.49	64-dan/bah[1)]
313.15	794.50 ± 0.30	0.28	55-kus[1)]	300.00	803.00 ± 1.50	–1.39	64-dan/bah[1)]
323.15	786.80 ± 0.30	0.50	55-kus[1)]	320.00	787.00 ± 1.50	–1.82	64-dan/bah[1)]
333.15	778.90 ± 0.30	0.72	55-kus[1)]	340.00	770.00 ± 1.50	–2.47	64-dan/bah[1)]
343.15	770.70 ± 0.40	0.89	55-kus[1)]	360.00	752.00 ± 1.50	–3.04	64-dan/bah[1)]
353.15	762.20 ± 0.40	1.05	55-kus(X)	380.00	732.00 ± 1.50	–4.23	64-dan/bah[1)]
213.15	867.70 ± 0.50	–2.81	58-cos/bow[1)]	400.00	711.00 ± 2.00	–4.71	64-dan/bah[1)]
233.15	852.80 ± 0.50	–1.96	58-cos/bow[1)]	420.00	690.00 ± 2.00	–3.20	64-dan/bah[1)]
253.15	838.20 ± 0.50	–1.36	58-cos/bow[1)]	440.00	663.00 ± 2.00	–4.01	64-dan/bah[1)]
273.15	823.40 ± 0.50	–1.21	58-cos/bow[1)]	460.00	637.00 ± 2.00	–1.11	64-dan/bah[1)]
293.15	808.60 ± 0.40	–1.00	58-cos/bow[1)]	480.00	610.00 ± 2.00	2.40	64-dan/bah[1)]
313.15	793.60 ± 0.40	–0.62	58-cos/bow[1)]	500.00	577.00 ± 2.00	3.52	64-dan/bah[1)]
333.15	779.00 ± 0.50	0.82	58-cos/bow[1)]	520.00	540.00 ± 2.00	9.39	64-dan/bah(X)
353.15	761.60 ± 0.50	0.45	58-cos/bow[1)]	540.00	485.00 ± 2.50	15.67	64-dan/bah(X)
373.15	743.40 ± 0.50	0.55	58-cos/bow(X)	560.00	380.00 ± 6.00	29.03	64-dan/bah[1)]
393.15	723.70 ± 0.50	0.75	58-cos/bow(X)	513.15	548.30 ± 3.00	1.56	64-ska/kay[1)]
413.15	701.60 ± 0.60	0.45	58-cos/bow[1)]	518.15	537.70 ± 3.00	2.55	64-ska/kay[1)]
433.15	678.70 ± 0.60	2.26	58-cos/bow(X)	523.15	526.20 ± 3.00	3.67	64-ska/kay[1)]
443.15	654.50 ± 0.60	–8.07	58-cos/bow(X)	528.15	514.20 ± 3.00	5.52	64-ska/kay[1)]
303.15	802.00 ± 0.30	0.03	58-lin/van[1)]	533.15	501.30 ± 3.00	7.95	64-ska/kay[1)]
328.15	782.90 ± 0.30	0.63	58-lin/van[1)]	538.15	486.80 ± 3.00	10.58	64-ska/kay[1)]
348.15	767.00 ± 0.40	1.48	58-lin/van(X)	543.15	470.00 ± 3.00	13.20	64-ska/kay[1)]
368.15	748.20 ± 0.40	0.64	58-lin/van(X)	548.15	450.00 ± 3.00	15.71	64-ska/kay(X)
273.15	824.59 ± 0.10	-0.02	59-mck/ski(X)	553.15	426.10 ± 3.00	19.00	64-ska/kay(X)
293.15	809.60 ± 0.30	0.00	63-amb/tow[1)]	558.15	391.60 ± 3.00	21.09	64-ska/kay(X)
440.08	669.40 ± 0.50	2.50	63-amb/tow(X)	273.15	826.00 ± 0.50	1.39	66-efr[1)]
451.72	653.80 ± 0.50	3.54	63-amb/tow(X)	293.15	810.00 ± 0.50	0.40	66-efr[1)]
462.05	640.10 ± 0.50	5.04	63-amb/tow(X)	313.15	795.00 ± 0.50	0.78	66-efr[1)]
469.47	627.90 ± 0.50	3.99	63-amb/tow(X)	333.15	779.00 ± 1.00	0.82	66-efr[1)]
475.77	618.20 ± 0.50	3.96	63-amb/tow(X)	353.15	762.00 ± 1.00	0.85	66-efr[1)]
481.90	608.70 ± 1.00	4.13	63-amb/tow(X)	373.15	743.00 ± 1.00	0.15	66-efr[1)]
489.78	595.00 ± 1.00	3.40	63-amb/tow(X)	393.15	723.00 ± 1.00	0.05	66-efr[1)]
499.72	576.40 ± 1.00	2.40	63-amb/tow(X)	413.15	702.00 ± 1.00	0.85	66-efr[1)]
508.23	559.20 ± 1.00	1.89	63-amb/tow(X)	433.15	676.00 ± 1.50	-0.44	66-efr[1)]
515.59	543.70 ± 1.00	2.50	63-amb/tow(X)	453.15	697.00 ± 1.50	48.82	66-efr[1)]
521.20	531.40 ± 1.00	3.81	63-amb/tow(X)	473.15	617.00 ± 1.50	–1.29	66-efr[1)]
530.10	508.00 ± 1.00	5.10	63-amb/tow(X)	493.15	582.00 ± 2.00	–3.82	66-efr[1)]
534.05	489.60 ± 1.00	–0.82	63-amb/tow(X)	513.15	541.00 ± 2.00	–5.74	66-efr[1)]

[1)] Not included in Fig. 1.

cont.

1-Butanol (cont.)

Table 2. (cont.)

$\frac{T}{K}$	$\frac{\rho_{exp} \pm 2\sigma_{est}}{kg \cdot m^{-3}}$	$\frac{\rho_{exp} - \rho_{calc}}{kg \cdot m^{-3}}$	Ref. (Symbol in Fig. 1)	$\frac{T}{K}$	$\frac{\rho_{exp} \pm 2\sigma_{est}}{kg \cdot m^{-3}}$	$\frac{\rho_{exp} - \rho_{calc}}{kg \cdot m^{-3}}$	Ref. (Symbol in Fig. 1)
533.15	490.00 ± 2.00	–3.35	66-efr(×)	400.00	714.96 ± 0.20	–0.75	76-hal/ell(×)
543.15	457.00 ± 2.00	0.20	66-efr(×)	420.00	692.25 ± 0.20	–0.95	76-hal/ell[1)]
553.15	408.00 ± 3.00	0.90	66-efr(×)	440.00	667.32 ± 0.20	0.31	76-hal/ell(×)
559.15	343.00 ± 3.00	–17.71	66-efr(×)	460.00	639.77 ± 0.25	1.66	76-hal/ell(×)
560.15	314.00 ± 5.00	–35.06	66-efr[1)]	470.00	624.85 ± 0.25	1.75	76-hal/ell(×)
186.15	891.20 ± 0.70	–2.00	67-kom/man[1)]	480.00	608.98 ± 0.30	1.38	76-hal/ell(×)
193.15	885.40 ± 0.70	–1.72	67-kom/man[1)]	490.00	592.03 ± 0.30	0.80	76-hal/ell(×)
203.15	877.50 ± 0.60	–1.19	67-kom/man[1)]	213.15	869.20 ± 0.50	–1.31	76-kat/nit(×)
213.15	869.60 ± 0.60	–0.91	67-kom/man[1)]	233.15	853.90 ± 0.50	–0.86	76-kat/nit[1)]
223.15	862.20 ± 0.50	–0.35	67-kom/man[1)]	253.15	839.00 ± 0.50	–0.56	76-kat/nit[1)]
233.15	854.70 ± 0.50	–0.06	67-kom/man[1)]	273.15	824.30 ± 0.50	–0.31	76-kat/nit[1)]
243.15	847.10 ± 0.50	–0.01	67-kom/man[1)]	303.15	802.09 ± 0.10	0.12	78-red/nai-1[1)]
253.15	839.50 ± 0.40	–0.06	67-kom/man(×)	313.15	794.51 ± 0.10	0.29	78-red/nai-1(×)
263.15	831.80 ± 0.40	–0.27	67-kom/man(×)	298.15	805.84 ± 0.15	0.04	79-dia/tar[1)]
273.15	824.30 ± 0.40	–0.31	67-kom/man[1)]	308.15	798.02 ± 0.15	–0.10	79-dia/tar[1)]
283.15	816.60 ± 0.30	–0.53	67-kom/man[1)]	318.15	790.18 ± 0.15	–0.11	79-dia/tar(×)
293.15	809.10 ± 0.30	–0.50	67-kom/man[1)]	333.15	778.10 ± 0.15	–0.08	79-dia/tar(×)
298.15	805.40 ± 0.30	–0.40	67-kom/man[1)]	298.15	805.83 ± 0.02	0.03	79-kiy/ben(□)
303.15	801.95 ± 0.04	–0.02	74-rao/nai-1(Δ)	303.15	802.01 ± 0.05	0.04	81-nar/dha(◆)
293.15	809.56 ± 0.20	–0.04	76-hal/cll[1)]	283.15	817.02 ± 0.10	–0.11	86-hei/sch(×)
298.15	805.75 ± 0.20	–0.05	76-hal/ell[1)]	298.15	805.76 ± 0.10	–0.04	86-hei/sch[1)]
303.15	801.91 ± 0.20	–0.06	76-hal/ell[1)]	313.15	794.22 ± 0.10	–0.00	86-hei/sch(×)
320.00	788.73 ± 0.20	–0.09	76-hal/ell(×)	298.15	805.74 ± 0.02	–0.06	86-oga/mur(○)
340.00	772.32 ± 0.20	–0.15	76-hal/ell(×)	298.15	805.74 ± 0.05	–0.06	87-oga/mur(∇)
360.00	754.76 ± 0.20	–0.28	76-hal/ell(×)	303.15	801.91 ± 0.15	–0.06	87-pik[1)]
380.00	735.73 ± 0.20	–0.50	76-hal/ell(×)	333.15	778.22 ± 0.15	0.04	87-pik(×)

[1)] Not included in Fig. 1.

Further references: [1864-lan, 1871-lie/ros, 1880-bru-1, 1881-pri/han, 1883-sch-3, 1884-zan, 1886-tra, 1893-eyk-1, 1898-kah, 00-loo, 06-car/fer, 08-dor/dvo, 08-ric/mat, 13-atk/wal, 14-low, 15-pea, 19-eyk, 19-ort/jon, 19-rei/ral, 21-bru/cre, 21-rei/hic, 22-her/sch, 23-pop, 23-wil/smi, 24-ter, 25-nor/ash, 25-pal/con, 25-per, 26-mat, 26-mun, 27-cla/rob, 27-ver/coo, 28-llo/bro, 29-ber, 29-jon, 29-mah/das, 29-pre, 30-bil/fis-1, 31-smy/wal, 31-sto/hul, 32-ern/lit, 33-azi/bha, 33-but/tho, 33-nev/jat, 33-tre/wat, 33-vos/con, 35-bra/fel, 35-bru/fur, 35-but/ram, 35-hen, 35-kef/mcl, 36-tom, 39-all/lin, 39-lar/hun, 41-hus/age, 42-mul, 42-sny/gil, 43-bru/bog, 44-was/str, 45-add, 46-kre/now, 48-con/elv, 48-jon/bow, 48-laz, 48-vog-2, 48-wei, 49-tsc/ric, 49-tsv/mar, 50-mum/phi, 50-pic/zie, 50-sac/sau, 51-sew, 51-tei/gor, 52-coo, 52-dun/was, 52-von, 53-ani, 53-mcc/jon, 53-par/cha, 54-jon/mcc, 54-pur/bow, 54-skr/mur, 55-bou/cle, 55-dan/col, 55-sin/she, 56-rus/ame, 56-tor-1, 57-mur/las, 57-rao/rao, 57-rom, 58-ano-5, 58-hol/len, 58-lin/tua, 59-ale, 59-ell/raz, 60-cop/fin, 60-tje, 61-bel/shu-1, 62-bro/smi, 62-par/mis, 63-gol/bag, 63-hov/sea, 63-man/she, 63-man/she-1, 63-mcc/lai, 63-sub/rao-1, 64-sta/kor, 65-fin/kid, 65-vij/des, 66-vij/des, 67-dei, 67-gol/per, 67-mur/rao, 67-vij/des-1, 68-ano, 69-fin/cop, 69-smi/kur, 69-sub/nag, 70-gal, 70-kat/kon, 70-kri/kom-1, 71-kat/lob, 72-bon/pik, 73-dak/rao, 73-dak/vee, 73-daw/new, 73-khi/ale, 73-svo/ves, 74-dut/mat, 74-rao/nai, 75-kub/tan, 75-mat/fer]

cont.

[76-bul/pro, 76-kow/kas, 76-red/nai, 77-rat/sal, 77-tre/ben, 78-dap/don, 78-sac/pes, 79-cha/ses-1, 79-jim/paz, 79-kum/pra, 79-sah/hay, 80-arc/bla, 80-fuk/ogi, 80-sue/mul, 81-nai/nai, 81-sjo/dyh, 82-ber/rog-1, 82-kar/red, 82-nai/nai, 82-ort, 82-ven/dha, 83-fuk/ogi, 83-pik-1, 83-pik-2, 83-rau/ste, 83-tri, 84-ber/pen, 84-sak/nak, 85-fer/pin, 85-mat/ben, 85-mat/ben-1, 85-ort/paz-1, 85-sar/paz, 86-ash/sri, 86-gat/woo, 86-gou/tom, 86-lep/mat, 86-mah/daw, 86-miy/hay, 86-san/sha, 86-tan/toy, 87-ber-5, 87-ber-7, 87-fer/ber, 87-kri/cho, 87-kub/tan, 87-nag, 87-ogi/ara, 88-fer/lap, 88-nag-2, 89-ala/sal, 89-mac/fra, 89-nao/sur, 89-ort/sus, 89-vij/nai, 90-cha/kat, 90-sri/nai, 90-vij/nai, 91-cab/bel, 91-fen/wan, 91-gar/her, 91-ram/muk, 92-ban/san, 92-gra/san, 92-tan/mur, 93-ami/ara, 93-ami/rai, 93-sus/ort, 94-kim/lee, 94-pap/pan-1, 94-yu /tsa-1, 95-fra/jim, 95-fra/men, 95-red/ram-1, 96-bha/mak, 96-dej/gon-1, 96-dom/rod, 96-elb, 96-gon/ort, 96-nik/jad, 96-nik/mah, 98-ami/ban, 98-ami/pat-1, 98-nik/shi].

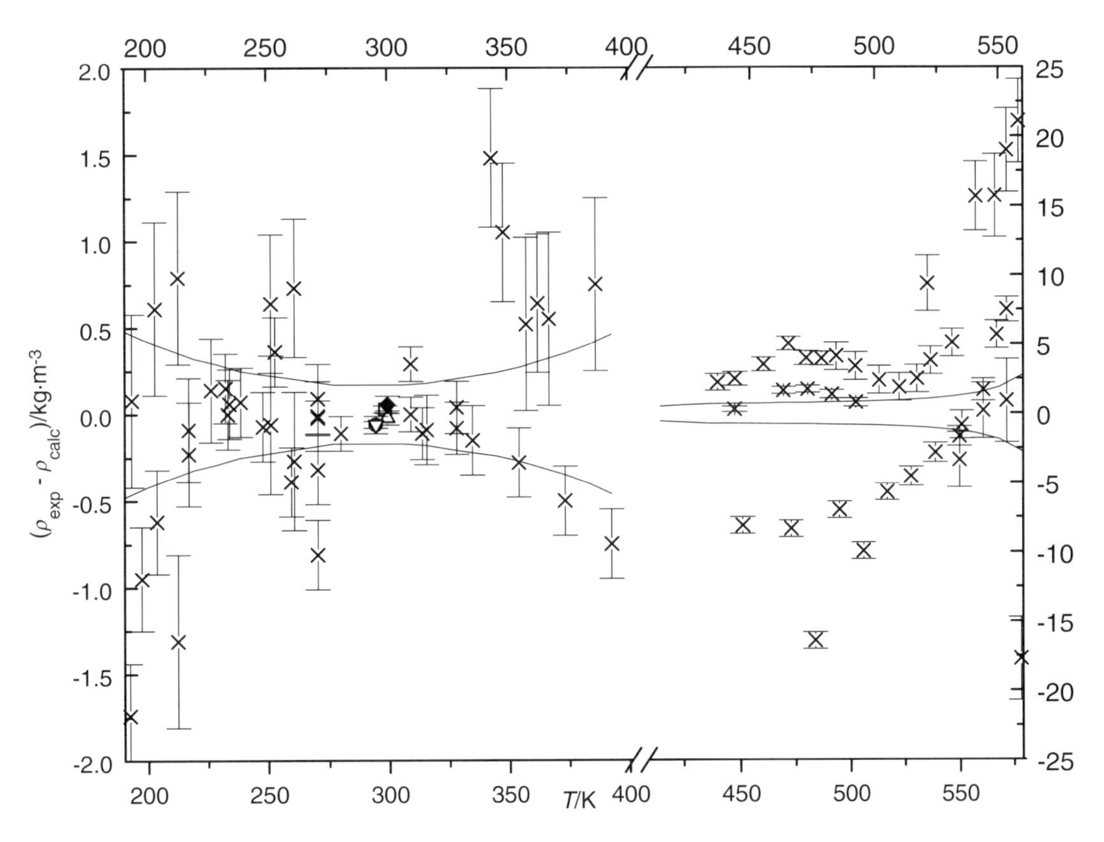

Fig. 1. The symbols show the deviation of the calculated from the experimental values from Table 2. The curves above and below the zero line indicate the calculated error region of the recommended values given in Table 3. The error bars represent the experimental errors. (Error bars smaller than the symbols are omitted for clarity of the figure.)

cont.

1-Butanol (cont.)

Table 3. Recommended values (fit to the reliable experimental values according to the equations $\rho = A + BT + CT^2 + DT^3 + \dots$ or $\rho = [1 + 1.75(1 - T/T_c)^{1/3} + 0.75(1 - T/T_c)][\rho_c + A(T_c - T) + B(T_c - T)^2 + C(T_c - T)^3 + D(T_c - T)^4])$.

$\frac{T}{K}$	$\frac{\rho \pm \sigma_{fit}}{kg \cdot m^{-3}}$	$\frac{T}{K}$	$\frac{\rho \pm \sigma_{fit}}{kg \cdot m^{-3}}$	$\frac{T}{K}$	$\frac{\rho \pm \sigma_{fit}}{kg \cdot m^{-3}}$
190.00	889.84 ± 0.48	310.00	796.68 ± 0.17	450.00	652.75 ± 0.73
200.00	881.32 ± 0.42	320.00	788.82 ± 0.18	460.00	638.11 ± 0.75
210.00	873.06 ± 0.37	330.00	780.76 ± 0.20	470.00	623.10 ± 0.77
220.00	865.03 ± 0.32	340.00	772.47 ± 0.22	480.00	607.60 ± 0.80
230.00	857.19 ± 0.29	350.00	763.91 ± 0.24	490.00	591.23 ± 0.83
240.00	849.51 ± 0.25	360.00	755.04 ± 0.27	500.00	573.48 ± 0.88
250.00	841.93 ± 0.23	370.00	745.83 ± 0.31	510.00	553.60 ± 0.95
260.00	834.43 ± 0.21	380.00	736.23 ± 0.35	520.00	530.61 ± 1.05
270.00	826.96 ± 0.19	390.00	726.20 ± 0.40	530.00	503.20 ± 1.20
280.00	819.49 ± 0.17	400.00	715.71 ± 0.46	540.00	469.33 ± 1.43
290.00	811.98 ± 0.17	410.00	704.73 ± 0.53	550.00	424.91 ± 1.86
293.15	809.60 ± 0.17	420.00	693.20 ± 0.62	560.00	350.97 ± 2.78
298.15	805.80 ± 0.17	430.00	680.65 ± 0.70		
300.00	804.39 ± 0.17	440.00	667.01 ± 0.71		

***d*-2-Butanol** **[4221-99-2]** **$C_4H_{10}O$** **MW = 74.12** **6**

Table 1. Coefficients of the polynomial expansion equation.
Standard deviations (see introduction):
$\sigma_{c,w} = 1.8025 \cdot 10^{-1}$ (combined temperature ranges, weighted),
$\sigma_{c,uw} = 4.7605 \cdot 10^{-2}$ (combined temperature ranges, unweighted).

Coefficient	T = 273.15 to 348.95 K $\rho = A + BT + CT^2 + DT^3 + \dots$
A	$8.60370 \cdot 10^{2}$
B	$4.37482 \cdot 10^{-1}$
C	$-2.11090 \cdot 10^{-3}$

Table 2. Experimental values with uncertainties and deviation from calculated values.

$\frac{T}{K}$	$\frac{\rho_{exp} \pm 2\sigma_{est}}{kg \cdot m^{-3}}$	$\frac{\rho_{exp} - \rho_{calc}}{kg \cdot m^{-3}}$	Ref. (Symbol in Fig. 1)	$\frac{T}{K}$	$\frac{\rho_{exp} \pm 2\sigma_{est}}{kg \cdot m^{-3}}$	$\frac{\rho_{exp} - \rho_{calc}}{kg \cdot m^{-3}}$	Ref. (Symbol in Fig. 1)
285.85	812.90 ± 0.60	−0.04	14-smi(○)	341.35	763.70 ± 0.70	−0.04	14-smi(○)
290.95	808.80 ± 0.60	−0.16	14-smi(○)	342.75	762.20 ± 0.70	−0.13	14-smi(○)
293.95	806.40 ± 0.60	−0.17	14-smi(○)	345.25	759.80 ± 0.70	0.00	14-smi(○)
295.95	804.70 ± 0.60	−0.26	14-smi(○)	347.45	757.50 ± 0.70	−0.04	14-smi(○)
313.45	790.40 ± 0.60	0.30	14-smi(○)	348.95	755.90 ± 0.70	−0.09	14-smi(○)
315.05	788.90 ± 0.60	0.22	14-smi(○)	273.15	822.71 ± 0.50	0.34	28-tim/mar(□)
317.85	786.40 ± 0.60	0.24	14-smi(○)	288.15	810.87 ± 0.50	−0.29	28-tim/mar(□)
320.35	783.90 ± 0.60	0.01	14-smi(○)	303.15	798.96 ± 0.50	−0.04	28-tim/mar(□)
322.15	782.40 ± 0.60	0.17	14-smi(○)				

[1)] Not included in Fig. 1.

cont.

Further references: [48-kor/pat].

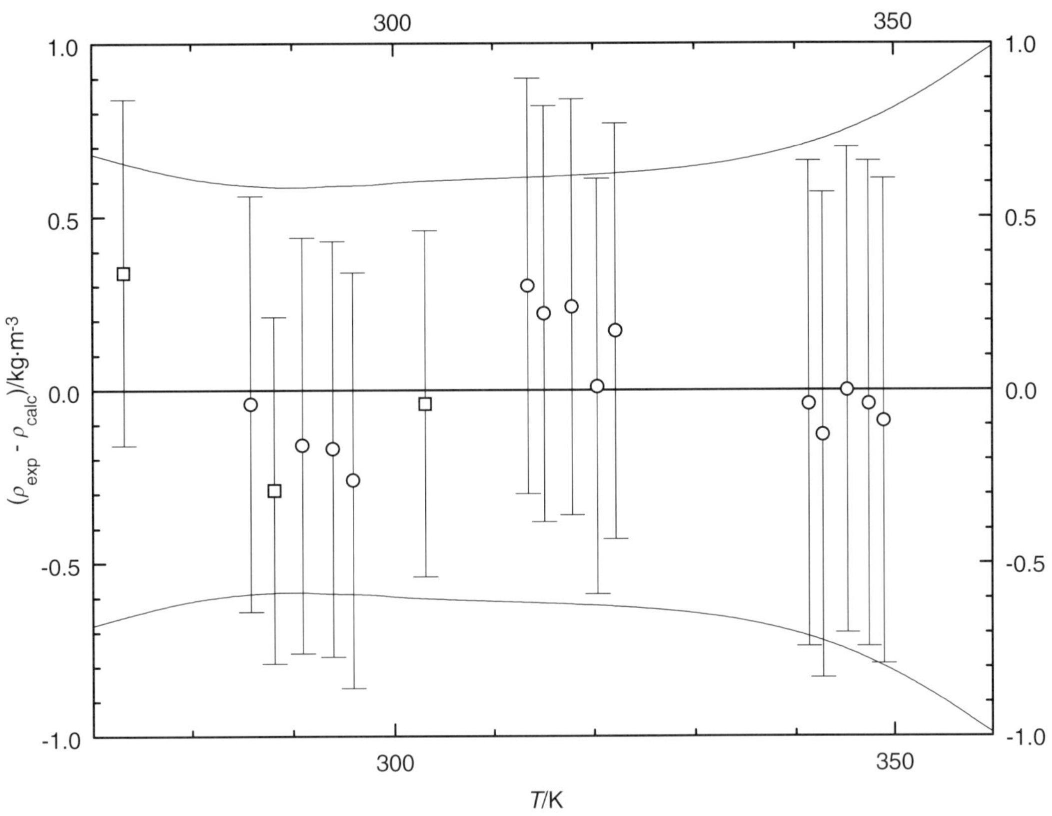

Fig. 1. The symbols show the deviation of the calculated from the experimental values from Table 2. The curves above and below the zero line indicate the calculated error region of the recommended values given in Table 3. The error bars represent the experimental errors. (Error bars smaller than the symbols are omitted for clarity of the figure.)

Table 3. Recommended values (fit to the reliable experimental values according to the equations
$\rho = A + BT + CT^2 + DT^3 + \ldots$ or $\rho = [1 + 1.75(1 - T/T_c)^{1/3} + 0.75(1 - T/T_c)][\rho_c + A(T_c - T) + B(T_c - T)^2 + C(T_c - T)^3 + D(T_c - T)^4]$).

$\frac{T}{K}$	$\frac{\rho \pm \sigma_{fit}}{kg \cdot m^{-3}}$	$\frac{T}{K}$	$\frac{\rho \pm \sigma_{fit}}{kg \cdot m^{-3}}$	$\frac{T}{K}$	$\frac{\rho \pm \sigma_{fit}}{kg \cdot m^{-3}}$
270.00	824.61 ± 0.68	298.15	803.16 ± 0.59	330.00	774.86 ± 0.64
280.00	817.37 ± 0.60	300.00	801.63 ± 0.60	340.00	765.09 ± 0.69
290.00	809.71 ± 0.58	310.00	793.13 ± 0.61	350.00	754.90 ± 0.80
293.15	807.21 ± 0.59	320.00	784.21 ± 0.62	360.00	744.29 ± 0.99

l-2-Butanol **[14898-79-4]** $C_4H_{10}O$ **MW = 74.12** **7**

Table 1. Experimental value with uncertainty.

$\frac{T}{K}$	$\frac{\rho_{exp} \pm 2\sigma_{est}}{kg \cdot m^{-3}}$	Ref.
298.15	804.2 ± 0.5	51-ler/luc

(*RS*)-2-Butanol **[78-92-2]** **$C_4H_{10}O$** **MW = 74.12** **8**

T_c = 536.18 K [89-tej/lee]
ρ_c = 276.00 kg·m^{-3} [89-tej/lee]

Table 1. Coefficients for the polynomial expansion equations. Standard deviations (see introduction): σ_ℓ = 2.1049 (low temperature range), $\sigma_{c,w}$ = (1.9422 combined temperature ranges, weighted), $\sigma_{c,uw} = 8.1750 \cdot 10^{-1}$ (combined temperature ranges, unweighted).

Coefficient	T = 200.00 to 410.00 K $\rho = A + BT + CT^2 + DT^3 + \ldots$	T = 410.00 to 536.18 K $\rho = [1 + 1.75(1 - T/T_c)^{1/3} + 0.75(1 - T/T_c)]$ $[\rho_c + A(T_c - T) + B(T_c - T)^2 + C(T_c - T)^3$ $+ D(T_c - T)^4]$
A	$1.16218 \cdot 10^{3}$	$4.61818 \cdot 10^{-1}$
B	-2.35615	$1.78871 \cdot 10^{-3}$
C	$6.45589 \cdot 10^{-3}$	$-6.92823 \cdot 10^{-5}$
D	$-8.71602 \cdot 10^{-6}$	$3.07819 \cdot 10^{-7}$

Table 2. Experimental values with uncertainties and deviation from calculated values.

T/K	$\rho_{exp} \pm 2\sigma_{est}$ / kg·m^{-3}	$\rho_{exp} - \rho_{calc}$ / kg·m^{-3}	Ref. (Symbol in Fig. 1)	T/K	$\rho_{exp} \pm 2\sigma_{est}$ / kg·m^{-3}	$\rho_{exp} - \rho_{calc}$ / kg·m^{-3}	Ref. (Symbol in Fig. 1)
273.15	822.61 ± 0.20	−0.04	22-tim(◆)	433.15	667.70 ± 0.60	18.32	58-cos/bow(X)
273.15	822.73 ± 0.15	0.08	28-tim/mar(Δ)	453.15	636.50 ± 0.60	16.34	58-cos/bow(X)
288.15	810.86 ± 0.15	0.10	28-tim/mar(Δ)	293.15	812.90 ± 0.30	6.20	63-amb/tow [1)]
303.15	798.93 ± 0.15	0.54	28-tim/mar(Δ)	407.23	694.40 ± 0.50	9.71	63-amb/tow [1)]
293.15	806.50 ± 0.20	−0.20	29-ber(X)	424.02	670.10 ± 0.50	8.62	63-amb/tow(X)
298.15	802.50 ± 0.15	−0.08	45-kol/bur(∇)	444.77	637.70 ± 0.50	4.59	63-amb/tow(X)
293.15	807.30 ± 0.40	0.60	55-kus [1)]	453.55	623.30 ± 0.50	3.79	63-amb/tow(X)
298.15	803.30 ± 0.40	0.72	55-kus [1)]	461.35	609.20 ± 0.50	3.07	63-amb/tow(X)
303.15	799.10 ± 0.40	0.71	55-kus [1)]	473.25	587.10 ± 1.00	4.22	63-amb/tow(X)
313.15	790.60 ± 0.40	0.82	55-kus [1)]	478.56	575.80 ± 1.00	4.53	63-amb/tow(X)
323.15	781.70 ± 0.40	0.87	55-kus(X)	494.70	539.00 ± 2.00	8.35	63-amb/tow(X)
333.15	772.30 ± 0.50	0.82	55-kus [1)]	504.39	513.20 ± 3.00	11.41	63-amb/tow(X)
343.15	762.20 ± 0.50	0.52	55-kus(X)	302.75	798.80 ± 0.40	0.07	63-tho/mea [1)]
353.15	752.10 ± 0.50	0.73	55-kus(X)	318.25	785.70 ± 0.40	0.44	63-tho/mea(X)
213.15	866.80 ± 0.60	−2.07	58-cos/bow(X)	332.35	772.10 ± 0.40	−0.14	63-tho/mea(X)
233.15	851.70 ± 0.60	−1.62	58-cos/bow(X)	347.75	756.70 ± 0.50	−0.30	63-tho/mea(X)
253.15	837.20 ± 0.60	−0.85	58-cos/bow(X)	362.15	740.90 ± 0.50	−0.73	63-tho/mea(X)
273.15	822.10 ± 0.50	−0.55	58-cos/bow [1)]	200.00	875.00 ± 1.50	−4.46	64-dan/bah(X)
293.15	806.00 ± 0.50	−0.70	58-cos/bow [1)]	220.00	862.00 ± 1.50	−1.49	64-dan/bah(X)
313.15	789.40 ± 0.50	−0.38	58-cos/bow [1)]	240.00	846.00 ± 1.50	−2.08	64-dan/bah(X)
333.15	771.40 ± 0.50	−0.08	58-cos/bow [1)]	260.00	830.00 ± 1.50	−2.81	64-dan/bah(X)
353.15	753.90 ± 0.50	2.53	58-cos/bow(X)	280.00	814.00 ± 1.50	−3.27	64-dan/bah [1)]
373.15	735.60 ± 0.60	6.56	58-cos/bow(X)	300.00	798.00 ± 1.50	−3.04	64-dan/bah [1)]
393.15	716.90 ± 0.60	12.83	58-cos/bow [1)]	320.00	781.00 ± 1.50	−2.69	64-dan/bah [1)]
413.15	695.50 ± 0.60	19.33	58-cos/bow(X)	340.00	761.00 ± 1.50	−3.82	64-dan/bah [1)]

[1)] Not included in Fig. 1.

cont.

Table 2. (cont.)

$\frac{T}{K}$	$\frac{\rho_{exp} \pm 2\sigma_{est}}{kg \cdot m^{-3}}$	$\frac{\rho_{exp} - \rho_{calc}}{kg \cdot m^{-3}}$	Ref. (Symbol in Fig. 1)	$\frac{T}{K}$	$\frac{\rho_{exp} \pm 2\sigma_{est}}{kg \cdot m^{-3}}$	$\frac{\rho_{exp} - \rho_{calc}}{kg \cdot m^{-3}}$	Ref. (Symbol in Fig. 1)
360.00	742.00 ± 1.50	−2.00	64-dan/bah[1)]	420.00	667.41 ± 0.20	0.60	76-hal/ell(×)
380.00	717.00 ± 1.50	−3.81	64-dan/bah[1)]	440.00	637.44 ± 0.20	−2.54	76-hal/ell(×)
400.00	693.00 ± 1.50	−1.84	64-dan/bah[1)]	460.00	604.02 ± 0.25	−4.52	76-hal/ell(×)
420.00	667.00 ± 1.70	0.19	64-dan/bah[1)]	483.83	557.48 ± 0.30	−1.44	76-hal/ell(×)
440.00	638.00 ± 1.70	−1.98	64-dan/bah[1)]	490.00	543.63 ± 0.30	0.27	76-hal/ell(×)
460.00	607.00 ± 2.00	−1.54	64-dan/bah[1)]	293.15	807.70 ± 0.50	1.00	81-kor/kov[1)]
480.00	572.00 ± 2.00	4.02	64-dan/bah(×)	293.15	807.80 ± 0.50	1.10	81-kor/kov[1)]
500.00	532.00 ± 2.00	16.66	64-dan/bah(×)	353.15	755.80 ± 0.50	4.43	81-kor/kov(×)
520.00	465.00 ± 3.00	20.36	64-dan/bah(×)	353.15	755.80 ± 0.50	4.43	81-kor/kov(×)
293.15	806.61 ± 0.20	−0.09	76-hal/ell(×)	298.15	802.50 ± 0.30	−0.08	85-ogi/ara[1)]
298.15	802.44 ± 0.20	−0.14	76-hal/ell[1)]	313.15	790.00 ± 0.30	0.22	85-ogi/ara(×)
303.15	798.25 ± 0.20	−0.14	76-hal/ell(×)	293.15	806.66 ± 0.10	−0.04	86-cha/lam(○)
320.00	783.42 ± 0.20	−0.27	76-hal/ell(×)	313.15	789.66 ± 0.10	−0.12	86-cha/lam(○)
340.00	764.17 ± 0.20	−0.65	76-hal/ell(×)	323.15	780.46 ± 0.15	−0.37	86-cha/lam(○)
360.00	742.98 ± 0.20	−1.02	76-hal/ell(×)	333.15	770.78 ± 0.15	−0.70	86-cha/lam(○)
380.00	719.83 ± 0.20	−0.98	76-hal/ell(×)	298.15	802.49 ± 0.06	−0.09	88-oka/oga(□)
400.00	694.75 ± 0.20	−0.09	76-hal/ell(×)				

[1)] Not included in Fig. 1.

Further references: [1869-lie, 06-car/fer, 11-pic/ken, 13-pic/ken, 14-smi, 14-vav, 16-wil/bru, 19-beh, 19-eyk, 21-bru/cre, 21-rei/hic, 23-bru, 23-clo/joh, 25-fai, 25-nor/ash, 26-mun, 33-huc/ack, 33-nev/jat, 35-but/ram, 35-mah-1, 39-all/lin, 39-bar/atk, 39-lar/hun, 42-boe/han-1, 48-wei, 49-ber/ped, 50-pic/zie, 51-ami/wei, 51-bot, 52-coo, 52-dun/was, 56-rus/ame, 62-bro/smi, 63-mcc/lai, 65-fin/kid, 67-gol/per, 68-ano, 68-eva/lin, 69-bro/foc, 78-sac/pes, 80-cha/ses, 80-kas/izy, 80-rig/ube, 83-fuk/ogi, 85-rao/red, 86-kar/cam, 87-kri/cho, 88-nag-2, 89-nao/sur, 92-tan/mur, 94-ben/car, 96-bha/mak, 96-dej/gon-1, 98-fen/cho].

Table 3. Recommended values (fit to the reliable experimental values according to the equations $\rho = A + BT + CT^2 + DT^3 + \ldots$ or $\rho = [1 + 1.75(1 - T/T_c)^{1/3} + 0.75(1 - T/T_c)][\rho_c + A(T_c - T) + B(T_c - T)^2 + C(T_c - T)^3 + D(T_c - T)^4]$).

$\frac{T}{K}$	$\frac{\rho \pm \sigma_{fit}}{kg \cdot m^{-3}}$	$\frac{T}{K}$	$\frac{\rho \pm \sigma_{fit}}{kg \cdot m^{-3}}$	$\frac{T}{K}$	$\frac{\rho \pm \sigma_{fit}}{kg \cdot m^{-3}}$
200.00	879.46 ± 1.58	300.00	801.04 ± 0.30	420.00	666.81 ± 0.60
210.00	871.38 ± 1.33	310.00	792.53 ± 0.29	430.00	653.58 ± 0.62
220.00	863.49 ± 1.11	320.00	783.69 ± 0.29	440.00	639.98 ± 0.64
230.00	855.74 ± 0.92	330.00	774.47 ± 0.29	450.00	625.17 ± 0.66
240.00	848.08 ± 0.76	340.00	764.82 ± 0.31	460.00	608.54 ± 0.70
250.00	840.45 ± 0.63	350.00	754.68 ± 0.33	470.00	589.60 ± 0.74
260.00	832.81 ± 0.52	360.00	744.00 ± 0.35	480.00	567.98 ± 0.81
270.00	825.10 ± 0.44	370.00	732.73 ± 0.38	490.00	543.36 ± 0.90
280.00	817.27 ± 0.38	380.00	720.81 ± 0.41	500.00	515.34 ± 1.05
290.00	809.27 ± 0.33	390.00	708.20 ± 0.44	510.00	483.12 ± 1.31
293.15	806.70 ± 0.32	400.00	694.84 ± 0.47	520.00	444.64 ± 1.82
298.15	802.58 ± 0.31	410.00	680.68 ± 0.51	530.00	391.57 ± 3.20

cont.

(*RS*)-2-Butanol (cont.)

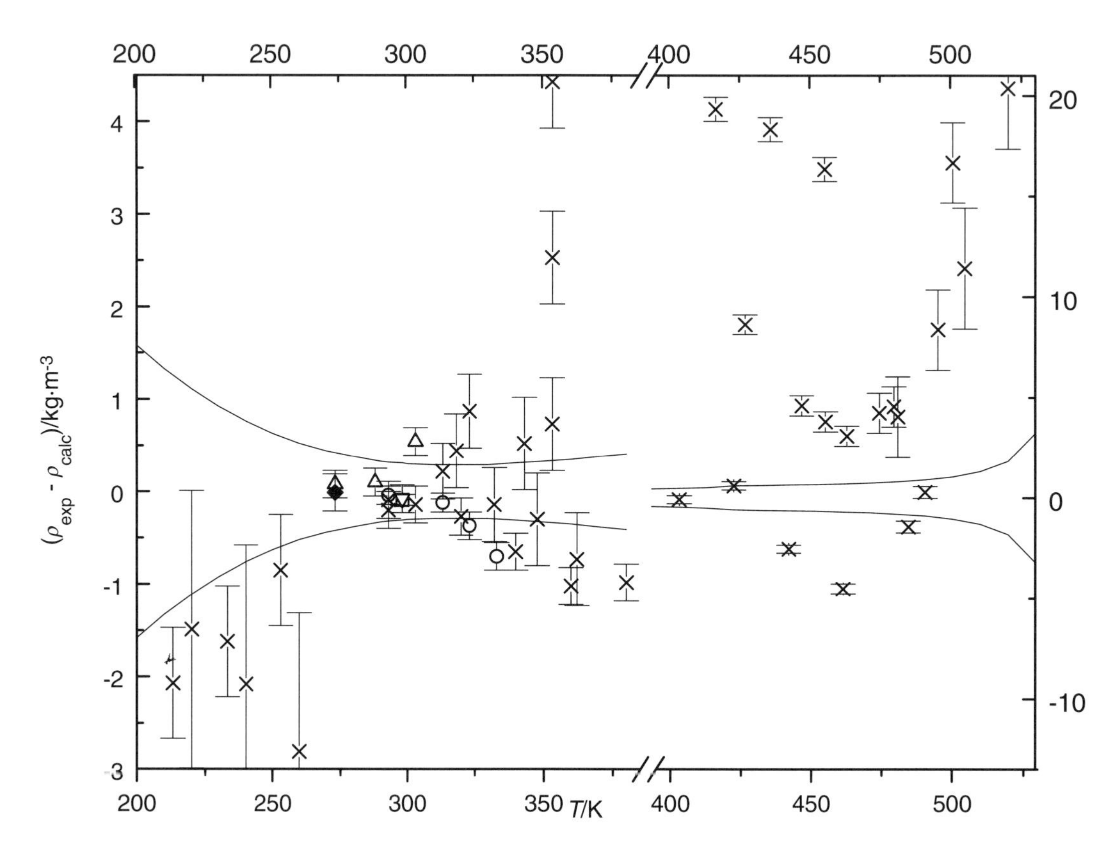

Fig. 1. The symbols show the deviation of the calculated from the experimental values from Table 2. The curves above and below the zero line indicate the calculated error region of the recommended values given in Table 3. The error bars represent the experimental errors. (Error bars smaller than the symbols are omitted for clarity of the figure.)

2-Methyl-1-propanol **[78-83-1]** **$C_4H_{10}O$** **MW = 74.12** **9**

T_c = 547.74 K [63-amb/bro]
ρ_c = 272.20 kg·m^{-3} [63-amb/bro]

Table 1. Coefficients for the polynomial expansion equations. Standard deviations (see introduction): σ_ℓ = 1.9234 (low temperature range), $\sigma_{c,w}$ = (1.4721 combined temperature ranges, weighted), $\sigma_{c,uw}$ = 8.2141 · 10^{-1} (combined temperature ranges, unweighted).

Coefficient	T = 213.15 to 440.00 K $\rho = A + BT + CT^2 + DT^3 + \ldots$	T = 440.00 to 547.74 K $\rho = [1 + 1.75(1 - T/T_c)^{1/3} + 0.75(1 - T/T_c)]$ $[\rho_c + A(T_c - T) + B(T_c - T)^2 + C(T_c - T)^3$ $+ D(T_c - T)^4]$
A	1.16796 · 10^3	8.96238 · 10^{-1}
B	−2.41227	−1.28076 · 10^{-2}
C	6.24427 · 10^{-3}	9.68824 · 10^{-5}
D	−7.77901 · 10^{-6}	−2.95139 · 10^{-7}

cont.

Table 2. Experimental values with uncertainties and deviation from calculated values.

$\frac{T}{K}$	$\frac{\rho_{exp} \pm 2\sigma_{est}}{kg \cdot m^{-3}}$	$\frac{\rho_{exp} - \rho_{calc}}{kg \cdot m^{-3}}$	Ref. (Symbol in Fig. 1)	$\frac{T}{K}$	$\frac{\rho_{exp} \pm 2\sigma_{est}}{kg \cdot m^{-3}}$	$\frac{\rho_{exp} - \rho_{calc}}{kg \cdot m^{-3}}$	Ref. (Symbol in Fig. 1)
273.15	816.22 ± 0.70	−0.18	1881-ber[1)]	273.15	819.60 ± 0.60	3.20	22-mic(×)
280.71	810.50 ± 0.70	−0.28	1881-ber[1)]	273.15	817.00 ± 0.60	0.60	26-han(×)
281.64	809.80 ± 0.70	−0.29	1881-ber[1)]	380.15	726.00 ± 0.60	0.04	27-arb-2(×)
287.65	805.20 ± 0.70	−0.39	1881-ber[1)]	380.15	726.00 ± 0.60	0.04	27-arb-2(×)
290.95	802.70 ± 0.70	−0.40	1881-ber[1)]	273.15	817.03 ± 0.60	0.63	28-tim/mar(×)
292.15	801.80 ± 0.70	−0.40	1881-ber[1)]	288.15	805.74 ± 0.60	0.53	28-tim/mar[1)]
292.25	801.70 ± 0.70	−0.42	1881-ber[1)]	303.15	794.35 ± 0.60	0.54	28–tim/mar[1)]
303.86	792.70 ± 0.70	−0.56	1881-ber[1)]	303.15	795.00 ± 0.45	1.19	33-azi/bha[1)]
304.40	792.30 ± 0.70	−0.54	1881-ber[1)]	313.15	788.90 ± 0.45	2.89	33-azi/bha[1)]
310.44	787.20 ± 0.70	−0.94	1881-ber[1)]	323.15	778.80 ± 0.45	0.81	33-azi/bha(×)
319.71	780.00 ± 0.70	−0.78	1881-ber[1)]	333.15	770.70 ± 0.45	0.98	33-azi/bha(×)
319.86	779.90 ± 0.70	−0.75	1881-ber[1)]	348.15	754.50 ± 0.45	−2.22	33-azi/bha(×)
333.85	768.00 ± 0.70	−1.13	1881-ber[1)]	273.15	817.20 ± 0.30	0.80	45-alb/was(◆)
342.12	760.80 ± 1.00	−1.24	1881-ber[1)]	298.15	798.09 ± 0.30	0.45	45-alb/was[1)]
354.01	749.70 ± 1.00	−1.72	1881-ber[1)]	293.15	802.10 ± 0.60	0.66	48-vog-2[1)]
372.43	731.80 ± 1.00	−2.02	1881-ber(×)	315.85	785.20 ± 0.60	1.34	48-vog-2[1)]
372.63	731.50 ± 1.00	−2.12	1881-ber(×)	334.65	769.90 ± 0.60	1.45	48-vog-2(×)
375.12	729.50 ± 1.00	−1.61	1881-ber(×)	358.15	750.10 ± 0.60	2.51	48-vog-2(×)
379.75	726.50 ± 1.00	0.12	1883-sch-3(×)	493.15	555.50 ± 0.50	−2.98	55-kay/don(×)
379.55	726.50 ± 1.00	−0.08	1884-sch-6(×)	503.15	533.80 ± 0.50	−1.35	55-kay/don(×)
283.15	809.90 ± 0.50	0.94	1890-gar[1)]	513.15	503.20 ± 0.50	−5.19	55-kay/don(×)
293.15	801.90 ± 0.50	0.46	1890-gar[1)]	523.15	478.90 ± 0.60	2.36	55-kay/don(×)
303.15	794.00 ± 0.50	0.19	1890-gar[1)]	533.15	441.20 ± 2.00	4.89	55-kay/don(×)
313.15	786.50 ± 0.50	0.49	1890-gar[1)]	543.15	283.00 ± 3.00	−92.91	55-kay/don[1)]
323.15	779.20 ± 0.50	1.21	1890-gar(×)	547.74	269.00 ± 5.00	−3.20	55-kay/don(×)
273.15	816.96 ± 0.60	0.56	02-you/for(×)	293.15	802.30 ± 0.50	0.86	55-kus[1)]
289.50	804.59 ± 0.60	0.39	02-you/for[1)]	298.15	798.30 ± 0.50	0.66	55-kus[1)]
289.40	806.78 ± 0.50	2.51	12-sch-3[1)]	303.15	794.30 ± 0.50	0.49	55-kus[1)]
304.95	794.21 ± 0.50	1.79	12-sch-3[1)]	313.15	786.30 ± 0.50	0.29	55-kus[1)]
326.15	776.49 ± 0.50	0.95	12-sch-3(×)	323.15	778.50 ± 0.50	0.51	55-kus(×)
347.80	758.27 ± 0.70	1.24	12-sch-3(×)	333.15	770.20 ± 0.50	0.48	55-kus(×)
289.40	806.78 ± 0.50	2.51	15-sch-1[1)]	343.15	761.60 ± 0.50	0.46	55-kus(×)
304.95	794.21 ± 0.50	1.79	15-sch-1[1)]	353.15	752.70 ± 0.50	0.49	55-kus(×)
326.15	776.49 ± 0.50	0.95	15-sch-1(×)	293.15	801.60 ± 0.50	0.16	56-tor-1[1)]
347.80	758.27 ± 0.70	1.24	15-sch-1(×)	313.15	786.20 ± 0.50	0.19	56-tor-1[1)]
201.65	885.00 ± 1.50	13.35	17-jae[1)]	333.15	769.60 ± 0.50	−0.12	56-tor-1(×)
261.15	828.00 ± 1.50	2.70	17-jae(×)	213.15	861.30 ± 0.60	−0.85	58-cos/bow(×)
273.45	817.00 ± 1.50	0.82	17-jae[1)]	233.15	846.40 ± 0.60	0.02	58-cos/bow(×)
283.55	807.00 ± 1.50	−1.66	17-jae[1)]	253.15	831.40 ± 0.60	0.14	58-cos/bow(×)
298.25	794.00 ± 1.50	−3.57	17-jae[1)]	273.15	816.70 ± 0.50	0.30	58-cos/bow(×)
308.25	785.00 ± 1.50	−4.85	17-jae[1)]	293.15	802.10 ± 0.50	0.66	58-cos/bow[1)]
322.85	771.00 ± 1.50	−7.24	17-jae[1)]	313.15	785.80 ± 0.50	−0.21	58-cos/bow[1)]
342.75	753.00 ± 1.50	−8.49	17-jae[1)]	333.15	769.20 ± 0.50	−0.52	58-cos/bow(×)
374.15	731.00 ± 1.50	−1.09	17-jae[1)]	353.15	751.00 ± 0.50	−1.21	58-cos/bow(×)

[1)] Not included in Fig. 1.

cont.

2-Methyl-1-propanol (cont.)

Table 2. (cont.)

$\frac{T}{K}$	$\frac{\rho_{exp} \pm 2\sigma_{est}}{kg \cdot m^{-3}}$	$\frac{\rho_{exp} - \rho_{calc}}{kg \cdot m^{-3}}$	Ref. (Symbol in Fig. 1)	$\frac{T}{K}$	$\frac{\rho_{exp} \pm 2\sigma_{est}}{kg \cdot m^{-3}}$	$\frac{\rho_{exp} - \rho_{calc}}{kg \cdot m^{-3}}$	Ref. (Symbol in Fig. 1)
373.15	731.30 ± 0.60	−1.80	58-cos/bow(×)	280.00	810.00 ± 1.50	−1.31	64-dan/bah[1)]
393.15	711.50 ± 0.60	−0.52	58-cos/bow(×)	300.00	795.00 ± 1.50	−1.23	64-dan/bah[1)]
413.15	689.00 ± 0.60	0.41	58-cos/bow(×)	320.00	778.00 ± 1.50	−2.54	64-dan/bah[1)]
433.15	664.30 ± 0.60	1.85	58-cos/bow(×)	340.00	762.00 ± 1.50	−1.88	64-dan/bah[1)]
293.15	802.20 ± 0.40	0.76	63-amb/tow[1)]	360.00	743.00 ± 1.50	−2.86	64-dan/bah[1)]
421.16	667.50 ± 0.50	−10.97	63-amb/tow[1)]	380.00	722.00 ± 1.50	−4.12	64-dan/bah[1)]
434.40	659.40 ± 0.50	−1.32	63-amb/tow(×)	400.00	700.00 ± 1.70	−4.28	64-dan/bah(×)
454.31	629.70 ± 0.50	−1.22	63-amb/tow(×)	420.00	676.00 ± 1.70	−3.96	64-dan/bah[1)]
459.55	621.10 ± 0.50	−1.21	63-amb/tow(×)	440.00	650.00 ± 2.00	−2.80	64-dan/bah[1)]
467.04	611.10 ± 0.50	1.65	63-amb/tow(×)	460.00	620.00 ± 2.00	−1.55	64-dan/bah[1)]
476.88	594.30 ± 1.00	2.78	63-amb/tow(×)	480.00	590.00 ± 2.00	4.46	64-dan/bah(×)
487.09	574.60 ± 1.00	3.23	63-amb/tow(×)	500.00	553.00 ± 2.00	10.18	64-dan/bah(×)
497.04	553.90 ± 1.00	4.16	63-amb/tow(×)	520.00	510.00 ± 2.50	22.74	64-dan/bah(×)
502.46	541.60 ± 1.00	4.74	63-amb/tow(×)	540.00	440.00 ± 6.00	40.84	64-dan/bah(×)
509.32	523.90 ± 1.00	4.77	63-amb/tow(×)	298.15	794.35 ± 0.15	−3.29	80-kas/izy(Δ)
515.30	508.80 ± 1.00	6.76	63-amb/tow(×)	303.15	794.39 ± 0.20	0.58	86-mou/nai(∇)
521.85	488.80 ± 2.00	7.75	63-amb/tow(×)	298.15	797.87 ± 0.06	0.23	88-oka/oga(□)
526.06	474.30 ± 3.00	8.38	63-amb/tow(×)	298.15	797.81 ± 0.10	0.17	96-dom/rod(○)
535.44	431.30 ± 5.00	6.14	63-amb/tow(×)	298.15	798.30 ± 0.40	0.66	98-nik/shi[1)]
200.00	868.00 ± 1.50	−5.04	64-dan/bah[1)]	303.15	794.31 ± 0.40	0.50	98-nik/shi[1)]
220.00	855.00 ± 1.50	−1.65	64-dan/bah(×)	308.15	790.26 ± 0.40	0.33	98-nik/shi[1)]
240.00	840.00 ± 1.50	−1.15	64-dan/bah(×)	313.15	786.23 ± 0.40	0.22	98-nik/shi(×)
260.00	825.00 ± 1.50	−1.16	64-dan/bah(×)				

[1)] Not included in Fig. 1.

Further references: [1872-lin/von-2, 1872-pie/puc, 1884-gla, 1884-per, 1891-jah, 1891-sch/kos, 1892-lan/jah, 1894-jah/mol, 1898-kah, 08-dor/dvo, 11-dor, 13-rom, 14-eng/tur, 14-gas, 14-wor, 15-pea, 15-ric/coo, 16-wro/rei, 19-eyk, 21-bru/cre, 21-rei/hic, 23-pop, 23-wil/smi, 25-nor/ash, 26-mat, 26-mun, 31-sto/hul, 32-byl, 33-huc/ack, 33-tre/wat, 33-vos/con, 34-smi, 35-but/ram, 35-mah-1, 39-all/lin, 39-lar/hun, 42-sny/gil, 43-boh, 46-fri/sto, 47-sho/pri, 48-wei, 50-cro/van, 50-mum/phi, 50-pic/zie, 51-hau, 51-tei/gor, 52-coo, 52-dun/was, 53-ani, 53-ani-1, 53-hag/dec, 53-par/cha, 55-dan/col, 56-rus/ame, 58-ano-5, 62-bro/smi, 63-mcc/lai, 67-gol/per, 67-nat/rao, 68-ano, 69-bro/foc, 69-smi/kur, 70-sus/hol, 82-ber/rog-1, 82-kar/red, 82-ven/dha, 83-fuk/ogi, 83-lin, 84-ber/pen, 85-mat/ben, 85-mat/ben-1, 85-rao/red, 87-ber-1, 87-kri/cho, 88-nag, 88-nag-1, 90-mal/rao, 92-kum/sre, 93-ami/rai, 95-red/ram-1, 96-bha/mak, 96-nik/jad, 96-nik/mah, 98-art/dom, 98-sen].

cont.

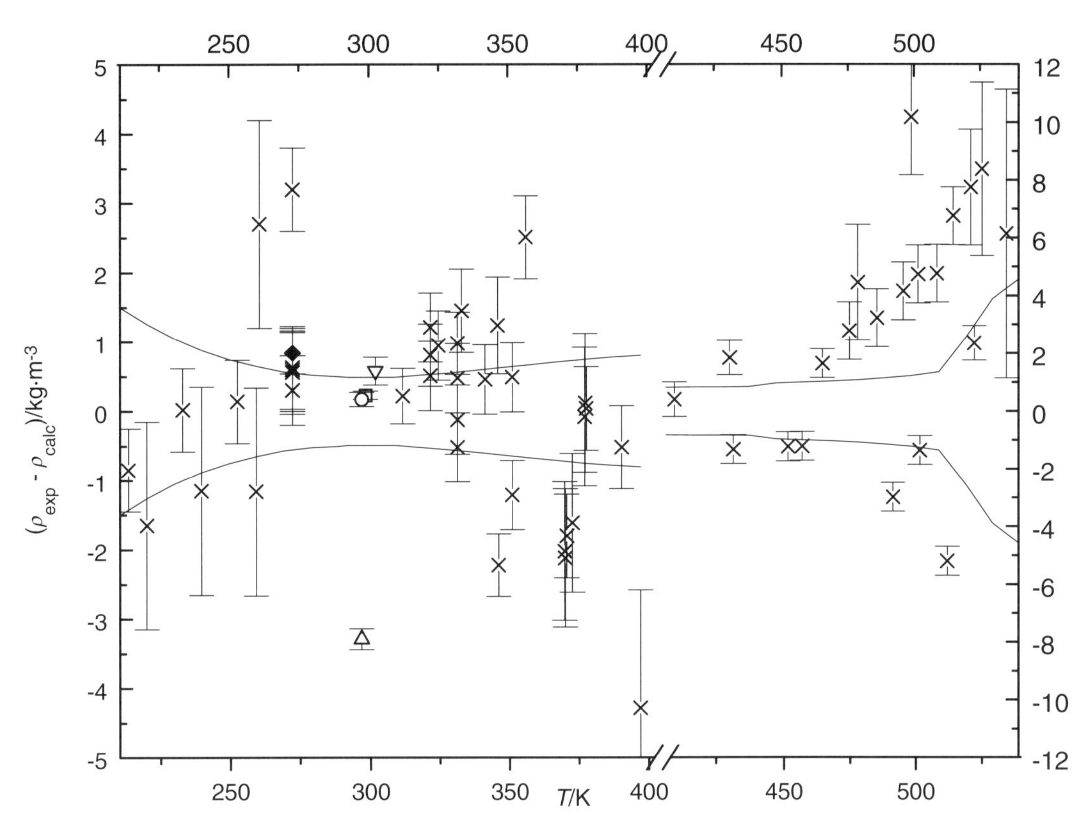

Fig. 1. The symbols show the deviation of the calculated from the experimental values from Table 2. The curves above and below the zero line indicate the calculated error region of the recommended values given in Table 3. The error bars represent the experimental errors. (Error bars smaller than the symbols are omitted for clarity of the figure.)

Table 3. Recommended values (fit to the reliable experimental values according to the equations $\rho = A + BT + CT^2 + DT^3 + \ldots$ or $\rho = [1 + 1.75(1 - T/T_c)^{1/3} + 0.75(1 - T/T_c)][\rho_c + A(T_c - T) + B(T_c - T)^2 + C(T_c - T)^3 + D(T_c - T)^4]$).

$\frac{T}{\text{K}}$	$\frac{\rho \pm \sigma_{\text{fit}}}{\text{kg} \cdot \text{m}^{-3}}$	$\frac{T}{\text{K}}$	$\frac{\rho \pm \sigma_{\text{fit}}}{\text{kg} \cdot \text{m}^{-3}}$	$\frac{T}{\text{K}}$	$\frac{\rho \pm \sigma_{\text{fit}}}{\text{kg} \cdot \text{m}^{-3}}$
210.00	864.71 ± 1.50	310.00	788.48 ± 0.49	430.00	666.76 ± 0.83
220.00	856.65 ± 1.26	320.00	780.54 ± 0.52	440.00	652.80 ± 0.84
230.00	848.81 ± 1.05	330.00	772.36 ± 0.55	450.00	637.78 ± 0.97
240.00	841.15 ± 0.88	340.00	763.88 ± 0.59	460.00	621.55 ± 1.00
250.00	833.61 ± 0.75	350.00	755.06 ± 0.63	470.00	604.19 ± 1.03
260.00	826.16 ± 0.65	360.00	745.86 ± 0.67	480.00	585.54 ± 1.08
270.00	818.74 ± 0.57	370.00	736.23 ± 0.71	490.00	565.28 ± 1.14
280.00	811.31 ± 0.52	380.00	726.12 ± 0.75	500.00	542.82 ± 1.22
290.00	803.82 ± 0.50	390.00	715.48 ± 0.78	510.00	517.27 ± 1.35
293.15	801.44 ± 0.49	400.00	704.28 ± 0.80	520.00	487.26 ± 2.55
298.15	797.64 ± 0.49	410.00	692.45 ± 0.82	530.00	450.24 ± 3.89
300.00	796.23 ± 0.49	420.00	679.96 ± 0.83	540.00	399.16 ± 4.57

2-Methyl-2-propanol **[75-65-0]** **$C_4H_{10}O$** **MW = 74.12** **10**

$T_c = 506.20$ K [89-tej/lee]
$\rho_c = 270.00$ kg·m^{-3} [89-tej/lee]

Table 1. Coefficients for the polynomial expansion equations. Standard deviations (see introduction): $\sigma_\ell = 4.8919$ (low temperature range), $\sigma_{c,w} = (1.7560$ combined temperature ranges, weighted), $\sigma_{c,uw} = 4.6459 \cdot 10^{-1}$ (combined temperature ranges, unweighted).

Coefficient	$T = 273.15$ to 400.00 K $\rho = A + BT + CT^2 + DT^3 + \ldots$	$T = 400.00$ to 506.20 K $\rho = [1 + 1.75(1 - T/T_c)^{1/3} + 0.75(1 - T/T_c)]$ $[\rho_c + A(T_c - T) + B(T_c - T)^2 + C(T_c - T)^3$ $+ D(T_c - T)^4]$
A	$9.06345 \cdot 10^2$	1.32387
B	$1.53910 \cdot 10^{-1}$	$-2.89194 \cdot 10^{-2}$
C	$-1.93019 \cdot 10^{-3}$	$2.93882 \cdot 10^{-4}$
D		$-1.07285 \cdot 10^{-6}$

Table 2. Experimental values with uncertainties and deviation from calculated values.

$\frac{T}{K}$	$\frac{\rho_{exp} \pm 2\sigma_{est}}{kg \cdot m^{-3}}$	$\frac{\rho_{exp} - \rho_{calc}}{kg \cdot m^{-3}}$	Ref. (Symbol in Fig. 1)	$\frac{T}{K}$	$\frac{\rho_{exp} \pm 2\sigma_{est}}{kg \cdot m^{-3}}$	$\frac{\rho_{exp} - \rho_{calc}}{kg \cdot m^{-3}}$	Ref. (Symbol in Fig. 1)
273.15	807.40 ± 1.00	3.03	1872-but(×)	447.23	575.40 ± 0.50	2.89	63-amb/tow(×)
303.15	775.40 ± 1.00	−0.22	1872-but[1)]	456.20	556.10 ± 0.50	2.63	63-amb/tow(×)
303.15	777.50 ± 0.50	1.88	31-smy/dor-1[1)]	465.81	533.00 ± 0.50	2.08	63-amb/tow(×)
323.15	756.30 ± 0.50	1.78	31-smy/dor-1[1)]	473.52	512.80 ± 1.00	2.70	63-amb/tow(×)
343.15	734.30 ± 0.50	2.42	31-smy/dor-1(×)	480.56	490.50 ± 1.00	2.68	63-amb/tow(×)
293.15	786.68 ± 0.50	1.09	34-tim/del(◆)	300.00	776.00 ± 1.50	−2.80	64-dan/bah[1)]
303.15	776.18 ± 0.50	0.56	34-tim/del[1)]	320.00	756.00 ± 1.50	−1.94	64-dan/bah[1)]
308.15	770.88 ± 0.50	0.39	34-tim/del[1)]	340.00	736.00 ± 1.50	0.46	64-dan/bah[1)]
313.15	765.58 ± 0.50	0.32	34-tim/del[1)]	360.00	712.00 ± 1.50	0.40	64-dan/bah[1)]
298.15	781.60 ± 0.40	0.95	55-kus[1)]	380.00	686.00 ± 1.50	−0.11	64-dan/bah[1)]
303.15	776.60 ± 0.40	0.98	55-kus[1)]	400.00	654.00 ± 1.50	−5.08	64-dan/bah[1)]
313.15	766.50 ± 0.40	1.24	55-kus[1)]	420.00	618.00 ± 1.70	−7.60	64-dan/bah[1)]
323.15	755.90 ± 0.40	1.38	55-kus[1)]	440.00	580.00 ± 1.70	−7.09	64-dan/bah[1)]
333.15	741.50 ± 0.50	−1.89	55-kus[1)]	460.00	536.00 ± 2.00	−8.90	64-dan/bah[1)]
343.15	734.20 ± 0.50	2.32	55-kus(×)	480.00	476.00 ± 2.50	−13.74	64-dan/bah[1)]
353.15	722.60 ± 0.50	2.63	55-kus(×)	500.00	380.00 ± 6.00	−11.52	64-dan/bah(×)
313.15	765.20 ± 0.50	−0.06	58-cos/bow[1)]	300.65	778.10 ± 0.12	−0.05	83-hal/gun(∇)
333.15	742.90 ± 0.50	−0.49	58-cos/bow[1)]	303.15	775.43 ± 0.12	−0.19	83-hal/gun(∇)
353.15	721.20 ± 0.50	1.23	58-cos/bow(×)	308.15	770.23 ± 0.12	−0.26	83-hal/gun(∇)
373.15	697.40 ± 0.50	2.38	58-cos/bow(×)	313.15	757.60 ± 0.15	−7.66	83-hal/gun(∇)
393.15	671.50 ± 0.60	2.99	58-cos/bow(×)	340.00	735.05 ± 0.15	−0.49	83-hal/gun(∇)
413.15	645.00 ± 0.60	6.98	58-cos/bow(×)	360.00	710.82 ± 0.17	−0.78	83-hal/gun(∇)
300.15	779.20 ± 0.40	0.55	63-amb/tow[1)]	380.00	684.41 ± 0.17	−1.70	83-hal/gun(∇)
419.82	628.10 ± 0.50	2.16	63-amb/tow(×)	400.00	655.47 ± 0.17	−3.61	83-hal/gun(∇)
433.61	603.50 ± 0.50	3.85	63-amb/tow(×)	420.00	623.36 ± 0.20	−2.24	83-hal/gun(∇)

[1)] Not included in Fig. 1.

cont.

Table 2. (cont.)

$\frac{T}{K}$	$\frac{\rho_{exp} \pm 2\sigma_{est}}{kg \cdot m^{-3}}$	$\frac{\rho_{exp} - \rho_{calc}}{kg \cdot m^{-3}}$	Ref. (Symbol in Fig. 1)	$\frac{T}{K}$	$\frac{\rho_{exp} \pm 2\sigma_{est}}{kg \cdot m^{-3}}$	$\frac{\rho_{exp} - \rho_{calc}}{kg \cdot m^{-3}}$	Ref. (Symbol in Fig. 1)
440.00	587.07 ± 0.20	−0.02	83-hal/gun(∇)	308.15	769.97 ± 0.10	−0.52	88-kim/mar(□)
450.00	566.74 ± 0.20	−0.03	83-hal/gun(∇)	313.15	764.69 ± 0.10	−0.57	88-kim/mar(□)
460.00	544.38 ± 0.20	−0.52	83-hal/gun(∇)	318.15	759.37 ± 0.10	−0.57	88-kim/mar(□)
465.00	532.21 ± 0.20	−0.73	83-hal/gun(∇)	323.15	754.01 ± 0.10	−0.51	88-kim/mar(□)
303.15	775.46 ± 0.15	−0.16	87-pik[1)]	328.15	748.61 ± 0.10	−0.39	88-kim/mar(□)
333.15	743.18 ± 0.15	−0.21	87-pik(Δ)	299.15	779.48 ± 0.15	−0.17	88-oka/oga(○)
303.15	775.21 ± 0.10	−0.41	88-kim/mar(□)				

[1)] Not included in Fig. 1.

Further references: [1872-lin/von-2, 1880-bru-1, 1884-per, 1893-tho/jon, 02-you/for, 06-car/fer, 08-ric/mat, 11-dor-1, 19-beh, 19-eyk, 25-nor/ash, 26-mun, 29-pre, 29-swa, 33-gin/her, 33-nev/jat, 35-hen, 36-ipa/cor, 36-spe, 39-owe/qua, 44-ira, 46-sim/was, 48-wei, 49-dre/mar, 50-pic/zie, 52-dun/was, 54-wes/aud, 55-bou/cle, 55-wes, 56-rus/ame, 62-bro/smi, 63-mcc/lai, 69-bro/foc, 70-sus/hol, 71-des/bha-1, 76-tri/kri, 77-gov/and-1, 80-kas/izy, 83-pik-2, 85-rao/red, 87-kri/cho, 88-cac/cos, 95-red/ram-1, 96-bha/mak, 96-nik/mah, 98-nik/shi].

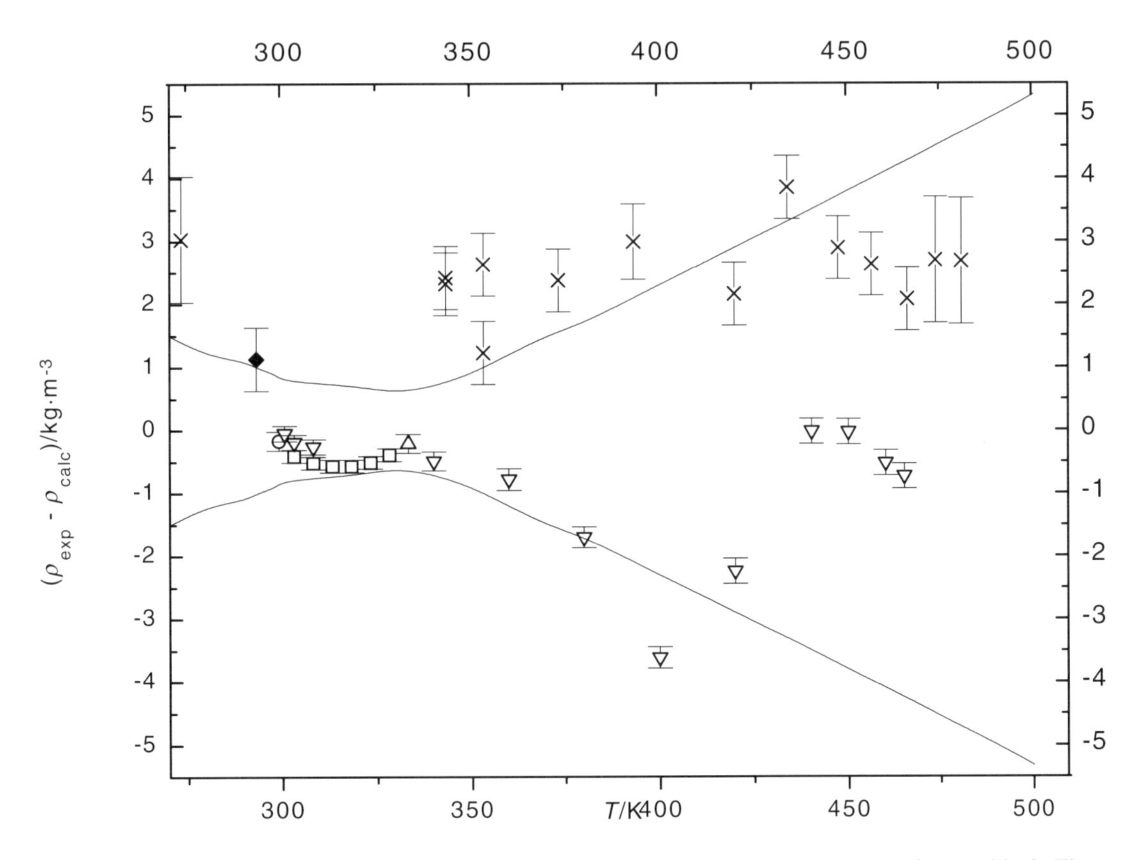

Fig. 1. The symbols show the deviation of the calculated from the experimental values from Table 2. The curves above and below the zero line indicate the calculated error region of the recommended values given in Table 3. The error bars represent the experimental errors. (Error bars smaller than the symbols are omitted for clarity of the figure.)

cont.

2-Methyl-2-propanol (cont.)

Table 3. Recommended values (fit to the reliable experimental values according to the equations $\rho = A + BT + CT^2 + DT^3 + \ldots$ or $\rho = [1 + 1.75(1 - T/T_c)^{1/3} + 0.75(1 - T/T_c)][\rho_c + A(T_c - T) + B(T_c - T)^2 + C(T_c - T)^3 + D(T_c - T)^4])$.

$\frac{T}{\text{K}}$	$\frac{\rho \pm \sigma_{fit}}{\text{kg} \cdot \text{m}^{-3}}$	$\frac{T}{\text{K}}$	$\frac{\rho \pm \sigma_{fit}}{\text{kg} \cdot \text{m}^{-3}}$	$\frac{T}{\text{K}}$	$\frac{\rho \pm \sigma_{fit}}{\text{kg} \cdot \text{m}^{-3}}$
270.00	807.19 ± 1.50	340.00	735.54 ± 0.72	430.00	606.64 ± 3.20
280.00	798.11 ± 1.20	350.00	723.77 ± 0.90	440.00	587.09 ± 3.53
290.00	788.65 ± 1.13	360.00	711.60 ± 1.22	450.00	566.77 ± 3.80
293.15	785.59 ± 1.00	370.00	699.05 ± 1.51	460.00	544.90 ± 4.10
298.15	780.65 ± 0.91	380.00	686.11 ± 1.74	470.00	519.99 ± 4.40
300.00	778.80 ± 0.80	390.00	672.79 ± 2.00	480.00	489.74 ± 4.70
310.00	768.57 ± 0.75	400.00	659.08 ± 2.30	490.00	450.24 ± 5.00
320.00	757.94 ± 0.70	410.00	643.47 ± 2.60	500.00	391.52 ± 5.32
330.00	746.94 ± 0.60	420.00	625.60 ± 2.90		

2.1.2 Alkanols, C_5 - C_6

2,2-Dimethyl-1-propanol **[75-84-3]** **$C_5H_{12}O$** **MW = 88.15** **11**

Table 1. Experimental value with uncertainty.

$\frac{T}{K}$	$\frac{\rho_{exp} \pm 2\sigma_{est}}{kg \cdot m^{-3}}$	Ref.
303.15	799.0 ± 0.3	74-mye/cle

2-Methyl-1-butanol **[137-32-6]** **$C_5H_{12}O$** **MW = 88.15** **12**

Table 1. Coefficients of the polynomial expansion equation.
Standard deviations (see introduction):
$\sigma_{c,w} = 7.9216 \cdot 10^{-1}$ (combined temperature ranges, weighted),
$\sigma_{c,uw} = 1.8175 \cdot 10^{-1}$ (combined temperature ranges, unweighted).

Coefficient	T = 293.15 to 390.65 K $\rho = A + BT + CT^2 + DT^3 + \ldots$
A	$1.00094 \cdot 10^{3}$
B	$-4.27416 \cdot 10^{-1}$
C	$-6.53533 \cdot 10^{-4}$

Table 2. Experimental values with uncertainties and deviation from calculated values.

$\frac{T}{K}$	$\frac{\rho_{exp} \pm 2\sigma_{est}}{kg \cdot m^{-3}}$	$\frac{\rho_{exp} - \rho_{calc}}{kg \cdot m^{-3}}$	Ref. (Symbol in Fig. 1)	$\frac{T}{K}$	$\frac{\rho_{exp} \pm 2\sigma_{est}}{kg \cdot m^{-3}}$	$\frac{\rho_{exp} - \rho_{calc}}{kg \cdot m^{-3}}$	Ref. (Symbol in Fig. 1)
298.15	815.20 ± 0.50	−0.21	27-nor/cor(○)	367.95	754.90 ± 0.50	−0.29	63-tho/mea(∇)
296.15	816.00 ± 1.00	−1.04	31-lev/mar-4(×)	378.15	745.10 ± 0.55	−0.75	63-tho/mea(∇)
293.15	819.30 ± 0.60	−0.18	37-bra(◆)	390.65	734.30 ± 0.55	0.07	63-tho/mea(∇)
293.15	818.44 ± 1.00	−1.04	38-whi/ole(×)	293.15	818.80 ± 0.50	−0.68	68-ano(Δ)
308.15	807.40 ± 1.00	0.23	46-haf/lov(×)	298.15	814.80 ± 0.30	−0.61	82-dap/don(□)
293.15	820.00 ± 1.00	0.52	51-lyu/ter(×)	293.15	819.80 ± 1.00	0.32	83-fuk/ogi(×)
293.15	819.80 ± 1.00	0.32	52-coo(×)	298.15	815.90 ± 1.00	0.49	83-fuk/ogi(×)
298.15	815.00 ± 1.00	−0.41	56-ike/kep(×)	303.15	812.40 ± 1.00	1.10	83-fuk/ogi(×)
302.65	811.10 ± 0.50	−0.62	63-tho/mea(∇)	313.15	804.60 ± 1.50	1.60	83-fuk/ogi(×)
324.35	792.80 ± 0.50	−0.75	63-tho/mea(∇)	323.15	796.80 ± 1.50	2.23	83-fuk/ogi(×)
337.95	781.40 ± 0.50	−0.45	63-tho/mea(∇)	333.15	788.80 ± 1.50	2.79	83-fuk/ogi[1)]
353.05	768.70 ± 0.50	0.12	63-tho/mea(∇)	343.15	780.10 ± 1.50	2.79	83-fuk/ogi[1)]

[1)] Not included in Fig. 1.

Further references: [1884-per, 01-mar/mck, 16-wil/bru, 33-tre/wat, 35-gui, 37-gin/bau, 39-bar/atk, 48-bro/bro, 48-wei, 50-pic/zie, 52-hel, 63-mcc/lai]

cont.

2-Methyl-1-butanol (cont.)

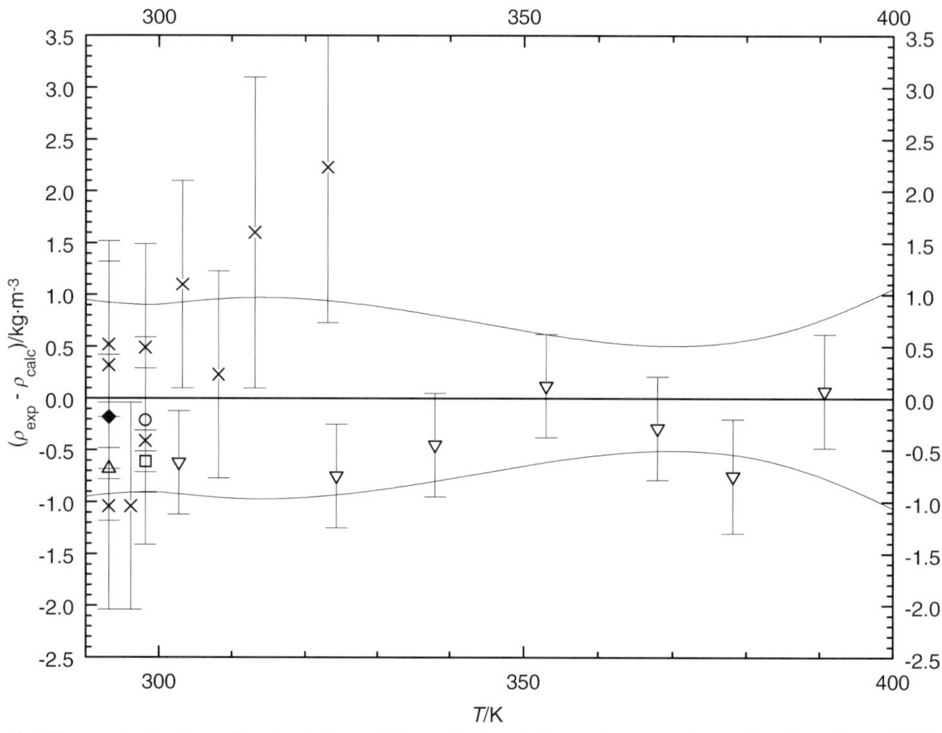

Fig. 1. The symbols show the deviation of the calculated from the experimental values from Table 2. The curves above and below the zero line indicate the calculated error region of the recommended values given in Table 3. The error bars represent the experimental errors. (Error bars smaller than the symbols are omitted for clarity of the figure.)

Table 3. Recommended values (fit to the reliable experimental values according to the equations
$$\rho = A + BT + CT^2 + DT^3 + \ldots \text{ or } \rho = [1 + 1.75(1 - T/T_c)^{1/3} + 0.75(1 - T/T_c)][\rho_c + A(T_c - T) + B(T_c - T)^2 + C(T_c - T)^3 + D(T_c - T)^4]\,).$$

$\frac{T}{K}$	$\frac{\rho \pm \sigma_{fit}}{kg \cdot m^{-3}}$	$\frac{T}{K}$	$\frac{\rho \pm \sigma_{fit}}{kg \cdot m^{-3}}$	$\frac{T}{K}$	$\frac{\rho \pm \sigma_{fit}}{kg \cdot m^{-3}}$
290.00	822.02 ± 0.95	320.00	797.24 ± 0.97	370.00	753.32 ± 0.49
293.15	819.48 ± 0.92	330.00	788.72 ± 0.89	380.00	744.15 ± 0.54
298.15	815.41 ± 0.90	340.00	780.07 ± 0.78	390.00	734.84 ± 0.73
300.00	813.89 ± 0.90	350.00	771.28 ± 0.65	400.00	725.40 ± 1.06
310.00	805.63 ± 0.98	360.00	762.37 ± 0.54		

(*S*)-(-)-2-Methyl-1-butanol **[1565-80-6]** **$C_5H_{12}O$** **MW = 88.15** **13**

Table 1. Fit with estimated *B* coefficient for 7 accepted points. Deviation $\sigma_w = 0.464$.

Coefficient	$\rho = A + BT$
A	1047.24
B	−0.780

cont.

Table 2. Experimental values with uncertainties and deviation from calculated values.

$\frac{T}{K}$	$\frac{\rho_{exp} \pm 2\sigma_{est}}{kg \cdot m^{-3}}$	$\frac{\rho_{exp} - \rho_{calc}}{kg \cdot m^{-3}}$	Ref.	$\frac{T}{K}$	$\frac{\rho_{exp} \pm 2\sigma_{est}}{kg \cdot m^{-3}}$	$\frac{\rho_{exp} - \rho_{calc}}{kg \cdot m^{-3}}$	Ref.
293.15	818.9 ± 0.5	0.32	38-whi/ole	293.15	818.3 ± 1.0	−0.29	38-whi/ole-1
293.15	818.8 ± 0.6	0.18	38-whi/ole-1	296.15	814.0 ± 1.5	−2.24	52-hel
300.65	813.0 ± 2.0	0.27	38-whi/ole-1	289.15	821.6 ± 0.5	−0.10	59-pin/lar
293.15	818.4 ± 1.0	−0.16	38-whi/ole-1				

Table 3. Recommended values.

$\frac{T}{K}$	$\frac{\rho_{exp} \pm 2\sigma_{est}}{kg \cdot m^{-3}}$
280.00	828.8 ± 1.6
290.00	821.0 ± 1.0
293.15	818.6 ± 1.0
298.15	814.7 ± 1.2
310.00	805.4 ± 2.1

2-Methyl-2-butanol [75-85-4] $C_5H_{12}O$ MW = 88.15 14

Table 1. Coefficients of the polynomial expansion equation.
Standard deviations (see introduction):
$\sigma_{c,w}$ = 1.1040 (combined temperature ranges, weighted),
$\sigma_{c,uw}$ = 2.1679 · 10^{-1} (combined temperature ranges, unweighted).

Coefficient	T = 268.15 to 453.15 K $\rho = A + BT + CT^2 + DT^3 + \ldots$
A	9.31063 · 10^{2}
B	6.74937 · 10^{-2}
C	−1.64419 · 10^{-3}

Table 2. Experimental values with uncertainties and deviation from calculated values.

$\frac{T}{K}$	$\frac{\rho_{exp} \pm 2\sigma_{est}}{kg \cdot m^{-3}}$	$\frac{\rho_{exp} - \rho_{calc}}{kg \cdot m^{-3}}$	Ref. (Symbol in Fig. 1)	$\frac{T}{K}$	$\frac{\rho_{exp} \pm 2\sigma_{est}}{kg \cdot m^{-3}}$	$\frac{\rho_{exp} - \rho_{calc}}{kg \cdot m^{-3}}$	Ref. (Symbol in Fig. 1)
374.75	724.10 ± 1.00	−1.35	1883-sch-3(×)	298.15	801.80 ± 2.00	−3.23	39-owe/qua[1)]
375.15	724.10 ± 1.00	−0.88	1884-sch-6(×)	308.15	792.30 ± 2.00	−3.44	39-owe/qua[1)]
273.15	827.00 ± 0.60	0.18	10-ric(×)	318.15	782.70 ± 2.00	−3.41	39-owe/qua[1)]
288.15	813.80 ± 0.50	−0.19	27-nor/reu(×)	328.15	773.10 ± 2.00	−3.06	39-owe/qua(×)
298.15	804.73 ± 0.50	−0.30	27-nor/reu[1)]	328.15	773.10 ± 2.00	−3.06	39-owe/qua[1)]
273.15	827.16 ± 0.60	0.34	32-tim/hen(×)	338.15	762.40 ± 0.50	−3.48	39-owe/qua[1)]
288.15	813.44 ± 0.60	−0.55	32-tim/hen(×)	273.15	825.00 ± 0.50	−1.82	58-cos/bow(×)
303.15	799.72 ± 0.60	−0.70	32-tim/hen[1)]	293.15	808.40 ± 0.50	−1.15	58-cos/bow[1)]
298.15	805.99 ± 0.60	0.96	35-but/ram(□)	313.15	790.00 ± 0.50	−0.96	58-cos/bow(×)
273.15	826.20 ± 2.00	−0.62	39-owe/qua[1)]	333.15	771.30 ± 0.50	0.24	58-cos/bow(×)
273.15	826.20 ± 0.50	−0.62	39-owe/qua(×)	353.15	750.40 ± 0.50	0.56	58-cos/bow(×)
298.15	801.80 ± 0.50	−3.23	39-owe/qua[1)]	373.15	728.10 ± 0.50	0.79	58-cos/bow(×)

[1)] Not included in Fig.

cont.

2-Methyl-2-butanol (cont.)

Table 2. (cont.)

$\frac{T}{\text{K}}$	$\frac{\rho_{exp} \pm 2\sigma_{est}}{\text{kg}\cdot\text{m}^{-3}}$	$\frac{\rho_{exp} - \rho_{calc}}{\text{kg}\cdot\text{m}^{-3}}$	Ref. (Symbol in Fig. 1)	$\frac{T}{\text{K}}$	$\frac{\rho_{exp} \pm 2\sigma_{est}}{\text{kg}\cdot\text{m}^{-3}}$	$\frac{\rho_{exp} - \rho_{calc}}{\text{kg}\cdot\text{m}^{-3}}$	Ref. (Symbol in Fig. 1)
393.15	704.50 ± 0.50	1.04	58-cos/bow(×)	298.15	806.90 ± 0.50	1.87	85-dap/don[1)]
413.15	679.40 ± 1.00	1.10	58-cos/bow(×)	308.15	796.20 ± 0.50	0.46	85-dap/don[1)]
433.15	653.00 ± 1.00	1.18	58-cos/bow(×)	298.15	804.37 ± 0.60	−0.66	85-tre/ben(Δ)
453.15	623.90 ± 1.00	−0.12	58-cos/bow(×)	298.15	805.00 ± 0.20	−0.03	86-rig/mar(O)
303.15	799.73 ± 0.25	−0.69	75-hsu/cle(∇)	303.15	800.60 ± 0.20	0.18	86-rig/mar(O)
298.15	806.70 ± 0.50	1.67	82-dap/don(◆)	308.15	796.00 ± 0.20	0.26	86-rig/mar(O)
268.15	833.20 ± 0.50	2.26	85-dap/don(×)	313.15	791.20 ± 0.20	0.24	86-rig/mar(O)
278.15	822.60 ± 0.50	-0.03	85-dap/don(×)				

[1)] Not included in Fig. 1.

Further references: [1878-wis, 1884-per, 1893-tho/jon, 19-eyk, 26-mun, 33-nev/jat, 35-mah-1, 36-par, 37-gin/bau, 48-wei, 50-pic/zie, 51-lev/fai, 52-coo, 55-soe/fre, 57-pet/sus, 58-pan/osi, 63-mcc/lai, 67-gol/per, 76-tri/kri, 88-cab/bar, 96-nik/jad, 98-pai/che, 98-sen].

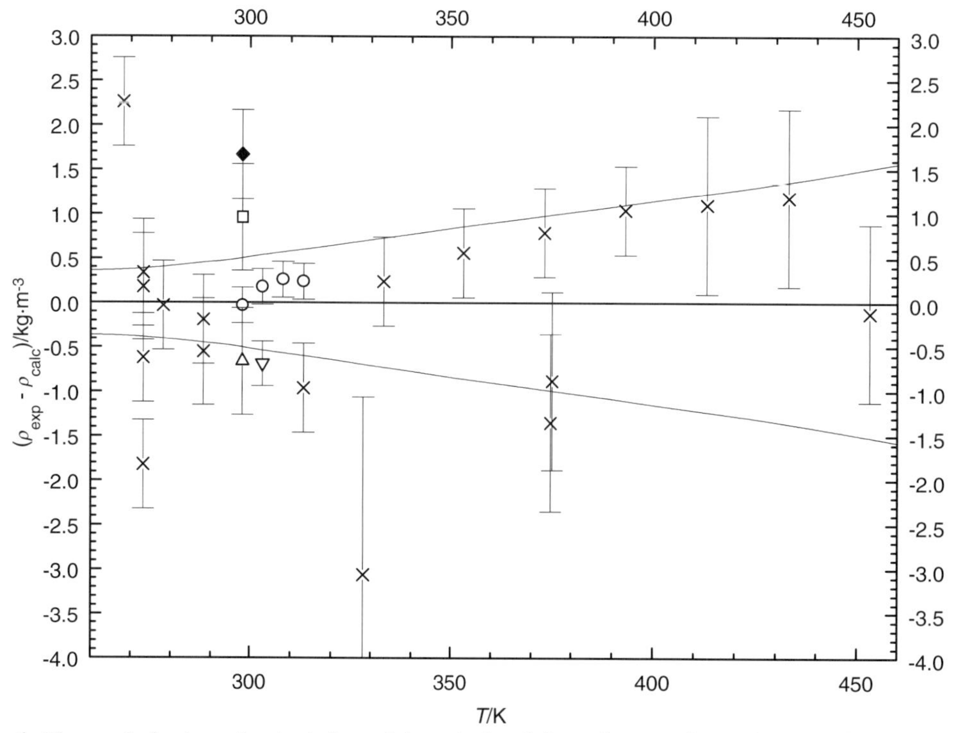

Fig. 1. The symbols show the deviation of the calculated from the experimental values from Table 2. The curves above and below the zero line indicate the calculated error region of the recommended values given in Table 3. The error bars represent the experimental errors. (Error bars smaller than the symbols are omitted for clarity of the figure.)

cont.

Table 3. Recommended values (fit to the reliable experimental values according to the equations $\rho = A + BT + CT^2 + DT^3 + \ldots$ or $\rho = [1 + 1.75(1 - T/T_c)^{1/3} + 0.75(1 - T/T_c)][\rho_c + A(T_c - T) + B(T_c - T)^2 + C(T_c - T)^3 + D(T_c - T)^4]$).

T/K	$\rho \pm \sigma_{fit}$/kg·m^{-3}	T/K	$\rho \pm \sigma_{fit}$/kg·m^{-3}	T/K	$\rho \pm \sigma_{fit}$/kg·m^{-3}
260.00	837.46 ± 0.36	320.00	784.30 ± 0.64	400.00	694.99 ± 1.15
270.00	829.43 ± 0.37	330.00	774.28 ± 0.71	410.00	682.35 ± 1.21
280.00	821.06 ± 0.41	340.00	763.94 ± 0.77	420.00	669.38 ± 1.27
290.00	812.36 ± 0.46	350.00	753.27 ± 0.84	430.00	656.07 ± 1.34
293.15	809.55 ± 0.47	360.00	742.27 ± 0.90	440.00	642.45 ± 1.41
298.15	805.03 ± 0.50	370.00	730.95 ± 0.96	450.00	628.49 ± 1.49
300.00	803.33 ± 0.52	380.00	719.29 ± 1.02	460.00	614.20 ± 1.57
310.00	793.98 ± 0.58	390.00	707.30 ± 1.08		

3-Methyl-1-butanol **[123-51-3]** **$C_5H_{12}O$** **MW = 88.15** **15**

Table 1. Coefficients of the polynomial expansion equation.
Standard deviations (see introduction):
$\sigma_{c,w} = 9.8281 \cdot 10^{-1}$ (combined temperature ranges, weighted),
$\sigma_{c,uw} = 2.2532 \cdot 10^{-1}$ (combined temperature ranges, unweighted).

Coefficient	T = 273.15 to 347.85 K $\rho = A + BT + CT^2 + DT^3 + \ldots$
A	$8.89389 \cdot 10^{2}$
B	$1.88877 \cdot 10^{-1}$
C	$-1.56150 \cdot 10^{-3}$

Table 2. Experimental values with uncertainties and deviation from calculated values.

T/K	$\rho_{exp} \pm 2\sigma_{est}$/kg·m^{-3}	$\rho_{exp} - \rho_{calc}$/kg·m^{-3}	Ref. (Symbol in Fig. 1)	T/K	$\rho_{exp} \pm 2\sigma_{est}$/kg·m^{-3}	$\rho_{exp} - \rho_{calc}$/kg·m^{-3}	Ref. (Symbol in Fig. 1)
285.45	815.83 ± 0.00	−0.24	11-sch[1)]	303.15	804.19 ± 0.30	1.04	85-rao/red(×)
307.15	799.79 ± 0.50	−0.30	11-sch(×)	303.15	801.77 ± 0.30	−1.38	86-mou/nai(×)
327.15	783.63 ± 0.50	−0.43	11-sch(×)	298.15	807.10 ± 0.20	0.20	86-rig/mar(□)
347.85	767.20 ± 0.50	1.05	11-sch(×)	303.15	802.70 ± 0.20	−0.45	86-rig/mar(□)
293.15	810.50 ± 0.50	−0.07	26-mat(×)	308.15	799.00 ± 0.20	−0.32	86-rig/mar(□)
293.15	810.20 ± 0.50	−0.37	28-mon(×)	313.15	794.90 ± 0.20	−0.51	86-rig/mar(□)
273.15	823.88 ± 0.50	−0.60	29-tim/hen(Δ)	303.15	801.78 ± 1.00	−1.37	92-kum/sre(×)
288.15	812.87 ± 1.00	−1.29	29-tim/hen[1)]	298.15	806.90 ± 0.20	0.00	92-tan/mur(×)
303.15	801.73 ± 1.00	−1.42	29-tim/hen[1)]	298.15	806.90 ± 0.20	0.00	93-ami/rai(∇)
293.15	808.60 ± 2.00	−1.97	56-tor-1[1)]	303.15	803.00 ± 0.20	−0.15	93-ami/rai(∇)
313.15	793.80 ± 2.00	−1.61	56-tor-1[1)]	308.15	799.30 ± 0.20	−0.02	93-ami/rai(∇)
333.15	778.10 ± 1.00	−0.90	56-tor-1(×)	298.15	807.10 ± 0.30	0.20	98-sen(×)
293.15	811.57 ± 0.50	1.00	81-joo/arl(◆)				

[1)] Not included in Fig. 1.

cont.

3-Methyl-1-butanol (cont.)

Further references: [1876-bal, 1884-per, 1884-sch-6, 1886-tra, 1891-sch/kos, 1893-tho/jon, 1898-kah, 04-bru/sch, 07-che-1, 08-ric/mat, 13-muc, 14-eng/tur, 16-wro/rei, 21-rei/hic, 23-pop, 26-han, 26-mun, 27-krc/wil, 27-nor/cor, 28-har-2, 28-llo/bro, 33-nev/jat, 35-but/ram, 35-cou/hop, 35-mah-1, 36-spe, 36-tom, 37-gin/bau, 41-hus/age, 42-boh, 42-mul, 43-boh, 45-add, 48-vog-2, 48-wei, 49-udo/kal, 50-pic/zie, 52-coo, 53-par/cha, 56-ike/kep, 57-ano-1, 57-pet/sus, 57-rom-1, 58-arn/was, 60-ter/kep, 63-mcc/lai, 63-raj/ran, 65-red/rao, 67-gol/per, 68-ana/rao, 68-ano, 71-abr/ber, 71-tha/rao, 79-sub/rao, 82-dap/don, 82-kar/red, 87-isl/qua].

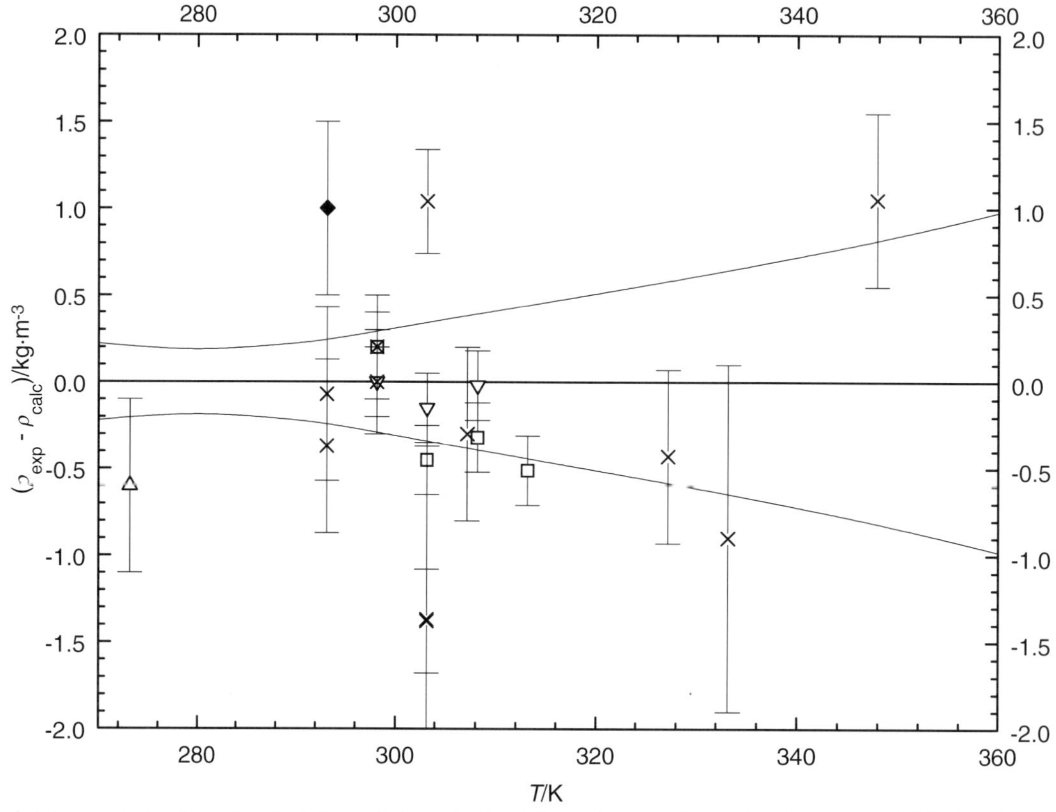

Fig. 1. The symbols show the deviation of the calculated from the experimental values from Table 2. The curves above and below the zero line indicate the calculated error region of the recommended values given in Table 3. The error bars represent the experimental errors. (Error bars smaller than the symbols are omitted for clarity of the figure.)

Table 3. Recommended values (fit to the reliable experimental values according to the equations

$$\rho = A + BT + CT^2 + DT^3 + \ldots \text{ or } \rho = [1 + 1.75(1 - T/T_c)^{1/3} + 0.75(1 - T/T_c)][\rho_c + A(T_c - T) + B(T_c - T)^2 + C(T_c - T)^3 + D(T_c - T)^4]).$$

$\frac{T}{K}$	$\frac{\rho \pm \sigma_{fit}}{kg \cdot m^{-3}}$	$\frac{T}{K}$	$\frac{\rho \pm \sigma_{fit}}{kg \cdot m^{-3}}$	$\frac{T}{K}$	$\frac{\rho \pm \sigma_{fit}}{kg \cdot m^{-3}}$
270.00	826.55 ± 0.22	298.15	806.90 ± 0.29	330.00	781.67 ± 0.61
280.00	819.85 ± 0.17	300.00	805.52 ± 0.31	340.00	773.10 ± 0.72
290.00	812.84 ± 0.22	310.00	797.88 ± 0.41	350.00	764.21 ± 0.84
293.15	810.57 ± 0.24	320.00	789.93 ± 0.51	360.00	755.01 ± 0.98

3-Methyl-2-butanol **[598-75-4]** **$C_5H_{12}O$** **MW = 88.15** **16**

Table 1. Coefficients of the polynomial expansion equation.
Standard deviations (see introduction):
$\sigma_{c,w} = 7.5344 \cdot 10^{-1}$ (combined temperature ranges, weighted),
$\sigma_{c,uw} = 2.8698 \cdot 10^{-1}$ (combined temperature ranges, unweighted).

Coefficient	T = 293.15 to 378.15 K $\rho = A + BT + CT^2 + DT^3 + \ldots$
A	$8.82581 \cdot 10^{2}$
B	$3.69971 \cdot 10^{-1}$
C	$-2.00997 \cdot 10^{-3}$

Table 2. Experimental values with uncertainties and deviation from calculated values.

$\frac{T}{\mathrm{K}}$	$\frac{\rho_{exp} \pm 2\sigma_{est}}{\mathrm{kg \cdot m^{-3}}}$	$\frac{\rho_{exp} - \rho_{calc}}{\mathrm{kg \cdot m^{-3}}}$	Ref. (Symbol in Fig. 1)	$\frac{T}{\mathrm{K}}$	$\frac{\rho_{exp} \pm 2\sigma_{est}}{\mathrm{kg \cdot m^{-3}}}$	$\frac{\rho_{exp} - \rho_{calc}}{\mathrm{kg \cdot m^{-3}}}$	Ref. (Symbol in Fig. 1)
293.15	818.00 ± 0.50	–0.31	12-pic/ken(□)	324.35	790.10 ± 0.50	–1.03	63-tho/mea(Δ)
293.15	818.20 ± 0.50	–0.11	50-pic/zie(O)	337.95	777.80 ± 0.50	–0.25	63-tho/mea(Δ)
293.15	820.00 ± 1.00	1.69	63-tho/mea(Δ)	353.05	759.80 ± 0.50	–2.87	63-tho/mea[1)]
298.15	815.00 ± 1.00	0.79	63-tho/mea(Δ)	367.95	746.10 ± 0.50	–0.49	63-tho/mea(Δ)
302.55	810.70 ± 0.50	0.17	63-tho/mea(Δ)	378.15	734.60 ± 0.55	–0.47	63-tho/mea(Δ)

[1)] Not included in Fig. 1.

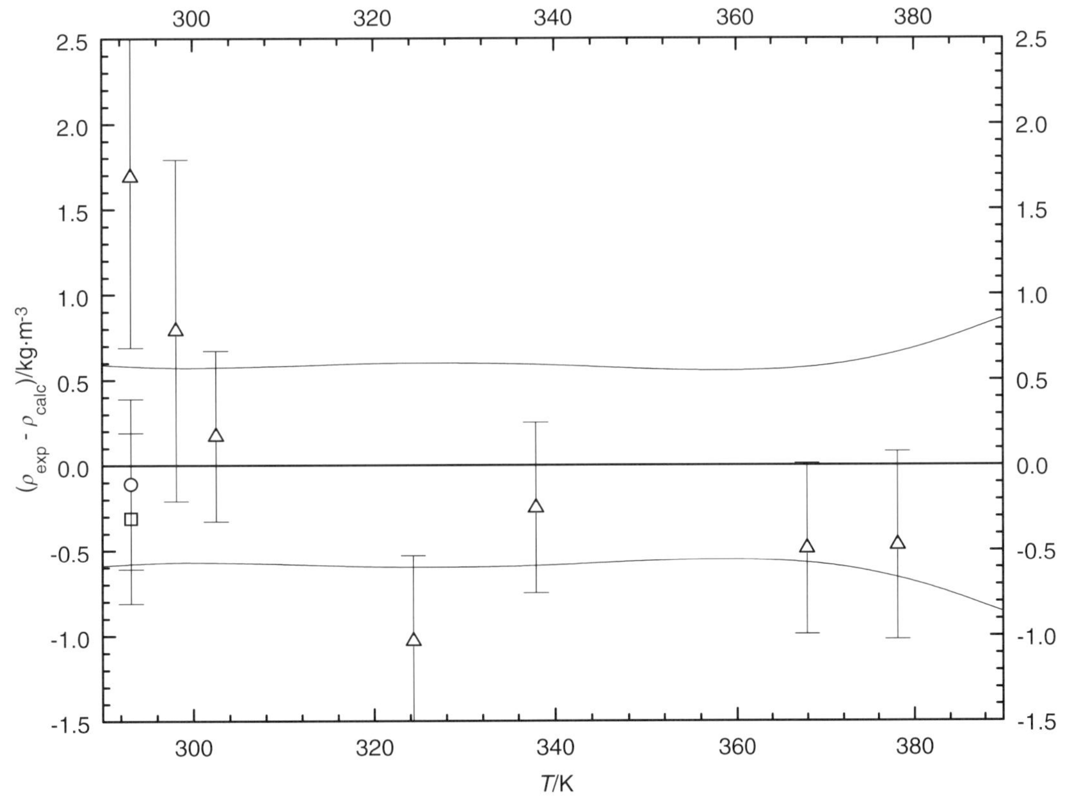

Fig. 1. The symbols show the deviation of the calculated from the experimental values from Table 2. The curves above and below the zero line indicate the calculated error region of the recommended values given in Table 3. The error bars represent the experimental errors. (Error bars smaller than the symbols are omitted for clarity of the figure.)

3-Methyl-2-butanol (cont.)

Further references: [1876-mun, 1878-win, 1878-wis, 33-ste, 37-gin/bau, 38-whi/joh, 48-mcm/rop, 52-coo, 63-mcc/lai].

Table 3. Recommended values (fit to the reliable experimental values according to the equations $\rho = A + BT + CT^2 + DT^3 + \ldots$ or $\rho = [1 + 1.75(1 - T/T_c)^{1/3} + 0.75(1 - T/T_c)][\rho_c + A(T_c - T) + B(T_c - T)^2 + C(T_c - T)^3 + D(T_c - T)^4]$).

$\frac{T}{K}$	$\frac{\rho \pm \sigma_{fit}}{kg \cdot m^{-3}}$	$\frac{T}{K}$	$\frac{\rho \pm \sigma_{fit}}{kg \cdot m^{-3}}$	$\frac{T}{K}$	$\frac{\rho \pm \sigma_{fit}}{kg \cdot m^{-3}}$
290.00	820.83 ± 0.59	320.00	795.15 ± 0.60	370.00	744.31 ± 0.57
293.15	818.31 ± 0.58	330.00	785.79 ± 0.60	380.00	732.93 ± 0.67
298.15	814.21 ± 0.57	340.00	776.02 ± 0.59	390.00	721.15 ± 0.86
300.00	812.68 ± 0.57	350.00	765.85 ± 0.56		
310.00	804.11 ± 0.58	360.00	755.28 ± 0.55		

(*S*)-3-Methyl-2-butanol [500029-41-4] $C_5H_{12}O$ MW = 88.15 17

Table 1. Experimental value with uncertainty.

$\frac{T}{K}$	$\frac{\rho_{exp} \pm 2\sigma_{est}}{kg \cdot m^{-3}}$	Ref.
298.15	810.0 ± 2.0	33-ste

1-Pentanol [71-41-0] $C_5H_{12}O$ MW = 88.15 18

T_c = 588.00 K [89-tej/lee]
ρ_c = 270.00 kg·m^{-3} [89-tej/lee]

Table 1. Coefficients for the polynomial expansion equations. Standard deviations (see introduction): $\sigma_\ell = 7.5696 \cdot 10^{-1}$ (low temperature range), $\sigma_{c,w} = (6.3480 \cdot 10^{-1}$ combined temperature ranges, weighted), $\sigma_{c,uw} = 3.3635 \cdot 10^{-1}$ (combined temperature ranges, unweighted).

Coefficient	T = 213.15 to 470.00 K $\rho = A + BT + CT^2 + DT^3 + \ldots$	T = 470.00 to 588.00 K $\rho = [1 + 1.75(1 - T/T_c)^{1/3} + 0.75(1 - T/T_c)]$ $[\rho_c + A(T_c - T) + B(T_c - T)^2 + C(T_c - T)^3$ $+ D(T_c - T)^4]$
A	$1.10034 \cdot 10^{3}$	$1.06991 \cdot 10^{0}$
B	$-1.58026 \cdot 10^{0}$	$-1.91438 \cdot 10^{-2}$
C	$3.34384 \cdot 10^{-3}$	$1.56345 \cdot 10^{-4}$
D	$-4.34342 \cdot 10^{-6}$	$-4.63597 \cdot 10^{-7}$

cont.

Table 2. Experimental values with uncertainties and deviation from calculated values.

$\frac{T}{K}$	$\frac{\rho_{exp} \pm 2\sigma_{est}}{kg \cdot m^{-3}}$	$\frac{\rho_{exp} - \rho_{calc}}{kg \cdot m^{-3}}$	Ref. (Symbol in Fig. 1)	$\frac{T}{K}$	$\frac{\rho_{exp} \pm 2\sigma_{est}}{kg \cdot m^{-3}}$	$\frac{\rho_{exp} - \rho_{calc}}{kg \cdot m^{-3}}$	Ref. (Symbol in Fig. 1)
273.15	829.60 ± 0.50	–0.06	26-han(×)	400.00	724.98 ± 0.20	–0.29	76-hal/ell(×)
273.15	829.61 ± 0.20	–0.05	32-ell/rei(×)	420.00	704.25 ± 0.20	–0.44	76-hal/ell(×)
298.15	811.59 ± 0.20	0.28	32-ell/rei[1)]	440.00	681.80 ± 0.25	–0.60	76-hal/ell(×)
213.15	873.00 ± 1.00	–0.36	58-cos/bow(×)	460.00	657.53 ± 0.25	–0.67	76-hal/ell[1)]
233.15	858.10 ± 1.00	–0.52	58-cos/bow(×)	470.00	644.61 ± 0.30	–0.71	76-hal/ell[1)]
253.15	843.00 ± 1.00	–1.12	58-cos/bow(×)	480.00	631.11 ± 0.30	–0.40	76-hal/ell(×)
273.15	827.80 ± 1.00	–1.86	58-cos/bow[1)]	490.00	616.92 ± 0.30	0.14	76-hal/ell(×)
293.15	813.30 ± 1.00	–1.72	58-cos/bow[1)]	288.15	818.20 ± 0.30	–0.51	78-dap/don(×)
313.15	798.10 ± 1.00	–1.91	58-cos/bow[1)]	298.15	810.90 ± 0.30	–0.41	78-dap/don[1)]
333.15	783.40 ± 1.00	–1.00	58-cos/bow[1)]	308.15	803.90 ± 0.30	0.09	78-dap/don[1)]
353.15	768.00 ± 1.00	0.00	58-cos/bow[1)]	318.15	796.00 ± 0.30	–0.17	78-dap/don[1)]
373.15	751.50 ± 1.00	0.91	58-cos/bow[1)]	303.15	807.58 ± 0.20	0.00	78-red/nai-1[1)]
393.15	735.60 ± 2.00	3.64	58-cos/bow(×)	313.15	799.92 ± 0.20	–0.09	78-red/nai-1(×)
413.15	718.40 ± 2.00	6.48	58-cos/bow[1)]	298.15	810.97 ± 0.15	–0.34	79-dia/tar(∇)
433.15	698.30 ± 2.00	8.06	58-cos/bow[1)]	308.15	803.61 ± 0.15	–0.20	79-dia/tar(∇)
453.15	675.40 ± 2.00	8.68	58-cos/bow[1)]	318.15	796.05 ± 0.15	–0.12	79-dia/tar(∇)
273.15	830.83 ± 0.10	1.17	59-mck/ski(Δ)	333.15	784.44 ± 0.15	0.04	79-dia/tar(∇)
273.15	831.00 ± 1.50	1.34	66-efr[1)]	278.15	825.34 ± 0.20	–0.68	83-dap/del(◆)
293.15	815.00 ± 1.50	–0.02	66-efr[1)]	288.15	818.21 ± 0.20	–0.50	83-dap/del(◆)
313.15	800.00 ± 1.50	–0.01	66-efr[1)]	298.15	811.34 ± 0.20	0.03	83-dap/del[1)]
333.15	780.00 ± 1.50	–4.40	66-efr[1)]	308.15	803.50 ± 0.20	–0.31	83-dap/del(◆)
353.15	784.00 ± 1.50	16.00	66-efr[1)]	318.15	795.83 ± 0.20	–0.34	83-dap/del(◆)
373.15	752.00 ± 1.50	1.41	66-efr[1)]	298.15	810.86 ± 0.10	–0.45	84-sak/nak(O)
393.15	732.00 ± 1.50	0.04	66-efr[1)]	268.15	839.00 ± 0.50	5.72	85-dap/don(×)
413.15	712.00 ± 1.50	0.08	66-efr(×)	278.15	825.30 ± 0.50	–0.72	85-dap/don(×)
433.15	690.00 ± 1.50	–0.24	66-efr(×)	298.15	810.90 ± 0.50	–0.41	85-dap/don[1)]
453.15	665.00 ± 1.50	–1.72	66-efr(×)	308.15	803.90 ± 0.50	0.09	85-dap/don[1)]
473.15	640.00 ± 1.50	–1.11	66-efr(×)	298.15	811.00 ± 0.20	–0.31	86-rig/mar[1)]
493.15	612.00 ± 1.50	–0.01	66-efr(×)	303.15	807.30 ± 0.20	–0.28	86-rig/mar[1)]
513.15	581.00 ± 1.50	0.43	66-efr(×)	308.15	803.90 ± 0.20	0.09	86-rig/mar(×)
533.15	545.00 ± 1.50	–0.47	66-efr(×)	313.15	799.90 ± 0.20	–0.11	86-rig/mar(×)
553.15	500.00 ± 2.00	–0.71	66-efr(×)	323.15	792.18 ± 0.10	–0.11	93-gar/ban-1(□)
563.15	471.00 ± 2.00	–0.02	66-efr(×)	328.15	788.92 ± 0.10	0.55	93-gar/ban-1(□)
568.15	454.00 ± 2.00	0.84	66-efr(×)	333.15	784.39 ± 0.10	-0.01	93-gar/ban-1(□)
573.15	434.00 ± 3.00	1.63	66-efr(×)	338.15	780.59 ± 0.10	0.21	93-gar/ban-1(□)
578.15	410.00 ± 3.00	2.82	66-efr(×)	343.15	776.19 ± 0.10	–0.12	93-gar/ban-1(□)
583.15	365.00 ± 5.00	–8.61	66-efr(×)	348.15	771.92 ± 0.10	–0.27	93-gar/ban-1(□)
584.15	350.00 ± 5.00	–14.84	66-efr(×)	353.15	768.08 ± 0.10	0.08	93-gar/ban-1(□)
585.15	324.00 ± 5.00	–30.71	66-efr[1)]	358.15	763.64 ± 0.10	–0.11	93-gar/ban-1(□)
293.15	814.45 ± 0.20	–0.57	76-hal/ell(×)	363.15	759.36 ± 0.10	–0.07	93-gar/ban-1(□)
298.15	810.80 ± 0.20	–0.51	76-hal/ell[1)]	373.15	750.46 ± 0.10	–0.13	93-gar/ban-1(□)
303.15	807.12 ± 0.20	–0.46	76-hal/ell[1)]	278.15	825.34 ± 0.40	–0.68	94-rom/pel(×)
320.00	794.46 ± 0.20	–0.28	76-hal/ell(×)	288.15	818.10 ± 0.40	–0.61	94-rom/pel[1)]
340.00	778.81 ± 0.20	–0.07	76-hal/ell(×)	298.15	810.87 ± 0.40	–0.44	94-rom/pel[1)]
360.00	762.11 ± 0.20	–0.05	76-hal/ell(×)	308.15	803.40 ± 0.40	–0.41	94-rom/pel[1)]
380.00	744.25 ± 0.20	-0.11	76-hal/ell(×)				

[1)] Not included in Fig. 1..

cont.

1-Pentanol (cont.)

Further references: [1854-kop, 1864-lan, 1871-lie/ros-3, 1880-bru-1, 1883-sch-3, 1884-zan, 1886-gar, 1892-lan/jah, 1894-jah/mol, 13-kis-3, 13-rom, 23-pop, 24-bou, 24-bus-1, 24-lie, 27-ver/coo, 28-llo/bro, 29-sim, 30-err/she, 30-she, 32-tim/hen, 33-but/tho, 35-but/ram, 39-lar/hun, 45-add, 48-jon/bow, 48-lad/smi, 48-vog-2, 48-wei, 49-tsc/ric, 50-mum/phi, 50-pic/zie, 50-sac/sau, 52-coo, 52-von, 54-dub/luf, 59-ale, 59-lis/kor, 63-hov/sea, 63-tho/mea, 67-gol/per, 68-ano, 70-mye/cle, 73-khi/ale, 73-min/rue, 74-rao/nai, 74-rao/nai-1, 75-mat/fer, 76-kow/kas, 76-red/nai, 78-ast, 78-paz/gar, 78-tre/ben, 79-gyl/apa, 79-kiy/ben, 81-joo/arl, 81-nai/nai, 81-sjo/dyh, 82-kar/red, 82-nai/nai, 82-ort; 82-ven/dha, 83-fuk/ogi, 85-fer/pin, 85-ort/paz-1, 85-sar/paz, 86-ash/sri, 86-ort/paz-1, 86-san/sha, 86-tan/toy, 87-fer/ber, 89-ala/sal, 89-ort/sus, 89-vij/nai, 90-sri/nai, 90-vij/nai, 91-gar/her, 91-ram/muk, 91-yos/kat, 93-ami/ara, 93-ami/rai, 94-kim/lee, 94-kum/nai, 94-kum/nai-1, 94-ven/ven, 94-yu /tsa-1, 95-org/igl, 95-red/ram-1, 96-elb, 98-ami/ban, 98-ami/pat-1, 98-pai/che].

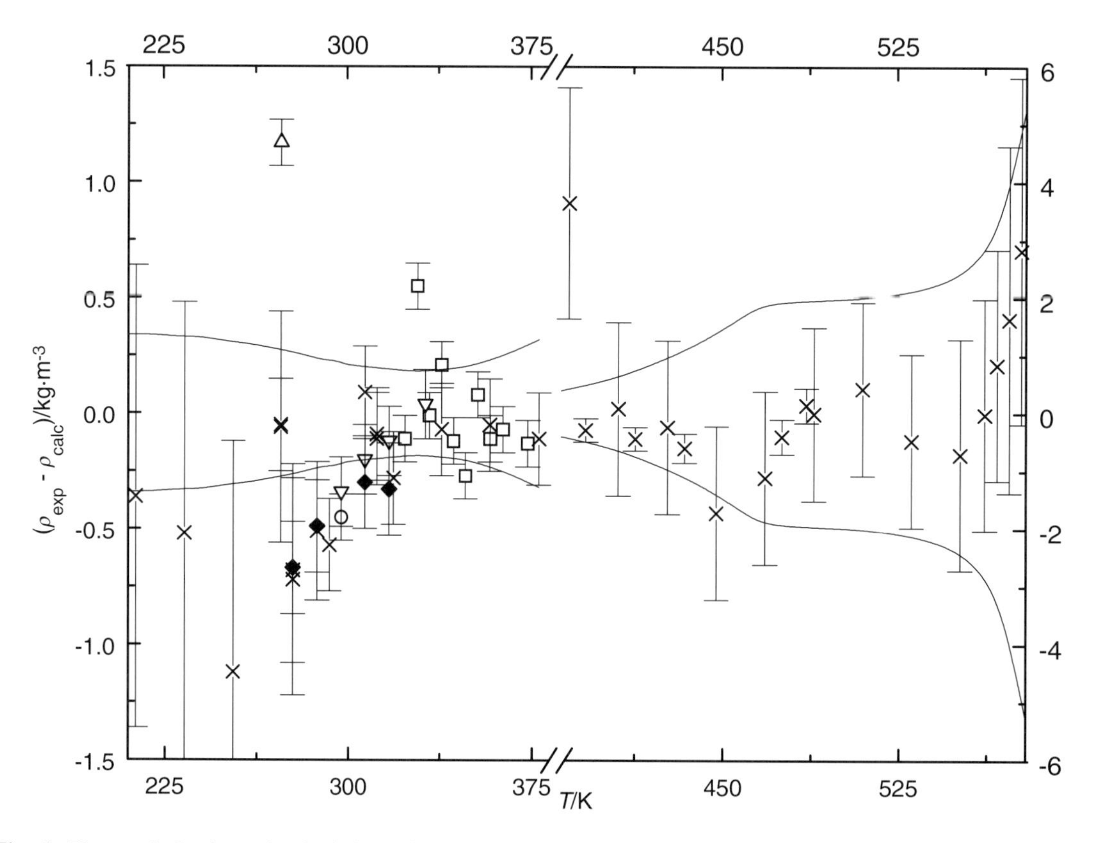

Fig. 1. The symbols show the deviation of the calculated from the experimental values from Table 2. The curves above and below the zero line indicate the calculated error region of the recommended values given in Table 3. The error bars represent the experimental errors. (Error bars smaller than the symbols are omitted for clarity of the figure.)

cont.

1-Pentanol (cont.)

Table 3. Recommended values (fit to the reliable experimental values according to the equations $\rho = A + BT + CT^2 + DT^3 + \ldots$ or $\rho = [1 + 1.75(1 - T/T_c)^{1/3} + 0.75(1 - T/T_c)][\rho_c + A(T_c - T) + B(T_c - T)^2 + C(T_c - T)^3 + D(T_c - T)^4]$).

$\frac{T}{K}$	$\frac{\rho \pm \sigma_{fit}}{kg \cdot m^{-3}}$	$\frac{T}{K}$	$\frac{\rho \pm \sigma_{fit}}{kg \cdot m^{-3}}$	$\frac{T}{K}$	$\frac{\rho \pm \sigma_{fit}}{kg \cdot m^{-3}}$
210.00	875.72 ± 0.34	330.00	786.91 ± 0.18	470.00	645.32 ± 1.86
220.00	868.27 ± 0.34	340.00	778.88 ± 0.19	480.00	631.51 ± 1.93
230.00	860.92 ± 0.33	350.00	770.64 ± 0.20	490.00	616.78 ± 1.95
240.00	853.64 ± 0.33	360.00	762.16 ± 0.23	500.00	601.48 ± 1.97
250.00	846.40 ± 0.31	370.00	753.41 ± 0.27	510.00	585.68 ± 1.99
260.00	839.17 ± 0.30	380.00	744.36 ± 0.32	520.00	569.16 ± 2.03
270.00	831.94 ± 0.28	390.00	734.99 ± 0.40	530.00	551.43 ± 2.09
280.00	824.68 ± 0.26	400.00	725.27 ± 0.49	540.00	531.69 ± 2.18
290.00	817.35 ± 0.23	410.00	715.18 ± 0.60	550.00	508.81 ± 2.33
293.15	815.02 ± 0.23	420.00	704.69 ± 0.74	560.00	481.13 ± 2.62
298.15	811.31 ± 0.22	430.00	693.77 ± 0.91	570.00	445.87 ± 3.28
300.00	809.93 ± 0.21	440.00	682.40 ± 1.10	580.00	396.14 ± 5.29
310.00	802.41 ± 0.20	450.00	670.55 ± 1.32		
320.00	794.74 ± 0.19	460.00	658.20 ± 1.57		

***d*-2-Pentanol** **[31087-44-2]** **$C_5H_{12}O$** **MW = 88.15** **19**

Table 1. Fit with estimated B coefficient for 2 accepted points. Deviation $\sigma_w = 0.785$.

Coefficient	$\rho = A + BT$
A	1039.28
B	−0.780

Table 2. Experimental values with uncertainties and deviation from calculated values.

$\frac{T}{K}$	$\frac{\rho_{exp} \pm 2\sigma_{est}}{kg \cdot m^{-3}}$	$\frac{\rho_{exp} - \rho_{calc}}{kg \cdot m^{-3}}$	Ref.
293.15	810.1 ± 1.0	−0.52	43-bra
298.15	807.9 ± 1.5	1.18	56-ike/kep
298.15	816.1 ± 3.0	9.38	60-ter/kep[1]

[1] Not included in calculation of linear coefficients.

Table 3. Recommended values.

$\frac{T}{K}$	$\frac{\rho_{exp} \pm 2\sigma_{est}}{kg \cdot m^{-3}}$
290.00	813.1 ± 1.2
293.15	810.6 ± 1.1
298.15	806.7 ± 1.1

2-Pentanol **[6032-29-7]** **$C_5H_{12}O$** **MW = 88.15** **20**

Table 1. Coefficients of the polynomial expansion equation.
Standard deviations (see introduction):
$\sigma_{c,w} = 5.1481 \cdot 10^{-1}$ (combined temperature ranges, weighted),
$\sigma_{c,uw} = 9.2505 \cdot 10^{-2}$ (combined temperature ranges, unweighted).

Coefficient	T = 233.15 to 381.35 K $\rho = A + BT + CT^2 + DT^3 + \ldots$
A	$9.37043 \cdot 10^{2}$
B	$-5.69541 \cdot 10^{-2}$
C	$-1.29071 \cdot 10^{-3}$

Table 2. Experimental values with uncertainties and deviation from calculated values.

T/K	$\rho_{exp} \pm 2\sigma_{est}$ / kg·m^{-3}	$\rho_{exp} - \rho_{calc}$ / kg·m^{-3}	Ref. (Symbol in Fig. 1)	T/K	$\rho_{exp} \pm 2\sigma_{est}$ / kg·m^{-3}	$\rho_{exp} - \rho_{calc}$ / kg·m^{-3}	Ref. (Symbol in Fig. 1)
273.15	824.79 ± 0.50	−0.39	32-ell/rei(×)	293.25	806.90 ± 0.50	−2.45	63-tho/mea[1)]
298.15	805.26 ± 0.50	−0.07	32-ell/rei[1)]	303.25	801.50 ± 0.50	0.42	63-tho/mea[1)]
273.15	824.68 ± 0.50	−0.50	32-tim/hen(×)	314.45	792.30 ± 0.50	0.79	63-tho/mea(×)
288.15	813.17 ± 0.50	−0.29	32-tim/hen(×)	323.95	783.60 ± 0.60	0.46	63-tho/mea(×)
303.15	801.18 ± 0.50	0.02	32-tim/hen[1)]	347.75	761.70 ± 0.60	0.55	63-tho/mea(×)
298.15	805.25 ± 0.30	−0.08	35-but/ram(◆)	366.65	742.80 ± 0.60	0.15	63-tho/mea(×)
293.15	808.80 ± 0.40	−0.63	50-pic/zie(×)	381.35	727.40 ± 0.70	−0.22	63-tho/mea(×)
293.15	809.70 ± 0.30	0.27	52-coo(○)	298.15	805.60 ± 0.30	0.27	82-dap/don(∇)
293.15	809.80 ± 0.40	0.37	54-pom/foo-1(×)	298.15	805.16 ± 0.30	−0.17	86-ort/paz-1(Δ)
233.15	854.90 ± 1.00	1.30	58-cos/bow(×)	303.15	800.97 ± 0.30	0.19	86-ort/paz-1(Δ)
253.15	840.20 ± 0.60	0.29	58-cos/bow(×)	308.15	796.97 ± 0.30	0.04	86-ort/paz-1(Δ)
273.15	824.60 ± 0.60	−0.58	58-cos/bow(×)	318.15	788.26 ± 0.30	−0.02	86-ort/paz-1(Δ)
293.15	808.90 ± 0.60	−0.53	58-cos/bow[1)]	323.15	783.75 ± 0.30	−0.10	86-ort/paz-1(Δ)
313.15	791.90 ± 0.60	−0.74	58-cos/bow[1)]	298.15	805.50 ± 0.20	0.17	86-rig/mar(□)
333.15	774.20 ± 0.60	−0.61	58-cos/bow(×)	303.15	801.00 ± 0.20	−0.16	86-rig/mar(□)
353.15	755.10 ± 0.70	−0.86	58-cos/bow(×)	308.15	796.90 ± 0.20	−0.03	86-rig/mar(□)
373.15	730.40 ± 0.70	−5.67	58-cos/bow[1)]	313.15	792.80 ± 0.20	0.16	86-rig/mar(□)

[1)] Not included in Fig. 1.

Table 3. Recommended values (fit to the reliable experimental values according to the equations
$\rho = A + BT + CT^2 + DT^3 + \ldots$ or $\rho = [1 + 1.75(1 - T/T_c)^{1/3} + 0.75(1 - T/T_c)][\rho_c + A(T_c - T) + B(T_c - T)^2 + C(T_c - T)^3 + D(T_c - T)^4]$).

T/K	$\rho \pm \sigma_{fit}$ / kg·m^{-3}	T/K	$\rho \pm \sigma_{fit}$ / kg·m^{-3}	T/K	$\rho \pm \sigma_{fit}$ / kg·m^{-3}
230.00	855.66 ± 1.21	293.15	809.43 ± 0.35	350.00	759.00 ± 0.58
240.00	849.03 ± 0.96	298.15	805.33 ± 0.34	360.00	749.26 ± 0.66
250.00	842.13 ± 0.75	300.00	803.79 ± 0.34	370.00	739.27 ± 0.74
260.00	834.98 ± 0.59	310.00	795.35 ± 0.36	380.00	729.02 ± 0.82
270.00	827.57 ± 0.47	320.00	786.65 ± 0.40	390.00	718.51 ± 0.89
280.00	819.90 ± 0.40	330.00	777.69 ± 0.45		
290.00	811.98 ± 0.36	340.00	768.47 ± 0.51		

cont.

2-Pentanol (cont.)

Further references: [1876-lin, 11-pic/ken, 14-vav, 19-eyk, 23-bru, 23-clo/joh, 25-fai, 26-mun, 27-nor/cor, 30-err/she, 36-nor/has, 37-gin/bau, 56-goe/mcc, 63-mcc/lai].

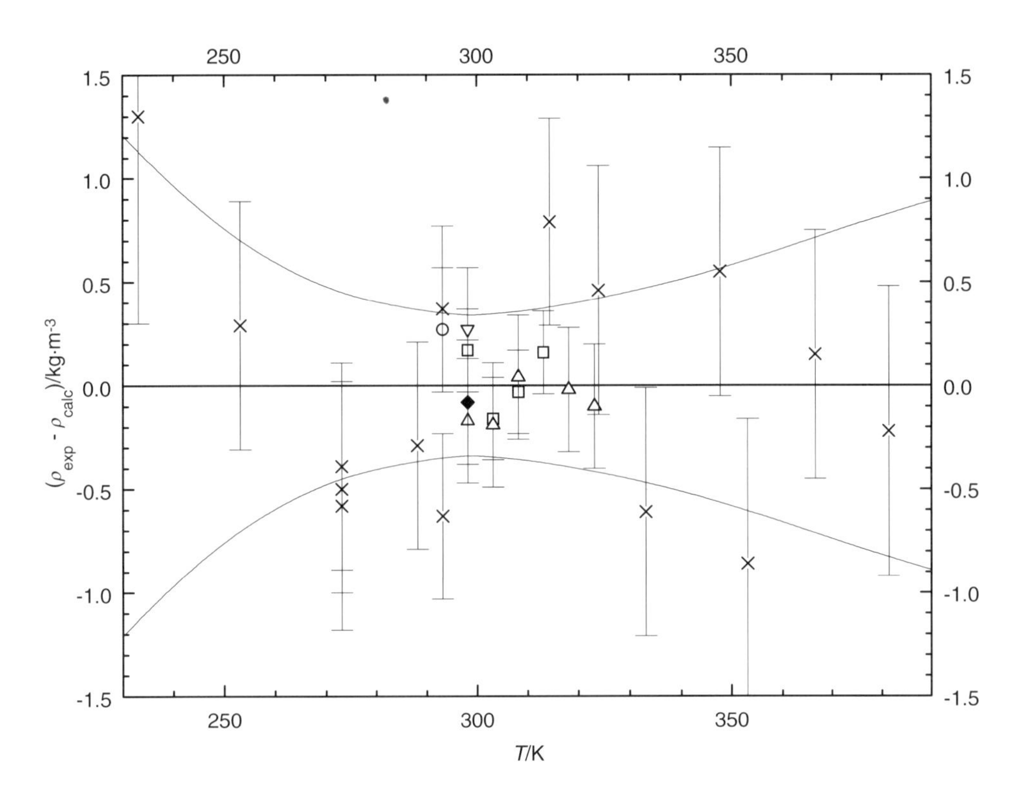

Fig. 1. The symbols show the deviation of the calculated from the experimental values from Table 2. The curves above and below the zero line indicate the calculated error region of the recommended values given in Table 3. The error bars represent the experimental errors. (Error bars smaller than the symbols are omitted for clarity of the figure.)

3-Pentanol **[584-02-1]** $C_5H_{12}O$ **MW = 88.15** **21**

Table 1. Coefficients of the polynomial expansion equation.
Standard deviations (see introduction):
$\sigma_{c,w}$ = 1.0029 (combined temperature ranges, weighted),
$\sigma_{c,uw}$ = 1.4273 · 10^{-1} (combined temperature ranges, unweighted).

Coefficient	T = 273.15 to 381.35 K $\rho = A + BT + CT^2 + DT^3 + \dots$
A	9.59951 · 10^{2}
B	−9.37135 · 10^{-2}
C	−1.30516 · 10^{-3}

cont.

3-Pentanol (cont.)

Table 2. Experimental values with uncertainties and deviation from calculated values.

$\frac{T}{\mathrm{K}}$	$\frac{\rho_{exp} \pm 2\sigma_{est}}{\mathrm{kg \cdot m^{-3}}}$	$\frac{\rho_{exp} - \rho_{calc}}{\mathrm{kg \cdot m^{-3}}}$	Ref. (Symbol in Fig. 1)	$\frac{T}{\mathrm{K}}$	$\frac{\rho_{exp} \pm 2\sigma_{est}}{\mathrm{kg \cdot m^{-3}}}$	$\frac{\rho_{exp} - \rho_{calc}}{\mathrm{kg \cdot m^{-3}}}$	Ref. (Symbol in Fig. 1)
298.15	815.40 ± 0.50	−0.59	23-bru(□)	292.55	820.80 ± 0.40	−0.03	63-tho/mea(X)
273.15	836.79 ± 1.00	−0.18	32-tim/hen(X)	303.05	811.90 ± 0.40	0.21	63-tho/mea(X)
288.15	824.64 ± 1.00	0.06	32-tim/hen(X)	314.45	801.50 ± 0.40	0.07	63-tho/mea(X)
303.15	811.78 ± 1.00	0.18	32-tim/hen(X)	323.97	792.80 ± 0.50	0.19	63-tho/mea(X)
303.15	812.20 ± 1.00	0.60	33-nev/jat(X)	347.75	769.80 ± 0.50	0.27	63-tho/mea(X)
293.15	820.20 ± 0.40	−0.12	50-lad/smi(Δ)	366.65	750.10 ± 0.50	−0.04	63-tho/mea(X)
293.15	820.30 ± 0.50	−0.02	50-pic/zie(X)	381.35	733.60 ± 0.60	−0.81	63-tho/mea(X)
293.15	820.70 ± 0.40	0.38	54-pom/foo-1(∇)	298.15	815.20 ± 0.30	−0.79	82-dap/don(O)
298.15	815.00 ± 0.50	−0.99	63-mcc/lai(X)				

Further references: [1875-wag/say, 13-pic/ken, 14-vav, 19-eyk, 27-nor/cor, 30-err/she, 35-les/lom, 36-spe, 37-gin/bau, 37-she/mat, 40-whi/sur, 48-wei, 52-coo, 53-ano-1, 56-goe/mcc, 56-shu/bel].

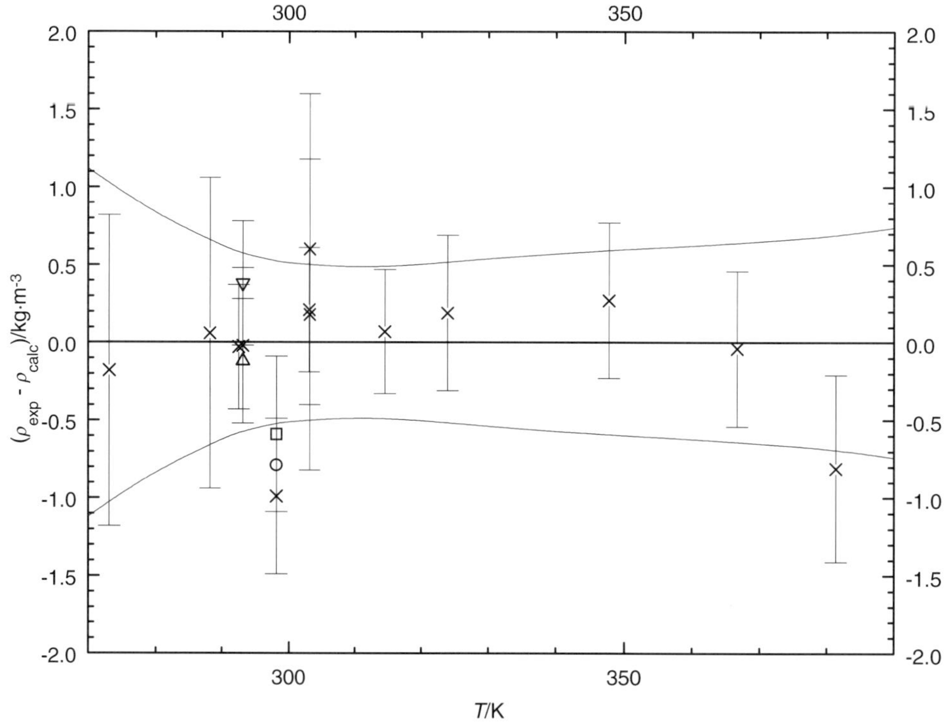

Fig. 1. The symbols show the deviation of the calculated from the experimental values from Table 2. The curves above and below the zero line indicate the calculated error region of the recommended values given in Table 3. The error bars represent the experimental errors. (Error bars smaller than the symbols are omitted for clarity of the figure.)

cont.

Table 3. Recommended values (fit to the reliable experimental values according to the equations $\rho = A + BT + CT^2 + DT^3 + \ldots$ or $\rho = [1 + 1.75(1 - T/T_c)^{1/3} + 0.75(1 - T/T_c)][\rho_c + A(T_c - T) + B(T_c - T)^2 + C(T_c - T)^3 + D(T_c - T)^4]$).

T/K	$\rho \pm \sigma_{fit}$/kg·m^{-3}	T/K	$\rho \pm \sigma_{fit}$/kg·m^{-3}	T/K	$\rho \pm \sigma_{fit}$/kg·m^{-3}
270.00	839.50 ± 1.12	300.00	814.37 ± 0.51	350.00	767.27 ± 0.60
280.00	831.39 ± 0.82	310.00	805.47 ± 0.48	360.00	757.07 ± 0.62
290.00	823.01 ± 0.62	320.00	796.31 ± 0.50	370.00	746.60 ± 0.65
293.15	820.32 ± 0.57	330.00	786.89 ± 0.54	380.00	735.87 ± 0.68
298.15	815.99 ± 0.52	340.00	777.21 ± 0.57	390.00	724.89 ± 0.74

2,2-Dimethyl-1-butanol [1185-33-7] $C_6H_{14}O$ MW = 102.18 22

Table 1. Coefficients of the polynomial expansion equation.
Standard deviations (see introduction):
$\sigma_{c,w} = 3.0286 \cdot 10^{-1}$ (combined temperature ranges, weighted),
$\sigma_{c,uw} = 9.4262 \cdot 10^{-2}$ (combined temperature ranges, unweighted).

Coefficient	T = 278.15 to 408.15 K $\rho = A + BT + CT^2 + DT^3 + \ldots$
A	$9.76530 \cdot 10^2$
B	$-2.16929 \cdot 10^{-1}$
C	$-9.82641 \cdot 10^{-4}$

Table 2. Experimental values with uncertainties and deviation from calculated values.

T/K	$\rho_{exp} \pm 2\sigma_{est}$/kg·m^{-3}	$\rho_{exp} - \rho_{calc}$/kg·m^{-3}	Ref. (Symbol in Fig. 1)	T/K	$\rho_{exp} \pm 2\sigma_{est}$/kg·m^{-3}	$\rho_{exp} - \rho_{calc}$/kg·m^{-3}	Ref. (Symbol in Fig. 1)
278.15	841.01 ± 0.40	0.84	40-hov/lan-2(□)	368.15	763.86 ± 0.40	0.37	40-hov/lan-2(□)
288.15	832.37 ± 0.40	–0.06	40-hov/lan-2(□)	378.15	754.30 ± 0.40	0.32	40-hov/lan-2(□)
298.15	824.27 ± 0.40	–0.23	40-hov/lan-2(□)	388.15	744.38 ± 0.40	0.10	40-hov/lan-2(□)
308.15	816.20 ± 0.40	–0.18	40-hov/lan-2(□)	398.15	734.38 ± 0.40	–0.01	40-hov/lan-2(□)
318.15	807.90 ± 0.40	–0.15	40-hov/lan-2(□)	408.15	724.02 ± 0.40	–0.28	40-hov/lan-2(□)
328.15	799.47 ± 0.40	–0.06	40-hov/lan-2(□)	293.15	828.30 ± 0.60	–0.19	50-pic/zie(O)
338.15	790.97 ± 0.40	0.16	40-hov/lan-2(□)	293.15	827.50 ± 1.00	–0.99	64-blo/hag(Δ)
348.15	782.03 ± 0.40	0.13	40-hov/lan-2(□)	293.15	828.50 ± 1.00	0.01	65-shu/puz(∇)
358.15	773.02 ± 0.40	0.23	40-hov/lan-2(□)				

Table 3. Recommended values (fit to the reliable experimental values according to the equations $\rho = A + BT + CT^2 + DT^3 + \ldots$ or $\rho = [1 + 1.75(1 - T/T_c)^{1/3} + 0.75(1 - T/T_c)][\rho_c + A(T_c - T) + B(T_c - T)^2 + C(T_c - T)^3 + D(T_c - T)^4]$).

T/K	$\rho \pm \sigma_{fit}$/kg·m^{-3}	T/K	$\rho \pm \sigma_{fit}$/kg·m^{-3}	T/K	$\rho \pm \sigma_{fit}$/kg·m^{-3}
270.00	846.32 ± 0.76	310.00	814.85 ± 0.52	370.00	761.74 ± 0.40
280.00	838.75 ± 0.68	320.00	806.49 ± 0.49	380.00	752.20 ± 0.40
290.00	830.98 ± 0.61	330.00	797.93 ± 0.46	390.00	742.47 ± 0.41
293.15	828.49 ± 0.59	340.00	789.18 ± 0.45	400.00	732.54 ± 0.44
298.15	824.50 ± 0.56	350.00	780.23 ± 0.43	410.00	722.41 ± 0.50
300.00	823.01 ± 0.56	360.00	771.09 ± 0.41	420.00	712.08 ± 0.57

cont.

2,2-Dimethyl-1-butanol (cont.)

Further references: [04-bou/bla-2, 38-gin/web].

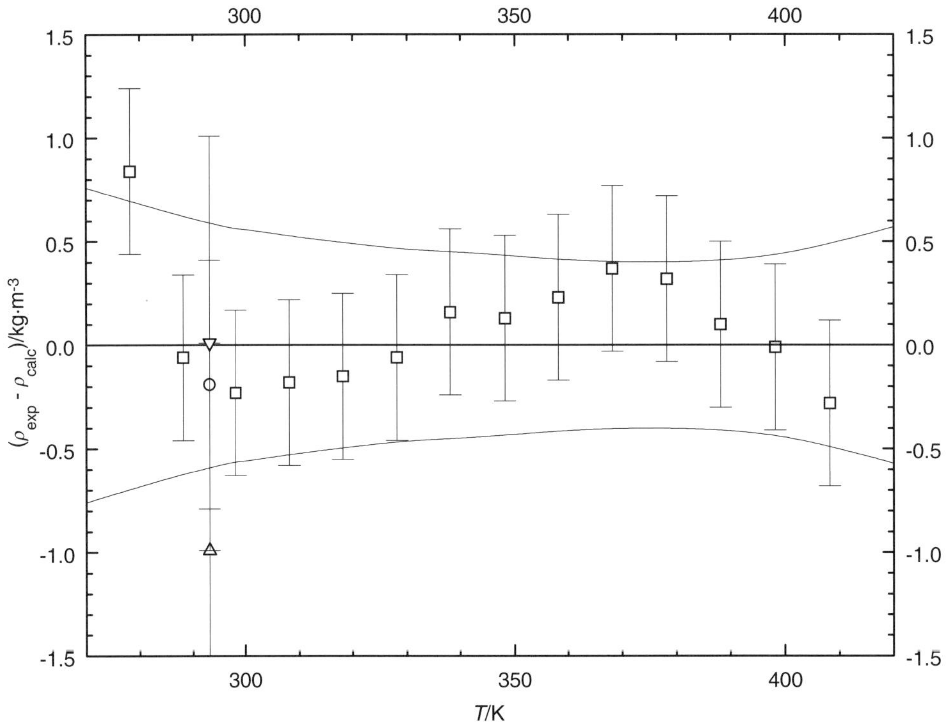

Fig. 1. The symbols show the deviation of the calculated from the experimental values from Table 2. The curves above and below the zero line indicate the calculated error region of the recommended values given in Table 3. The error bars represent the experimental errors. (Error bars smaller than the symbols are omitted for clarity of the figure.)

2,3-Dimethyl-1-butanol **[19550-30-2]** **$C_6H_{14}O$** **MW = 102.18** **23**

Table 1. Experimental and recommended values with uncertainties.

$\frac{T}{\mathrm{K}}$	$\frac{\rho_{exp} \pm 2\sigma_{est}}{\mathrm{kg \cdot m^{-3}}}$	Ref.	$\frac{T}{\mathrm{K}}$	$\frac{\rho_{exp} \pm 2\sigma_{est}}{\mathrm{kg \cdot m^{-3}}}$	Ref.
298.15	823.7 ± 2.0	59-tsu/hay[1)]	293.15	830.1 ± 0.6	50-pic/zie
293.65	829.7 ± 0.6	13-gor	293.15	830.6 ± 1.0	52-coo
298.15	823.0 ± 2.0	35-lev/mar[1)]	293.15	830.0 ± 0.6	Recommended

[1)] Not included in calculation of recommended value.

(*R*)-2,3-Dimethyl-1-butanol **[15019-27-9]** **$C_6H_{14}O$** **MW = 102.18** **24**

Table 1. Experimental value with uncertainty.

$\frac{T}{\mathrm{K}}$	$\frac{\rho_{exp} \pm 2\sigma_{est}}{\mathrm{kg \cdot m^{-3}}}$	Ref.
298.15	823.7 ± 0.8	59-tsu/hay

2,3-Dimethyl-2-butanol **[594-60-5]** **$C_6H_{14}O$** **MW = 102.18** **25**

Table 1. Coefficients of the polynomial expansion equation.
Standard deviations (see introduction):
$\sigma_{c,w} = 6.3478 \cdot 10^{-1}$ (combined temperature ranges, weighted),
$\sigma_{c,uw} = 1.4692 \cdot 10^{-1}$ (combined temperature ranges, unweighted).

Coefficient	T = 273.15 to 388.15 K $\rho = A + BT + CT^2 + DT^3 + \ldots$
A	$9.39911 \cdot 10^{2}$
B	$6.32414 \cdot 10^{-2}$
C	$-1.58086 \cdot 10^{-3}$

Table 2. Experimental values with uncertainties and deviation from calculated values.

$\frac{T}{\mathrm{K}}$	$\frac{\rho_{exp} \pm 2\sigma_{est}}{\mathrm{kg \cdot m^{-3}}}$	$\frac{\rho_{exp} - \rho_{calc}}{\mathrm{kg \cdot m^{-3}}}$	Ref. (Symbol in Fig. 1)	$\frac{T}{\mathrm{K}}$	$\frac{\rho_{exp} \pm 2\sigma_{est}}{\mathrm{kg \cdot m^{-3}}}$	$\frac{\rho_{exp} - \rho_{calc}}{\mathrm{kg \cdot m^{-3}}}$	Ref. (Symbol in Fig. 1)
273.15	838.60 ± 1.00	–0.64	1879-pav(∇)	348.15	770.02 ± 0.20	–0.29	41-hov/lan(□)
292.15	821.90 ± 1.00	–1.56	1879-pav(∇)	358.15	759.67 ± 0.20	–0.11	41-hov/lan(□)
278.15	835.11 ± 0.20	–0.08	41-hov/lan(□)	368.15	748.92 ± 0.20	–0.01	41-hov/lan(□)
288.15	826.76 ± 0.20	–0.11	41-hov/lan(□)	378.15	738.04 ± 0.20	0.27	41-hov/lan(□)
298.15	818.59 ± 0.20	0.35	41-hov/lan(□)	388.15	726.36 ± 0.20	0.07	41-hov/lan(□)
308.15	809.51 ± 0.20	0.22	41-hov/lan(□)	293.15	823.60 ± 0.40	1.00	47-how/mea(○)
318.15	799.84 ± 0.20	–0.18	41-hov/lan(□)	298.15	819.30 ± 0.40	1.06	47-how/mea(○)
328.15	790.56 ± 0.20	0.13	41-hov/lan(□)	293.15	822.30 ± 0.60	–0.30	50-pic/zie(Δ)
338.15	780.70 ± 0.20	0.17	41-hov/lan(□)				

Further references: [1873-fri/sil, 52-lev/tan, 52-ove/ber].

Table 3. Recommended values (fit to the reliable experimental values according to the equations $\rho = A + BT + CT^2 + DT^3 + \ldots$ or $\rho = [1 + 1.75(1 - T/T_c)^{1/3} + 0.75(1 - T/T_c)][\rho_c + A(T_c - T) + B(T_c - T)^2 + C(T_c - T)^3 + D(T_c - T)^4]$).

$\frac{T}{\mathrm{K}}$	$\frac{\rho \pm \sigma_{fit}}{\mathrm{kg \cdot m^{-3}}}$	$\frac{T}{\mathrm{K}}$	$\frac{\rho \pm \sigma_{fit}}{\mathrm{kg \cdot m^{-3}}}$	$\frac{T}{\mathrm{K}}$	$\frac{\rho \pm \sigma_{fit}}{\mathrm{kg \cdot m^{-3}}}$
270.00	841.74 ± 0.73	310.00	807.60 ± 0.31	370.00	746.89 ± 0.19
280.00	833.68 ± 0.60	320.00	798.27 ± 0.25	380.00	735.67 ± 0.24
290.00	825.30 ± 0.48	330.00	788.62 ± 0.21	390.00	724.13 ± 0.31
293.15	822.60 ± 0.45	340.00	778.67 ± 0.18	400.00	712.27 ± 0.40
298.15	818.24 ± 0.40	350.00	768.39 ± 0.17		
300.00	816.61 ± 0.38	360.00	757.80 ± 0.17		

cont.

2,3-Dimethyl-2-butanol (cont.)

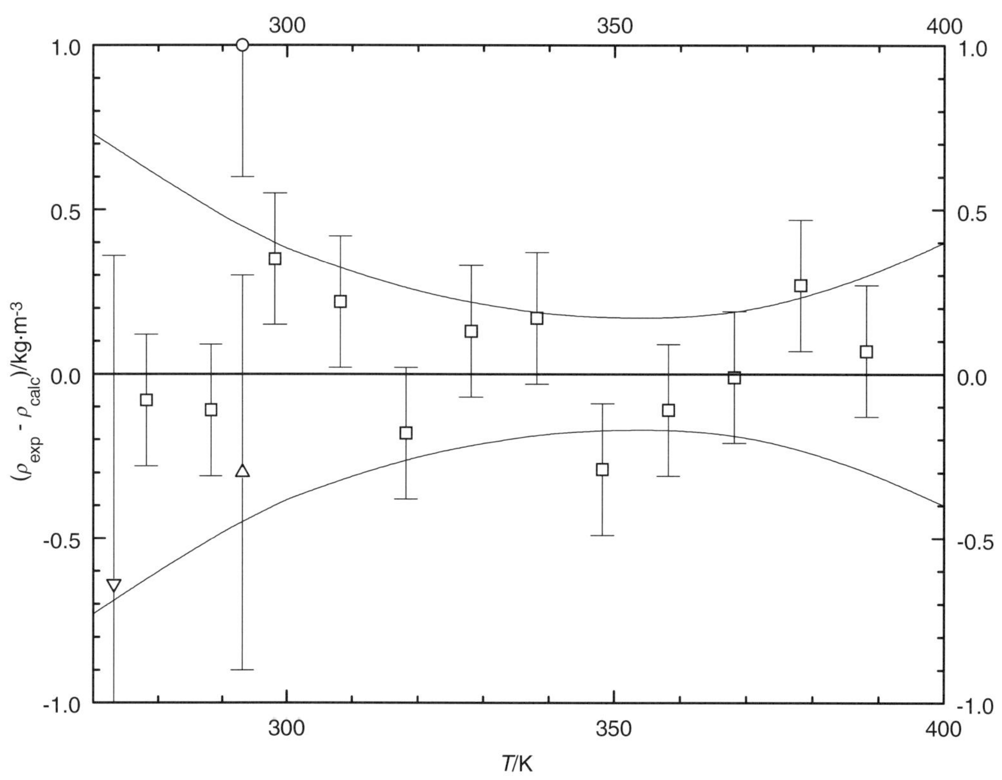

Fig. 1. The symbols show the deviation of the calculated from the experimental values from Table 2. The curves above and below the zero line indicate the calculated error region of the recommended values given in Table 3. The error bars represent the experimental errors. (Error bars smaller than the symbols are omitted for clarity of the figure.)

3,3-Dimethyl-1-butanol **[624-95-3]** **$C_6H_{14}O$** **MW = 102.18** **26**

Table 1. Experimental values with uncertainties.

$\frac{T}{K}$	$\frac{\rho_{exp} \pm 2\sigma_{est}}{kg \cdot m^{-3}}$	Ref.
297.15	814.0 ± 2.0	41-hus/age
288.15	844.0 ± 10.0	49-mal/vol
298.15	809.7 ± 2.0	56-sar/new

3,3-Dimethyl-2-butanol **[464-07-3]** **$C_6H_{14}O$** **MW = 102.18** **27**

Table 1. Experimental and recommended values with uncertainties.

$\frac{T}{K}$	$\frac{\rho_{exp} \pm 2\sigma_{est}}{kg \cdot m^{-3}}$	Ref.	$\frac{T}{K}$	$\frac{\rho_{exp} \pm 2\sigma_{est}}{kg \cdot m^{-3}}$	Ref.
293.15	818.5 ± 1.0	14-low	298.15	812.2 ± 2.0	56-sar/new[1)]
298.15	810.0 ± 3.0	33-ste[1)]	293.15	816.7 ± 1.0	50-pic/zie
298.15	815.7 ± 2.0	38-gin/web[1)]	293.15	817.9 ± 1.2	Recommended
293.15	818.4 ± 1.0	50-mos/lac			

[1)] Not included in calculation of recommended value.

(*S*)-3,3-Dimethyl-2-butanol **[1517-67-5]** **$C_6H_{14}O$** **MW = 102.18** **28**

Table 1. Experimental value with uncertainty.

T / K	$\rho_{exp} \pm 2\sigma_{est}$ / $kg \cdot m^{-3}$	Ref.
298.15	810.0 ± 2.0	33-ste

2-Ethyl-1-butanol **[97-95-0]** **$C_6H_{14}O$** **MW = 102.18** **29**

Table 1. Coefficients of the polynomial expansion equation.
Standard deviations (see introduction):
$\sigma_{c,w} = 2.6167 \cdot 10^{-1}$ (combined temperature ranges, weighted),
$\sigma_{c,uw} = 5.9806 \cdot 10^{-2}$ (combined temperature ranges, unweighted).

Coefficient	T = 288.15 to 418.15 K $\rho = A + BT + CT^2 + DT^3 + \ldots$
A	$9.36526 \cdot 10^{2}$
B	$1.79278 \cdot 10^{-2}$
C	$-1.26674 \cdot 10^{-3}$

Table 2. Experimental values with uncertainties and deviation from calculated values.

T / K	$\rho_{exp} \pm 2\sigma_{est}$ / $kg \cdot m^{-3}$	$\rho_{exp} - \rho_{calc}$ / $kg \cdot m^{-3}$	Ref. (Symbol in Fig. 1)	T / K	$\rho_{exp} \pm 2\sigma_{est}$ / $kg \cdot m^{-3}$	$\rho_{exp} - \rho_{calc}$ / $kg \cdot m^{-3}$	Ref. (Symbol in Fig. 1)
293.15	832.60 ± 1.00	–0.32	38-whi/kar(×)	368.15	771.39 ± 0.50	–0.05	40-hov/lan-2(Δ)
293.15	832.80 ± 0.50	–0.12	39-gol/tay(◆)	378.15	762.31 ± 0.50	0.15	40-hov/lan-2(Δ)
278.15	845.04 ± 0.50	1.53	40-hov/lan-2[1)]	388.15	752.47 ± 0.50	–0.17	40-hov/lan-2(Δ)
288.15	837.33 ± 0.50	0.82	40-hov/lan-2(Δ)	398.15	742.87 ± 0.50	0.01	40-hov/lan-2(Δ)
298.15	829.53 ± 0.50	0.26	40-hov/lan-2(Δ)	408.15	732.74 ± 0.50	–0.08	40-hov/lan-2(Δ)
308.15	822.03 ± 0.50	0.27	40-hov/lan-2(Δ)	418.15	722.29 ± 0.50	–0.24	40-hov/lan-2(Δ)
318.15	814.11 ± 0.50	0.10	40-hov/lan-2(Δ)	293.15	833.10 ± 0.60	0.18	52-coo(×)
328.15	805.93 ± 0.50	–0.07	40-hov/lan-2(Δ)	293.15	832.90 ± 0.50	–0.02	68-ano(∇)
338.15	797.38 ± 0.50	–0.36	40-hov/lan-2(Δ)	298.15	829.11 ± 0.20	–0.16	84-kim/ben(□)
348.15	789.18 ± 0.50	–0.05	40-hov/lan-2(Δ)	298.15	829.07 ± 0.20	–0.20	85-ort/paz-2(○)
358.15	780.51 ± 0.50	0.05	40-hov/lan-2(Δ)				

[1)] Not included in Fig. 1.

Table 3. Recommended values (fit to the reliable experimental values according to the equations
$\rho = A + BT + CT^2 + DT^3 + \ldots$ or $\rho = [1 + 1.75(1 - T/T_c)^{1/3} + 0.75(1 - T/T_c)][\rho_c + A(T_c - T) + B(T_c - T)^2 + C(T_c - T)^3 + D(T_c - T)^4]$).

T / K	$\rho \pm \sigma_{fit}$ / $kg \cdot m^{-3}$	T / K	$\rho \pm \sigma_{fit}$ / $kg \cdot m^{-3}$	T / K	$\rho \pm \sigma_{fit}$ / $kg \cdot m^{-3}$
280.00	842.23 ± 0.55	320.00	812.55 ± 0.51	380.00	760.42 ± 0.51
290.00	835.19 ± 0.52	330.00	804.49 ± 0.52	390.00	750.85 ± 0.51
293.15	832.92 ± 0.52	340.00	796.19 ± 0.52	400.00	741.02 ± 0.52
298.15	829.27 ± 0.51	350.00	787.62 ± 0.52	410.00	730.94 ± 0.55
300.00	827.90 ± 0.51	360.00	778.81 ± 0.52	420.00	720.60 ± 0.60
310.00	820.35 ± 0.51	370.00	769.74 ± 0.52	430.00	710.01 ± 0.68

cont.

2-Ethyl-1-butanol (cont.)

Further references: [48-wei, 50-pic/zie, 53-ano-1, 58-ano-5, 58-cos/bow].

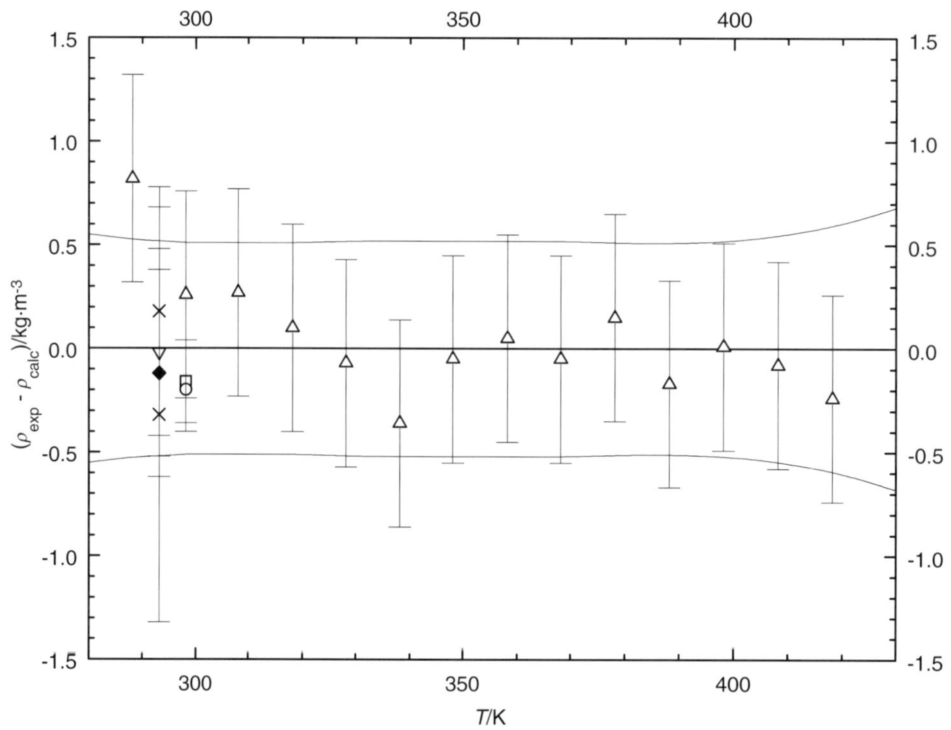

Fig. 1. The symbols show the deviation of the calculated from the experimental values from Table 2. The curves above and below the zero line indicate the calculated error region of the recommended values given in Table 3. The error bars represent the experimental errors. (Error bars smaller than the symbols are omitted for clarity of the figure.)

1-Hexanol **[111-27-3]** **$C_6H_{14}O$** **MW = 102.18** **30**

T_c = 611.00 K [89-tej/lee]
ρ_c = 268.00 kg·m^{-3} [89-tej/lee]

Table 1. Coefficients for the polynomial expansion equations. Standard deviations (see introduction): $\sigma_\ell = 7.2858 \cdot 10^{-1}$ (low temperature range), $\sigma_{c,w}$ = (4.6480 · 10^{-1} combined temperature ranges, weighted), $\sigma_{c,uw} = 2.7537 \cdot 10^{-1}$ (combined temperature ranges, unweighted).

Coefficient	T = 235.72 to 500.00 K $\rho = A + BT + CT^2 + DT^3 + \ldots$	T = 500.00 to 611.00 K $\rho = [1 + 1.75(1 - T/T_c)^{1/3} + 0.75(1 - T/T_c)]$ $[\rho_c + A(T_c - T) + B(T_c - T)^2 + C(T_c - T)^3$ $+ D(T_c - T)^4]$
A	$1.01544 \cdot 10^{3}$	$8.35316 \cdot 10^{-1}$
B	$-7.78875 \cdot 10^{-1}$	$-1.13272 \cdot 10^{-2}$
C	$8.80075 \cdot 10^{-4}$	$8.40546 \cdot 10^{-5}$
D	$-1.73223 \cdot 10^{-6}$	$-2.64908 \cdot 10^{-7}$

cont.

Table 2. Experimental values with uncertainties and deviation from calculated values.

$\frac{T}{K}$	$\frac{\rho_{exp} \pm 2\sigma_{est}}{kg \cdot m^{-3}}$	$\frac{\rho_{exp} - \rho_{calc}}{kg \cdot m^{-3}}$	Ref. (Symbol in Fig. 1)	$\frac{T}{K}$	$\frac{\rho_{exp} \pm 2\sigma_{est}}{kg \cdot m^{-3}}$	$\frac{\rho_{exp} - \rho_{calc}}{kg \cdot m^{-3}}$	Ref. (Symbol in Fig. 1)
273.15	833.34 ± 0.20	0.29	32-ell/rei(×)	553.15	545.00 ± 1.50	–0.12	66-efr(×)
298.15	815.79 ± 0.20	0.25	32-ell/rei[1)]	563.15	524.00 ± 1.50	0.10	66-efr(×)
278.15	829.83 ± 0.50	0.23	38-hov/lan[1)]	573.15	500.00 ± 1.50	0.24	66-efr(×)
288.15	822.67 ± 0.50	0.04	38-hov/lan[1)]	583.15	473.00 ± 2.00	1.53	66-efr(×)
298.15	815.54 ± 0.50	0.00	38-hov/lan[1)]	588.15	457.00 ± 2.00	1.89	66-efr(×)
308.15	808.34 ± 0.50	0.03	38-hov/lan[1)]	593.15	436.00 ± 2.00	–0.66	66-efr(×)
318.15	801.03 ± 0.50	0.09	38-hov/lan[1)]	598.15	410.00 ± 3.00	–5.23	66-efr(×)
328.15	793.56 ± 0.50	0.15	38-hov/lan[1)]	603.15	385.00 ± 5.00	–3.80	66-efr(×)
338.15	785.34 ± 0.50	–0.37	38-hov/lan[1)]	605.15	370.00 ± 5.00	–5.74	66-efr(×)
348.15	778.20 ± 0.50	0.35	38-hov/lan[1)]	607.15	347.00 ± 5.00	–12.95	66-efr[1)]
358.15	770.22 ± 0.50	0.43	38-hov/lan[1)]	235.72	859.24 ± 0.16	1.19	73-fin(×)
368.15	762.12 ± 0.50	0.58	38-hov/lan[1)]	249.28	849.58 ± 0.16	0.45	73-fin(×)
378.15	753.39 ± 0.60	0.31	38-hov/lan(×)	262.88	840.02 ± 0.16	–0.02	73-fin(×)
388.15	744.60 ± 0.60	0.19	38-hov/lan(×)	273.00	832.97 ± 0.16	–0.18	73-fin(×)
398.15	735.48 ± 0.60	–0.03	38-hov/lan(×)	286.10	823.72 ± 0.16	–0.35	73-fin(×)
408.15	726.02 ± 0.60	–0.35	38-hov/lan(×)	293.12	818.88 ± 0.16	–0.24	73-fin[1)]
418.15	716.16 ± 0.60	–0.82	38-hov/lan(×)	293.14	818.80 ± 0.16	–0.31	73-fin[1)]
432.15	702.09 ± 0.60	–1.31	38-hov/lan(×)	303.16	811.63 ± 0.16	–0.30	73-fin[1)]
253.15	850.30 ± 1.00	3.74	58-cos/bow[1)]	313.13	804.42 ± 0.16	–0.24	73-fin[1)]
273.15	835.90 ± 1.00	2.85	58-cos/bow[1)]	323.08	797.04 ± 0.16	–0.20	73-fin[1)]
293.15	821.70 ± 1.00	2.60	58-cos/bow[1)]	333.14	789.42 ± 0.16	–0.17	73-fin(×)
313.15	807.00 ± 1.00	2.36	58-cos/bow[1)]	303.15	811.95 ± 0.04	0.01	74-rao/nai-1(□)
333.15	792.40 ± 1.00	2.82	58-cos/bow[1)]	293.15	818.75 ± 0.10	–0.35	78-jel/leo(◆)
353.15	776.60 ± 1.00	2.76	58-cos/bow[1)]	298.15	815.34 ± 0.10	–0.20	78-jel/leo[1)]
373.15	759.90 ± 1.00	2.56	58-cos/bow[1)]	303.15	811.62 ± 0.10	–0.32	78-jel/leo(◆)
393.15	743.20 ± 1.00	3.21	58-cos/bow(×)	308.15	808.02 ± 0.10	–0.29	78-jel/leo(◆)
413.15	724.70 ± 1.00	2.99	58-cos/bow(×)	303.15	811.94 ± 0.10	–0.00	78-red/nai-1(∇)
433.15	705.60 ± 1.50	3.19	58-cos/bow(×)	313.15	804.48 ± 0.10	–0.16	78-red/nai-1(∇)
453.15	685.00 ± 1.50	2.98	58-cos/bow(×)	290.10	820.00 ± 0.82	–1.26	79-gyl/apa[1)]
473.15	663.80 ± 1.50	3.35	58-cos/bow(×)	305.90	807.50 ± 0.81	–2.45	79-gyl/apa[1)]
493.15	640.60 ± 1.50	2.98	58-cos/bow(×)	350.80	743.40 ± 0.74	–32.33	79-gyl/apa[1)]
513.15	616.30 ± 1.50	3.97	58-cos/bow(×)	380.90	744.00 ± 0.74	–6.72	79-gyl/apa[1)]
273.15	832.82 ± 0.10	–0.23	59-mck/ski(Δ)	422.80	712.30 ± 0.71	–0.23	79-gyl/apa(×)
273.15	834.00 ± 1.50	0.95	66-efr[1)]	283.15	822.14 ± 0.15	–3.99	86-hei/sch(×)
293.15	819.00 ± 1.50	–0.10	66-efr[1)]	298.15	815.12 ± 0.15	–0.42	86-hei/sch[1)]
313.15	804.00 ± 1.50	–0.64	66-efr[1)]	308.15	807.93 ± 0.15	–0.38	86-hei/sch[1)]
333.15	788.00 ± 1.50	–1.58	66-efr[1)]	323.15	796.86 ± 0.15	–0.33	86-hei/sch(×)
353.15	773.00 ± 1.50	–0.84	66-efr[1)]	323.15	797.34 ± 0.10	0.15	93-gar/ban-1(O)
373.15	757.00 ± 1.50	–0.34	66-efr[1)]	328.15	794.09 ± 0.10	0.68	93-gar/ban-1(O)
393.15	739.00 ± 1.50	–0.99	66-efr[1)]	333.15	789.67 ± 0.10	0.09	93-gar/ban-1(O)
413.15	720.00 ± 1.50	–1.71	66-efr[1)]	338.15	786.02 ± 0.10	0.31	93-gar/ban-1(O)
433.15	700.00 ± 1.50	–2.41	66-efr(×)	343.15	781.78 ± 0.10	–0.02	93-gar/ban-1(O)
453.15	679.00 ± 1.50	–3.02	66-efr(×)	348.15	777.68 ± 0.10	–0.17	93-gar/ban-1(O)
473.15	657.00 ± 1.50	–3.45	66-efr(×)	353.15	773.89 ± 0.10	0.05	93-gar/ban-1(O)
493.15	634.00 ± 1.50	–3.62	66-efr(×)	358.15	769.75 ± 0.10	–0.04	93-gar/ban-1(O)
513.15	610.00 ± 1.50	–2.33	66-efr(×)	363.15	765.47 ± 0.10	–0.22	93-gar/ban-1(O)
533.15	580.00 ± 1.50	–1.60	66-efr(×)	373.15	757.03 ± 0.10	–0.31	93-gar/ban-1(O)

[1)] Not included in Fig. 1.

cont.

1-Hexanol (cont.)

Further references: [1877-lie/jan, 1883-fre, 1884-zan, 1886-gar, 18-bro/hum, 19-beh, 27-nor/cor, 27-ver/coo, 29-kel-3, 29-mah/das, 30-wal/ros, 33-but/tho, 35-bil/gis, 35-but/ram, 39-gol/tay, 41-hus/age, 45-add, 48-jon/bow, 48-lad/smi, 48-vog-2, 48-wei, 50-gor, 50-mum/phi, 50-pic/zie, 50-sac/sau, 52-coo, 52-eri-1, 53-ano-1, 53-par/cha, 54-naz/kak-2, 56-shu/bel, 57-col/fal, 58-ano-5, 58-lin/tua, 62-bro/smi, 63-nik, 68-ano, 68-pfl/pop, 68-rao/chi, 69-kat/pat, 70-mye/cle, 70-puz/bul, 74-moo/wel, 74-rao/nai, 75-hsu/cle, 75-mat/fer, 76-kow/kas, 76-red/nai, 78-ast, 78-dap/don, 78-paz/gar, 78-tre/ben, 79-dia/tar, 79-kiy/ben, 80-tre/ben, 81-han/hal, 81-kim/ben, 81-kor/kov, 81-nai/nai, 81-sjo/dyh, 82-nai/nai, 82-ort, 82-sin/sin, 82-ven/dha, 83-fuk/ogi, 83-kim/ben, 84-bra/pin, 84-ort/ang, 84-sak/nak, 85-fer/pin, 85-ort, 85-ort/paz-1, 85-sar/paz, 85-sin/sin, 85-sin/sin-1, 86-ash/sri, 86-dew/meh, 86-sin/sin, 86-tan/toy, 86-wag/hei, 87-dew/meh, 88-ort/gar, 89-ala/sal, 89-dew/gup, 89-mat/mak-1, 89-sin/sin, 89-vij/nai, 90-vij/nai, 91-fen/wan, 91-ram/muk, 92-lie/sen-1, 93-ami/ara, 93-ami/rai, 93-yan/mae, 94-kim/lee, 94-kum/nai, 94-kum/nai-1, 94-ven/ven, 94-yu /tsa, 94-yu /tsa-1, 95-fra/jim, 95-fra/men, 96-elb, 98-ami/ban, 98-ami/pat-1].

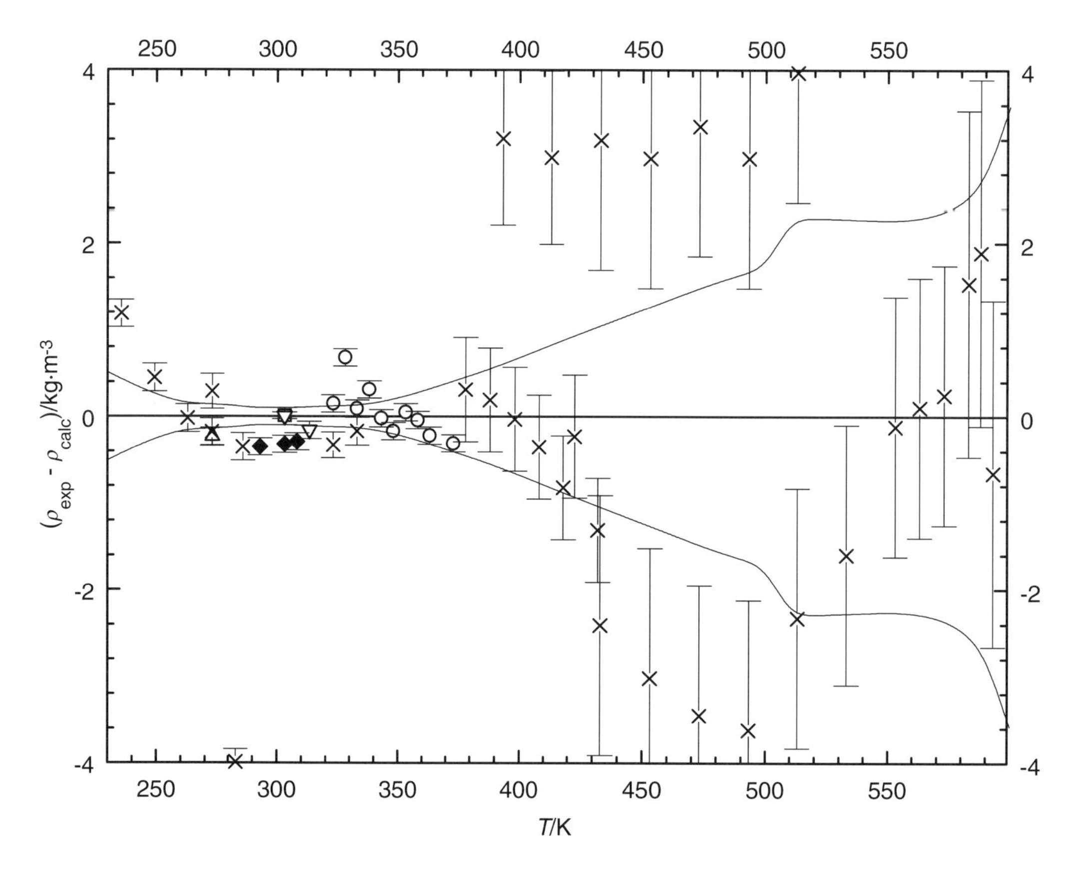

Fig. 1. The symbols show the deviation of the calculated from the experimental values from Table 2. The curves above and below the zero line indicate the calculated error region of the recommended values given in Table 3. The error bars represent the experimental errors. (Error bars smaller than the symbols are omitted for clarity of the figure.)

cont.

Table 3. Recommended values (fit to the reliable experimental values according to the equations $\rho = A + BT + CT^2 + DT^3 + \ldots$ or $\rho = [1 + 1.75(1 - T/T_c)^{1/3} + 0.75(1 - T/T_c)][\rho_c + A(T_c - T) + B(T_c - T)^2 + C(T_c - T)^3 + D(T_c - T)^4]$).

T / K	$\rho \pm \sigma_{fit}$ / kg·m^{-3}	T / K	$\rho \pm \sigma_{fit}$ / kg·m^{-3}	T / K	$\rho \pm \sigma_{fit}$ / kg·m^{-3}
230.00	861.78 ± 0.51	350.00	776.37 ± 0.21	490.00	641.30 ± 1.64
240.00	855.25 ± 0.37	360.00	768.28 ± 0.29	500.00	629.49 ± 1.73
250.00	848.66 ± 0.25	370.00	759.99 ± 0.38	510.00	616.68 ± 2.28
260.00	841.98 ± 0.16	380.00	751.50 ± 0.47	520.00	602.39 ± 2.29
270.00	835.20 ± 0.14	390.00	742.78 ± 0.57	530.00	586.79 ± 2.28
280.00	828.32 ± 0.13	400.00	733.84 ± 0.68	540.00	569.84 ± 2.27
290.00	821.33 ± 0.10	410.00	724.65 ± 0.79	550.00	551.33 ± 2.26
293.15	819.10 ± 0.10	420.00	715.22 ± 0.90	560.00	530.85 ± 2.27
298.15	815.54 ± 0.10	430.00	705.52 ± 1.01	570.00	507.74 ± 2.32
300.00	814.21 ± 0.10	440.00	695.56 ± 1.12	580.00	480.94 ± 2.44
310.00	806.96 ± 0.10	450.00	685.31 ± 1.23	590.00	448.56 ± 2.72
320.00	799.56 ± 0.12	460.00	674.77 ± 1.34	600.00	406.21 ± 3.58
330.00	792.00 ± 0.12	470.00	663.93 ± 1.45	610.00	324.59 ± 9.90
340.00	784.27 ± 0.15	480.00	652.78 ± 1.55		

2-Hexanol **[626-93-7]** **$C_6H_{14}O$** **MW = 102.18** **31**

Table 1. Coefficients of the polynomial expansion equation.
Standard deviations (see introduction):
$\sigma_{c,w} = 4.6999 \cdot 10^{-1}$ (combined temperature ranges, weighted),
$\sigma_{c,uw} = 8.6236 \cdot 10^{-2}$ (combined temperature ranges, unweighted).

Coefficient	T = 273.15 to 413.15 K $\rho = A + BT + CT^2 + DT^3 + \ldots$
A	$9.26903 \cdot 10^{2}$
B	$-3.13386 \cdot 10^{-3}$
C	$-1.30510 \cdot 10^{-3}$

Table 2. Experimental values with uncertainties and deviation from calculated values.

T / K	$\rho_{exp} \pm 2\sigma_{est}$ / kg·m^{-3}	$\rho_{exp} - \rho_{calc}$ / kg·m^{-3}	Ref. (Symbol in Fig. 1)	T / K	$\rho_{exp} \pm 2\sigma_{est}$ / kg·m^{-3}	$\rho_{exp} - \rho_{calc}$ / kg·m^{-3}	Ref. (Symbol in Fig. 1)
273.15	828.70 ± 0.60	0.03	08-zel/prz(X)	298.15	810.34 ± 0.50	0.39	38-hov/lan(X)
293.15	814.10 ± 0.60	0.27	08-zel/prz(X)	308.15	802.21 ± 0.50	0.20	38-hov/lan(X)
293.15	815.00 ± 0.60	1.17	11-pic/ken(X)	318.15	794.01 ± 0.50	0.20	38-hov/lan(X)
371.65	744.40 ± 1.00	−1.07	12-pic/ken(X)	328.15	785.49 ± 0.50	0.15	38-hov/lan(X)
393.15	726.30 ± 1.00	2.35	12-pic/ken[1)]	338.15	776.74 ± 0.50	0.13	38-hov/lan(X)
273.15	828.30 ± 0.50	−0.37	26-van(X)	348.15	767.76 ± 0.50	0.14	38-hov/lan(X)
288.15	817.10 ± 0.50	−0.54	26-van(X)	358.15	758.01 ± 0.50	-0.36	38-hov/lan(X)
298.15	809.75 ± 0.20	−0.20	32-ell/rei(O)	368.15	749.26 ± 0.50	0.40	38-hov/lan(X)
298.15	810.80 ± 0.50	0.85	38-gin/web(◆)	378.15	738.95 ± 0.60	−0.14	38-hov/lan(X)
278.15	825.80 ± 0.50	0.74	38-hov/lan(X)	388.15	728.80 ± 0.60	−0.26	38-hov/lan(X)
288.15	818.12 ± 0.50	0.48	38-hov/lan(X)	398.15	718.43 ± 0.60	−0.34	38-hov/lan(X)

[1)] Not included in Fig. 1

cont.

2-Hexanol (cont.)

Table 2. (cont.)

$\frac{T}{K}$	$\frac{\rho_{exp} \pm 2\sigma_{est}}{kg \cdot m^{-3}}$	$\frac{\rho_{exp} - \rho_{calc}}{kg \cdot m^{-3}}$	Ref. (Symbol in Fig. 1)	$\frac{T}{K}$	$\frac{\rho_{exp} \pm 2\sigma_{est}}{kg \cdot m^{-3}}$	$\frac{\rho_{exp} - \rho_{calc}}{kg \cdot m^{-3}}$	Ref. (Symbol in Fig. 1)
408.15	707.72 ± 0.60	−0.49	38-hov/lan(X)	300.15	808.05 ± 0.50	−0.34	85-ort(X)
413.15	702.22 ± 0.60	−0.62	38-hov/lan(X)	302.15	806.46 ± 0.50	−0.35	85-ort(X)
293.15	815.00 ± 0.60	1.17	42-air/bal(X)	304.15	804.85 ± 0.50	−0.37	85-ort(X)
293.15	814.20 ± 0.60	0.37	50-pic/zie(X)	306.15	803.27 ± 0.50	−0.35	85-ort(X)
293.15	814.30 ± 0.50	0.47	52-coo(∇)	308.15	801.89 ± 0.50	−0.12	85-ort(X)
293.15	813.56 ± 0.50	−0.27	85-ort(X)	298.15	809.69 ± 0.10	−0.26	86-ort/paz(□)
298.15	809.61 ± 0.50	−0.34	85-ort(X)	298.15	809.61 ± 0.50	−0.34	94-tar/jun(Δ)

Further references: [23-clo/joh, 27-nor/cor, 35-les/lom, 39-bar/atk].

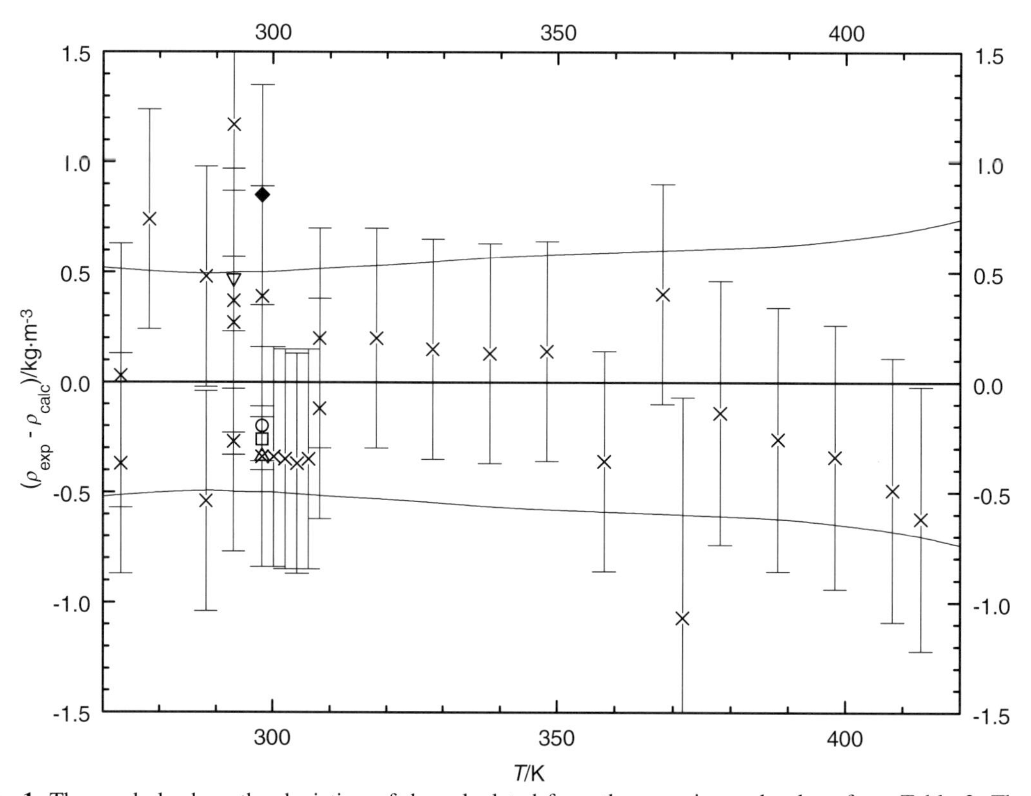

Fig. 1. The symbols show the deviation of the calculated from the experimental values from Table 2. The curves above and below the zero line indicate the calculated error region of the recommended values given in Table 3. The error bars represent the experimental errors. (Error bars smaller than the symbols are omitted for clarity of the figure.)

cont.

Table 3. Recommended values (fit to the reliable experimental values according to the equations
$\rho = A + BT + CT^2 + DT^3 + \ldots$ or $\rho = [1 + 1.75(1 - T/T_c)^{1/3} + 0.75(1 - T/T_c)][\rho_c + A(T_c - T) + B(T_c - T)^2 + C(T_c - T)^3 + D(T_c - T)^4]$).

$\frac{T}{K}$	$\frac{\rho \pm \sigma_{fit}}{kg \cdot m^{-3}}$	$\frac{T}{K}$	$\frac{\rho \pm \sigma_{fit}}{kg \cdot m^{-3}}$	$\frac{T}{K}$	$\frac{\rho \pm \sigma_{fit}}{kg \cdot m^{-3}}$
270.00	830.92 ± 0.52	310.00	800.51 ± 0.52	370.00	747.08 ± 0.60
280.00	823.71 ± 0.50	320.00	792.26 ± 0.53	380.00	737.26 ± 0.61
290.00	816.24 ± 0.49	330.00	783.74 ± 0.55	390.00	727.18 ± 0.62
293.15	813.83 ± 0.50	340.00	774.97 ± 0.57	400.00	716.83 ± 0.65
298.15	809.95 ± 0.50	350.00	765.93 ± 0.58	410.00	706.23 ± 0.68
300.00	808.50 ± 0.50	360.00	756.63 ± 0.59	420.00	695.37 ± 0.74

(*RS*)-2-Hexanol **[20281-86-1]** **$C_6H_{14}O$** **MW = 102.18** **32**

Table 1. Experimental value with uncertainty.

$\frac{T}{K}$	$\frac{\rho_{exp} \pm 2\sigma_{est}}{kg \cdot m^{-3}}$	Ref.
298.15	810.0 ± 0.2	88-tan/luo

3-Hexanol **[623-37-0]** **$C_6H_{14}O$** **MW = 102.18** **33**

Table 1. Coefficients of the polynomial expansion equation.
Standard deviations (see introduction):
$\sigma_{c,w} = 3.6198 \cdot 10^{-1}$ (combined temperature ranges, weighted),
$\sigma_{c,uw} = 7.2518 \cdot 10^{-2}$ (combined temperature ranges, unweighted).

Coefficient	T = 273.15 to 409.15 K $\rho = A + BT + CT^2 + DT^3 + \ldots$
A	$9.66285 \cdot 10^{2}$
B	$-1.77415 \cdot 10^{-1}$
C	$-1.11253 \cdot 10^{-3}$

Table 2. Experimental values with uncertainties and deviation from calculated values.

$\frac{T}{K}$	$\frac{\rho_{exp} \pm 2\sigma_{est}}{kg \cdot m^{-3}}$	$\frac{\rho_{exp} - \rho_{calc}}{kg \cdot m^{-3}}$	Ref. (Symbol in Fig. 1)	$\frac{T}{K}$	$\frac{\rho_{exp} \pm 2\sigma_{est}}{kg \cdot m^{-3}}$	$\frac{\rho_{exp} - \rho_{calc}}{kg \cdot m^{-3}}$	Ref. (Symbol in Fig. 1)
273.15	833.50 ± 1.00	−1.32	1875-lie[1)]	338.15	779.06 ± 0.50	−0.02	38-hov/lan(○)
293.15	818.80 ± 1.00	0.13	1875-lie(×)	348.15	769.72 ± 0.50	0.05	38-hov/lan(○)
273.15	833.50 ± 1.00	−1.32	1875-lie/vol[1)]	358.15	759.69 ± 0.50	−0.35	38-hov/lan(○)
293.15	818.80 ± 1.00	0.13	1875-lie/vol(×)	368.15	750.40 ± 0.50	0.22	38-hov/lan(○)
273.15	834.31 ± 1.00	−0.51	1876-dec(∇)	378.15	740.13 ± 0.50	0.02	38-hov/lan(○)
293.15	818.23 ± 1.00	−0.44	1876-dec(∇)	388.15	730.34 ± 0.50	0.53	38-hov/lan(○)
298.15	814.30 ± 0.50	−0.19	38-gin/web(Δ)	398.15	719.18 ± 0.50	−0.11	38-hov/lan(○)
278.15	830.96 ± 0.50	0.10	38-hov/lan(○)	409.15	707.06 ± 0.50	−0.39	38-hov/lan(○)
288.15	822.73 ± 0.50	−0.06	38-hov/lan(○)	293.15	819.30 ± 1.00	0.63	42-air/bal(◆)
298.15	814.26 ± 0.50	−0.23	38-hov/lan(○)	293.15	819.50 ± 1.00	0.83	44-hen/mat(×)
308.15	805.80 ± 0.50	−0.17	38-hov/lan(○)	293.15	818.60 ± 0.50	−0.07	50-pic/zie(□)
318.15	796.97 ± 0.50	−0.26	38-hov/lan(○)	293.15	819.00 ± 1.00	0.33	52-coo(×)
328.15	788.10 ± 0.50	−0.17	38-hov/lan(○)				

[1)] Not included in Fig. 1.

cont.

3-Hexanol (cont.)

Further references: [13-pic/ken, 34-les/lom, 35-les/lom, 39-spi/tin-1, 54-naz/kak-3, 56-shu/bel].

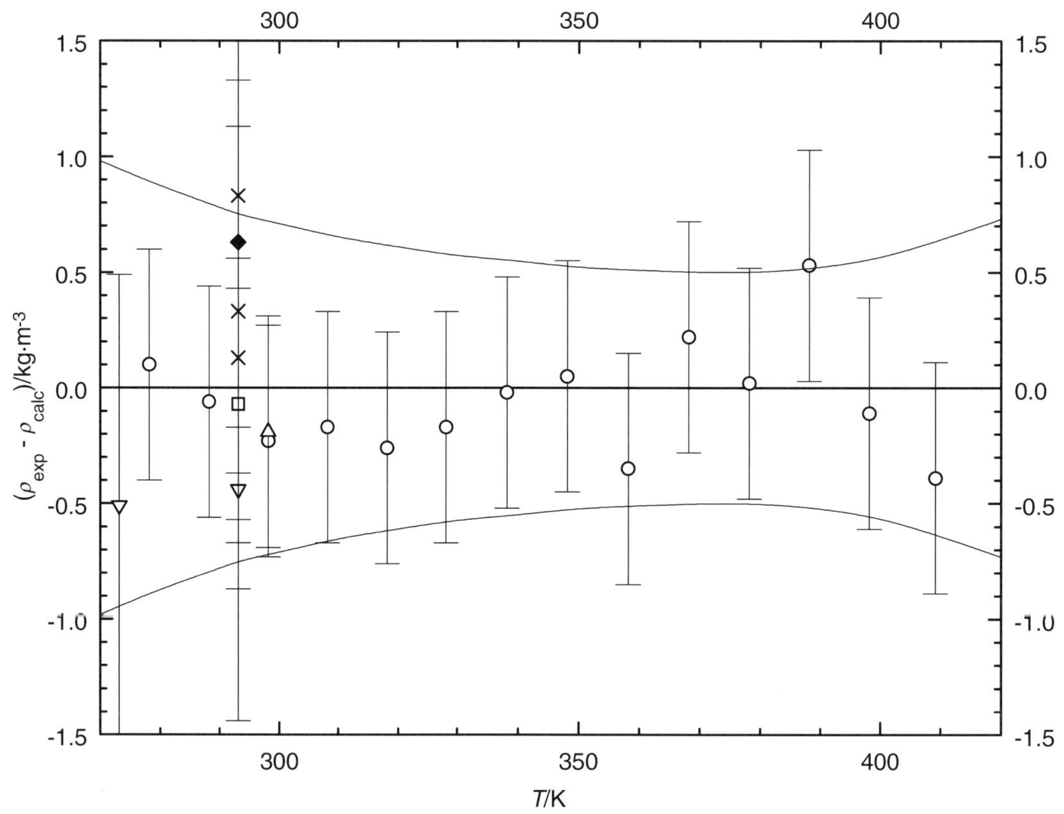

Fig. 1. The symbols show the deviation of the calculated from the experimental values from Table 2. The curves above and below the zero line indicate the calculated error region of the recommended values given in Table 3. The error bars represent the experimental errors. (Error bars smaller than the symbols are omitted for clarity of the figure.)

Table 3. Recommended values (fit to the reliable experimental values according to the equations
$\rho = A + BT + CT^2 + DT^3 + \ldots$ or $\rho = [1 + 1.75(1 - T/T_c)^{1/3} + 0.75(1 - T/T_c)][\rho_c + A(T_c - T) + B(T_c - T)^2 + C(T_c - T)^3 + D(T_c - T)^4]$).

$\frac{T}{K}$	$\frac{\rho \pm \sigma_{fit}}{kg \cdot m^{-3}}$	$\frac{T}{K}$	$\frac{\rho \pm \sigma_{fit}}{kg \cdot m^{-3}}$	$\frac{T}{K}$	$\frac{\rho \pm \sigma_{fit}}{kg \cdot m^{-3}}$
270.00	837.28 ± 0.98	310.00	804.37 ± 0.65	370.00	748.34 ± 0.50
280.00	829.39 ± 0.87	320.00	795.59 ± 0.61	380.00	738.22 ± 0.50
290.00	821.27 ± 0.78	330.00	786.58 ± 0.57	390.00	727.88 ± 0.52
293.15	818.67 ± 0.75	340.00	777.36 ± 0.55	400.00	717.31 ± 0.56
298.15	814.49 ± 0.72	350.00	767.91 ± 0.52	410.00	706.53 ± 0.64
300.00	812.93 ± 0.71	360.00	758.23 ± 0.51	420.00	695.52 ± 0.73

2-Methyl-1-pentanol **[105-30-6]** **$C_6H_{14}O$** **MW = 102.18** **34**

Table 1. Coefficients of the polynomial expansion equation.
Standard deviations (see introduction):
$\sigma_{c,w} = 3.6425 \cdot 10^{-1}$ (combined temperature ranges, weighted),
$\sigma_{c,uw} = 7.9675 \cdot 10^{-2}$ (combined temperature ranges, unweighted).

Coefficient	T = 273.15 to 408.15 K $\rho = A + BT + CT^2 + DT^3 + \ldots$
A	$9.25662 \cdot 10^{2}$
B	$1.79466 \cdot 10^{-2}$
C	$-1.24362 \cdot 10^{-3}$

Table 2. Experimental values with uncertainties and deviation from calculated values.

$\frac{T}{K}$	$\frac{\rho_{exp} \pm 2\sigma_{est}}{kg \cdot m^{-3}}$	$\frac{\rho_{exp} - \rho_{calc}}{kg \cdot m^{-3}}$	Ref. (Symbol in Fig. 1)	$\frac{T}{K}$	$\frac{\rho_{exp} \pm 2\sigma_{est}}{kg \cdot m^{-3}}$	$\frac{\rho_{exp} - \rho_{calc}}{kg \cdot m^{-3}}$	Ref. (Symbol in Fig. 1)
273.15	839.50 ± 1.50	1.72	1883-lie/zei[1)]	338.15	789.50 ± 0.50	−0.03	38-hov/lan(Δ)
273.15	837.40 ± 1.50	−0.38	1883-lie/zei(∇)	348.15	781.19 ± 0.50	0.02	38-hov/lan(Δ)
290.75	824.60 ± 1.50	−1.15	1883-lie/zei(∇)	358.15	772.69 ± 0.50	0.12	38-hov/lan(Δ)
296.85	821.40 ± 1.50	−0.00	1883-lie/zei(∇)	368.15	764.13 ± 0.50	0.41	38-hov/lan(Δ)
278.15	834.70 ± 0.50	0.26	38-hov/lan(Δ)	378.15	754.77 ± 0.50	0.16	38-hov/lan(Δ)
288.15	827.99 ± 0.50	0.42	38-hov/lan(Δ)	388.15	745.34 ± 0.50	0.08	38-hov/lan(Δ)
298.15	820.63 ± 0.50	0.17	38-hov/lan(Δ)	398.15	735.69 ± 0.50	0.03	38-hov/lan(Δ)
308.15	813.07 ± 0.50	−0.03	38-hov/lan(Δ)	408.15	725.74 ± 0.50	−0.08	38-hov/lan(Δ)
318.15	805.48 ± 0.50	−0.01	38-hov/lan(Δ)	293.15	824.30 ± 0.50	0.25	50-pic/zie(○)
328.15	797.55 ± 0.50	−0.08	38-hov/lan(Δ)	293.15	824.00 ± 0.50	−0.05	68-ano(□)

[1)] Not included in Fig. 1.

Further references: [24-ter, 27-nor/cor, 36-oli, 51-hau, 52-coo, 53-ano-1, 56-ano-1, 84-bra/pin, 85-ort, 94-tar/jun].

Table 3. Recommended values (fit to the reliable experimental values according to the equations
$\rho = A + BT + CT^2 + DT^3 + \ldots$ or $\rho = [1 + 1.75(1 - T/T_c)^{1/3} + 0.75(1 - T/T_c)][\rho_c + A(T_c - T) + B(T_c - T)^2 + C(T_c - T)^3 + D(T_c - T)^4])$.

$\frac{T}{K}$	$\frac{\rho \pm \sigma_{fit}}{kg \cdot m^{-3}}$	$\frac{T}{K}$	$\frac{\rho \pm \sigma_{fit}}{kg \cdot m^{-3}}$	$\frac{T}{K}$	$\frac{\rho \pm \sigma_{fit}}{kg \cdot m^{-3}}$
270.00	839.85 ± 1.24	310.00	811.71 ± 0.60	370.00	762.05 ± 0.53
280.00	833.19 ± 1.01	320.00	804.06 ± 0.55	380.00	752.90 ± 0.54
290.00	826.28 ± 0.82	330.00	796.15 ± 0.52	390.00	743.51 ± 0.55
293.15	824.05 ± 0.78	340.00	788.00 ± 0.51	400.00	733.86 ± 0.56
298.15	820.46 ± 0.71	350.00	779.60 ± 0.51	410.00	723.97 ± 0.59
300.00	819.12 ± 0.69	360.00	770.95 ± 0.52	420.00	713.82 ± 0.63

cont.

2-Methyl-1-pentanol (cont.)

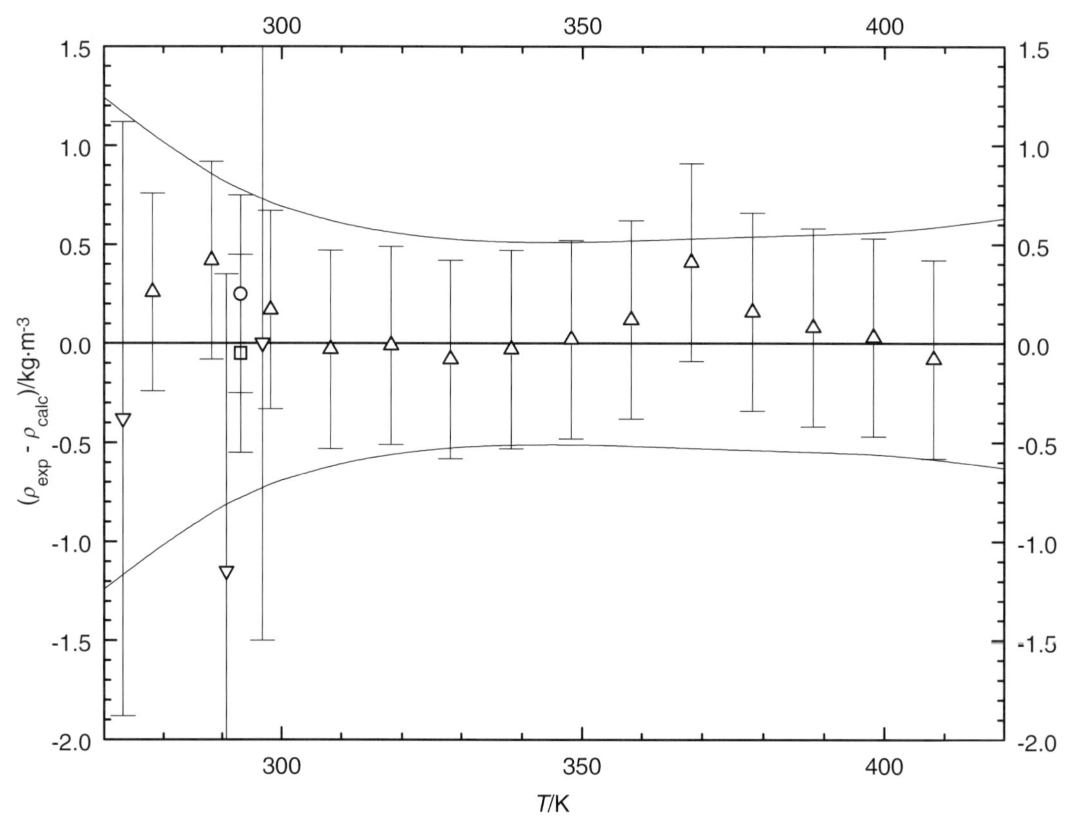

Fig. 1. The symbols show the deviation of the calculated from the experimental values from Table 2. The curves above and below the zero line indicate the calculated error region of the recommended values given in Table 3. The error bars represent the experimental errors. (Error bars smaller than the symbols are omitted for clarity of the figure.)

2-Methyl-2-pentanol **[590-36-3]** **$C_6H_{14}O$** **MW = 102.18** **35**

Table 1. Coefficients of the polynomial expansion equation.
Standard deviations (see introduction):
$\sigma_{c,w} = 2.6940 \cdot 10^{-1}$ (combined temperature ranges, weighted),
$\sigma_{c,uw} = 1.0135 \cdot 10^{-1}$ (combined temperature ranges, unweighted).

Coefficient	T = 278.15 to 358.15 K $\rho = A + BT + CT^2 + DT^3 + \ldots$
A	$9.36813 \cdot 10^{2}$
B	$-1.65836 \cdot 10^{-2}$
C	$-1.37719 \cdot 10^{-3}$

cont.

Table 2. Experimental values with uncertainties and deviation from calculated values.

$\frac{T}{K}$	$\frac{\rho_{exp} \pm 2\sigma_{est}}{kg \cdot m^{-3}}$	$\frac{\rho_{exp} - \rho_{calc}}{kg \cdot m^{-3}}$	Ref. (Symbol in Fig. 1)	$\frac{T}{K}$	$\frac{\rho_{exp} \pm 2\sigma_{est}}{kg \cdot m^{-3}}$	$\frac{\rho_{exp} - \rho_{calc}}{kg \cdot m^{-3}}$	Ref. (Symbol in Fig. 1)
278.15	826.00 ± 0.50	0.35	33-hov/lan(□)	328.15	783.02 ± 0.50	–0.05	33-hov/lan(□)
288.15	817.10 ± 0.50	–0.59	33-hov/lan(□)	338.15	773.72 ± 0.50	–0.01	33-hov/lan(□)
298.15	809.68 ± 0.50	0.23	33-hov/lan(□)	348.15	764.32 ± 0.50	0.21	33-hov/lan(□)
308.15	801.10 ± 0.50	0.17	33-hov/lan(□)	358.15	754.11 ± 0.50	–0.11	33-hov/lan(□)
318.15	792.33 ± 0.50	0.19	33-hov/lan(□)	293.15	813.40 ± 1.00	–0.20	50-pic/zie(○)

Further references: [36-nor/has, 38-gin/web, 38-whi/joh, 39-owe/qua, 52-lev/tan, 85-ort, 94-tar/jun].

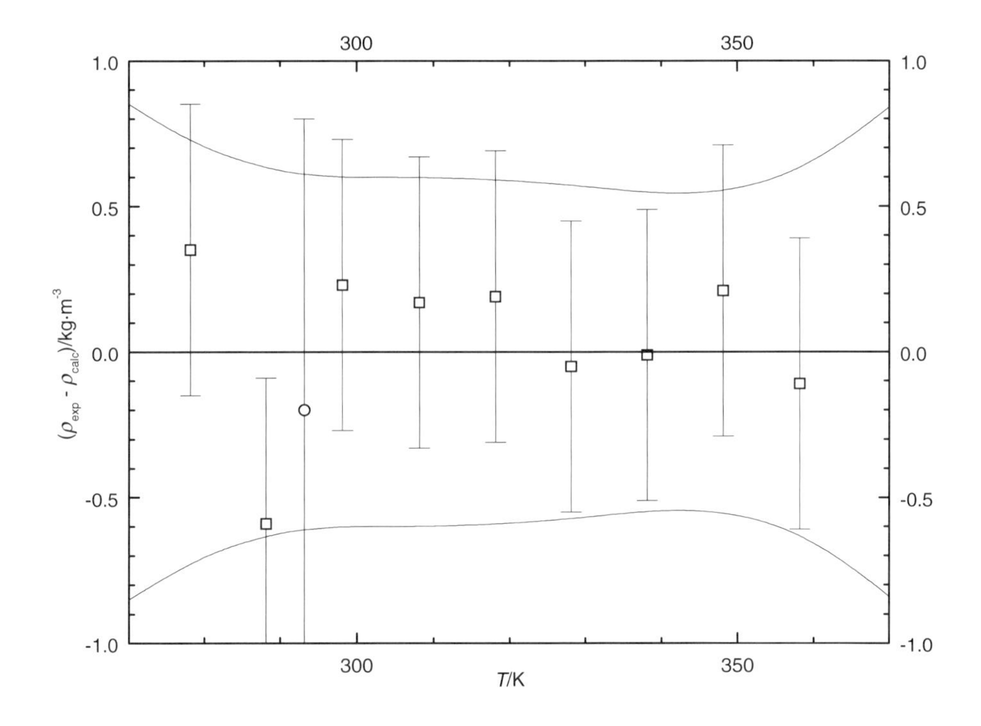

Fig. 1. The symbols show the deviation of the calculated from the experimental values from Table 2. The curves above and below the zero line indicate the calculated error region of the recommended values given in Table 3. The error bars represent the experimental errors. (Error bars smaller than the symbols are omitted for clarity of the figure.)

Table 3. Recommended values (fit to the reliable experimental values according to the equations

$$\rho = A + BT + CT^2 + DT^3 + \dots \text{ or } \rho = [1 + 1.75(1 - T/T_c)^{1/3} + 0.75(1 - T/T_c)][\rho_c + A(T_c - T) + B(T_c - T)^2 + C(T_c - T)^3 + D(T_c - T)^4]\,).$$

$\frac{T}{K}$	$\frac{\rho \pm \sigma_{fit}}{kg \cdot m^{-3}}$	$\frac{T}{K}$	$\frac{\rho \pm \sigma_{fit}}{kg \cdot m^{-3}}$	$\frac{T}{K}$	$\frac{\rho \pm \sigma_{fit}}{kg \cdot m^{-3}}$
270.00	831.94 ± 0.85	300.00	807.89 ± 0.60	350.00	762.30 ± 0.55
280.00	824.20 ± 0.69	310.00	799.32 ± 0.60	360.00	752.36 ± 0.64
290.00	816.18 ± 0.62	320.00	790.48 ± 0.59	370.00	742.14 ± 0.84
293.15	813.60 ± 0.61	330.00	781.36 ± 0.57		
298.15	809.45 ± 0.60	340.00	771.97 ± 0.54		

2-Methyl-3-pentanol **[565-67-3]** **$C_6H_{14}O$** **MW = 102.18** **36**

Table 1. Coefficients of the polynomial expansion equation.
Standard deviations (see introduction):
$\sigma_{c,w} = 6.8420 \cdot 10^{-1}$ (combined temperature ranges, weighted),
$\sigma_{c,uw} = 1.7852 \cdot 10^{-1}$ (combined temperature ranges, unweighted).

Coefficient	T = 278.15 to 398.15 K $\rho = A + BT + CT^2 + DT^3 + \ldots$
A	$9.13614 \cdot 10^{2}$
B	$2.23285 \cdot 10^{-1}$
C	$-1.80054 \cdot 10^{-3}$

Table 2. Experimental values with uncertainties and deviation from calculated values.

$\frac{T}{\text{K}}$	$\frac{\rho_{exp} \pm 2\sigma_{est}}{\text{kg}\cdot\text{m}^{-3}}$	$\frac{\rho_{exp} - \rho_{calc}}{\text{kg}\cdot\text{m}^{-3}}$	Ref. (Symbol in Fig. 1)	$\frac{T}{\text{K}}$	$\frac{\rho_{exp} \pm 2\sigma_{est}}{\text{kg}\cdot\text{m}^{-3}}$	$\frac{\rho_{exp} - \rho_{calc}}{\text{kg}\cdot\text{m}^{-3}}$	Ref. (Symbol in Fig. 1)
293.15	824.70 ± 1.00	0.36	12-pic/ken(X)	338.15	783.78 ± 0.60	0.55	40-hov/lan-1(O)
288.95	826.90 ± 1.00	−0.90	28-kro/see(X)	348.15	773.60 ± 0.60	0.49	40-hov/lan-1(O)
298.15	819.30 ± 1.00	−0.83	36-nor/has(Δ)	358.15	762.93 ± 0.60	0.30	40-hov/lan-1(O)
298.15	818.60 ± 1.00	−1.53	38-gin/web(X)	368.15	751.71 ± 0.60	−0.07	40-hov/lan-1(O)
293.15	824.00 ± 1.00	−0.34	38-whi/joh(∇)	378.15	739.85 ± 0.60	−0.73	40-hov/lan-1(O)
278.15	837.11 ± 0.60	0.69	40-hov/lan-1(O)	388.15	727.85 ± 0.60	−1.16	40-hov/lan-1(O)
288.15	829.09 ± 0.60	0.64	40-hov/lan-1(O)	398.15	718.58 ± 0.60	1.49	40-hov/lan-1(O)
298.15	820.62 ± 0.60	0.49	40-hov/lan-1(O)	293.15	823.00 ± 1.00	−1.34	44-hen/mat(X)
308.15	811.80 ± 0.60	0.35	40-hov/lan-1(O)	293.15	824.90 ± 1.00	0.56	50-pic/zie(◆)
318.15	802.98 ± 0.60	0.58	40-hov/lan-1(O)	298.15	820.08 ± 0.40	−0.05	85-ort/paz-2(□)
328.15	793.44 ± 0.60	0.44	40-hov/lan-1(O)				

Further references: [19-beh].

Table 3. Recommended values (fit to the reliable experimental values according to the equations $\rho = A + BT + CT^2 + DT^3 + \ldots$ or $\rho = [1 + 1.75(1 - T/T_c)^{1/3} + 0.75(1 - T/T_c)][\rho_c + A(T_c - T) + B(T_c - T)^2 + C(T_c - T)^3 + D(T_c - T)^4]$).

$\frac{T}{\text{K}}$	$\frac{\rho \pm \sigma_{fit}}{\text{kg}\cdot\text{m}^{-3}}$	$\frac{T}{\text{K}}$	$\frac{\rho \pm \sigma_{fit}}{\text{kg}\cdot\text{m}^{-3}}$	$\frac{T}{\text{K}}$	$\frac{\rho \pm \sigma_{fit}}{\text{kg}\cdot\text{m}^{-3}}$
270.00	842.64 ± 0.92	310.00	809.80 ± 0.77	370.00	749.74 ± 0.58
280.00	834.97 ± 0.87	320.00	800.69 ± 0.73	380.00	738.46 ± 0.60
290.00	826.94 ± 0.83	330.00	791.22 ± 0.69	390.00	726.83 ± 0.68
293.15	824.34 ± 0.82	340.00	781.39 ± 0.65	400.00	714.84 ± 0.80
298.15	820.13 ± 0.80	350.00	771.20 ± 0.62	410.00	702.49 ± 0.99
300.00	818.55 ± 0.80	360.00	760.65 ± 0.59		

cont.

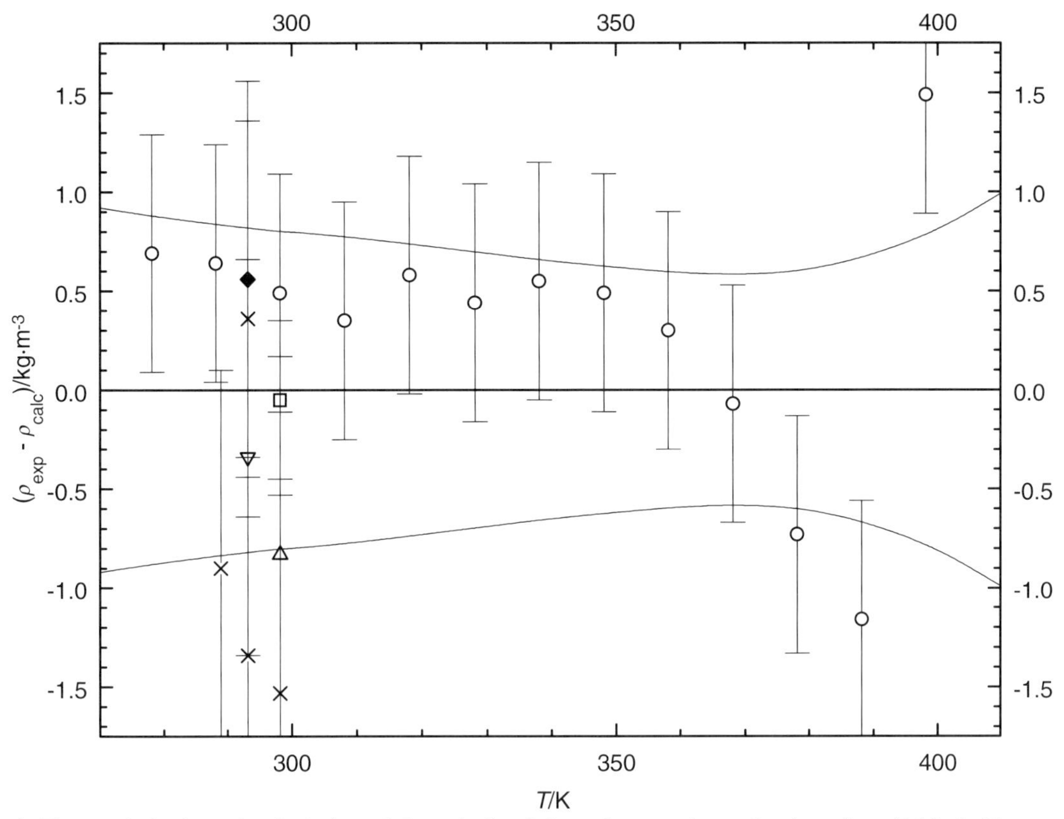

Fig. 1. The symbols show the deviation of the calculated from the experimental values from Table 2. The curves above and below the zero line indicate the calculated error region of the recommended values given in Table 3. The error bars represent the experimental errors. (Error bars smaller than the symbols are omitted for clarity of the figure.)

3-Methyl-1-pentanol **[589-35-5]** **$C_6H_{14}O$** **MW = 102.18** **37**

Table 1. Coefficients of the polynomial expansion equation.
Standard deviations (see introduction):
$\sigma_{c,w} = 8.2830 \cdot 10^{-1}$ (combined temperature ranges, weighted),
$\sigma_{c,uw} = 2.1862 \cdot 10^{-1}$ (combined temperature ranges, unweighted).

Coefficient	T = 278.15 to 423.15 K $\rho = A + BT + CT^2 + DT^3 + \ldots$
A	$9.12528 \cdot 10^{2}$
B	$7.80681 \cdot 10^{-2}$
C	$-1.29987 \cdot 10^{-3}$

cont.

3-Methyl-1-pentanol (cont.)

Table 2. Experimental values with uncertainties and deviation from calculated values.

$\frac{T}{\mathrm{K}}$	$\frac{\rho_{exp} \pm 2\sigma_{est}}{\mathrm{kg \cdot m^{-3}}}$	$\frac{\rho_{exp} - \rho_{calc}}{\mathrm{kg \cdot m^{-3}}}$	Ref. (Symbol in Fig. 1)	$\frac{T}{\mathrm{K}}$	$\frac{\rho_{exp} \pm 2\sigma_{est}}{\mathrm{kg \cdot m^{-3}}}$	$\frac{\rho_{exp} - \rho_{calc}}{\mathrm{kg \cdot m^{-3}}}$	Ref. (Symbol in Fig. 1)
293.65	826.20 ± 2.00	2.84	08-har[1)]	338.15	789.62 ± 0.60	−0.67	40-hov/lan(∇)
302.15	819.30 ± 2.00	1.85	08-har(×)	348.15	781.71 ± 0.60	−0.44	40-hov/lan(∇)
313.15	810.70 ± 2.00	1.19	08-har(×)	358.15	773.50 ± 0.60	−0.25	40-hov/lan(∇)
333.15	796.60 ± 2.00	2.33	08-har(×)	368.15	765.18 ± 0.60	0.09	40-hov/lan(∇)
353.15	781.20 ± 2.00	3.22	08-har[1)]	378.15	756.64 ± 0.60	0.47	40-hov/lan(∇)
298.15	820.50 ± 0.50	0.25	27-nor/cor(○)	388.15	747.82 ± 0.60	0.83	40-hov/lan(∇)
293.15	824.20 ± 0.50	0.49	36-oli(Δ)	398.15	737.26 ± 0.60	−0.29	40-hov/lan(∇)
278.15	833.26 ± 0.60	−0.42	40-hov/lan(∇)	408.15	727.23 ± 0.60	−0.62	40-hov/lan(∇)
288.15	826.00 ± 0.60	−1.09	40-hov/lan(∇)	418.15	717.48 ± 0.60	−0.41	40-hov/lan(∇)
298.15	818.68 ± 0.60	−1.57	40-hov/lan(∇)	423.15	712.58 ± 0.60	−0.23	40-hov/lan(∇)
308.15	811.58 ± 0.60	−1.57	40-hov/lan(∇)	293.15	824.10 ± 0.60	0.39	52-coo(◆)
318.15	804.82 ± 0.60	−0.97	40-hov/lan(∇)	298.15	821.70 ± 0.50	1.45	59-pin/lar(□)
328.15	797.38 ± 0.60	−0.79	40-hov/lan(∇)				

[1)] Not included in Fig. 1.

Further references: [12-ipa, 41-boh, 41-hus/age, 43-boh, 50-pic/zie].

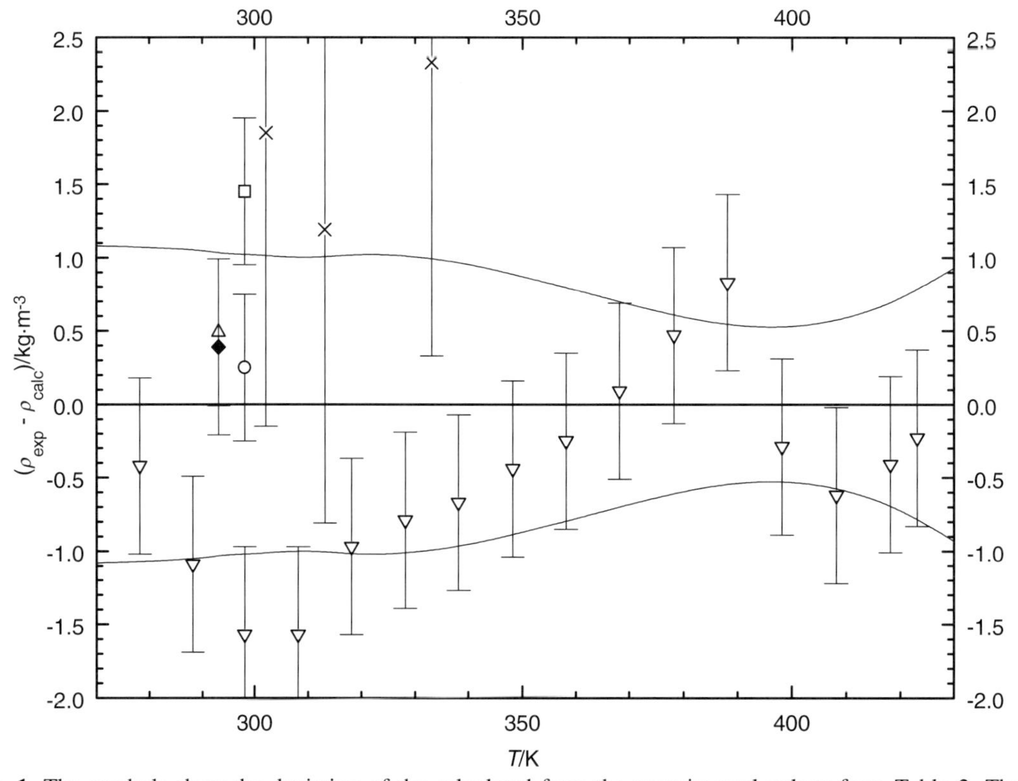

Fig. 1. The symbols show the deviation of the calculated from the experimental values from Table 2. The curves above and below the zero line indicate the calculated error region of the recommended values given in Table 3. The error bars represent the experimental errors. (Error bars smaller than the symbols are omitted for clarity of the figure.)

cont.

Table 3. Recommended values (fit to the reliable experimental values according to the equations $\rho = A + BT + CT^2 + DT^3 + \ldots$ or $\rho = [1 + 1.75(1 - T/T_c)^{1/3} + 0.75(1 - T/T_c)][\rho_c + A(T_c - T) + B(T_c - T)^2 + C(T_c - T)^3 + D(T_c - T)^4]$).

T / K	$\rho \pm \sigma_{fit}$ / $kg \cdot m^{-3}$	T / K	$\rho \pm \sigma_{fit}$ / $kg \cdot m^{-3}$	T / K	$\rho \pm \sigma_{fit}$ / $kg \cdot m^{-3}$
270.00	838.85 ± 1.08	320.00	804.40 ± 1.03	390.00	745.26 ± 0.53
280.00	832.48 ± 1.07	330.00	796.73 ± 1.01	400.00	735.78 ± 0.52
290.00	825.85 ± 1.05	340.00	788.81 ± 0.96	410.00	726.03 ± 0.58
293.15	823.71 ± 1.03	350.00	780.62 ± 0.87	420.00	716.02 ± 0.71
298.15	820.25 ± 1.02	360.00	772.17 ± 0.78	430.00	705.75 ± 0.93
300.00	818.96 ± 1.02	370.00	763.46 ± 0.68		
310.00	811.81 ± 0.99	380.00	754.49 ± 0.59		

(*R*)-3-Methyl-1-pentanol [70224-28-1] $C_6H_{14}O$ MW = 102.18 38

Table 1. Fit with estimated *B* coefficient for 2 accepted points. Deviation $\sigma_w = 0.020$.

Coefficient	$\rho = A + BT$
A	1060.34
B	−0.800

Table 2. Experimental values with uncertainties and deviation from calculated values.

T / K	$\rho_{exp} \pm 2\sigma_{est}$ / $kg \cdot m^{-3}$	$\rho_{exp} - \rho_{calc}$ / $kg \cdot m^{-3}$	Ref.
292.95	826.0 ± 0.5	0.02	59-pin/lar
298.15	821.8 ± 0.5	−0.02	59-pin/lar

Table 3. Recommended values.

T / K	$\rho_{exp} \pm 2\sigma_{est}$ / $kg \cdot m^{-3}$
290.00	828.3 ± 0.5
293.15	825.8 ± 0.5
298.15	821.8 ± 0.5

3-Methyl-2-pentanol [565-60-6] $C_6H_{14}O$ MW = 102.18 39

Table 1. Coefficients of the polynomial expansion equation. Standard deviations (see introduction):
$\sigma_{c,w} = 3.8542 \cdot 10^{-1}$ (combined temperature ranges, weighted),
$\sigma_{c,uw} = 1.6563 \cdot 10^{-1}$ (combined temperature ranges, unweighted).

Coefficient	T = 278.15 to 398.15 K $\rho = A + BT + CT^2 + DT^3 + \ldots$
A	$9.72247 \cdot 10^{2}$
B	$-1.48421 \cdot 10^{-1}$
C	$-1.16182 \cdot 10^{-3}$

cont.

3-Methyl-2-pentanol (cont.)

Table 2. Experimental values with uncertainties and deviation from calculated values.

$\frac{T}{K}$	$\frac{\rho_{exp} \pm 2\sigma_{est}}{kg \cdot m^{-3}}$	$\frac{\rho_{exp} - \rho_{calc}}{kg \cdot m^{-3}}$	Ref. (Symbol in Fig. 1)	$\frac{T}{K}$	$\frac{\rho_{exp} \pm 2\sigma_{est}}{kg \cdot m^{-3}}$	$\frac{\rho_{exp} - \rho_{calc}}{kg \cdot m^{-3}}$	Ref. (Symbol in Fig. 1)
291.15	830.70 ± 1.00	0.15	1883-wis(∇)	318.15	807.50 ± 0.50	0.07	41-hov/lan(□)
298.15	824.00 ± 1.00	−0.72	36-nor/has(Δ)	328.15	798.53 ± 0.50	0.10	41-hov/lan(□)
298.15	826.40 ± 1.00	1.68	36-nor/has(Δ)	338.15	789.32 ± 0.50	0.11	41-hov/lan(□)
298.15	823.50 ± 1.00	−1.22	36-nor/has(Δ)	348.15	779.86 ± 0.50	0.11	41-hov/lan(□)
298.15	823.10 ± 1.00	−1.62	38-gin/web(O)	358.15	770.28 ± 0.50	0.22	41-hov/lan(□)
278.15	841.51 ± 0.50	0.43	41-hov/lan(□)	368.15	760.33 ± 0.50	0.19	41-hov/lan(□)
288.15	833.26 ± 0.50	0.25	41-hov/lan(□)	378.15	750.17 ± 0.50	0.19	41-hov/lan(□)
298.15	824.73 ± 0.50	0.01	41-hov/lan(□)	388.15	739.63 ± 0.50	0.03	41-hov/lan(□)
308.15	816.31 ± 0.50	0.12	41-hov/lan(□)	398.15	728.86 ± 0.50	−0.12	41-hov/lan(□)

Further references: [01-zel/zel, 50-pic/zie, 84-bra/pin, 85-ort, 94-tar/jun].

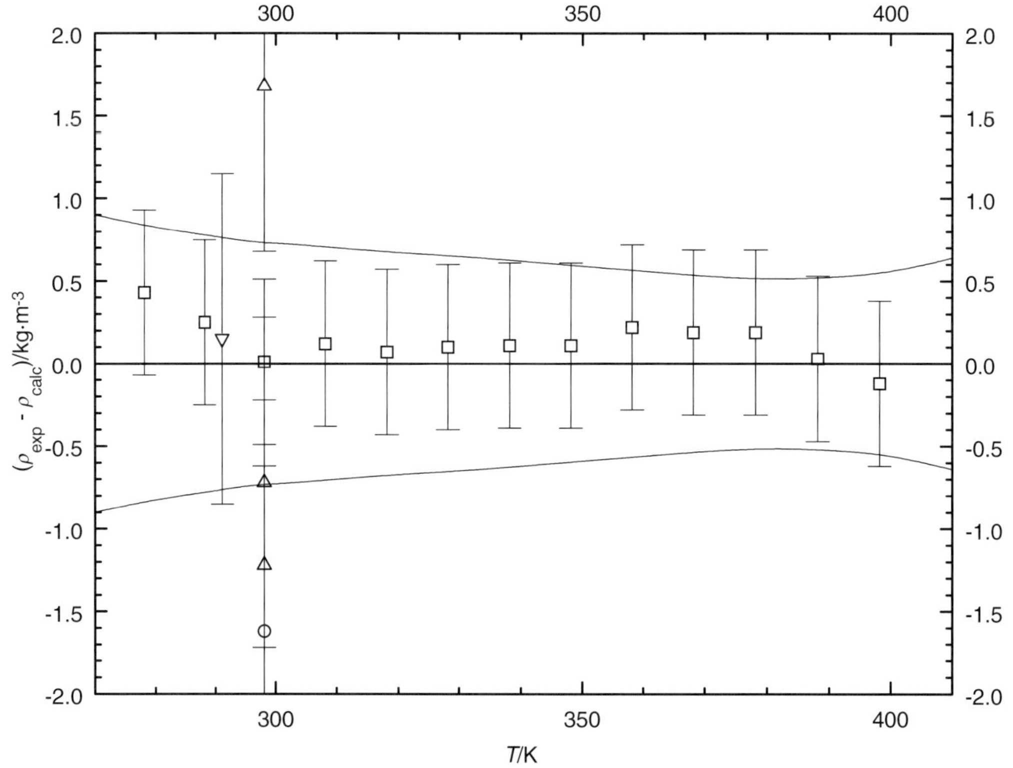

Fig. 1. The symbols show the deviation of the calculated from the experimental values from Table 2. The curves above and below the zero line indicate the calculated error region of the recommended values given in Table 3. The error bars represent the experimental errors. (Error bars smaller than the symbols are omitted for clarity of the figure.)

cont.

Table 3. Recommended values (fit to the reliable experimental values according to the equations $\rho = A + BT + CT^2 + DT^3 + \ldots$ or $\rho = [1 + 1.75(1 - T/T_c)^{1/3} + 0.75(1 - T/T_c)][\rho_c + A(T_c - T) + B(T_c - T)^2 + C(T_c - T)^3 + D(T_c - T)^4]$).

$\frac{T}{K}$	$\frac{\rho \pm \sigma_{fit}}{kg \cdot m^{-3}}$	$\frac{T}{K}$	$\frac{\rho \pm \sigma_{fit}}{kg \cdot m^{-3}}$	$\frac{T}{K}$	$\frac{\rho \pm \sigma_{fit}}{kg \cdot m^{-3}}$
270.00	847.48 ± 0.90	310.00	814.59 ± 0.70	370.00	758.28 ± 0.53
280.00	839.60 ± 0.82	320.00	805.78 ± 0.67	380.00	748.08 ± 0.51
290.00	831.50 ± 0.77	330.00	796.75 ± 0.65	390.00	737.65 ± 0.52
293.15	828.89 ± 0.75	340.00	787.48 ± 0.62	400.00	726.99 ± 0.55
298.15	824.72 ± 0.73	350.00	777.98 ± 0.59	410.00	716.09 ± 0.64
300.00	823.16 ± 0.73	360.00	768.24 ± 0.56		

3-Methyl-3-pentanol **[77-74-7]** **$C_6H_{14}O$** **MW = 102.18** **40**

Table 1. Coefficients of the polynomial expansion equation.
Standard deviations (see introduction):
$\sigma_{c,w} = 5.7626 \cdot 10^{-1}$ (combined temperature ranges, weighted),
$\sigma_{c,uw} = 1.1588 \cdot 10^{-1}$ (combined temperature ranges, unweighted).

Coefficient	T = 273.15 to 338.15 K $\rho = A + BT + CT^2 + DT^3 + \ldots$
A	$9.06284 \cdot 10^{2}$
B	$3.36695 \cdot 10^{-1}$
C	$-2.05718 \cdot 10^{-3}$

Table 2. Experimental values with uncertainties and deviation from calculated values.

$\frac{T}{K}$	$\frac{\rho_{exp} \pm 2\sigma_{est}}{kg \cdot m^{-3}}$	$\frac{\rho_{exp} - \rho_{calc}}{kg \cdot m^{-3}}$	Ref. (Symbol in Fig. 1)	$\frac{T}{K}$	$\frac{\rho_{exp} \pm 2\sigma_{est}}{kg \cdot m^{-3}}$	$\frac{\rho_{exp} - \rho_{calc}}{kg \cdot m^{-3}}$	Ref. (Symbol in Fig. 1)
273.15	845.20 ± 0.80	0.44	21-par/sim(✕)	298.15	823.00 ± 0.40	−0.80	84-bra/pin(Δ)
298.15	824.20 ± 0.50	0.40	38-gin/web(∇)	293.15	827.96 ± 0.50	−0.24	85-ort(✕)
273.15	844.50 ± 1.00	−0.26	39-owe/qua(✕)	298.15	823.63 ± 0.50	−0.17	85-ort(✕)
298.15	821.50 ± 1.00	−2.30	39-owe/qua[1)]	300.15	821.84 ± 0.50	−0.17	85-ort(✕)
308.15	812.20 ± 1.00	−2.49	39-owe/qua[1)]	302.15	819.99 ± 0.50	−0.22	85-ort(✕)
318.15	803.10 ± 1.00	−2.08	39-owe/qua[1)]	304.15	818.32 ± 0.50	−0.07	85-ort(✕)
328.15	794.00 ± 1.00	−1.25	39-owe/qua(✕)	306.15	816.50 ± 0.50	−0.05	85-ort(✕)
338.15	784.80 ± 1.00	−0.11	39-owe/qua(✕)	308.15	814.77 ± 0.50	0.08	85-ort(✕)
293.15	828.60 ± 0.40	0.40	47-how/mea(○)	283.15	837.20 ± 0.50	0.51	88-cac/cos(◆)
298.15	824.30 ± 0.40	0.50	47-how/mea(○)	298.15	824.50 ± 0.50	0.70	88-cac/cos(◆)
273.15	844.00 ± 2.00	−0.76	48-zal(✕)	313.15	810.54 ± 0.50	0.55	88-cac/cos(◆)
280.15	837.10 ± 2.00	−2.05	48-zal[1)]	323.15	800.95 ± 0.50	0.69	88-cac/cos(◆)
293.15	827.90 ± 2.00	−0.30	48-zal[1)]	298.15	823.63 ± 0.30	−0.17	94-tar/jun(□)

[1)] Not included in Fig. 1.

Further references: [1887-ref, 19-eyk, 27-nor/cor, 35-sav, 36-nor/has, 40-hov/lan-1, 50-pic/zie, 52-van, 55-soe/fre, 56-sok/fed].

cont.

3-Methyl-3-pentanol (cont.)

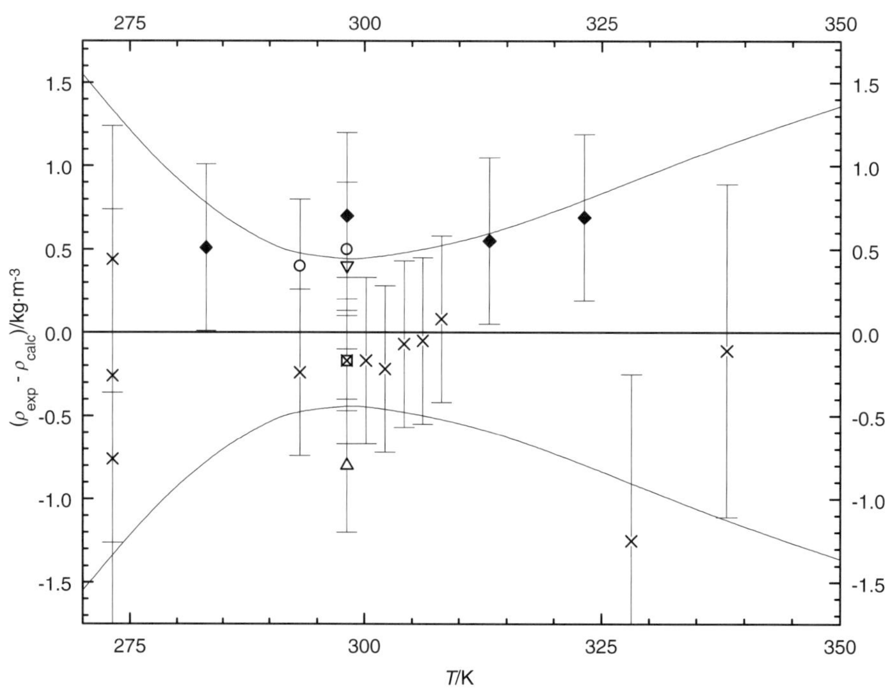

Fig. 1. The symbols show the deviation of the calculated from the experimental values from Table 2. The curves above and below the zero line indicate the calculated error region of the recommended values given in Table 3. The error bars represent the experimental errors. (Error bars smaller than the symbols are omitted for clarity of the figure.)

Table 3. Recommended values (fit to the reliable experimental values according to the equations $\rho = A + BT + CT^2 + DT^3 + \dots$ or $\rho = [1 + 1.75(1 - T/T_c)^{1/3} + 0.75(1 - T/T_c)][\rho_c + A(T_c - T) + B(T_c - T)^2 + C(T_c - T)^3 + D(T_c - T)^4]$).

$\frac{T}{K}$	$\frac{\rho \pm \sigma_{fit}}{kg \cdot m^{-3}}$	$\frac{T}{K}$	$\frac{\rho \pm \sigma_{fit}}{kg \cdot m^{-3}}$	$\frac{T}{K}$	$\frac{\rho \pm \sigma_{fit}}{kg \cdot m^{-3}}$
270.00	847.22 ± 1.55	298.15	823.80 ± 0.44	330.00	793.37 ± 0.95
280.00	839.28 ± 0.87	300.00	822.15 ± 0.44	340.00	782.95 ± 1.17
290.00	830.92 ± 0.52	310.00	812.96 ± 0.53	350.00	772.12 ± 1.36
293.15	828.20 ± 0.47	320.00	803.37 ± 0.72		

4-Methyl-1-pentanol **[626-89-1]** **$C_6H_{14}O$** **MW = 102.18** **41**

Table 1. Coefficients of the polynomial expansion equation.
Standard deviations (see introduction):
$\sigma_{c,w} = 3.9343 \cdot 10^{-1}$ (combined temperature ranges, weighted),
$\sigma_{c,uw} = 8.2317 \cdot 10^{-2}$ (combined temperature ranges, unweighted).

Coefficient	T = 273.15 to 423.15 K $\rho = A + BT + CT^2 + DT^3 + \dots$
A	$9.17775 \cdot 10^{2}$
B	$-2.40364 \cdot 10^{-2}$
C	$-1.13490 \cdot 10^{-3}$

cont.

Table 2. Experimental values with uncertainties and deviation from calculated values.

$\frac{T}{K}$	$\frac{\rho_{exp} \pm 2\sigma_{est}}{kg \cdot m^{-3}}$	$\frac{\rho_{exp} - \rho_{calc}}{kg \cdot m^{-3}}$	Ref. (Symbol in Fig. 1)	$\frac{T}{K}$	$\frac{\rho_{exp} \pm 2\sigma_{est}}{kg \cdot m^{-3}}$	$\frac{\rho_{exp} - \rho_{calc}}{kg \cdot m^{-3}}$	Ref. (Symbol in Fig. 1)
273.15	827.00 ± 1.00	0.47	19-beh(∇)	358.15	763.76 ± 0.50	0.17	40-hov/lan(O)
291.65	814.00 ± 1.00	−0.23	19-beh(∇)	368.15	755.24 ± 0.50	0.13	40-hov/lan(O)
293.15	813.10 ± 0.60	−0.10	36-oli(Δ)	378.15	746.81 ± 0.60	0.41	40-hov/lan(O)
278.15	823.74 ± 0.50	0.45	40-hov/lan(O)	388.15	737.78 ± 0.60	0.32	40-hov/lan(O)
288.15	816.45 ± 0.50	−0.17	40-hov/lan(O)	398.15	728.48 ± 0.60	0.18	40-hov/lan(O)
298.15	809.38 ± 0.50	−0.34	40-hov/lan(O)	408.15	718.78 ± 0.60	−0.13	40-hov/lan(O)
308.15	802.44 ± 0.50	−0.16	40-hov/lan(O)	418.15	708.98 ± 0.60	−0.31	40-hov/lan(O)
318.15	794.91 ± 0.50	−0.34	40-hov/lan(O)	423.15	703.90 ± 0.60	−0.49	40-hov/lan(O)
328.15	787.40 ± 0.50	−0.28	40-hov/lan(O)	293.15	812.90 ± 0.50	−0.30	50-pic/zie(□)
338.15	779.66 ± 0.50	−0.22	40-hov/lan(O)	293.15	814.20 ± 1.50	1.00	52-coo(◆)
348.15	771.78 ± 0.50	−0.07	40-hov/lan(O)				

Further references: [02-gri/tis, 27-nor/cor, 41-hus/age].

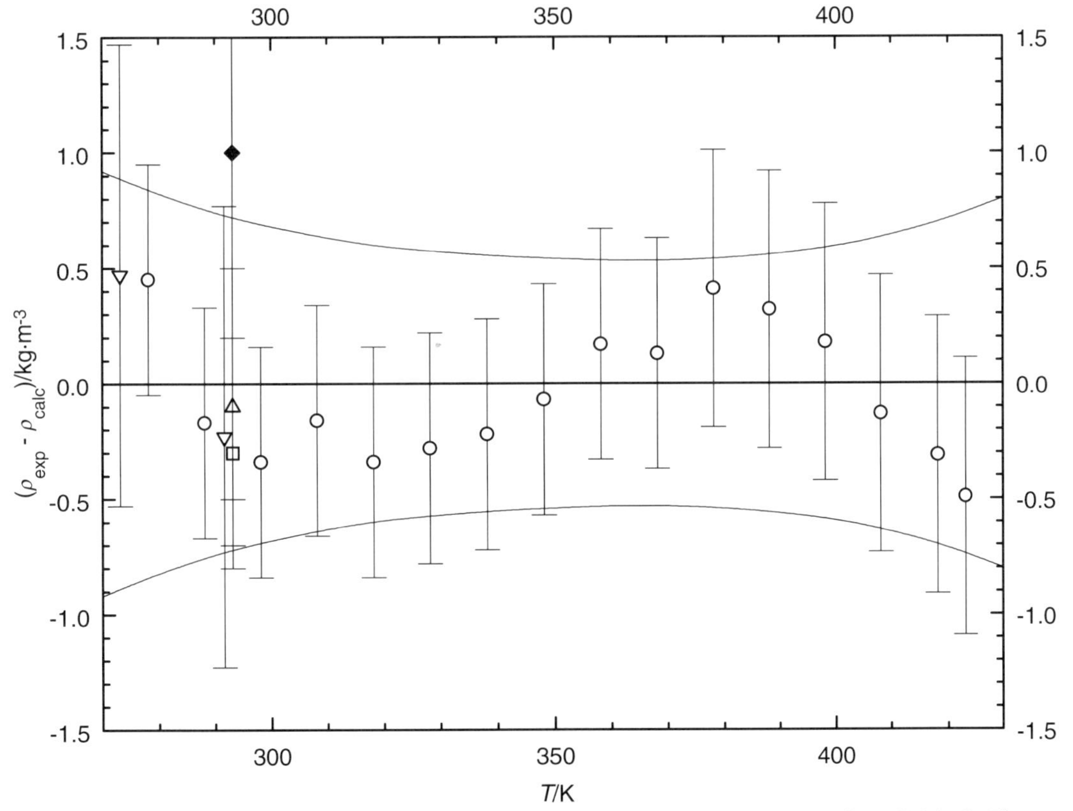

Fig. 1. The symbols show the deviation of the calculated from the experimental values from Table 2. The curves above and below the zero line indicate the calculated error region of the recommended values given in Table 3. The error bars represent the experimental errors. (Error bars smaller than the symbols are omitted for clarity of the figure.)

cont.

4-Methyl-1-pentanol (cont.)

Table 3. Recommended values (fit to the reliable experimental values according to the equations $\rho = A + BT + CT^2 + DT^3 + \ldots$ or $\rho = [1 + 1.75(1 - T/T_c)^{1/3} + 0.75(1 - T/T_c)][\rho_c + A(T_c - T) + B(T_c - T)^2 + C(T_c - T)^3 + D(T_c - T)^4])$.

T/K	$\rho \pm \sigma_{fit}$/kg·m^{-3}	T/K	$\rho \pm \sigma_{fit}$/kg·m^{-3}	T/K	$\rho \pm \sigma_{fit}$/kg·m^{-3}
270.00	828.55 ± 0.92	320.00	793.87 ± 0.59	390.00	735.78 ± 0.56
280.00	822.07 ± 0.82	330.00	786.25 ± 0.57	400.00	726.58 ± 0.59
290.00	815.36 ± 0.74	340.00	778.41 ± 0.55	410.00	717.14 ± 0.64
293.15	813.20 ± 0.72	350.00	770.34 ± 0.54	420.00	707.48 ± 0.71
298.15	809.72 ± 0.69	360.00	762.04 ± 0.53	430.00	697.60 ± 0.80
300.00	808.42 ± 0.68	370.00	753.51 ± 0.53		
310.00	801.26 ± 0.63	380.00	744.76 ± 0.54		

4-Methyl-2-pentanol **[108-11-2]** **$C_6H_{14}O$** **MW = 102.18** **42**

Table 1. Coefficients of the polynomial expansion equation.
Standard deviations (see introduction):
$\sigma_{c,w} = 3.2132 \cdot 10^{-1}$ (combined temperature ranges, weighted),
$\sigma_{c,uw} = 4.5436 \cdot 10^{-2}$ (combined temperature ranges, unweighted).

Coefficient	T = 273.15 to 405.15 K $\rho = A + BT + CT^2 + DT^3 + \ldots$
A	$9.40784 \cdot 10^2$
B	$-1.09111 \cdot 10^{-1}$
C	$-1.18327 \cdot 10^{-3}$

Table 2. Experimental values with uncertainties and deviation from calculated values.

T/K	$\rho_{exp} \pm 2\sigma_{est}$/kg·m^{-3}	$\rho_{exp} - \rho_{calc}$/kg·m^{-3}	Ref. (Symbol in Fig. 1)	T/K	$\rho_{exp} \pm 2\sigma_{est}$/kg·m^{-3}	$\rho_{exp} - \rho_{calc}$/kg·m^{-3}	Ref. (Symbol in Fig. 1)
273.15	823.00 ± 1.00	0.30	09-gue(✕)	398.15	710.12 ± 0.50	0.35	38-hov/lan(∇)
298.15	803.40 ± 0.60	0.33	38-gin/web(✕)	405.15	702.10 ± 0.50	−0.25	38-hov/lan(∇)
278.15	819.07 ± 0.50	0.18	38-hov/lan(∇)	293.15	807.50 ± 0.50	0.39	44-hen/mat(✕)
288.15	811.18 ± 0.50	0.08	38-hov/lan(∇)	292.65	807.50 ± 0.60	−0.01	49-ken/str(✕)
298.15	803.04 ± 0.50	−0.03	38-hov/lan(∇)	293.15	806.90 ± 0.50	−0.21	68-ano(◆)
308.15	794.82 ± 0.50	0.02	38-hov/lan(∇)	298.15	802.72 ± 0.40	−0.35	84-bra/pin(Δ)
318.15	786.37 ± 0.50	0.07	38-hov/lan(∇)	293.15	806.99 ± 0.40	−0.12	85-ort(○)
328.15	777.69 ± 0.50	0.13	38-hov/lan(∇)	298.15	802.98 ± 0.40	−0.09	85-ort(○)
338.15	768.95 ± 0.50	0.36	38-hov/lan(∇)	300.15	801.29 ± 0.40	−0.14	85-ort(○)
348.15	759.87 ± 0.50	0.49	38-hov/lan(∇)	302.15	799.64 ± 0.40	−0.15	85-ort(○)
358.15	749.60 ± 0.50	−0.33	38-hov/lan(∇)	304.15	798.01 ± 0.40	−0.13	85-ort(○)
368.15	739.95 ± 0.50	−0.29	38-hov/lan(∇)	306.15	796.38 ± 0.40	−0.09	85-ort(○)
378.15	730.12 ± 0.50	−0.20	38-hov/lan(∇)	308.15	794.69 ± 0.40	−0.11	85-ort(○)
388.15	720.04 ± 0.50	−0.12	38-hov/lan(∇)	298.15	802.98 ± 0.30	−0.09	94-tar/jun(□)

cont.

Further references: [11-pic/ken, 12-gue, 14-vav, 19-beh, 21-bru/cre, 23-bru, 33-van, 38-whi/joh, 39-dup/dar, 47-tuo/guy, 48-wei, 50-pic/zie, 53-ano-1, 58-ano-5, 58-rao/ram].

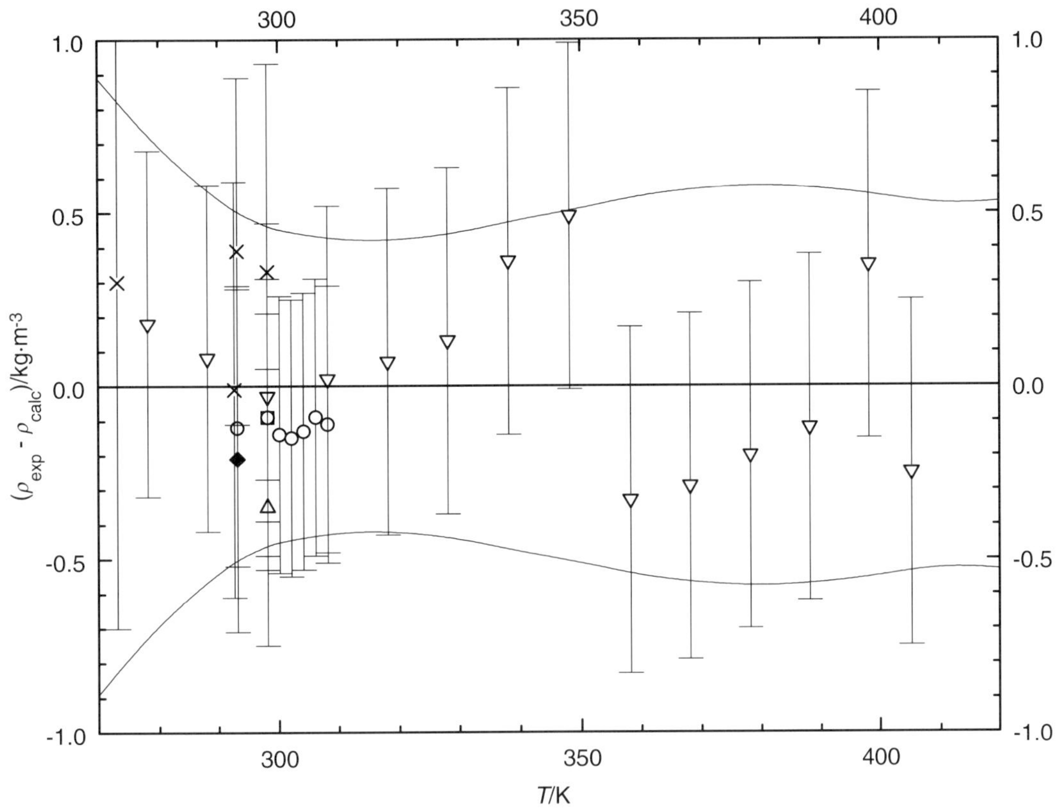

Fig. 1. The symbols show the deviation of the calculated from the experimental values from Table 2. The curves above and below the zero line indicate the calculated error region of the recommended values given in Table 3. The error bars represent the experimental errors. (Error bars smaller than the symbols are omitted for clarity of the figure.)

Table 3. Recommended values (fit to the reliable experimental values according to the equations
$\rho = A + BT + CT^2 + DT^3 + \ldots$ or $\rho = [1 + 1.75(1 - T/T_c)^{1/3} + 0.75(1 - T/T_c)][\rho_c + A(T_c - T) + B(T_c - T)^2 + C(T_c - T)^3 + D(T_c - T)^4]$).

$\frac{T}{K}$	$\frac{\rho \pm \sigma_{fit}}{kg \cdot m^{-3}}$	$\frac{T}{K}$	$\frac{\rho \pm \sigma_{fit}}{kg \cdot m^{-3}}$	$\frac{T}{K}$	$\frac{\rho \pm \sigma_{fit}}{kg \cdot m^{-3}}$
270.00	825.06 ± 0.89	310.00	793.25 ± 0.42	370.00	738.42 ± 0.57
280.00	817.46 ± 0.68	320.00	784.70 ± 0.42	380.00	728.46 ± 0.58
290.00	809.63 ± 0.54	330.00	775.92 ± 0.44	390.00	718.26 ± 0.57
293.15	807.11 ± 0.50	340.00	766.90 ± 0.48	400.00	707.82 ± 0.55
298.15	803.07 ± 0.46	350.00	757.64 ± 0.51	410.00	697.14 ± 0.52
300.00	801.56 ± 0.45	360.00	748.15 ± 0.55	420.00	686.23 ± 0.53

2.1.3 Alkanols, C_7

2,2-Dimethyl-1-pentanol **[2370-12-9]** **$C_7H_{16}O$** **MW = 116.2** **43**

Table 1. Experimental value with uncertainty.

$\frac{T}{K}$	$\frac{\rho_{exp} \pm 2\sigma_{est}}{kg \cdot m^{-3}}$	Ref.
293.15	837.9 ± 1.0	65-shu/puz

2,2-Dimethyl-3-pentanol **[3970-62-5]** **$C_7H_{16}O$** **MW = 116.2** **44**

Table 1. Fit with estimated B coefficient for 3 accepted points. Deviation $\sigma_w = 0.493$.

Coefficient	$\rho = A + BT$
A	1060.34
B	–0.800

Table 2. Experimental values with uncertainties and deviation from calculated values.

$\frac{T}{K}$	$\frac{\rho_{exp} \pm 2\sigma_{est}}{kg \cdot m^{-3}}$	$\frac{\rho_{exp} - \rho_{calc}}{kg \cdot m^{-3}}$	Ref.	$\frac{T}{K}$	$\frac{\rho_{exp} \pm 2\sigma_{est}}{kg \cdot m^{-3}}$	$\frac{\rho_{exp} - \rho_{calc}}{kg \cdot m^{-3}}$	Ref.
293.15	824.6 ± 2.0	–1.22	13-fav [1)]	293.15	825.4 ± 0.6	–0.42	49-boo/gre
293.15	824.0 ± 2.0	–1.82	29-edg/cal [1)]	293.15	826.4 ± 1.0	0.58	50-pic/zie
298.15	822.4 ± 1.0	0.58	38-gin/hau	293.15	828.1 ± 2.0	2.28	55-gay/cau [1)]

[1)] Not included in calculation of linear coefficients.

Table 3. Recommended values.

$\frac{T}{K}$	$\frac{\rho_{exp} \pm 2\sigma_{est}}{kg \cdot m^{-3}}$
290.00	828.3 ± 0.8
293.15	825.8 ± 0.6
298.15	821.8 ± 0.7

2,3-Dimethyl-1-pentanol **[10143-23-4]** **$C_7H_{16}O$** **MW = 116.2** **45**

Table 1. Experimental value with uncertainty.

$\frac{T}{K}$	$\frac{\rho_{exp} \pm 2\sigma_{est}}{kg \cdot m^{-3}}$	Ref.
296.15	836.0 ± 2.0	31-lev/mar-2

2,3-Dimethyl-2-pentanol **[4911-70-0]** **$C_7H_{16}O$** **MW = 116.2** **46**

Table 1. Fit with estimated B coefficient for 5 accepted points. Deviation $\sigma_w = 0.857$.

Coefficient	$\rho = A + BT$
A	1060.84
B	–0.780

cont.

2,3-Dimethyl-2-pentanol (cont.)

Table 2. Experimental values with uncertainties and deviation from calculated values.

$\frac{T}{K}$	$\frac{\rho_{exp} \pm 2\sigma_{est}}{kg \cdot m^{-3}}$	$\frac{\rho_{exp} - \rho_{calc}}{kg \cdot m^{-3}}$	Ref.	$\frac{T}{K}$	$\frac{\rho_{exp} \pm 2\sigma_{est}}{kg \cdot m^{-3}}$	$\frac{\rho_{exp} - \rho_{calc}}{kg \cdot m^{-3}}$	Ref.
293.15	803.0 ± 5.0	−29.18	29-edg/cal[1)]	298.15	830.7 ± 2.0	2.42	38-gin/hau
298.15	828.5 ± 1.0	0.22	36-nor/has	293.15	832.4 ± 1.0	0.22	43-jam
298.15	827.6 ± 1.0	−0.68	36-nor/has	293.15	830.7 ± 2.0	−1.48	50-pic/zie

[1)] Not included in calculation of linear coefficients.

Table 3. Recommended values.

$\frac{T}{K}$	$\frac{\rho_{exp} \pm 2\sigma_{est}}{kg \cdot m^{-3}}$
290.00	834.6 ± 1.5
293.15	832.2 ± 1.4
298.15	828.3 ± 1.4

l-2,3-Dimethyl-2-pentanol **[28357-68-8]** **$C_7H_{16}O$** **MW = 116.2** **47**

Table 1. Experimental value with uncertainty.

$\frac{T}{K}$	$\frac{\rho_{exp} \pm 2\sigma_{est}}{kg \cdot m^{-3}}$	Ref.
293.15	836.0 ± 1.0	57-luk/lan

2,3-Dimethyl-3-pentanol **[595-41-5]** **$C_7H_{16}O$** **MW = 116.2** **48**

Table 1. Fit with estimated *B* coefficient for 6 accepted points. Deviation $\sigma_w = 0.901$.

Coefficient	$\rho = A + BT$
A	1075.79
B	−0.800

Table 2. Experimental values with uncertainties and deviation from calculated values.

$\frac{T}{K}$	$\frac{\rho_{exp} \pm 2\sigma_{est}}{kg \cdot m^{-3}}$	$\frac{\rho_{exp} - \rho_{calc}}{kg \cdot m^{-3}}$	Ref.	$\frac{T}{K}$	$\frac{\rho_{exp} \pm 2\sigma_{est}}{kg \cdot m^{-3}}$	$\frac{\rho_{exp} - \rho_{calc}}{kg \cdot m^{-3}}$	Ref.
273.15	858.6 ± 1.0	1.33	21-par/sim	298.15	838.2 ± 1.0	0.93	36-nor/has
293.15	841.5 ± 1.0	0.23	21-par/sim	298.15	836.5 ± 1.0	−0.77	38-gin/hau
293.15	833.0 ± 4.0	−8.27	33-whi/eve[1)]	293.15	840.2 ± 1.0	−1.07	50-pic/zie
298.15	836.6 ± 1.0	−0.67	36-nor/has	293.15	832.0 ± 4.0	−9.27	56-sok/fed[1)]

[1)] Not included in calculation of linear coefficients.

Table 3. Recommended values.

$\frac{T}{K}$	$\frac{\rho_{exp} \pm 2\sigma_{est}}{kg \cdot m^{-3}}$	$\frac{T}{K}$	$\frac{\rho_{exp} \pm 2\sigma_{est}}{kg \cdot m^{-3}}$
270.00	859.8 ± 1.6	293.15	841.3 ± 1.1
280.00	851.8 ± 1.3	298.15	837.3 ± 1.2
290.00	843.8 ± 1.1		

2,4-Dimethyl-1-pentanol **[6305-71-1]** **$C_7H_{16}O$** **MW = 116.2** **49**

Table 1. Fit with estimated B coefficient for 2 accepted points. Deviation $\sigma_w = 0.500$.

Coefficient	$\rho = A + BT$
A	1054.02
B	–0.800

Table 2. Experimental values with uncertainties and deviation from calculated values.

$\frac{T}{\mathrm{K}}$	$\frac{\rho_{exp} \pm 2\sigma_{est}}{\mathrm{kg \cdot m^{-3}}}$	$\frac{\rho_{exp} - \rho_{calc}}{\mathrm{kg \cdot m^{-3}}}$	Ref.
293.15	793.0 ± 15.0	–26.50	31-chu/mar[1)]
298.15	821.0 ± 5.0	5.50	32-mor/har[1)]
298.15	816.0 ± 2.0	0.50	35-lev/mar
293.15	819.0 ± 2.0	–0.50	39-gol/tay

[1)] Not included in calculation of linear coefficients.

Table 3. Recommended values.

$\frac{T}{\mathrm{K}}$	$\frac{\rho_{exp} \pm 2\sigma_{est}}{\mathrm{kg \cdot m^{-3}}}$
290.00	822.0 ± 1.2
293.15	819.5 ± 1.1
298.15	815.5 ± 1.1

2,4-Dimethyl-2-pentanol **[625-06-9]** **$C_7H_{16}O$** **MW = 116.2** **50**

Table 1. Fit with estimated B coefficient for 7 accepted points. Deviation $\sigma_w = 0.680$.

Coefficient	$\rho = A + BT$
A	1034.29
B	–0.760

Table 2. Experimental values with uncertainties and deviation from calculated values.

$\frac{T}{\mathrm{K}}$	$\frac{\rho_{exp} \pm 2\sigma_{est}}{\mathrm{kg \cdot m^{-3}}}$	$\frac{\rho_{exp} - \rho_{calc}}{\mathrm{kg \cdot m^{-3}}}$	Ref.	$\frac{T}{\mathrm{K}}$	$\frac{\rho_{exp} \pm 2\sigma_{est}}{\mathrm{kg \cdot m^{-3}}}$	$\frac{\rho_{exp} - \rho_{calc}}{\mathrm{kg \cdot m^{-3}}}$	Ref.
293.15	815.7 ± 3.0	4.20	09-kho[1)]	293.15	811.0 ± 1.0	–0.50	38-whi/joh
273.15	832.5 ± 3.0	5.80	09-kho[1)]	293.15	812.2 ± 1.0	0.70	50-pic/zie
293.15	812.2 ± 1.0	0.70	24-cha/deg	293.15	811.9 ± 1.0	0.40	57-pet/sus
293.15	810.3 ± 1.0	–1.20	31-deg	293.15	811.9 ± 1.0	0.40	59-pet/zak-1
288.15	814.8 ± 1.0	–0.50	31-deg	293.15	814.6 ± 2.0	3.10	66-are/tav[1)]
298.15	810.0 ± 2.0	2.30	38-gin/hau[1)]				

[1)] Not included in calculation of linear coefficients.

cont.

2,4-Dimethyl-2-pentanol (cont.)

Table 3. Recommended values.

$\frac{T}{\text{K}}$	$\frac{\rho_{exp} \pm 2\sigma_{est}}{\text{kg} \cdot \text{m}^{-3}}$
280.00	821.5 ± 1.5
290.00	813.9 ± 0.9
293.15	811.5 ± 0.9
298.15	807.7 ± 1.1

2,4-Dimethyl-3-pentanol **[600-36-2]** $C_7H_{16}O$ **MW = 116.2** **51**

Table 1. Coefficients of the polynomial expansion equation.
Standard deviations (see introduction):
$\sigma_{c,w}$ = 1.0054 (combined temperature ranges, weighted),
$\sigma_{c,uw} = 2.2149 \cdot 10^{-1}$ (combined temperature ranges, unweighted).

Coefficient	T = 290.15 to 394.75 K $\rho = A + BT + CT^2 + DT^3 + \ldots$
A	$1.11751 \cdot 10^3$
B	$-9.81289 \cdot 10^{-1}$

Table 2. Experimental values with uncertainties and deviation from calculated values.

$\frac{T}{\text{K}}$	$\frac{\rho_{exp} \pm 2\sigma_{est}}{\text{kg} \cdot \text{m}^{-3}}$	$\frac{\rho_{exp} - \rho_{calc}}{\text{kg} \cdot \text{m}^{-3}}$	Ref. (Symbol in Fig. 1)	$\frac{T}{\text{K}}$	$\frac{\rho_{exp} \pm 2\sigma_{est}}{\text{kg} \cdot \text{m}^{-3}}$	$\frac{\rho_{exp} - \rho_{calc}}{\text{kg} \cdot \text{m}^{-3}}$	Ref. (Symbol in Fig. 1)
290.15	832.30 ± 1.00	−0.49	1876-mun(×)	293.15	830.20 ± 0.50	0.36	53-raz/old(□)
293.15	828.80 ± 1.00	−1.04	1891-pol(×)	306.95	815.30 ± 1.00	−1.00	63-tho/mea(∇)
298.15	825.40 ± 1.00	0.46	38-gin/hau(◆)	316.55	808.10 ± 1.00	1.22	63-tho/mea(∇)
293.15	831.00 ± 1.00	1.16	38-whi/joh(×)	335.85	789.30 ± 1.00	1.36	63-tho/mea(∇)
293.15	828.80 ± 0.60	−1.04	39-gol/tay(Δ)	353.55	771.60 ± 1.00	1.03	63-tho/mea(∇)
303.15	819.20 ± 2.00	−0.83	48-wei(×)	365.75	758.70 ± 1.00	0.10	63-tho/mea(∇)
293.15	829.40 ± 1.00	−0.44	50-pic/zie(×)	378.65	745.50 ± 1.00	−0.44	63-tho/mea(∇)
293.15	830.40 ± 0.50	0.56	52-coo(○)	394.75	729.20 ± 1.00	−0.94	63-tho/mea(∇)

Further references: [44-hus/awu, 49-naz/pin].

Table 3. Recommended values (fit to the reliable experimental values according to the equations $\rho = A + BT + CT^2 + DT^3 + \ldots$ or $\rho = [1 + 1.75(1 - T/T_c)^{1/3} + 0.75(1 - T/T_c)][\rho_c + A(T_c - T) + B(T_c - T)^2 + C(T_c - T)^3 + D(T_c - T)^4])$.

$\frac{T}{\text{K}}$	$\frac{\rho \pm \sigma_{fit}}{\text{kg} \cdot \text{m}^{-3}}$	$\frac{T}{\text{K}}$	$\frac{\rho \pm \sigma_{fit}}{\text{kg} \cdot \text{m}^{-3}}$	$\frac{T}{\text{K}}$	$\frac{\rho \pm \sigma_{fit}}{\text{kg} \cdot \text{m}^{-3}}$
290.00	832.93 ± 0.90	320.00	803.50 ± 0.91	370.00	754.43 ± 1.03
293.15	829.84 ± 0.90	330.00	793.68 ± 0.93	380.00	744.62 ± 1.06
298.15	824.94 ± 0.90	340.00	783.87 ± 0.95	390.00	734.81 ± 1.10
300.00	823.12 ± 0.90	350.00	774.06 ± 0.97	400.00	724.99 ± 1.14
310.00	813.31 ± 0.90	360.00	764.24 ± 1.00		

cont.

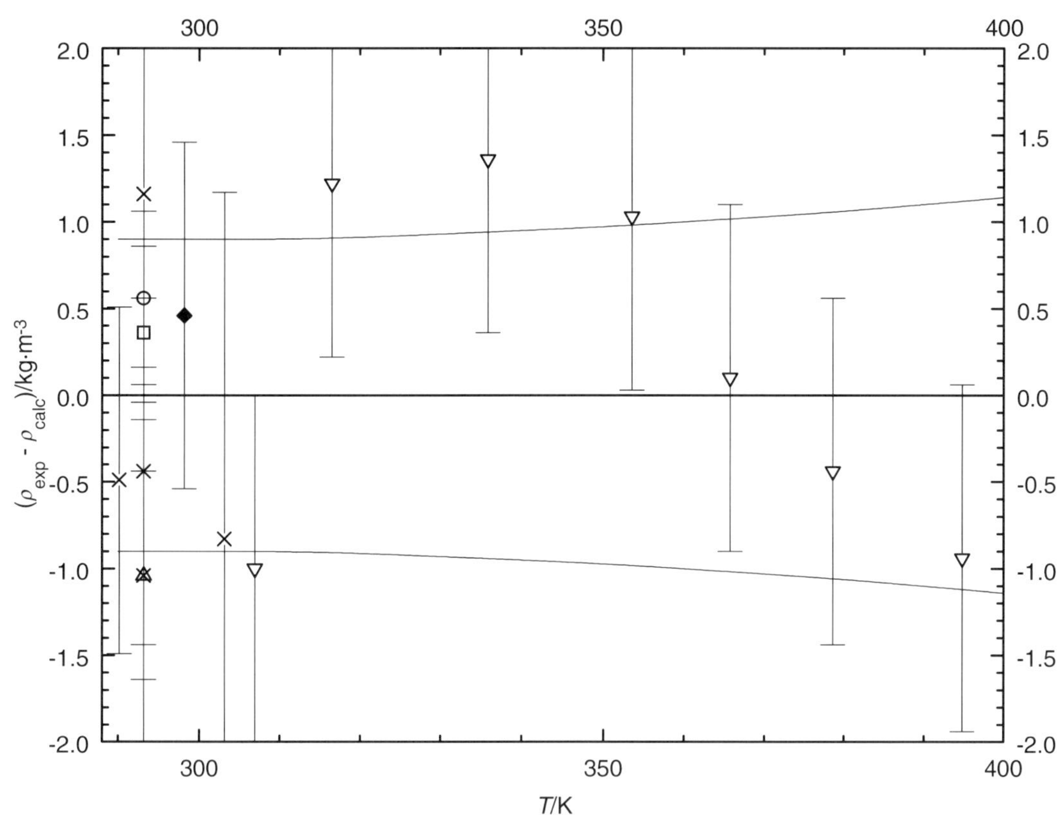

Fig. 1. The symbols show the deviation of the calculated from the experimental values from Table 2. The curves above and below the zero line indicate the calculated error region of the recommended values given in Table 3. The error bars represent the experimental errors. (Error bars smaller than the symbols are omitted for clarity of the figure.)

3,3-Dimethyl-1-pentanol **[19264-94-9]** **$C_7H_{16}O$** **MW = 116.2** **52**

Table 1. Experimental value with uncertainty.

$\frac{T}{K}$	$\frac{\rho_{exp} \pm 2\sigma_{est}}{kg \cdot m^{-3}}$	Ref.
293.15	832.0 ± 1.0	45-sch

3,3-Dimethyl-2-pentanol **[19781-24-9]** **$C_7H_{16}O$** **MW = 116.2** **53**

Table 1. Experimental value with uncertainty.

$\frac{T}{K}$	$\frac{\rho_{exp} \pm 2\sigma_{est}}{kg \cdot m^{-3}}$	Ref.
293.15	827.0 ± 1.5	29-edg/cal

3,4-Dimethyl-1-pentanol **[6570-87-2]** **$C_7H_{16}O$** **MW = 116.2** **54**

Table 1. Experimental values with uncertainties.

T/K	$\rho_{exp} \pm 2\sigma_{est}$/kg·m^{-3}	Ref.
297.15	819.0 ± 2.0	41-hus/age
293.15	834.5 ± 2.0	47-det/cra
325.15	803.5 ± 2.0	60-tsu/kis

3,4-Dimethyl-2-pentanol **[64502-86-9]** **$C_7H_{16}O$** **MW = 116.2** **55**

Table 1. Experimental value with uncertainty.

T/K	$\rho_{exp} \pm 2\sigma_{est}$/kg·m^{-3}	Ref.
294.15	836.0 ± 2.0	19-wil/hat

4,4-Dimethyl-1-pentanol **[3121-79-7]** **$C_7H_{16}O$** **MW = 116.2** **56**

Table 1. Experimental and recommended values with uncertainties.

T/K	$\rho_{exp} \pm 2\sigma_{est}$/kg·m^{-3}	Ref.
290.15	820.0 ± 2.0	49-mal/vol[1)]
293.15	815.2 ± 1.0	31-hom
293.15	815.0 ± 1.0	33-whi/hom-1
293.15	815.1 ± 1.0	Recommended

[1)] Not included in calculation of recommended value.

4,4-Dimethyl-2-pentanol **[6144-93-0]** **$C_7H_{16}O$** **MW = 116.2** **57**

Table 1. Experimental and recommended values with uncertainties.

T/K	$\rho_{exp} \pm 2\sigma_{est}$/kg·m^{-3}	Ref.	T/K	$\rho_{exp} \pm 2\sigma_{est}$/kg·m^{-3}	Ref.
293.15	811.5 ± 1.0	33-whi/hom	293.15	811.3 ± 1.0	52-coo
293.15	812.0 ± 1.0	33-whi/hom-1	293.15	813.3 ± 1.0	55-gay/cau
293.15	811.5 ± 1.0	33-whi/kru	293.15	811.9 ± 1.2	Recommended

2-Ethyl-2-methyl-1-butanol **[18371-13-6]** **$C_7H_{16}O$** **MW = 116.2** **58**

Table 1. Fit with estimated B coefficient for 2 accepted points. Deviation σ_w = 1.550.

Coefficient	$\rho = A + BT$
A	1055.31
B	−0.780

cont.

Table 2. Experimental values with uncertainties and deviation from calculated values.

$\frac{T}{K}$	$\frac{\rho_{exp} \pm 2\sigma_{est}}{kg \cdot m^{-3}}$	$\frac{\rho_{exp} - \rho_{calc}}{kg \cdot m^{-3}}$	Ref.
293.15	828.2 ± 1.0	1.55	25-fav/zal
273.15	840.7 ± 1.0	−1.55	25-fav/zal

Table 3. Recommended values.

$\frac{T}{K}$	$\frac{\rho_{exp} \pm 2\sigma_{est}}{kg \cdot m^{-3}}$
270.00	844.7 ± 1.9
280.00	836.9 ± 1.8
290.00	829.1 ± 1.8
293.15	826.7 ± 1.9
298.15	822.7 ± 1.9

2-Ethyl-3-methyl-1-butanol [32444-34-1] $C_7H_{16}O$ MW = 116.2 59

Table 1. Experimental and recommended values with uncertainties.

$\frac{T}{K}$	$\frac{\rho_{exp} \pm 2\sigma_{est}}{kg \cdot m^{-3}}$	Ref.
298.15	832.7 ± 1.0	56-sar/new
298.15	832.6 ± 0.8	59-tsu/hay
298.15	832.6 ± 1.0	60-tsu/kis
298.15	832.6 ± 0.8	Recommended

2-Ethyl-1-pentanol [27522-11-8] $C_7H_{16}O$ MW = 116.2 60

Table 1. Experimental and recommended values with uncertainties.

$\frac{T}{K}$	$\frac{\rho_{exp} \pm 2\sigma_{est}}{kg \cdot m^{-3}}$	Ref.
298.15	832.0 ± 3.0	32-mor/har[1)]
298.15	828.0 ± 2.0	36-lev/rot-1
298.15	829.6 ± 2.0	50-ada/van
298.15	828.8 ± 2.1	Recommended

[1)] Not included in calculation of recommended value.

3-Ethyl-2-pentanol [609-27-8] $C_7H_{16}O$ MW = 116.2 61

Table 1. Fit with estimated B coefficient for 2 accepted points. Deviation $\sigma_w = 0.200$.

Coefficient	$\rho = A + BT$
A	1071.82
B	−0.800

cont.

3-Ethyl-2-pentanol (cont.)

Table 2. Experimental values with uncertainties and deviation from calculated values.

$\frac{T}{K}$	$\frac{\rho_{exp} \pm 2\sigma_{est}}{kg \cdot m^{-3}}$	$\frac{\rho_{exp} - \rho_{calc}}{kg \cdot m^{-3}}$	Ref.
273.15	853.1 ± 2.0	−0.20	07-fou/tif
298.15	833.5 ± 2.0	0.20	29-luc

Table 3. Recommended values.

$\frac{T}{K}$	$\frac{\rho_{exp} \pm 2\sigma_{est}}{kg \cdot m^{-3}}$
270.00	855.8 ± 2.4
280.00	847.8 ± 1.9
290.00	839.8 ± 1.9
293.15	837.3 ± 2.0
298.15	833.3 ± 2.2

3-Ethyl-3-pentanol **[597-49-9]** **$C_7H_{16}O$** **MW = 116.2** **62**

Table 1. Coefficients of the polynomial expansion equation.
Standard deviations (see introduction):
$\sigma_{c,w} = 5.5395 \cdot 10^{-1}$ (combined temperature ranges, weighted),
$\sigma_{c,uw} = 3.5871 \cdot 10^{-1}$ (combined temperature ranges, unweighted).

Coefficient	T = 273.15 to 338.15 K $\rho = A + BT + CT^2 + DT^3 + \ldots$
A	$8.65425 \cdot 10^{2}$
B	$7.69225 \cdot 10^{-1}$
C	$-2.87450 \cdot 10^{-3}$

Table 2. Experimental values with uncertainties and deviation from calculated values.

$\frac{T}{K}$	$\frac{\rho_{exp} \pm 2\sigma_{est}}{kg \cdot m^{-3}}$	$\frac{\rho_{exp} - \rho_{calc}}{kg \cdot m^{-3}}$	Ref. (Symbol in Fig. 1)	$\frac{T}{K}$	$\frac{\rho_{exp} \pm 2\sigma_{est}}{kg \cdot m^{-3}}$	$\frac{\rho_{exp} - \rho_{calc}}{kg \cdot m^{-3}}$	Ref. (Symbol in Fig. 1)
287.05	849.00 ± 2.00	−0.38	19-eyk(×)	338.15	798.20 ± 1.00	1.35	39-owe/qua(Δ)
298.15	841.30 ± 1.00	2.05	29-luc(∇)	293.15	842.00 ± 2.00	−1.90	53-gru/ost(×)
295.54	840.70 ± 2.00	−0.99	32-boe/wil(◆)	273.15	861.18 ± 0.40	0.11	55-tim/hen(□)
273.15	860.30 ± 1.00	−0.77	39-owe/qua[1)]	288.15	848.29 ± 0.40	−0.12	55-tim/hen(□)
298.15	836.70 ± 1.00	−2.55	39-owe/qua[1)]	303.15	834.83 ± 0.40	0.38	55-tim/hen(□)
308.15	827.30 ± 1.00	−2.21	39-owe/qua[1)]	293.15	845.09 ± 2.00	1.19	56-lib/lap-1(×)
318.15	817.50 ± 1.00	−1.70	39-owe/qua(Δ)	298.15	839.56 ± 0.40	0.31	88-cac/cos(○)
328.15	808.00 ± 1.00	−0.31	39-owe/qua(Δ)				

[1)] Not included in Fig. 1.

Further references: [29-edg/cal, 50-pic/zie, 55-soe/fre, 62-and/kuk-1].

cont.

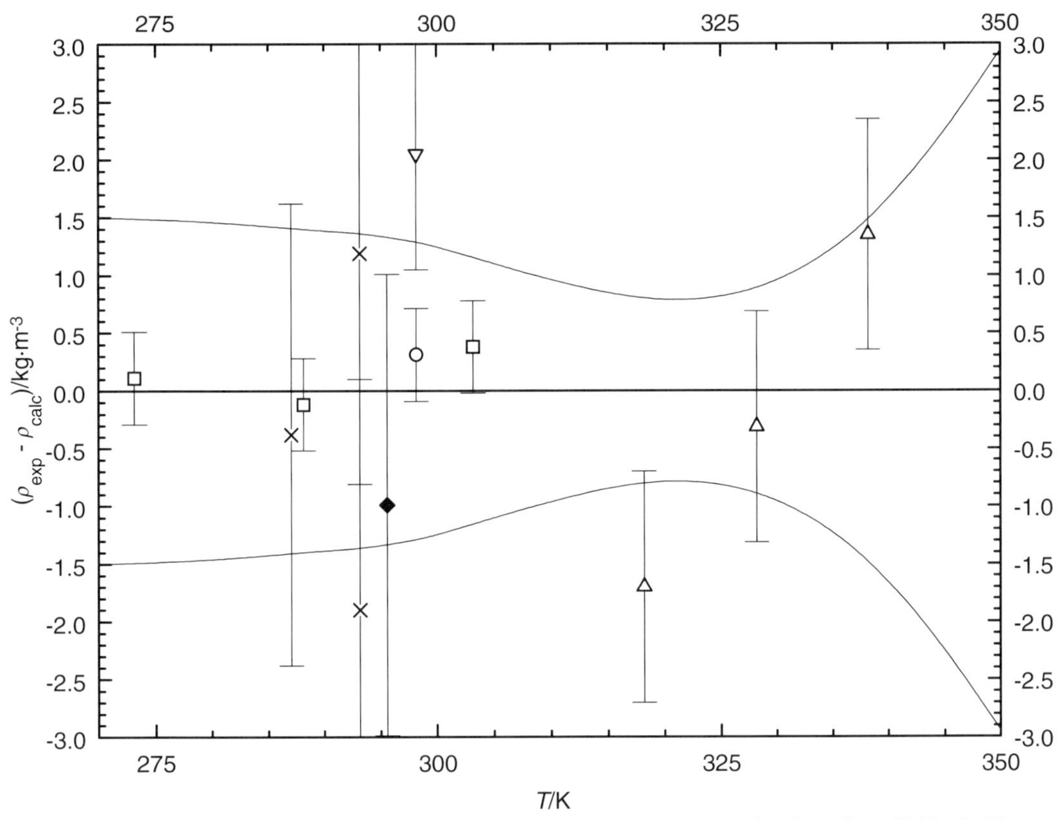

Fig. 1. The symbols show the deviation of the calculated from the experimental values from Table 2. The curves above and below the zero line indicate the calculated error region of the recommended values given in Table 3. The error bars represent the experimental errors. (Error bars smaller than the symbols are omitted for clarity of the figure.)

Table 3. Recommended values (fit to the reliable experimental values according to the equations $\rho = A + BT + CT^2 + DT^3 + \dots$ or $\rho = [1 + 1.75(1 - T/T_c)^{1/3} + 0.75(1 - T/T_c)][\rho_c + A(T_c - T) + B(T_c - T)^2 + C(T_c - T)^3 + D(T_c - T)^4]$).

$\frac{T}{K}$	$\frac{\rho \pm \sigma_{fit}}{kg \cdot m^{-3}}$	$\frac{T}{K}$	$\frac{\rho \pm \sigma_{fit}}{kg \cdot m^{-3}}$	$\frac{T}{K}$	$\frac{\rho \pm \sigma_{fit}}{kg \cdot m^{-3}}$
270.00	863.57 ± 1.50	298.15	839.25 ± 1.29	330.00	806.24 ± 0.87
280.00	855.45 ± 1.47	300.00	837.49 ± 1.25	340.00	794.67 ± 1.54
290.00	846.76 ± 1.38	310.00	827.65 ± 0.95	350.00	782.53 ± 2.94
293.15	843.90 ± 1.37	320.00	817.23 ± 0.73		

1-Heptanol **[111-70-6]** **$C_7H_{16}O$** **MW = 116.2** **63**

T_c = 632.50 K [89-tej/lee]
ρ_c = 267.00 kg·m^{-3} [89-tej/lee]

Table 1. Coefficients for the polynomial expansion equations. Standard deviations (see introduction): $\sigma_t = 7.2440 \cdot 10^{-1}$ (low temperature range), $\sigma_{c,w} = (2.0233 \cdot 10^{-1}$ combined temperature ranges, weighted), $\sigma_{c,uw} = 3.3667 \cdot 10^{-1}$ (combined temperature ranges, unweighted).

Coefficient	T = 273.15 to 510.00 K $\rho = A + BT + CT^2 + DT^3 + \ldots$	T = 510.00 to 632.50 K $\rho = [1 + 1.75(1 - T/T_c)^{1/3} + 0.75(1 - T/T_c)]$ $[\rho_c + A(T_c - T) + B(T_c - T)^2 + C(T_c - T)^3$ $+ D(T_c - T)^4]$
A	$9.98526 \cdot 10^{2}$	$9.10034 \cdot 10^{-1}$
B	$-6.23212 \cdot 10^{-1}$	$-1.32340 \cdot 10^{-2}$
C	$4.74025 \cdot 10^{-4}$	$8.23480 \cdot 10^{-5}$
D	$-1.34624 \cdot 10^{-6}$	$-1.83963 \cdot 10^{-7}$

Table 2. Experimental values with uncertainties and deviation from calculated values.

T/K	$\rho_{exp} \pm 2\sigma_{est}$ / kg·m^{-3}	$\rho_{exp} - \rho_{calc}$ / kg·m^{-3}	Ref. (Symbol in Fig. 1)	T/K	$\rho_{exp} \pm 2\sigma_{est}$ / kg·m^{-3}	$\rho_{exp} - \rho_{calc}$ / kg·m^{-3}	Ref. (Symbol in Fig. 1)
273.15	836.20 ± 0.50	−0.03	31-def(×)	533.15	600.00 ± 1.50	2.80	66-efr(×)
273.15	836.20 ± 1.00	−0.03	31-def[1)]	553.15	574.00 ± 1.50	3.18	66-efr(×)
288.15	825.99 ± 0.50	−0.11	31-def[1)]	573.15	540.00 ± 1.50	0.23	66-efr(×)
288.15	825.99 ± 1.00	−0.11	31-def[1)]	593.15	496.00 ± 1.50	−3.22	66-efr(×)
303.15	815.81 ± 0.50	0.15	31-def[1)]	603.15	470.00 ± 1.50	−2.90	66-efr(×)
303.15	818.81 ± 1.00	3.15	31-def[1)]	613.15	440.00 ± 2.00	−0.03	66-efr(×)
273.15	836.32 ± 0.20	0.09	32-ell/rei(◆)	618.15	420.00 ± 2.00	0.20	66-efr(×)
298.15	819.13 ± 0.20	−0.04	32-ell/rei[1)]	623.15	400.00 ± 5.00	4.68	66-efr(×)
273.15	836.04 ± 0.30	−0.19	35-bil/gis(×)	628.15	370.00 ± 5.00	7.80	66-efr(×)
288.15	825.80 ± 0.30	−0.30	35-bil/gis(×)	298.15	819.30 ± 0.25	0.13	76-kow/kas[1)]
303.15	815.57 ± 0.30	−0.09	35-bil/gis[1)]	298.15	819.29 ± 0.25	0.12	76-kow/kas[1)]
273.15	836.50 ± 0.10	0.27	59-mck/ski(Δ)	313.15	810.47 ± 0.25	1.96	76-kow/kas(×)
273.15	839.00 ± 1.50	2.77	66-efr[1)]	298.15	819.42 ± 0.20	0.25	79-dia/tar[1)]
293.15	824.00 ± 1.50	1.35	66-efr[1)]	308.15	812.28 ± 0.20	0.18	79-dia/tar(×)
313.15	809.00 ± 1.50	0.49	66-efr[1)]	318.15	805.10 ± 0.20	0.22	79-dia/tar(×)
333.15	794.00 ± 1.50	0.26	66-efr[1)]	333.15	794.05 ± 0.20	0.31	79-dia/tar(×)
353.15	778.00 ± 1.50	−0.26	66-efr[1)]	293.15	822.10 ± 0.30	−0.55	82-ort[1)]
373.15	762.00 ± 1.50	−0.03	66-efr[1)]	298.15	819.50 ± 0.30	0.33	82-ort[1)]
393.15	745.00 ± 1.50	0.03	66-efr[1)]	303.15	815.70 ± 0.30	0.04	82-ort[1)]
413.15	726.00 ± 1.50	−1.02	66-efr(×)	308.15	812.10 ± 0.30	−0.00	82-ort[1)]
433.15	707.00 ± 1.50	−1.11	66-efr(×)	313.15	811.20 ± 0.30	2.69	82-ort[1)]
453.15	688.00 ± 1.50	−0.19	66-efr(×)	318.15	804.80 ± 0.30	−0.08	82-ort(×)
473.15	669.00 ± 1.50	1.83	66-efr(×)	298.15	818.24 ± 0.10	−0.93	84-sak/nak(○)
493.15	648.00 ± 1.50	2.99	66-efr[1)]	303.15	815.72 ± 0.20	0.06	89-vij/nai(∇)
513.15	626.00 ± 1.50	4.36	66-efr(×)	323.15	801.09 ± 0.10	−0.12	93-gar/ban-1(□)

[1)] Not included in Fig. 1.

cont.

Table 2. (cont.)

$\frac{T}{K}$	$\frac{\rho_{exp} \pm 2\sigma_{est}}{kg \cdot m^{-3}}$	$\frac{\rho_{exp} - \rho_{calc}}{kg \cdot m^{-3}}$	Ref. (Symbol in Fig. 1)	$\frac{T}{K}$	$\frac{\rho_{exp} \pm 2\sigma_{est}}{kg \cdot m^{-3}}$	$\frac{\rho_{exp} - \rho_{calc}}{kg \cdot m^{-3}}$	Ref. (Symbol in Fig. 1)
328.15	797.68 ± 0.10	0.19	93-gar/ban-1(□)	363.15	770.31 ± 0.10	0.06	93-gar/ban-1(□)
333.15	793.62 ± 0.10	–0.12	93-gar/ban-1(□)	373.15	762.00 ± 0.10	–0.03	93-gar/ban-1(□)
338.15	790.17 ± 0.10	0.23	93-gar/ban-1(□)	278.15	832.91 ± 0.40	0.03	94-rom/pel(×)
343.15	786.06 ± 0.10	–0.03	93-gar/ban-1(□)	288.15	825.96 ± 0.40	–0.14	94-rom/pel[1)]
348.15	781.91 ± 0.10	–0.29	93-gar/ban-1(□)	298.15	819.06 ± 0.40	–0.11	94-rom/pel[1)]
353.15	778.40 ± 0.10	0.14	93-gar/ban-1(□)	308.15	811.99 ± 0.40	–0.11	94-rom/pel[1)]
358.15	774.25 ± 0.10	–0.03	93-gar/ban-1(□)				

[1)] Not included in Fig. 1.

Further references: [1877-cro, 1877-cro-1, 1880-bru-1, 1884-per, 1884-zan, 1886-gar, 1890-gar, 1893-eyk, 06-car/fer, 09-fal-1, 14-low, 18-lev/tay, 19-beh, 19-eyk, 25-fai-1, 27-nor/cor, 27-ver/coo, 28-har-2, 30-bin/for, 30-err/she, 30-she, 33-ano, 33-but/tho, 34-bur/adk, 34-car/jon, 35-but/ram, 35-mah-1, 37-oli, 41-hus/age, 42-mul, 44-app/dob, 45-add, 48-jon/bow, 48-vog-2, 49-pra/dra, 50-mum/phi, 50-pic/zie, 50-sac/sau, 52-coo, 52-eri-1, 66-rob/edm, 67-gol/per, 70-puz/bul, 73-min/rue, 74-moo/wel, 75-mat/fer, 81-sjo/dyh, 82-ven/dha, 83-fuk/ogi, 83-rau/ste, 85-fer/pin, 85-ogi/ara, 85-ort-1, 85-ort/paz-1, 85-sar/paz, 85-zhu, 86-ash/sri, 86-dew/meh, 87-dew/meh, 88-ort/gar, 89-ala/sal, 89-dew/gup, 90-vij/nai, 91-ram/muk, 93-ami/ara, 93-yan/mae, 94-kum/nai, 94-kum/nai-1, 94-yu /tsa-1, 95-cas/cal, 96-elb].

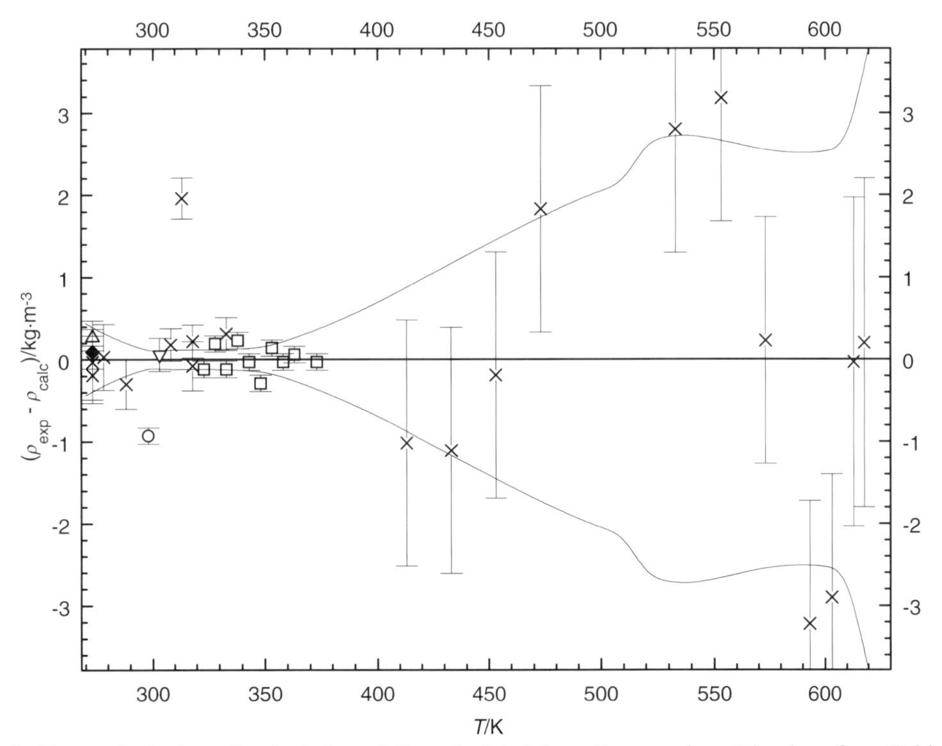

Fig. 1. The symbols show the deviation of the calculated from the experimental values from Table 2. The curves above and below the zero line indicate the calculated error region of the recommended values given in Table 3. The error bars represent the experimental errors. (Error bars smaller than the symbols are omitted for clarity of the figure.)

cont.

1-Heptanol (cont.)

Table 3. Recommended values (fit to the reliable experimental values according to the equations $\rho = A + BT + CT^2 + DT^3 + \ldots$ or $\rho = [1 + 1.75(1 - T/T_c)^{1/3} + 0.75(1 - T/T_c)][\rho_c + A(T_c - T) + B(T_c - T)^2 + C(T_c - T)^3 + D(T_c - T)^4]$).

$\frac{T}{K}$	$\frac{\rho \pm \sigma_{fit}}{kg \cdot m^{-3}}$	$\frac{T}{K}$	$\frac{\rho \pm \sigma_{fit}}{kg \cdot m^{-3}}$	$\frac{T}{K}$	$\frac{\rho \pm \sigma_{fit}}{kg \cdot m^{-3}}$
270.00	838.32 ± 0.44	380.00	756.28 ± 0.44	510.00	625.40 ± 2.14
280.00	831.64 ± 0.29	390.00	747.72 ± 0.56	520.00	613.39 ± 2.64
290.00	824.83 ± 0.18	400.00	738.93 ± 0.69	530.00	601.13 ± 2.72
293.15	822.65 ± 0.15	410.00	729.91 ± 0.83	540.00	588.48 ± 2.73
298.15	819.17 ± 0.11	420.00	720.65 ± 0.98	550.00	575.19 ± 2.68
300.00	817.88 ± 0.11	430.00	711.16 ± 1.12	560.00	560.90 ± 2.62
310.00	810.78 ± 0.12	440.00	701.41 ± 1.27	570.00	545.14 ± 2.56
320.00	803.52 ± 0.12	450.00	691.39 ± 1.41	580.00	527.28 ± 2.52
330.00	796.11 ± 0.12	460.00	681.11 ± 1.55	590.00	506.52 ± 2.51
340.00	788.52 ± 0.13	470.00	670.56 ± 1.69	600.00	481.77 ± 2.52
350.00	780.75 ± 0.15	480.00	659.72 ± 1.82	610.00	451.29 ± 2.59
360.00	772.79 ± 0.23	490.00	648.58 ± 1.94	620.00	411.37 ± 3.78
370.00	764.64 ± 0.33	500.00	637.15 ± 2.05	630.00	344.48 ± 5.46

2-Heptanol **[543-49-7]** **$C_7H_{16}O$** **MW = 116.2** **64**

Table 1. Coefficients of the polynomial expansion equation.
Standard deviations (see introduction):
$\sigma_{c,w} = 8.0415 \cdot 10^{-1}$ (combined temperature ranges, weighted),
$\sigma_{c,uw} = 1.7660 \cdot 10^{-1}$ (combined temperature ranges, unweighted).

Coefficient	T = 273.15 to 421.45 K $\rho = A + BT + CT^2 + DT^3 + \ldots$
A	$9.15825 \cdot 10^2$
B	$4.92881 \cdot 10^{-2}$
C	$-1.31078 \cdot 10^{-3}$

Table 2. Experimental values with uncertainties and deviation from calculated values.

$\frac{T}{K}$	$\frac{\rho_{exp} \pm 2\sigma_{est}}{kg \cdot m^{-3}}$	$\frac{\rho_{exp} - \rho_{calc}}{kg \cdot m^{-3}}$	Ref. (Symbol in Fig. 1)	$\frac{T}{K}$	$\frac{\rho_{exp} \pm 2\sigma_{est}}{kg \cdot m^{-3}}$	$\frac{\rho_{exp} - \rho_{calc}}{kg \cdot m^{-3}}$	Ref. (Symbol in Fig. 1)
293.15	818.50 ± 1.00	0.87	11-pic/ken(×)	293.15	816.70 ± 1.00	−0.93	52-pom(×)
383.15	741.70 ± 1.00	−0.58	12-pic/ken(◆)	298.15	812.90 ± 1.00	−1.10	52-pom(×)
295.15	815.30 ± 0.60	−0.89	30-err/she(∇)	303.25	811.80 ± 1.00	1.57	63-tho/mea(×)
273.15	831.50 ± 1.00	0.01	30-she(×)	321.15	797.80 ± 1.00	1.34	63-tho/mea(×)
288.15	820.30 ± 1.00	−0.89	30-she(×)	332.35	788.00 ± 1.00	0.58	63-tho/mea(×)
293.15	816.70 ± 1.00	−0.93	30-she(×)	348.05	774.80 ± 1.00	0.61	63-tho/mea(×)
273.15	832.01 ± 0.50	0.52	32-ell/rei(○)	362.45	761.50 ± 1.00	0.01	63-tho/mea(×)
298.15	813.38 ± 0.50	−0.62	32-ell/rei(○)	376.75	748.50 ± 1.00	0.16	63-tho/mea(×)
303.15	809.80 ± 0.50	−0.51	48-wei(Δ)	393.15	732.20 ± 1.00	−0.40	63-tho/mea(×)
293.15	819.00 ± 1.50	1.37	50-pic/zie(×)	408.45	717.10 ± 1.00	−0.18	63-tho/mea(×)
293.15	817.90 ± 0.50	0.27	52-coo(□)	421.45	703.50 ± 1.00	−0.28	63-tho/mea(×)

cont.

Further references: [09-hen, 09-mas, 13-tho, 49-mal/vol, 62-cuv/nor].

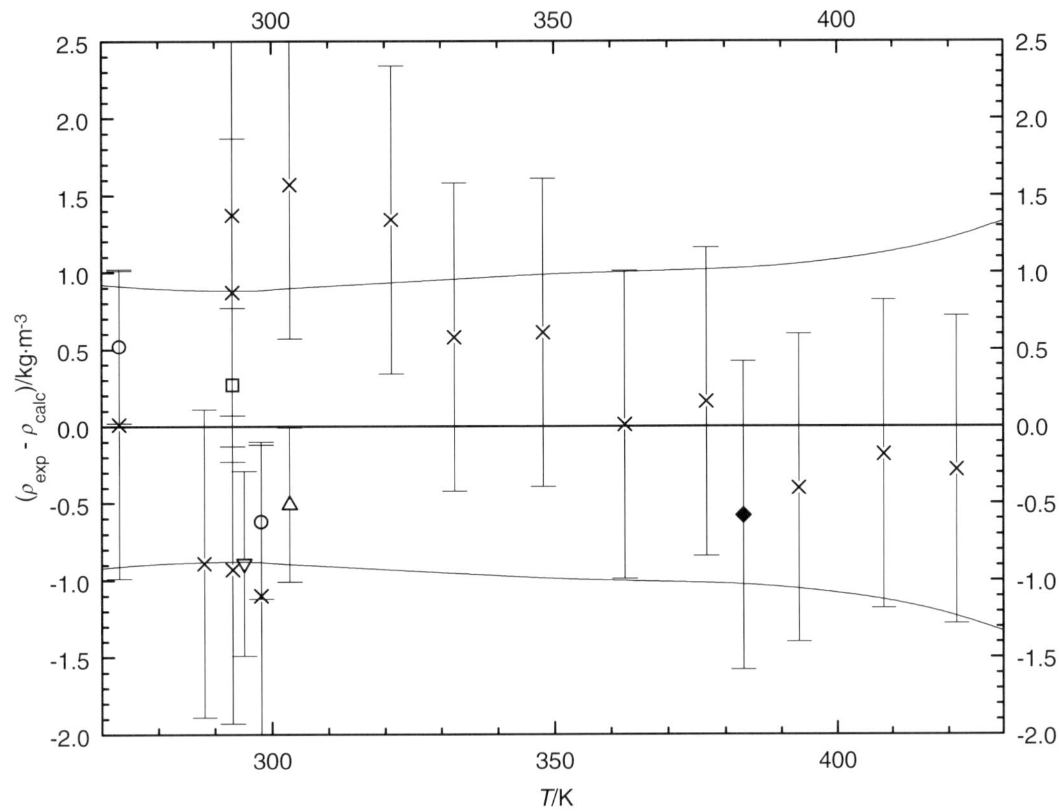

Fig. 1. The symbols show the deviation of the calculated from the experimental values from Table 2. The curves above and below the zero line indicate the calculated error region of the recommended values given in Table 3. The error bars represent the experimental errors. (Error bars smaller than the symbols are omitted for clarity of the figure.)

Table 3. Recommended values (fit to the reliable experimental values according to the equations
$\rho = A + BT + CT^2 + DT^3 + \ldots$ or $\rho = [1 + 1.75(1 - T/T_c)^{1/3} + 0.75(1 - T/T_c)][\rho_c + A(T_c - T) + B(T_c - T)^2 + C(T_c - T)^3 + D(T_c - T)^4]$).

$\frac{T}{\mathrm{K}}$	$\frac{\rho \pm \sigma_{fit}}{\mathrm{kg \cdot m^{-3}}}$	$\frac{T}{\mathrm{K}}$	$\frac{\rho \pm \sigma_{fit}}{\mathrm{kg \cdot m^{-3}}}$	$\frac{T}{\mathrm{K}}$	$\frac{\rho \pm \sigma_{fit}}{\mathrm{kg \cdot m^{-3}}}$
270.00	833.58 ± 0.92	320.00	797.37 ± 0.93	390.00	735.68 ± 1.04
280.00	826.86 ± 0.89	330.00	789.35 ± 0.95	400.00	725.82 ± 1.08
290.00	819.88 ± 0.88	340.00	781.06 ± 0.97	410.00	715.69 ± 1.13
293.15	817.63 ± 0.88	350.00	772.51 ± 0.99	420.00	705.30 ± 1.21
298.15	814.00 ± 0.88	360.00	763.69 ± 1.00	430.00	694.66 ± 1.33
300.00	812.64 ± 0.89	370.00	754.62 ± 1.01		
310.00	805.14 ± 0.91	380.00	745.28 ± 1.02		

3-Heptanol **[589-82-2]** **$C_7H_{16}O$** **MW = 116.2** **65**

Table 1. Coefficients of the polynomial expansion equation.
Standard deviations (see introduction):
$\sigma_{c,w} = 5.5943 \cdot 10^{-1}$ (combined temperature ranges, weighted),
$\sigma_{c,uw} = 2.3013 \cdot 10^{-1}$ (combined temperature ranges, unweighted).

Coefficient	T = 273.15 to 416.55 K $\rho = A + BT + CT^2 + DT^3 + \ldots$
A	$9.65814 \cdot 10^{2}$
B	$-1.88467 \cdot 10^{-1}$
C	$-1.04313 \cdot 10^{-3}$

Table 2. Experimental values with uncertainties and deviation from calculated values.

$\frac{T}{\text{K}}$	$\frac{\rho_{exp} \pm 2\sigma_{est}}{\text{kg}\cdot\text{m}^{-3}}$	$\frac{\rho_{exp} - \rho_{calc}}{\text{kg}\cdot\text{m}^{-3}}$	Ref. (Symbol in Fig. 1)	$\frac{T}{\text{K}}$	$\frac{\rho_{exp} \pm 2\sigma_{est}}{\text{kg}\cdot\text{m}^{-3}}$	$\frac{\rho_{exp} - \rho_{calc}}{\text{kg}\cdot\text{m}^{-3}}$	Ref. (Symbol in Fig. 1)
298.15	815.90 ± 2.00	−1.00	28-dil/luc(✕)	306.95	810.00 ± 1.00	0.32	63-tho/mea(∇)
295.15	819.40 ± 0.50	0.08	30-err/she(○)	316.55	802.10 ± 1.00	0.47	63-tho/mea(∇)
273.15	837.00 ± 0.50	0.49	30-she(□)	335.95	785.80 ± 1.00	1.03	63-tho/mea(∇)
288.15	825.10 ± 0.50	0.20	30-she(□)	353.55	767.40 ± 1.00	−1.39	63-tho/mea(∇)
293.15	821.00 ± 0.50	0.08	30-she(□)	365.75	758.30 ± 1.00	0.96	63-tho/mea(∇)
298.15	816.50 ± 1.00	−0.40	50-ada/van(✕)	378.65	745.90 ± 1.00	1.01	63-tho/mea(∇)
293.15	820.40 ± 1.00	−0.52	50-pic/zie(◆)	405.55	716.60 ± 1.00	−1.22	63-tho/mea(∇)
293.15	821.10 ± 0.50	0.18	52-coo(Δ)	416.55	708.00 ± 1.00	1.69	63-tho/mea(∇)
293.15	820.90 ± 1.00	−0.02	53-ano-1(✕)	298.15	814.92 ± 2.00	−1.98	88-tan/luo(✕)

Further references: [13-pic/ken, 61-tis/sta].

Table 3. Recommended values (fit to the reliable experimental values according to the equations $\rho = A + BT + CT^2 + DT^3 + \ldots$ or $\rho = [1 + 1.75(1 - T/T_c)^{1/3} + 0.75(1 - T/T_c)][\rho_c + A(T_c - T) + B(T_c - T)^2 + C(T_c - T)^3 + D(T_c - T)^4]$).

$\frac{T}{\text{K}}$	$\frac{\rho \pm \sigma_{fit}}{\text{kg}\cdot\text{m}^{-3}}$	$\frac{T}{\text{K}}$	$\frac{\rho \pm \sigma_{fit}}{\text{kg}\cdot\text{m}^{-3}}$	$\frac{T}{\text{K}}$	$\frac{\rho \pm \sigma_{fit}}{\text{kg}\cdot\text{m}^{-3}}$
270.00	838.88 ± 0.91	320.00	798.69 ± 0.91	390.00	733.65 ± 1.03
280.00	831.26 ± 0.88	330.00	790.02 ± 0.93	400.00	723.53 ± 1.07
290.00	823.43 ± 0.87	340.00	781.15 ± 0.95	410.00	713.19 ± 1.14
293.15	820.92 ± 0.87	350.00	772.07 ± 0.96	420.00	702.65 ± 1.23
298.15	816.90 ± 0.87	360.00	762.78 ± 0.97	430.00	691.90 ± 1.36
300.00	815.39 ± 0.87	370.00	753.28 ± 0.99		
310.00	807.14 ± 0.89	380.00	743.57 ± 1.00		

cont.

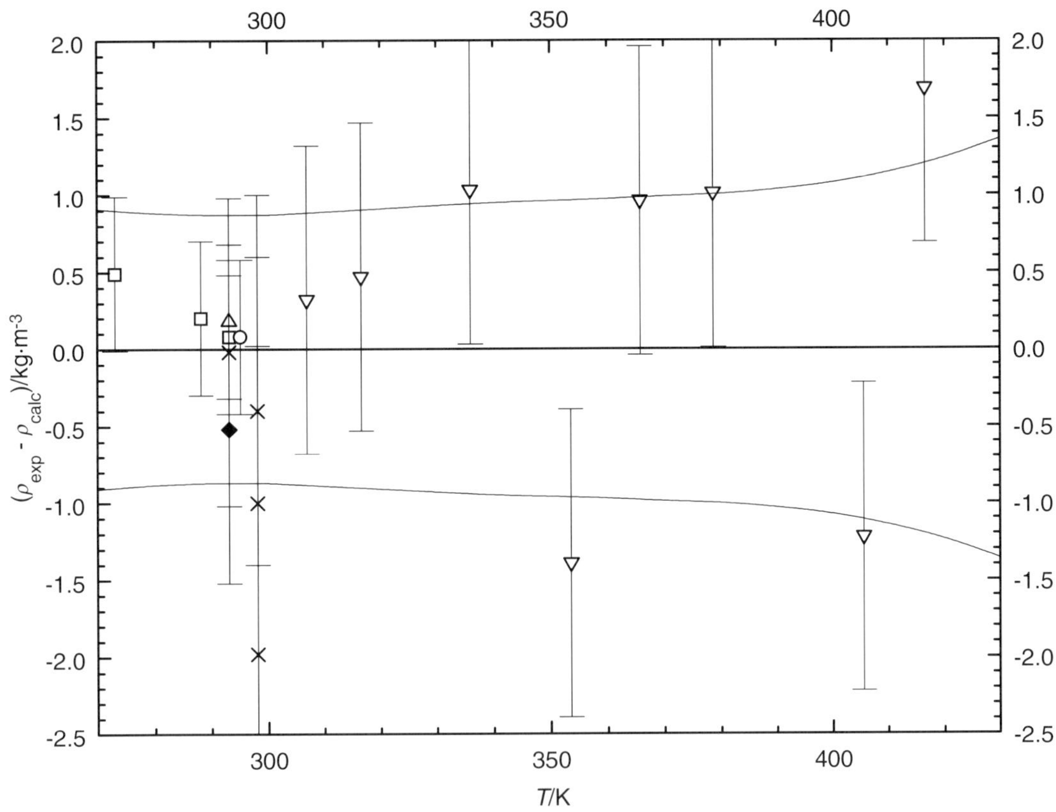

Fig. 1. The symbols show the deviation of the calculated from the experimental values from Table 2. The curves above and below the zero line indicate the calculated error region of the recommended values given in Table 3. The error bars represent the experimental errors. (Error bars smaller than the symbols are omitted for clarity of the figure.)

4-Heptanol **[589-55-9]** **$C_7H_{16}O$** **MW = 116.2** **66**

Table 1. Coefficients of the polynomial expansion equation.
Standard deviations (see introduction):
$\sigma_{c,w} = 1.1547$ (combined temperature ranges, weighted),
$\sigma_{c,uw} = 2.3522 \cdot 10^{-1}$ (combined temperature ranges, unweighted).

Coefficient	$T = 273.15$ to 416.05 K $\rho = A + BT + CT^2 + DT^3 + \ldots$
A	$9.10927 \cdot 10^{2}$
B	$9.60428 \cdot 10^{-2}$
C	$-1.40301 \cdot 10^{-3}$

cont.

4-Heptanol (cont.)

Table 2. Experimental values with uncertainties and deviation from calculated values.

$\frac{T}{K}$	$\frac{\rho_{exp} \pm 2\sigma_{est}}{kg \cdot m^{-3}}$	$\frac{\rho_{exp} - \rho_{calc}}{kg \cdot m^{-3}}$	Ref. (Symbol in Fig. 1)	$\frac{T}{K}$	$\frac{\rho_{exp} \pm 2\sigma_{est}}{kg \cdot m^{-3}}$	$\frac{\rho_{exp} - \rho_{calc}}{kg \cdot m^{-3}}$	Ref. (Symbol in Fig. 1)
288.15	822.00 ± 1.00	−0.11	14-vav(×)	293.15	817.20 ± 2.00	−1.31	52-coo(×)
293.15	820.00 ± 2.00	1.49	19-beh(×)	293.15	819.60 ± 2.00	1.09	56-shu/bel(×)
290.85	819.30 ± 2.00	−0.88	19-eyk(×)	303.05	811.40 ± 1.00	0.22	63-tho/mea(∇)
298.15	812.90 ± 0.30	−1.94	23-bru(□)	312.95	804.60 ± 1.00	1.02	63-tho/mea(∇)
298.15	815.60 ± 1.00	0.76	27-nor/cor(◆)	327.95	791.90 ± 1.00	0.37	63-tho/mea(∇)
295.15	817.00 ± 0.40	−0.05	30-err/she(○)	349.15	773.40 ± 1.00	−0.03	63-tho/mea(∇)
273.15	833.50 ± 0.60	1.02	30-she(Δ)	364.65	759.10 ± 1.00	−0.29	63-tho/mea(∇)
288.15	821.90 ± 0.60	−0.21	30-she(Δ)	381.45	742.90 ± 1.00	−0.52	63-tho/mea(∇)
293.15	818.30 ± 0.60	−0.21	30-she(Δ)	406.25	717.30 ± 1.00	−1.09	63-tho/mea(∇)
293.15	820.00 ± 2.00	1.49	39-gol/tay(×)	416.05	706.70 ± 1.00	−1.33	63-tho/mea(∇)

Further references: [28-dil/luc, 36-tuo, 42-boe/han, 50-pic/zie].

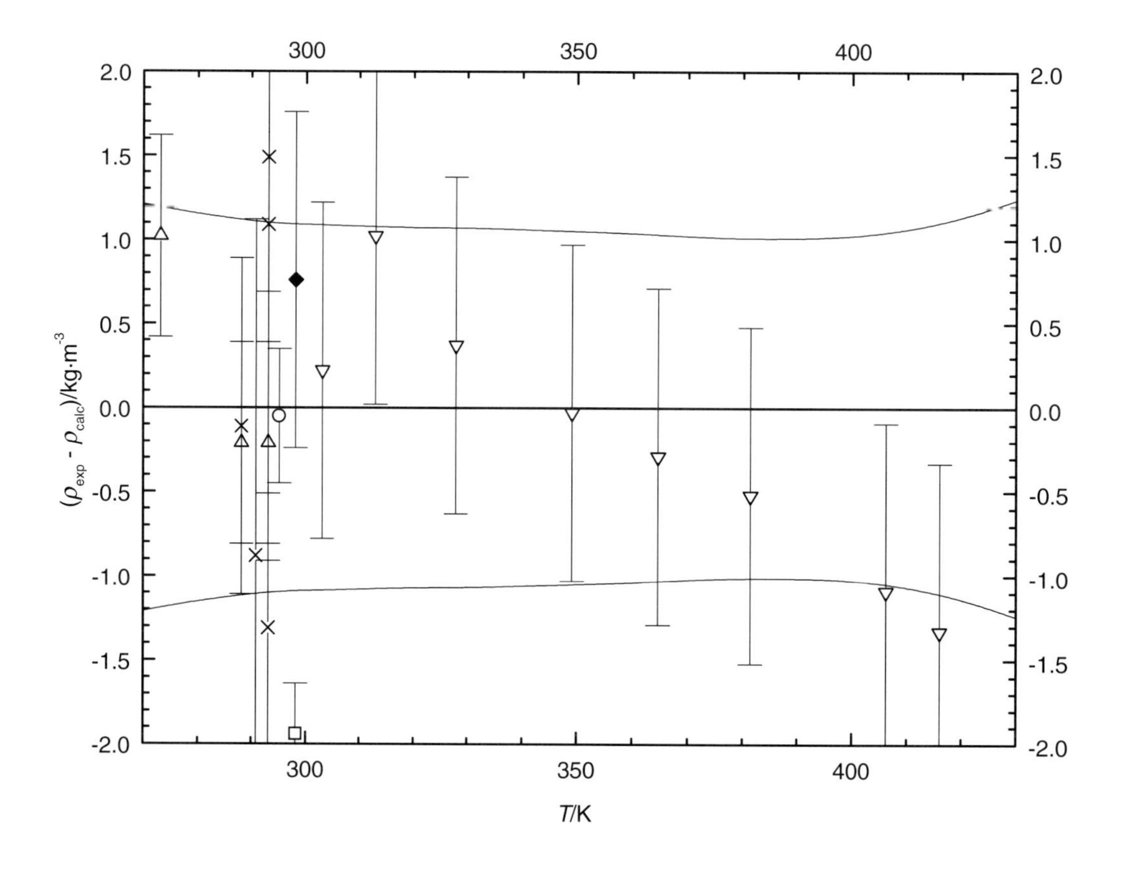

Fig. 1. The symbols show the deviation of the calculated from the experimental values from Table 2. The curves above and below the zero line indicate the calculated error region of the recommended values given in Table 3. The error bars represent the experimental errors. (Error bars smaller than the symbols are omitted for clarity of the figure.)

cont.

Table 3. Recommended values (fit to the reliable experimental values according to the equations $\rho = A + BT + CT^2 + DT^3 + \ldots$ or $\rho = [1 + 1.75(1 - T/T_c)^{1/3} + 0.75(1 - T/T_c)][\rho_c + A(T_c - T) + B(T_c - T)^2 + C(T_c - T)^3 + D(T_c - T)^4]$).

$\frac{T}{\text{K}}$	$\frac{\rho \pm \sigma_{fit}}{\text{kg} \cdot \text{m}^{-3}}$	$\frac{T}{\text{K}}$	$\frac{\rho \pm \sigma_{fit}}{\text{kg} \cdot \text{m}^{-3}}$	$\frac{T}{\text{K}}$	$\frac{\rho \pm \sigma_{fit}}{\text{kg} \cdot \text{m}^{-3}}$
270.00	834.58 ± 1.21	320.00	797.99 ± 1.07	390.00	734.99 ± 1.01
280.00	827.82 ± 1.15	330.00	789.83 ± 1.07	400.00	724.86 ± 1.02
290.00	820.79 ± 1.11	340.00	781.39 ± 1.06	410.00	714.46 ± 1.06
293.15	818.51 ± 1.10	350.00	772.67 ± 1.05	420.00	703.77 ± 1.13
298.15	814.84 ± 1.09	360.00	763.67 ± 1.04	430.00	692.81 ± 1.24
300.00	813.47 ± 1.09	370.00	754.39 ± 1.02		
310.00	805.87 ± 1.08	380.00	744.83 ± 1.01		

2-Methyl-1-hexanol **[624-22-6]** **$C_7H_{16}O$** **MW = 116.2** **67**

Table 1. Fit with estimated B coefficient for 3 accepted points. Deviation $\sigma_w = 0.583$.

Coefficient	$\rho = A + BT$
A	1054.84
B	−0.780

Table 2. Experimental values with uncertainties and deviation from calculated values.

$\frac{T}{\text{K}}$	$\frac{\rho_{exp} \pm 2\sigma_{est}}{\text{kg} \cdot \text{m}^{-3}}$	$\frac{\rho_{exp} - \rho_{calc}}{\text{kg} \cdot \text{m}^{-3}}$	Ref.
286.15	831.3 ± 1.0	−0.34	08-zel/prz
293.15	827.0 ± 1.0	0.82	08-zel/prz
293.15	825.7 ± 1.0	−0.48	50-pic/zie
293.15	829.0 ± 2.0	2.82	54-naz/kak-1[1)]

[1)] Not included in calculation of linear coefficients.

Table 3. Recommended values.

$\frac{T}{\text{K}}$	$\frac{\rho_{exp} \pm 2\sigma_{est}}{\text{kg} \cdot \text{m}^{-3}}$	$\frac{T}{\text{K}}$	$\frac{\rho_{exp} \pm 2\sigma_{est}}{\text{kg} \cdot \text{m}^{-3}}$
280.00	836.4 ± 1.4	293.15	826.2 ± 0.9
290.00	828.6 ± 0.9	298.15	822.3 ± 1.2

2-Methyl-2-hexanol **[625-23-0]** **$C_7H_{16}O$** **MW = 116.2** **68**

Table 1. Coefficients of the polynomial expansion equation. Standard deviations (see introduction):
$\sigma_{c,w} = 4.8651 \cdot 10^{-1}$ (combined temperature ranges, weighted),
$\sigma_{c,uw} = 1.2332 \cdot 10^{-1}$ (combined temperature ranges, unweighted).

Coefficient	T = 273.15 to 388.25 K $\rho = A + BT + CT^2 + DT^3 + \ldots$
A	$9.60129 \cdot 10^{2}$
B	$-1.73590 \cdot 10^{-1}$
C	$-1.10338 \cdot 10^{-3}$

cont.

2-Methyl-2-hexanol (cont.)

Table 2. Experimental values with uncertainties and deviation from calculated values.

$\frac{T}{K}$	$\frac{\rho_{exp} \pm 2\sigma_{est}}{kg \cdot m^{-3}}$	$\frac{\rho_{exp} - \rho_{calc}}{kg \cdot m^{-3}}$	Ref. (Symbol in Fig. 1)	$\frac{T}{K}$	$\frac{\rho_{exp} \pm 2\sigma_{est}}{kg \cdot m^{-3}}$	$\frac{\rho_{exp} - \rho_{calc}}{kg \cdot m^{-3}}$	Ref. (Symbol in Fig. 1)
298.15	809.30 ± 1.00	−0.99	38-gin/hau(×)	293.15	814.60 ± 1.00	0.18	54-naz/kak-2(×)
293.15	814.60 ± 0.70	0.18	38-whi/ore(×)	293.15	815.00 ± 1.00	0.58	54-naz/kak-2(×)
273.15	831.10 ± 1.00	0.71	39-owe/qua(□)	293.15	815.00 ± 1.00	0.58	57-pet/sus(×)
298.15	809.70 ± 1.00	−0.59	39-owe/qua(□)	293.15	815.30 ± 1.00	0.88	59-pet/zak-1(×)
308.15	801.40 ± 1.00	−0.46	39-owe/qua(□)	303.05	806.40 ± 0.50	0.21	63-tho/mea(Δ)
318.15	792.90 ± 1.00	−0.32	39-owe/qua(□)	311.95	798.00 ± 0.50	−0.60	63-tho/mea(Δ)
328.15	784.20 ± 1.00	−0.15	39-owe/qua(□)	327.75	785.20 ± 0.50	0.49	63-tho/mea(Δ)
338.15	776.00 ± 1.00	0.74	39-owe/qua(□)	349.05	765.90 ± 0.50	0.79	63-tho/mea(Δ)
293.15	814.20 ± 0.50	−0.22	50-pic/zie(○)	364.75	749.80 ± 0.50	−0.22	63-tho/mea(Δ)
293.15	813.60 ± 0.70	−0.82	52-lev/tan(◆)	380.55	734.20 ± 0.55	−0.08	63-tho/mea(Δ)
298.15	809.80 ± 0.60	−0.49	53-sut(∇)	388.25	726.00 ± 0.55	−0.41	63-tho/mea(Δ)

Further references: [09-hen, 29-edg/cal, 33-whi/woo, 38-whi/joh].

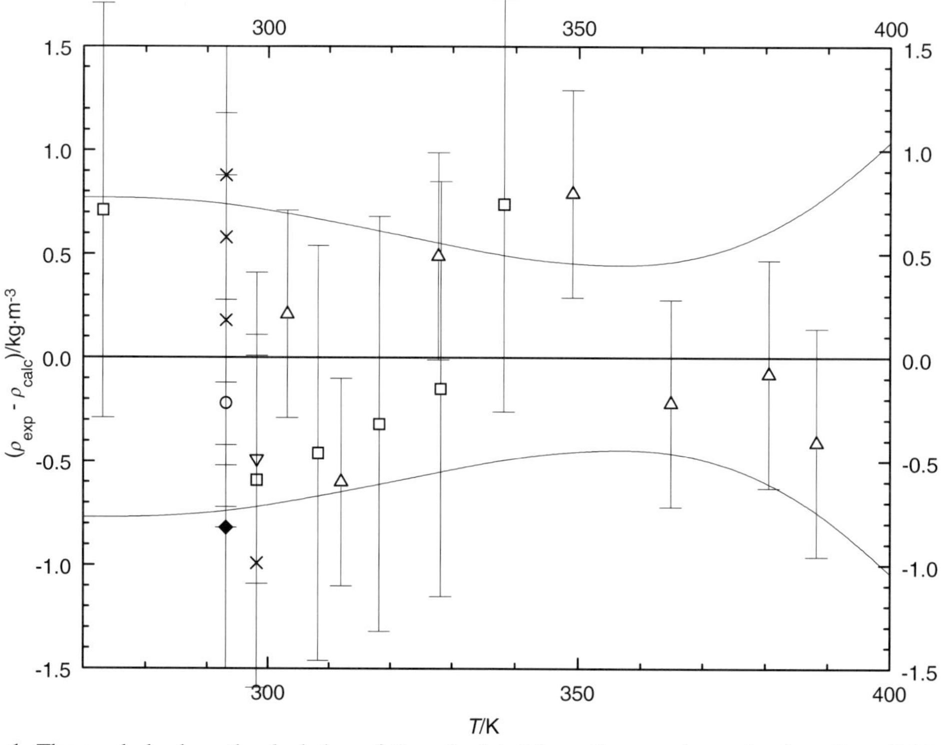

Fig. 1. The symbols show the deviation of the calculated from the experimental values from Table 2. The curves above and below the zero line indicate the calculated error region of the recommended values given in Table 3. The error bars represent the experimental errors. (Error bars smaller than the symbols are omitted for clarity of the figure.)

cont.

Table 3. Recommended values (fit to the reliable experimental values according to the equations $\rho = A + BT + CT^2 + DT^3 + \ldots$ or $\rho = [1 + 1.75(1 - T/T_c)^{1/3} + 0.75(1 - T/T_c)][\rho_c + A(T_c - T) + B(T_c - T)^2 + C(T_c - T)^3 + D(T_c - T)^4])$.

$\frac{T}{K}$	$\frac{\rho \pm \sigma_{fit}}{kg \cdot m^{-3}}$	$\frac{T}{K}$	$\frac{\rho \pm \sigma_{fit}}{kg \cdot m^{-3}}$	$\frac{T}{K}$	$\frac{\rho \pm \sigma_{fit}}{kg \cdot m^{-3}}$
270.00	832.82 ± 0.77	310.00	800.28 ± 0.66	370.00	744.85 ± 0.48
280.00	825.02 ± 0.77	320.00	791.59 ± 0.60	380.00	734.84 ± 0.59
290.00	816.99 ± 0.75	330.00	782.69 ± 0.54	390.00	724.60 ± 0.77
293.15	814.42 ± 0.74	340.00	773.56 ± 0.48	400.00	714.15 ± 1.04
298.15	810.29 ± 0.72	350.00	764.21 ± 0.45		
300.00	808.75 ± 0.71	360.00	754.64 ± 0.44		

2-Methyl-3-hexanol **[617-29-8]** **$C_7H_{16}O$** **MW = 116.2** **69**

Table 1. Fit with estimated B coefficient for 6 accepted points. Deviation $\sigma_w = 0.661$.

Coefficient	$\rho = A + BT$
A	1052.32
B	−0.780

Table 2. Experimental values with uncertainties and deviation from calculated values.

$\frac{T}{K}$	$\frac{\rho_{exp} \pm 2\sigma_{est}}{kg \cdot m^{-3}}$	$\frac{\rho_{exp} - \rho_{calc}}{kg \cdot m^{-3}}$	Ref.	$\frac{T}{K}$	$\frac{\rho_{exp} \pm 2\sigma_{est}}{kg \cdot m^{-3}}$	$\frac{\rho_{exp} - \rho_{calc}}{kg \cdot m^{-3}}$	Ref.
290.15	821.0 ± 6.0	−5.01	06-mus[1)]	284.15	830.3 ± 2.0	−0.39	48-mal/kon
293.15	825.0 ± 2.0	1.33	12-pic/ken	293.15	823.9 ± 0.6	0.23	50-pic/zie
293.15	822.0 ± 2.0	−1.67	38-whi/joh	289.15	825.8 ± 2.0	−0.99	54-naz/kak-1
293.15	822.8 ± 2.0	−0.87	43-geo				

[1)] Not included in calculation of linear coefficients.

Table 3. Recommended values.

$\frac{T}{K}$	$\frac{\rho_{exp} \pm 2\sigma_{est}}{kg \cdot m^{-3}}$	$\frac{T}{K}$	$\frac{\rho_{exp} \pm 2\sigma_{est}}{kg \cdot m^{-3}}$
280.00	833.9 ± 1.9	293.15	823.7 ± 1.5
290.00	826.1 ± 1.5	298.15	819.8 ± 1.6

3-Methyl-1-hexanol **[13231-81-7]** **$C_7H_{16}O$** **MW = 116.2** **70**

Table 1. Experimental and recommended values with uncertainties.

$\frac{T}{K}$	$\frac{\rho_{exp} \pm 2\sigma_{est}}{kg \cdot m^{-3}}$	Ref.	$\frac{T}{K}$	$\frac{\rho_{exp} \pm 2\sigma_{est}}{kg \cdot m^{-3}}$	Ref.
293.15	825.8 ± 1.0	24-dew/wec	297.15	817.0 ± 5.0	41-hus/age[1)]
298.15	824.5 ± 2.0	27-nor/cor[1)]	293.15	825.8 ± 1.0	50-pic/zie
302.15	820.8 ± 2.0	31-lev/mar-5[1)]	293.15	825.8 ± 1.0	Recommended

[1)] Not included in calculation of recommended value.

3-Methyl-2-hexanol **[2313-65-7]** $C_7H_{16}O$ **MW = 116.2** **71**

Table 1. Experimental value with uncertainty.

T/K	$\rho_{exp} \pm 2\sigma_{est}$/kg·m^{-3}	Ref.
298.15	822.0 ± 0.8	12-bje-0

3-Methyl-3-hexanol **[597-96-6]** $C_7H_{16}O$ **MW = 116.2** **72**

Table 1. Coefficients of the polynomial expansion equation.
Standard deviations (see introduction):
$\sigma_{c,w} = 8.4017 \cdot 10^{-1}$ (combined temperature ranges, weighted),
$\sigma_{c,uw} = 2.5702 \cdot 10^{-1}$ (combined temperature ranges, unweighted).

Coefficient	T = 273.15 to 338.15 K $\rho = A + BT + CT^2 + DT^3 + \ldots$
A	$1.09518 \cdot 10^{3}$
B	$-9.22352 \cdot 10^{-1}$

Table 2. Experimental values with uncertainties and deviation from calculated values.

T/K	$\rho_{exp} \pm 2\sigma_{est}$/kg·m^{-3}	$\rho_{exp} - \rho_{calc}$/kg·m^{-3}	Ref. (Symbol in Fig. 1)	T/K	$\rho_{exp} \pm 2\sigma_{est}$/kg·m^{-3}	$\rho_{exp} - \rho_{calc}$/kg·m^{-3}	Ref. (Symbol in Fig. 1)
296.75	819.90 ± 2.00	−1.57	19-eyk(×)	318.15	801.60 ± 1.00	−0.13	39-owe/qua(∇)
288.15	829.60 ± 1.00	0.20	25-deg(Δ)	328.15	792.50 ± 1.00	−0.01	39-owe/qua(∇)
293.15	825.40 ± 1.00	0.61	25-deg(Δ)	338.15	783.50 ± 1.00	0.21	39-owe/qua(∇)
298.15	820.20 ± 1.00	0.02	38-gin/hau(×)	290.15	829.40 ± 1.00	1.84	46-shi(◆)
273.15	841.50 ± 1.00	−1.74	39-owe/qua(∇)	293.15	825.40 ± 0.60	0.61	50-pic/zie(□)
298.15	819.80 ± 1.00	−0.38	39-owe/qua(∇)	293.15	825.40 ± 0.60	0.61	55-soe/fre(○)
308.15	810.70 ± 1.00	−0.26	39-owe/qua(∇)				

Further references: [07-fou/tif, 14-hal, 29-edg/cal].

Table 3. Recommended values (fit to the reliable experimental values according to the equations $\rho = A + BT + CT^2 + DT^3 + \ldots$ or $\rho = [1 + 1.75(1 - T/T_c)^{1/3} + 0.75(1 - T/T_c)][\rho_c + A(T_c - T) + B(T_c - T)^2 + C(T_c - T)^3 + D(T_c - T)^4]$).

T/K	$\rho \pm \sigma_{fit}$/kg·m^{-3}	T/K	$\rho \pm \sigma_{fit}$/kg·m^{-3}	T/K	$\rho \pm \sigma_{fit}$/kg·m^{-3}
270.00	846.14 ± 1.12	298.15	820.18 ± 1.01	330.00	790.80 ± 1.13
280.00	836.92 ± 1.06	300.00	818.47 ± 1.01	340.00	781.58 ± 1.20
290.00	827.70 ± 1.03	310.00	809.25 ± 1.03	350.00	772.36 ± 1.30
293.15	824.79 ± 1.02	320.00	800.03 ± 1.07		

cont.

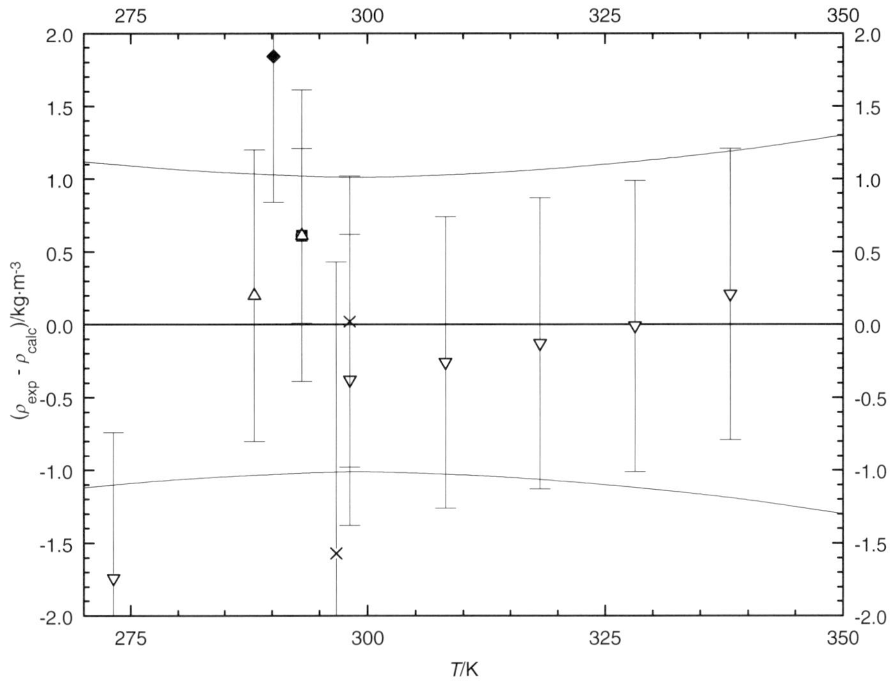

Fig. 1. The symbols show the deviation of the calculated from the experimental values from Table 2. The curves above and below the zero line indicate the calculated error region of the recommended values given in Table 3. The error bars represent the experimental errors. (Error bars smaller than the symbols are omitted for clarity of the figure.)

4-Methyl-1-hexanol **[818-49-5]** **$C_7H_{16}O$** **MW = 116.2** **73**

Table 1. Fit with estimated B coefficient for 3 accepted points. Deviation $\sigma_w = 0.104$.

Coefficient	$\rho = A + BT$
A	1052.63
B	−0.780

Table 2. Experimental values with uncertainties and deviation from calculated values.

$\frac{T}{\text{K}}$	$\frac{\rho_{exp} \pm 2\sigma_{est}}{\text{kg} \cdot \text{m}^{-3}}$	$\frac{\rho_{exp} - \rho_{calc}}{\text{kg} \cdot \text{m}^{-3}}$	Ref.
293.15	823.9 ± 1.0	−0.07	24-dew/wec
297.15	821.0 ± 1.0	0.15	41-hus/age
293.15	823.9 ± 1.0	−0.07	50-pic/zie

Table 3. Recommended values.

$\frac{T}{\text{K}}$	$\frac{\rho_{exp} \pm 2\sigma_{est}}{\text{kg} \cdot \text{m}^{-3}}$
290.00	826.4 ± 0.9
293.15	824.0 ± 0.9
298.15	820.1 ± 0.9

4-Methyl-2-hexanol **[2313-61-3]** **$C_7H_{16}O$** **MW = 116.2** **74**

Table 1. Experimental values with uncertainties.

T / K	$\rho_{exp} \pm 2\sigma_{est}$ / $kg \cdot m^{-3}$	Ref.
298.15	817.7 ± 2.0	30-dav/dix
295.15	816.0 ± 2.0	31-lev/mar-4
293.15	810.0 ± 8.0	53-ner/hen

5-Methyl-1-hexanol **[627-98-5]** **$C_7H_{16}O$** **MW = 116.2** **75**

Table 1. Experimental and recommended values with uncertainties.

T / K	$\rho_{exp} \pm 2\sigma_{est}$ / $kg \cdot m^{-3}$	Ref.
298.15	819.2 ± 6.0	16-lev/all[1)]
297.15	819.0 ± 5.0	41-hus/age[1)]
293.15	822.6 ± 3.0	54-naz/kak-2[1)]
293.15	815.7 ± 1.0	52-coo
293.15	815.7 ± 1.0	Recommended

[1)] Not included in calculation of recommended value.

5-Methyl-2-hexanol **[627-59-8]** **$C_7H_{16}O$** **MW = 116.2** **76**

Table 1. Fit with estimated B coefficient for 4 accepted points. Deviation σ_w = 1.301.

Coefficient	$\rho = A + BT$
A	1049.27
B	−0.800

Table 2. Experimental values with uncertainties and deviation from calculated values.

T / K	$\rho_{exp} \pm 2\sigma_{est}$ / $kg \cdot m^{-3}$	$\rho_{exp} - \rho_{calc}$ / $kg \cdot m^{-3}$	Ref.	T / K	$\rho_{exp} \pm 2\sigma_{est}$ / $kg \cdot m^{-3}$	$\rho_{exp} - \rho_{calc}$ / $kg \cdot m^{-3}$	Ref.
290.65	818.5 ± 2.0	1.75	1878-roh	293.15	814.0 ± 2.0	−0.75	38-whi/joh
290.65	817.4 ± 2.0	0.65	1878-roh	277.15	822.0 ± 4.0	−5.55	49-mal/vol[1)]
293.15	813.1 ± 2.0	−1.65	36-tuo				

[1)] Not included in calculation of linear coefficients.

Table 3. Recommended values.

T / K	$\rho_{exp} \pm 2\sigma_{est}$ / $kg \cdot m^{-3}$
290.00	817.3 ± 1.9
293.15	814.7 ± 1.9
298.15	810.7 ± 2.0

5-Methyl-3-hexanol **[623-55-2]** **$C_7H_{16}O$** **MW = 116.2** **77**

Table 1. Experimental value with uncertainty.

$\frac{T}{K}$	$\frac{\rho_{exp} \pm 2\sigma_{est}}{kg \cdot m^{-3}}$	Ref.
293.15	833.1 ± 1.0	57-shu/bel

2,2,3-Trimethyl-1-butanol **[55505-23-2]** **$C_7H_{16}O$** **MW = 116.2** **78**

Table 1. Experimental value with uncertainty.

$\frac{T}{K}$	$\frac{\rho_{exp} \pm 2\sigma_{est}}{kg \cdot m^{-3}}$	Ref.
293.15	846.6 ± 1.0	41-gle

2,3,3-Trimethyl-1-butanol **[36794-64-6]** **$C_7H_{16}O$** **MW = 116.2** **79**

Table 1. Experimental values with uncertainties.

$\frac{T}{K}$	$\frac{\rho_{exp} \pm 2\sigma_{est}}{kg \cdot m^{-3}}$	Ref.
298.15	823.8 ± 2.0	56-sar/new
293.15	837.2 ± 2.0	66-far/per

2,3,3-Trimethyl-2-butanol **[594-83-2]** **$C_7H_{16}O$** **MW = 116.2** **80**

Table 1. Fit with estimated B coefficient for 4 accepted points. Deviation $\sigma_w = 0.081$.

Coefficient	$\rho = A + BT$
A	1116.79
B	–0.950

Table 2. Experimental values with uncertainties and deviation from calculated values.

$\frac{T}{K}$	$\frac{\rho_{exp} \pm 2\sigma_{est}}{kg \cdot m^{-3}}$	$\frac{\rho_{exp} - \rho_{calc}}{kg \cdot m^{-3}}$	Ref.	$\frac{T}{K}$	$\frac{\rho_{exp} \pm 2\sigma_{est}}{kg \cdot m^{-3}}$	$\frac{\rho_{exp} - \rho_{calc}}{kg \cdot m^{-3}}$	Ref.
298.15	838.0 ± 3.0	4.46	38-gin/hau[1)]	298.15	833.5 ± 0.5	–0.03	88-sip/wie
298.15	833.5 ± 0.5	–0.03	85-wie/sip	303.15	828.9 ± 0.5	0.14	88-sip/wie
293.15	838.2 ± 0.5	–0.07	88-sip/wie	308.15	819.2 ± 2.0	–4.87	88-sip/wie[1)]

[1)] Not included in calculation of linear coefficients.

Table 3. Recommended values.

$\frac{T}{K}$	$\frac{\rho_{exp} \pm 2\sigma_{est}}{kg \cdot m^{-3}}$
290.00	841.3 ± 0.5
293.15	838.3 ± 0.4
298.15	833.5 ± 0.3
310.00	822.3 ± 0.7

2.1.4 Alkanols, C_8

3,3-Dimethyl-2-ethyl-1-butanol **[66576-56-5]** **$C_8H_{18}O$** **MW = 130.23** **81**

Table 1. Experimental value with uncertainty.

$\frac{T}{K}$	$\frac{\rho_{exp} \pm 2\sigma_{est}}{kg \cdot m^{-3}}$	Ref.
298.15	842.5 ± 1.0	56-sar/new

2,2-Dimethyl-1-hexanol **[2370-13-0]** **$C_8H_{18}O$** **MW = 130.23** **82**

Table 1. Experimental values with uncertainties.

$\frac{T}{K}$	$\frac{\rho_{exp} \pm 2\sigma_{est}}{kg \cdot m^{-3}}$	Ref.
293.15	826.5 ± 2.0	64-blo/hag
293.15	839.8 ± 2.0	65-shu/puz

2,2-Dimethyl-3-hexanol **[4209-90-9]** **$C_8H_{18}O$** **MW = 130.23** **83**

Table 1. Experimental and recommended values with uncertainties.

$\frac{T}{K}$	$\frac{\rho_{exp} \pm 2\sigma_{est}}{kg \cdot m^{-3}}$	Ref.
290.15	830.0 ± 5.0	21-ler[1)]
293.15	834.2 ± 2.0	44-hen/mat
293.15	834.2 ± 2.0	Recommended

[1)] Not included in calculation of recommended value.

2,3-Dimethyl-2-hexanol **[19550-03-9]** **$C_8H_{18}O$** **MW = 130.23** **84**

Table 1. Experimental and recommended values with uncertainties.

$\frac{T}{K}$	$\frac{\rho_{exp} \pm 2\sigma_{est}}{kg \cdot m^{-3}}$	Ref.
298.15	831.0 ± 2.0	43-ste/gre[1)]
293.15	836.5 ± 1.0	41-hus/gui
293.15	836.5 ± 1.0	Recommended

[1)] Not included in calculation of recommended value.

2,3-Dimethyl-3-hexanol **[4166-46-5]** **$C_8H_{18}O$** **MW = 130.23** **85**

Table 1. Experimental value with uncertainty.

$\frac{T}{K}$	$\frac{\rho_{exp} \pm 2\sigma_{est}}{kg \cdot m^{-3}}$	Ref.
293.15	837.1 ± 2.0	26-sta

2,4-Dimethyl-2-hexanol **[42328-76-7** **$C_8H_{18}O$** **MW = 130.23** **86**

Table 1. Experimental and recommended values with uncertainties.

T / K	$\rho_{exp} \pm 2\sigma_{est}$ / kg·m^{-3}	Ref.
294.15	827.0 ± 10.0	31-lev/mar-2[1)]
293.15	809.9 ± 2.0	41-hus/gui[1)]
301.15	806.5 ± 1.0	60-tha/vas
301.15	806.5 ± 1.0	Recommended

[1)] Not included in calculation of recommended value.

2,5-Dimethyl-1-hexanol **[6886-16-4]** **$C_8H_{18}O$** **MW = 130.23** **87**

Table 1. Coefficients of the polynomial expansion equation. Standard deviations (see introduction):
$\sigma_{c,w} = 1.3629 \cdot 10^{-1}$ (combined temperature ranges, weighted),
$\sigma_{c,uw} = 3.8771 \cdot 10^{-2}$ (combined temperature ranges, unweighted).

Coefficient	T = 273.15 to 373.15 K $\rho = A + BT + CT^2 + DT^3 + \ldots$
A	$9.26323 \cdot 10^{2}$
B	$-1.60894 \cdot 10^{-3}$
C	$-1.13959 \cdot 10^{-3}$

Table 2. Experimental values with uncertainties and deviation from calculated values.

T / K	$\rho_{exp} \pm 2\sigma_{est}$ / kg·m^{-3}	$\rho_{exp} - \rho_{calc}$ / kg·m^{-3}	Ref. (Symbol in Fig. 1)	T / K	$\rho_{exp} \pm 2\sigma_{est}$ / kg·m^{-3}	$\rho_{exp} - \rho_{calc}$ / kg·m^{-3}	Ref. (Symbol in Fig. 1)
273.15	841.00 ± 1.50	0.14	1879-car(□)	313.15	814.00 ± 1.50	−0.07	1879-car(□)
285.15	833.00 ± 1.50	−0.20	1879-car(□)	323.15	807.00 ± 1.50	0.20	1879-car(□)
293.15	828.00 ± 1.50	0.08	1879-car(□)	373.15	767.00 ± 2.00	−0.05	1879-car(□)
303.15	821.00 ± 1.50	−0.11	1879-car(□)				

Table 3. Recommended values (fit to the reliable experimental values according to the equations $\rho = A + BT + CT^2 + DT^3 + \ldots$ or $\rho = [1 + 1.75(1 - T/T_c)^{1/3} + 0.75(1 - T/T_c)][\rho_c + A(T_c - T) + B(T_c - T)^2 + C(T_c - T)^3 + D(T_c - T)^4]$).

T / K	$\rho \pm \sigma_{fit}$ / kg·m^{-3}	T / K	$\rho \pm \sigma_{fit}$ / kg·m^{-3}	T / K	$\rho \pm \sigma_{fit}$ / kg·m^{-3}
270.00	842.81 ± 1.75	300.00	823.28 ± 1.53	350.00	786.16 ± 1.77
280.00	836.53 ± 1.59	310.00	816.31 ± 1.55	360.00	778.05 ± 1.95
290.00	830.02 ± 1.53	320.00	809.11 ± 1.58	370.00	769.72 ± 2.23
293.15	827.92 ± 1.53	330.00	801.69 ± 1.62	380.00	761.15 ± 2.64
298.15	824.54 ± 1.53	340.00	794.04 ± 1.67		

cont.

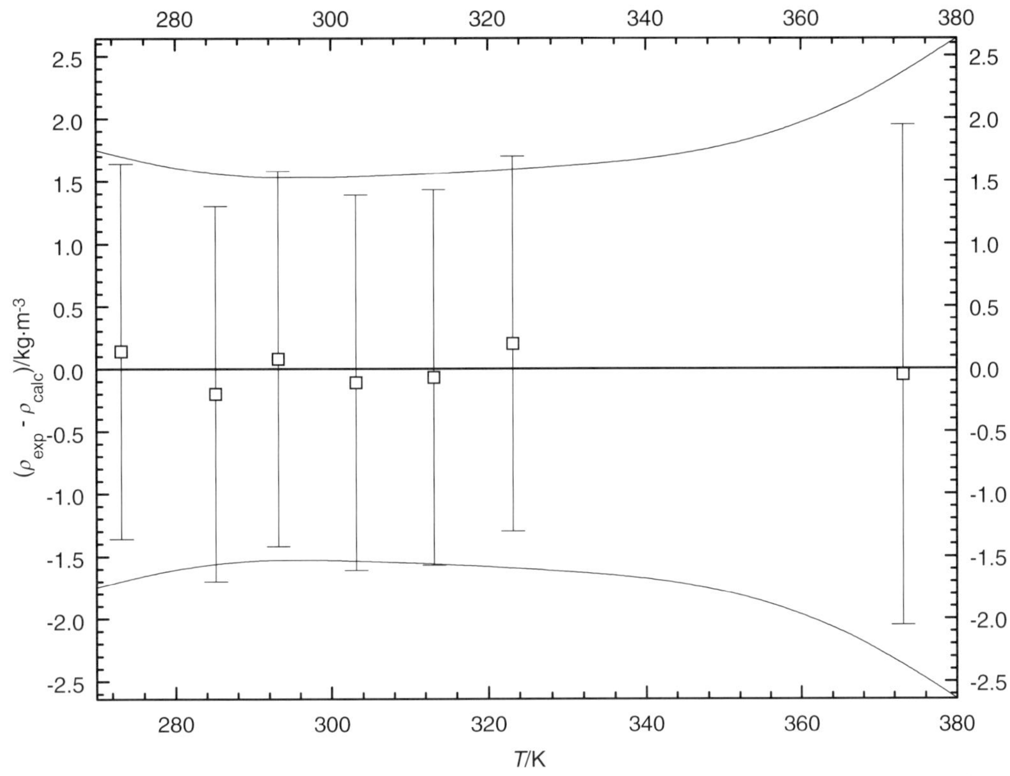

Fig. 1. The symbols show the deviation of the calculated from the experimental values from Table 2. The curves above and below the zero line indicate the calculated error region of the recommended values given in Table 3. The error bars represent the experimental errors. (Error bars smaller than the symbols are omitted for clarity of the figure.)

2,5-Dimethyl-2-hexanol **[3730-60-7]** **$C_8H_{18}O$** **MW = 130.23** **88**

Table 1. Fit with estimated *B* coefficient for 2 accepted points. Deviation $\sigma_w = 0.390$.

Coefficient	$\rho = A + BT$
A	1044.07
B	–0.780

Table 2. Experimental values with uncertainties and deviation from calculated values.

$\frac{T}{K}$	$\frac{\rho_{exp} \pm 2\sigma_{est}}{kg \cdot m^{-3}}$	$\frac{\rho_{exp} - \rho_{calc}}{kg \cdot m^{-3}}$	Ref.
293.15	822.8 ± 5.0	7.39	02-kon[1)]
293.15	810.5 ± 3.0	–4.91	33-mey/tuo[1)]
293.15	815.8 ± 1.0	0.39	41-hus/gui
302.15	808.0 ± 1.0	–0.39	56-woo/vio

[1)] Not included in calculation of linear coefficients.

cont.

2,5-Dimethyl-2-hexanol (cont.)

Table 3. Recommended values.

$\frac{T}{K}$	$\frac{\rho_{exp} \pm 2\sigma_{est}}{kg \cdot m^{-3}}$
290.00	817.9 ± 0.7
293.15	815.4 ± 0.6
298.15	811.5 ± 0.6
310.00	802.3 ± 0.9

2,5-Dimethyl-3-hexanol **[19550-07-3]** **$C_8H_{18}O$** **MW = 130.23** **89**

Table 1. Coefficients of the polynomial expansion equation.
Standard deviations (see introduction):
$\sigma_{c,w} = 8.1519 \cdot 10^{-1}$ (combined temperature ranges, weighted),
$\sigma_{c,uw} = 3.4359 \cdot 10^{-1}$ (combined temperature ranges, unweighted).

Coefficient	T = 273.15 to 323.15 K $\rho = A + BT + CT^2 + DT^3 + \ldots$
A	$8.97260 \cdot 10^{2}$
B	$2.17973 \cdot 10^{-1}$
C	$-1.67297 \cdot 10^{-3}$

Table 2. Experimental values with uncertainties and deviation from calculated values.

$\frac{T}{K}$	$\frac{\rho_{exp} \pm 2\sigma_{est}}{kg \cdot m^{-3}}$	$\frac{\rho_{exp} - \rho_{calc}}{kg \cdot m^{-3}}$	Ref. (Symbol in Fig. 1)	$\frac{T}{K}$	$\frac{\rho_{exp} \pm 2\sigma_{est}}{kg \cdot m^{-3}}$	$\frac{\rho_{exp} - \rho_{calc}}{kg \cdot m^{-3}}$	Ref. (Symbol in Fig. 1)
288.15	820.00 ± 1.50	−1.16	1879-car(Δ)	303.15	811.00 ± 1.50	1.41	00-car-1(O)
293.15	817.00 ± 2.00	−0.39	1879-car(Δ)	313.15	801.00 ± 2.00	−0.46	00-car-1(O)
298.15	814.00 ± 2.00	0.47	1879-car(Δ)	323.15	793.00 ± 2.00	0.00	00-car-1(O)
303.15	811.00 ± 1.50	1.41	1879-car(Δ)	273.15	832.40 ± 1.00	0.42	13-fav(□)
313.15	801.00 ± 2.00	−0.46	1879-car(Δ)	293.15	822.10 ± 4.00	4.71	13-fav[1)]
323.15	793.00 ± 2.00	0.00	1879-car(Δ)	293.15	819.50 ± 2.00	2.11	43-geo(∇)
288.15	820.00 ± 1.50	−1.16	00-car-1(O)	293.15	815.20 ± 3.00	−2.19	53-sok(◆)

[1)] Not included in Fig. 1.

Further references: [12-mic, 36-tuo, 48-mal/kon].

Table 3. Recommended values (fit to the reliable experimental values according to the equations
$\rho = A + BT + CT^2 + DT^3 + \ldots$ or $\rho = [1 + 1.75(1 - T/T_c)^{1/3} + 0.75(1 - T/T_c)][\rho_c + A(T_c - T) + B(T_c - T)^2 + C(T_c - T)^3 + D(T_c - T)^4])$.

$\frac{T}{K}$	$\frac{\rho \pm \sigma_{fit}}{kg \cdot m^{-3}}$	$\frac{T}{K}$	$\frac{\rho \pm \sigma_{fit}}{kg \cdot m^{-3}}$	$\frac{T}{K}$	$\frac{\rho \pm \sigma_{fit}}{kg \cdot m^{-3}}$
270.00	834.15 ± 2.07	293.15	817.39 ± 1.99	310.00	804.06 ± 1.86
280.00	827.13 ± 2.05	298.15	813.53 ± 1.97	320.00	795.70 ± 2.10
290.00	819.78 ± 2.03	300.00	812.08 ± 1.96	330.00	787.00 ± 3.06

cont.

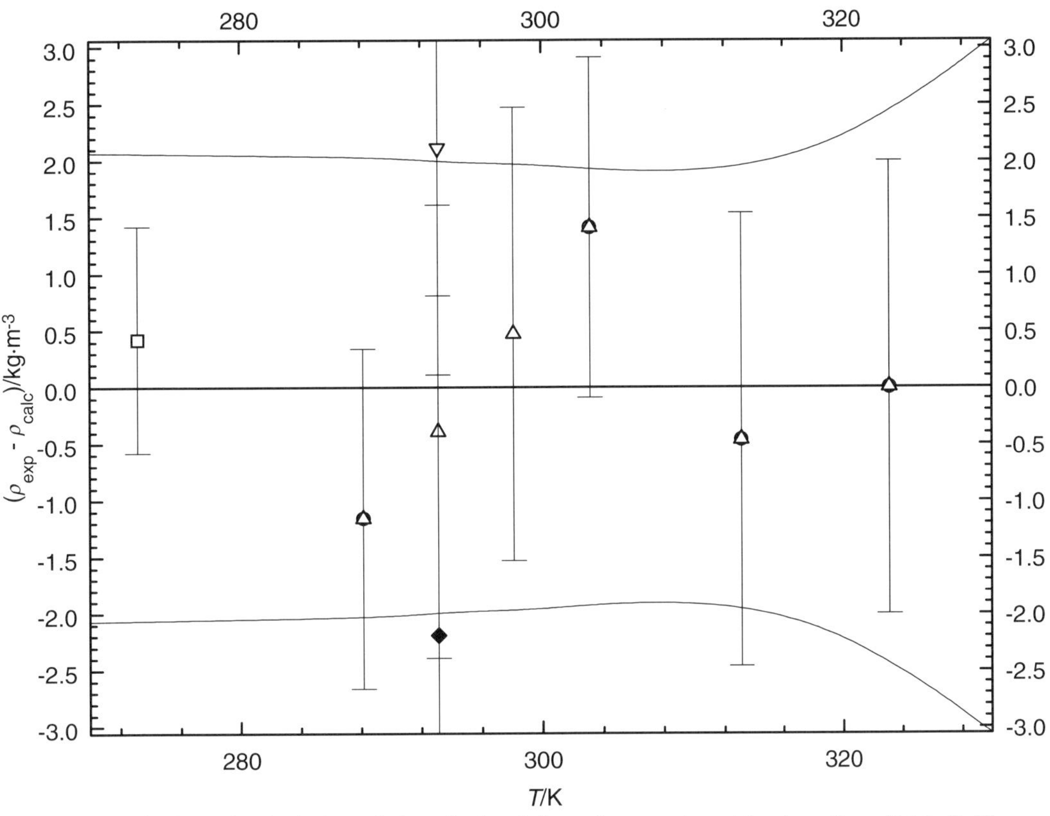

Fig. 1. The symbols show the deviation of the calculated from the experimental values from Table 2. The curves above and below the zero line indicate the calculated error region of the recommended values given in Table 3. The error bars represent the experimental errors. (Error bars smaller than the symbols are omitted for clarity of the figure.)

3,3-Dimethyl-1-hexanol **[10524-70-6]** **$C_8H_{18}O$** **MW = 130.23** **90**

Table 1. Experimental value with uncertainty.

$\frac{T}{K}$	$\frac{\rho_{exp} \pm 2\sigma_{est}}{kg \cdot m^{-3}}$	Ref.
293.15	839.0 ± 2.0	64-sta-1

3,3-Dimethyl-2-hexanol **[22025-20-3]** **$C_8H_{18}O$** **MW = 130.23** **91**

Table 1. Experimental and recommended values with uncertainties.

$\frac{T}{K}$	$\frac{\rho_{exp} \pm 2\sigma_{est}}{kg \cdot m^{-3}}$	Ref.
293.15	845.7 ± 1.0	57-bol/ego-1
273.15	880.8 ± 10.	57-bol/ego-1 [1)]
293.15	845.9 ± 1.0	58-ego
293.15	845.9 ± 1.0	Recommended

[1)] Not included in calculation of recommended value.

3,4-Dimethyl-3-hexanol **[19550-08-4]** **$C_8H_{18}O$** **MW = 130.23** **92**

Table 1. Experimental value with uncertainty.

T / K	$\rho_{exp} \pm 2\sigma_{est}$ / $kg \cdot m^{-3}$	Ref.
298.15	834.5 ± 2.0	48-hus/goe

3,5-Dimethyl-1-hexanol **[13501-73-0]** **$C_8H_{18}O$** **MW = 130.23** **93**

Table 1. Experimental value with uncertainty.

T / K	$\rho_{exp} \pm 2\sigma_{est}$ / $kg \cdot m^{-3}$	Ref.
293.15	828.2 ± 2.0	59-hos/nis

3,5-Dimethyl-3-hexanol **[4209-91-0]** **$C_8H_{18}O$** **MW = 130.23** **94**

Table 2. Experimental and recommended values with uncertainties.

T / K	$\rho_{exp} \pm 2\sigma_{est}$ / $kg \cdot m^{-3}$	Ref.	T / K	$\rho_{exp} \pm 2\sigma_{est}$ / $kg \cdot m^{-3}$	Ref.
288.15	830.0 ± 3.0	09-bod/tab[1)]	293.15	826.1 ± 2.0	59-pet/zak
293.15	822.8 ± 3.0	33-mey/tuo[1)]	293.15	826.1 ± 2.0	59-pet/zak-1
288.15	830.0 ± 3.0	50-doe/zei[1)]	293.15	827.2 + 2.0	61-sok/she
298.15	827.0 ± 3.0	56-woo/vio[1)]	293.15	827.0 ± 2.0	Recommended
293.15	828.8 ± 2.0	58-pan/osi			

[1)] Not included in calculation of recommended value.

4,4-Dimethyl-3-hexanol **[19550-09-5]** **$C_8H_{18}O$** **MW = 130.23** **95**

Table 1. Experimental value with uncertainty.

T / K	$\rho_{exp} \pm 2\sigma_{est}$ / $kg \cdot m^{-3}$	Ref.
293.15	834.1 ± 2.0	51-lev/fai

5,5-Dimethyl-3-hexanol **[66576-31-6]** **$C_8H_{18}O$** **MW = 130.23** **96**

Table 1. Experimental value with uncertainty.

T / K	$\rho_{exp} \pm 2\sigma_{est}$ / $kg \cdot m^{-3}$	Ref.
293.15	833.9 ± 1.0	41-whi/whi

2-Ethyl-1-hexanol **[104-76-7]** **$C_8H_{18}O$** **MW = 130.23** **97**

Table 1. Coefficients of the polynomial expansion equation.
Standard deviations (see introduction):
$\sigma_{c,w} = 8.7629 \cdot 10^{-1}$ (combined temperature ranges, weighted),
$\sigma_{c,uw} = 1.8894 \cdot 10^{-1}$ (combined temperature ranges, unweighted).

Coefficient	T = 273.15 to 373.64 K $\rho = A + BT + CT^2 + DT^3 + \ldots$
A	$1.00464 \cdot 10^{3}$
B	$-4.24121 \cdot 10^{-1}$
C	$-5.58140 \cdot 10^{-4}$

Table 2. Experimental values with uncertainties and deviation from calculated values.

$\frac{T}{\mathrm{K}}$	$\frac{\rho_{exp} \pm 2\sigma_{est}}{\mathrm{kg \cdot m^{-3}}}$	$\frac{\rho_{exp} - \rho_{calc}}{\mathrm{kg \cdot m^{-3}}}$	Ref. (Symbol in Fig. 1)	$\frac{T}{\mathrm{K}}$	$\frac{\rho_{exp} \pm 2\sigma_{est}}{\mathrm{kg \cdot m^{-3}}}$	$\frac{\rho_{exp} - \rho_{calc}}{\mathrm{kg \cdot m^{-3}}}$	Ref. (Symbol in Fig. 1)
273.15	848.30 ± 2.00	1.15	01-gue-1(×)	293.15	832.60 ± 0.40	0.26	56-ano-1(○)
273.19	846.38 ± 1.00	–0.74	30-bin/dar(×)	293.15	832.50 ± 0.50	0.16	58-ano-5(Δ)
283.15	838.93 ± 1.00	–0.87	30-bin/dar(×)	293.15	832.60 ± 0.40	0.26	61-dyk/sep(□)
293.15	831.60 ± 1.00	–0.74	30-bin/dar[1)]	293.15	832.60 ± 0.50	0.26	68-ano(◆)
303.15	824.13 ± 1.00	–0.64	30-bin/dar(×)	293.15	832.10 ± 1.00	–0.24	85-chi/lin[1)]
313.15	816.79 ± 1.00	–0.30	30-bin/dar(×)	298.15	827.50 ± 1.00	–1.07	85-chi/lin(×)
333.15	801.15 ± 1.00	–0.25	30-bin/dar(×)	303.15	823.10 ± 1.00	–1.67	85-chi/lin(×)
353.15	785.61 ± 1.00	0.36	30-bin/dar(×)	308.15	818.50 ± 0.00	–2.45	85-chi/lin[1)]
373.64	768.40 ± 1.00	0.15	30-bin/dar(×)	313.15	812.40 ± 0.00	–4.69	85-chi/lin[1)]
298.15	829.30 ± 1.00	0.73	36-lev/rot-1(×)	318.15	807.30 ± 0.00	–5.91	85-chi/lin[1)]
292.15	834.20 ± 2.00	1.11	39-ken/pla[1)]	323.15	801.80 ± 0.00	–7.50	85-chi/lin[1)]
293.15	833.40 ± 2.00	1.06	39-ken/pla[1)]	328.15	795.20 ± 0.00	–10.16	85-chi/lin[1)]
310.15	820.50 ± 2.00	1.09	39-ken/pla(×)	333.15	790.70 ± 0.00	–10.70	85-chi/lin[1)]
317.15	815.00 ± 2.00	1.01	39-ken/pla(×)	298.15	828.70 ± 0.50	0.13	88-cab/bar(∇)
367.15	777.00 ± 2.00	3.31	39-ken/pla[1)]				

[1)] Not included in Fig. 1.

Further references: [22-lev/tay-1, 34-von/man, 37-mas, 38-mas, 39-gol/tay, 51-hau, 58-kut/lyu, 58-lyu, 58-lyu/bel, 60-tje, 63-lyu/mer].

Table 3. Recommended values (fit to the reliable experimental values according to the equations
$\rho = A + BT + CT^2 + DT^3 + \ldots$ or $\rho = [1 + 1.75(1 - T/T_c)^{1/3} + 0.75(1 - T/T_c)][\rho_c + A(T_c - T) + B(T_c - T)^2 + C(T_c - T)^3 + D(T_c - T)^4]$).

$\frac{T}{\mathrm{K}}$	$\frac{\rho \pm \sigma_{fit}}{\mathrm{kg \cdot m^{-3}}}$	$\frac{T}{\mathrm{K}}$	$\frac{\rho \pm \sigma_{fit}}{\mathrm{kg \cdot m^{-3}}}$	$\frac{T}{\mathrm{K}}$	$\frac{\rho \pm \sigma_{fit}}{\mathrm{kg \cdot m^{-3}}}$
270.00	849.44 ± 1.21	300.00	827.17 ± 1.02	350.00	787.82 ± 1.15
280.00	842.13 ± 1.06	310.00	819.52 ± 1.05	360.00	779.62 ± 1.19
290.00	834.70 ± 1.01	320.00	811.77 ± 1.09	370.00	771.30 ± 1.29
293.15	832.34 ± 1.01	330.00	803.90 ± 1.11	380.00	762.88 ± 1.48
298.15	828.57 ± 1.01	340.00	795.92 ± 1.13		

cont.

2-Ethyl-1-hexanol (cont.)

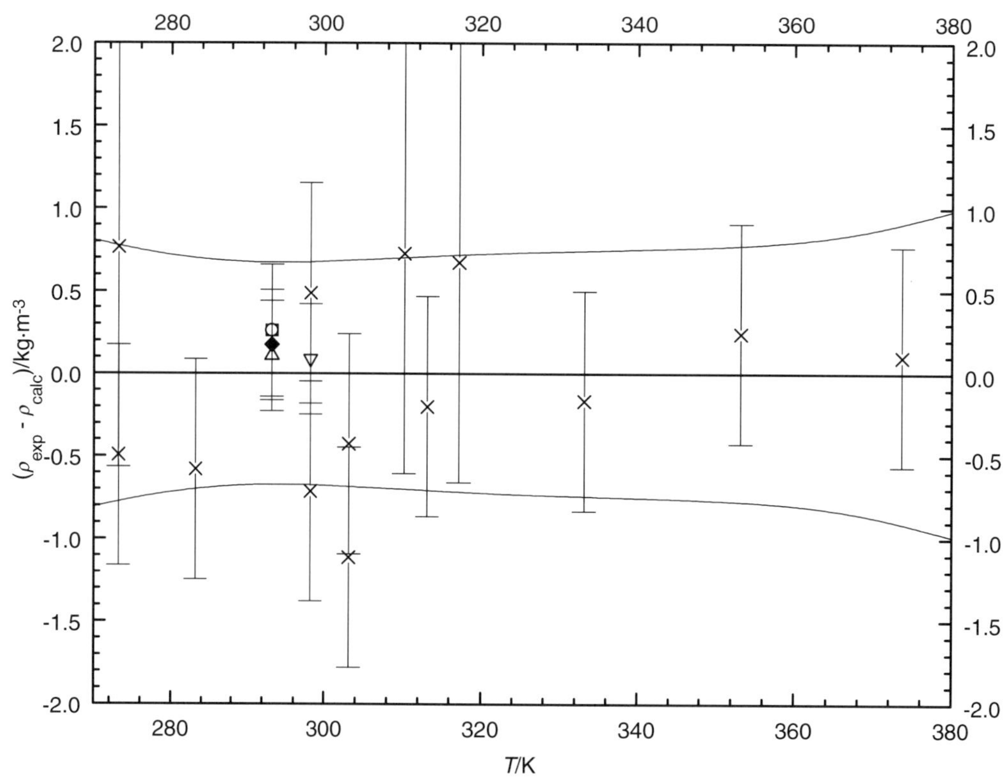

Fig. 1. The symbols show the deviation of the calculated from the experimental values from Table 2. The curves above and below the zero line indicate the calculated error region of the recommended values given in Table 3. The error bars represent the experimental errors. (Error bars smaller than the symbols are omitted for clarity of the figure.)

3-Ethyl-1-hexanol **[41065-95-6]** **$C_8H_{18}O$** **MW = 130.23** **98**

Table 1. Experimental value with uncertainty.

$\frac{T}{\mathrm{K}}$	$\frac{\rho_{exp} \pm 2\sigma_{est}}{\mathrm{kg \cdot m^{-3}}}$	Ref.
301.15	829.0 ± 3.0	31-lev/mar-3

3-Ethyl-3-hexanol **[597-76-2]** **$C_8H_{18}O$** **MW = 130.23** **99**

Table 1. Coefficients of the polynomial expansion equation.
Standard deviations (see introduction):
$\sigma_{c,w} = 8.6240 \cdot 10^{-1}$ (combined temperature ranges, weighted),
$\sigma_{c,uw} = 4.3902 \cdot 10^{-1}$ (combined temperature ranges, unweighted).

Coefficient	T = 273.15 to 338.15 K $\rho = A + BT + CT^2 + DT^3 + \ldots$
A	$8.03283 \cdot 10^2$
B	1.10512
C	$-3.35884 \cdot 10^{-3}$

cont.

Table 2. Experimental values with uncertainties and deviation from calculated values.

$\frac{T}{\text{K}}$	$\frac{\rho_{exp} \pm 2\sigma_{est}}{\text{kg}\cdot\text{m}^{-3}}$	$\frac{\rho_{exp} - \rho_{calc}}{\text{kg}\cdot\text{m}^{-3}}$	Ref. (Symbol in Fig. 1)	$\frac{T}{\text{K}}$	$\frac{\rho_{exp} \pm 2\sigma_{est}}{\text{kg}\cdot\text{m}^{-3}}$	$\frac{\rho_{exp} - \rho_{calc}}{\text{kg}\cdot\text{m}^{-3}}$	Ref. (Symbol in Fig. 1)
295.15	836.50 ± 1.00	−0.36	14-hal(Δ)	308.15	824.60 ± 1.00	−0.28	39-owe/qua(□)
288.10	843.40 ± 2.00	0.52	19-eyk(◆)	318.15	815.60 ± 1.00	0.70	39-owe/qua(□)
293.15	837.30 ± 2.00	−1.30	22-bae(∇)	328.15	806.70 ± 1.00	2.46	39-owe/qua(□)
273.15	854.60 ± 1.00	0.06	39-owe/qua(□)	338.15	790.80 ± 1.00	−2.11	39-owe/qua(□)
298.15	833.30 ± 1.00	−0.90	39-owe/qua(□)	293.15	839.80 ± 1.00	1.20	55-ano(O)

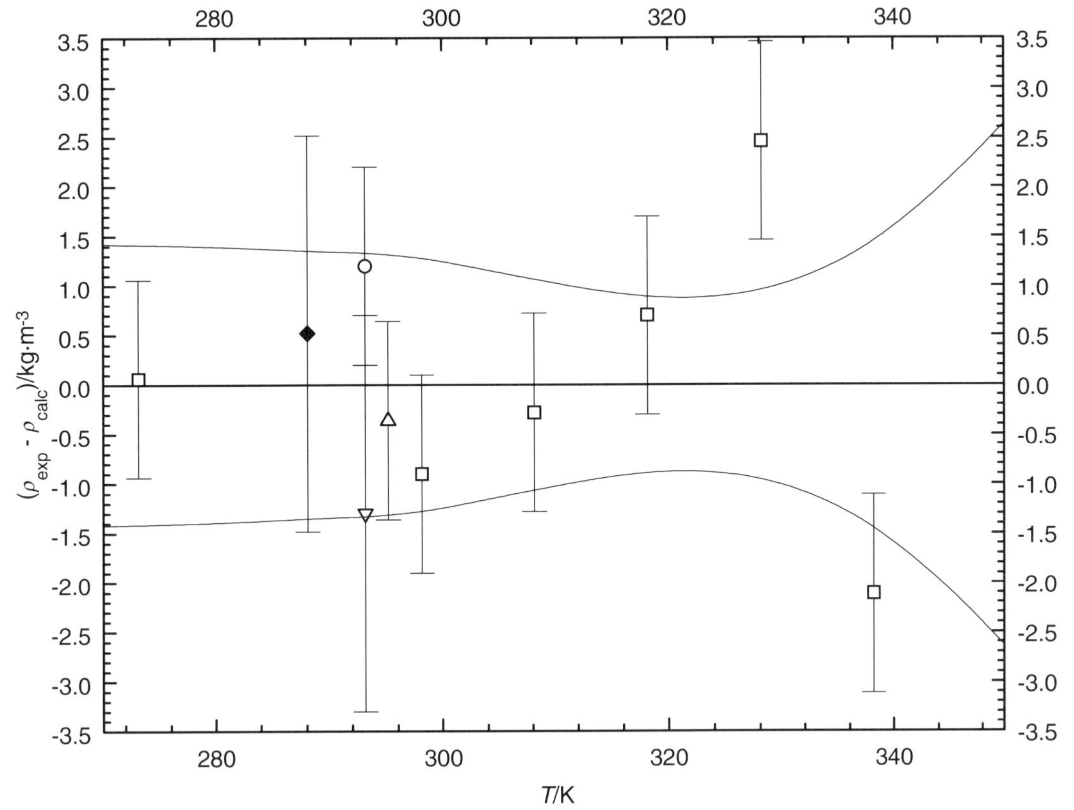

Fig. 1. The symbols show the deviation of the calculated from the experimental values from Table 2. The curves above and below the zero line indicate the calculated error region of the recommended values given in Table 3. The error bars represent the experimental errors. (Error bars smaller than the symbols are omitted for clarity of the figure.)

Table 3. Recommended values (fit to the reliable experimental values according to the equations
$\rho = A + BT + CT^2 + DT^3 + \ldots$ or $\rho = [1 + 1.75(1 - T/T_c)^{1/3} + 0.75(1 - T/T_c)][\rho_c + A(T_c - T) + B(T_c - T)^2 + C(T_c - T)^3 + D(T_c - T)^4]$).

$\frac{T}{\text{K}}$	$\frac{\rho \pm \sigma_{fit}}{\text{kg}\cdot\text{m}^{-3}}$	$\frac{T}{\text{K}}$	$\frac{\rho \pm \sigma_{fit}}{\text{kg}\cdot\text{m}^{-3}}$	$\frac{T}{\text{K}}$	$\frac{\rho \pm \sigma_{fit}}{\text{kg}\cdot\text{m}^{-3}}$
270.00	856.81 ± 1.42	298.15	834.20 ± 1.28	330.00	802.19 ± 0.94
280.00	849.38 ± 1.40	300.00	832.52 ± 1.25	340.00	790.74 ± 1.50
290.00	841.29 ± 1.34	310.00	823.09 ± 1.02	350.00	778.62 ± 2.63
293.15	838.60 ± 1.34	320.00	812.98 ± 0.83		

4-Ethyl-3-hexanol **[19780-44-0]** **$C_8H_{18}O$** **MW = 130.23** **100**

Table 1. Experimental value with uncertainty.

$\frac{T}{K}$	$\frac{\rho_{exp} \pm 2\sigma_{est}}{kg \cdot m^{-3}}$	Ref.
273.15	835.0 ± 3.0	07-fou/tif

2-Ethyl-4-methyl-1-pentanol **[106-67-2]** **$C_8H_{18}O$** **MW = 130.23** **101**

Table 1. Experimental values with uncertainties.

$\frac{T}{K}$	$\frac{\rho_{exp} \pm 2\sigma_{est}}{kg \cdot m^{-3}}$	Ref.
293.15	830.0 ± 2.0	58-ano-3
293.15	827.3 ± 0.6	61-dyk/sep

3-Ethyl-2-methyl-2-pentanol **[19780-63-3]** **$C_8H_{18}O$** **MW = 130.23** **102**

Table 1. Fit with estimated B coefficient for 2 accepted points. Deviation $\sigma_w = 0.150$.

Coefficient	$\rho = A + BT$
A	1067.01
B	−0.780

Table 2. Experimental values with uncertainties and deviation from calculated values.

$\frac{T}{K}$	$\frac{\rho_{exp} \pm 2\sigma_{est}}{kg \cdot m^{-3}}$	$\frac{\rho_{exp} - \rho_{calc}}{kg \cdot m^{-3}}$	Ref.
293.15	838.2 ± 1.0	−0.15	41-hus/gui
298.15	834.6 ± 1.	0.15	54-ski/flo

Table 3. Recommended values.

$\frac{T}{K}$	$\frac{\rho_{exp} \pm 2\sigma_{est}}{kg \cdot m^{-3}}$
290.00	840.8 ± 0.9
293.15	838.4 ± 0.9
298.15	834.5 ± 0.9

3-Ethyl-2-methyl-3-pentanol **[597-05-7]** **$C_8H_{18}O$** **MW = 130.23** **103**

Table 1. Fit with estimated B coefficient for 2 accepted points. Deviation $\sigma_w = 0.200$.

Coefficient	$\rho = A + BT$
A	1070.08
B	−0.820

cont.

Table 2. Experimental values with uncertainties and deviation from calculated values.

$\frac{T}{K}$	$\frac{\rho_{exp} \pm 2\sigma_{est}}{kg \cdot m^{-3}}$	$\frac{\rho_{exp} - \rho_{calc}}{kg \cdot m^{-3}}$	Ref.
273.15	846.3 ± 2.0	0.20	1891-gri/paw
293.15	829.5 ± 2.0	–0.20	1891-gri/paw

Table 3. Recommended values.

$\frac{T}{K}$	$\frac{\rho_{exp} \pm 2\sigma_{est}}{kg \cdot m^{-3}}$
270.00	848.7 ± 2.2
280.00	840.5 ± 1.8
290.00	832.3 ± 1.9
293.15	829.7 ± 2.1
298.15	825.6 ± 2.3

3-Ethyl-3-methyl-2-pentanol [66576-22-5] $C_8H_{18}O$ MW = 130.23 104

Table 1. Experimental values with uncertainties.

$\frac{T}{K}$	$\frac{\rho_{exp} \pm 2\sigma_{est}}{kg \cdot m^{-3}}$	Ref.
293.15	857.6 ± 1.0	57-bol/ego
293.15	857.6 ± 1.0	58-ego

2-Methyl-1-heptanol [60435-70-3] $C_8H_{18}O$ MW = 130.23 105

Table 1. Coefficients of the polynomial expansion equation.
Standard deviations (see introduction):
$\sigma_{c,w} = 2.1063 \cdot 10^{-1}$ (combined temperature ranges, weighted),
$\sigma_{c,uw} = 9.7608 \cdot 10^{-2}$ (combined temperature ranges, unweighted).

Coefficient	T = 273.15 to 372.82 K $\rho = A + BT + CT^2 + DT^3 + \ldots$
A	$9.49326 \cdot 10^{2}$
B	$-2.66559 \cdot 10^{-1}$
C	$-8.01222 \cdot 10^{-4}$

Table 2. Experimental values with uncertainties and deviation from calculated values.

$\frac{T}{K}$	$\frac{\rho_{exp} \pm 2\sigma_{est}}{kg \cdot m^{-3}}$	$\frac{\rho_{exp} - \rho_{calc}}{kg \cdot m^{-3}}$	Ref. (Symbol in Fig. 1)	$\frac{T}{K}$	$\frac{\rho_{exp} \pm 2\sigma_{est}}{kg \cdot m^{-3}}$	$\frac{\rho_{exp} - \rho_{calc}}{kg \cdot m^{-3}}$	Ref. (Symbol in Fig. 1)
273.23	816.93 ± 1.00	0.25	30-bin/dar(○)	353.15	755.69 ± 1.00	0.42	30-bin/dar(○)
283.15	809.19 ± 1.00	–0.42	30-bin/dar(○)	372.82	738.39 ± 1.00	–0.19	30-bin/dar(○)
293.15	801.99 ± 1.00	–0.34	30-bin/dar(○)	273.15	816.80 ± 0.60	0.06	41-dor/gla(□)
303.15	795.42 ± 1.00	0.53	30-bin/dar(○)	293.15	802.30 ± 0.60	–0.03	41-dor/gla(□)
313.15	787.03 ± 1.00	–0.25	30-bin/dar(○)	298.15	798.70 ± 0.60	0.07	41-dor/gla(□)
333.15	771.49 ± 1.00	–0.11	30-bin/dar(○)				

[1] Not included in Fig. 1.

cont.

2-Methyl-1-heptanol (cont.)

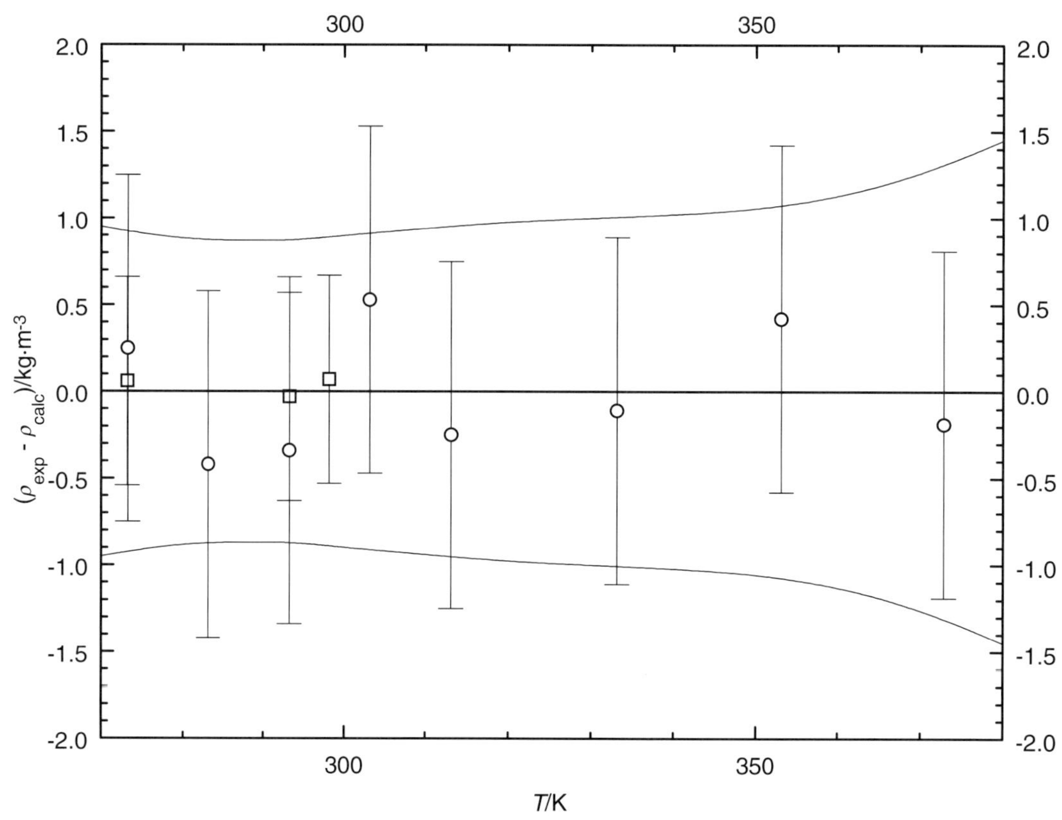

Fig. 1. The symbols show the deviation of the calculated from the experimental values from Table 2. The curves above and below the zero line indicate the calculated error region of the recommended values given in Table 3. The error bars represent the experimental errors. (Error bars smaller than the symbols are omitted for clarity of the figure.)

Table 3. Recommended values (fit to the reliable experimental values according to the equations
$\rho = A + BT + CT^2 + DT^3 + \ldots$ or $\rho = [1 + 1.75(1 - T/T_c)^{1/3} + 0.75(1 - T/T_c)][\rho_c + A(T_c - T) + B(T_c - T)^2 + C(T_c - T)^3 + D(T_c - T)^4])$.

$\frac{T}{K}$	$\frac{\rho \pm \sigma_{fit}}{kg \cdot m^{-3}}$	$\frac{T}{K}$	$\frac{\rho \pm \sigma_{fit}}{kg \cdot m^{-3}}$	$\frac{T}{K}$	$\frac{\rho \pm \sigma_{fit}}{kg \cdot m^{-3}}$
270.00	818.95 ± 0.95	300.00	797.25 ± 0.90	350.00	757.88 ± 1.05
280.00	811.87 ± 0.87	310.00	789.70 ± 0.94	360.00	749.53 ± 1.12
290.00	804.64 ± 0.87	320.00	781.98 ± 0.98	370.00	741.01 ± 1.25
293.15	802.33 ± 0.87	330.00	774.11 ± 1.00	380.00	732.34 ± 1.45
298.15	798.63 ± 0.89	340.00	766.07 ± 1.02		

2-Methyl-2-heptanol **[625-25-2]** **$C_8H_{18}O$** **MW = 130.23** **106**

Table 1. Coefficients of the polynomial expansion equation.
Standard deviations (see introduction):
$\sigma_{c,w} = 7.1719 \cdot 10^{-1}$ (combined temperature ranges, weighted),
$\sigma_{c,uw} = 2.6066 \cdot 10^{-1}$ (combined temperature ranges, unweighted).

Coefficient	$T = 273.15$ to 372.77 K $\rho = A + BT + CT^2 + DT^3 + \ldots$
A	$9.54283 \cdot 10^{2}$
B	$-1.78842 \cdot 10^{-1}$
C	$-1.09478 \cdot 10^{-3}$

Table 2. Experimental values with uncertainties and deviation from calculated values.

$\frac{T}{\text{K}}$	$\frac{\rho_{exp} \pm 2\sigma_{est}}{\text{kg} \cdot \text{m}^{-3}}$	$\frac{\rho_{exp} - \rho_{calc}}{\text{kg} \cdot \text{m}^{-3}}$	Ref. (Symbol in Fig. 1)	$\frac{T}{\text{K}}$	$\frac{\rho_{exp} \pm 2\sigma_{est}}{\text{kg} \cdot \text{m}^{-3}}$	$\frac{\rho_{exp} - \rho_{calc}}{\text{kg} \cdot \text{m}^{-3}}$	Ref. (Symbol in Fig. 1)
273.19	822.98 ± 1.00	−0.74	30-bin/dar(○)	333.15	773.28 ± 1.00	0.09	30-bin/dar(○)
283.15	814.66 ± 1.00	−1.21	30-bin/dar(○)	353.15	755.17 ± 1.00	0.58	30-bin/dar(○)
293.15	807.30 ± 1.00	−0.47	30-bin/dar(○)	372.77	735.84 ± 1.00	0.35	30-bin/dar(○)
303.15	798.91 ± 1.00	−0.55	30-bin/dar(○)	273.15	824.50 ± 0.60	0.75	41-dor/gla(□)
313.15	790.76 ± 1.00	−0.16	30-bin/dar(○)	298.15	805.00 ± 0.60	1.36	41-dor/gla(□)

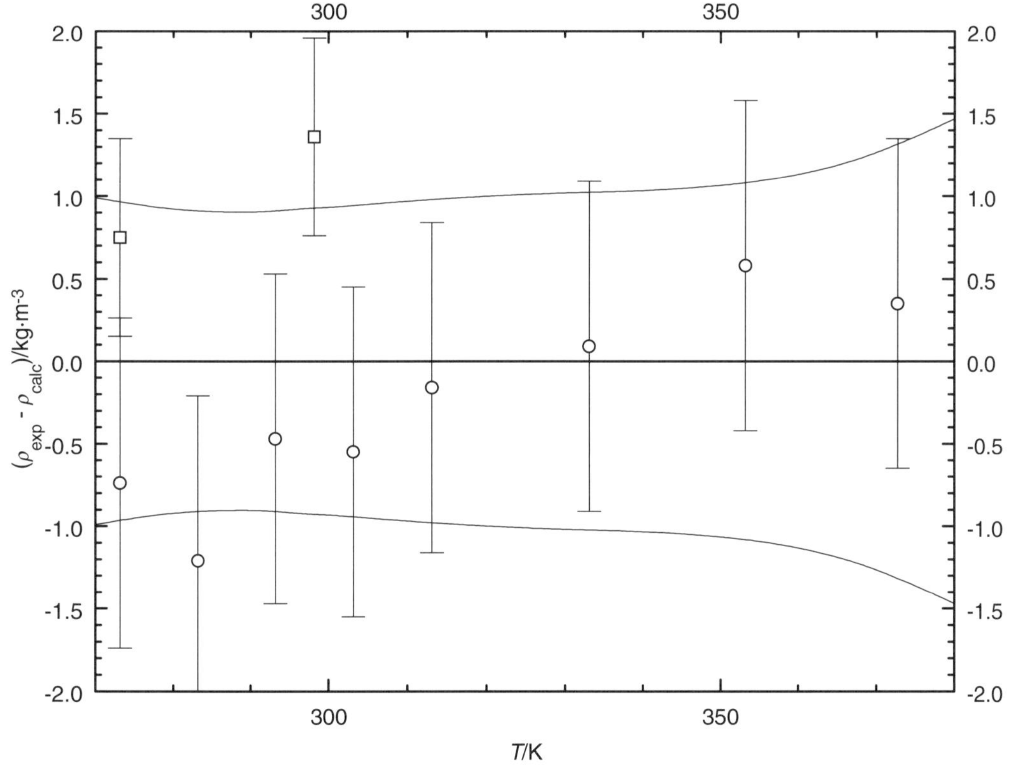

Fig. 1. The symbols show the deviation of the calculated from the experimental values from Table 2. The curves above and below the zero line indicate the calculated error region of the recommended values given in Table 3. The error bars represent the experimental errors. (Error bars smaller than the symbols are omitted for clarity of the figure.)

cont.

2-Methyl-2-heptanol (cont.)

Further references: [06-mus, 33-whi/wil, 41-hus/gui, 44-qua/sma, 57-col/fal].

Table 3. Recommended values (fit to the reliable experimental values according to the equations $\rho = A + BT + CT^2 + DT^3 + \ldots$ or $\rho = [1 + 1.75(1 - T/T_c)^{1/3} + 0.75(1 - T/T_c)][\rho_c + A(T_c - T) + B(T_c - T)^2 + C(T_c - T)^3 + D(T_c - T)^4]$).

$\frac{T}{K}$	$\frac{\rho \pm \sigma_{fit}}{kg \cdot m^{-3}}$	$\frac{T}{K}$	$\frac{\rho \pm \sigma_{fit}}{kg \cdot m^{-3}}$	$\frac{T}{K}$	$\frac{\rho \pm \sigma_{fit}}{kg \cdot m^{-3}}$
270.00	826.19 ± 0.99	300.00	802.10 ± 0.93	350.00	757.58 ± 1.06
280.00	818.38 ± 0.91	310.00	793.63 ± 0.97	360.00	748.02 ± 1.12
290.00	810.35 ± 0.90	320.00	784.95 ± 1.00	370.00	738.24 ± 1.25
293.15	807.77 ± 0.91	330.00	776.04 ± 1.02	380.00	728.24 ± 1.47
298.15	803.64 ± 0.93	340.00	766.92 ± 1.03		

2-Methyl-3-heptanol **[18720-62-2]** **$C_8H_{18}O$** **MW = 130.23** **107**

Table 1. Coefficients of the polynomial expansion equation.
Standard deviations (see introduction):
$\sigma_{c,w} = 2.3940 \cdot 10^{-1}$ (combined temperature ranges, weighted),
$\sigma_{c,uw} = 1.0583 \cdot 10^{-1}$ (combined temperature ranges, unweighted).

Coefficient	T = 273.15 to 373.52 K $\rho = A + BT + CT^2 + DT^3 + \ldots$
A	$9.51750 \cdot 10^{2}$
B	$-8.19206 \cdot 10^{-2}$
C	$-1.19870 \cdot 10^{-3}$

Table 2. Experimental values with uncertainties and deviation from calculated values.

$\frac{T}{K}$	$\frac{\rho_{exp} \pm 2\sigma_{est}}{kg \cdot m^{-3}}$	$\frac{\rho_{exp} - \rho_{calc}}{kg \cdot m^{-3}}$	Ref. (Symbol in Fig. 1)	$\frac{T}{K}$	$\frac{\rho_{exp} \pm 2\sigma_{est}}{kg \cdot m^{-3}}$	$\frac{\rho_{exp} - \rho_{calc}}{kg \cdot m^{-3}}$	Ref. (Symbol in Fig. 1)
293.15	825.00 ± 1.00	0.28	06-mus(X)	333.15	791.26 ± 1.00	–0.16	30-bin/dar(◆)
293.15	823.50 ± 1.00	–1.22	12-pic/ken(Δ)	353.15	773.57 ± 1.00	0.25	30-bin/dar(◆)
298.15	821.00 ± 1.00	0.23	13-tho(X)	373.52	753.98 ± 1.00	0.07	30-bin/dar(◆)
273.21	840.19 ± 1.00	0.30	30-bin/dar(◆)	273.15	840.00 ± 0.60	0.06	41-dor/gla(□)
283.15	832.09 ± 1.00	–0.36	30-bin/dar(◆)	298.15	821.00 ± 0.60	0.23	41-dor/gla(□)
293.15	824.67 ± 1.00	–0.05	30-bin/dar(◆)	293.15	825.10 ± 1.00	0.38	43-geo(∇)
303.15	816.66 ± 1.00	–0.10	30-bin/dar(◆)	284.15	831.60 ± 1.00	–0.09	48-mal/kon(O)
313.15	808.73 ± 1.00	0.18	30-bin/dar(◆)				

Table 3. Recommended values (fit to the reliable experimental values according to the equations $\rho = A + BT + CT^2 + DT^3 + \ldots$ or $\rho = [1 + 1.75(1 - T/T_c)^{1/3} + 0.75(1 - T/T_c)][\rho_c + A(T_c - T) + B(T_c - T)^2 + C(T_c - T)^3 + D(T_c - T)^4]$).

$\frac{T}{K}$	$\frac{\rho \pm \sigma_{fit}}{kg \cdot m^{-3}}$	$\frac{T}{K}$	$\frac{\rho \pm \sigma_{fit}}{kg \cdot m^{-3}}$	$\frac{T}{K}$	$\frac{\rho \pm \sigma_{fit}}{kg \cdot m^{-3}}$
270.00	842.25 ± 1.31	300.00	819.29 ± 1.21	350.00	776.24 ± 1.10
280.00	834.83 ± 1.34	310.00	811.16 ± 1.09	360.00	766.91 ± 1.46
290.00	827.18 ± 1.30	320.00	802.79 ± 0.98	370.00	757.34 ± 2.04
293.15	824.72 ± 1.28	330.00	794.18 ± 0.91	380.00	747.53 ± 2.85
298.15	820.77 ± 1.23	340.00	785.33 ± 0.93		

cont.

Further references: [14-wal-1, 33-bri, 36-tuo].

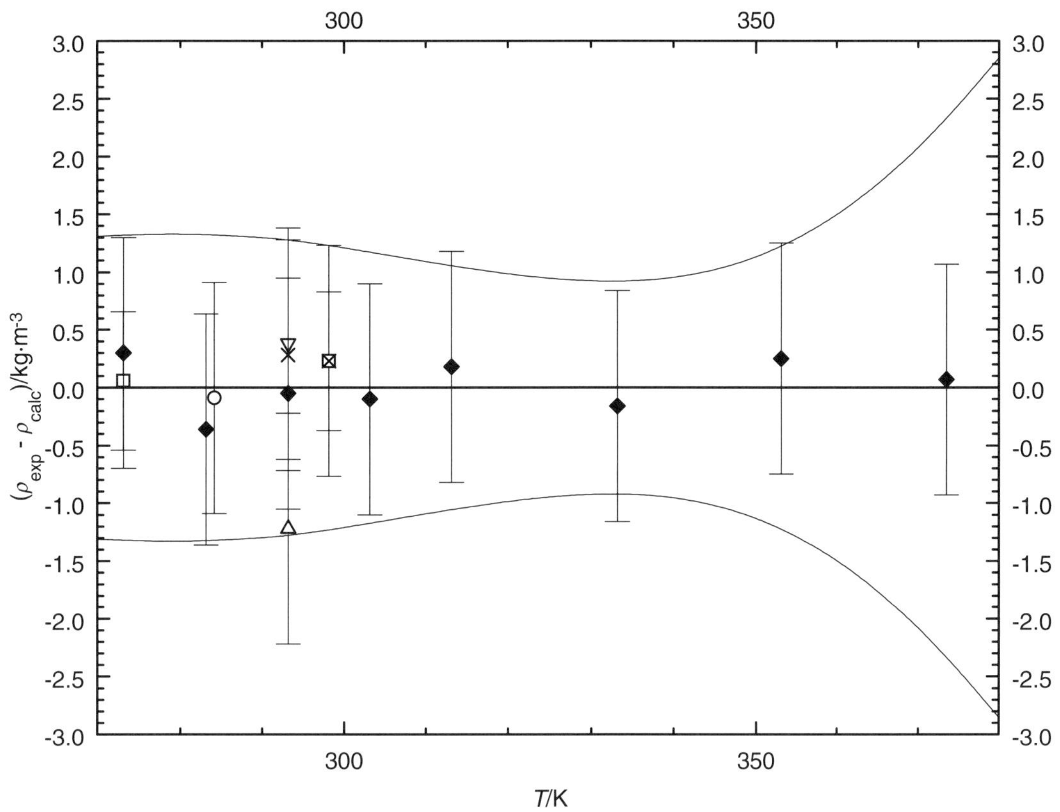

Fig. 1. The symbols show the deviation of the calculated from the experimental values from Table 2. The curves above and below the zero line indicate the calculated error region of the recommended values given in Table 3. The error bars represent the experimental errors. (Error bars smaller than the symbols are omitted for clarity of the figure.)

2-Methyl-4-heptanol [21570-35-4] $C_8H_{18}O$ MW = 130.23 108

Table 1. Coefficients of the polynomial expansion equation.
Standard deviations (see introduction):
$\sigma_{c,w} = 1.1397 \cdot 10^{-1}$ (combined temperature ranges, weighted),
$\sigma_{c,uw} = 5.7770 \cdot 10^{-2}$ (combined temperature ranges, unweighted).

Coefficient	T = 273.15 to 373.54 K $\rho = A + BT + CT^2 + DT^3 + \ldots$
A	$9.33558 \cdot 10^{2}$
B	$-7.11104 \cdot 10^{-2}$
C	$-1.15359 \cdot 10^{-3}$

cont.

2-Methyl-4-heptanol (cont.)

Table 2. Experimental values with uncertainties and deviation from calculated values.

$\frac{T}{K}$	$\frac{\rho_{exp} \pm 2\sigma_{est}}{kg \cdot m^{-3}}$	$\frac{\rho_{exp} - \rho_{calc}}{kg \cdot m^{-3}}$	Ref. (Symbol in Fig. 1)	$\frac{T}{K}$	$\frac{\rho_{exp} \pm 2\sigma_{est}}{kg \cdot m^{-3}}$	$\frac{\rho_{exp} - \rho_{calc}}{kg \cdot m^{-3}}$	Ref. (Symbol in Fig. 1)
273.23	828.36 ± 1.00	0.35	30-bin/dar(○)	353.15	764.82 ± 1.00	0.24	30-bin/dar(○)
283.15	820.75 ± 1.00	–0.19	30-bin/dar(○)	373.54	745.82 ± 1.00	–0.21	30-bin/dar(○)
293.15	813.47 ± 1.00	–0.11	30-bin/dar(○)	273.15	828.10 ± 0.60	0.04	41-dor/gla(□)
303.15	805.80 ± 1.00	–0.19	30-bin/dar(○)	293.15	813.50 ± 0.60	–0.08	41-dor/gla(□)
313.15	798.28 ± 1.00	0.11	30-bin/dar(○)	298.15	809.80 ± 0.60	–0.01	41-dor/gla(□)
333.15	781.86 ± 1.00	0.03	30-bin/dar(○)				

Further references: [06-mus, 36-tuo, 57-shu/bel].

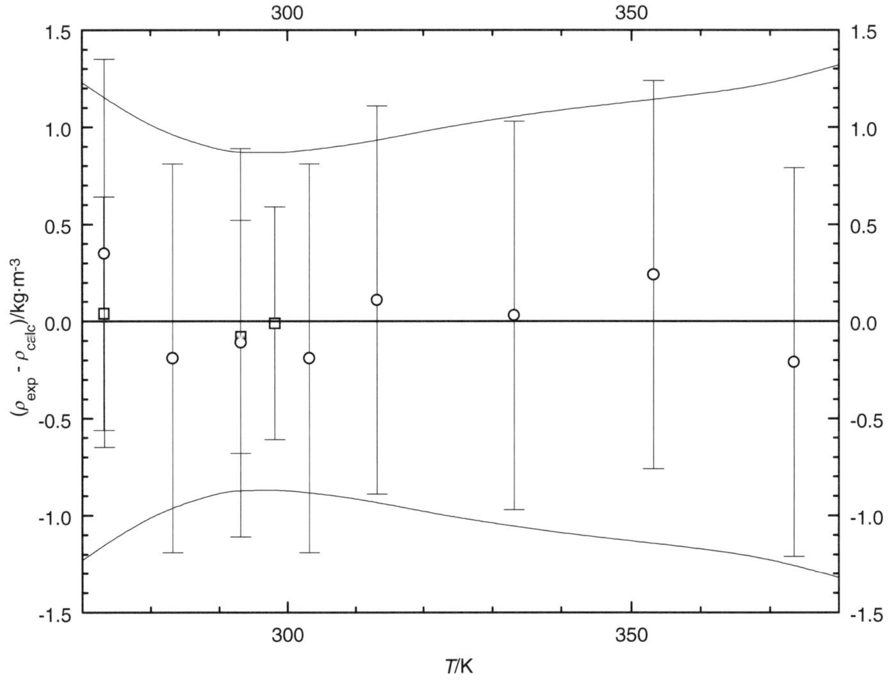

Fig. 1. The symbols show the deviation of the calculated from the experimental values from Table 2. The curves above and below the zero line indicate the calculated error region of the recommended values given in Table 3. The error bars represent the experimental errors. (Error bars smaller than the symbols are omitted for clarity of the figure.)

Table 3. Recommended values (fit to the reliable experimental values according to the equations
$\rho = A + BT + CT^2 + DT^3 + \ldots$ or $\rho = [1 + 1.75(1 - T/T_c)^{1/3} + 0.75(1 - T/T_c)][\rho_c + A(T_c - T) + B(T_c - T)^2 + C(T_c - T)^3 + D(T_c - T)^4]$).

$\frac{T}{K}$	$\frac{\rho \pm \sigma_{fit}}{kg \cdot m^{-3}}$	$\frac{T}{K}$	$\frac{\rho \pm \sigma_{fit}}{kg \cdot m^{-3}}$	$\frac{T}{K}$	$\frac{\rho \pm \sigma_{fit}}{kg \cdot m^{-3}}$
270.00	830.26 ± 1.23	300.00	808.40 ± 0.87	350.00	767.35 ± 1.13
280.00	823.21 ± 0.99	310.00	800.65 ± 0.91	360.00	758.45 ± 1.17
290.00	815.92 ± 0.88	320.00	792.68 ± 0.98	370.00	749.32 ± 1.22
293.15	813.58 ± 0.87	330.00	784.47 ± 1.04	380.00	739.96 ± 1.32
298.15	809.81 ± 0.87	340.00	776.03 ± 1.09		

3-Methyl-1-heptanol **[1070-32-2]** **$C_8H_{18}O$** **MW = 130.23** **109**

Table 1. Coefficients of the polynomial expansion equation.
Standard deviations (see introduction):
$\sigma_{c,w} = 7.0802 \cdot 10^{-2}$ (combined temperature ranges, weighted),
$\sigma_{c,uw} = 3.3702 \cdot 10^{-2}$ (combined temperature ranges, unweighted).

Coefficient	T = 273.15 to 373.62 K $\rho = A + BT + CT^2 + DT^3 + \ldots$
A	$9.62326 \cdot 10^2$
B	$-4.50758 \cdot 10^{-1}$
C	$-4.89103 \cdot 10^{-4}$

Table 2. Experimental values with uncertainties and deviation from calculated values.

$\frac{T}{\text{K}}$	$\frac{\rho_{exp} \pm 2\sigma_{est}}{\text{kg} \cdot \text{m}^{-3}}$	$\frac{\rho_{exp} - \rho_{calc}}{\text{kg} \cdot \text{m}^{-3}}$	Ref. (Symbol in Fig. 1)	$\frac{T}{\text{K}}$	$\frac{\rho_{exp} \pm 2\sigma_{est}}{\text{kg} \cdot \text{m}^{-3}}$	$\frac{\rho_{exp} - \rho_{calc}}{\text{kg} \cdot \text{m}^{-3}}$	Ref. (Symbol in Fig. 1)
273.19	802.83 ± 1.00	0.15	30-bin/dar(○)	333.15	757.81 ± 1.00	−0.06	30-bin/dar(○)
283.15	795.29 ± 1.00	−0.19	30-bin/dar(○)	353.15	742.17 ± 1.00	0.03	30-bin/dar(○)
293.15	788.15 ± 1.00	−0.00	30-bin/dar(○)	373.62	725.64 ± 1.00	0.00	30-bin/dar(○)
303.15	780.64 ± 1.00	−0.09	30-bin/dar(○)	273.15	802.70 ± 0.60	−0.01	41-dor/gla(□)
313.15	773.34 ± 1.00	0.13	30-bin/dar(○)	298.15	784.50 ± 0.60	0.05	41-dor/gla(□)

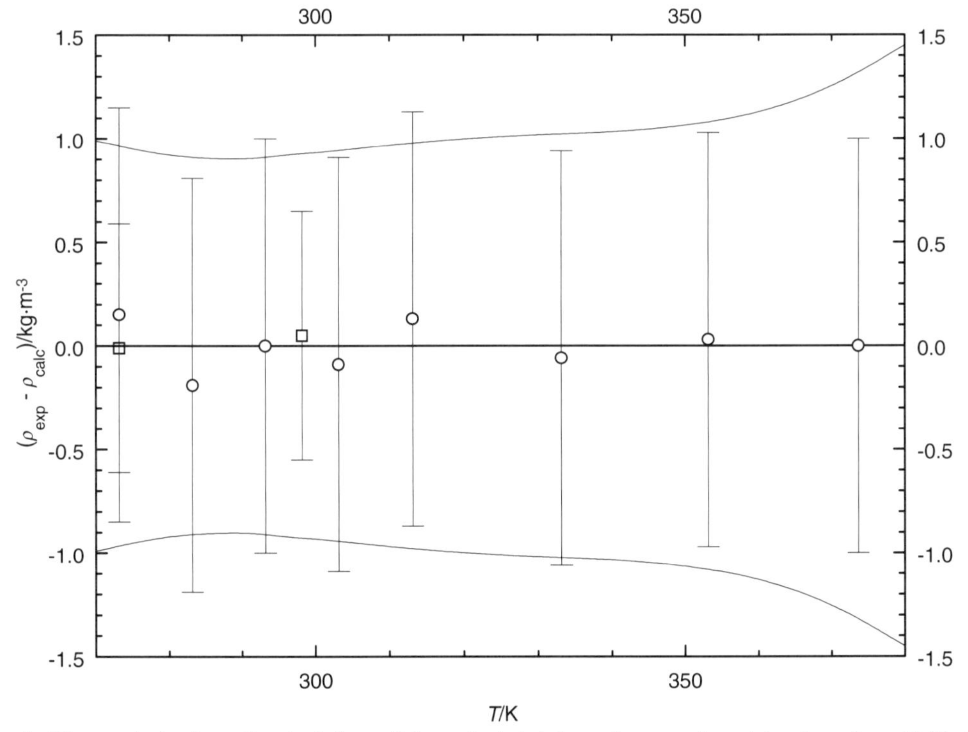

Fig. 1. The symbols show the deviation of the calculated from the experimental values from Table 2. The curves above and below the zero line indicate the calculated error region of the recommended values given in Table 3. The error bars represent the experimental errors. (Error bars smaller than the symbols are omitted for clarity of the figure.)

cont.

3-Methyl-1-heptanol (cont.)

Further references: [31-lev/mar-5, 33-bri, 41-hus/age, 56-lic/dur].

Table 3. Recommended values (fit to the reliable experimental values according to the equations $\rho = A + BT + CT^2 + DT^3 + \ldots$ or $\rho = [1 + 1.75(1 - T/T_c)^{1/3} + 0.75(1 - T/T_c)][\rho_c + A(T_c - T) + B(T_c - T)^2 + C(T_c - T)^3 + D(T_c - T)^4]$).

T/K	$\rho \pm \sigma_{fit}$ / $kg \cdot m^{-3}$	T/K	$\rho \pm \sigma_{fit}$ / $kg \cdot m^{-3}$	T/K	$\rho \pm \sigma_{fit}$ / $kg \cdot m^{-3}$
270.00	804.97 ± 0.99	300.00	783.08 ± 0.93	350.00	744.65 ± 1.06
280.00	797.77 ± 0.91	310.00	775.59 ± 0.97	360.00	736.67 ± 1.12
290.00	790.47 ± 0.90	320.00	768.00 ± 1.00	370.00	728.59 ± 1.24
293.15	788.15 ± 0.91	330.00	760.31 ± 1.02	380.00	720.41 ± 1.45
298.15	784.45 ± 0.93	340.00	752.53 ± 1.03		

3-Methyl-2-heptanol **[31367-46-1]** **$C_8H_{18}O$** **MW = 130.23** **110**

Table 1. Fit with estimated B coefficient for 3 accepted points. Deviation $\sigma_w = 0.094$.

Coefficient	$\rho = A + BT$
A	1056.15
B	−0.800

Table 2. Experimental values with uncertainties and deviation from calculated values.

T/K	$\rho_{exp} \pm 2\sigma_{est}$ / $kg \cdot m^{-3}$	$\rho_{exp} - \rho_{calc}$ / $kg \cdot m^{-3}$	Ref.	T/K	$\rho_{exp} \pm 2\sigma_{est}$ / $kg \cdot m^{-3}$	$\rho_{exp} - \rho_{calc}$ / $kg \cdot m^{-3}$	Ref.
286.15	827.2 ± 2.00	−0.03	24-pow[1)]	303.15	781.5 ± 10.00	−32.14	30-bin/dar[1)]
273.15	837.5 ± 1.00	−0.13	41-dor/gla	313.15	773.7 ± 10.00	−31.88	30-bin/dar[1)]
298.15	817.7 ± 1.00	0.07	41-dor/gla	333.15	756.0 ± 10.00	−33.60	30-bin/dar[1)]
293.15	821.7 ± 1.00	0.07	41-dor/gla	353.15	738.4 ± 10.00	−35.19	30-bin/dar[1)]
283.15	797.6 ± 10.00	−32.05	30-bin/dar[1)]	373.44	723.5 ± 10.00	−33.92	30-bin/dar[1)]
293.15	790.0 ± 10.00	−31.62	30-bin/dar[1)]	273.20	806.3 ± 10.00	−31.27	30-bin/dar[1)]

[1)] Not included in calculation of linear coefficients.

Table 3. Recommended values.

T/K	$\rho_{exp} \pm 2\sigma_{est}$ / $kg \cdot m^{-3}$
270.00	840.2 ± 0.9
280.00	832.2 ± 0.5
290.00	824.2 ± 0.3
293.15	821.6 ± 0.3
298.15	817.6 ± 0.6

3-Methyl-3-heptanol **[5582-82-1]** **$C_8H_{18}O$** **MW = 130.23** **111**

Table 1. Coefficients of the polynomial expansion equation.
Standard deviations (see introduction):
$\sigma_{c,w} = 9.7126 \cdot 10^{-1}$ (combined temperature ranges, weighted),
$\sigma_{c,uw} = 2.4649 \cdot 10^{-1}$ (combined temperature ranges, unweighted).

Coefficient	T = 273.15 to 373.54 K $\rho = A + BT + CT^2 + DT^3 + \ldots$
A	$1.00165 \cdot 10^{3}$
B	$-3.47181 \cdot 10^{-1}$
C	$-8.23886 \cdot 10^{-4}$

Table 2. Experimental values with uncertainties and deviation from calculated values.

$\frac{T}{\text{K}}$	$\frac{\rho_{exp} \pm 2\sigma_{est}}{\text{kg} \cdot \text{m}^{-3}}$	$\frac{\rho_{exp} - \rho_{calc}}{\text{kg} \cdot \text{m}^{-3}}$	Ref. (Symbol in Fig. 1)	$\frac{T}{\text{K}}$	$\frac{\rho_{exp} \pm 2\sigma_{est}}{\text{kg} \cdot \text{m}^{-3}}$	$\frac{\rho_{exp} - \rho_{calc}}{\text{kg} \cdot \text{m}^{-3}}$	Ref. (Symbol in Fig. 1)
273.23	845.31 ± 1.00	0.03	30-bin/dar(Δ)	293.15	828.20 ± 2.00	−0.87	33-whi/woo(×)
283.15	837.17 ± 1.00	−0.12	30-bin/dar(Δ)	298.15	824.90 ± 2.00	0.00	33-whi/woo(×)
293.15	829.19 ± 1.00	0.12	30-bin/dar(Δ)	273.15	847.10 ± 1.00	1.75	41-dor/gla(∇)
303.15	821.02 ± 1.00	0.33	30-bin/dar(Δ)	298.15	827.10 ± 1.00	2.20	41-dor/gla(∇)
313.15	812.61 ± 1.00	0.47	30-bin/dar(Δ)	298.15	825.20 ± 1.00	0.30	44-qua/sma(○)
333.15	794.47 ± 1.00	−0.07	30-bin/dar(Δ)	318.15	808.30 ± 1.00	0.50	44-qua/sma(○)
353.15	776.10 ± 1.00	−0.19	30-bin/dar(Δ)	328.15	799.90 ± 1.00	0.90	44-qua/sma(○)
373.54	757.17 ± 1.00	0.17	30-bin/dar(Δ)	298.15	823.00 ± 2.00	−1.90	61-wib/fos(×)
273.15	844.60 ± 2.00	−0.75	30-van(◆)	293.15	826.90 ± 1.00	−2.17	66-are/tav(□)
288.15	832.50 ± 2.00	−0.70	30-van(◆)				

Further references: [02-kon, 65-col/des].

Table 3. Recommended values (fit to the reliable experimental values according to the equations
$\rho = A + BT + CT^2 + DT^3 + \ldots$ or $\rho = [1 + 1.75(1 - T/T_c)^{1/3} + 0.75(1 - T/T_c)][\rho_c + A(T_c - T) + B(T_c - T)^2 + C(T_c - T)^3 + D(T_c - T)^4]$).

$\frac{T}{\text{K}}$	$\frac{\rho \pm \sigma_{fit}}{\text{kg} \cdot \text{m}^{-3}}$	$\frac{T}{\text{K}}$	$\frac{\rho \pm \sigma_{fit}}{\text{kg} \cdot \text{m}^{-3}}$	$\frac{T}{\text{K}}$	$\frac{\rho \pm \sigma_{fit}}{\text{kg} \cdot \text{m}^{-3}}$
270.00	847.85 ± 1.56	300.00	823.35 ± 1.30	350.00	779.21 ± 1.09
280.00	839.85 ± 1.43	310.00	814.85 ± 1.26	360.00	769.89 ± 1.10
290.00	831.68 ± 1.35	320.00	806.19 ± 1.22	370.00	760.40 ± 1.18
293.15	829.07 ± 1.33	330.00	797.36 ± 1.17	380.00	750.75 ± 1.35
298.15	824.90 ± 1.31	340.00	788.37 ± 1.13		

cont.

3-Methyl-3-heptanol (cont.)

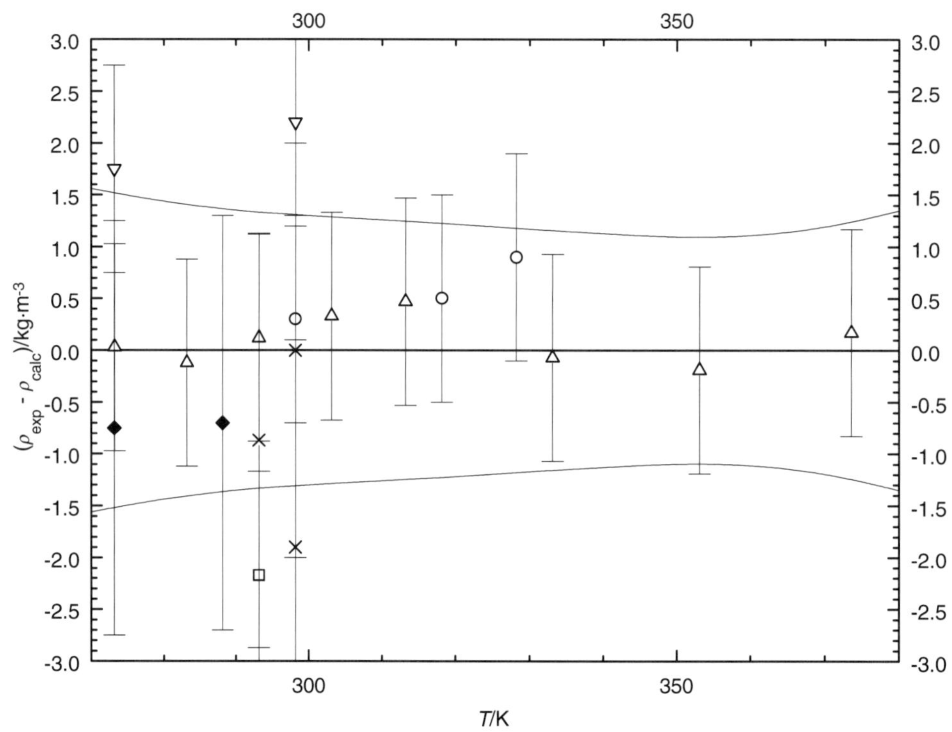

Fig. 1. The symbols show the deviation of the calculated from the experimental values from Table 2. The curves above and below the zero line indicate the calculated error region of the recommended values given in Table 3. The error bars represent the experimental errors. (Error bars smaller than the symbols are omitted for clarity of the figure.)

3-Methyl-4-heptanol **[1838-73-9]** **$C_8H_{18}O$** **MW = 130.23** **112**

Table 1. Coefficients of the polynomial expansion equation.
Standard deviations (see introduction):
$\sigma_{c,w} = 5.5178 \cdot 10^{-1}$ (combined temperature ranges, weighted),
$\sigma_{c,uw} = 2.7388 \cdot 10^{-1}$ (combined temperature ranges, unweighted).

Coefficient	T = 273.15 to 373.12 K $\rho = A + BT + CT^2 + DT^3 + \ldots$
A	$1.08833 \cdot 10^3$
B	$-8.57031 \cdot 10^{-1}$

Table 2. Experimental values with uncertainties and deviation from calculated values.

$\frac{T}{\text{K}}$	$\frac{\rho_{exp} \pm 2\sigma_{est}}{\text{kg}\cdot\text{m}^{-3}}$	$\frac{\rho_{exp} - \rho_{calc}}{\text{kg}\cdot\text{m}^{-3}}$	Ref. (Symbol in Fig. 1)	$\frac{T}{\text{K}}$	$\frac{\rho_{exp} \pm 2\sigma_{est}}{\text{kg}\cdot\text{m}^{-3}}$	$\frac{\rho_{exp} - \rho_{calc}}{\text{kg}\cdot\text{m}^{-3}}$	Ref. (Symbol in Fig. 1)
273.20	853.53 ± 1.00	−0.66	30-bin/dar(O)	353.15	785.61 ± 1.00	−0.06	30-bin/dar(O)
283.15	844.81 ± 1.00	−0.85	30-bin/dar(O)	373.12	767.40 ± 1.00	−1.16	30-bin/dar(O)
293.15	836.96 ± 1.00	−0.13	30-bin/dar(O)	273.15	853.70 ± 0.60	−0.53	41-dor/gla(□)
303.15	828.57 ± 1.00	0.05	30-bin/dar(O)	298.15	833.50 ± 0.60	0.69	41-dor/gla(□)
313.15	820.34 ± 1.00	0.39	30-bin/dar(O)	298.15	835.00 ± 2.00	2.19	65-col/des(Δ)
333.15	802.89 ± 1.00	0.08	30-bin/dar(O)				

cont.

Further references: [33-bri].

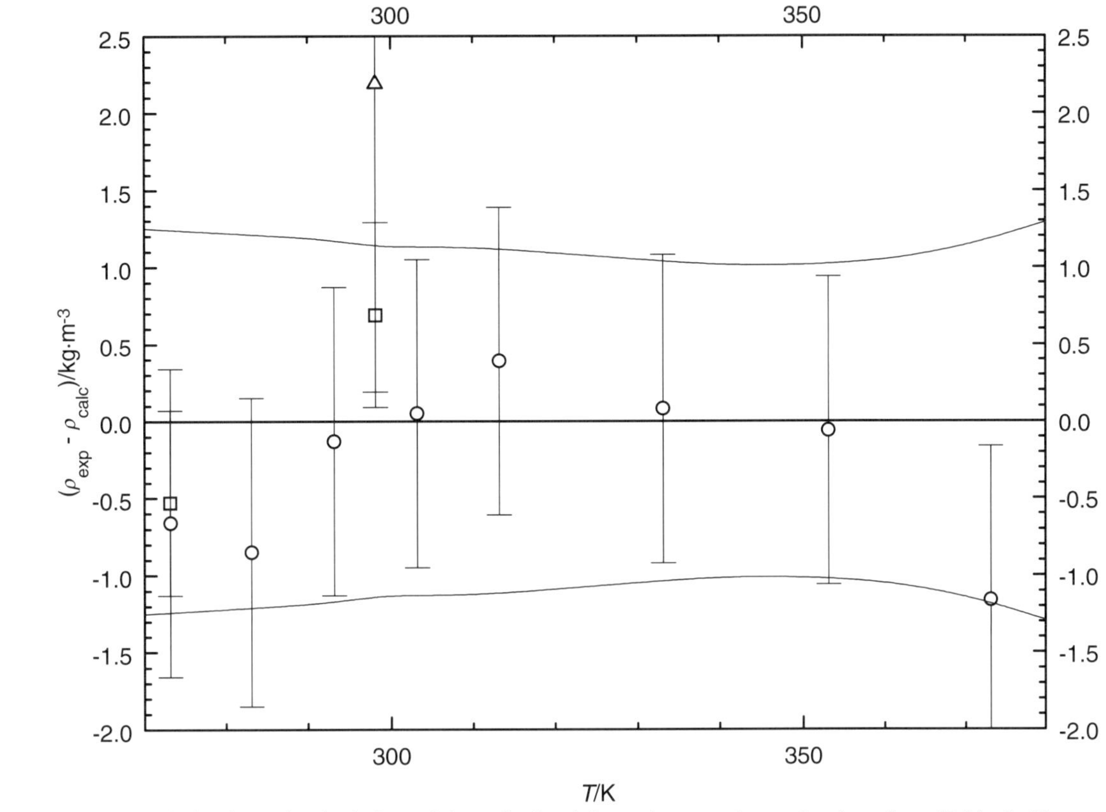

Fig. 1. The symbols show the deviation of the calculated from the experimental values from Table 2. The curves above and below the zero line indicate the calculated error region of the recommended values given in Table 3. The error bars represent the experimental errors. (Error bars smaller than the symbols are omitted for clarity of the figure.)

Table 3. Recommended values (fit to the reliable experimental values according to the equations $\rho = A + BT + CT^2 + DT^3 + \ldots$ or $\rho = [1 + 1.75(1 - T/T_c)^{1/3} + 0.75(1 - T/T_c)][\rho_c + A(T_c - T) + B(T_c - T)^2 + C(T_c - T)^3 + D(T_c - T)^4]$).

$\frac{T}{K}$	$\frac{\rho \pm \sigma_{fit}}{kg \cdot m^{-3}}$	$\frac{T}{K}$	$\frac{\rho \pm \sigma_{fit}}{kg \cdot m^{-3}}$	$\frac{T}{K}$	$\frac{\rho \pm \sigma_{fit}}{kg \cdot m^{-3}}$
270.00	856.93 ± 1.25	300.00	831.22 ± 1.13	350.00	788.37 ± 1.01
280.00	848.36 ± 1.22	310.00	822.65 ± 1.13	360.00	779.80 ± 1.04
290.00	839.79 ± 1.19	320.00	814.08 ± 1.09	370.00	771.23 ± 1.13
293.15	837.09 ± 1.17	330.00	805.51 ± 1.05	380.00	762.66 ± 1.29
298.15	832.81 ± 1.14	340.00	796.94 ± 1.01		

4-Methyl-1-heptanol **[817-91-4]** **$C_8H_{18}O$** **MW = 130.23** **113**

Table 1. Coefficients of the polynomial expansion equation.
Standard deviations (see introduction):
$\sigma_{c,w} = 5.6667 \cdot 10^{-1}$ (combined temperature ranges, weighted),
$\sigma_{c,uw} = 1.8889 \cdot 10^{-1}$ (combined temperature ranges, unweighted).

Coefficient	T = 273.15 to 373.35 K $\rho = A + BT + CT^2 + DT^3 + \ldots$
A	$1.03650 \cdot 10^{3}$
B	$-7.73026 \cdot 10^{-1}$

Table 2. Experimental values with uncertainties and deviation from calculated values.

$\frac{T}{\text{K}}$	$\frac{\rho_{exp} \pm 2\sigma_{est}}{\text{kg} \cdot \text{m}^{-3}}$	$\frac{\rho_{exp} - \rho_{calc}}{\text{kg} \cdot \text{m}^{-3}}$	Ref. (Symbol in Fig. 1)	$\frac{T}{\text{K}}$	$\frac{\rho_{exp} \pm 2\sigma_{est}}{\text{kg} \cdot \text{m}^{-3}}$	$\frac{\rho_{exp} - \rho_{calc}}{\text{kg} \cdot \text{m}^{-3}}$	Ref. (Symbol in Fig. 1)
273.27	824.88 ± 1.00	−0.38	30-bin/dar(□)	333.15	779.42 ± 1.00	0.45	30-bin/dar(□)
283.15	817.13 ± 1.00	−0.49	30-bin/dar(□)	353.15	763.77 ± 1.00	0.26	30-bin/dar(□)
293.15	810.24 ± 1.00	0.35	30-bin/dar(□)	373.35	746.88 ± 1.00	−1.01	30-bin/dar(□)
303.15	802.57 ± 1.00	0.41	30-bin/dar(□)	273.15	824.60 ± 1.00	−0.75	41-dor/gla(O)
313.15	795.10 ± 1.00	0.67	30-bin/dar(□)	298.15	806.50 ± 1.00	0.48	41-dor/gla(O)

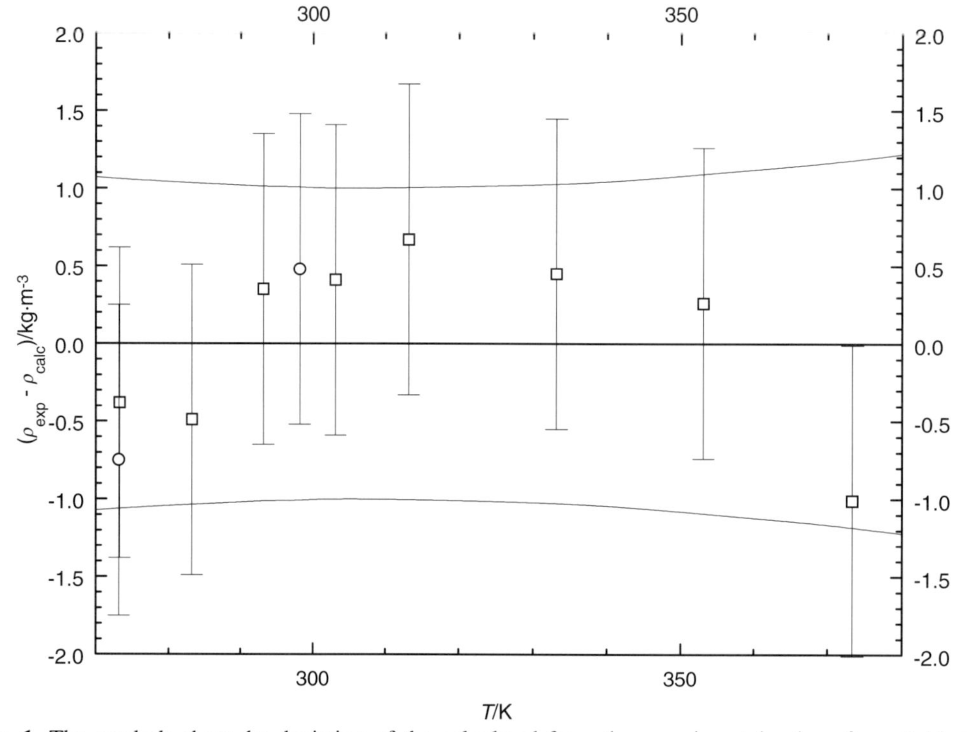

Fig. 1. The symbols show the deviation of the calculated from the experimental values from Table 2. The curves above and below the zero line indicate the calculated error region of the recommended values given in Table 3. The error bars represent the experimental errors. (Error bars smaller than the symbols are omitted for clarity of the figure.)

cont.

Table 3. Recommended values (fit to the reliable experimental values according to the equations $\rho = A + BT + CT^2 + DT^3 + \ldots$ or $\rho = [1 + 1.75(1 - T/T_c)^{1/3} + 0.75(1 - T/T_c)][\rho_c + A(T_c - T) + B(T_c - T)^2 + C(T_c - T)^3 + D(T_c - T)^4]$).

T/K	$\rho \pm \sigma_{fit}$ / kg·m^{-3}	T/K	$\rho \pm \sigma_{fit}$ / kg·m^{-3}	T/K	$\rho \pm \sigma_{fit}$ / kg·m^{-3}
270.00	827.78 ± 1.07	300.00	804.59 ± 1.00	350.00	765.94 ± 1.08
280.00	820.05 ± 1.04	310.00	796.86 ± 1.00	360.00	758.21 ± 1.12
290.00	812.32 ± 1.02	320.00	789.13 ± 1.01	370.00	750.48 ± 1.16
293.15	809.89 ± 1.01	330.00	781.40 ± 1.02	380.00	742.75 ± 1.22
298.15	806.02 ± 1.01	340.00	773.67 ± 1.04		

4-Methyl-2-heptanol **[56298-90-9]** **$C_8H_{18}O$** **MW = 130.23** **114**

Table 1. Coefficients of the polynomial expansion equation.
Standard deviations (see introduction):
$\sigma_{c,w} = 1.0278 \cdot 10^{-1}$ (combined temperature ranges, weighted),
$\sigma_{c,uw} = 5.1187 \cdot 10^{-2}$ (combined temperature ranges, unweighted).

Coefficient	T = 273.15 to 373.13 K $\rho = A + BT + CT^2 + DT^3 + \ldots$
A	$9.32872 \cdot 10^{2}$
B	$-1.40659 \cdot 10^{-1}$
C	$-1.03498 \cdot 10^{-3}$

Table 2. Experimental values with uncertainties and deviation from calculated values.

T/K	$\rho_{exp} \pm 2\sigma_{est}$ / kg·m^{-3}	$\rho_{exp} - \rho_{calc}$ / kg·m^{-3}	Ref. (Symbol in Fig. 1)	T/K	$\rho_{exp} \pm 2\sigma_{est}$ / kg·m^{-3}	$\rho_{exp} - \rho_{calc}$ / kg·m^{-3}	Ref. (Symbol in Fig. 1)
273.20	817.33 ± 1.00	0.13	30-bin/dar(○)	333.15	771.31 ± 1.00	0.17	30-bin/dar(○)
283.15	809.78 ± 1.00	−0.29	30-bin/dar(○)	353.15	755.23 ± 1.00	1.11	30-bin/dar[1)]
293.15	802.57 ± 1.00	−0.13	30-bin/dar(○)	373.13	736.27 ± 1.00	−0.02	30-bin/dar(○)
303.15	795.04 ± 1.00	−0.08	30-bin/dar(○)	273.15	817.30 ± 0.60	0.07	41-dor/gla(□)
313.15	787.40 ± 1.00	0.07	30-bin/dar(○)	298.15	799.00 ± 0.60	0.07	41-dor/gla(□)

[1)] Not included in Fig. 1.

Table 3. Recommended values (fit to the reliable experimental values according to the equations $\rho = A + BT + CT^2 + DT^3 + \ldots$ or $\rho = [1 + 1.75(1 - T/T_c)^{1/3} + 0.75(1 - T/T_c)][\rho_c + A(T_c - T) + B(T_c - T)^2 + C(T_c - T)^3 + D(T_c - T)^4]$).

T/K	$\rho \pm \sigma_{fit}$ / kg·m^{-3}	T/K	$\rho \pm \sigma_{fit}$ / kg·m^{-3}	T/K	$\rho \pm \sigma_{fit}$ / kg·m^{-3}
270.00	819.44 ± 0.99	300.00	797.53 ± 0.93	350.00	756.86 ± 1.09
280.00	812.35 ± 0.91	310.00	789.81 ± 0.97	360.00	748.10 ± 1.17
290.00	805.04 ± 0.90	320.00	781.88 ± 1.01	370.00	739.14 ± 1.31
293.15	802.70 ± 0.91	330.00	773.75 ± 1.03	380.00	729.97 ± 1.52
298.15	798.93 ± 0.92	340.00	765.40 ± 1.05		

cont.

4-Methyl-2-heptanol (cont.)

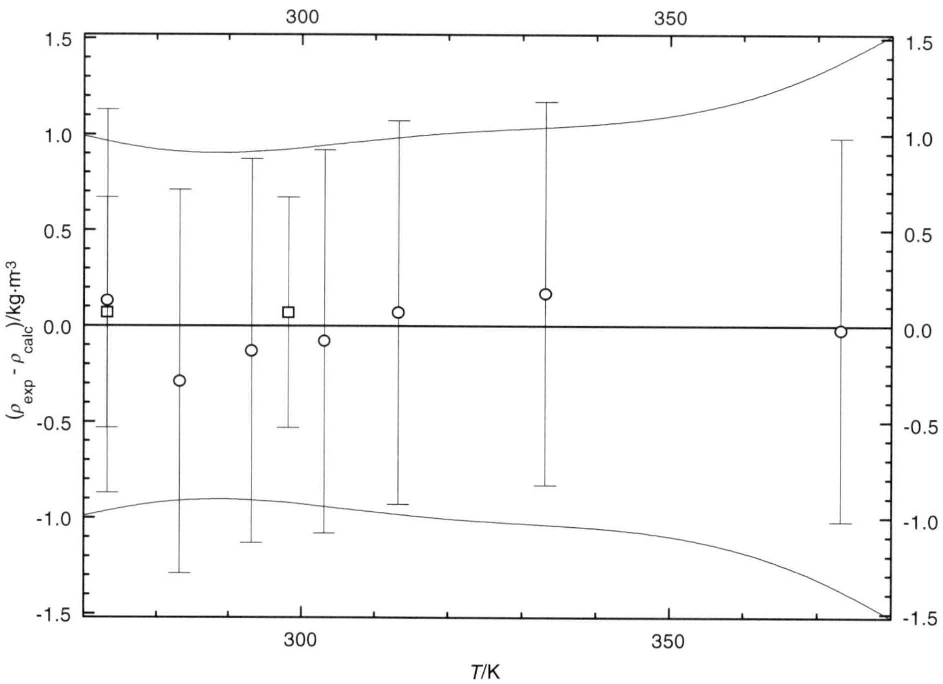

Fig. 1. The symbols show the deviation of the calculated from the experimental values from Table 2. The curves above and below the zero line indicate the calculated error region of the recommended values given in Table 3. The error bars represent the experimental errors. (Error bars smaller than the symbols are omitted for clarity of the figure.)

4-Methyl-3-heptanol **[14979-39-6]** **$C_8H_{18}O$** **MW = 130.23** **115**

Table 1. Coefficients of the polynomial expansion equation.
Standard deviations (see introduction):
$\sigma_{c,w} = 2.8281 \cdot 10^{-1}$ (combined temperature ranges, weighted),
$\sigma_{c,uw} = 1.0734 \cdot 10^{-1}$ (combined temperature ranges, unweighted).

Coefficient	T = 273.15 to 373.56 K $\rho = A + BT + CT^2 + DT^3 + \ldots$
A	$9.85558 \cdot 10^{2}$
B	$-4.61789 \cdot 10^{-1}$
C	$-6.04236 \cdot 10^{-4}$

Table 2. Experimental values with uncertainties and deviation from calculated values.

T/K	$\rho_{exp} \pm 2\sigma_{est}$ / kg·m^{-3}	$\rho_{exp} - \rho_{calc}$ / kg·m^{-3}	Ref. (Symbol in Fig. 1)	T/K	$\rho_{exp} \pm 2\sigma_{est}$ / kg·m^{-3}	$\rho_{exp} - \rho_{calc}$ / kg·m^{-3}	Ref. (Symbol in Fig. 1)
273.18	814.33 ± 1.00	0.02	30-bin/dar(O)	353.15	747.50 ± 1.00	0.38	30-bin/dar(O)
283.15	805.67 ± 1.00	−0.69	30-bin/dar(O)	373.56	728.86 ± 1.00	0.13	30-bin/dar(O)
293.15	798.02 ± 1.00	−0.24	30-bin/dar(O)	273.15	814.50 ± 0.60	0.16	41-dor/gla(□)
303.15	789.76 ± 1.00	−0.28	30-bin/dar(O)	293.15	798.60 ± 0.60	0.34	41-dor/gla(□)
313.15	781.62 ± 1.00	−0.08	30-bin/dar(O)	298.15	794.60 ± 0.60	0.44	41-dor/gla(□)
333.15	764.47 ± 1.00	−0.18	30-bin/dar(O)				

cont.

Further references: [12-bje].

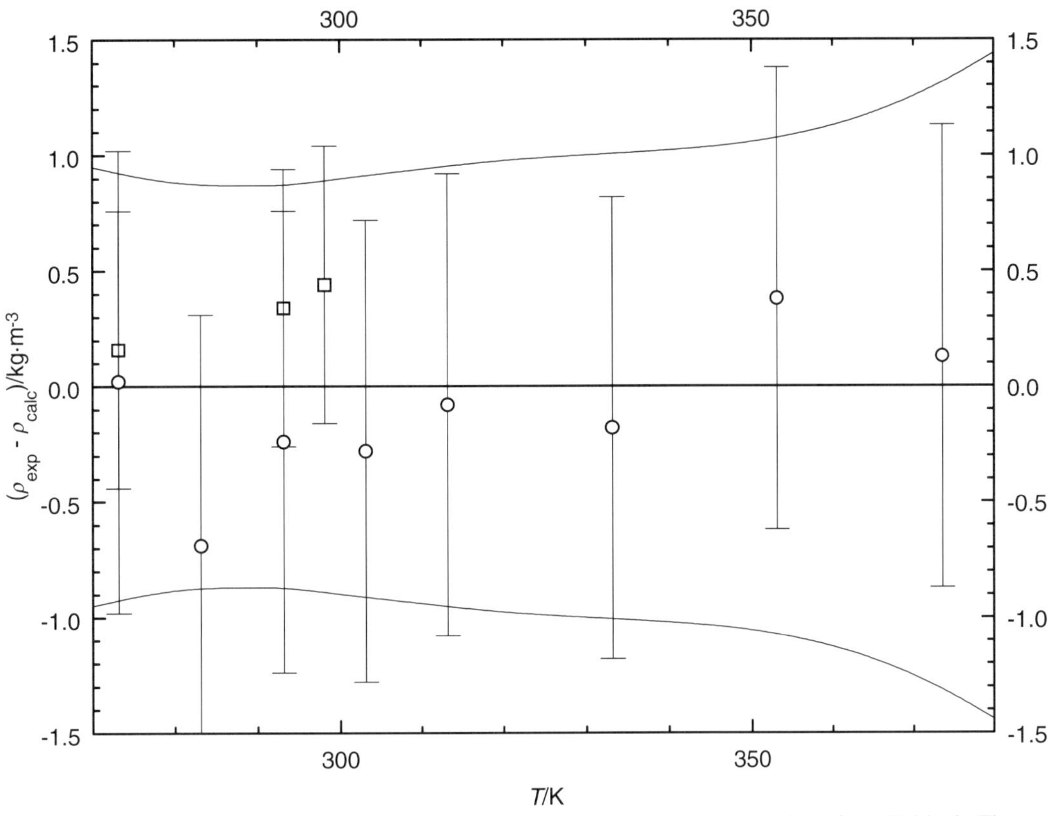

Fig. 1. The symbols show the deviation of the calculated from the experimental values from Table 2. The curves above and below the zero line indicate the calculated error region of the recommended values given in Table 3. The error bars represent the experimental errors. (Error bars smaller than the symbols are omitted for clarity of the figure.)

Table 3. Recommended values (fit to the reliable experimental values according to the equations
$\rho = A + BT + CT^2 + DT^3 + \ldots$ or $\rho = [1 + 1.75(1 - T/T_c)^{1/3} + 0.75(1 - T/T_c)][\rho_c + A(T_c - T) + B(T_c - T)^2 + C(T_c - T)^3 + D(T_c - T)^4]$).

$\frac{T}{\text{K}}$	$\frac{\rho \pm \sigma_{fit}}{\text{kg} \cdot \text{m}^{-3}}$	$\frac{T}{\text{K}}$	$\frac{\rho \pm \sigma_{fit}}{\text{kg} \cdot \text{m}^{-3}}$	$\frac{T}{\text{K}}$	$\frac{\rho \pm \sigma_{fit}}{\text{kg} \cdot \text{m}^{-3}}$
270.00	816.83 ± 0.95	300.00	792.64 ± 0.90	350.00	749.91 ± 1.05
280.00	808.89 ± 0.87	310.00	784.34 ± 0.94	360.00	741.01 ± 1.12
290.00	800.82 ± 0.87	320.00	775.91 ± 0.98	370.00	731.98 ± 1.24
293.15	798.26 ± 0.87	330.00	767.37 ± 1.00	380.00	722.83 ± 1.44
298.15	794.16 ± 0.89	340.00	758.70 ± 1.02		

4-Methyl-4-heptanol **[598-01-6]** **$C_8H_{18}O$** **MW = 130.23** **116**

Table 1. Coefficients of the polynomial expansion equation.
Standard deviations (see introduction):
$\sigma_{c,w} = 5.8580 \cdot 10^{-1}$ (combined temperature ranges, weighted),
$\sigma_{c,uw} = 1.7570 \cdot 10^{-1}$ (combined temperature ranges, unweighted).

Coefficient	T = 273.15 to 373.42 K $\rho = A + BT + CT^2 + DT^3 + \ldots$
A	$9.96959 \cdot 10^{2}$
B	$-3.50068 \cdot 10^{-1}$
C	$-8.24303 \cdot 10^{-4}$

Table 2. Experimental values with uncertainties and deviation from calculated values.

$\frac{T}{\mathrm{K}}$	$\frac{\rho_{exp} \pm 2\sigma_{est}}{\mathrm{kg \cdot m^{-3}}}$	$\frac{\rho_{exp} - \rho_{calc}}{\mathrm{kg \cdot m^{-3}}}$	Ref. (Symbol in Fig. 1)	$\frac{T}{\mathrm{K}}$	$\frac{\rho_{exp} \pm 2\sigma_{est}}{\mathrm{kg \cdot m^{-3}}}$	$\frac{\rho_{exp} - \rho_{calc}}{\mathrm{kg \cdot m^{-3}}}$	Ref. (Symbol in Fig. 1)
289.85	826.90 ± 2.00	0.66	19-eyk(◆)	298.15	818.30 ± 1.00	−1.01	39-owe/qua(○)
273.26	840.12 ± 1.00	0.37	30-bin/dar(Δ)	308.15	809.60 ± 1.00	−1.21	39-owe/qua(○)
283.15	831.67 ± 1.00	−0.08	30-bin/dar(Δ)	318.15	801.00 ± 1.00	−1.15	39-owe/qua(○)
293.15	823.99 ± 1.00	0.49	30-bin/dar(Δ)	328.15	792.50 ± 1.00	−0.82	39-owe/qua(○)
303.15	815.79 ± 1.00	0.71	30-bin/dar(Δ)	338.15	783.90 ± 1.00	−0.43	39-owe/qua(○)
313.15	807.56 ± 1.00	1.06	30-bin/dar(Δ)	273.15	839.70 ± 0.60	−0.14	41-dor/gla(□)
333.15	789.45 ± 1.00	0.60	30-bin/dar(Δ)	293.15	824.10 ± 0.60	0.60	41-dor/gla(□)
353.15	771.13 ± 1.00	0.60	30-bin/dar(Δ)	298.15	820.20 ± 0.60	0.89	41-dor/gla(□)
373.42	751.48 ± 1.00	0.19	30-bin/dar(Δ)	293.15	822.70 ± 1.00	−0.80	59-yur/bel(∇)
273.15	839.30 ± 1.00	−0.54	39-owe/qua(○)				

Table 3. Recommended values (fit to the reliable experimental values according to the equations $\rho = A + BT + CT^2 + DT^3 + \ldots$ or $\rho = [1 + 1.75(1 - T/T_c)^{1/3} + 0.75(1 - T/T_c)][\rho_c + A(T_c - T) + B(T_c - T)^2 + C(T_c - T)^3 + D(T_c - T)^4]$).

$\frac{T}{\mathrm{K}}$	$\frac{\rho \pm \sigma_{fit}}{\mathrm{kg \cdot m^{-3}}}$	$\frac{T}{\mathrm{K}}$	$\frac{\rho \pm \sigma_{fit}}{\mathrm{kg \cdot m^{-3}}}$	$\frac{T}{\mathrm{K}}$	$\frac{\rho \pm \sigma_{fit}}{\mathrm{kg \cdot m^{-3}}}$
270.00	842.35 ± 1.07	300.00	817.75 ± 1.00	350.00	773.46 ± 1.06
280.00	834.31 ± 1.01	310.00	809.22 ± 1.01	360.00	764.10 ± 1.13
290.00	826.12 ± 0.99	320.00	800.53 ± 1.02	370.00	754.59 ± 1.24
293.15	823.50 ± 0.99	330.00	791.67 ± 1.03	380.00	744.90 ± 1.43
298.15	819.31 ± 1.00	340.00	782.65 ± 1.04		

cont.

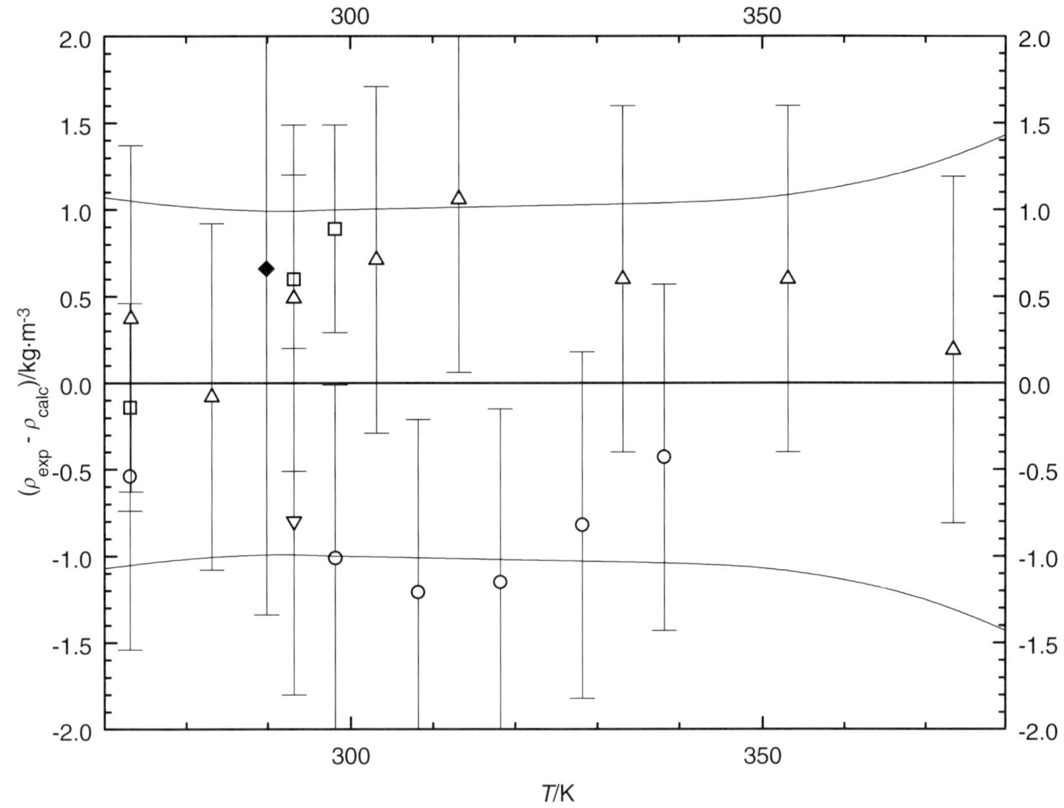

Fig. 1. The symbols show the deviation of the calculated from the experimental values from Table 2. The curves above and below the zero line indicate the calculated error region of the recommended values given in Table 3. The error bars represent the experimental errors. (Error bars smaller than the symbols are omitted for clarity of the figure.)

5-Methyl-1-heptanol **[7212-53-5]** **$C_8H_{18}O$** **MW = 130.23** **117**

Table 1. Coefficients of the polynomial expansion equation.
Standard deviations (see introduction):
$\sigma_{c,w} = 2.1134 \cdot 10^{-1}$ (combined temperature ranges, weighted),
$\sigma_{c,uw} = 9.5153 \cdot 10^{-2}$ (combined temperature ranges, unweighted).

Coefficient	T = 273.15 to 373.22 K $\rho = A + BT + CT^2 + DT^3 + \ldots$
A	$9.74236 \cdot 10^{2}$
B	$-3.44128 \cdot 10^{-1}$
C	$-6.33911 \cdot 10^{-4}$

cont.

5-Methyl-1-heptanol (cont.)

Table 2. Experimental values with uncertainties and deviation from calculated values.

$\frac{T}{K}$	$\frac{\rho_{exp} \pm 2\sigma_{est}}{kg \cdot m^{-3}}$	$\frac{\rho_{exp} - \rho_{calc}}{kg \cdot m^{-3}}$	Ref. (Symbol in Fig. 1)	$\frac{T}{K}$	$\frac{\rho_{exp} \pm 2\sigma_{est}}{kg \cdot m^{-3}}$	$\frac{\rho_{exp} - \rho_{calc}}{kg \cdot m^{-3}}$	Ref. (Symbol in Fig. 1)
273.32	833.33 ± 1.00	0.51	30-bin/dar(○)	353.15	774.23 ± 1.00	0.58	30-bin/dar(○)
283.15	825.76 ± 1.00	–0.21	30-bin/dar(○)	373.22	757.06 ± 1.00	–0.44	30-bin/dar(○)
293.15	818.80 ± 1.00	–0.08	30-bin/dar(○)	273.15	832.90 ± 0.60	–0.04	41-dor/gla(□)
303.15	811.56 ± 1.00	–0.10	30-bin/dar(○)	293.15	818.70 ± 0.60	–0.18	41-dor/gla(□)
313.15	804.38 ± 1.00	0.07	30-bin/dar(○)	298.15	815.20 ± 0.60	–0.08	41-dor/gla(□)
333.15	789.20 ± 1.00	–0.03	30-bin/dar(○)				

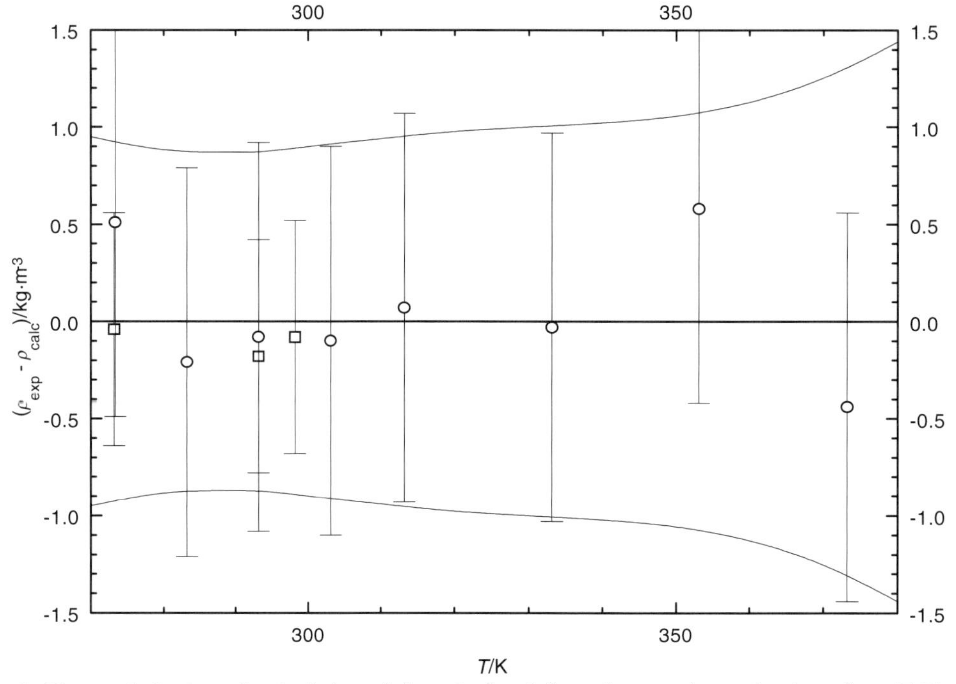

Fig. 1. The symbols show the deviation of the calculated from the experimental values from Table 2. The curves above and below the zero line indicate the calculated error region of the recommended values given in Table 3. The error bars represent the experimental errors. (Error bars smaller than the symbols are omitted for clarity of the figure.)

Table 3. Recommended values (fit to the reliable experimental values according to the equations $\rho = A + BT + CT^2 + DT^3 + \ldots$ or $\rho = [1 + 1.75(1 - T/T_c)^{1/3} + 0.75(1 - T/T_c)][\rho_c + A(T_c - T) + B(T_c - T)^2 + C(T_c - T)^3 + D(T_c - T)^4]$).

$\frac{T}{K}$	$\frac{\rho \pm \sigma_{fit}}{kg \cdot m^{-3}}$	$\frac{T}{K}$	$\frac{\rho \pm \sigma_{fit}}{kg \cdot m^{-3}}$	$\frac{T}{K}$	$\frac{\rho \pm \sigma_{fit}}{kg \cdot m^{-3}}$
270.00	835.11 ± 0.95	300.00	813.94 ± 0.90	350.00	776.14 ± 1.05
280.00	828.18 ± 0.87	310.00	806.64 ± 0.94	360.00	768.19 ± 1.12
290.00	821.13 ± 0.87	320.00	799.20 ± 0.98	370.00	760.13 ± 1.24
293.15	818.88 ± 0.87	330.00	791.64 ± 1.00	380.00	751.93 ± 1.44
298.15	815.28 ± 0.89	340.00	783.95 ± 1.02		

L(+)-5-Methyl-1-heptanol **[500006-90-6]** **$C_8H_{18}O$** **MW = 130.23** **118**

Table 1. Experimental value with uncertainty.

T/K	$\rho_{exp} \pm 2\sigma_{est}$ / $kg \cdot m^{-3}$	Ref.
298.15	825.9 ± 0.6	62-lar/sal

5-Methyl-2-heptanol **[54630-50-1]** **$C_8H_{18}O$** **MW = 130.23** **119**

Table 1. Coefficients of the polynomial expansion equation.
Standard deviations (see introduction):
$\sigma_{c,w} = 3.6541 \cdot 10^{-1}$ (combined temperature ranges, weighted),
$\sigma_{c,uw} = 1.7122 \cdot 10^{-1}$ (combined temperature ranges, unweighted).

Coefficient	T = 273.15 to 373.50 K $\rho = A + BT + CT^2 + DT^3 + \ldots$
A	$8.98516 \cdot 10^{2}$
B	$2.01122 \cdot 10^{-1}$
C	$-1.67280 \cdot 10^{-3}$

Table 2. Experimental values with uncertainties and deviation from calculated values.

T/K	$\rho_{exp} \pm 2\sigma_{est}$ / $kg \cdot m^{-3}$	$\rho_{exp} - \rho_{calc}$ / $kg \cdot m^{-3}$	Ref. (Symbol in Fig. 1)	T/K	$\rho_{exp} \pm 2\sigma_{est}$ / $kg \cdot m^{-3}$	$\rho_{exp} - \rho_{calc}$ / $kg \cdot m^{-3}$	Ref. (Symbol in Fig. 1)
273.20	828.78 ± 1.00	0.17	30-bin/dar(O)	353.15	764.06 ± 1.00	3.14	30-bin/dar[1)]
283.15	820.55 ± 1.00	-0.80	30-bin/dar(O)	373.50	740.14 ± 1.00	–0.13	30-bin/dar(O)
293.15	813.34 ± 1.00	–0.38	30-bin/dar(O)	273.15	829.00 ± 0.60	0.36	41-dor/gla(□)
303.15	805.22 ± 1.00	–0.54	30-bin/dar(O)	293.15	813.80 ± 0.60	0.08	41-dor/gla(□)
313.15	797.45 ± 1.00	–0.01	30-bin/dar(O)	298.15	810.00 ± 0.60	0.22	41-dor/gla(□)
333.15	780.88 ± 1.00	1.02	30-bin/dar(O)				

[1)] Not included in Fig. 1.

Table 3. Recommended values (fit to the reliable experimental values according to the equations $\rho = A + BT + CT^2 + DT^3 + \ldots$ or $\rho = [1 + 1.75(1 - T/T_c)^{1/3} + 0.75(1 - T/T_c)][\rho_c + A(T_c - T) + B(T_c - T)^2 + C(T_c - T)^3 + D(T_c - T)^4]$).

T/K	$\rho \pm \sigma_{fit}$ / $kg \cdot m^{-3}$	T/K	$\rho \pm \sigma_{fit}$ / $kg \cdot m^{-3}$	T/K	$\rho \pm \sigma_{fit}$ / $kg \cdot m^{-3}$
270.00	830.87 ± 0.95	300.00	808.30 ± 0.90	350.00	763.99 ± 1.08
280.00	823.68 ± 0.87	310.00	800.11 ± 0.94	360.00	754.12 ± 1.16
290.00	816.16 ± 0.86	320.00	791.58 ± 0.98	370.00	743.92 ± 1.30
293.15	813.72 ± 0.87	330.00	782.72 ± 1.01	380.00	733.39 ± 1.50
298.15	809.78 ± 0.89	340.00	773.52 ± 1.04		

cont.

5-Methyl-2-heptanol (cont.)

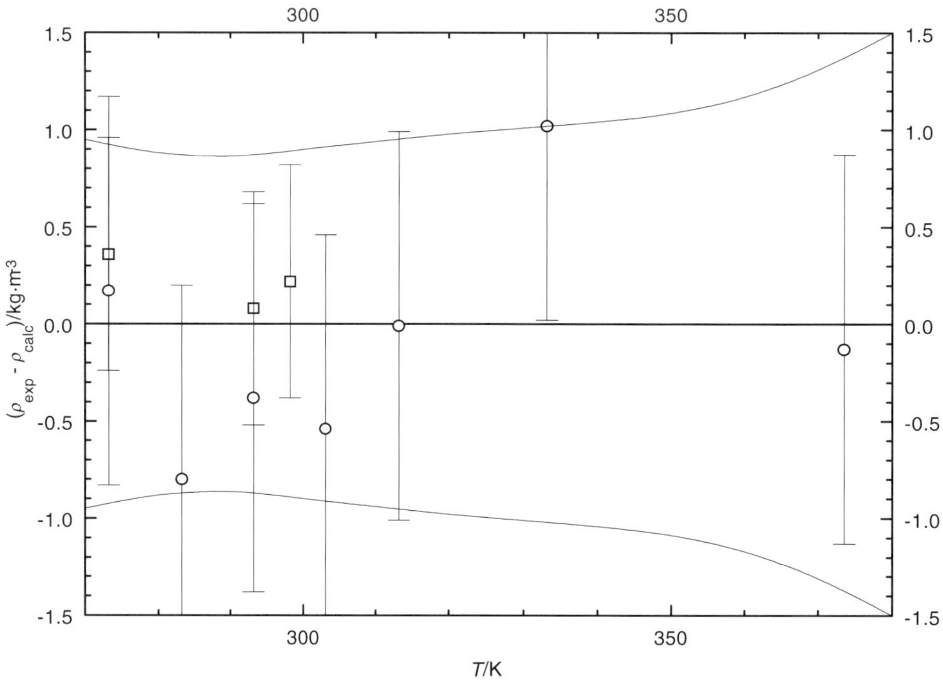

Fig. 1. The symbols show the deviation of the calculated from the experimental values from Table 2. The curves above and below the zero line indicate the calculated error region of the recommended values given in Table 3. The error bars represent the experimental errors. (Error bars smaller than the symbols are omitted for clarity of the figure.)

5-Methyl-3-heptanol **[18720-65-5]** **$C_8H_{18}O$** **MW = 130.23** **120**

Table 1. Coefficients of the polynomial expansion equation.
Standard deviations (see introduction):
$\sigma_{c,w} = 9.4864 \cdot 10^{-1}$ (combined temperature ranges, weighted),
$\sigma_{c,uw} = 2.9572 \cdot 10^{-1}$ (combined temperature ranges, unweighted).

Coefficient	T = 273.15 to 373.20 K $\rho = A + BT + CT^2 + DT^3 + \ldots$
A	$1.07627 \cdot 10^3$
B	$-8.79191 \cdot 10^{-1}$

Table 2. Experimental values with uncertainties and deviation from calculated values.

T/K	$\rho_{exp} \pm 2\sigma_{est}$/kg·m^{-3}	$\rho_{exp} - \rho_{calc}$/kg·m^{-3}	Ref. (Symbol in Fig. 1)	T/K	$\rho_{exp} \pm 2\sigma_{est}$/kg·m^{-3}	$\rho_{exp} - \rho_{calc}$/kg·m^{-3}	Ref. (Symbol in Fig. 1)
273.19	834.52 ± 1.00	−1.57	30-bin/dar(O)	353.15	766.11 ± 1.00	0.32	30-bin/dar(O)
283.15	826.04 ± 1.00	−1.29	30-bin/dar(O)	373.20	748.00 ± 1.00	−0.16	30-bin/dar(O)
293.15	817.86 ± 1.00	−0.68	30-bin/dar(O)	296.15	816.00 ± 2.00	0.10	31-lev/mar-4(Δ)
303.15	809.59 ± 1.00	−0.15	30-bin/dar(O)	273.15	836.70 ± 0.60	0.58	41-dor/gla(□)
313.15	801.28 ± 1.00	0.33	30-bin/dar(O)	298.15	816.20 ± 0.60	2.06	41-dor/gla(□)
333.15	783.82 ± 1.00	0.45	30-bin/dar(O)				

cont.

Further references: [12-gue, 31-pow/sec, 59-mac/bar].

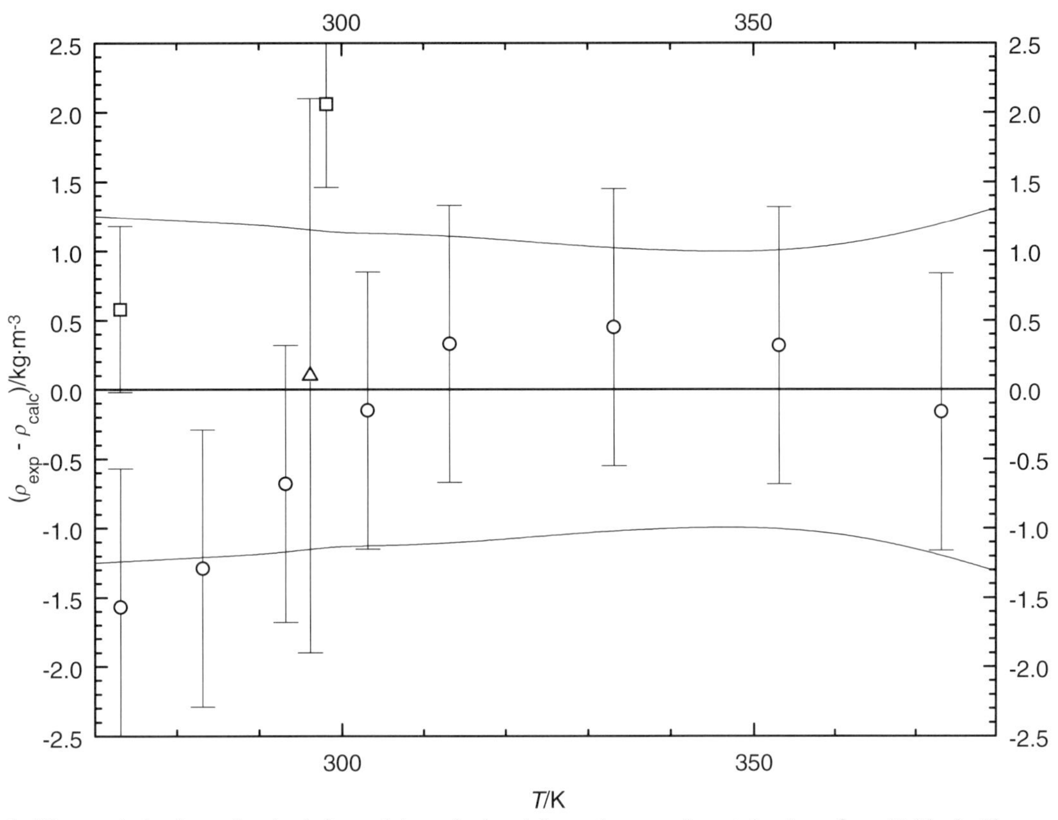

Fig. 1. The symbols show the deviation of the calculated from the experimental values from Table 2. The curves above and below the zero line indicate the calculated error region of the recommended values given in Table 3. The error bars represent the experimental errors. (Error bars smaller than the symbols are omitted for clarity of the figure.)

Table 3. Recommended values (fit to the reliable experimental values according to the equations $\rho = A + BT + CT^2 + DT^3 + \ldots$ or $\rho = [1 + 1.75(1 - T/T_c)^{1/3} + 0.75(1 - T/T_c)][\rho_c + A(T_c - T) + B(T_c - T)^2 + C(T_c - T)^3 + D(T_c - T)^4]$).

$\frac{T}{K}$	$\frac{\rho \pm \sigma_{fit}}{kg \cdot m^{-3}}$	$\frac{T}{K}$	$\frac{\rho \pm \sigma_{fit}}{kg \cdot m^{-3}}$	$\frac{T}{K}$	$\frac{\rho \pm \sigma_{fit}}{kg \cdot m^{-3}}$
270.00	838.89 ± 1.25	300.00	812.51 ± 1.13	350.00	768.55 ± 0.99
280.00	830.10 ± 1.22	310.00	803.72 ± 1.12	360.00	759.76 ± 1.03
290.00	821.31 ± 1.19	320.00	794.93 ± 1.08	370.00	750.97 ± 1.14
293.15	818.54 ± 1.17	330.00	786.14 ± 1.03	380.00	742.18 ± 1.31
298.15	814.14 ± 1.14	340.00	777.35 ± 1.00		

6-Methyl-1-heptanol **[1653-40-3]** **$C_8H_{18}O$** **MW = 130.23** **121**

Table 1. Coefficients of the polynomial expansion equation.
Standard deviations (see introduction):
$\sigma_{c,w} = 1.1940 \cdot 10^{-1}$ (combined temperature ranges, weighted),
$\sigma_{c,uw} = 5.7073 \cdot 10^{-2}$ (combined temperature ranges, unweighted).

Coefficient	$T = 273.15$ to 373.57 K $\rho = A + BT + CT^2 + DT^3 + \dots$
A	$9.65307 \cdot 10^{2}$
B	$-2.92230 \cdot 10^{-1}$
C	$-6.81907 \cdot 10^{-4}$

Table 2. Experimental values with uncertainties and deviation from calculated values.

$\frac{T}{\text{K}}$	$\frac{\rho_{exp} \pm 2\sigma_{est}}{\text{kg} \cdot \text{m}^{-3}}$	$\frac{\rho_{exp} - \rho_{calc}}{\text{kg} \cdot \text{m}^{-3}}$	Ref. (Symbol in Fig. 1)	$\frac{T}{\text{K}}$	$\frac{\rho_{exp} \pm 2\sigma_{est}}{\text{kg} \cdot \text{m}^{-3}}$	$\frac{\rho_{exp} - \rho_{calc}}{\text{kg} \cdot \text{m}^{-3}}$	Ref. (Symbol in Fig. 1)
273.37	839.14 ± 3.00	4.68	30-bin/dar[1)]	353.15	777.42 ± 1.00	0.36	30-bin/dar(○)
283.15	827.61 ± 1.00	−0.28	30-bin/dar(○)	373.57	760.86 ± 1.00	−0.12	30-bin/dar(○)
293.15	820.95 ± 1.00	−0.09	30-bin/dar(○)	273.15	834.70 ± 0.60	0.09	41-dor/gla(□)
303.15	814.07 ± 1.00	0.02	30-bin/dar(○)	293.15	821.10 ± 0.60	0.06	41-dor/gla(□)
313.15	806.97 ± 1.00	0.04	30-bin/dar(○)	298.15	817.60 ± 0.60	0.04	41-dor/gla(□)
333.15	792.14 ± 1.00	−0.13	30-bin/dar(○)				

[1)] Not included in Fig. 1.

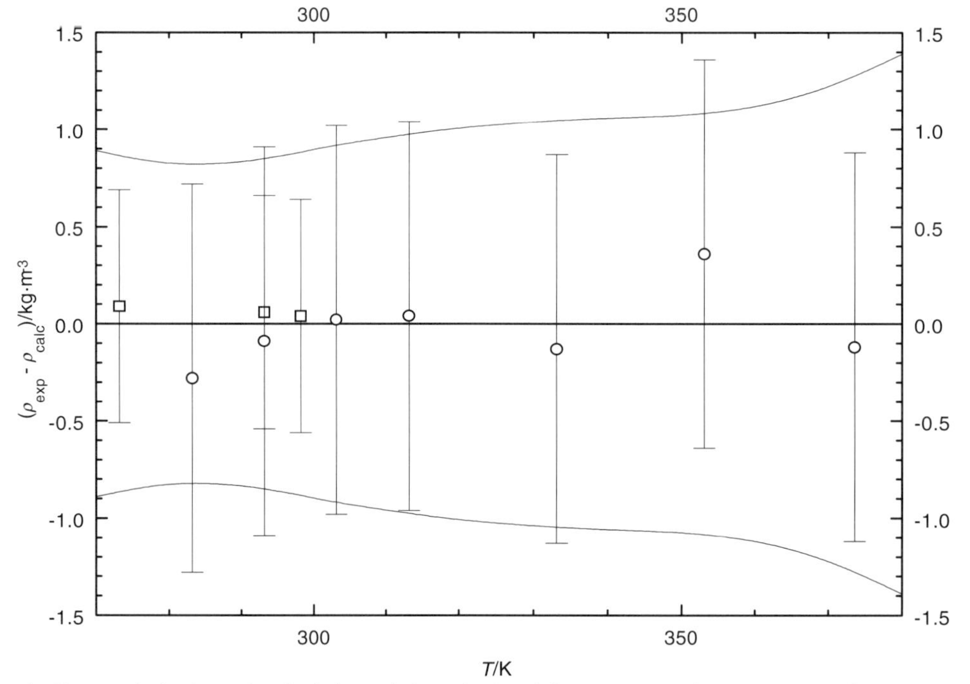

Fig. 1. The symbols show the deviation of the calculated from the experimental values from Table 2. The curves above and below the zero line indicate the calculated error region of the recommended values given in Table 3. The error bars represent the experimental errors. (Error bars smaller than the symbols are omitted for clarity of the figure.)

cont.

Further references: [16-lev/all].

Table 3. Recommended values (fit to the reliable experimental values according to the equations $\rho = A + BT + CT^2 + DT^3 + \ldots$ or $\rho = [1 + 1.75(1 - T/T_c)^{1/3} + 0.75(1 - T/T_c)][\rho_c + A(T_c - T) + B(T_c - T)^2 + C(T_c - T)^3 + D(T_c - T)^4]$).

$\frac{T}{K}$	$\frac{\rho \pm \sigma_{fit}}{kg \cdot m^{-3}}$	$\frac{T}{K}$	$\frac{\rho \pm \sigma_{fit}}{kg \cdot m^{-3}}$	$\frac{T}{K}$	$\frac{\rho \pm \sigma_{fit}}{kg \cdot m^{-3}}$
270.00	836.69 ± 0.89	300.00	816.27 ± 0.90	350.00	779.49 ± 1.07
280.00	830.02 ± 0.81	310.00	809.18 ± 0.96	360.00	771.73 ± 1.11
290.00	823.21 ± 0.83	320.00	801.97 ± 1.01	370.00	763.83 ± 1.21
293.15	821.04 ± 0.85	330.00	794.61 ± 1.04	380.00	755.79 ± 1.39
298.15	817.56 ± 0.88	340.00	787.12 ± 1.06		

6-Methyl-2-heptanol **[4730-22-7]** **$C_8H_{18}O$** **MW = 130.23** **122**

Table 1. Coefficients of the polynomial expansion equation.
Standard deviations (see introduction):
$\sigma_{c,w} = 1.6961 \cdot 10^{-1}$ (combined temperature ranges, weighted),
$\sigma_{c,uw} = 8.6825 \cdot 10^{-2}$ (combined temperature ranges, unweighted).

Coefficient	T = 273.15 to 373.37 K $\rho = A + BT + CT^2 + DT^3 + \ldots$
A	$9.63426 \cdot 10^{2}$
B	$-3.06269 \cdot 10^{-1}$
C	$-7.74629 \cdot 10^{-4}$

Table 2. Experimental values with uncertainties and deviation from calculated values.

$\frac{T}{K}$	$\frac{\rho_{exp} \pm 2\sigma_{est}}{kg \cdot m^{-3}}$	$\frac{\rho_{exp} - \rho_{calc}}{kg \cdot m^{-3}}$	Ref. (Symbol in Fig. 1)	$\frac{T}{K}$	$\frac{\rho_{exp} \pm 2\sigma_{est}}{kg \cdot m^{-3}}$	$\frac{\rho_{exp} - \rho_{calc}}{kg \cdot m^{-3}}$	Ref. (Symbol in Fig. 1)
284.15	814.10 ± 2.00	0.24	28-esc(Δ)	333.15	775.43 ± 1.00	0.01	30-bin/dar(○)
273.17	822.37 ± 1.00	0.41	30-bin/dar(○)	353.15	759.07 ± 1.00	0.41	30-bin/dar(○)
283.15	814.46 ± 1.00	−0.14	30-bin/dar(○)	373.37	740.47 ± 1.00	−0.62	30-bin/dar(○)
293.15	807.04 ± 1.00	−0.03	30-bin/dar(○)	273.15	821.90 ± 0.60	−0.07	41-dor/gla(□)
303.15	799.17 ± 1.00	−0.22	30-bin/dar(○)	293.15	807.10 ± 0.60	0.03	41-dor/gla(□)
313.15	791.39 ± 1.00	−0.17	30-bin/dar(○)	298.15	803.40 ± 0.60	0.15	41-dor/gla(□)

Table 3. Recommended values (fit to the reliable experimental values according to the equations $\rho = A + BT + CT^2 + DT^3 + \ldots$ or $\rho = [1 + 1.75(1 - T/T_c)^{1/3} + 0.75(1 - T/T_c)][\rho_c + A(T_c - T) + B(T_c - T)^2 + C(T_c - T)^3 + D(T_c - T)^4]$).

$\frac{T}{K}$	$\frac{\rho \pm \sigma_{fit}}{kg \cdot m^{-3}}$	$\frac{T}{K}$	$\frac{\rho \pm \sigma_{fit}}{kg \cdot m^{-3}}$	$\frac{T}{K}$	$\frac{\rho \pm \sigma_{fit}}{kg \cdot m^{-3}}$
270.00	824.26 ± 1.03	300.00	801.83 ± 1.03	350.00	761.34 ± 1.12
280.00	816.94 ± 1.05	310.00	794.04 ± 0.98	360.00	752.78 ± 1.27
290.00	809.46 ± 1.06	320.00	786.10 ± 0.93	370.00	744.06 ± 1.33
293.15	807.07 ± 1.05	330.00	778.00 ± 0.90	380.00	735.19 ± 1.47
298.15	803.25 ± 1.04	340.00	769.75 ± 0.95		

cont.

6-Methyl-2-heptanol (cont.)

Further references: [08-bue, 08-bue-1, 12-gue-2, 14-wal-1, 52-her/zao, 53-her/zao, 62-mir/fed].

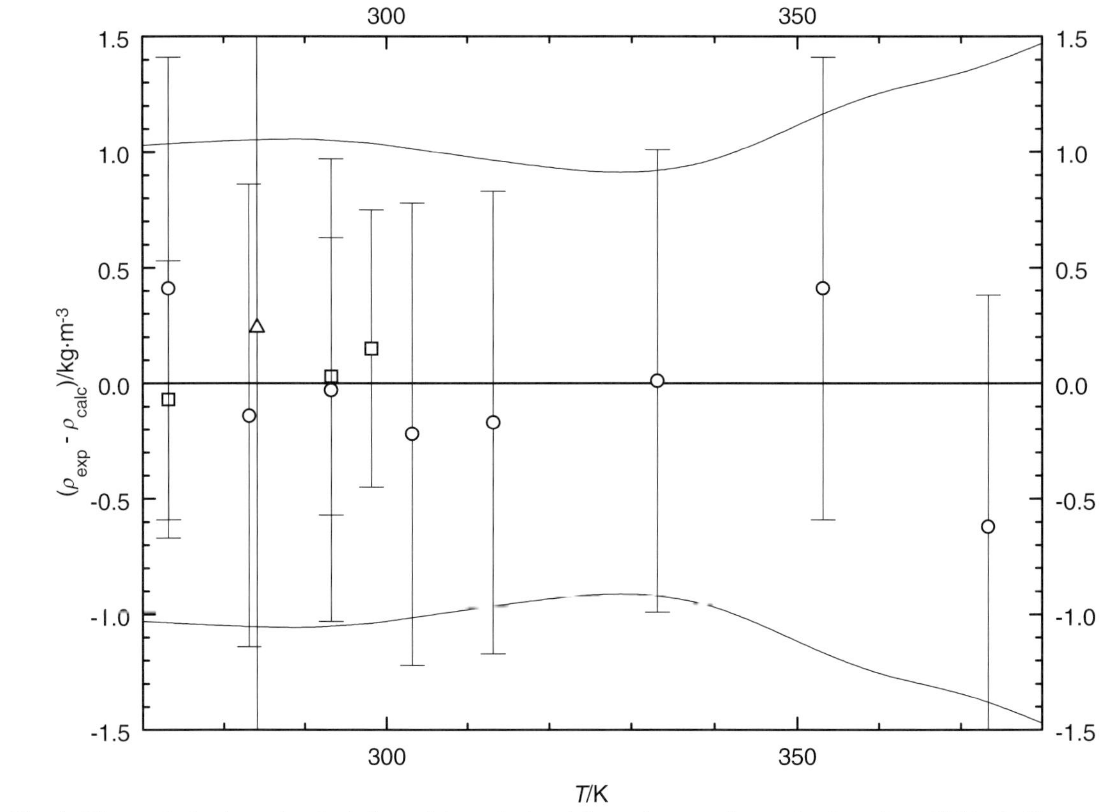

Fig. 1. The symbols show the deviation of the calculated from the experimental values from Table 2. The curves above and below the zero line indicate the calculated error region of the recommended values given in Table 3. The error bars represent the experimental errors. (Error bars smaller than the symbols are omitted for clarity of the figure.)

6-Methyl-3-heptanol [18720-66-6] $C_8H_{18}O$ MW = 130.23 123

Table 1. Coefficients of the polynomial expansion equation.
Standard deviations (see introduction):
$\sigma_{c,w} = 7.5441 \cdot 10^{-1}$ (combined temperature ranges, weighted),
$\sigma_{c,uw} = 3.0306 \cdot 10^{-1}$ (combined temperature ranges, unweighted).

Coefficient	T = 273.15 to 373.32 K $\rho = A + BT + CT^2 + DT^3 + \dots$
A	$1.02123 \cdot 10^3$
B	$-8.20546 \cdot 10^{-1}$

cont.

Table 2. Experimental values with uncertainties and deviation from calculated values.

$\frac{T}{\mathrm{K}}$	$\frac{\rho_{exp} \pm 2\sigma_{est}}{\mathrm{kg \cdot m^{-3}}}$	$\frac{\rho_{exp} - \rho_{calc}}{\mathrm{kg \cdot m^{-3}}}$	Ref. (Symbol in Fig. 1)	$\frac{T}{\mathrm{K}}$	$\frac{\rho_{exp} \pm 2\sigma_{est}}{\mathrm{kg \cdot m^{-3}}}$	$\frac{\rho_{exp} - \rho_{calc}}{\mathrm{kg \cdot m^{-3}}}$	Ref. (Symbol in Fig. 1)
273.42	796.69 ± 1.00	−0.19	30-bin/dar(O)	353.15	731.42 ± 1.00	−0.04	30-bin/dar(O)
283.15	788.58 ± 1.00	−0.32	30-bin/dar(O)	373.32	712.81 ± 1.00	−2.10	30-bin/dar(O)
293.15	781.43 ± 1.00	0.74	30-bin/dar(O)	273.15	796.00 ± 0.60	−1.10	41-dor/gla(□)
303.15	773.46 ± 1.00	0.97	30-bin/dar(O)	293.15	780.50 ± 0.60	−0.19	41-dor/gla(□)
313.15	765.81 ± 1.00	1.53	30-bin/dar(O)	298.15	776.60 ± 0.60	0.01	41-dor/gla(□)
333.15	748.56 ± 1.00	0.69	30-bin/dar(O)				

Further references: [08-bue-1, 33-bri, 44-hen/mat].

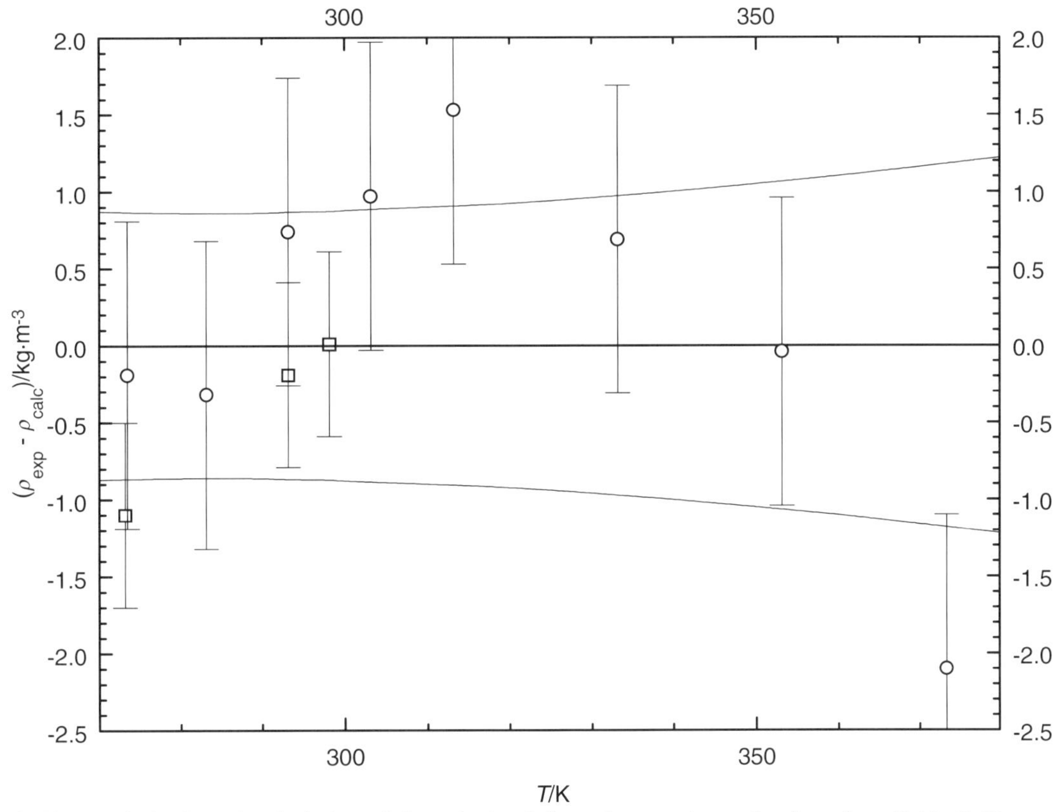

Fig. 1. The symbols show the deviation of the calculated from the experimental values from Table 2. The curves above and below the zero line indicate the calculated error region of the recommended values given in Table 3. The error bars represent the experimental errors. (Error bars smaller than the symbols are omitted for clarity of the figure.)

cont.

6-Methyl-3-heptanol (cont.)

Table 3. Recommended values (fit to the reliable experimental values according to the equations $\rho = A + BT + CT^2 + DT^3 + \dots$ or $\rho = [1 + 1.75(1 - T/T_c)^{1/3} + 0.75(1 - T/T_c)][\rho_c + A(T_c - T) + B(T_c - T)^2 + C(T_c - T)^3 + D(T_c - T)^4]$).

$\frac{T}{K}$	$\frac{\rho \pm \sigma_{fit}}{kg \cdot m^{-3}}$	$\frac{T}{K}$	$\frac{\rho \pm \sigma_{fit}}{kg \cdot m^{-3}}$	$\frac{T}{K}$	$\frac{\rho \pm \sigma_{fit}}{kg \cdot m^{-3}}$
270.00	799.69 ± 0.87	300.00	775.07 ± 0.88	350.00	734.04 ± 1.05
280.00	791.48 ± 0.86	310.00	766.87 ± 0.90	360.00	725.84 ± 1.10
290.00	783.28 ± 0.86	320.00	758.66 ± 0.92	370.00	717.63 ± 1.16
293.15	780.69 ± 0.87	330.00	750.45 ± 0.96	380.00	709.43 ± 1.22
298.15	776.59 ± 0.87	340.00	742.25 ± 1.00		

3-Methyl-2-(1-methylethyl)-1-butanol [18593-92-5] $C_8H_{18}O$ MW = 130.23 124

Table 1. Experimental value with uncertainty.

$\frac{T}{K}$	$\frac{\rho_{exp} \pm 2\sigma_{est}}{kg \cdot m^{-3}}$	Ref.
298.15	842.5 ± 1.0	56-sar/new

1-Octanol [111-87-5] $C_8H_{18}O$ MW = 130.23 125

T_c = 652.60 K [89-tej/lee] ρ_c = 266.00 kg·m^{-3} [89-tej/lee]

Table 1. Coefficients for the polynomial expansion equations. Standard deviations (see introduction): $\sigma_\ell = 3.5218 \cdot 10^{-1}$ (low temperature range), $\sigma_{c,w} = (2.6366 \cdot 10^{-1}$ combined temperature ranges, weighted), $\sigma_{c,uw} = 3.0252 \cdot 10^{-1}$ (combined temperature ranges, unweighted).

Coefficient	T = 258.08 to 530.00 K $\rho = A + BT + CT^2 + DT^3 + \dots$	T = 530.00 to 652.60 K $\rho = [1 + 1.75(1 - T/T_c)^{1/3} + 0.75(1 - T/T_c)]$ $[\rho_c + A(T_c - T) + B(T_c - T)^2 + C(T_c - T)^3 + D(T_c - T)^4]$
A	$1.02020 \cdot 10^3$	1.11884
B	$-7.81865 \cdot 10^{-1}$	$-1.68799 \cdot 10^{-2}$
C	$8.70573 \cdot 10^{-4}$	$8.96699 \cdot 10^{-5}$
D	$-1.61006 \cdot 10^{-6}$	$-1.31742 \cdot 10^{-7}$

Table 2. Experimental values with uncertainties and deviation from calculated values.

$\frac{T}{K}$	$\frac{\rho_{exp} \pm 2\sigma_{est}}{kg \cdot m^{-3}}$	$\frac{\rho_{exp} - \rho_{calc}}{kg \cdot m^{-3}}$	Ref. (Symbol in Fig. 1)	$\frac{T}{K}$	$\frac{\rho_{exp} \pm 2\sigma_{est}}{kg \cdot m^{-3}}$	$\frac{\rho_{exp} - \rho_{calc}}{kg \cdot m^{-3}}$	Ref. (Symbol in Fig. 1)
	crystal			303.15	818.60 ± 0.60	0.27	29-smy/sto[1)]
78.15	1023.0 ± 3.0		30-bil/fis-1	303.15	818.60 ± 0.40	0.27	29-smy/sto[1)]
194.15	1001.0 ± 3.0		30-bil/fis-1	313.15	811.50 ± 0.40	0.21	29-smy/sto[1)]
	liquid			323.15	804.20 ± 0.50	0.08	29-smy/sto[1)]
263.15	846.10 ± 0.50	0.70	29-smy/sto(×)	333.15	797.00 ± 0.50	0.19	29-smy/sto[1)]
273.15	839.10 ± 0.50	0.33	29-smy/sto[1)]	273.15	838.46 ± 0.20	−0.31	32-ell/rei(×)
283.15	832.20 ± 0.50	0.14	29-smy/sto[1)]	298.15	821.35 ± 0.20	−0.45	32-ell/rei[1)]
293.15	825.30 ± 0.40	0.05	29-smy/sto[1)]	273.15	837.50 ± 0.50	−1.27	58-cos/bow[1)]
293.15	825.30 ± 0.60	0.05	29-smy/sto[1)]	293.15	822.70 ± 0.52	−2.55	58-cos/bow[1)]

[1)] Not included in Fig. 1.

cont.

Table 2. (cont.)

T/K	$\rho_{exp} \pm 2\sigma_{est}$ / $kg \cdot m^{-3}$	$\rho_{exp} - \rho_{calc}$ / $kg \cdot m^{-3}$	Ref. (Symbol in Fig. 1)	T/K	$\rho_{exp} \pm 2\sigma_{est}$ / $kg \cdot m^{-3}$	$\rho_{exp} - \rho_{calc}$ / $kg \cdot m^{-3}$	Ref. (Symbol in Fig. 1)
313.15	809.10 ± 0.54	−2.19	58-cos/bow[1]	323.07	804.15 ± 0.16	−0.03	73-fin(×)
333.15	794.80 ± 0.56	−2.01	58-cos/bow[1]	333.12	796.87 ± 0.16	0.04	73-fin(×)
353.15	780.40 ± 0.58	−1.34	58-cos/bow[1]	333.18	796.85 ± 0.16	0.06	73-fin(×)
373.15	764.30 ± 0.60	−1.71	58-cos/bow[1]	293.15	824.99 ± 0.25	−0.26	76-hal/ell[1]
393.15	747.70 ± 0.62	−1.83	58-cos/bow(×)	298.15	821.57 ± 0.25	−0.23	76-hal/ell[1]
413.15	730.50 ± 0.64	−1.73	58-cos/bow(×)	303.15	818.11 ± 0.25	−0.22	76-hal/ell[1]
433.15	713.10 ± 0.66	−0.92	58-cos/bow(×)	320.00	806.27 ± 0.25	−0.12	76-hal/ell[1]
453.15	695.10 ± 0.68	0.26	58-cos/bow(×)	340.00	791.73 ± 0.25	0.01	76-hal/ell[1]
473.15	675.60 ± 0.70	0.99	58-cos/bow(×)	360.00	776.47 ± 0.25	0.04	76-hal/ell(×)
493.15	655.30 ± 0.72	2.06	58-cos/bow(×)	380.00	760.46 ± 0.25	0.01	76-hal/ell(×)
513.15	635.10 ± 1.50	4.43	58-cos/bow[1]	400.00	743.64 ± 0.30	−0.06	76-hal/ell(×)
533.15	613.70 ± 1.50	6.81	58-cos/bow(×)	420.00	725.96 ± 0.30	−0.14	76-hal/ell(×)
553.15	590.00 ± 2.00	4.73	58-cos/bow(×)	440.00	707.32 ± 0.30	−0.25	76-hal/ell(×)
273.15	838.93 ± 0.10	0.16	59-mck/ski(Δ)	460.00	687.82 ± 0.30	−0.22	76-hal/ell(×)
273.15	843.00 ± 1.50	4.23	66-efr[1]	470.00	677.71 ± 0.30	−0.16	76-hal/ell(×)
293.15	827.00 ± 1.50	1.75	66-efr[1]	480.00	667.39 ± 0.35	−0.03	76-hal/ell(×)
313.15	812.00 ± 1.50	0.71	66-efr[1]	490.00	656.77 ± 0.35	0.08	76-hal/ell(×)
333.15	797.00 ± 1.50	0.19	66-efr[1]	293.15	825.23 ± 0.10	−0.02	78-jel/leo(◆)
353.15	782.00 ± 1.50	0.26	66-efr[1]	298.15	821.87 ± 0.10	0.07	78-jel/leo(◆)
373.15	767.00 ± 1.50	0.99	66-efr[1]	303.15	818.30 ± 0.10	−0.03	78-jel/leo(◆)
393.15	750.00 ± 1.50	0.47	66-efr[1]	298.15	822.60 ± 0.20	0.80	79-dia/tar[1]
413.15	733.00 ± 1.50	0.77	66-efr[1]	308.15	815.52 ± 0.20	0.70	79-dia/tar(×)
433.15	714.00 ± 1.50	−0.02	66-efr[1]	318.15	808.42 ± 0.20	0.70	79-dia/tar(×)
453.15	696.00 ± 1.50	1.16	66-efr[1]	333.15	797.58 ± 0.20	0.77	79-dia/tar[1]
473.15	677.00 ± 1.50	2.39	66-efr[1]	298.15	821.79 ± 0.02	−0.01	79-kiy/ben(□)
493.15	657.00 ± 1.50	3.76	66-efr[1]	298.15	820.85 ± 0.10	−0.95	84-sak/nak(∇)
513.15	637.00 ± 1.50	6.33	66-efr[1]	283.15	831.83 ± 0.20	−0.23	86-hei/sch(×)
533.15	615.00 ± 1.50	8.11	66-efr(×)	298.15	821.59 ± 0.20	−0.21	86-hei/sch[1]
553.15	592.00 ± 1.50	6.73	66-efr(×)	313.15	811.14 ± 0.20	−0.15	86-hei/sch(×)
573.15	565.00 ± 1.50	0.79	66-efr(×)	323.15	804.18 ± 0.10	0.06	93-gar/ban-1(○)
593.15	534.00 ± 1.50	−4.26	66-efr(×)	328.15	801.15 ± 0.10	0.67	93-gar/ban-1(○)
613.15	500.00 ± 2.00	−0.69	66-efr(×)	333.15	796.80 ± 0.10	−0.01	93-gar/ban-1(○)
633.15	444.00 ± 2.00	2.66	66-efr(×)	338.15	793.35 ± 0.10	0.25	93-gar/ban-1(○)
643.15	400.00 ± 3.00	4.51	66-efr(×)	343.15	789.36 ± 0.10	0.00	93-gar/ban-1(○)
258.08	849.12 ± 0.17	0.40	73-fin(×)	348.15	785.43 ± 0.10	−0.14	93-gar/ban-1(○)
258.33	848.86 ± 0.17	0.30	73-fin(×)	353.15	781.89 ± 0.10	0.15	93-gar/ban-1(○)
286.14	829.91 ± 0.17	−0.12	73-fin(×)	358.15	777.90 ± 0.10	0.02	93-gar/ban-1(○)
293.15	825.10 ± 0.17	−0.15	73-fin[1]	363.15	773.73 ± 0.10	−0.24	93-gar/ban-1(○)
303.15	818.18 ± 0.16	−0.15	73-fin[1]	373.15	765.97 ± 0.10	−0.04	93-gar/ban-1(○)
313.15	811.19 ± 0.16	−0.10	73-fin(×)				

[1]) Not included in Fig. 1.

Further references: [1869-zin, 1884-per, 1884-zan, 1886-gar, 1890-gar, 13-har, 14-low, 19-beh, 19-eyk, 27-ver/coo, 29-mah/das, 31-def, 32-kom/tal, 33-ano, 33-but/tho, 33-nev/jat, 35-but/ram, 35-saw, 37-oli, 41-dor/gla, 41-hus/age, 45-add, 48-jon/bow, 48-vog-2, 48-wei, 49-dre/mar, 49-tsv/mar, 50-mum/phi, 50-sac/sau, 52-coo, 52-eri-1, 53-par/cha, 56-goe/mcc, 56-goe/mcc-1, 62-bro/smi, 62-gei/fru, 62-par/mis, 64-blo/hag, 67-gol/per, 69-kat/pat, 73-min/rue, 75-mat/fer, 78-ast, 78-tre/ben, 80-sue/mul,] cont.

1-Octanol (cont.)

Further references (cont.)

[81-sjo/dyh, 82-aww/pet, 82-kat/wat, 82-ort, 82-ven/dha, 83-rau/ste, 85-fer/pin, 85-ort/paz-1, 85-sar/paz, 85-zhu, 86-ash/sri, 86-dew/meh, 86-wag/hei, 87-dew/meh, 88-ort/gar, 89-ala/sal, 89-dew/gup, 89-mat/mak-1, 89-vij/nai, 90-vij/nai, 91-ram/muk, 92-lie/sen-1, 93-yan/mae, 94-yu /tsa-1, 95-arc/bla, 95-cas/cal, 95-fra/jim, 95-fra/men, 95-sen/say, 96-elb].

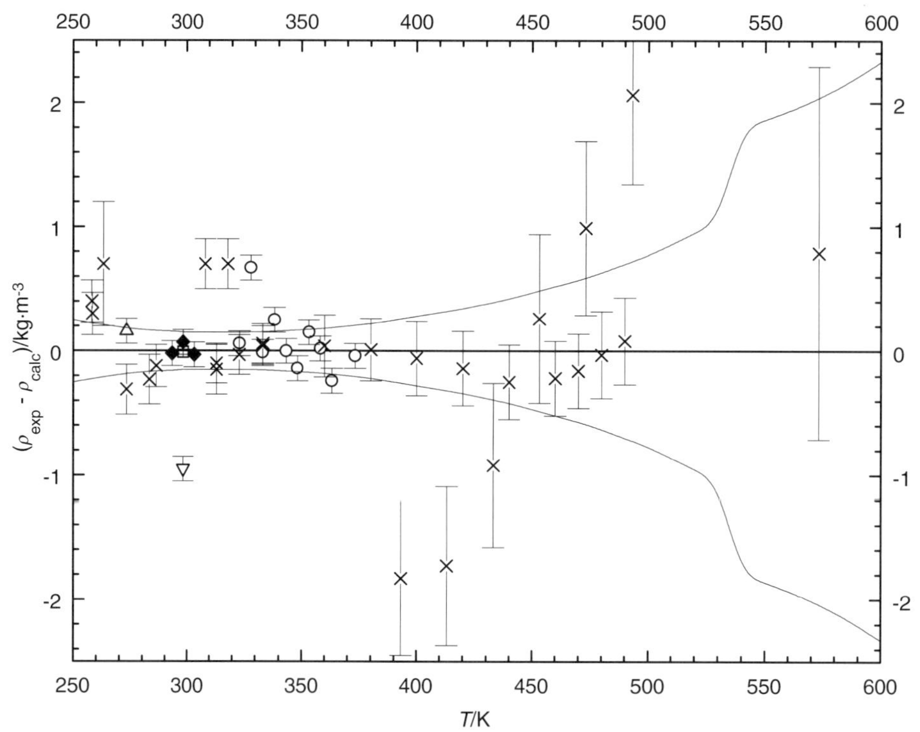

Fig. 1. The symbols show the deviation of the calculated from the experimental values from Table 2. The curves above and below the zero line indicate the calculated error region of the recommended values given in Table 3. The error bars represent the experimental errors. (Error bars smaller than the symbols are omitted for clarity of the figure.)

Table 3. Recommended values (fit to the reliable experimental values according to the equations
$\rho = A + BT + CT^2 + DT^3 + \ldots$ or $\rho = [1 + 1.75(1 - T/T_c)^{1/3} + 0.75(1 - T/T_c)][\rho_c + A(T_c - T) + B(T_c - T)^2 + C(T_c - T)^3 + D(T_c - T)^4]$).

$\frac{T}{K}$	$\frac{\rho \pm \sigma_{fit}}{kg \cdot m^{-3}}$	$\frac{T}{K}$	$\frac{\rho \pm \sigma_{fit}}{kg \cdot m^{-3}}$	$\frac{T}{K}$	$\frac{\rho \pm \sigma_{fit}}{kg \cdot m^{-3}}$
250.00	853.99 ± 0.25	298.15	821.80 ± 0.15	350.00	784.16 ± 0.17
260.00	847.47 ± 0.22	300.00	820.52 ± 0.15	360.00	776.43 ± 0.18
270.00	840.87 ± 0.19	310.00	813.52 ± 0.15	370.00	768.54 ± 0.20
280.00	834.18 ± 0.17	320.00	806.39 ± 0.15	380.00	760.45 ± 0.22
290.00	827.40 ± 0.16	330.00	799.13 ± 0.15	390.00	752.18 ± 0.25
293.15	825.25 ± 0.16	340.00	791.72 ± 0.16	400.00	743.70 ± 0.28

cont.

Table 3. (cont.)

T / K	$\rho \pm \sigma_{fit}$ / $kg \cdot m^{-3}$	T / K	$\rho \pm \sigma_{fit}$ / $kg \cdot m^{-3}$	T / K	$\rho \pm \sigma_{fit}$ / $kg \cdot m^{-3}$
410.00	735.01 ± 0.31	500.00	645.65 ± 0.77	590.00	542.93 ± 2.21
420.00	726.10 ± 0.34	510.00	634.31 ± 0.86	600.00	527.10 ± 2.33
430.00	716.95 ± 0.38	520.00	622.64 ± 0.95	610.00	507.71 ± 2.47
440.00	707.57 ± 0.42	530.00	610.65 ± 1.05	620.00	483.56 ± 2.64
450.00	697.93 ± 0.47	540.00	599.13 ± 1.80	630.00	452.84 ± 3.85
460.00	688.04 ± 0.52	550.00	588.51 ± 1.86	640.00	411.94 ± 4.10
470.00	677.87 ± 0.57	560.00	578.25 ± 1.93	650.00	344.17 ± 5.40
480.00	667.42 ± 0.63	570.00	567.70 ± 2.01		
490.00	656.69 ± 0.70	580.00	556.18 ± 2.10		

2-Octanol **[123-96-6]** **$C_8H_{18}O$** **MW = 130.23** **126**

Table 1. Coefficients of the polynomial expansion equation.
Standard deviations (see introduction):
$\sigma_{c,w} = 1.9507 \cdot 10^{-1}$ (combined temperature ranges, weighted),
$\sigma_{c,uw} = 1.1763 \cdot 10^{-1}$ (combined temperature ranges, unweighted).

Coefficient	T = 273.15 to 452.15 K $\rho = A + BT + CT^2 + DT^3 + \ldots$
A	$9.55354 \cdot 10^{2}$
B	$-1.73168 \cdot 10^{-1}$
C	$-9.75788 \cdot 10^{-4}$

Table 2. Experimental values with uncertainties and deviation from calculated values.

T / K	$\rho_{exp} \pm 2\sigma_{est}$ / $kg \cdot m^{-3}$	$\rho_{exp} - \rho_{calc}$ / $kg \cdot m^{-3}$	Ref. (Symbol in Fig. 1)	T / K	$\rho_{exp} \pm 2\sigma_{est}$ / $kg \cdot m^{-3}$	$\rho_{exp} - \rho_{calc}$ / $kg \cdot m^{-3}$	Ref. (Symbol in Fig. 1)
452.15	678.10 ± 1.00	0.53	1883-sch-3(×)	293.15	820.60 ± 0.50	−0.13	62-gei/fru(∇)
452.15	678.20 ± 1.00	0.63	1883-sch-3(×)	293.15	820.00 ± 0.40	−0.73	90-bar/paz(○)
273.15	835.30 ± 0.60	0.05	41-dor/gla(Δ)	298.15	817.10 ± 0.40	0.12	93-ami/rai(□)
298.15	817.00 ± 0.60	0.02	41-dor/gla(Δ)	303.15	813.00 ± 0.40	−0.18	93-ami/rai(□)
293.15	820.50 ± 0.50	−0.23	46-bra(×)	308.15	809.20 ± 0.40	−0.13	93-ami/rai(□)
293.15	820.80 ± 0.50	0.07	52-coo(◆)				

Further references: [1863-gla/dal, 1880-bru-1, 1884-per, 11-pic/ken, 19-beh, 19-eyk, 33-whi/her, 36-par, 37-zep, 42-mul, 45-add, 50-naz/fis, 53-ani-1, 57-tra/bat, 70-puz/bul].

cont.

2-Octanol (cont.)

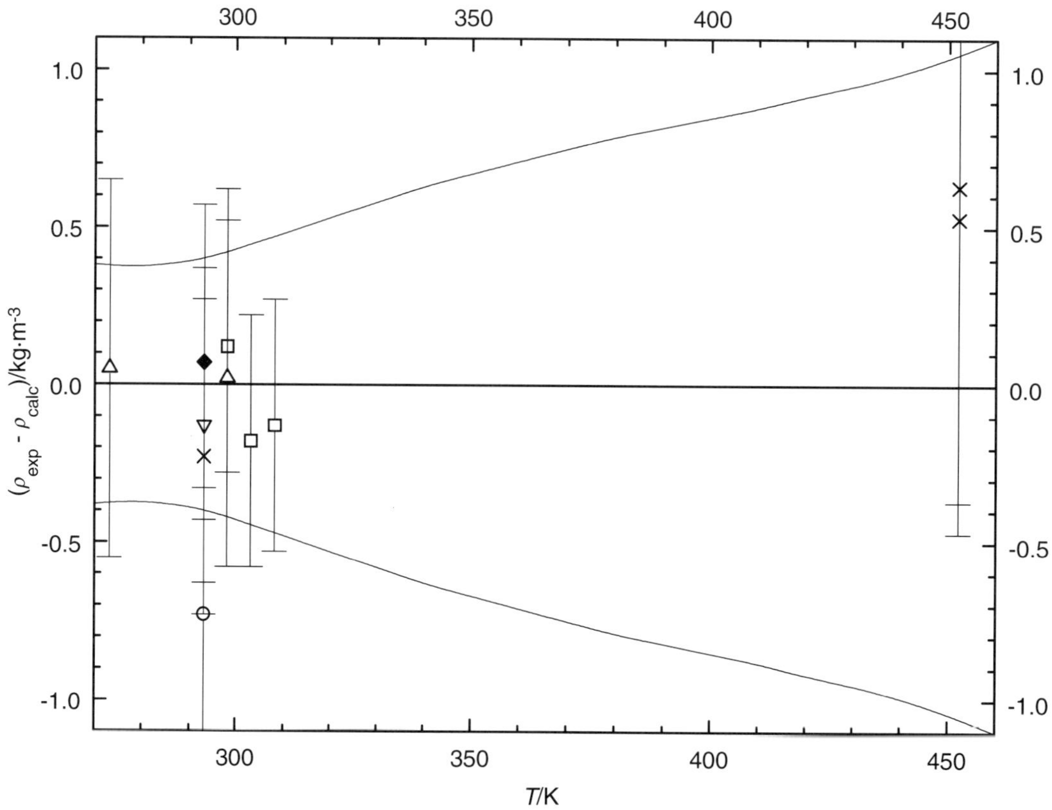

Fig. 1. The symbols show the deviation of the calculated from the experimental values from Table 2. The curves above and below the zero line indicate the calculated error region of the recommended values given in Table 3. The error bars represent the experimental errors. (Error bars smaller than the symbols are omitted for clarity of the figure.)

Table 3. Recommended values (fit to the reliable experimental values according to the equations $\rho = A + BT + CT^2 + DT^3 + \ldots$ or $\rho = [1 + 1.75(1 - T/T_c)^{1/3} + 0.75(1 - T/T_c)][\rho_c + A(T_c - T) + B(T_c - T)^2 + C(T_c - T)^3 + D(T_c - T)^4]$).

$\frac{T}{\text{K}}$	$\frac{\rho\pm\sigma_{\text{fit}}}{\text{kg}\cdot\text{m}^{-3}}$	$\frac{T}{\text{K}}$	$\frac{\rho\pm\sigma_{\text{fit}}}{\text{kg}\cdot\text{m}^{-3}}$	$\frac{T}{\text{K}}$	$\frac{\rho\pm\sigma_{\text{fit}}}{\text{kg}\cdot\text{m}^{-3}}$
270.00	837.46 ± 0.38	330.00	791.95 ± 0.58	410.00	720.33 ± 0.88
280.00	830.37 ± 0.37	340.00	783.68 ± 0.63	420.00	710.49 ± 0.92
290.00	823.07 ± 0.39	350.00	775.21 ± 0.67	430.00	700.47 ± 0.95
293.15	820.73 ± 0.40	360.00	766.55 ± 0.71	440.00	690.25 ± 0.99
298.15	816.98 ± 0.42	370.00	757.70 ± 0.75	450.00	679.83 ± 1.04
300.00	815.58 ± 0.43	380.00	748.65 ± 0.79	460.00	669.22 ± 1.10
310.00	807.90 ± 0.48	390.00	739.40 ± 0.82		
320.00	800.02 ± 0.53	400.00	729.96 ± 0.85		

(*R*)-(-)-2-Octanol **[5978-70-1]** **$C_8H_{18}O$** **MW = 130.23** **127**

Table 1. Experimental and recommended values with uncertainties.

$\frac{T}{K}$	$\frac{\rho_{exp} \pm 2\sigma_{est}}{kg \cdot m^{-3}}$	Ref.
298.15	815.2 ± 1.0	55-ber/sch
298.15	816.0 ± 2.0	56-goe/mcc
298.15	815.4 ± 1.0	Recommended

(*RS*)-2-Octanol **[4128-31-8]** **$C_8H_{18}O$** **MW = 130.23** **128**

Table 1. Fit with estimated *B* coefficient for 3 accepted points. Deviation $\sigma_w = 0.189$.

Coefficient	$\rho = A + BT$
A	1043.06
B	–0.760

Table 2. Experimental values with uncertainties and deviation from calculated values.

$\frac{T}{K}$	$\frac{\rho_{exp} \pm 2\sigma_{est}}{kg \cdot m^{-3}}$	$\frac{\rho_{exp} - \rho_{calc}}{kg \cdot m^{-3}}$	Ref.
293.15	820.2 ± 1.0	–0.07	39-cop/gos
293.15	820.2 ± 1.0	–0.07	47-kor/lic
298.15	817.0 ± 2.0	0.53	56-goe/mcc

Table 3. Recommended values.

$\frac{T}{K}$	$\frac{\rho_{exp} \pm 2\sigma_{est}}{kg \cdot m^{-3}}$
290.00	822.7 ± 1.3
293.15	820.3 ± 1.2
298.15	816.5 ± 1.3

(*S*)-(+)-2-Octanol **[6169-06-8]** **$C_8H_{18}O$** **MW = 130.23** **129**

Table 1. Coefficients of the polynomial expansion equation.
Standard deviations (see introduction):
$\sigma_{c,w} = 4.6275 \cdot 10^{-1}$ (combined temperature ranges, weighted),
$\sigma_{c,uw} = 1.8230 \cdot 10^{-1}$ (combined temperature ranges, unweighted).

Coefficient	T = 273.15 to 355.65 K $\rho = A + BT + CT^2 + DT^3 + \ldots$
A	$1.03721 \cdot 10^{3}$
B	$-7.40112 \cdot 10^{-1}$

cont.

(*S*)-(+)-2-Octanol (cont.)

Table 2. Experimental values with uncertainties and deviation from calculated values.

$\frac{T}{K}$	$\frac{\rho_{exp} \pm 2\sigma_{est}}{kg \cdot m^{-3}}$	$\frac{\rho_{exp} - \rho_{calc}}{kg \cdot m^{-3}}$	Ref. (Symbol in Fig. 1)	$\frac{T}{K}$	$\frac{\rho_{exp} \pm 2\sigma_{est}}{kg \cdot m^{-3}}$	$\frac{\rho_{exp} - \rho_{calc}}{kg \cdot m^{-3}}$	Ref. (Symbol in Fig. 1)
290.15	822.90 ± 2.00	0.43	07-pic/ken(×)	320.45	799.70 ± 1.00	−0.34	37-pat/hol(○)
293.15	822.10 ± 2.00	1.85	07-pic/ken(×)	333.45	790.20 ± 1.00	−0.22	37-pat/hol(○)
273.15	835.03 ± 2.00	−0.02	32-ell/rei(◆)	346.75	780.30 ± 1.00	−0.28	37-pat/hol(○)
298.15	823.01 ± 2.00	6.47	32-ell/rei[1]	355.65	773.90 ± 1.00	−0.09	37-pat/hol(○)
273.15	834.10 ± 1.00	−0.95	37-pat/hol(○)	293.15	820.10 ± 1.00	−0.15	39-cop/gos(Δ)
290.95	821.60 ± 1.00	−0.27	37-pat/hol(○)	293.15	820.10 ± 1.00	−0.15	47-kor/lic(∇)
305.65	810.70 ± 1.00	−0.29	37-pat/hol(○)	298.15	817.00 ± 1.00	0.46	56-goe/mcc(□)

[1] Not included in Fig. 1.

Further references: [12-pic/ken].

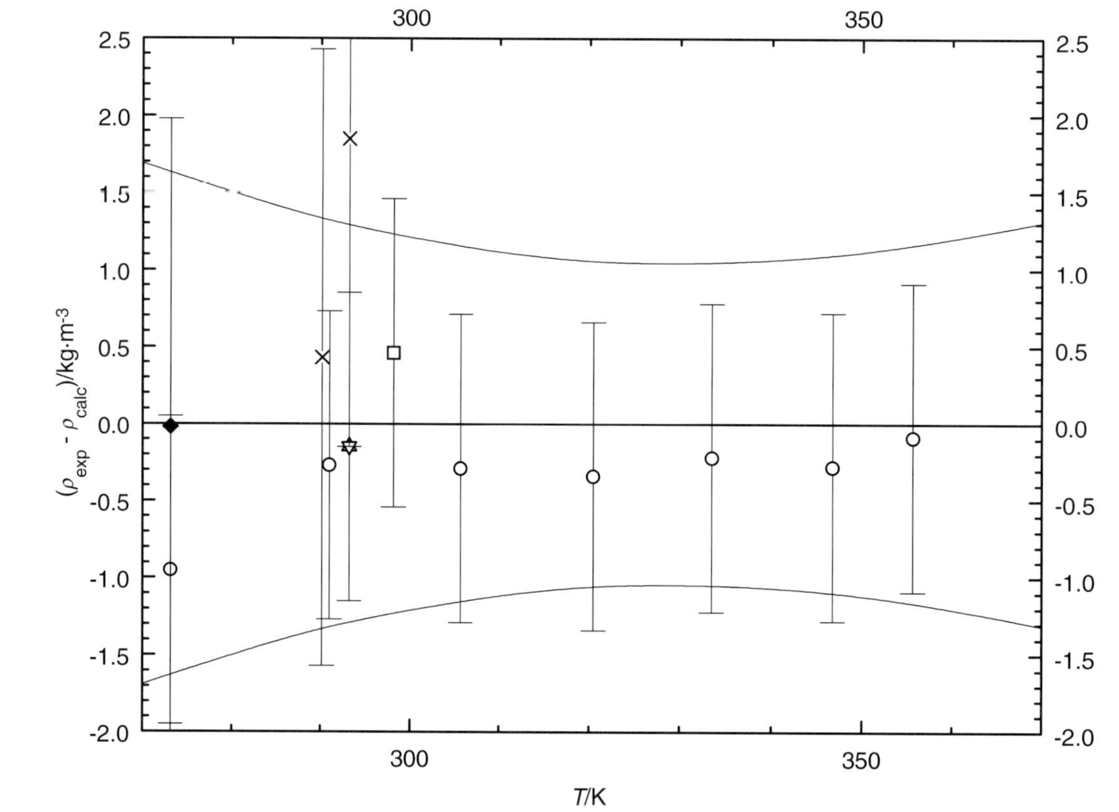

Fig. 1. The symbols show the deviation of the calculated from the experimental values from Table 2. The curves above and below the zero line indicate the calculated error region of the recommended values given in Table 3. The error bars represent the experimental errors. (Error bars smaller than the symbols are omitted for clarity of the figure.)

cont.

Table 3. Recommended values (fit to the reliable experimental values according to the equations $\rho = A + BT + CT^2 + DT^3 + \ldots$ or $\rho = [1 + 1.75(1 - T/T_c)^{1/3} + 0.75(1 - T/T_c)][\rho_c + A(T_c - T) + B(T_c - T)^2 + C(T_c - T)^3 + D(T_c - T)^4]$).

$\frac{T}{K}$	$\frac{\rho \pm \sigma_{fit}}{kg \cdot m^{-3}}$	$\frac{T}{K}$	$\frac{\rho \pm \sigma_{fit}}{kg \cdot m^{-3}}$	$\frac{T}{K}$	$\frac{\rho \pm \sigma_{fit}}{kg \cdot m^{-3}}$
270.00	837.38 ± 1.69	300.00	815.18 ± 1.21	350.00	778.17 ± 1.11
280.00	829.98 ± 1.50	310.00	807.77 ± 1.11	360.00	770.77 ± 1.20
290.00	822.58 ± 1.33	320.00	800.37 ± 1.05	370.00	763.37 ± 1.31
293.15	820.25 ± 1.29	330.00	792.97 ± 1.04		
298.15	816.54 ± 1.23	340.00	785.57 ± 1.06		

3-Octanol **[589-98-0]** **$C_8H_{18}O$** **MW = 130.23** **130**

Table 1. Coefficients of the polynomial expansion equation.
Standard deviations (see introduction):
$\sigma_{c,w} = 5.3499 \cdot 10^{-1}$ (combined temperature ranges, weighted),
$\sigma_{c,uw} = 4.6865 \cdot 10^{-1}$ (combined temperature ranges, unweighted).

Coefficient	T = 273.15 to 373.01 K $\rho = A + BT + CT^2 + DT^3 + \ldots$
A	$9.61167 \cdot 10^{2}$
B	$-1.85752 \cdot 10^{-1}$
C	$-9.98691 \cdot 10^{-4}$

Table 2. Experimental values with uncertainties and deviation from calculated values.

$\frac{T}{K}$	$\frac{\rho_{exp} \pm 2\sigma_{est}}{kg \cdot m^{-3}}$	$\frac{\rho_{exp} - \rho_{calc}}{kg \cdot m^{-3}}$	Ref. (Symbol in Fig. 1)	$\frac{T}{K}$	$\frac{\rho_{exp} \pm 2\sigma_{est}}{kg \cdot m^{-3}}$	$\frac{\rho_{exp} - \rho_{calc}}{kg \cdot m^{-3}}$	Ref. (Symbol in Fig. 1)
293.15	824.70 ± 3.00	3.81	13-pic/ken(∇)	333.15	787.53 ± 1.00	−0.91	30-bin/dar(○)
353.15	771.10 ± 3.00	0.08	13-pic/ken(∇)	353.15	770.12 ± 1.00	−0.90	30-bin/dar(○)
273.32	835.49 ± 1.00	−0.30	30-bin/dar(○)	373.01	751.09 ± 1.00	−1.84	30-bin/dar(○)
283.15	827.47 ± 1.00	−1.03	30-bin/dar(○)	273.15	836.10 ± 0.60	0.18	41-dor/gla(□)
293.15	820.88 ± 1.00	−0.01	30-bin/dar(○)	298.15	816.00 ± 0.60	−1.01	41-dor/gla(□)
303.15	812.61 ± 1.00	−0.47	30-bin/dar(○)	293.15	823.90 ± 2.00	3.01	69-nav/tul(Δ)
313.15	804.44 ± 1.00	−0.62	30-bin/dar(○)				

Table 3. Recommended values (fit to the reliable experimental values according to the equations $\rho = A + BT + CT^2 + DT^3 + \ldots$ or $\rho = [1 + 1.75(1 - T/T_c)^{1/3} + 0.75(1 - T/T_c)][\rho_c + A(T_c - T) + B(T_c - T)^2 + C(T_c - T)^3 + D(T_c - T)^4]$).

$\frac{T}{K}$	$\frac{\rho \pm \sigma_{fit}}{kg \cdot m^{-3}}$	$\frac{T}{K}$	$\frac{\rho \pm \sigma_{fit}}{kg \cdot m^{-3}}$	$\frac{T}{K}$	$\frac{\rho \pm \sigma_{fit}}{kg \cdot m^{-3}}$
270.00	838.21 ± 1.95	300.00	815.56 ± 1.72	350.00	773.81 ± 1.55
280.00	830.86 ± 1.90	310.00	807.61 ± 1.38	360.00	764.87 ± 1.57
290.00	823.31 ± 1.82	320.00	799.46 ± 1.49	370.00	755.72 ± 1.59
293.15	820.89 ± 1.74	330.00	791.11 ± 1.55	380.00	746.37 ± 1.67
298.15	817.01 ± 1.73	340.00	782.56 ± 1.55		

cont.

3-Octanol (cont.)

Further references: [56-woo/vio, 62-gei/fru].

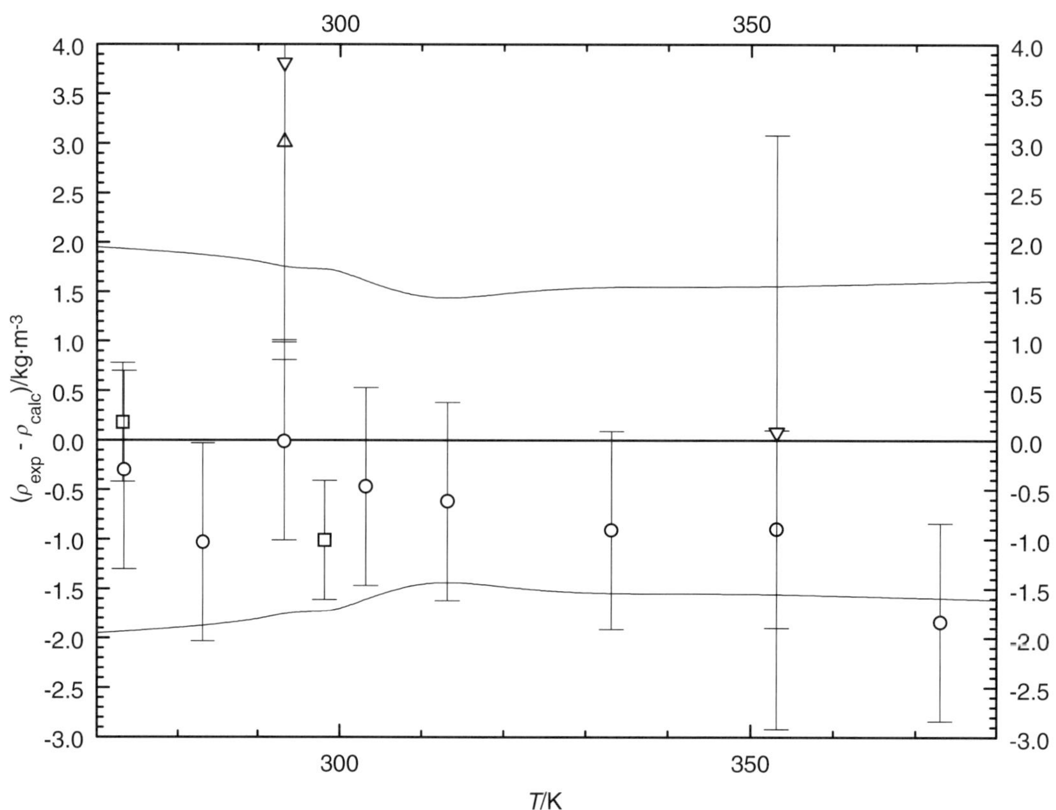

Fig. 1. The symbols show the deviation of the calculated from the experimental values from Table 2. The curves above and below the zero line indicate the calculated error region of the recommended values given in Table 3. The error bars represent the experimental errors. (Error bars smaller than the symbols are omitted for clarity of the figure.)

(*RS*)-3-Octanol **[20296-29-1]** **$C_8H_{18}O$** **MW = 130.23** **131**

Table 1. Experimental value with uncertainty.

$\frac{T}{\mathrm{K}}$	$\frac{\rho_{\mathrm{exp}} \pm 2\sigma_{\mathrm{est}}}{\mathrm{kg \cdot m^{-3}}}$	Ref.
295.35	817.7 ± 0.8	33-bri

(*S*)-(+)-3-Octanol **[22658-92-0]** **$C_8H_{18}O$** **MW = 130.23** **132**

Table 1. Experimental value with uncertainty.

$\frac{T}{\mathrm{K}}$	$\frac{\rho_{\mathrm{exp}} \pm 2\sigma_{\mathrm{est}}}{\mathrm{kg \cdot m^{-3}}}$	Ref.
293.15	834.4 ± 1.0	57-shu/bel

4-Octanol **[589-62-8]** **$C_8H_{18}O$** **MW = 130.23** **133**

Table 1. Coefficients of the polynomial expansion equation.
Standard deviations (see introduction):
$\sigma_{c,w} = 2.5809 \cdot 10^{-1}$ (combined temperature ranges, weighted),
$\sigma_{c,uw} = 1.7589 \cdot 10^{-1}$ (combined temperature ranges, unweighted).

Coefficient	$T = 273.15$ to 373.32 K $\rho = A + BT + CT^2 + DT^3 + \ldots$
A	$9.65289 \cdot 10^{2}$
B	$-2.26785 \cdot 10^{-1}$
C	$-9.21832 \cdot 10^{-4}$

Table 2. Experimental values with uncertainties and deviation from calculated values.

$\frac{T}{\text{K}}$	$\frac{\rho_{exp} \pm 2\sigma_{est}}{\text{kg} \cdot \text{m}^{-3}}$	$\frac{\rho_{exp} - \rho_{calc}}{\text{kg} \cdot \text{m}^{-3}}$	Ref. (Symbol in Fig. 1)	$\frac{T}{\text{K}}$	$\frac{\rho_{exp} \pm 2\sigma_{est}}{\text{kg} \cdot \text{m}^{-3}}$	$\frac{\rho_{exp} - \rho_{calc}}{\text{kg} \cdot \text{m}^{-3}}$	Ref. (Symbol in Fig. 1)
273.29	834.86 ± 1.00	0.40	30-bin/dar(○)	353.15	770.59 ± 1.00	0.36	30-bin/dar(○)
283.15	827.27 ± 1.00	0.10	30-bin/dar(○)	373.32	752.16 ± 1.00	0.01	30-bin/dar(○)
293.15	819.94 ± 1.00	0.35	30-bin/dar(○)	295.15	818.00 ± 2.00	−0.05	31-lev/mar-2(Δ)
303.15	811.89 ± 1.00	0.07	30-bin/dar(○)	293.15	817.80 ± 2.00	−1.79	36-tuo(∇)
313.15	804.05 ± 1.00	0.18	30-bin/dar(○)	273.15	834.60 ± 0.60	0.04	41-dor/gla(□)
333.15	787.59 ± 1.00	0.17	30-bin/dar(○)	298.15	815.90 ± 0.60	0.17	41-dor/gla(□)

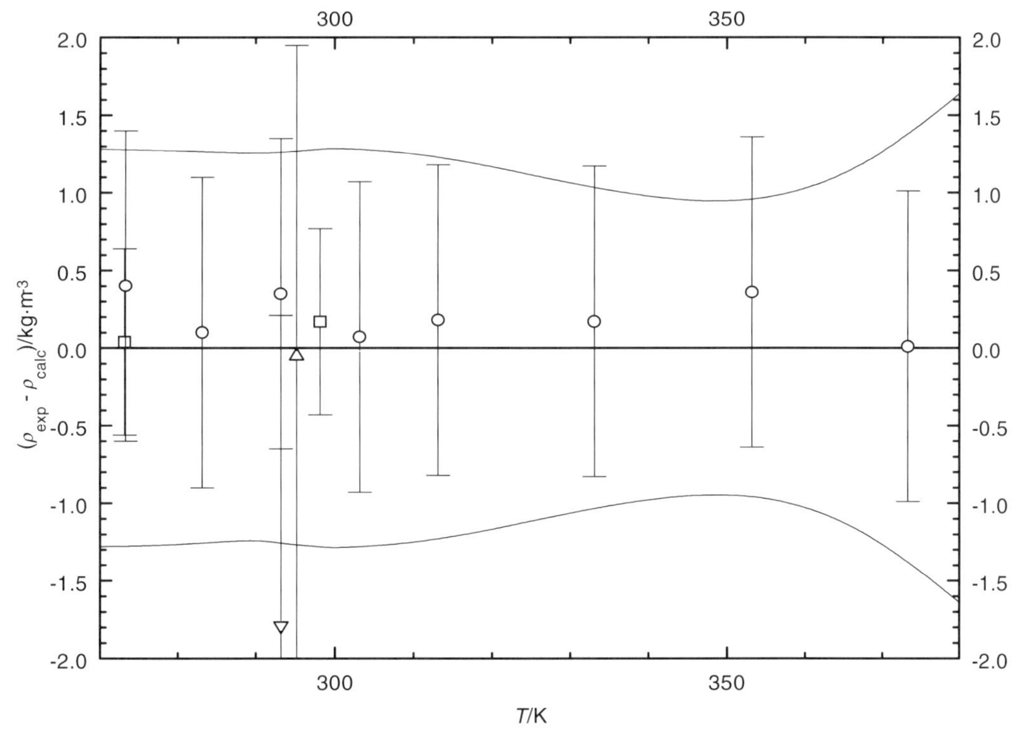

Fig. 1. The symbols show the deviation of the calculated from the experimental values from Table 2. The curves above and below the zero line indicate the calculated error region of the recommended values given in Table 3. The error bars represent the experimental errors. (Error bars smaller than the symbols are omitted for clarity of the figure.)

cont.

4-Octanol (cont.)

Further references: [06-bou/loc, 50-naz/fis, 62-gei/fru].

Table 3. Recommended values (fit to the reliable experimental values according to the equations $\rho = A + BT + CT^2 + DT^3 + \ldots$ or $\rho = [1 + 1.75(1 - T/T_c)^{1/3} + 0.75(1 - T/T_c)][\rho_c + A(T_c - T) + B(T_c - T)^2 + C(T_c - T)^3 + D(T_c - T)^4])$.

T/K	$\rho \pm \sigma_{fit}$ / $kg \cdot m^{-3}$	T/K	$\rho \pm \sigma_{fit}$ / $kg \cdot m^{-3}$	T/K	$\rho \pm \sigma_{fit}$ / $kg \cdot m^{-3}$
270.00	836.86 ± 1.28	300.00	814.29 ± 1.29	350.00	772.99 ± 0.93
280.00	829.52 ± 1.27	310.00	806.40 ± 1.26	360.00	764.18 ± 1.00
290.00	822.00 ± 1.25	320.00	798.32 ± 1.17	370.00	755.18 ± 1.24
293.15	819.59 ± 1.26	330.00	790.06 ± 1.06	380.00	746.00 ± 1.64
298.15	815.73 ± 1.28	340.00	781.62 ± 0.97		

2-Propyl-1-pentanol **[58175-57-8]** **$C_8H_{18}O$** **MW = 130.23** **134**

Table 1. Experimental value with uncertainty.

T/K	$\rho_{exp} \pm 2\sigma_{est}$ / $kg \cdot m^{-3}$	Ref.
294.15	900.0 ± 10.0	58-wie/thu

2,2,3-Trimethyl-3-pentanol **[7294-05-5]** **$C_8H_{18}O$** **MW = 130.23** **135**

Table 1. Fit with estimated B coefficient for 5 accepted points. Deviation $\sigma_w = 0.823$.

Coefficient	$\rho = A + BT$
A	1075.47
B	−0.780

Table 2. Experimental values with uncertainties and deviation from calculated values.

T/K	$\rho_{exp} \pm 2\sigma_{est}$ / $kg \cdot m^{-3}$	$\rho_{exp} - \rho_{calc}$ / $kg \cdot m^{-3}$	Ref.	T/K	$\rho_{exp} \pm 2\sigma_{est}$ / $kg \cdot m^{-3}$	$\rho_{exp} - \rho_{calc}$ / $kg \cdot m^{-3}$	Ref.
293.15	849.0 ± 2.00	2.19	33-whi/lau	298.15	842.0 ± 1.00	−0.91	39-gin/col
298.15	842.3 ± 1.00	−0.61	36-nor/has	293.15	831.0 ± 10.00	−15.81	57-pet/sus [1)]
298.15	843.4 ± 1.00	0.49	36-nor/has	293.15	847.3 ± 1.00	0.49	62-gei/fru

[1)] Not included in calculation of linear coefficients.

Table 3. Recommended values.

T/K	$\rho_{exp} \pm 2\sigma_{est}$ / $kg \cdot m^{-3}$
290.00	849.3 ± 1.4
293.15	846.8 ± 1.3
298.15	842.9 ± 1.2

2,2,4-Trimethyl-1-pentanol **[123-44-4]** **$C_8H_{18}O$** **MW = 130.23** **136**

Table 1. Experimental and recommended values with uncertainties.

$\frac{T}{K}$	$\frac{\rho_{exp} \pm 2\sigma_{est}}{kg \cdot m^{-3}}$	Ref.
293.15	823.2 ± 10.0	25-ter[1)]
293.15	838.4 ± 1.0	58-ano-3
293.15	838.4 ± 1.0	Recommended

[1)] Not included in calculation of recommended value.

2,2,4-Trimethyl-3-pentanol **[5162-48-1]** **$C_8H_{18}O$** **MW = 130.23** **137**

Table 1. Experimental values with uncertainties.

$\frac{T}{K}$	$\frac{\rho_{exp} \pm 2\sigma_{est}}{kg \cdot m^{-3}}$	Ref.
293.15	832.2 ± 1.0	51-smi/cre
293.15	832.6 ± 1.0	62-gei/fru
293.15	832.4 ± 1.0	Recommended

2,3,3-Trimethyl-2-pentanol **[23171-85-9]** **$C_8H_{18}O$** **MW = 130.23** **138**

Table 1. Experimental values with uncertainties.

$\frac{T}{K}$	$\frac{\rho_{exp} \pm 2\sigma_{est}}{kg \cdot m^{-3}}$	Ref.
293.15	861.0 ± 2.0	33-whi/lau
298.15	815.1 ± 5.0	36-nor/has
293.15	851.7 ± 2.0	41-hus/gui

2,3,4-Trimethyl-1-pentanol **[6570-88-3]** **$C_8H_{18}O$** **MW = 130.23** **139**

Table 1. Experimental value with uncertainty.

$\frac{T}{K}$	$\frac{\rho_{exp} \pm 2\sigma_{est}}{kg \cdot m^{-3}}$	Ref.
293.15	849.8 ± 1.0	58-per/can

2,3,4-Trimethyl-2-pentanol **[66576-26-9]** **$C_8H_{18}O$** **MW = 130.23** **140**

Table 1. Experimental values with uncertainties.

$\frac{T}{K}$	$\frac{\rho_{exp} \pm 2\sigma_{est}}{kg \cdot m^{-3}}$	Ref.
293.15	808.1 ± 2.0	39-hus/gui
293.15	808.0 ± 2.0	41-hus/gui
293.15	843.2 ± 2.0	48-hus/kra

2,3,4-Trimethyl-3-pentanol **[3054-92-0]** **$C_8H_{18}O$** **MW = 130.23** **141**

Table 1. Experimental value with uncertainty.

$\frac{T}{K}$	$\frac{\rho_{exp} \pm 2\sigma_{est}}{kg \cdot m^{-3}}$	Ref.
293.15	849.2 ± 2.0	26-sta

2,4,4-Trimethyl-1-pentanol **[16325-63-6]** **$C_8H_{18}O$** **MW = 130.23** **142**

Table 1. Experimental value with uncertainty.

$\frac{T}{K}$	$\frac{\rho_{exp} \pm 2\sigma_{est}}{kg \cdot m^{-3}}$	Ref.
293.15	833.0 ± 2.0	40-sut

2,4,4-Trimethyl-2-pentanol **[690-37-9]** **$C_8H_{18}O$** **MW = 130.23** **143**

Table 1. Experimental and recommended values with uncertainties.

$\frac{T}{K}$	$\frac{\rho_{exp} \pm 2\sigma_{est}}{kg \cdot m^{-3}}$	Ref.
296.65	830.9 ± 6.0	48-rit[1)]
293.15	822.5 ± 2.0	41-hus/gui
293.15	825.0 ± 2.0	41-whi/wil
293.15	823.7 ± 2.2	Recommended

[1)] Not included in calculation of recommended value.

3,3,4-Trimethyl-2-pentanol **[19411-41-7]** **$C_8H_{18}O$** **MW = 130.23** **144**

Table 1. Experimental and recommended values with uncertainties.

$\frac{T}{K}$	$\frac{\rho_{exp} \pm 2\sigma_{est}}{kg \cdot m^{-3}}$	Ref.
293.15	812.8 ± 4.0	08-bue[1)]
293.15	855.7 ± 1.0	55-mes/pet
293.15	855.7 ± 1.0	Recommended

[1)] Not included in calculation of recommended value.

3,4,4-Trimethyl-2-pentanol **[10575-56-1]** **$C_8H_{18}O$** **MW = 130.23** **145**

Table 1. Experimental value with uncertainty.

$\frac{T}{K}$	$\frac{\rho_{exp} \pm 2\sigma_{est}}{kg \cdot m^{-3}}$	Ref.
293.15	840.8 ± 1.0	41-whe

2.1.5 Alkanols, C_9

2,2-Dimethyl-3-ethyl-3-pentanol **[66793-96-2]** **$C_9H_{20}O$** **MW = 144.26** **146**

Table 1. Fit with estimated *B* coefficient for 2 accepted points. Deviation $\sigma_w = 0.250$.

Coefficient	$\rho = A + BT$
A	1097.33
B	−0.820

Table 2. Experimental values with uncertainties and deviation from calculated values.

$\frac{T}{K}$	$\frac{\rho_{exp} \pm 2\sigma_{est}}{kg \cdot m^{-3}}$	$\frac{\rho_{exp} - \rho_{calc}}{kg \cdot m^{-3}}$	Ref.
293.15	852.4 ± 3.0	−4.55	38-whi/mey[1)]
293.15	857.2 ± 1.0	0.25	47-how/mea
298.15	852.6 ± 1.0	−0.25	47-how/mea

[1)] Not included in calculation of linear coefficients.

Table 3. Recommended values.

$\frac{T}{K}$	$\frac{\rho_{exp} \pm 2\sigma_{est}}{kg \cdot m^{-3}}$
290.00	859.5 ± 0.9
293.15	857.0 ± 0.7
298.15	852.9 ± 0.7

2,4-Dimethyl-2-ethyl-1-pentanol **[66793-98-4]** **$C_9H_{20}O$** **MW = 144.26** **147**

Table 1. Experimental value with uncertainty.

$\frac{T}{K}$	$\frac{\rho_{exp} \pm 2\sigma_{est}}{kg \cdot m^{-3}}$	Ref.
298.15	837.0 ± 1.5	52-doe/far

2,4-Dimethyl-3-ethyl-3-pentanol **[3970-59-0]** **$C_9H_{20}O$** **MW = 144.26** **148**

Table 1. Fit with estimated *B* coefficient for 4 accepted points. Deviation $\sigma_w = 0.542$.

Coefficient	$\rho = A + BT$
A	1081.51
B	−0.760

cont.

2,4-Dimethyl-3-ethyl-3-pentanol (cont.)

Table 2. Experimental values with uncertainties and deviation from calculated values.

$\frac{T}{K}$	$\frac{\rho_{exp} \pm 2\sigma_{est}}{kg \cdot m^{-3}}$	$\frac{\rho_{exp} - \rho_{calc}}{kg \cdot m^{-3}}$	Ref.	$\frac{T}{K}$	$\frac{\rho_{exp} \pm 2\sigma_{est}}{kg \cdot m^{-3}}$	$\frac{\rho_{exp} - \rho_{calc}}{kg \cdot m^{-3}}$	Ref.
273.15	877.3 ± 2.0	3.38	26-sta[1)]	293.15	858.8 ± 0.6	0.08	47-how/mea
293.15	860.8 ± 2.0	2.08	26-sta[1)]	298.15	854.3 ± 0.6	−0.62	47-how/mea
303.15	853.2 ± 2.0	2.08	26-sta[1)]	293.15	858.8 ± 0.6	0.08	51-smi/cre
293.15	860.0 ± 1.0	1.28	43-geo				

[1)] Not included in calculation of linear coefficients.

Table 3. Recommended values.

$\frac{T}{K}$	$\frac{\rho_{exp} \pm 2\sigma_{est}}{kg \cdot m^{-3}}$
290.00	861.1 ± 0.8
293.15	858.7 ± 0.7
298.15	854.9 ± 0.7

2,2-Dimethyl-3-heptanol **[19549-70-3]** **$C_9H_{20}O$** **MW = 144.26** **149**

Table 1. Fit with estimated *B* coefficient for 2 accepted points. Deviation σ_w = 0.020.

Coefficient	$\rho = A + BT$
A	1050.33
B	−0.760

Table 2. Experimental values with uncertainties and deviation from calculated values.

$\frac{T}{K}$	$\frac{\rho_{exp} \pm 2\sigma_{est}}{kg \cdot m^{-3}}$	$\frac{\rho_{exp} - \rho_{calc}}{kg \cdot m^{-3}}$	Ref.
295.15	826.0 ± 1.0	−0.02	49-col/lag
299.15	823.0 ± 1.0	0.02	59-fol/wel

Table 3. Recommended values.

$\frac{T}{K}$	$\frac{\rho_{exp} \pm 2\sigma_{est}}{kg \cdot m^{-3}}$
290.00	829.9 ± 1.1
293.15	827.5 ± 1.0
298.15	823.7 ± 0.9

2,3-Dimethyl-2-heptanol **[66794-00-1]** **$C_9H_{20}O$** **MW = 144.26** **150**

Table 1. Experimental value with uncertainty.

$\frac{T}{K}$	$\frac{\rho_{exp} \pm 2\sigma_{est}}{kg \cdot m^{-3}}$	Ref.
293.15	839.6 ± 1.0	48-naz/tor

2,3-Dimethyl-3-heptanol **[19549-71-4]** **$C_9H_{20}O$** **MW = 144.26** **151**

Table 1. Experimental values with uncertainties.

$\frac{T}{K}$	$\frac{\rho_{exp} \pm 2\sigma_{est}}{kg \cdot m^{-3}}$	Ref.
293.15	839.5 ± 2.0	33-whi/eve
294.15	838.3 ± 2.0	33-whi/eve
293.15	834.9 ± 3.0	50-naz/bak

2,4-Dimethyl-2-heptanol **[65822-93-7]** **$C_9H_{20}O$** **MW = 144.26** **152**

Table 1. Experimental value with uncertainty.

$\frac{T}{K}$	$\frac{\rho_{exp} \pm 2\sigma_{est}}{kg \cdot m^{-3}}$	Ref.
298.15	828.0 ± 1.0	59-col/gau

2,4-Dimethyl-4-heptanol **[19549-77-0]** **$C_9H_{20}O$** **MW = 144.26** **153**

Table 1. Experimental and recommended values with uncertainties.

$\frac{T}{K}$	$\frac{\rho_{exp} \pm 2\sigma_{est}}{kg \cdot m^{-3}}$	Ref.	$\frac{T}{K}$	$\frac{\rho_{exp} \pm 2\sigma_{est}}{kg \cdot m^{-3}}$	Ref.
293.15	826.0 ± 1.0	09-bod/tab	293.15	824.2 ± 1.0	59-pet/zak
293.15	821.5 ± 3.0	33-mey/tuo[1)]	293.15	824.2 ± 1.0	59-pet/zak-1
293.15	825.4 ± 1.0	42-hen/all	293.15	824.8 ± 1.0	Recommended
293.15	824.2 ± 1.0	54-naz/kak-3			

[1)] Not included in calculation of recommended value.

2,5-Dimethyl-2-heptanol **[1561-18-8]** **$C_9H_{20}O$** **MW = 144.26** **154**

Table 1. Experimental value with uncertainty.

$\frac{T}{K}$	$\frac{\rho_{exp} \pm 2\sigma_{est}}{kg \cdot m^{-3}}$	Ref.
295.15	830.0 ± 2.0	31-lev/mar-2

2,6-Dimethyl-2-heptanol **[13254-34-7]** **$C_9H_{20}O$** **MW = 144.26** **155**

Table 1. Experimental values with uncertainties.

$\frac{T}{K}$	$\frac{\rho_{exp} \pm 2\sigma_{est}}{kg \cdot m^{-3}}$	Ref.
293.15	818.6 ± 2.0	26-pas/zam
283.65	816.2 ± 2.0	28-esc

2,6-Dimethyl-3-heptanol **[19549-73-6]** **$C_9H_{20}O$** **MW = 144.26** **156**

Table 1. Experimental and recommended values with uncertainties.

$\frac{T}{K}$	$\frac{\rho_{exp} \pm 2\sigma_{est}}{kg \cdot m^{-3}}$	Ref.
293.15	821.2 ± 4.0	12-mic[1)]
293.15	814.8 ± 2.0	36-tuo
293.15	814.8 ± 2.0	Recommended

[1)] Not included in calculation of recommended value.

2,6-Dimethyl-4-heptanol **[108-82-7]** **$C_9H_{20}O$** **MW =144.26** **157**

Table 1. Fit with estimated *B* coefficient for 8 accepted points. Deviation σ_w = 0.427.

Coefficient	$\rho = A + BT$
A	1033.02
B	–0.760

Table 2. Experimental values with uncertainties and deviation from calculated values.

$\frac{T}{K}$	$\frac{\rho_{exp} \pm 2\sigma_{est}}{kg \cdot m^{-3}}$	$\frac{\rho_{exp} - \rho_{calc}}{kg \cdot m^{-3}}$	Ref.	$\frac{T}{K}$	$\frac{\rho_{exp} \pm 2\sigma_{est}}{kg \cdot m^{-3}}$	$\frac{\rho_{exp} - \rho_{calc}}{kg \cdot m^{-3}}$	Ref.
294.15	809.0 ± 1.0	–0.47	14-vav	333.15	779.9 ± 0.4	0.05	47-str/gab
293.15	811.4 ± 1.0	1.17	36-tuo	293.15	810.6 ± 1.0	0.37	53-ano-5
293.15	809.7 ± 0.4	–0.56	47-str/gab	293.15	810.7 ± 1.0	0.47	58-ano-5
313.15	795.2 ± 0.4	0.17	47-str/gab	293.15	810.7 ± 1.0	0.47	68-ano

Table 3. Recommended values.

$\frac{T}{K}$	$\frac{\rho_{exp} \pm 2\sigma_{est}}{kg \cdot m^{-3}}$	$\frac{T}{K}$	$\frac{\rho_{exp} \pm 2\sigma_{est}}{kg \cdot m^{-3}}$	$\frac{T}{K}$	$\frac{\rho_{exp} \pm 2\sigma_{est}}{kg \cdot m^{-3}}$
290.00	812.6 ± 1.3	310.00	797.4 ± 0.8	330.00	782.2 ± 1.3
293.15	810.2 ± 1.1	320.00	789.8 ± 1.0	340.00	774.6 ± 1.8
298.15	806.4 ± 1.0				

3,5-Dimethyl-3-heptanol **[19549-74-7]** **$C_9H_{20}O$** **MW = 144.26** **158**

Table 1. Fit with estimated *B* coefficient for 2 accepted points. Deviation σ_w = 1.930.

Coefficient	$\rho = A + BT$
A	1066.31
B	–0.820

Table 2. Experimental values with uncertainties and deviation from calculated values.

$\frac{T}{K}$	$\frac{\rho_{exp} \pm 2\sigma_{est}}{kg \cdot m^{-3}}$	$\frac{\rho_{exp} - \rho_{calc}}{kg \cdot m^{-3}}$	Ref.
298.15	817.7 ± 3.0	–4.13	30-dav/dix[1)]
297.15	834.0 ± 8.0	11.35	48-bro/bro[1)]
301.15	831.0 ± 8.0	11.63	48-bro/bro[1)]
293.15	824.0 ± 2.0	–1.93	59-mac/bar
301.15	821.3 ± 2.0	1.93	60-tha/vas

[1)] Not included in calculation of linear coefficients.

Table 3. Recommended values.

$\frac{T}{K}$	$\frac{\rho_{exp} \pm 2\sigma_{est}}{kg \cdot m^{-3}}$
290.00	828.5 ± 2.1
293.15	825.9 ± 2.1
298.15	821.8 ± 2.1
310.00	812.1 ± 2.2

3,5-Dimethyl-4-heptanol **[19549-79-2]** **$C_9H_{20}O$** **MW = 144.26** **159**

Table 1. Experimental and recommended values with uncertainties.

$\frac{T}{K}$	$\frac{\rho_{exp} \pm 2\sigma_{est}}{kg \cdot m^{-3}}$	Ref.
291.15	836.0 ± 10.0	23-vav/iva[1)]
293.15	859.2 ± 1.0	52-lev/shu
293.15	859.2 ± 1.0	Recommended

[1)] Not included in calculation of recommended value.

3,6-Dimethyl-1-heptanol **[1573-33-7]** **$C_9H_{20}O$** **MW = 144.26** **160**

Table 1. Experimental value with uncertainty.

$\frac{T}{K}$	$\frac{\rho_{exp} \pm 2\sigma_{est}}{kg \cdot m^{-3}}$	Ref.
300.15	823.0 ± 2.0	32-lev/mar

3,6-Dimethyl-3-heptanol **[1573-28-0]** **$C_9H_{20}O$** **MW = 144.26** **161**

Table 1. Experimental value with uncertainty.

$\frac{T}{K}$	$\frac{\rho_{exp} \pm 2\sigma_{est}}{kg \cdot m^{-3}}$	Ref.
289.15	828.5 ± 2.0	04-kon

4,6-Dimethyl-2-heptanol **[51079-52-8]** **$C_9H_{20}O$** **MW = 144.26** **162**

Table 1. Experimental and recommended values with uncertainties.

$\frac{T}{K}$	$\frac{\rho_{exp} \pm 2\sigma_{est}}{kg \cdot m^{-3}}$	Ref.
273.15	878.7 ± 2.0	09-gue[1)]
273.15	880.1 ± 2.0	12-gue[1)]
298.15	822.0 ± 1.5	64-hin/dre
298.15	822.0 ± 1.5	Recommended

[1)] Not included in calculation of recommended value.

5,6-Dimethyl-2-heptanol **[58795-24-7]** **$C_9H_{20}O$** **MW = 144.26** **163**

Table 1. Experimental value with uncertainty.

$\frac{T}{K}$	$\frac{\rho_{exp} \pm 2\sigma_{est}}{kg \cdot m^{-3}}$	Ref.
291.15	835.0 ± 2.0	11-wal

6,6-Dimethyl-1-heptanol **[65769-10-0]** $C_9H_{20}O$ **MW = 144.26** **164**

Table 1. Experimental value with uncertainty.

T / K	$\rho_{exp} \pm 2\sigma_{est}$ / $kg \cdot m^{-3}$	Ref.
293.15	843.9 ± 1.0	56-gol/kon

3-Ethyl-1-heptanol **[3525-25-5]** $C_9H_{20}O$ **MW = 144.26** **165**

Table 1. Experimental value with uncertainty.

T / K	$\rho_{exp} \pm 2\sigma_{est}$ / $kg \cdot m^{-3}$	Ref.
296.15	834.0 ± 2.0	31-lev/mar-3

3-Ethyl-3-heptanol **[19780-41-7]** $C_9H_{20}O$ **MW = 144.26** **166**

Table 1. Coefficients of the polynomial expansion equation.
Standard deviations (see introduction):
$\sigma_{c,w} = 3.4886 \cdot 10^{-1}$ (combined temperature ranges, weighted),
$\sigma_{c,uw} = 1.4242 \cdot 10^{-1}$ (combined temperature ranges, unweighted).

Coefficient	T = 273.15 to 338.15 K $\rho = A + BT + CT^2 + DT^3 + \ldots$
A	$1.07793 \cdot 10^{3}$
B	$-8.33559 \cdot 10^{-1}$

Table 2. Experimental values with uncertainties and deviation from calculated values.

T / K	$\rho_{exp} \pm 2\sigma_{est}$ / $kg \cdot m^{-3}$	$\rho_{exp} - \rho_{calc}$ / $kg \cdot m^{-3}$	Ref. (Symbol in Fig. 1)	T / K	$\rho_{exp} \pm 2\sigma_{est}$ / $kg \cdot m^{-3}$	$\rho_{exp} - \rho_{calc}$ / $kg \cdot m^{-3}$	Ref. (Symbol in Fig. 1)
293.15	832.90 ± 1.00	−0.67	38-whi/ore(□)	318.15	812.60 ± 1.00	−0.14	39-owe/qua(O)
273.15	850.50 ± 1.00	0.25	39-owe/qua(O)	328.15	804.30 ± 1.00	−0.10	39-owe/qua(O)
298.15	829.90 ± 1.00	0.49	39-owe/qua(O)	338.15	796.30 ± 1.00	0.24	39-owe/qua(O)
308.15	821.00 ± 1.00	−0.07	39-owe/qua(O)				

Further references: [29-con/bla, 33-whi/woo].

Table 3. Recommended values (fit to the reliable experimental values according to the equations $\rho = A + BT + CT^2 + DT^3 + \ldots$ or $\rho = [1 + 1.75(1 - T/T_c)^{1/3} + 0.75(1 - T/T_c)][\rho_c + A(T_c - T) + B(T_c - T)^2 + C(T_c - T)^3 + D(T_c - T)^4])$.

T / K	$\rho \pm \sigma_{fit}$ / $kg \cdot m^{-3}$	T / K	$\rho \pm \sigma_{fit}$ / $kg \cdot m^{-3}$	T / K	$\rho \pm \sigma_{fit}$ / $kg \cdot m^{-3}$
270.00	852.87 ± 1.22	298.15	829.41 ± 1.02	330.00	802.86 ± 1.08
280.00	844.54 ± 1.13	300.00	827.86 ± 1.01	340.00	794.52 ± 1.16
290.00	836.20 ± 1.05	310.00	819.53 ± 1.00	350.00	786.19 ± 1.26
293.15	833.57 ± 1.04	320.00	811.19 ± 1.02		

cont.

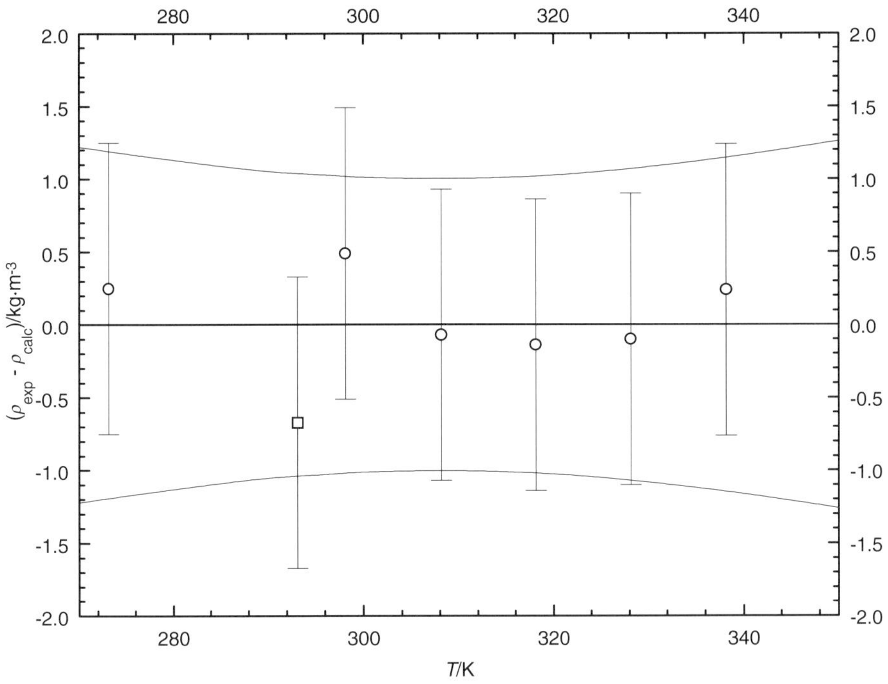

Fig. 1. The symbols show the deviation of the calculated from the experimental values from Table 2. The curves above and below the zero line indicate the calculated error region of the recommended values given in Table 3. The error bars represent the experimental errors. (Error bars smaller than the symbols are omitted for clarity of the figure.)

5-Ethyl-1-heptanol **[998-65-2]** **$C_9H_{20}O$** **MW = 144.26** **167**

Table 1. Experimental value with uncertainty.

$\frac{T}{K}$	$\frac{\rho_{exp} \pm 2\sigma_{est}}{kg \cdot m^{-3}}$	Ref.
298.15	848.0 ± 1.0	62-col/des

2-Ethyl-3-methyl-1-hexanol **[66794-04-5]** **$C_9H_{20}O$** **MW = 144.26** **168**

Table 1. Experimental value with uncertainty.

$\frac{T}{K}$	$\frac{\rho_{exp} \pm 2\sigma_{est}}{kg \cdot m^{-3}}$	Ref.
298.15	835.8 ± 1.0	32-con/adk

2-Ethyl-4-methyl-1-hexanol **[66794-06-7]** **$C_9H_{20}O$** **MW = 144.26** **169**

Table 1. Experimental value with uncertainty.

$\frac{T}{K}$	$\frac{\rho_{exp} \pm 2\sigma_{est}}{kg \cdot m^{-3}}$	Ref.
293.15	828.8 ± 1.0	58-hag/hud

2-Ethyl-5-methyl-1-hexanol **[66794-07-8]** **$C_9H_{20}O$** **MW = 144.26** **170**

Table 1. Experimental value with uncertainty.

$\frac{T}{K}$	$\frac{\rho_{exp} \pm 2\sigma_{est}}{kg \cdot m^{-3}}$	Ref.
298.15	820.8 ± 1.0	32-con/adk

3-Ethyl-2-methyl-1-hexanol **[66794-01-2]** **$C_9H_{20}O$** **MW = 144.26** **171**

Table 1. Experimental and recommended values with uncertainties.

$\frac{T}{K}$	$\frac{\rho_{exp} \pm 2\sigma_{est}}{kg \cdot m^{-3}}$	Ref.
295.15	848.0 ± 2.0	54-naz/kak-4[1)]
293.15	851.2 ± 1.0	54-naz/kak-4
293.15	851.2 ± 1.0	Recommended

[1)] Not included in calculation of recommended value.

3-Ethyl-2-methyl-2-hexanol **[66794-02-3]** **$C_9H_{20}O$** **MW = 144.26** **172**

Table 1. Experimental value with uncertainty.

$\frac{T}{K}$	$\frac{\rho_{exp} \pm 2\sigma_{est}}{kg \cdot m^{-3}}$	Ref.
298.15	833.4 ± 1.0	54-ski/flo

3-Ethyl-2-methyl-3-hexanol **[66794-03-4]** **$C_9H_{20}O$** **MW = 144.26** **173**

Table 1. Experimental and recommended values with uncertainties.

$\frac{T}{K}$	$\frac{\rho_{exp} \pm 2\sigma_{est}}{kg \cdot m^{-3}}$	Ref.
293.15	850.6 ± 2.0	26-sta[1)]
295.15	845.2 ± 2.0	54-naz/kak-4[1)]
298.15	844.5 ± 1.0	54-ski/flo
298.15	844.5 ± 1.0	Recommended

[1)] Not included in calculation of recommended value.

3-Ethyl-4-methyl-3-hexanol **[51200-80-7]** **$C_9H_{20}O$** **MW = 144.26** **174**

Table 1. Experimental value with uncertainty.

$\frac{T}{K}$	$\frac{\rho_{exp} \pm 2\sigma_{est}}{kg \cdot m^{-3}}$	Ref.
288.15	861.0 ± 1.0	44-pre/zal

3-Ethyl-5-methyl-3-hexanol **[597-77-3]** **$C_9H_{20}O$** **MW = 144.26** **175**

Table 1. Experimental value with uncertainty.

$\frac{T}{K}$	$\frac{\rho_{exp} \pm 2\sigma_{est}}{kg \cdot m^{-3}}$	Ref.
295.15	839.6 ± 2.0	14-hal

4-Ethyl-2-methyl-3-hexanol **[33943-21-4]** **$C_9H_{20}O$** **MW = 144.26** **176**

Table 1. Experimental value with uncertainty.

$\frac{T}{K}$	$\frac{\rho_{exp} \pm 2\sigma_{est}}{kg \cdot m^{-3}}$	Ref.
293.15	827.5 ± 1.0	43-geo

4-Ethyl-3-methyl-3-hexanol **[66794-05-6]** **$C_9H_{20}O$** **MW = 144.26** **177**

Table 1. Experimental value with uncertainty.

$\frac{T}{K}$	$\frac{\rho_{exp} \pm 2\sigma_{est}}{kg \cdot m^{-3}}$	Ref.
298.15	899.4 ± 1.0	54-ski/flo

1-Ethyl-1-propyl-1-butanol **[597-90-0]** **$C_9H_{20}O$** **MW = 144.26** **178**

Table 1. Fit with estimated B coefficient for 4 accepted points. Deviation $\sigma_w = 0.464$.

Coefficient	$\rho = A + BT$
A	1074.08
B	−0.820

Table 2. Experimental values with uncertainties and deviation from calculated values.

$\frac{T}{K}$	$\frac{\rho_{exp} \pm 2\sigma_{est}}{kg \cdot m^{-3}}$	$\frac{\rho_{exp} - \rho_{calc}}{kg \cdot m^{-3}}$	Ref.	$\frac{T}{K}$	$\frac{\rho_{exp} \pm 2\sigma_{est}}{kg \cdot m^{-3}}$	$\frac{\rho_{exp} - \rho_{calc}}{kg \cdot m^{-3}}$	Ref.
289.15	838.4 ± 2.0	1.42	19-eyk[1)]	298.15	829.9 ± 1.0	0.30	39-owe/qua
293.15	833.7 ± 1.5	0.00	26-sta	328.15	804.3 ± 1.0	−0.70	39-owe/qua
273.15	850.5 ± 1.0	0.40	39-owe/qua	293.15	836.4 ± 2.0	2.70	54-naz/kak-3[1)]

[1)] Not included in calculation of linear coefficients.

Table 3. Recommended values.

$\frac{T}{K}$	$\frac{\rho_{exp} \pm 2\sigma_{est}}{kg \cdot m^{-3}}$	$\frac{T}{K}$	$\frac{\rho_{exp} \pm 2\sigma_{est}}{kg \cdot m^{-3}}$	$\frac{T}{K}$	$\frac{\rho_{exp} \pm 2\sigma_{est}}{kg \cdot m^{-3}}$
270.00	852.7 ± 1.9	293.15	833.7 ± 0.9	320.00	811.7 ± 1.5
280.00	844.5 ± 1.4	298.15	829.6 ± 0.8	330.00	803.5 ± 2.0
290.00	836.3 ± 1.0	310.00	819.9 ± 1.1		

2-Methyl-1-octanol **[818-81-5]** **$C_9H_{20}O$** **MW = 144.26** **179**

Table 1. Experimental value with uncertainty.

T/K	$\rho_{exp} \pm 2\sigma_{est}$/kg·m^{-3}	Ref.
277.15	841.8 ± 1.0	04-bou/bla-2

2-Methyl-2-octanol **[628-44-4]** **$C_9H_{20}O$** **MW = 144.26** **180**

Table 1. Coefficients of the polynomial expansion equation. Standard deviations (see introduction):
$\sigma_{c,w}$ = 1.2926 (combined temperature ranges, weighted),
$\sigma_{c,uw}$ = 8.1386 · 10^{-1} (combined temperature ranges, unweighted).

Coefficient	T = 273.15 to 338.15 K $\rho = A + BT + CT^2 + DT^3 + \ldots$
A	1.05740 · 10^{3}
B	−8.13336 · 10^{-1}

Table 2. Experimental values with uncertainties and deviation from calculated values.

T/K	$\rho_{exp} \pm 2\sigma_{est}$/kg·m^{-3}	$\rho_{exp} - \rho_{calc}$/kg·m^{-3}	Ref. (Symbol in Fig. 1)	T/K	$\rho_{exp} \pm 2\sigma_{est}$/kg·m^{-3}	$\rho_{exp} - \rho_{calc}$/kg·m^{-3}	Ref. (Symbol in Fig. 1)
293.15	821.00 ± 2.00	2.02	38-whi/ore(○)	318.15	797.70 ± 1.00	−0.94	39-owe/qua(□)
273.15	833.60 ± 1.00	−1.64	39-owe/qua(□)	328.15	789.70 ± 1.00	−0.81	39-owe/qua(□)
298.15	813.40 ± 1.00	1.51	39-owe/qua(□)	338.15	781.60 ± 1.00	0.77	39-owe/qua(□)
308.15	805.50 ± 1.00	−1.27	39-owe/qua(□)	293.15	823.90 ± 4.00	4.92	56-tar/tai(Δ)

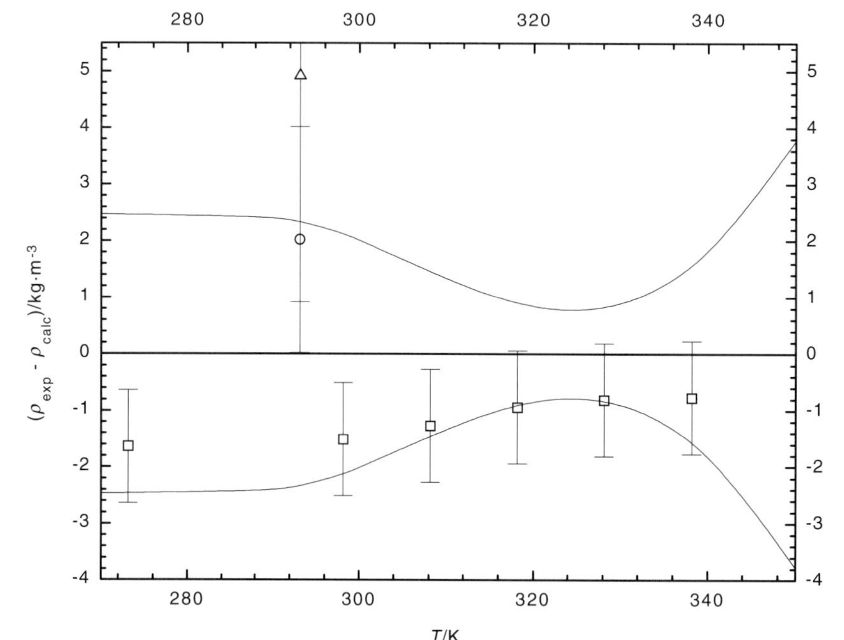

Fig. 1. The symbols show the deviation of the calculated from the experimental values from Table 2. The curves above and below the zero line indicate the calculated error region of the recommended values given in Table 3. The error bars represent the experimental errors. (Error bars smaller than the symbols are omitted for clarity of the figure.)

cont.

Further references: [59-her].

Table 3. Recommended values (fit to the reliable experimental values according to the equations $\rho = A + BT + CT^2 + DT^3 + \dots$ or $\rho = [1 + 1.75(1 - T/T_c)^{1/3} + 0.75(1 - T/T_c)][\rho_c + A(T_c - T) + B(T_c - T)^2 + C(T_c - T)^3 + D(T_c - T)^4]$).

T/K	$\rho \pm \sigma_{fit}$/kg·m^{-3}	T/K	$\rho \pm \sigma_{fit}$/kg·m^{-3}	T/K	$\rho \pm \sigma_{fit}$/kg·m^{-3}
270.00	837.80 ± 2.47	298.15	814.91 ± 2.13	330.00	789.00 ± 0.76
280.00	829.67 ± 2.45	300.00	813.40 ± 2.02	340.00	780.87 ± 1.60
290.00	821.54 ± 2.42	310.00	805.27 ± 1.32	350.00	772.74 ± 3.78
293.15	818.98 ± 2.35	320.00	797.14 ± 0.75		

2-Methyl-3-octanol **[26533-34-6]** **$C_9H_{20}O$** **MW = 144.26** **181**

Table 1. Experimental and recommended values with uncertainties.

T/K	$\rho_{exp} \pm 2\sigma_{est}$/kg·m^{-3}	Ref.
293.15	827.0 ± 3.0	12-pic/ken[1)]
293.15	833.1 ± 2.0	31-lev/mar-4
293.15	833.1 ± 2.0	Recommended

[1)] Not included in calculation of recommended value.

2-Methyl-4-octanol **[40575-41-5]** **$C_9H_{20}O$** **MW = 144.26** **182**

Table 1. Experimental and recommended values with uncertainties.

T/K	$\rho_{exp} \pm 2\sigma_{est}$/kg·m^{-3}	Ref.
293.15	815.0 ± 3.0	06-mal-1[1)]
293.15	813.0 ± 4.0	36-tuo[1)]
298.15	815.0 ± 2.0	54-dub/luf
298.15	815.0 ± 2.0	Recommended

[1)] Not included in calculation of recommended value.

3-Methyl-1-octanol **[38514-02-2]** **$C_9H_{20}O$** **MW = 144.26** **183**

Table 1. Experimental values with uncertainties.

T/K	$\rho_{exp} \pm 2\sigma_{est}$/kg·m^{-3}	Ref.
297.15	827.0 ± 2.0	31-lev/mar-1
297.15	827.0 ± 2.0	31-lev/mar-5

3-Methyl-2-octanol **[27644-49-1]** **$C_9H_{20}O$** **MW = 144.26** **184**

Table 1. Experimental value with uncertainty.

T / K	$\rho_{exp} \pm 2\sigma_{est}$ / $kg \cdot m^{-3}$	Ref.
300.15	831.0 ± 2.0	33-pow/mur

3-Methyl-3-octanol **[5340-36-3]** **$C_9H_{20}O$** **MW = 144.26** **185**

Table 1. Fit with estimated B coefficient for 6 accepted points. Deviation σ_w = 0.605.

Coefficient	$\rho = A + BT$
A	1072.01
B	−0.820

Table 2. Experimental values with uncertainties and deviation from calculated values.

T / K	$\rho_{exp} \pm 2\sigma_{est}$ / $kg \cdot m^{-3}$	$\rho_{exp} - \rho_{calc}$ / $kg \cdot m^{-3}$	Ref.	T / K	$\rho_{exp} \pm 2\sigma_{est}$ / $kg \cdot m^{-3}$	$\rho_{exp} - \rho_{calc}$ / $kg \cdot m^{-3}$	Ref.
298.15	825.8 ± 2.0	−1.73	33-whi/wil	328.15	802.9 ± 1.0	−0.03	44-qua/sma
293.15	821.6 ± 8.0	−10.03	34-gre[1)]	293.15	833.1 ± 2.0	1.47	54-pom/foo-1
298.15	827.5 ± 1.0	−0.03	44-qua/sma	293.15	832.2 ± 2.0	0.57	61-sok/she
318.15	811.1 ± 1.0	−0.03	44-qua/sma				

[1)] Not included in calculation of linear coefficients.

Table 3. Recommended values.

T / K	$\rho_{exp} \pm 2\sigma_{est}$ / $kg \cdot m^{-3}$	T / K	$\rho_{exp} \pm 2\sigma_{est}$ / $kg \cdot m^{-3}$	T / K	$\rho_{exp} \pm 2\sigma_{est}$ / $kg \cdot m^{-3}$
290.00	834.2 ± 2.5	298.15	827.5 ± 1.8	320.00	809.6 ± 1.6
293.15	831.6 ± 2.2	310.00	817.8 ± 1.3	330.00	801.4 ± 2.3

3-Methyl-4-octanol **[26533-35-7]** **$C_9H_{20}O$** **MW = 144.26** **186**

Table 1. Experimental value with uncertainty.

T / K	$\rho_{exp} \pm 2\sigma_{est}$ / $kg \cdot m^{-3}$	Ref.
291.15	834.0 ± 2.0	23-vav/iva

4-Methyl-1-octanol **[38514-03-3]** **$C_9H_{20}O$** **MW = 144.26** **187**

Table 1. Experimental value with uncertainty.

T / K	$\rho_{exp} \pm 2\sigma_{est}$ / $kg \cdot m^{-3}$	Ref.
300.65	820.0 ± 2.0	31-lev/mar-5

4-Methyl-3-octanol **[66793-80-4]** **$C_9H_{20}O$** **MW = 144.26** **188**

Table 1. Experimental value with uncertainty.

T / K	$\rho_{exp} \pm 2\sigma_{est}$ / $kg \cdot m^{-3}$	Ref.
298.15	843.7 ± 2.0	34-gre

4-Methyl-4-octanol **[23418-37-3]** $C_9H_{20}O$ **MW = 144.26** **189**

Table 1. Fit with estimated B coefficient for 4 accepted points. Deviation $\sigma_w = 0.328$.

Coefficient	$\rho = A + BT$
A	1069.04
B	−0.820

Table 2. Experimental values with uncertainties and deviation from calculated values.

T/K	$\rho_{exp} \pm 2\sigma_{est}$ / $kg \cdot m^{-3}$	$\rho_{exp} - \rho_{calc}$ / $kg \cdot m^{-3}$	Ref.	T/K	$\rho_{exp} \pm 2\sigma_{est}$ / $kg \cdot m^{-3}$	$\rho_{exp} - \rho_{calc}$ / $kg \cdot m^{-3}$	Ref.
293.15	826.7 ± 2.0	−1.96	33-whi/woo[1)]	328.15	799.9 ± 1.0	−0.06	44-qua/sma
298.15	823.7 ± 2.0	−0.86	33-whi/woo[1)]	293.15	828.4 ± 0.6	−0.26	54-pom/foo-1
298.15	825.2 ± 1.0	0.64	44-qua/sma	293.15	822.7 ± 4.0	−5.96	59-yur/bel-1[1)]
318.15	808.3 ± 1.0	0.14	44-qua/sma				

[1)] Not included in calculation of linear coefficients.

Table 3. Recommended values.

T/K	$\rho_{exp} \pm 2\sigma_{est}$ / $kg \cdot m^{-3}$	T/K	$\rho_{exp} \pm 2\sigma_{est}$ / $kg \cdot m^{-3}$	T/K	$\rho_{exp} \pm 2\sigma_{est}$ / $kg \cdot m^{-3}$
290.00	831.2 ± 0.9	298.15	824.6 ± 0.6	320.00	806.6 ± 1.0
293.15	828.7 ± 0.8	310.00	814.8 ± 0.6	330.00	798.4 ± 1.4

5-Methyl-1-octanol **[38514-04-4]** $C_9H_{20}O$ **MW = 144.26** **190**

Table 1. Experimental value with uncertainty.

T/K	$\rho_{exp} \pm 2\sigma_{est}$ / $kg \cdot m^{-3}$	Ref.
297.15	828.0 ± 2.0	33-lev/mar-1

5-Methyl-2-octanol **[66793-81-5]** $C_9H_{20}O$ **MW = 144.26** **191**

Table 1. Experimental value with uncertainty.

T/K	$\rho_{exp} \pm 2\sigma_{est}$ / $kg \cdot m^{-3}$	Ref.
298.15	821.0 ± 2.0	31-lev/mar-4

5-Methyl-4-octanol **[59734-23-5]** $C_9H_{20}O$ **MW = 144.26** **192**

Table 1. Experimental value with uncertainty.

T/K	$\rho_{exp} \pm 2\sigma_{est}$ / $kg \cdot m^{-3}$	Ref.
298.15	815.6 ± 0.8	12-bje

(*S*)-6-Methyl-1-octanol **[110453-78-6]** **$C_9H_{20}O$** **MW = 144.26** **193**

Table 1. Experimental value with uncertainty.

$\frac{T}{K}$	$\frac{\rho_{exp} \pm 2\sigma_{est}}{kg \cdot m^{-3}}$	Ref.
297.15	828.0 ± 2.0	33-lev/mar-1

6-Methyl-3-octanol **[40225-75-0]** **$C_9H_{20}O$** **MW = 144.26** **194**

Table 1. Experimental value with uncertainty.

$\frac{T}{K}$	$\frac{\rho_{exp} \pm 2\sigma_{est}}{kg \cdot m^{-3}}$	Ref.
301.15	832.0 ± 1.0	36 -pow/bal -0

6-Methyl-4-octanol **[66793-82-6]** **$C_9H_{20}O$** **MW = 144.26** **195**

Table 1. Experimental value with uncertainty.

$\frac{T}{K}$	$\frac{\rho_{exp} \pm 2\sigma_{est}}{kg \cdot m^{-3}}$	Ref.
296.15	822.0 ± 2.0	31-lev/mar-4

7-Methyl-1-octanol **[2430-22-0]** **$C_9H_{20}O$** **MW = 144.26** **196**

Table 1. Experimental value with uncertainty.

$\frac{T}{K}$	$\frac{\rho_{exp} \pm 2\sigma_{est}}{kg \cdot m^{-3}}$	Ref.
298.15	826.0 ± 2.0	16-lev/all

7-Methyl-3-octanol **[66793-84-8]** **$C_9H_{20}O$** **MW = 144.26** **197**

Table 1. Experimental value with uncertainty.

$\frac{T}{K}$	$\frac{\rho_{exp} \pm 2\sigma_{est}}{kg \cdot m^{-3}}$	Ref.
285.15	840.2 ± 2.0	25-tho/kah

7-Methyl-4-octanol **[33933-77-6]** **$C_9H_{20}O$** **MW = 144.26** **198**

Table 1. Experimental value with uncertainty.

$\frac{T}{K}$	$\frac{\rho_{exp} \pm 2\sigma_{est}}{kg \cdot m^{-3}}$	Ref.
293.15	813.6 ± 1.0	36-tuo

4-Methyl-2-propyl-1-pentanol **[54004-41-0]** **$C_9H_{20}O$** **MW = 144.26** **199**

Table 1. Experimental value with uncertainty.

$\frac{T}{K}$	$\frac{\rho_{exp} \pm 2\sigma_{est}}{kg \cdot m^{-3}}$	Ref.
293.15	825.6 ± 1.0	58-hag/hud

1-Nonanol **[143-08-8]** **$C_9H_{20}O$** **MW = 144.26** **200**

T_c = 670.50 K [89-tej/lee] ρ_c = 264.00 kg·m^{-3} [89-tej/lee]

Table 1. Coefficients for the polynomial expansion equations. Standard deviations (see introduction): $\sigma_t = 9.8705 \cdot 10^{-1}$ (low temperature range), $\sigma_{c,w} = (5.4570 \cdot 10^{-1}$ combined temperature ranges, weighted), $\sigma_{c,uw} = 4.0529 \cdot 10^{-1}$ (combined temperature ranges, unweighted).

Coefficient	T = 273.15 to 540.00 K $\rho = A + BT + CT^2 + DT^3 + \ldots$	T = 540.00 to 670.50 K $\rho = [1 + 1.75(1 - T/T_c)^{1/3} + 0.75(1 - T/T_c)]$ $[\rho_c + A(T_c - T) + B(T_c - T)^2 + C(T_c - T)^3 + D(T_c - T)^4]$
A	$9.49744 \cdot 10^{2}$	2.41155
B	$-7.07347 \cdot 10^{-2}$	$-5.85206 \cdot 10^{-2}$
C	$-1.43841 \cdot 10^{-3}$	$5.31779 \cdot 10^{-4}$
D	$8.99651 \cdot 10^{-7}$	$-1.62241 \cdot 10^{-6}$

Table 2. Experimental values with uncertainties and deviation from calculated values.

T / K	$\rho_{exp} \pm 2\sigma_{est}$ / kg·m^{-3}	$\rho_{exp} - \rho_{calc}$ / kg·m^{-3}	Ref. (Symbol in Fig. 1)	T / K	$\rho_{exp} \pm 2\sigma_{est}$ / kg·m^{-3}	$\rho_{exp} - \rho_{calc}$ / kg·m^{-3}	Ref. (Symbol in Fig. 1)
273.15	841.50 ± 1.00	0.06	1886-kra(×)	288.15	831.00 ± 1.70	−0.45	71-gol/dob[1)]
273.15	841.50 ± 0.60	0.06	1886-kra(×)	288.45	831.00 ± 1.70	−0.25	71-gol/dob[1)]
283.15	834.60 ± 0.60	−0.22	1886-kra(×)	293.15	827.80 ± 1.70	−0.26	71-gol/dob[1)]
283.15	834.60 ± 1.00	−0.22	1886-kra[1)]	298.85	822.30 ± 1.60	−1.85	71-gol/dob[1)]
293.15	827.90 ± 0.60	−0.16	1886-kra[1)]	298.95	822.30 ± 1.60	−1.78	71-gol/dob[1)]
293.15	827.90 ± 1.00	−0.16	1886-kra[1)]	303.15	821.80 ± 1.60	0.63	71-gol/dob[1)]
273.15	840.00 ± 0.30	−1.44	32-ell/rei(∇)	313.15	815.50 ± 1.60	1.33	71-gol/dob[1)]
298.15	823.01 ± 0.30	−1.62	32-ell/rei(∇)	322.85	809.50 ± 1.60	2.25	71-gol/dob[1)]
273.15	827.00 ± 1.50	−14.44	66-efr[1)]	323.15	809.20 ± 1.60	2.16	71-gol/dob[1)]
293.15	827.00 ± 1.50	−1.06	66-efr[1)]	323.45	809.10 ± 1.60	2.28	71-gol/dob[1)]
313.15	813.00 ± 1.50	−1.17	66-efr[1)]	333.15	802.80 ± 1.60	3.00	71-gol/dob[1)]
333.15	800.00 ± 1.50	0.20	66-efr[1)]	343.15	796.10 ± 1.60	3.65	71-gol/dob[1)]
353.15	786.00 ± 1.50	1.00	66-efr[1)]	347.75	793.20 ± 1.60	4.17	71-gol/dob[1)]
373.15	771.00 ± 1.50	1.19	66-efr[1)]	347.75	793.10 ± 1.60	4.07	71-gol/dob[1)]
393.15	755.00 ± 1.50	0.73	66-efr(×)	347.85	792.90 ± 1.60	3.94	71-gol/dob[1)]
413.15	739.00 ± 1.50	0.56	66-efr(×)	353.15	789.20 ± 1.60	4.20	71-gol/dob[1)]
433.15	722.00 ± 1.50	−0.34	66-efr(×)	363.15	782.20 ± 1.60	4.75	71-gol/dob[1)]
453.15	704.00 ± 1.50	−2.03	66-efr(×)	369.95	777.50 ± 1.60	5.24	71-gol/dob[1)]
473.15	686.00 ± 1.50	−3.55	66-efr(×)	370.25	777.20 ± 1.60	5.17	71-gol/dob[1)]
493.15	667.00 ± 1.50	−5.94	66-efr(×)	370.35	776.80 ± 1.60	4.84	71-gol/dob[1)]
513.15	648.00 ± 1.50	−8.24	66-efr(×)	373.15	775.00 ± 1.60	5.19	71-gol/dob[1)]
533.15	629.00 ± 1.50	−10.50	66-efr[1)]	383.15	767.10 ± 1.50	5.02	71-gol/dob(×)
553.15	607.00 ± 1.50	−10.86	66-efr(×)	393.15	759.40 ± 1.50	5.13	71-gol/dob(×)
573.15	585.00 ± 1.50	−1.41	66-efr(×)	403.15	751.20 ± 1.50	4.81	71-gol/dob(×)
593.15	560.00 ± 1.50	0.90	66-efr(×)	413.15	742.80 ± 1.50	4.36	71-gol/dob(×)
613.15	541.00 ± 1.50	4.05	66-efr(×)	423.15	734.10 ± 1.50	3.68	71-gol/dob(×)
633.15	500.00 ± 2.00	−7.88	66-efr(×)	424.35	733.20 ± 1.50	3.75	71-gol/dob(×)
653.15	450.00 ± 2.00	2.95	66-efr(×)	424.75	732.80 ± 1.50	3.67	71-gol/dob(×)
663.15	420.00 ± 3.00	30.57	66-efr[1)]	424.85	732.30 ± 1.50	3.25	71-gol/dob(×)

[1)] Not included in Fig. 1.

cont.

1-Nonanol (cont.)

Table 2. Experimental values with uncertainties and deviation from calculated values.

T/K	$\rho_{exp} \pm 2\sigma_{est}$ / $kg \cdot m^{-3}$	$\rho_{exp} - \rho_{calc}$ / $kg \cdot m^{-3}$	Ref. (Symbol in Fig. 1)	T/K	$\rho_{exp} \pm 2\sigma_{est}$ / $kg \cdot m^{-3}$	$\rho_{exp} - \rho_{calc}$ / $kg \cdot m^{-3}$	Ref. (Symbol in Fig. 1)
433.15	725.00 ± 1.50	2.66	71-gol/dob(✕)	298.15	824.60 ± 0.30	−0.03	82-ort(◆)
443.15	715.50 ± 1.40	1.29	71-gol/dob(✕)	303.15	821.60 ± 0.30	0.43	82-ort(◆)
453.15	705.80 ± 1.40	−0.23	71-gol/dob(✕)	308.15	818.10 ± 0.30	0.41	82-ort(◆)
463.15	695.00 ± 1.40	−2.81	71-gol/dob(✕)	313.15	814.70 ± 0.30	0.53	82-ort(◆)
473.15	685.00 ± 1.40	−4.55	71-gol/dob(✕)	318.15	811.00 ± 0.30	0.38	82-ort(◆)
473.15	685.80 ± 1.40	−3.75	71-gol/dob(✕)	288.15	831.00 ± 0.40	−0.45	83-rau/ste(✕)
474.45	683.10 ± 1.40	−5.38	71-gol/dob(✕)	298.15	824.40 ± 0.40	−0.23	83-rau/ste[1)]
483.15	673.60 ± 1.30	−7.66	71-gol/dob[1)]	308.15	817.70 ± 0.40	0.01	83-rau/ste[1)]
300.45	823.39 ± 0.40	0.34	73-nay/kud[1)]	318.15	811.10 ± 0.50	0.48	83-rau/ste[1)]
310.27	816.77 ± 0.40	0.57	73-nay/kud(✕)	328.15	804.30 ± 0.50	0.87	83-rau/ste[1)]
320.89	809.11 ± 0.40	0.45	73-nay/kud(✕)	298.15	824.72 ± 0.20	0.09	85-fer/pin(□)
330.83	802.40 ± 0.40	0.91	73-nay/kud[1)]	323.15	806.92 ± 0.10	−0.12	93-gar/ban-1(✕)
340.81	794.57 ± 0.40	0.39	73-nay/kud[1)]	328.15	803.98 ± 0.10	0.55	93-gar/ban-1(✕)
350.02	788.06 ± 0.40	0.72	73-nay/kud[1)]	333.15	799.70 ± 0.10	−0.10	93-gar/ban-1(✕)
359.78	780.30 ± 0.40	0.30	73-nay/kud[1)]	338.15	796.32 ± 0.10	0.19	93-gar/ban-1(✕)
369.70	772.24 ± 0.40	−0.21	73-nay/kud(✕)	343.15	792.34 ± 0.10	−0.11	93-gar/ban-1(✕)
293.15	828.10 ± 0.30	0.04	76-hon/sin(Δ)	348.15	788.36 ± 0.10	−0.37	93-gar/ban-1(✕)
298.15	824.47 ± 0.20	−0.16	79-dia/tar(O)	353.15	785.08 ± 0.10	0.08	93-gar/ban-1(✕)
308.15	817.58 ± 0.20	−0.11	79-dia/tar(O)	358.15	781.10 ± 0.10	−0.13	93-gar/ban-1(✕)
318.15	810.62 ± 0.20	0.00	79-dia/tar(O)	363.15	777.09 ± 0.10	−0.36	93-gar/ban-1(✕)
333.15	799.91 ± 0.20	0.11	79-dia/tar(O)	373.15	769.58 ± 0.10	−0.23	93-gar/ban-1(✕)
293.15	827.70 ± 0.30	−0.36	82-ort(◆)				

[1)] Not included in Fig. 1.

Table 3. Recommended values (fit to the reliable experimental values according to the equations $\rho = A + BT + CT^2 + DT^3 + \ldots$ or $\rho = [1 + 1.75(1 - T/T_c)^{1/3} + 0.75(1 - T/T_c)][\rho_c + A(T_c - T) + B(T_c - T)^2 + C(T_c - T)^3 + D(T_c - T)^4]$).

T/K	$\rho \pm \sigma_{fit}$ / $kg \cdot m^{-3}$	T/K	$\rho \pm \sigma_{fit}$ / $kg \cdot m^{-3}$	T/K	$\rho \pm \sigma_{fit}$ / $kg \cdot m^{-3}$
270.00	843.49 ± 0.59	400.00	748.88 ± 0.95	550.00	622.37 ± 3.08
280.00	836.92 ± 0.45	410.00	740.95 ± 1.08	560.00	607.28 ± 3.32
290.00	830.20 ± 0.35	420.00	732.95 ± 1.21	570.00	591.31 ± 3.56
293.15	828.06 ± 0.33	430.00	724.89 ± 1.34	580.00	576.25 ± 3.71
298.15	824.63 ± 0.30	440.00	716.78 ± 1.46	590.00	562.92 ± 3.96
300.00	823.36 ± 0.29	450.00	708.62 ± 1.58	600.00	551.28 ± 4.23
310.00	816.39 ± 0.26	460.00	700.41 ± 1.69	610.00	540.44 ± 4.41
320.00	809.30 ± 0.26	470.00	692.16 ± 1.79	620.00	528.74 ± 4.61
330.00	802.09 ± 0.29	480.00	683.88 ± 1.88	630.00	513.76 ± 4.83
340.00	794.77 ± 0.33	490.00	675.56 ± 1.96	640.00	492.33 ± 5.29
350.00	787.35 ± 0.40	500.00	667.23 ± 2.03	650.00	460.27 ± 5.49
360.00	779.84 ± 0.49	510.00	658.88 ± 2.09	660.00	410.90 ± 6.75
370.00	772.22 ± 0.59	520.00	650.51 ± 2.14	670.00	307.42 ± 8.10
380.00	764.52 ± 0.70	530.00	642.14 ± 2.40		
390.00	756.74 ± 0.82	540.00	633.77 ± 2.76		

cont.

Further references: [00-ste, 19-beh, 27-ver/coo, 29-mah/das, 33-ano, 37-oli, 42-mul, 48-vog-2, 50-sac/sau, 52-coo, 52-eri-1, 57-gol/kon, 81-sjo/dyh, 85-ort/paz-1, 88-ort/gar, 90-klo/pal, 93-ami/rai, 93-yan/mae, 94-yu /tsa-1].

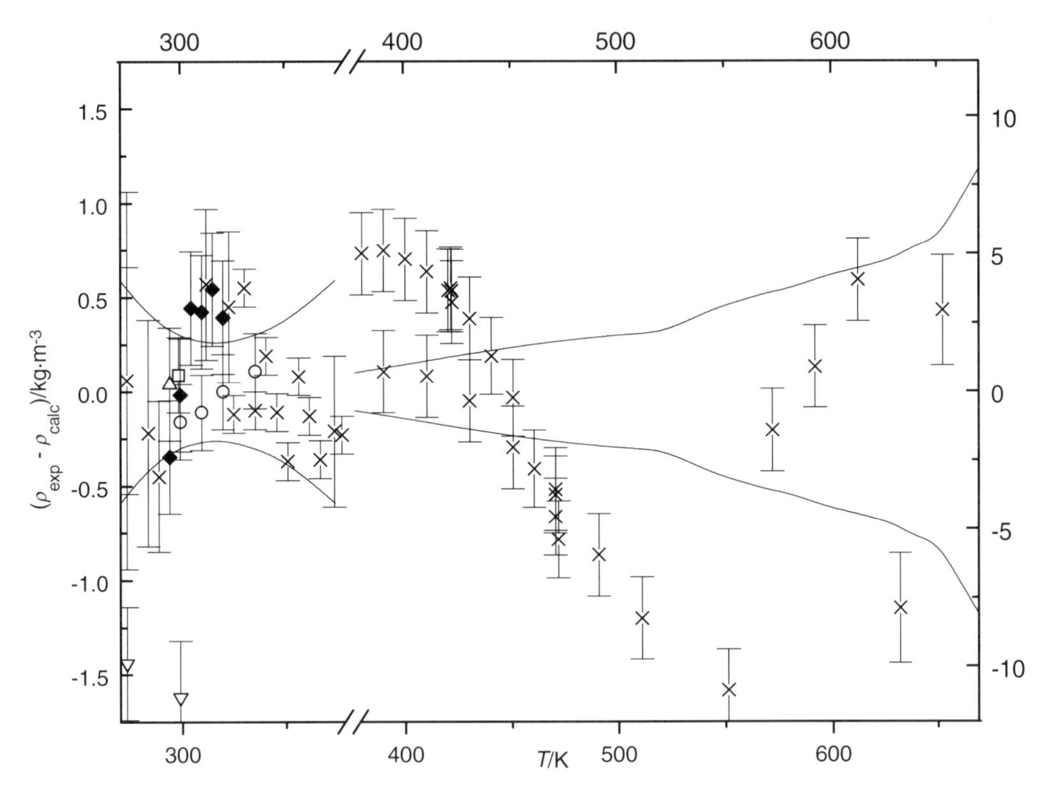

Fig. 1. The symbols show the deviation of the calculated from the experimental values from Table 2. The curves above and below the zero line indicate the calculated error region of the recommended values given in Table 3. The error bars represent the experimental errors. (Error bars smaller than the symbols are omitted for clarity of the figure.)

2-Nonanol **[628-99-9]** **$C_9H_{20}O$** **MW = 144.26** **201**

Table 1. Fit with estimated B coefficient for 4 accepted points. Deviation $\sigma_w = 0.974$.

Coefficient	$\rho = A + BT$
A	1045.94
B	−0.760

Table 2. Experimental values with uncertainties and deviation from calculated values.

T/K	$\rho_{exp} \pm 2\sigma_{est}$ / $kg \cdot m^{-3}$	$\rho_{exp} - \rho_{calc}$ / $kg \cdot m^{-3}$	Ref.	T/K	$\rho_{exp} \pm 2\sigma_{est}$ / $kg \cdot m^{-3}$	$\rho_{exp} - \rho_{calc}$ / $kg \cdot m^{-3}$	Ref.
293.15	847.1 ± 20.0	23.93	06-van[1)]	403.15	730.4 ± 3.0	−9.15	12-pic/ken[1)]
273.15	839.9 ± 1.0	1.55	09-mas	273.15	837.2 ± 1.0	−1.14	32-ell/rei
298.15	823.0 ± 4.0	3.65	10-hal/las[1)]	298.15	819.1 ± 1.0	−0.27	32-ell/rei
293.15	823.0 ± 1.0	−0.15	11-pic/ken				

[1)] Not included in calculation of linear coefficients.

cont.

2-Nonanol (cont.)

Table 3. Recommended values.

$\frac{T}{K}$	$\frac{\rho_{exp} \pm 2\sigma_{est}}{kg \cdot m^{-3}}$	$\frac{T}{K}$	$\frac{\rho_{exp} \pm 2\sigma_{est}}{kg \cdot m^{-3}}$
270.00	840.7 ± 1.8	293.15	823.1 ± 1.4
280.00	833.1 ± 1.2	298.15	819.3 ± 1.8
290.00	825.5 ± 1.2		

3-Nonanol **[624-51-1]** $C_9H_{20}O$ **MW = 144.26** **202**

Table 1. Fit with estimated *B* coefficient for 2 accepted points. Deviation $\sigma_w = 1.150$.

Coefficient	$\rho = A + BT$
A	1068.03
B	−0.820

Table 2. Experimental values with uncertainties and deviation from calculated values.

$\frac{T}{K}$	$\frac{\rho_{exp} \pm 2\sigma_{est}}{kg \cdot m^{-3}}$	$\frac{\rho_{exp} - \rho_{calc}}{kg \cdot m^{-3}}$	Ref.
293.15	826.5 ± 2.0	−1.15	13-pic/ken
353.15	779.6 ± 2.0	1.15	13-pic/ken
293.15	819.6 ± 4.0	−8.05	36-tuo[1)]

[1)] Not included in calculation of linear coefficients.

Table 3. Recommended values.

$\frac{T}{K}$	$\frac{\rho_{exp} \pm 2\sigma_{est}}{kg \cdot m^{-3}}$	$\frac{T}{K}$	$\frac{\rho_{exp} \pm 2\sigma_{est}}{kg \cdot m^{-3}}$	$\frac{T}{K}$	$\frac{\rho_{exp} \pm 2\sigma_{est}}{kg \cdot m^{-3}}$
290.00	830.2 ± 2.9	310.00	813.8 ± 1.9	340.00	789.2 ± 2.0
293.15	827.6 ± 2.7	320.00	805.6 ± 1.7	350.00	781.0 ± 2.5
298.15	823.5 ± 2.4	330.00	797.4 ± 1.7	360.00	772.8 ± 3.1

4-Nonanol **[5932-79-6]** $C_9H_{20}O$ **MW = 144.26** **203**

Table 1. Experimental value with uncertainty.

$\frac{T}{K}$	$\frac{\rho_{exp} \pm 2\sigma_{est}}{kg \cdot m^{-3}}$	Ref.
293.15	826.3 ± 1.0	57-shu/bel

5-nonanol **[623-93-8]** $C_9H_{20}O$ **MW = 144.26** **204**

Table 1. Fit with estimated *B* coefficient for 4 accepted points. Deviation $\sigma_w = 0.293$.

Coefficient	$\rho = A + BT$
A	1044.87
B	−0.760

cont.

Table 2. Experimental values with uncertainties and deviation from calculated values.

$\frac{T}{K}$	$\frac{\rho_{exp} \pm 2\sigma_{est}}{kg \cdot m^{-3}}$	$\frac{\rho_{exp} - \rho_{calc}}{kg \cdot m^{-3}}$	Ref.	$\frac{T}{K}$	$\frac{\rho_{exp} \pm 2\sigma_{est}}{kg \cdot m^{-3}}$	$\frac{\rho_{exp} - \rho_{calc}}{kg \cdot m^{-3}}$	Ref.
293.15	823.0 ± 1.0	0.92	06-mal-1	293.15	825.7 ± 2.0	3.62	47-tuo/guy[1)]
289.15	824.4 ± 2.0	–0.72	19-eyk[1)]	293.15	822.0 ± 0.4	–0.08	50-mea/foo
291.15	823.0 ± 1.0	–0.60	23-vav/iva	298.15	818.3 ± 0.4	0.02	50-mea/foo
293.65	835.6 ± 10.0	13.90	42-boe/han[1)]				

[1)] Not included in calculation of linear coefficients.

Table 3. Recommended values.

$\frac{T}{K}$	$\frac{\rho_{exp} \pm 2\sigma_{est}}{kg \cdot m^{-3}}$
290.00	824.5 ± 0.5
293.15	822.1 ± 0.5
298.15	818.3 ± 0.5

2,2,3,4-Tetramethyl-3-pentanol **[29772-39-2]** **$C_9H_{20}O$** **MW = 144.26** **205**

Table 1. Fit with estimated *B* coefficient for 5 accepted points. Deviation $\sigma_w = 0.468$.

Coefficient	$\rho = A + BT$
A	1096.61
B	–0.820

Table 2. Experimental values with uncertainties and deviation from calculated values.

$\frac{T}{K}$	$\frac{\rho_{exp} \pm 2\sigma_{est}}{kg \cdot m^{-3}}$	$\frac{\rho_{exp} - \rho_{calc}}{kg \cdot m^{-3}}$	Ref.	$\frac{T}{K}$	$\frac{\rho_{exp} \pm 2\sigma_{est}}{kg \cdot m^{-3}}$	$\frac{\rho_{exp} - \rho_{calc}}{kg \cdot m^{-3}}$	Ref.
293.15	856.4 ± 1.0	0.18	29-con/bla	293.15	856.5 ± 0.6	0.28	47-how/mea
293.15	856.0 ± 1.0	–0.22	33-whi/lau	298.15	852.3 ± 0.6	0.18	47-how/mea
288.15	862.2 ± 2.0	1.88	36-naz[1)]	293.15	855.0 ± 1.0	–1.22	48-cad/foo

[1)] Not included in calculation of linear coefficients.

Table 3. Recommended values.

$\frac{T}{K}$	$\frac{\rho_{exp} \pm 2\sigma_{est}}{kg \cdot m^{-3}}$
290.00	858.8 ± 0.8
293.15	856.2 ± 0.8
298.15	852.1 ± 0.8

2,2,3-Trimethyl-3-hexanol **[5340-41-0]** $C_9H_{20}O$ **MW = 144.26** **206**

Table 1. Experimental and recommended values with uncertainties.

T/K	$\rho_{exp} \pm 2\sigma_{est}$ / $kg \cdot m^{-3}$	Ref.
293.15	846.4 ± 1.0	43-geo
293.15	848.5 ± 2.0	48-ruo[1)]
293.15	846.3 ± 1.0	55-pet/sus
293.15	846.0 ± 1.0	61-mar/pet
293.15	846.2 ± 1.0	Recommended

[1)] Not included in calculation of recommended value.

2,3,4-Trimethyl-2-hexanol **[21102-13-6]** $C_9H_{20}O$ **MW = 144.26** **207**

Table 1. Experimental value with uncertainty.

T/K	$\rho_{exp} \pm 2\sigma_{est}$ / $kg \cdot m^{-3}$	Ref.
288.15	835.3 ± 1.0	35-col-1

2,3,5-Trimethyl-3-hexanol **[65927-60-8]** $C_9H_{20}O$ **MW = 144.26** **208**

Table 1. Experimental and recommended values with uncertainties.

T/K	$\rho_{exp} \pm 2\sigma_{est}$ / $kg \cdot m^{-3}$	Ref.
293.15	825.6 ± 3.0	33-mey/tuo[1)]
293.15	831.2 ± 2.0	59-pet/zak-1
293.15	831.2 ± 2.0	Recommended

[1)] Not included in calculation of recommended value.

2,4,4-Trimethyl-2-hexanol **[66793-91-7]** $C_9H_{20}O$ **MW = 144.26** **209**

Table 1. Experimental value with uncertainty.

T/K	$\rho_{exp} \pm 2\sigma_{est}$ / $kg \cdot m^{-3}$	Ref.
293.15	847.5 ± 1.0	40-mos

2,4,4-Trimethyl-3-hexanol **[66793-92-8]** **$C_9H_{20}O$** **MW = 144.26** **210**

Table 1. Experimental value with uncertainty.

$\frac{T}{K}$	$\frac{\rho_{exp} \pm 2\sigma_{est}}{kg \cdot m^{-3}}$	Ref.
293.15	848.8 ± 1.0	48-ruo

2,5,5-Trimethyl-3-hexanol **[66793-72-4]** **$C_9H_{20}O$** **MW = 144.26** **211**

Table 1. Experimental value with uncertainty.

$\frac{T}{K}$	$\frac{\rho_{exp} \pm 2\sigma_{est}}{kg \cdot m^{-3}}$	Ref.
293.15	825.0 ± 1.0	42-whi/for

3,4,4-Trimethyl-3-hexanol **[66793-74-6]** **$C_9H_{20}O$** **MW = 144.26** **212**

Table 1. Experimental value with uncertainty.

$\frac{T}{K}$	$\frac{\rho_{exp} \pm 2\sigma_{est}}{kg \cdot m^{-3}}$	Ref.
294.15	832.3 ± 2.0	06-kon/mil

3,5,5-Trimethyl-1-hexanol **[3452-97-9]** **$C_9H_{20}O$** **MW = 144.26** **213**

Table 1. Experimental value with uncertainty.

$\frac{T}{K}$	$\frac{\rho_{exp} \pm 2\sigma_{est}}{kg \cdot m^{-3}}$	Ref.
298.15	823.6 ± 1.0	49-bru

3,5,5-Trimethyl-3-hexanol **[66810-87-5]** **$C_9H_{20}O$** **MW = 144.26** **214**

Table 1. Experimental value with uncertainty.

$\frac{T}{K}$	$\frac{\rho_{exp} \pm 2\sigma_{est}}{kg \cdot m^{-3}}$	Ref.
293.15	835.0 ± 0.5	47-how/mea

4,5,5-Trimethyl-1-hexanol **[66793-75-7]** **$C_9H_{20}O$** **MW = 144.26** **215**

Table 1. Experimental value with uncertainty.

$\frac{T}{K}$	$\frac{\rho_{exp} \pm 2\sigma_{est}}{kg \cdot m^{-3}}$	Ref.
293.15	846.0 ± 1.0	61-mar/pet

2.1.6 Alkanols, C_{10}

2-Butyl-1-hexanol **[2768-15-2]** **$C_{10}H_{22}O$** **MW = 158.28** **216**

Table 1. Experimental value with uncertainty.

$\frac{T}{K}$	$\frac{\rho_{exp} \pm 2\sigma_{est}}{kg \cdot m^{-3}}$	Ref.
289.15	836.0 ± 3.0	16-lev/all

1-Decanol **[112-30-1]** **$C_{10}H_{22}O$** **MW = 158.28** **217**

T_c = 687.10 K [89-tej/lee] ρ_c = 264.00 kg·m^{-3} [89-tej/lee]

Table 1. Coefficients for the polynomial expansion equations. Standard deviations (see introduction): $\sigma_\ell = 4.3461 \cdot 10^{-1}$ (low temperature range), $\sigma_{c,w} = (2.6342 \cdot 10^{-1}$ combined temperature ranges, weighted), $\sigma_{c,uw} = 3.3635 \cdot 10^{-1}$ (combined temperature ranges, unweighted).

Coefficient	T = 279.63 to 550.00 K $\rho = A + BT + CT^2 + DT^3 + \ldots$	T = 550.00 to 687.10 K $\rho = [1 + 1.75(1 - T/T_c)^{1/3} + 0.75(1 - T/T_c)]$ $[\rho_c + A(T_c - T) + B(T_c - T)^2 + C(T_c - T)^3 + D(T_c - T)^4]$
A	$1.01735 \cdot 10^{3}$	1.30964
B	$-6.79957 \cdot 10^{-1}$	$-1.59285 \cdot 10^{-2}$
C	$3.98783 \cdot 10^{-4}$	$5.30923 \cdot 10^{-5}$
D	$-8.88242 \cdot 10^{-7}$	$7.66905 \cdot 10^{-9}$

Table 2. Experimental values with uncertainties and deviation from calculated values.

$\frac{T}{K}$	$\frac{\rho_{exp} \pm 2\sigma_{est}}{kg \cdot m^{-3}}$	$\frac{\rho_{exp} - \rho_{calc}}{kg \cdot m^{-3}}$	Ref. (Symbol in Fig. 1)	$\frac{T}{K}$	$\frac{\rho_{exp} \pm 2\sigma_{est}}{kg \cdot m^{-3}}$	$\frac{\rho_{exp} - \rho_{calc}}{kg \cdot m^{-3}}$	Ref. (Symbol in Fig. 1)
280.15	838.90 ± 1.00	0.27	1883-kra[1)]	413.15	738.70 ± 0.64	−3.16	58-cos/bow(×)
293.15	829.70 ± 1.00	−0.22	1883-kra[1)]	433.15	722.40 ± 0.66	−3.06	58-cos/bow(×)
371.85	773.40 ± 1.00	−0.58	1883-kra(×)	453.15	705.30 ± 0.68	−3.17	58-cos/bow(×)
293.10	830.10 ± 0.29	0.15	55-kus[1)]	473.15	688.30 ± 0.70	−2.52	58-cos/bow(×)
298.10	826.90 ± 0.30	0.33	55-kus[1)]	493.15	670.80 ± 0.72	−1.69	58-cos/bow(×)
303.10	823.50 ± 0.30	0.34	55-kus[1)]	513.15	653.50 ± 1.00	0.08	58-cos/bow(×)
313.10	816.80 ± 0.32	0.51	55-kus[1)]	533.15	635.00 ± 1.00	1.42	58-cos/bow(×)
323.10	810.00 ± 0.34	0.67	55-kus[1)]	553.15	612.70 ± 1.00	−0.35	58-cos/bow(×)
333.10	802.70 ± 0.36	0.42	55-kus[1)]	293.15	829.70 ± 1.50	−0.22	66-efr[1)]
343.10	795.40 ± 0.38	0.27	55-kus(×)	313.15	816.20 ± 1.50	−0.05	66-efr[1)]
353.10	788.10 ± 0.39	0.22	55-kus(×)	333.15	802.20 ± 1.50	−0.04	66-efr[1)]
358.10	784.20 ± 0.40	−0.01	55-kus(×)	353.15	787.60 ± 1.50	−0.24	66-efr[1)]
293.15	826.00 ± 0.52	−3.92	58-cos/bow[1)]	373.15	772.40 ± 1.50	−0.60	66-efr[1)]
313.15	812.70 ± 0.54	−3.55	58-cos/bow[1)]	393.15	757.20 ± 1.50	−0.49	66-efr(×)
333.15	799.80 ± 0.56	−2.44	58-cos/bow[1)]	413.15	741.40 ± 1.50	−0.46	66-efr(×)
353.15	785.80 ± 0.58	−2.04	58-cos/bow[1)]	433.15	727.50 ± 1.50	2.04	66-efr[1)]
373.15	770.50 ± 0.60	−2.50	58-cos/bow(×)	453.15	707.20 ± 1.50	−1.27	66-efr[1)]
393.15	755.00 ± 0.62	−2.69	58-cos/bow(×)	473.15	694.00 ± 1.50	3.18	66-efr[1)]

[1)] Not included in Fig. 1.

cont.

1-Decanol (cont.)

Table 2. Experimental values with uncertainties and deviation from calculated values.

$\frac{T}{K}$	$\frac{\rho_{exp} \pm 2\sigma_{est}}{kg \cdot m^{-3}}$	$\frac{\rho_{exp} - \rho_{calc}}{kg \cdot m^{-3}}$	Ref. (Symbol in Fig. 1)	$\frac{T}{K}$	$\frac{\rho_{exp} \pm 2\sigma_{est}}{kg \cdot m^{-3}}$	$\frac{\rho_{exp} - \rho_{calc}}{kg \cdot m^{-3}}$	Ref. (Symbol in Fig. 1)
493.15	675.00 ± 1.50	2.51	66-efr[1]	303.15	823.20 ± 0.33	0.07	89-mat/mak-1[1]
513.15	659.00 ± 1.50	5.58	66-efr(×)	308.15	819.70 ± 0.33	–0.00	89-mat/mak-1[1]
533.15	643.00 ± 1.50	9.42	66-efr[1]	313.15	816.30 ± 0.33	0.05	89-mat/mak-1[1]
553.15	624.00 ± 1.50	10.95	66-efr(×)	318.15	812.80 ± 0.33	0.01	89-mat/mak-1[1]
573.15	605.00 ± 1.50	7.91	66-efr(×)	323.15	809.30 ± 0.32	0.01	89-mat/mak-1[1]
593.15	584.00 ± 1.50	0.66	66-efr(×)	323.15	809.40 ± 0.49	0.11	89-mat/mak-1[1]
613.15	564.00 ± 1.50	–1.22	66-efr(×)	328.15	805.90 ± 0.32	0.12	89-mat/mak-1[1]
633.15	530.00 ± 2.00	–6.50	66-efr(×)	333.15	802.30 ± 0.32	0.06	89-mat/mak-1[1]
653.15	496.00 ± 2.00	5.35	66-efr(×)	338.15	798.70 ± 0.32	0.02	89-mat/mak-1(×)
673.15	450.00 ± 3.00	33.09	66-efr[1]	343.15	795.10 ± 0.32	0.01	89-mat/mak-1(×)
279.63	838.86 ± 0.17	–0.12	73-fin(∇)	348.15	791.50 ± 0.47	0.02	89-mat/mak-1(×)
279.85	838.77 ± 0.17	–0.06	73-fin(∇)	348.15	791.40 ± 0.32	–0.08	89-mat/mak-1(×)
281.69	837.53 ± 0.17	–0.08	73-fin(∇)	283.15	834.30 ± 0.83	–2.33	90-apa/gyl[1]
286.12	834.51 ± 0.17	–0.13	73-fin(∇)	293.15	828.40 ± 0.83	–1.52	90-apa/gyl[1]
286.15	834.45 ± 0.17	–0.17	73-fin(∇)	303.15	822.60 ± 0.82	–0.53	90-apa/gyl[1]
293.15	829.73 ± 0.17	–0.19	73-fin[1]	323.15	811.10 ± 0.81	1.81	90-apa/gyl[1]
293.16	829.73 ± 0.17	–0.18	73-fin[1]	343.15	798.20 ± 0.80	3.11	90-apa/gyl[1]
303.14	822.94 ± 0.16	–0.19	73-fin[1]	363.15	784.10 ± 0.78	3.62	90-apa/gyl(×)
313.13	816.07 ± 0.16	–0.20	73-fin(∇)	383.15	769.40 ± 0.77	3.99	90-apa/gyl(×)
313.15	816.06 ± 0.16	–0.19	73-fin(∇)	403.15	753.50 ± 0.75	3.66	90-apa/gyl(×)
323.08	809.17 ± 0.16	–0.17	73-fin(∇)	423.15	737.20 ± 0.74	3.47	90-apa/gyl(×)
323.09	809.15 ± 0.16	–0.19	73-fin(∇)	433.15	728.00 ± 0.73	2.54	90-apa/gyl(×)
333.14	802.07 ± 0.16	–0.18	73-fin(∇)	443.15	719.20 ± 0.72	2.16	90-apa/gyl(×)
333.16	802.04 ± 0.16	–0.20	73-fin(∇)	453.15	710.10 ± 0.71	1.63	90-apa/gyl(×)
293.15	829.69 ± 0.10	–0.23	78-jel/leo(Δ)	463.15	700.60 ± 0.70	0.87	90-apa/gyl(×)
298.15	826.40 ± 0.10	–0.13	78-jel/leo(Δ)	473.15	690.10 ± 0.69	–0.72	90-apa/gyl(×)
303.15	823.04 ± 0.10	–0.09	78-jel/leo(Δ)	483.15	679.60 ± 0.68	–2.14	90-apa/gyl(×)
308.15	819.61 ± 0.10	–0.09	78-jel/leo(Δ)	493.15	669.10 ± 0.67	–3.39	90-apa/gyl(×)
298.15	826.23 ± 0.20	–0.30	79-dia/tar[1]	288.15	833.90 ± 0.60	0.62	92-lie/sen-1[1]
308.15	819.46 ± 0.20	–0.24	79-dia/tar[1]	293.15	830.20 ± 0.60	0.28	92-lie/sen-1[1]
318.15	812.58 ± 0.20	–0.21	79-dia/tar(◆)	298.15	827.00 ± 0.60	0.47	92-lie/sen-1[1]
333.15	802.03 ± 0.20	–0.21	79-dia/tar(◆)	303.15	823.70 ± 0.60	0.57	92-lie/sen-1[1]
298.15	826.57 ± 0.02	0.04	79-kiy/ben(×)	308.15	820.10 ± 0.60	0.40	92-lie/sen-1[1]
298.15	826.53 ± 0.02	–0.00	81-ben/han(□)	313.15	816.40 ± 0.60	0.15	92-lie/sen-1[1]
298.15	826.57 ± 0.10	0.04	85-kum/ben(O)	318.15	812.90 ± 0.60	0.11	92-lie/sen-1[1]
283.15	836.51 ± 0.20	–0.12	86-hei/sch(×)	323.15	809.50 ± 0.60	0.21	92-lie/sen-1[1]
298.15	826.40 ± 0.20	–0.13	86-hei/sch[1]	328.15	805.80 ± 0.60	0.02	92-lie/sen-1[1]
313.15	816.11 ± 0.20	–0.14	86-hei/sch(×)	333.15	802.40 ± 0.60	0.16	92-lie/sen-1[1]
293.15	829.82 ± 0.25	–0.10	86-wag/hei[1]	338.15	798.80 ± 0.60	0.12	92-lie/sen-1[1]
298.15	826.44 ± 0.25	–0.09	86-wag/hei[1]	343.15	795.20 ± 0.60	0.11	92-lie/sen-1[1]
333.15	802.02 ± 0.25	–0.22	86-wag/hei(×)	348.15	791.60 ± 0.60	0.12	92-lie/sen-1[1]
298.15	826.60 ± 0.33	0.07	89-mat/mak-1[1]	353.15	788.00 ± 0.60	0.16	92-lie/sen-1(×)
298.15	826.80 ± 0.50	0.27	89-mat/mak-1[1]				

[1] Not included in Fig. 1.

cont.

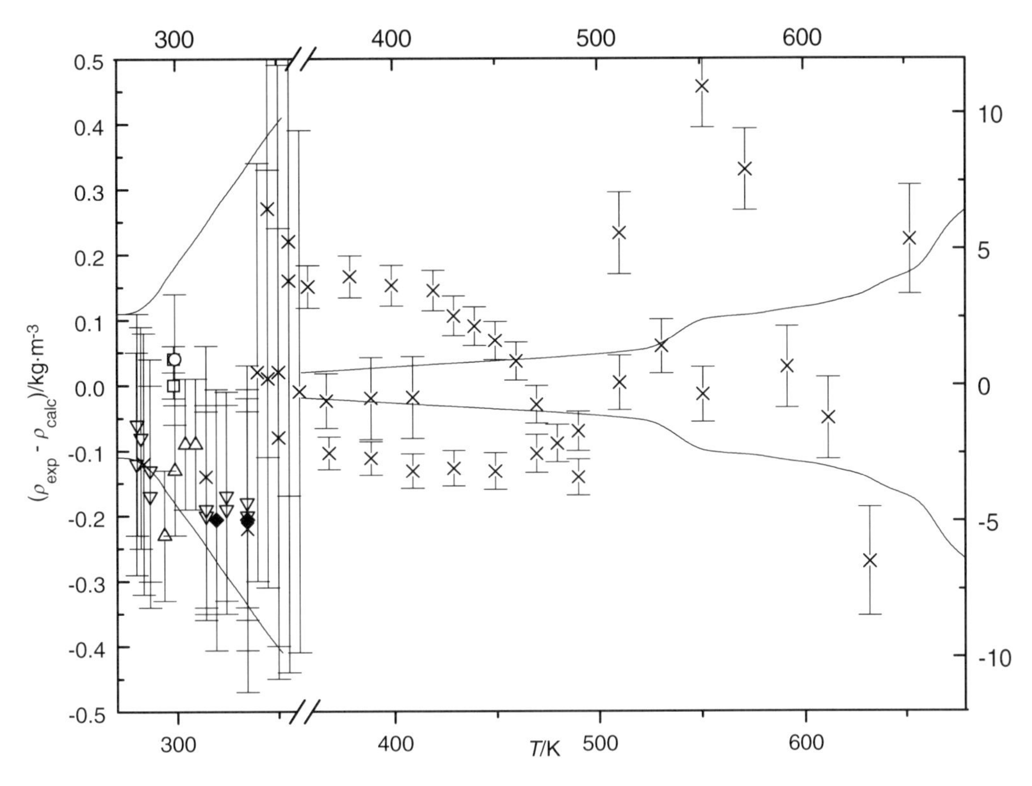

Fig. 1. The symbols show the deviation of the calculated from the experimental values from Table 2. The curves above and below the zero line indicate the calculated error region of the recommended values given in Table 3. The error bars represent the experimental errors. (Error bars smaller than the symbols are omitted for clarity of the figure.)

Table 3. Recommended values (fit to the reliable experimental values according to the equations
$\rho = A + BT + CT^2 + DT^3 + \dots$ or $\rho = [1 + 1.75(1 - T/T_c)^{1/3} + 0.75(1 - T/T_c)][\rho_c + A(T_c - T) + B(T_c - T)^2 + C(T_c - T)^3 + D(T_c - T)^4]$).

T/K	$\rho \pm \sigma_{fit}$/kg·m^{-3}	T/K	$\rho \pm \sigma_{fit}$/kg·m^{-3}	T/K	$\rho \pm \sigma_{fit}$/kg·m^{-3}
270.00	845.35 ± 0.11	400.00	752.33 ± 0.64	550.00	616.23 ± 2.38
280.00	838.73 ± 0.11	410.00	744.39 ± 0.69	560.00	606.90 ± 2.49
290.00	832.04 ± 0.14	420.00	736.31 ± 0.74	570.00	599.28 ± 2.54
293.15	829.92 ± 0.16	430.00	728.09 ± 0.78	580.00	592.47 ± 2.60
298.15	826.53 ± 0.18	440.00	719.71 ± 0.82	590.00	585.63 ± 2.76
300.00	825.27 ± 0.19	450.00	711.19 ± 0.87	600.00	577.94 ± 2.84
310.00	818.43 ± 0.23	460.00	702.50 ± 0.91	610.00	568.62 ± 2.94
320.00	811.50 ± 0.28	470.00	693.64 ± 0.95	620.00	556.89 ± 3.16
330.00	804.47 ± 0.32	480.00	684.62 ± 1.00	630.00	541.98 ± 3.30
340.00	797.36 ± 0.37	490.00	675.42 ± 1.05	640.00	523.10 ± 3.79
350.00	790.14 ± 0.41	500.00	666.04 ± 1.11	650.00	499.33 ± 4.03
360.00	782.81 ± 0.46	510.00	656.47 ± 1.17	660.00	469.47 ± 4.35
370.00	775.37 ± 0.51	520.00	646.71 ± 1.25	670.00	431.35 ± 5.79
380.00	767.81 ± 0.55	530.00	636.76 ± 1.34	680.00	378.50 ± 6.41
390.00	760.14 ± 0.60	540.00	626.60 ± 1.85		

cont.

1-Decanol (cont.)

Further references: [09-sch-1, 27-ver/coo, 48-jon/bow, 48-vog-2, 48-wei, 50-sac/sau, 56-rat/cur, 68-pfl/pop, 68-sin/ben, 69-smi/kur, 78-ast, 78-tre/ben, 81-kiy/ben, 81-sjo/dyh, 81-tre/kiy, 82-ort, 82-tre/han, 83-gop/rao, 83-rau/ste, 84-kum/ben, 85-fer/pin, 85-ort/paz-1, 86-dew/meh, 87-dew/meh, 88-ort/gar, 89-dew/gup, 93-yan/mae, 94-yu /tsa-1].

2-Decanol **[1120-06-5]** $C_{10}H_{22}O$ **MW = 158.28** **218**

Table 1. Experimental and recommended values with uncertainties.

$\frac{T}{\text{K}}$	$\frac{\rho_{exp} \pm 2\sigma_{est}}{\text{kg} \cdot \text{m}^{-3}}$	Ref.
293.15	825.0 ± 1.0	11-pic/ken
293.15	824.9 ± 1.0	59-asi/gei
293.15	825.6 ± 1.0	60-kor/pet
293.15	825.2 ± 1.0	Recommended

3-Decanol **[1565-81-7]** $C_{10}H_{22}O$ **MW = 158.28** **219**

Table 1. Fit with estimated B coefficient for 3 accepted points. Deviation $\sigma_w = 0.283$.

Coefficient	$\rho = A + BT$
A	1055.46
B	−0.780

Table 2. Experimental values with uncertainties and deviation from calculated values.

$\frac{T}{\text{K}}$	$\frac{\rho_{exp} \pm 2\sigma_{est}}{\text{kg} \cdot \text{m}^{-3}}$	$\frac{\rho_{exp} - \rho_{calc}}{\text{kg} \cdot \text{m}^{-3}}$	Ref.
293.15	827.2 ± 1.0	0.40	13-pic/ken
353.15	779.8 ± 1.0	−0.20	13-pic/ken
293.15	826.6 ± 1.0	−0.20	59-asi/gei

Table 3. Recommended values.

$\frac{T}{\text{K}}$	$\frac{\rho_{exp} \pm 2\sigma_{est}}{\text{kg} \cdot \text{m}^{-3}}$	$\frac{T}{\text{K}}$	$\frac{\rho_{exp} \pm 2\sigma_{est}}{\text{kg} \cdot \text{m}^{-3}}$	$\frac{T}{\text{K}}$	$\frac{\rho_{exp} \pm 2\sigma_{est}}{\text{kg} \cdot \text{m}^{-3}}$
290.00	829.3 ± 1.5	310.00	813.7 ± 1.0	340.00	790.3 ± 1.6
293.15	826.8 ± 1.4	320.00	805.9 ± 1.0	350.00	782.5 ± 2.1
298.15	822.9 ± 1.2	330.00	798.1 ± 1.3	360.00	774.7 ± 2.5

4-Decanol **[2051-31-2]** $C_{10}H_{22}O$ **MW = 158.28** **220**

Table 1. Experimental and recommended values with uncertainties.

$\frac{T}{\text{K}}$	$\frac{\rho_{exp} \pm 2\sigma_{est}}{\text{kg} \cdot \text{m}^{-3}}$	Ref.
293.15	826.2 ± 2.0	57-pet/nef-1
293.15	823.8 ± 2.0	59-asi/gei
293.15	825.0 ± 2.2	Recommended

5-Decanol **[5205-34-5]** $C_{10}H_{22}O$ **MW = 158.28** **221**

Table 1. Fit with estimated B coefficient for 2 accepted points. Deviation $\sigma_w = 0.221$.

Coefficient	$\rho = A + BT$
A	1047.15
B	–0.760

Table 2. Experimental values with uncertainties and deviation from calculated values.

$\frac{T}{\text{K}}$	$\frac{\rho_{exp} \pm 2\sigma_{est}}{\text{kg} \cdot \text{m}^{-3}}$	$\frac{\rho_{exp} - \rho_{calc}}{\text{kg} \cdot \text{m}^{-3}}$	Ref.
293.15	827.6 ± 3.0	3.29	56-gol/kon[1)]
293.15	823.8 ± 1.0	–0.55	59-asi/gei
298.15	820.6 ± 0.4	0.09	88-cac/cos

[1)] Not included in calculation of linear coefficients.

Table 3. Recommended values.

$\frac{T}{\text{K}}$	$\frac{\rho_{exp} \pm 2\sigma_{est}}{\text{kg} \cdot \text{m}^{-3}}$
290.00	826.7 ± 0.6
293.15	824.4 ± 0.5
298.15	820.6 ± 0.5

2,2-Dimethyl-4-ethyl-3-hexanol **[66719-47-9]** $C_{10}H_{22}O$ **MW = 158.28** **222**

Table 1. Experimental value with uncertainty.

$\frac{T}{\text{K}}$	$\frac{\rho_{exp} \pm 2\sigma_{est}}{\text{kg} \cdot \text{m}^{-3}}$	Ref.
293.15	833.9 ± 1.0	41-whi/whi

2,4-Dimethyl-4-ethyl-3-hexanol **[66719-48-0]** $C_{10}H_{22}O$ **MW = 158.28** **223**

Table 1. Experimental value with uncertainty.

$\frac{T}{\text{K}}$	$\frac{\rho_{exp} \pm 2\sigma_{est}}{\text{kg} \cdot \text{m}^{-3}}$	Ref.
293.15	860.6 ± 2.0	53-sok

5,5-Dimethyl-2-ethyl-1-hexanol **[66719-49-1]** $C_{10}H_{22}O$ **MW = 158.28** **224**

Table 1. Experimental value with uncertainty.

$\frac{T}{\text{K}}$	$\frac{\rho_{exp} \pm 2\sigma_{est}}{\text{kg} \cdot \text{m}^{-3}}$	Ref.
293.15	838.6 ± 1.0	56-gol/kon

5,5-Dimethyl-3-ethyl-3-hexanol **[5340-62-5]** $C_{10}H_{22}O$ **MW = 158.28** **225**

Table 1. Experimental and recommended values with uncertainties.

T/K	$\rho_{exp} \pm 2\sigma_{est}$ / $kg \cdot m^{-3}$	Ref.
293.15	842.2 ± 1.0	34-sta
293.15	845.0 ± 2.0	42-whi/for
293.15	842.8 ± 1.4	Recommended

2,4-Dimethyl-3-(1-methylethyl)-3-pentanol **[51200-83-0]** $C_{10}H_{22}O$ **MW = 158.28** **226**

Table 1. Fit with estimated B coefficient for 2 accepted points. Deviation $\sigma_w = 0.050$.

Coefficient	$\rho = A + BT$
A	1097.67
B	−0.800

Table 2. Experimental values with uncertainties and deviation from calculated values.

T/K	$\rho_{exp} \pm 2\sigma_{est}$ / $kg \cdot m^{-3}$	$\rho_{exp} - \rho_{calc}$ / $kg \cdot m^{-3}$	Ref.
293.15	874.0 ± 2.0	10.85	44-you/rob[1)]
292.15	861.6 ± 2.0	−2.35	46-vav/col[1)]
293.15	863.2 ± 0.6	0.05	47-how/mea
298.15	859.1 ± 0.6	−0.05	47-how/mea

[1)] Not included in calculation of linear coefficients.

Table 3. Recommended values.

T/K	$\rho_{exp} \pm 2\sigma_{est}$ / $kg \cdot m^{-3}$
290.00	865.7 ± 0.4
293.15	863.1 ± 0.3
298.15	859.1 ± 0.3

2,4-Dimethyl-3-propyl-3-pentanol **[500001-19-4]** $C_{10}H_{22}O$ **MW = 158.28** **227**

Table 1. Fit with estimated B coefficient for 3 accepted points. Deviation $\sigma_w = 0.262$.

Coefficient	$\rho = A + BT$
A	1088.19
B	−0.800

cont.

Table 2. Experimental values with uncertainties and deviation from calculated values.

$\frac{T}{\mathrm{K}}$	$\frac{\rho_{exp} \pm 2\sigma_{est}}{\mathrm{kg \cdot m^{-3}}}$	$\frac{\rho_{exp} - \rho_{calc}}{\mathrm{kg \cdot m^{-3}}}$	Ref.
293.15	853.9 ± 1.0	0.23	26-sta
273.15	869.3 ± 1.0	–0.37	26-sta
303.15	845.8 ± 1.0	0.13	26-sta

Table 3. Recommended values.

$\frac{T}{\mathrm{K}}$	$\frac{\rho_{exp} \pm 2\sigma_{est}}{\mathrm{kg \cdot m^{-3}}}$	$\frac{T}{\mathrm{K}}$	$\frac{\rho_{exp} \pm 2\sigma_{est}}{\mathrm{kg \cdot m^{-3}}}$	$\frac{T}{\mathrm{K}}$	$\frac{\rho_{exp} \pm 2\sigma_{est}}{\mathrm{kg \cdot m^{-3}}}$
270.00	872.2 ± 2.2	290.00	856.2 ± 0.9	298.15	849.7 ± 1.3
280.00	864.2 ± 1.4	293.15	853.7 ± 1.0	310.00	840.2 ± 2.2

1,1-Dimethyl-1-octanol [10297-57-1] $C_{10}H_{22}O$ MW = 158.28 228

Table 1. Experimental value with uncertainty.

$\frac{T}{\mathrm{K}}$	$\frac{\rho_{exp} \pm 2\sigma_{est}}{\mathrm{kg \cdot m^{-3}}}$	Ref.
298.15	807.8 ± 1.0	64-tis/sta

2,2-Dimethyl-1-octanol [2370-14-1] $C_{10}H_{22}O$ MW = 158.28 229

Table 1. Experimental and recommended values with uncertainties.

$\frac{T}{\mathrm{K}}$	$\frac{\rho_{exp} \pm 2\sigma_{est}}{\mathrm{kg \cdot m^{-3}}}$	Ref.
293.15	830.0 ± 1.0	64-blo/hag
293.15	843.0 ± 2.0	65-shu/puz[1)]
293.15	830.0 ± 1.0	Recommended

[1)] Not included in calculation of recommended value.

2,2-Dimethyl-4-octanol [66719-52-6] $C_{10}H_{22}O$ MW = 158.28 230

Table 1. Experimental value with uncertainty.

$\frac{T}{\mathrm{K}}$	$\frac{\rho_{exp} \pm 2\sigma_{est}}{\mathrm{kg \cdot m^{-3}}}$	Ref.
293.15	821.2 ± 1.0	38-whi/pop-1

2,3-Dimethyl-3-octanol [19781-10-3] $C_{10}H_{22}O$ MW = 158.28 231

Table 1. Fit with estimated *B* coefficient for 2 accepted points. Deviation $\sigma_w = 0.150$.

Coefficient	$\rho = A + BT$
A	1069.53
B	–0.820

cont.

2,3-Dimethyl-3-octanol (cont.)

Table 2. Experimental values with uncertainties and deviation from calculated values.

$\frac{T}{K}$	$\frac{\rho_{exp} \pm 2\sigma_{est}}{kg \cdot m^{-3}}$	$\frac{\rho_{exp} - \rho_{calc}}{kg \cdot m^{-3}}$	Ref.
293.15	840.1 ± 6.0	10.95	33-whi/eve[1)]
293.15	829.3 ± 0.6	0.15	50-mea/foo
298.15	824.9 ± 0.6	–0.15	50-mea/foo

[1)] Not included in calculation of linear coefficients.

Table 3. Recommended values.

$\frac{T}{K}$	$\frac{\rho_{exp} \pm 2\sigma_{est}}{kg \cdot m^{-3}}$
290.00	831.7 ± 0.7
293.15	829.2 ± 0.5
298.15	825.1 ± 0.5

2,4-Dimethyl-2-octanol **[18675-20-2]** **$C_{10}H_{22}O$** **MW = 158.28** **232**

Table 1. Experimental value with uncertainty.

$\frac{T}{K}$	$\frac{\rho_{exp} \pm 2\sigma_{est}}{kg \cdot m^{-3}}$	Ref.
293.15	825.7 ± 2.0	48-naz/tor

2,4-Dimethyl-4-octanol **[33933-79-8]** **$C_{10}H_{22}O$** **MW = 158.28** **233**

Table 1. Experimental and recommended values with uncertainties.

$\frac{T}{K}$	$\frac{\rho_{exp} \pm 2\sigma_{est}}{kg \cdot m^{-3}}$	Ref.	$\frac{T}{K}$	$\frac{\rho_{exp} \pm 2\sigma_{est}}{kg \cdot m^{-3}}$	Ref.
293.15	823.8 ± 1.0	33-mey/tuo	293.15	822.7 ± 1.0	59-pet/zak
293.15	823.6 ± 1.0	58-sok/kra	293.15	822.7 ± 1.0	59-pet/zak-1
293.15	832.8 ± 6.0	59-nog/dza[1)]	293.15	823.2 ± 1.0	Recommended

[1)] Not included in calculation of recommended value.

2,5-Dimethyl-4-octanol **[66719-53-7]** **$C_{10}H_{22}O$** **MW = 158.28** **234**

Table 1. Experimental value with uncertainty.

$\frac{T}{K}$	$\frac{\rho_{exp} \pm 2\sigma_{est}}{kg \cdot m^{-3}}$	Ref.
298.15	821.5 ± 0.8	12-bje

2,6-Dimethyl-1-octanol **[62417-08-7]** **$C_{10}H_{22}O$** **MW = 158.28** **235**

Table 1. Experimental value with uncertainty.

$\frac{T}{K}$	$\frac{\rho_{exp} \pm 2\sigma_{est}}{kg \cdot m^{-3}}$	Ref.
291.15	830.0 ± 2.0	14-vav

2,6-Dimethyl-2-octanol **[18479-57-7]** **$C_{10}H_{22}O$** **MW = 158.28** **236**

Table 1. Experimental and recommended values with uncertainties.

T/K	$\rho_{exp} \pm 2\sigma_{est}$/kg·m^{-3}	Ref.
298.15	802.3 ± 10.0	53-sut[1)]
293.15	833.5 ± 5.0	57-naz/kra[1)]
293.15	826.0 ± 2.0	59-hou/lev[1)]
293.15	827.3 ± 1.0	64-ohl/sei
293.15	827.3 ± 1.0	Recommended

[1)] Not included in calculation of recommended value.

2,7-Dimethyl-1-octanol **[15250-22-3]** **$C_{10}H_{22}O$** **MW = 158.28** **237**

Table 1. Experimental value with uncertainty.

T/K	$\rho_{exp} \pm 2\sigma_{est}$/kg·m^{-3}	Ref.
288.15	830.3 ± 1.0	40-pal

2,7-Dimethyl-2-octanol **[42007-73-8]** **$C_{10}H_{22}O$** **MW = 158.28** **238**

Table 1. Coefficients of the polynomial expansion equation.
Standard deviations (see introduction):
$\sigma_{c,w} = 7.1829 \cdot 10^{-2}$ (combined temperature ranges, weighted),
$\sigma_{c,uw} = 1.9108 \cdot 10^{-2}$ (combined temperature ranges, unweighted).

Coefficient	T = 293.15 to 358.15 K $\rho = A + BT + CT^2 + DT^3 + \dots$
A	$1.06429 \cdot 10^{3}$
B	$-8.15665 \cdot 10^{-1}$

Table 2. Experimental values with uncertainties and deviation from calculated values.

T/K	$\rho_{exp} \pm 2\sigma_{est}$/kg·m^{-3}	$\rho_{exp} - \rho_{calc}$/kg·m^{-3}	Ref. (Symbol in Fig. 1)	T/K	$\rho_{exp} \pm 2\sigma_{est}$/kg·m^{-3}	$\rho_{exp} - \rho_{calc}$/kg·m^{-3}	Ref. (Symbol in Fig. 1)
293.15	825.10 ± 1.00	−0.08	55-kus(□)	333.15	792.60 ± 1.00	0.05	55-kus(□)
298.15	821.10 ± 1.00	0.00	55-kus(□)	343.15	784.50 ± 1.00	0.11	55-kus(□)
303.15	817.00 ± 1.00	−0.02	55-kus(□)	353.15	776.20 ± 1.00	−0.04	55-kus(□)
313.15	808.90 ± 1.00	0.04	55-kus(□)	358.15	772.10 ± 1.00	−0.06	55-kus(□)
323.15	800.70 ± 1.00	−0.01	55-kus(□)				

cont.

2,7-Dimethyl-2-octanol (cont.)

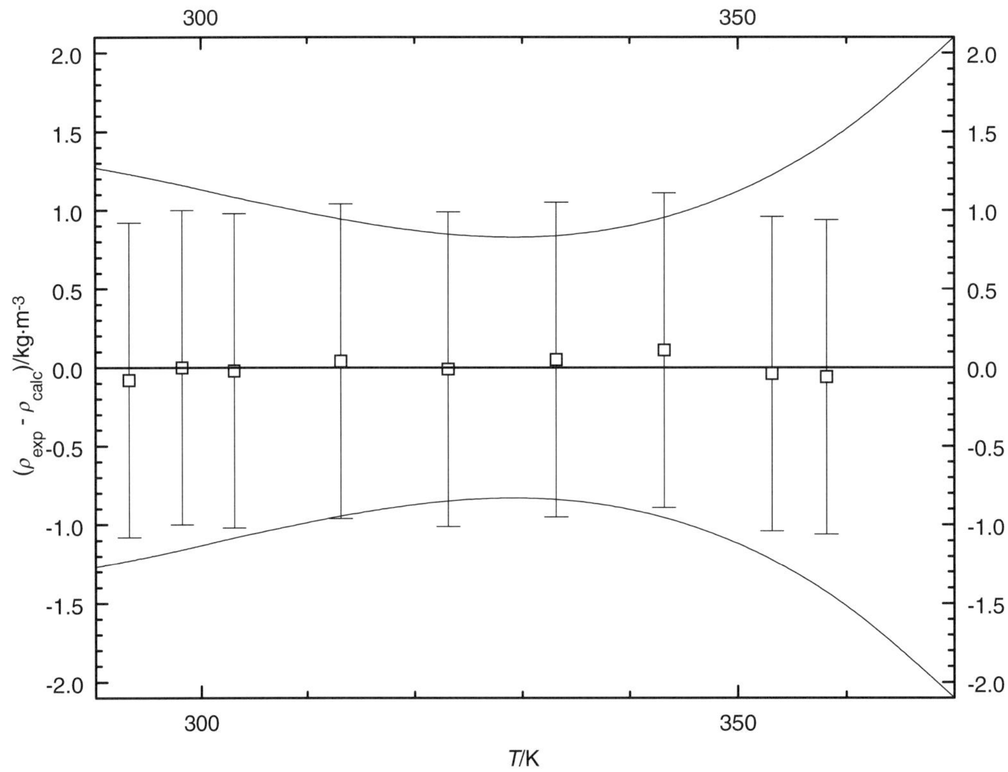

Fig. 1. The symbols show the deviation of the calculated from the experimental values from Table 2. The curves above and below the zero line indicate the calculated error region of the recommended values given in Table 3. The error bars represent the experimental errors. (Error bars smaller than the symbols are omitted for clarity of the figure.)

Table 3. Recommended values (fit to the reliable experimental values according to the equations $\rho = A + BT + CT^2 + DT^3 + \ldots$ or $\rho = [1 + 1.75(1 - T/T_c)^{1/3} + 0.75(1 - T/T_c)][\rho_c + A(T_c - T) + B(T_c - T)^2 + C(T_c - T)^3 + D(T_c - T)^4]$).

$\frac{T}{K}$	$\frac{\rho \pm \sigma_{fit}}{kg \cdot m^{-3}}$	$\frac{T}{K}$	$\frac{\rho \pm \sigma_{fit}}{kg \cdot m^{-3}}$	$\frac{T}{K}$	$\frac{\rho \pm \sigma_{fit}}{kg \cdot m^{-3}}$
290.00	827.75 ± 1.27	310.00	811.43 ± 0.98	350.00	778.81 ± 1.09
293.15	825.18 ± 1.23	320.00	803.28 ± 0.86	360.00	770.65 ± 1.48
298.15	821.10 ± 1.16	330.00	795.12 ± 0.81	370.00	762.49 ± 2.10
300.00	819.59 ± 1.13	340.00	786.96 ± 0.88		

2,7-Dimethyl-3-octanol **[66719-55-9]** **$C_{10}H_{22}O$** **MW = 158.28** **239**

Table 1. Experimental value with uncertainty.

$\frac{T}{K}$	$\frac{\rho_{exp} \pm 2\sigma_{est}}{kg \cdot m^{-3}}$	Ref.
293.15	815.2 ± 2.0	12-mic

2,7-Dimethyl-4-octanol **[19781-11-4]** **$C_{10}H_{22}O$** **MW = 158.28** **240**

Table 1. Experimental and recommended values with uncertainties.

$\frac{T}{K}$	$\frac{\rho_{exp} \pm 2\sigma_{est}}{kg \cdot m^{-3}}$	Ref.
293.15	814.0 ± 2.0	36-tuo[1)]
293.15	818.3 ± 1.0	44-pow/hag
293.15	818.3 ± 1.0	Recommended

[1)] Not included in calculation of recommended value.

3,5-Dimethyl-3-octanol **[56065-42-0]** **$C_{10}H_{22}O$** **MW = 158.28** **241**

Table 1. Experimental value with uncertainty.

$\frac{T}{K}$	$\frac{\rho_{exp} \pm 2\sigma_{est}}{kg \cdot m^{-3}}$	Ref.
298.15	837.0 ± 2.0	59-col/gau

3,6-Dimethyl-3-octanol **[151-19-9]** **$C_{10}H_{22}O$** **MW = 158.28** **242**

Table 1. Experimental value with uncertainty.

$\frac{T}{K}$	$\frac{\rho_{exp} \pm 2\sigma_{est}}{kg \cdot m^{-3}}$	Ref.
295.15	834.7 ± 2.0	13-dup

3,7-Dimethyl-1-octanol **[106-21-8]** **$C_{10}H_{22}O$** **MW = 158.28** **243**

Table 1. Experimental and recommended values with uncertainties.

$\frac{T}{K}$	$\frac{\rho_{exp} \pm 2\sigma_{est}}{kg \cdot m^{-3}}$	Ref.	$\frac{T}{K}$	$\frac{\rho_{exp} \pm 2\sigma_{est}}{kg \cdot m^{-3}}$	Ref.
291.15	838.0 ± 2.0	23-von/kai[1)]	296.15	828.1 ± 3.0	54-rin/ari[1)]
293.15	831.0 ± 3.0	24-lon/mar[1)]	293.15	840.2 ± 1.5	58-leb/kuk
284.15	837.4 ± 3.0	28-esc[1)]	293.15	828.6 ± 3.0	60-por/far[1)]
288.15	830.0 ± 6.0	40-pal[1)]	294.15	840.0 ± 2.0	61-yeh[1)]
293.15	835.7 ± 3.0	51-sor/suc[1)]	293.15	840.9 ± 1.5	63-yeh
291.15	837.0 ± 3.0	52-ino[1)]	293.15	840.6 ± 1.5	Recommended

[1)] Not included in calculation of recommended value.

3,7-Dimethyl-2-octanol **[15340-96-2]** **$C_{10}H_{22}O$** **MW = 158.28** **244**

Table 1. Experimental value with uncertainty.

$\frac{T}{K}$	$\frac{\rho_{exp} \pm 2\sigma_{est}}{kg \cdot m^{-3}}$	Ref.
293.15	829.1 ± 1.0	56-nav/des

3,7-Dimethyl-3-octanol **[78-69-3]** $C_{10}H_{22}O$ **MW = 158.28** **245**

Table 1. Experimental and recommended values with uncertainties.

$\frac{T}{K}$	$\frac{\rho_{exp} \pm 2\sigma_{est}}{kg \cdot m^{-3}}$	Ref.
293.15	828.0 ± 2.0	14-wal-1
298.15	864.7 ± 20.0	39-ste/mcn[1)]
292.15	833.9 ± 3.0	40-pal[1)]
293.15	829.5 ± 2.0	58-naz/gus
293.15	828.7 ± 2.1	Recommended

[1)] Not included in calculation of recommended value.

4,6-Dimethyl-4-octanol **[56065-43-1]** $C_{10}H_{22}O$ **MW = 158.28** **246**

Table 1. Experimental value with uncertainty.

$\frac{T}{K}$	$\frac{\rho_{exp} \pm 2\sigma_{est}}{kg \cdot m^{-3}}$	Ref.
301.15	825.8 ± 1.0	60-tha/vas

4,7-Dimethyl-4-octanol **[19781-13-6]** $C_{10}H_{22}O$ **MW = 158.28** **247**

Table 1. Experimental value with uncertainty.

$\frac{T}{K}$	$\frac{\rho_{exp} \pm 2\sigma_{est}}{kg \cdot m^{-3}}$	Ref.
273.15	842.1 ± 2.0	12-gue-1

3-Ethyl-2-methyl-3-heptanol **[66719-37-7]** $C_{10}H_{22}O$ **MW = 158.28** **248**

Table 1. Experimental value with uncertainty.

$\frac{T}{K}$	$\frac{\rho_{exp} \pm 2\sigma_{est}}{kg \cdot m^{-3}}$	Ref.
293.15	845.5 ± 2.0	15-wal-3

3-Ethyl-6-methyl-3-heptanol **[66719-40-2]** $C_{10}H_{22}O$ **MW = 158.28** **249**

Table 1. Fit with estimated *B* coefficient for 4 accepted points. Deviation $\sigma_w = 0.056$.

Coefficient	$\rho = A + BT$
A	1065.11
B	−0.780

Table 2. Experimental values with uncertainties and deviation from calculated values.

$\frac{T}{K}$	$\frac{\rho_{exp} \pm 2\sigma_{est}}{kg \cdot m^{-3}}$	$\frac{\rho_{exp} - \rho_{calc}}{kg \cdot m^{-3}}$	Ref.
273.15	852.0 ± 1.5	−0.06	04-gri
283.55	844.0 ± 1.5	0.06	04-gri
273.15	852.0 ± 1.5	−0.06	04-gri-1
283.55	844.0 ± 1.5	0.06	04-gri-1

cont.

Table 3. Recommended values.

$\frac{T}{\text{K}}$	$\frac{\rho_{exp} \pm 2\sigma_{est}}{\text{kg} \cdot \text{m}^{-3}}$
270.00	854.5 ± 1.5
280.00	846.7 ± 1.3
290.00	838.9 ± 1.6

5-Ethyl-4-methyl-3-heptanol [66731-94-0] $C_{10}H_{22}O$ MW = 158.28 250

Table 1. Experimental value with uncertainty.

$\frac{T}{\text{K}}$	$\frac{\rho_{exp} \pm 2\sigma_{est}}{\text{kg} \cdot \text{m}^{-3}}$	Ref.
290.15	865.0 ± 2.0	43-col/jol

l-3-Ethyl-1-octanol [900002-57-5] C10H22O MW = 158.28 251

Table 1. Experimental value with uncertainty.

$\frac{T}{\text{K}}$	$\frac{\rho_{exp} \pm 2\sigma_{est}}{\text{kg} \cdot \text{m}^{-3}}$	Ref.
295.15	833.0 ± 2.0	31-lev/mar-3

3-Ethyl-3-octanol [2051-32-3] $C_{10}H_{22}O$ MW = 158.28 252

Table 1. Experimental value with uncertainty.

$\frac{T}{\text{K}}$	$\frac{\rho_{exp} \pm 2\sigma_{est}}{\text{kg} \cdot \text{m}^{-3}}$	Ref.
298.15	836.1 ± 1.0	33-whi/wil

4-Ethyl-4-octanol [38395-42-5] $C_{10}H_{22}O$ MW = 158.28 253

Table 1. Coefficients of the polynomial expansion equation.
Standard deviations (see introduction):
$\sigma_{c,w} = 2.1329 \cdot 10^{-1}$ (combined temperature ranges, weighted),
$\sigma_{c,uw} = 9.5387 \cdot 10^{-2}$ (combined temperature ranges, unweighted).

Coefficient	T = 273.15 to 338.15 K $\rho = A + BT + CT^2 + DT^3 + \ldots$
A	$1.07866 \cdot 10^{3}$
B	$-8.43628 \cdot 10^{-1}$

Table 2. Experimental values with uncertainties and deviation from calculated values.

$\frac{T}{\text{K}}$	$\frac{\rho_{exp} \pm 2\sigma_{est}}{\text{kg} \cdot \text{m}^{-3}}$	$\frac{\rho_{exp} - \rho_{calc}}{\text{kg} \cdot \text{m}^{-3}}$	Ref. (Symbol in Fig. 1)	$\frac{T}{\text{K}}$	$\frac{\rho_{exp} \pm 2\sigma_{est}}{\text{kg} \cdot \text{m}^{-3}}$	$\frac{\rho_{exp} - \rho_{calc}}{\text{kg} \cdot \text{m}^{-3}}$	Ref. (Symbol in Fig. 1)
273.15	848.50 ± 1.00	0.28	39-owe/qua(×)	318.15	810.20 ± 1.00	−0.06	39-owe/qua(×)
298.15	826.80 ± 1.00	−0.33	39-owe/qua(×)	328.15	802.00 ± 1.00	0.18	39-owe/qua(×)
308.15	818.50 ± 1.00	−0.19	39-owe/qua(×)	338.15	793.50 ± 1.00	0.12	39-owe/qua(×)

cont.

4-Ethyl-4-octanol (cont.)

Further references: [33-whi/woo].

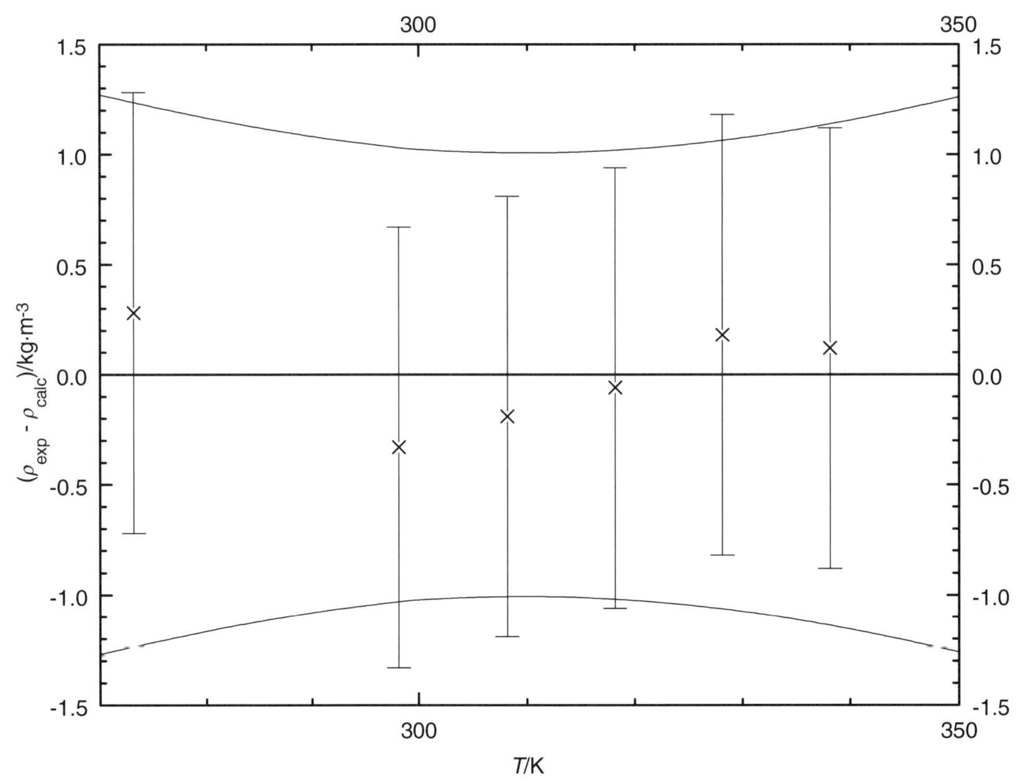

Fig. 1. The symbols show the deviation of the calculated from the experimental values from Table 2. The curves above and below the zero line indicate the calculated error region of the recommended values given in Table 3. The error bars represent the experimental errors. (Error bars smaller than the symbols are omitted for clarity of the figure.)

Table 3. Recommended values (fit to the reliable experimental values according to the equations
$\rho = A + BT + CT^2 + DT^3 + \ldots$ or $\rho = [1 + 1.75(1 - T/T_c)^{1/3} + 0.75(1 - T/T_c)][\rho_c + A(T_c - T) + B(T_c - T)^2 + C(T_c - T)^3 + D(T_c - T)^4]$).

$\frac{T}{\text{K}}$	$\frac{\rho \pm \sigma_{fit}}{\text{kg} \cdot \text{m}^{-3}}$	$\frac{T}{\text{K}}$	$\frac{\rho \pm \sigma_{fit}}{\text{kg} \cdot \text{m}^{-3}}$	$\frac{T}{\text{K}}$	$\frac{\rho \pm \sigma_{fit}}{\text{kg} \cdot \text{m}^{-3}}$
270.00	850.88 ± 1.27	298.15	827.13 ± 1.03	330.00	800.26 ± 1.07
280.00	842.44 ± 1.16	300.00	825.57 ± 1.02	340.00	791.82 ± 1.15
290.00	834.00 ± 1.08	310.00	817.13 ± 1.00	350.00	783.39 ± 1.26
293.15	831.35 ± 1.06	320.00	808.70 ± 1.02		

6-Ethyl-3-octanol **[19781-27-2]** $C_{10}H_{22}O$ **MW = 158.28** **254**

Table 1. Experimental value with uncertainty.

$\frac{T}{\text{K}}$	$\frac{\rho_{exp} \pm 2\sigma_{est}}{\text{kg} \cdot \text{m}^{-3}}$	Ref.
297.65	839.6 ± 0.5	36-pow/bal

3-Ethyl-2,2,4-trimethyl-3-pentanol [66256-41-5] $C_{10}H_{22}O$ MW = 158.28 255

Table 1. Experimental and recommended values with uncertainties.

T / K	$\rho_{exp} \pm 2\sigma_{est}$ / $kg \cdot m^{-3}$	Ref.
292.15	862.8 ± 1.5	36-naz[1)]
293.15	862.0 ± 1.0	51-smi/cre
293.15	862.8 ± 1.0	54-mos/cox
293.15	862.4 ± 1.0	Recommended

[1)] Not included in calculation of recommended value.

2-Methyl-3-(1-methylethyl)-3-hexanol [51200-81-8] $C_{10}H_{22}O$ MW = 158.28 256

Table 1. Experimental and recommended values with uncertainties.

T / K	$\rho_{exp} \pm 2\sigma_{est}$ / $kg \cdot m^{-3}$	Ref.
293.15	853.7 ± 2.0	26-sta
293.15	855.4 ± 3.0	43-geo[1)]
293.15	850.4 ± 1.0	48-cad/foo
293.15	851.1 ± 1.2	Recommended

[1)] Not included in calculation of recommended value.

5-Methyl-2-(1-methylethyl)-1-hexanol [2051-33-4] $C_{10}H_{22}O$ MW = 158.28 257

Table 1. Experimental and recommended values with uncertainties.

T / K	$\rho_{exp} \pm 2\sigma_{est}$ / $kg \cdot m^{-3}$	Ref.
293.15	832.2 ± 2.0	25-ter[1)]
290.15	836.6 ± 1.0	47-sch/sch
290.15	836.6 ± 1.0	Recommended

[1)] Not included in calculation of recommended value.

4-Methyl-2-(2-methylpropyl)-1-pentanol [22417-45-4] $C_{10}H_{22}O$ MW = 158.28 258

Table 1. Experimental values with uncertainties.

T / K	$\rho_{exp} \pm 2\sigma_{est}$ / $kg \cdot m^{-3}$	Ref.
277.15	846.0 ± 2.0	04-bou/bla
273.15	846.0 ± 2.0	10-fre

2-Methyl-1-nonanol [40589-14-8] $C_{10}H_{22}O$ MW = 158.28 259

Table 1. Fit with estimated B coefficient for 2 accepted points. Deviation σ_w = 0.350.

Coefficient	$\rho = A + BT$
A	1058.41
B	–0.780

cont.

2-Methyl-1-nonanol (cont.)

Table 2. Experimental values with uncertainties and deviation from calculated values.

$\frac{T}{K}$	$\frac{\rho_{exp} \pm 2\sigma_{est}}{kg \cdot m^{-3}}$	$\frac{\rho_{exp} - \rho_{calc}}{kg \cdot m^{-3}}$	Ref.
273.15	845.7 ± 2.0	0.35	02-gue
288.15	833.3 ± 2.0	–0.35	02-gue

Table 3. Recommended values.

$\frac{T}{K}$	$\frac{\rho_{exp} \pm 2\sigma_{est}}{kg \cdot m^{-3}}$
270.00	847.8 ± 2.1
280.00	840.0 ± 1.8
290.00	832.2 ± 2.0

2-Methyl-3-nonanol **[26533-33-5]** **$C_{10}H_{22}O$** **MW = 158.28** **260**

Table 1. Fit with estimated B coefficient for 4 accepted points. Deviation $\sigma_w = 0.309$.

Coefficient	$\rho = A + BT$
A	1051.17
B	–0.760

Table 2. Experimental values with uncertainties and deviation from calculated values.

$\frac{T}{K}$	$\frac{\rho_{exp} \pm 2\sigma_{est}}{kg \cdot m^{-3}}$	$\frac{\rho_{exp} - \rho_{calc}}{kg \cdot m^{-3}}$	Ref.
293.15	829.0 ± 1.0	0.63	12-pic/ken
293.15	828.7 ± 1.0	0.33	48-pet/old
293.15	828.1 ± 0.6	–0.27	50-mea/foo
298.15	824.5 ± 0.6	–0.07	50-mea/foo

Table 3. Recommended values.

$\frac{T}{K}$	$\frac{\rho_{exp} \pm 2\sigma_{est}}{kg \cdot m^{-3}}$
290.00	830.8 ± 0.8
293.15	828.4 ± 0.8
298.15	824.6 ± 0.8

2-Methyl-4-nonanol **[26533-31-3]** **$C_{10}H_{22}O$** **MW = 158.28** **261**

Table 1. Experimental value with uncertainty.

$\frac{T}{K}$	$\frac{\rho_{exp} \pm 2\sigma_{est}}{kg \cdot m^{-3}}$	Ref.
298.15	820.0 ± 2.0	54-dub/luf

2-Methyl-5-nonanol **[29843-62-7]** $C_{10}H_{22}O$ **MW = 158.28** **262**

Table 1. Experimental value with uncertainty.

T/K	$\rho_{exp} \pm 2\sigma_{est}$ / kg·m^{-3}	Ref.
293.15	821.5 ± 1.0	44-pow/hag

3-Methyl-1-nonanol **[22663-64-5]** $C_{10}H_{22}O$ **MW = 158.28** **263**

Table 1. Experimental and recommended values with uncertainties.

T/K	$\rho_{exp} \pm 2\sigma_{est}$ / kg·m^{-3}	Ref.
293.15	834.2 ± 3.0	22-lev/tay-1 [1)]
296.15	837.0 ± 2.0	31-lev/mar-5
296.15	837.0 ± 2.0	Recommended

[1)] Not included in calculation of recommended value.

(*R*)-3-Methyl-1-nonanol **[86414-45-1]** $C_{10}H_{22}O$ **MW = 158.28** **264**

Table 1. Experimental value with uncertainty.

T/K	$\rho_{exp} \pm 2\sigma_{est}$ / kg·m^{-3}	Ref.
293.15	829.0 ± 2.0	58-leg/ulr

3-Methyl-2-nonanol **[60671-32-1]** $C_{10}H_{22}O$ **MW = 158.28** **265**

Table 1. Experimental and recommended values with uncertainties.

T/K	$\rho_{exp} \pm 2\sigma_{est}$ / kg·m^{-3}	Ref.
295.15	845.0 ± 3.0	35-gre-2 [1)]
293.15	835.3 ± 2.0	57-pet/nef-1
293.15	835.3 ± 2.0	Recommended

[1)] Not included in calculation of recommended value.

3-Methyl-3-nonanol **[21078-72-8]** $C_{10}H_{22}O$ **MW = 158.28** **266**

Table 1. Coefficients of the polynomial expansion equation.
Standard deviations (see introduction):
$\sigma_{c,w} = 1.0876$ (combined temperature ranges, weighted),
$\sigma_{c,uw} = 6.1414 \cdot 10^{-1}$ (combined temperature ranges, unweighted).

Coefficient	T = 273.15 to 338.15 K $\rho = A + BT + CT^2 + DT^3 + \ldots$
A	$1.07303 \cdot 10^{3}$
B	$-8.27037 \cdot 10^{-1}$

cont.

3-Methyl-3-nonanol (cont.)

Table 2. Experimental values with uncertainties and deviation from calculated values.

$\frac{T}{K}$	$\frac{\rho_{exp} \pm 2\sigma_{est}}{kg \cdot m^{-3}}$	$\frac{\rho_{exp} - \rho_{calc}}{kg \cdot m^{-3}}$	Ref. (Symbol in Fig. 1)	$\frac{T}{K}$	$\frac{\rho_{exp} \pm 2\sigma_{est}}{kg \cdot m^{-3}}$	$\frac{\rho_{exp} - \rho_{calc}}{kg \cdot m^{-3}}$	Ref. (Symbol in Fig. 1)
296.15	829.00 ± 3.00	0.90	35-gre-2(∇)	318.15	809.20 ± 1.00	−0.70	39-owe/qua(O)
293.15	831.10 ± 1.00	0.52	38-whi/ore(□)	328.15	801.00 ± 1.00	−0.63	39-owe/qua(O)
273.15	845.50 ± 1.00	−1.62	39-owe/qua(O)	338.15	792.70 ± 1.00	−0.66	39-owe/qua(O)
298.15	825.10 ± 1.00	−1.34	39-owe/qua(O)	293.15	835.00 ± 3.00	4.42	61-sok/she(Δ)
308.15	817.30 ± 1.00	−0.87	39-owe/qua(O)				

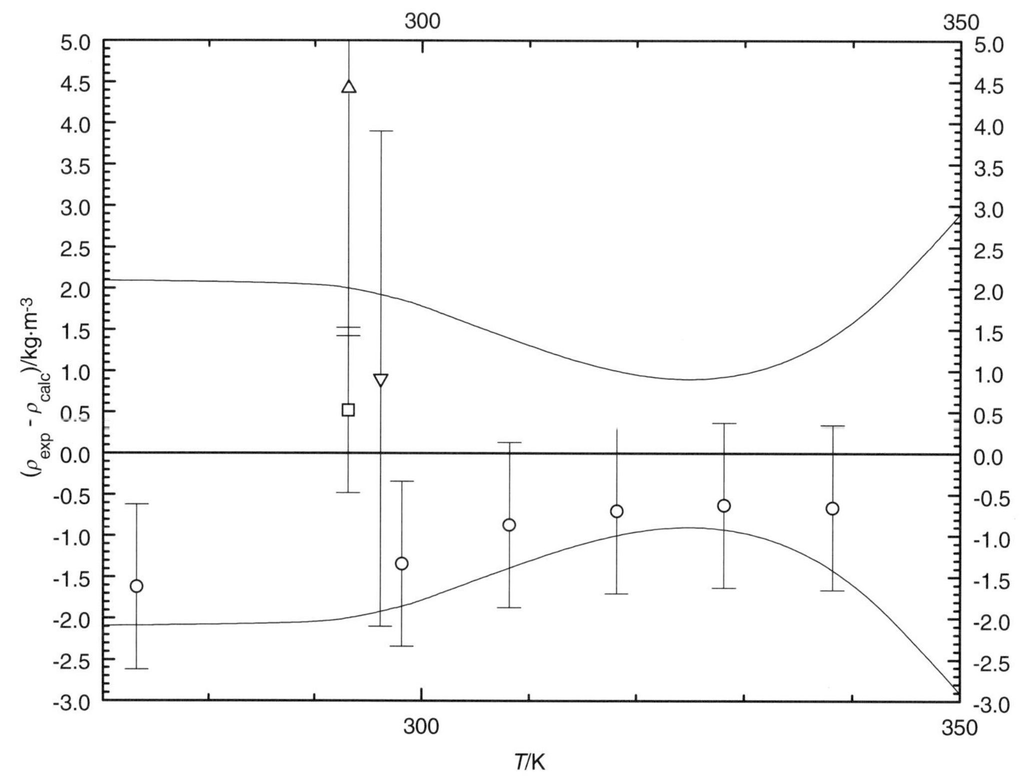

Fig. 1. The symbols show the deviation of the calculated from the experimental values from Table 2. The curves above and below the zero line indicate the calculated error region of the recommended values given in Table 3. The error bars represent the experimental errors. (Error bars smaller than the symbols are omitted for clarity of the figure.)

Table 3. Recommended values (fit to the reliable experimental values according to the equations $\rho = A + BT + CT^2 + DT^3 + \ldots$ or $\rho = [1 + 1.75(1 - T/T_c)^{1/3} + 0.75(1 - T/T_c)][\rho_c + A(T_c - T) + B(T_c - T)^2 + C(T_c - T)^3 + D(T_c - T)^4]$).

$\frac{T}{K}$	$\frac{\rho \pm \sigma_{fit}}{kg \cdot m^{-3}}$	$\frac{T}{K}$	$\frac{\rho \pm \sigma_{fit}}{kg \cdot m^{-3}}$	$\frac{T}{K}$	$\frac{\rho \pm \sigma_{fit}}{kg \cdot m^{-3}}$
270.00	849.73 ± 2.09	298.15	826.44 ± 1.86	330.00	800.10 ± 0.87
280.00	841.46 ± 2.08	300.00	824.91 ± 1.79	340.00	791.83 ± 1.44
290.00	833.19 ± 2.05	310.00	816.64 ± 1.29	350.00	783.56 ± 2.91
293.15	830.58 ± 2.01	320.00	808.37 ± 0.89		

_d_-3-Methyl-5-nonanol **[500021-26-1]** **$C_{10}H_{22}O$** **MW = 158.28** **267**

Table 1. Experimental value with uncertainty.

T/K	$\rho_{exp} \pm 2\sigma_{est}$ / $kg \cdot m^{-3}$	Ref.
298.15	821.0 ± 2.0	50-let/tra

4-Methyl-1-nonanol **[1489-47-0]** **$C_{10}H_{22}O$** **MW = 158.28** **268**

Table 1. Experimental value with uncertainty.

T/K	$\rho_{exp} \pm 2\sigma_{est}$ / $kg \cdot m^{-3}$	Ref.
300.15	826.0 ± 2.0	31-lev/mar-5

4-Methyl-4-nonanol **[23418-38-4]** **$C_{10}H_{22}O$** **MW = 158.28** **269**

Table 1. Fit with estimated B coefficient for 3 accepted points. Deviation $\sigma_w = 0.047$.

Coefficient	$\rho = A + BT$
A	1071.65
B	−0.820

Table 2. Experimental values with uncertainties and deviation from calculated values.

T/K	$\rho_{exp} \pm 2\sigma_{est}$ / $kg \cdot m^{-3}$	$\rho_{exp} - \rho_{calc}$ / $kg \cdot m^{-3}$	Ref.
298.15	824.5 ± 2.0	−2.67	33-whi/wil[1)]
298.15	827.1 ± 1.0	−0.07	44-qua/sma
318.15	810.8 ± 1.0	0.03	44-qua/sma
328.15	802.6 ± 1.0	0.03	44-qua/sma

[1)] Not included in calculation of linear coefficients.

Table 3. Recommended values.

T/K	$\rho_{exp} \pm 2\sigma_{est}$ / $kg \cdot m^{-3}$	T/K	$\rho_{exp} \pm 2\sigma_{est}$ / $kg \cdot m^{-3}$	T/K	$\rho_{exp} \pm 2\sigma_{est}$ / $kg \cdot m^{-3}$
290.00	833.8 ± 1.4	298.15	827.2 ± 1.1	320.00	809.2 ± 0.7
293.15	831.3 ± 1.3	310.00	817.4 ± 0.7	330.00	801.0 ± 1.0

5-Methyl-1-nonanol **[2768-16-3]** **$C_{10}H_{22}O$** **MW = 158.28** **270**

Table 1. Experimental value with uncertainty.

T/K	$\rho_{exp} \pm 2\sigma_{est}$ / $kg \cdot m^{-3}$	Ref.
297.15	831.0 ± 2.0	33-lev/mar-1

5-Methyl-4-nonanol **[66719-44-6]** $C_{10}H_{22}O$ **MW = 158.28** **271**

Table 1. Experimental value with uncertainty.

T / K	$\rho_{exp} \pm 2\sigma_{est}$ / $kg \cdot m^{-3}$	Ref.
300.15	826.0 ± 2.0	48-pow/nie

5-Methyl-5-nonanol **[33933-78-7]** $C_{10}H_{22}O$ **MW = 158.28** **272**

Table 1. Fit with estimated B coefficient for 5 accepted points. Deviation $\sigma_w = 0.264$.

Coefficient	$\rho = A + BT$
A	1063.90
B	−0.800

Table 2. Experimental values with uncertainties and deviation from calculated values.

T / K	$\rho_{exp} \pm 2\sigma_{est}$ / $kg \cdot m^{-3}$	$\rho_{exp} - \rho_{calc}$ / $kg \cdot m^{-3}$	Ref.	T / K	$\rho_{exp} \pm 2\sigma_{est}$ / $kg \cdot m^{-3}$	$\rho_{exp} - \rho_{calc}$ / $kg \cdot m^{-3}$	Ref.
293.15	829.0 ± 1.0	−0.38	33-whi/woo	328.15	801.3 ± 1.0	−0.08	44-qua/sma
298.15	825.3 ± 1.0	−0.08	33-whi/woo	318.15	809.5 ± 1.0	0.12	44-qua/sma
298.15	825.8 ± 1.0	0.42	44-qua/sma	293.15	830.5 ± 2.0	1.12	59-yur/bel[1)]

[1)] Not included in calculation of recommended value.

Table 3. Recommended values.

T / K	$\rho_{exp} \pm 2\sigma_{est}$ / $kg \cdot m^{-3}$	T / K	$\rho_{exp} \pm 2\sigma_{est}$ / $kg \cdot m^{-3}$	T / K	$\rho_{exp} \pm 2\sigma_{est}$ / $kg \cdot m^{-3}$
290.00	831.9 ± 1.2	298.15	825.4 ± 0.9	320.00	807.9 ± 1.0
293.15	829.4 ± 1.1	310.00	815.9 ± 0.8	330.00	799.9 ± 1.4

6-Methyl-2-nonanol **[66256-60-8]** $C_{10}H_{22}O$ **MW = 158.28** **273**

Table 1. Experimental value with uncertainty.

T / K	$\rho_{exp} \pm 2\sigma_{est}$ / $kg \cdot m^{-3}$	Ref.
293.15	833.2 ± 1.0	61-shv/pet

7-Methyl-1-nonanol **[33234-93-4]** $C_{10}H_{22}O$ **MW = 158.28** **274**

Table 1. Experimental value with uncertainty.

T / K	$\rho_{exp} \pm 2\sigma_{est}$ / $kg \cdot m^{-3}$	Ref.
298.15	828.2 ± 1.0	62-lar/sal

***L*(+)-7-Methyl-1-nonanol** **[500006-91-7]** $C_{10}H_{22}O$ **MW = 158.28** **275**

Table 1. Experimental value with uncertainty.

T / K	$\rho_{exp} \pm 2\sigma_{est}$ / $kg \cdot m^{-3}$	Ref.
298.15	828.2 ± 0.6	62-lar/sal

3-(1-Methylethyl)-1-heptanol **[38514-15-7]** $C_{10}H_{22}O$ **MW = 158.28** **276**

Table 1. Experimental value with uncertainty.

T / K	$\rho_{exp} \pm 2\sigma_{est}$ / $kg \cdot m^{-3}$	Ref.
303.15	834.1 ± 1.0	57-kit

4-(1-Methylethyl)-4-heptanol **[51200-82-9]** $C_{10}H_{22}O$ **MW = 158.28** **277**

Table 1. Fit with estimated B coefficient for 3 accepted points. Deviation $\sigma_w = 0.713$.

Coefficient	$\rho = A + BT$
A	1067.93
B	–0.760

Table 2. Experimental values with uncertainties and deviation from calculated values.

T / K	$\rho_{exp} \pm 2\sigma_{est}$ / $kg \cdot m^{-3}$	$\rho_{exp} - \rho_{calc}$ / $kg \cdot m^{-3}$	Ref.
293.15	844.9 ± 2.0	–0.23	26-sta
273.15	859.6 ± 2.0	–0.73	26-sta
303.15	838.5 ± 2.0	0.97	26-sta
293.15	847.1 ± 3.0	1.97	54-naz/kak-3[1)]

[1)] Not included in calculation of recommended value.

Table 3. Recommended values.

T / K	$\rho_{exp} \pm 2\sigma_{est}$ / $kg \cdot m^{-3}$	T / K	$\rho_{exp} \pm 2\sigma_{est}$ / $kg \cdot m^{-3}$	T / K	$\rho_{exp} \pm 2\sigma_{est}$ / $kg \cdot m^{-3}$
270.00	862.7 ± 2.5	290.00	847.5 ± 1.5	298.15	841.3 ± 1.7
280.00	855.1 ± 1.8	293.15	845.1 ± 1.6	310.00	832.3 ± 2.5

2-(1-Methylpropyl)-3-methyl-1-pentanol **[91717-78-1]** $C_{10}H_{22}O$ **MW = 158.28** **278**

Table 1. Experimental value with uncertainty.

T / K	$\rho_{exp} \pm 2\sigma_{est}$ / $kg \cdot m^{-3}$	Ref.
295.15	835.0 ± 2.0	62-thi

2-Propyl-1-heptanol **[10042-59-8]** $C_{10}H_{22}O$ **MW = 158.28** **279**

Table 1. Experimental value with uncertainty.

T / K	$\rho_{exp} \pm 2\sigma_{est}$ / $kg \cdot m^{-3}$	Ref.
293.15	832.2 ± 0.4	56-ano-4

4-Propyl-4-heptanol **[2198-72-3]** $C_{10}H_{22}O$ **MW = 158.28** **280**

Table 1. Fit with estimated *B* coefficient for 3 accepted points. Deviation $\sigma_w = 0.630$.

Coefficient	$\rho = A + BT$
A	1067.17
B	–0.800

Table 2. Experimental values with uncertainties and deviation from calculated values.

$\frac{T}{\text{K}}$	$\frac{\rho_{exp} \pm 2\sigma_{est}}{\text{kg} \cdot \text{m}^{-3}}$	$\frac{\rho_{exp} - \rho_{calc}}{\text{kg} \cdot \text{m}^{-3}}$	Ref.
294.15	833.7 ± 2.0	1.85	02-kon[1)]
290.35	835.0 ± 1.0	0.11	19-eyk
293.15	834.0 ± 1.0	1.35	61-mes/erz
298.15	828.3 ± 0.5	–0.37	88-cac/cos

[1)] Not included in calculation of linear coefficients.

Table 3. Recommended values.

$\frac{T}{\text{K}}$	$\frac{\rho_{exp} \pm 2\sigma_{est}}{\text{kg} \cdot \text{m}^{-3}}$
290.00	835.2 ± 1.0
293.15	832.6 ± 0.9
298.15	828.6 ± 0.9

2,2,3,4-Tetramethyl-3-hexanol **[66256-63-1]** $C_{10}H_{22}O$ **MW = 158.28** **281**

Table 1. Fit with estimated *B* coefficient for 2 accepted points. Deviation $\sigma_w = 0.080$.

Coefficient	$\rho = A + BT$
A	1104.69
B	–0.840

Table 2. Experimental values with uncertainties and deviation from calculated values.

$\frac{T}{\text{K}}$	$\frac{\rho_{exp} \pm 2\sigma_{est}}{\text{kg} \cdot \text{m}^{-3}}$	$\frac{\rho_{exp} - \rho_{calc}}{\text{kg} \cdot \text{m}^{-3}}$	Ref.
283.15	867.0 ± 2.0	0.16	37-naz
293.15	858.4 ± 1.0	–0.04	50-ste/coo

Table 3. Recommended values.

$\frac{T}{\text{K}}$	$\frac{\rho_{exp} \pm 2\sigma_{est}}{\text{kg} \cdot \text{m}^{-3}}$
280.00	869.5 ± 1.7
290.00	861.1 ± 1.3
293.15	858.4 ± 1.4
298.15	854.2 ± 1.5

2,2,3,5-Tetramethyl-3-hexanol [66256-64-2] $C_{10}H_{22}O$ MW = 158.28 282

Table 1. Experimental values with uncertainties.

$\frac{T}{K}$	$\frac{\rho_{exp} \pm 2\sigma_{est}}{kg \cdot m^{-3}}$	Ref.
293.15	839.3 ± 1.0	55-pet
293.15	839.3 ± 1.0	55-pet/sus

2,2,4,4-Tetramethyl-3-hexanol [66256-65-3] $C_{10}H_{22}O$ MW = 158.28 283

Table 1. Experimental value with uncertainty.

$\frac{T}{K}$	$\frac{\rho_{exp} \pm 2\sigma_{est}}{kg \cdot m^{-3}}$	Ref.
293.15	854.9 ± 1.0	34-sta

2,3,4,4-Tetramethyl-3-hexanol [66256-67-5] $C_{10}H_{22}O$ MW = 158.28 284

Table 1. Experimental and recommended values with uncertainties.

$\frac{T}{K}$	$\frac{\rho_{exp} \pm 2\sigma_{est}}{kg \cdot m^{-3}}$	Ref.
289.15	876.0 ± 2.0	37-naz[1)]
293.15	869.4 ± 3.0	40-mos[1)]
293.15	874.5 ± 1.0	50-ste/coo
293.15	874.5 ± 1.0	Recommended

[1)] Not included in calculation of recommended value.

2,3,5,5-Tetramethyl-3-hexanol [5396-09-8] $C_{10}H_{22}O$ MW = 158.28 285

Table 1. Experimental value with uncertainty.

$\frac{T}{K}$	$\frac{\rho_{exp} \pm 2\sigma_{est}}{kg \cdot m^{-3}}$	Ref.
293.15	837.8 ± 2.0	42-moe

3,4,4,5-Tetramethyl-3-hexanol [66256-39-1] $C_{10}H_{22}O$ MW = 158.28 286

Table 1. Experimental value with uncertainty.

$\frac{T}{K}$	$\frac{\rho_{exp} \pm 2\sigma_{est}}{kg \cdot m^{-3}}$	Ref.
293.15	874.2 ± 1.0	50-ste/coo

3,4,5,5-Tetramethyl-3-hexanol [66256-40-4] $C_{10}H_{22}O$ MW = 158.28 287

Table 1. Experimental value with uncertainty.

$\frac{T}{K}$	$\frac{\rho_{exp} \pm 2\sigma_{est}}{kg \cdot m^{-3}}$	Ref.
293.15	862.3 ± 1.0	50-ste/coo

2,2,3-Trimethyl-3-heptanol **[29772-40-5]** $C_{10}H_{22}O$ **MW = 158.28** **288**

Table 1. Experimental value with uncertainty.

$\frac{T}{K}$	$\frac{\rho_{exp} \pm 2\sigma_{est}}{kg \cdot m^{-3}}$	Ref.
293.15	848.7 ± 1.0	29-con/bla

2,2,4-Trimethyl-4-heptanol **[57233-31-5]** $C_{10}H_{22}O$ **MW = 158.28** **289**

Table 1. Experimental values with uncertainties.

$\frac{T}{K}$	$\frac{\rho_{exp} \pm 2\sigma_{est}}{kg \cdot m^{-3}}$	Ref.
293.15	833.0 ± 2.0	42-moe
293.15	833.0 ± 2.0	49-moe/whi

2,2,5-Trimethyl-4-heptanol **[66256-42-6]** $C_{10}H_{22}O$ **MW = 158.28** **290**

Table 1. Experimental value with uncertainty.

$\frac{T}{K}$	$\frac{\rho_{exp} \pm 2\sigma_{est}}{kg \cdot m^{-3}}$	Ref.
289.15	751.3 ± 1.0	39-pet/sum

2,2,6-Trimethyl-3-heptanol **[66256-43-7]** $C_{10}H_{22}O$ **MW = 158.28** **291**

Table 1. Experimental value with uncertainty.

$\frac{T}{K}$	$\frac{\rho_{exp} \pm 2\sigma_{est}}{kg \cdot m^{-3}}$	Ref.
293.15	823.6 ± 1.0	38-whi/mey

2,3,6-Trimethyl-3-heptanol **[58046-40-5]** $C_{10}H_{22}O$ **MW = 158.28** **292**

Table 1. Experimental value with uncertainty.

$\frac{T}{K}$	$\frac{\rho_{exp} \pm 2\sigma_{est}}{kg \cdot m^{-3}}$	Ref.
293.15	839.8 ± 1.0	63-tho/pal

2,4,6-Trimethyl-4-heptanol **[60836-07-9]** $C_{10}H_{22}O$ **MW = 158.28** **293**

Table 1. Experimental and recommended values with uncertainties.

$\frac{T}{K}$	$\frac{\rho_{exp} \pm 2\sigma_{est}}{kg \cdot m^{-3}}$	Ref.
294.15	823.0 ± 2.0	09-bod/tab[1)]
293.15	818.6 ± 3.0	33-mey/tuo[1)]
293.15	824.1 ± 1.0	59-pet/zak-1
293.15	824.1 ± 1.0	Recommended

[1)] Not included in calculation of recommended value.

2,5,5-Trimethyl-4-heptanol [66256-49-3] $C_{10}H_{22}O$ MW = 158.28 294

Table 1. Experimental value with uncertainty.

$\frac{T}{K}$	$\frac{\rho_{exp} \pm 2\sigma_{est}}{kg \cdot m^{-3}}$	Ref.
298.15	838.0 ± 2.0	57-tak/nak

2,5,6-Trimethyl-2-heptanol [66256-48-2] $C_{10}H_{22}O$ MW = 158.28 295

Table 1. Experimental value with uncertainty.

$\frac{T}{K}$	$\frac{\rho_{exp} \pm 2\sigma_{est}}{kg \cdot m^{-3}}$	Ref.
291.15	833.0 ± 2.0	11-wal

3,5,5-Trimethyl-3-heptanol [66256-50-6] $C_{10}H_{22}O$ MW = 158.28 296

Table 1. Experimental and recommended values with uncertainties.

$\frac{T}{K}$	$\frac{\rho_{exp} \pm 2\sigma_{est}}{kg \cdot m^{-3}}$	Ref.
293.15	851.3 ± 2.0	40-mos
293.15	851.3 ± 2.0	41-whi/mos
293.15	855.8 ± 1.0	50-ste/coo
293.15	854.3 ± 1.6	Recommended

4,6,6-Trimethyl-2-heptanol [51079-79-9] $C_{10}H_{22}O$ MW = 158.28 297

Table 1. Experimental value with uncertainty.

$\frac{T}{K}$	$\frac{\rho_{exp} \pm 2\sigma_{est}}{kg \cdot m^{-3}}$	Ref.
293.15	826.5 ± 2.0	46-dou-1

2.1.7 Alkanols, C_{11} - C_{12}

3-Butyl-2,4-dimethyl-3-pentanol **[900002-65-5]** **$C_{11}H_{24}O$** **MW = 172.31** **298**

Table 1. Experimental value with uncertainty.

$\frac{T}{K}$	$\frac{\rho_{exp} \pm 2\sigma_{est}}{kg \cdot m^{-3}}$	Ref.
293.15	860.0 ± 2.0	44-you/rob

3-Butyl-2-heptanol **[115667-95-3]** **$C_{11}H_{24}O$** **MW = 172.31** **299**

Table 1. Experimental value with uncertainty.

$\frac{T}{K}$	$\frac{\rho_{exp} \pm 2\sigma_{est}}{kg \cdot m^{-3}}$	Ref.
291.05	839.2 ± 1.0	25-hes/bap

2,6-Dimethyl-4-ethyl-4-heptanol **[54460-99-0]** **$C_{11}H_{24}O$** **MW = 172.31** **300**

Table 1. Experimental value with uncertainty.

$\frac{T}{K}$	$\frac{\rho_{exp} \pm 2\sigma_{est}}{kg \cdot m^{-3}}$	Ref.
298.15	825.2 ± 0.8	56-kru/cho

3,3-Dimethyl-5-ethyl-4-heptanol **[500000-48-6]** **$C_{11}H_{24}O$** **MW = 172.31** **301**

Table 1. Experimental value with uncertainty.

$\frac{T}{K}$	$\frac{\rho_{exp} \pm 2\sigma_{est}}{kg \cdot m^{-3}}$	Ref.
293.15	849.6 ± 1.0	41-whi/whi

l-2,4-Dimethyl-3-(1-methylethyl)-3-hexanol **[28357-71-3]** **$C_{11}H_{24}O$** **MW = 172.31** **302**

Table 1. Experimental value with uncertainty.

$\frac{T}{K}$	$\frac{\rho_{exp} \pm 2\sigma_{est}}{kg \cdot m^{-3}}$	Ref.
293.15	868.0 ± 2.0	44-you/rob

2,5-Dimethyl-3-(1-methylethyl)-3-hexanol **[57233-26-8]** **$C_{11}H_{24}O$** **MW = 172.31** **303**

Table 1. Experimental and recommended values with uncertainties.

$\frac{T}{K}$	$\frac{\rho_{exp} \pm 2\sigma_{est}}{kg \cdot m^{-3}}$	Ref.
293.15	847.0 ± 2.0	43-geo
293.15	848.6 ± 1.0	48-cad/foo
293.15	848.3 ± 1.0	Recommended

2,2-Dimethyl-1-nonanol **[14250-80-7]** $C_{11}H_{24}O$ **MW = 172.31** **304**

Table 1. Experimental value with uncertainty.

$\frac{T}{K}$	$\frac{\rho_{exp} \pm 2\sigma_{est}}{kg \cdot m^{-3}}$	Ref.
293.15	839.2 ± 1.0	65-shu/puz

2,2-Dimethyl-4-nonanol **[38206-58-5]** $C_{11}H_{24}O$ **MW = 172.31** **305**

Table 1. Experimental value with uncertainty.

$\frac{T}{K}$	$\frac{\rho_{exp} \pm 2\sigma_{est}}{kg \cdot m^{-3}}$	Ref.
293.15	822.5 ± 1.0	38-whi/pop-1

2,4-Dimethyl-4-nonanol **[74356-31-3]** $C_{11}H_{24}O$ **MW = 172.31** **306**

Table 1. Experimental value with uncertainty.

$\frac{T}{K}$	$\frac{\rho_{exp} \pm 2\sigma_{est}}{kg \cdot m^{-3}}$	Ref.
293.15	828.5 ± 2.0	59-pet/zak

2,6-Dimethyl-5-nonanol **[500001-10-5]** $C_{11}H_{24}O$ **MW = 172.31** **307**

Table 1. Experimental value with uncertainty.

$\frac{T}{K}$	$\frac{\rho_{exp} \pm 2\sigma_{est}}{kg \cdot m^{-3}}$	Ref.
298.15	821.6 ± 0.8	12-bje

3,5-Dimethyl-5-nonanol **[106593-61-7]** $C_{11}H_{24}O$ **MW = 172.31** **308**

Table 1. Experimental value with uncertainty.

$\frac{T}{K}$	$\frac{\rho_{exp} \pm 2\sigma_{est}}{kg \cdot m^{-3}}$	Ref.
301.15	826.4 ± 1.0	60-tha/vas

4,8-Dimethyl-1-nonanol **[33933-80-1]** $C_{11}H_{24}O$ **MW = 172.31** **309**

Table 1. Experimental values with uncertainties.

$\frac{T}{K}$	$\frac{\rho_{exp} \pm 2\sigma_{est}}{kg \cdot m^{-3}}$	Ref.
292.15	834.0 ± 2.0	23-von/kai
291.15	833.0 ± 2.0	23-von/kai

4,8-Dimethyl-4-nonanol **[91337-13-2]** **$C_{11}H_{24}O$** **MW = 172.31** **310**

Table 1. Experimental value with uncertainty.

T / K	$\rho_{exp} \pm 2\sigma_{est}$ / $kg \cdot m^{-3}$	Ref.
284.15	845.8 ± 2.0	28-esc

2,2-Dimethyl-3-propyl-3-hexanol **[900002-63-3]** **$C_{11}H_{24}O$** **MW = 172.31** **311**

Table 1. Experimental value with uncertainty.

T / K	$\rho_{exp} \pm 2\sigma_{est}$ / $kg \cdot m^{-3}$	Ref.
291.15	853.0 ± 2.0	21-ler

4,4-Dimethyl-3-(1,1-dimethylethyl)-1-pentanol **[79802-55-4]** **$C_{11}H_{24}O$** **MW = 172.31** **312**

Table 1. Experimental value with uncertainty.

T / K	$\rho_{exp} \pm 2\sigma_{est}$ / $kg \cdot m^{-3}$	Ref.
293.15	866.1 ± 2.0	60-pet/sok

5-Ethyl-2-nonanol **[103-08-2]** **$C_{11}H_{24}O$** **MW = 172.31** **313**

Table 1. Fit with estimated B coefficient for 2 accepted points. Deviation $\sigma_w = 0.150$.

Coefficient	$\rho = A + BT$
A	1063.31
B	–0.780

Table 2. Experimental values with uncertainties and deviation from calculated values.

T / K	$\rho_{exp} \pm 2\sigma_{est}$ / $kg \cdot m^{-3}$	$\rho_{exp} - \rho_{calc}$ / $kg \cdot m^{-3}$	Ref.
303.15	826.7 ± 1.0	–0.15	48-wei
293.15	834.8 ± 1.0	0.15	58-ano-13

Table 3. Recommended values.

T / K	$\rho_{exp} \pm 2\sigma_{est}$ / $kg \cdot m^{-3}$	T / K	$\rho_{exp} \pm 2\sigma_{est}$ / $kg \cdot m^{-3}$
290.00	837.1 ± 1.0	298.15	830.7 ± 0.9
293.15	834.7 ± 0.9	310.00	821.5 ± 1.1

5-Ethyl-3-nonanol **[19780-71-3]** **$C_{11}H_{24}O$** **MW = 172.31** **314**

Table 1. Experimental value with uncertainty.

T / K	$\rho_{exp} \pm 2\sigma_{est}$ / $kg \cdot m^{-3}$	Ref.
293.15	834.9 ± 1.0	58-ano-5

5-Ethyl-4-nonanol **[19780-73-5]** $C_{11}H_{24}O$ **MW = 172.31** **315**

Table 1. Experimental value with uncertainty.

T / K	$\rho_{exp} \pm 2\sigma_{est}$ / $kg \cdot m^{-3}$	Ref.
289.15	837.0 ± 2.0	48-pow/nie

5-Ethyl-5-nonanol **[5340-51-2]** $C_{11}H_{24}O$ **MW = 172.31** **316**

Table 1. Fit with estimated B coefficient for 2 accepted points. Deviation $\sigma_w = 0.150$.

Coefficient	$\rho = A + BT$
A	1037.09
B	−0.680

Table 2. Experimental values with uncertainties and deviation from calculated values.

T / K	$\rho_{exp} \pm 2\sigma_{est}$ / $kg \cdot m^{-3}$	$\rho_{exp} - \rho_{calc}$ / $kg \cdot m^{-3}$	Ref.
293.15	837.6 ± 1.0	−0.15	33-whi/woo
298.15	834.5 ± 1.0	0.15	33-whi/woo

Table 3. Recommended values.

T / K	$\rho_{exp} \pm 2\sigma_{est}$ / $kg \cdot m^{-3}$
290.00	839.9 ± 1.1
293.15	837.8 ± 0.9
298.15	834.4 ± 0.9

***d*-6-Ethyl-3-nonanol** **[900002-60-0]** $C_{11}H_{24}O$ **MW = 172.31** **317**

Table 1. Experimental value with uncertainty.

T / K	$\rho_{exp} \pm 2\sigma_{est}$ / $kg \cdot m^{-3}$	Ref.
295.15	830.0 ± 2.0	31-lev/mar-3

3-Ethyl-2,2,4,4-tetramethyl-3-pentanol **[32579-68-3]** $C_{11}H_{24}O$ **MW = 172.31** **318**

Table 1. Experimental value with uncertainty.

T / K	$\rho_{exp} \pm 2\sigma_{est}$ / $kg \cdot m^{-3}$	Ref.
293.15	811.9 ± 1.5	60-pet/sok

4-Ethyl-2,2,3-trimethyl-3-hexanol **[91337-10-9]** $C_{11}H_{24}O$ **MW = 172.31** **319**

Table 1. Experimental value with uncertainty.

T / K	$\rho_{exp} \pm 2\sigma_{est}$ / $kg \cdot m^{-3}$	Ref.
289.15	862.0 ± 2.0	37-naz

2-Methyl-4-decanol **[25564-57-2]** **$C_{11}H_{24}O$** **MW = 172.31** **320**

Table 1. Experimental value with uncertainty.

T / K	$\rho_{exp} \pm 2\sigma_{est}$ / $kg \cdot m^{-3}$	Ref.
298.15	816.8 ± 1.0	38-wer/bog

2-Methyl-5-decanol **[500001-13-8]** **$C_{11}H_{24}O$** **MW = 172.31** **321**

Table 1. Experimental value with uncertainty.

T / K	$\rho_{exp} \pm 2\sigma_{est}$ / $kg \cdot m^{-3}$	Ref.
293.15	826.6 ± 1.0	44-pow/hag

3-Methyl-2-decanol **[500001-16-1]** **$C_{11}H_{24}O$** **MW = 172.31** **322**

Table 1. Experimental value with uncertainty.

T / K	$\rho_{exp} \pm 2\sigma_{est}$ / $kg \cdot m^{-3}$	Ref.
296.15	834.0 ± 2.0	33-pow/mur

4-Methyl-3-decanol **[500001-14-9]** **$C_{11}H_{24}O$** **MW = 172.31** **323**

Table 1. Experimental value with uncertainty.

T / K	$\rho_{exp} \pm 2\sigma_{est}$ / $kg \cdot m^{-3}$	Ref.
296.15	829.0 ± 2.0	35-gre-2

4-Methyl-4-decanol **[26209-94-9]** **$C_{11}H_{24}O$** **MW = 172.31** **324**

Table 1. Coefficients of the polynomial expansion equation.
Standard deviations (see introduction):
$\sigma_{c,w} = 4.1396 \cdot 10^{-1}$ (combined temperature ranges, weighted),
$\sigma_{c,uw} = 2.6273 \cdot 10^{-1}$ (combined temperature ranges, unweighted).

Coefficient	T = 273.15 to 338.15 K $\rho = A + BT + CT^2 + DT^3 + \ldots$
A	$1.06090 \cdot 10^{3}$
B	$-7.94286 \cdot 10^{-1}$

Table 2. Experimental values with uncertainties and deviation from calculated values.

T / K	$\rho_{exp} \pm 2\sigma_{est}$ / $kg \cdot m^{-3}$	$\rho_{exp} - \rho_{calc}$ / $kg \cdot m^{-3}$	Ref. (Symbol in Fig. 1)	T / K	$\rho_{exp} \pm 2\sigma_{est}$ / $kg \cdot m^{-3}$	$\rho_{exp} - \rho_{calc}$ / $kg \cdot m^{-3}$	Ref. (Symbol in Fig. 1)
293.15	829.60 ± 1.00	1.54	38-whi/ore(□)	318.15	808.00 ± 0.50	−0.20	39-owe/qua(×)
273.15	843.70 ± 0.50	−0.24	39-owe/qua(×)	328.15	800.00 ± 0.50	−0.26	39-owe/qua(×)
298.15	823.60 ± 0.50	−0.49	39-owe/qua(×)	338.15	792.30 ± 0.50	−0.01	39-owe/qua(×)
308.15	815.80 ± 0.50	−0.34	39-owe/qua(×)				

[1)] Not included in Fig. 1.

cont.

4-Methyl-4-decanol (cont.)

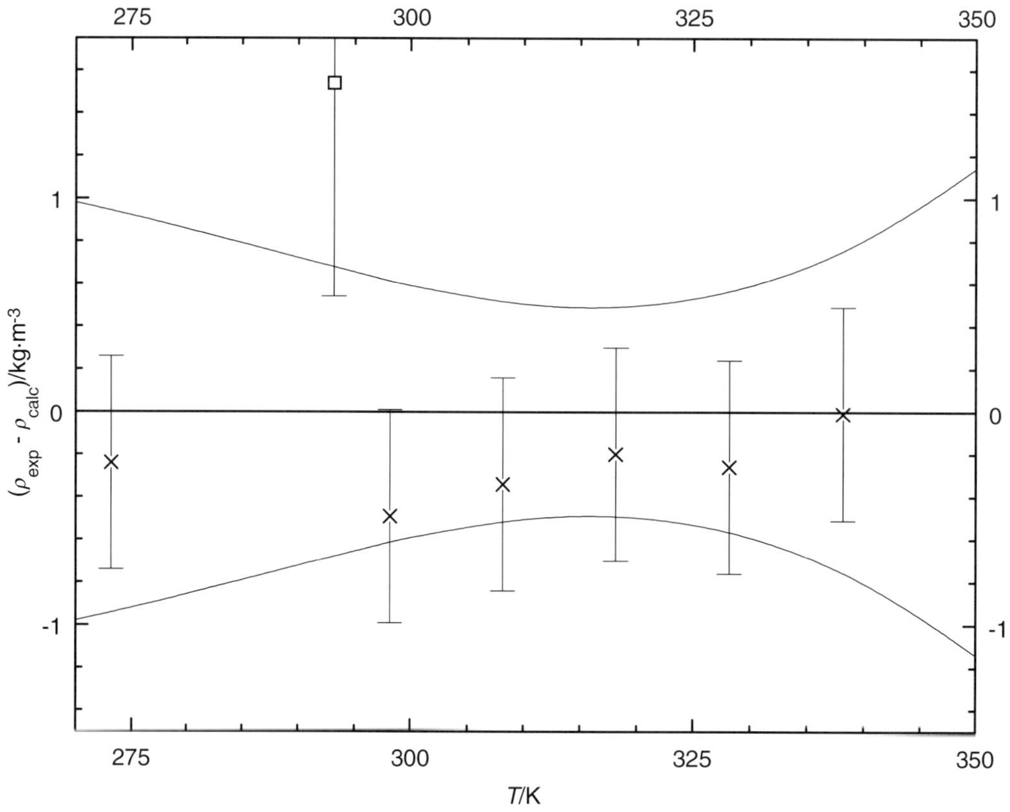

Fig. 1. The symbols show the deviation of the calculated from the experimental values from Table 2. The curves above and below the zero line indicate the calculated error region of the recommended values given in Table 3. The error bars represent the experimental errors. (Error bars smaller than the symbols are omitted for clarity of the figure.)

Table 3. Recommended values (fit to the reliable experimental values according to the equations
$\rho = A + BT + CT^2 + DT^3 + \ldots$ or $\rho = [1 + 1.75(1 - T/T_c)^{1/3} + 0.75(1 - T/T_c)][\rho_c + A(T_c - T) + B(T_c - T)^2 + C(T_c - T)^3 + D(T_c - T)^4]$).

$\frac{T}{K}$	$\frac{\rho \pm \sigma_{fit}}{kg \cdot m^{-3}}$	$\frac{T}{K}$	$\frac{\rho \pm \sigma_{fit}}{kg \cdot m^{-3}}$	$\frac{T}{K}$	$\frac{\rho \pm \sigma_{fit}}{kg \cdot m^{-3}}$
270.00	846.44 ± 0.98	298.15	824.09 ± 0.61	330.00	798.79 ± 0.57
280.00	838.50 ± 0.86	300.00	822.62 ± 0.59	340.00	790.84 ± 0.78
290.00	830.56 ± 0.72	310.00	814.67 ± 0.49	350.00	782.90 ± 1.14
293.15	828.06 ± 0.68	320.00	806.73 ± 0.48		

***L*-5-Methyl-1-decanol** **[500001-17-2]** **$C_{11}H_{24}O$** **MW = 172.31** **325**

Table 1. Experimental value with uncertainty.

$\frac{T}{K}$	$\frac{\rho_{exp} \pm 2\sigma_{est}}{kg \cdot m^{-3}}$	Ref.
297.15	840.0 ± 2.0	33-lev/mar-1

5-Methyl-5-decanol **[87258-26-2]** $C_{11}H_{24}O$ **MW = 172.31** **326**

Table 1. Experimental value with uncertainty.

T / K	$\rho_{exp} \pm 2\sigma_{est}$ / $kg \cdot m^{-3}$	Ref.
298.15	826.2 ± 1.0	33-whi/wil

***l*-6-Methyl-3-decanol** **[900002-59-7]** $C_{11}H_{24}O$ **MW = 172.31** **327**

Table 1. Experimental value with uncertainty.

T / K	$\rho_{exp} \pm 2\sigma_{est}$ / $kg \cdot m^{-3}$	Ref.
297.15	829.0 ± 2.0	31-lev/mar-4

2-Methyl-3-(1-methylethyl)-3-heptanol **[5340-35-2]** $C_{11}H_{24}O$ **MW = 172.31** **328**

Table 1. Experimental and recommended values with uncertainties.

T / K	$\rho_{exp} \pm 2\sigma_{est}$ / $kg \cdot m^{-3}$	Ref.
293.15	848.7 ± 1.0	29-con/bla
293.15	855.7 ± 3.0	43-geo[1)]
293.15	849.0 ± 1.0	49-naz/pin
293.15	848.9 ± 1.0	Recommended

[1)] Not included in calculation of recommended value.

5-Methyl-3-(2-methylpropyl)-2-hexanol **[900002-62-2]** $C_{11}H_{24}O$ **MW = 172.31** **329**

Table 1. Experimental value with uncertainty.

T / K	$\rho_{exp} \pm 2\sigma_{est}$ / $kg \cdot m^{-3}$	Ref.
277.15	844.0 ± 2.0	10-fre

2-Methyl-4-propyl-4-heptanol **[56065-39-5]** $C_{11}H_{24}O$ **MW = 172.31** **330**

Table 1. Experimental value with uncertainty.

T / K	$\rho_{exp} \pm 2\sigma_{est}$ / $kg \cdot m^{-3}$	Ref.
295.15	831.1 ± 2.0	14-hal

4-Methyl-2-propyl-1-hexanol **[66256-62-0]** $C_{11}H_{24}O$ **MW = 172.31** **331**

Table 1. Experimental value with uncertainty.

T / K	$\rho_{exp} \pm 2\sigma_{est}$ / $kg \cdot m^{-3}$	Ref.
293.15	828.6 ± 1.0	58-hag/hud

4-(1-Methylethyl)-4-octanol **[900002-61-1]** $C_{11}H_{24}O$ **MW = 172.31** **332**

Table 1. Experimental value with uncertainty.

$\frac{T}{K}$	$\frac{\rho_{exp} \pm 2\sigma_{est}}{kg \cdot m^{-3}}$	Ref.
293.15	840.0 ± 2.0	54-naz/kak-4

2,2,3,4,4-Pentamethyl-3-hexanol **[500000-99-7]** $C_{11}H_{24}O$ **MW = 172.31** **333**

Table 1. Experimental value with uncertainty.

$\frac{T}{K}$	$\frac{\rho_{exp} \pm 2\sigma_{est}}{kg \cdot m^{-3}}$	Ref.
289.15	886.0 ± 2.0	37-naz

2,2,3,4,5-Pentamethyl-3-hexanol **[500000-98-6]** $C_{11}H_{24}O$ **MW = 172.31** **334**

Table 1. Experimental value with uncertainty.

$\frac{T}{K}$	$\frac{\rho_{exp} \pm 2\sigma_{est}}{kg \cdot m^{-3}}$	Ref.
285.15	868.0 ± 2.0	37-naz

2,2,4,5,5-Pentamethyl-4-hexanol **[900002-64-4]** $C_{11}H_{24}O$ **MW = 172.31** **335**

Table 1. Experimental value with uncertainty.

$\frac{T}{K}$	$\frac{\rho_{exp} \pm 2\sigma_{est}}{kg \cdot m^{-3}}$	Ref.
293.15	841.0 ± 2.0	49-moe/whi

3,4,4,5,5-Pentamethyl-3-hexanol **[536-91-2]** $C_{11}H_{24}O$ **MW = 172.31** **336**

Table 1. Experimental value with uncertainty.

$\frac{T}{K}$	$\frac{\rho_{exp} \pm 2\sigma_{est}}{kg \cdot m^{-3}}$	Ref.
293.15	877.8 ± 2.0	60-pet/kao

3-Propyl-2-octanol **[500001-08-1]** $C_{11}H_{24}O$ **MW = 172.31** **337**

Table 1. Experimental value with uncertainty.

$\frac{T}{K}$	$\frac{\rho_{exp} \pm 2\sigma_{est}}{kg \cdot m^{-3}}$	Ref.
290.15	831.0 ± 2.0	12-gue-2

4-Propyl-4-octanol **[6632-94-6]** **$C_{11}H_{24}O$** **MW = 172.31** **338**

Table 1. Fit with estimated B coefficient for 2 accepted points. Deviation $\sigma_w = 0.100$.

Coefficient	$\rho = A + BT$
A	1034.54
B	–0.680

Table 2. Experimental values with uncertainties and deviation from calculated values.

$\frac{T}{\text{K}}$	$\frac{\rho_{exp} \pm 2\sigma_{est}}{\text{kg} \cdot \text{m}^{-3}}$	$\frac{\rho_{exp} - \rho_{calc}}{\text{kg} \cdot \text{m}^{-3}}$	Ref.
293.15	835.1 ± 1.0	–0.10	33-whi/woo
298.15	831.9 ± 1.0	0.10	33-whi/woo

Table 3. Recommended values.

$\frac{T}{\text{K}}$	$\frac{\rho_{exp} \pm 2\sigma_{est}}{\text{kg} \cdot \text{m}^{-3}}$
290.00	837.3 ± 1.1
293.15	835.2 ± 0.9
298.15	831.8 ± 0.9

2,2,3,4-Tetramethyl-3-heptanol **[91337-08-5]** **$C_{11}H_{24}O$** **MW = 172.31** **339**

Table 1. Experimental value with uncertainty.

$\frac{T}{\text{K}}$	$\frac{\rho_{exp} \pm 2\sigma_{est}}{\text{kg} \cdot \text{m}^{-3}}$	Ref.
273.15	866.0 ± 2.0	37-naz

2,2,3,6-Tetramethyl-3-heptanol **[106593-59-3]** **$C_{11}H_{24}O$** **MW = 172.31** **340**

Table 1. Experimental value with uncertainty.

$\frac{T}{\text{K}}$	$\frac{\rho_{exp} \pm 2\sigma_{est}}{\text{kg} \cdot \text{m}^{-3}}$	Ref.
293.15	850.6 ± 1.0	57-pet/sus

2,2,4,6-Tetramethyl-4-heptanol **[106593-60-6]** **$C_{11}H_{24}O$** **MW = 172.31** **341**

Table 1. Experimental value with uncertainty.

$\frac{T}{\text{K}}$	$\frac{\rho_{exp} \pm 2\sigma_{est}}{\text{kg} \cdot \text{m}^{-3}}$	Ref.
293.15	828.0 ± 2.0	49-moe/whi

2,3,4,4-Tetramethyl-3-heptanol **[91337-09-6]** **$C_{11}H_{24}O$** **MW = 172.31** **342**

Table 1. Experimental value with uncertainty.

$\frac{T}{\text{K}}$	$\frac{\rho_{exp} \pm 2\sigma_{est}}{\text{kg} \cdot \text{m}^{-3}}$	Ref.
288.15	874.0 ± 2.0	37-naz

2,2,4-Trimethyl-3-(1-methylethyl)-3-pentanol **[5457-41-0]** **$C_{11}H_{24}O$** **MW = 172.31** **343**

Table 1. Experimental and recommended values with uncertainties.

$\frac{T}{K}$	$\frac{\rho_{exp} \pm 2\sigma_{est}}{kg \cdot m^{-3}}$	Ref.
292.15	870.0 ± 2.0	46-vav/col[1)]
293.15	875.8 ± 1.0	51-smi/cre
293.15	875.8 ± 1.0	Recommended

[1)] Not included in calculation of recommended value.

2,2,4-Trimethyl-4-octanol **[5340-54-5]** **$C_{11}H_{24}O$** **MW = 172.31** **344**

Table 1. Experimental value with uncertainty.

$\frac{T}{K}$	$\frac{\rho_{exp} \pm 2\sigma_{est}}{kg \cdot m^{-3}}$	Ref.
293.15	835.0 ± 1.0	49-moe/whi

2,4,7-Trimethyl-4-octanol **[42842-13-7]** **$C_{11}H_{24}O$** **MW = 172.31** **345**

Table 1. Experimental value with uncertainty.

$\frac{T}{K}$	$\frac{\rho_{exp} \pm 2\sigma_{est}}{kg \cdot m^{-3}}$	Ref.
293.15	825.0 ± 1.0	33-mey/tuo

1-Undecanol **[112-42-5]** **$C_{11}H_{24}O$** **MW = 172.31** **346**

Table 1. Coefficients of the polynomial expansion equation. Standard deviations (see introduction):
$\sigma_{c,w} = 2.5713 \cdot 10^{-1}$ (combined temperature ranges, weighted),
$\sigma_{c,uw} = 1.0106 \cdot 10^{-1}$ (combined temperature ranges, unweighted).

Coefficient	T = 293.10 to 358.10 K $\rho = A + BT + CT^2 + DT^3 + \ldots$
A	$1.03691 \cdot 10^3$
B	$-6.95763 \cdot 10^{-1}$

Table 2. Experimental values with uncertainties and deviation from calculated values.

$\frac{T}{K}$	$\frac{\rho_{exp} \pm 2\sigma_{est}}{kg \cdot m^{-3}}$	$\frac{\rho_{exp} - \rho_{calc}}{kg \cdot m^{-3}}$	Ref. (Symbol in Fig. 1)	$\frac{T}{K}$	$\frac{\rho_{exp} \pm 2\sigma_{est}}{kg \cdot m^{-3}}$	$\frac{\rho_{exp} - \rho_{calc}}{kg \cdot m^{-3}}$	Ref. (Symbol in Fig. 1)
307.75	821.68 ± 1.00	−1.11	27-ver/coo(×)	343.10	798.30 ± 0.40	0.10	55-kus(○)
293.15	833.80 ± 1.00	0.85	42-mul(◆)	353.10	791.30 ± 0.40	0.06	55-kus(○)
297.45	829.80 ± 0.60	−0.16	50-sac/sau(▽)	358.10	787.70 ± 0.40	−0.06	55-kus(○)
293.10	833.20 ± 0.40	0.22	55-kus(○)	298.15	829.19 ± 0.30	−0.28	79-dia/tar(□)
298.10	829.80 ± 0.40	0.29	55-kus(○)	308.15	822.36 ± 0.30	−0.15	79-dia/tar(□)
303.10	826.30 ± 0.40	0.27	55-kus(○)	318.15	815.51 ± 0.30	−0.05	79-dia/tar(□)
313.10	819.40 ± 0.40	0.33	55-kus(○)	333.15	804.95 ± 0.30	−0.17	79-dia/tar(□)
323.10	812.40 ± 0.40	0.29	55-kus(○)	298.15	828.98 ± 0.60	−0.49	90-klo/pal(Δ)
333.10	805.20 ± 0.40	0.05	55-kus(○)				

[1)] Not included in Fig. 1.

cont.

Further references: [03-bla/gue, 29-mah/das, 48-vog-2, 56-gol/kon, 63-vil/gav].

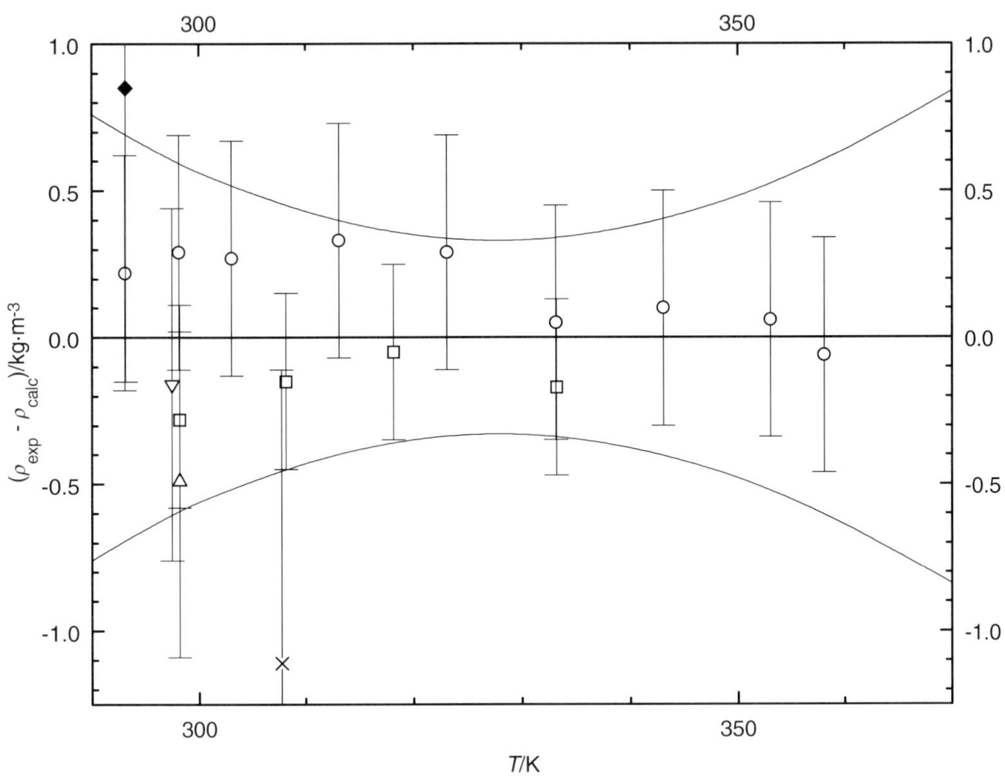

Fig. 1. The symbols show the deviation of the calculated from the experimental values from Table 2. The curves above and below the zero line indicate the calculated error region of the recommended values given in Table 3. The error bars represent the experimental errors. (Error bars smaller than the symbols are omitted for clarity of the figure.)

Table 3. Recommended values (fit to the reliable experimental values according to the equations $\rho = A + BT + CT^2 + DT^3 + \ldots$ or $\rho = [1 + 1.75(1 - T/T_c)^{1/3} + 0.75(1 - T/T_c)][\rho_c + A(T_c - T) + B(T_c - T)^2 + C(T_c - T)^3 + D(T_c - T)^4]$).

$\frac{T}{K}$	$\frac{\rho \pm \sigma_{fit}}{kg \cdot m^{-3}}$	$\frac{T}{K}$	$\frac{\rho \pm \sigma_{fit}}{kg \cdot m^{-3}}$	$\frac{T}{K}$	$\frac{\rho \pm \sigma_{fit}}{kg \cdot m^{-3}}$
290.00	835.14 ± 0.76	310.00	821.23 ± 0.42	350.00	793.40 ± 0.47
293.15	832.95 ± 0.69	320.00	814.27 ± 0.34	360.00	786.44 ± 0.63
298.15	829.47 ± 0.59	330.00	807.31 ± 0.32	370.00	779.48 ± 0.84
300.00	828.18 ± 0.56	340.00	800.35 ± 0.37		

2-Undecanol **[1653-30-1]** $C_{11}H_{24}O$ **MW = 172.31** **347**

Table 1. Fit with estimated B coefficient for 5 accepted points. Deviation $\sigma_w = 1.587$.

Coefficient	$\rho = A + BT$
A	1044.22
B	−0.740

cont.

2-Undecanol (cont.)

Table 2. Experimental values with uncertainties and deviation from calculated values.

$\frac{T}{K}$	$\frac{\rho_{exp} \pm 2\sigma_{est}}{kg \cdot m^{-3}}$	$\frac{\rho_{exp} - \rho_{calc}}{kg \cdot m^{-3}}$	Ref.	$\frac{T}{K}$	$\frac{\rho_{exp} \pm 2\sigma_{est}}{kg \cdot m^{-3}}$	$\frac{\rho_{exp} - \rho_{calc}}{kg \cdot m^{-3}}$	Ref.
292.15	826.8 ± 2.0	−1.23	1870-gie	293.15	827.0 ± 1.0	−0.29	11-pic/ken
291.15	826.3 ± 2.0	−2.47	03-tho/man	405.15	739.3 ± 3.0	−5.11	12-pic/ken[1)]
296.15	827.0 ± 2.0	1.93	10-hal/las	293.15	830.2 ± 2.0	2.91	42-mul

[1)] Not included in calculation of linear coefficients.

Table 3. Recommended values.

$\frac{T}{K}$	$\frac{\rho_{exp} \pm 2\sigma_{est}}{kg \cdot m^{-3}}$
290.00	829.6 ± 2.1
293.15	827.3 ± 2.1
298.15	823.6 ± 2.1

3-Undecanol **[6929-08-4]** **$C_{11}H_{24}O$** **MW = 172.31** **348**

Table 1. Experimental values with uncertainties.

$\frac{T}{K}$	$\frac{\rho_{exp} \pm 2\sigma_{est}}{kg \cdot m^{-3}}$	Ref.
293.15	829.5 ± 2.0	13-pic/ken
353.15	782.7 ± 2.0	13-pic/ken

4-Undecanol **[4272-06-4]** **$C_{11}H_{24}O$** **MW = 172.31** **349**

Table 1. Experimental value with uncertainty.

$\frac{T}{K}$	$\frac{\rho_{exp} \pm 2\sigma_{est}}{kg \cdot m^{-3}}$	Ref.
293.15	821.5 ± 1.0	65-das/mae

6-Undecanol **[23708-56-7]** **$C_{11}H_{24}O$** **MW = 172.31** **350**

Table 1. Experimental values with uncertainties.

$\frac{T}{K}$	$\frac{\rho_{exp} \pm 2\sigma_{est}}{kg \cdot m^{-3}}$	Ref.
293.15	833.4 ± 2.0	25-hes/bap
294.95	827.2 ± 2.0	25-hes/bap

3-Butyl-3-methyl-2-heptanol **[500001-77-4]** **$C_{12}H_{26}O$** **MW = 186.34** **351**

Table 1. Experimental value with uncertainty.

$\frac{T}{K}$	$\frac{\rho_{exp} \pm 2\sigma_{est}}{kg \cdot m^{-3}}$	Ref.
293.15	850.0 ± 2.0	33-whi/kru

2-Butyl-1-octanol **[3913-02-8]** $C_{12}H_{26}O$ **MW = 186.34** **352**

Table 1. Fit with estimated *B* coefficient for 2 accepted points. Deviation $\sigma_w = 0.255$.

Coefficient	$\rho = A + BT$
A	1051.19
B	–0.740

Table 2. Experimental values with uncertainties and deviation from calculated values.

$\frac{T}{K}$	$\frac{\rho_{exp} \pm 2\sigma_{est}}{kg \cdot m^{-3}}$	$\frac{\rho_{exp} - \rho_{calc}}{kg \cdot m^{-3}}$	Ref.
289.65	837.1 ± 1.0	0.25	38-mas
293.15	834.0 ± 1.0	–0.26	53-ano-15

Table 3. Recommended values.

$\frac{T}{K}$	$\frac{\rho_{exp} \pm 2\sigma_{est}}{kg \cdot m^{-3}}$
280.00	844.0 ± 1.1
290.00	836.6 ± 0.9
293.15	834.3 ± 0.9
298.15	830.6 ± 1.0

4-Butyl-1-octanol **[500001-73-0]** $C_{12}H_{26}O$ **MW = 186.34** **353**

Table 1. Experimental value with uncertainty.

$\frac{T}{K}$	$\frac{\rho_{exp} \pm 2\sigma_{est}}{kg \cdot m^{-3}}$	Ref.
289.15	841.0 ± 2.0	16-lev/all

2,4-Diethyl-1-octanol **[55514-25-5]** $C_{12}H_{26}O$ **MW = 186.34** **354**

Table 1. Experimental value with uncertainty.

$\frac{T}{K}$	$\frac{\rho_{exp} \pm 2\sigma_{est}}{kg \cdot m^{-3}}$	Ref.
293.15	840.0 ± 1.0	61-mil/ben

2,2-Dimethyl-1-decanol **[2370-15-2]** $C_{12}H_{26}O$ **MW = 186.34** **355**

Table 1. Experimental values with uncertainties.

$\frac{T}{K}$	$\frac{\rho_{exp} \pm 2\sigma_{est}}{kg \cdot m^{-3}}$	Ref.
293.15	834.5 ± 2.0	64-blo/hag
293.15	839.0 ± 2.0	65-shu/puz

2,4-Dimethyl-4-decanol **[106652-28-2]** **$C_{12}H_{26}O$** **MW = 186.34** **356**

Table 1. Experimental value with uncertainty.

$\frac{T}{K}$	$\frac{\rho_{exp} \pm 2\sigma_{est}}{kg \cdot m^{-3}}$	Ref.
293.15	828.4 ± 2.0	59-pet/zak

3,5-Dimethyl-5-decanol **[105900-70-7]** **$C_{12}H_{26}O$** **MW = 186.34** **357**

Table 1. Experimental value with uncertainty.

$\frac{T}{K}$	$\frac{\rho_{exp} \pm 2\sigma_{est}}{kg \cdot m^{-3}}$	Ref.
301.15	827.0 ± 1.0	60-tha/vas

5,9-Dimethyl-5-decanol **[900002-66-6]** **$C_{12}H_{26}O$** **MW = 186.34** **358**

Table 1. Experimental value with uncertainty.

$\frac{T}{K}$	$\frac{\rho_{exp} \pm 2\sigma_{est}}{kg \cdot m^{-3}}$	Ref.
284.15	852.6 ± 2.0	28-esc

6,6-Dimethyl-5-decanol **[500001-68-3]** **$C_{12}H_{26}O$** **MW = 186.34** **359**

Table 1. Experimental value with uncertainty.

$\frac{T}{K}$	$\frac{\rho_{exp} \pm 2\sigma_{est}}{kg \cdot m^{-3}}$	Ref.
293.15	845.0 ± 2.0	33-whi/kru

2,2-Dimethyl-3-(1,1-dimethylethyl)-3-hexanol **[32579-69-4]** **$C_{12}H_{26}O$** **MW = 186.34** **360**

Table 1. Experimental and recommended values with uncertainties.

$\frac{T}{K}$	$\frac{\rho_{exp} \pm 2\sigma_{est}}{kg \cdot m^{-3}}$	Ref.	$\frac{T}{K}$	$\frac{\rho_{exp} \pm 2\sigma_{est}}{kg \cdot m^{-3}}$	Ref.
293.15	859.5 ± 1.0	48-cad/foo	293.15	865.2 ± 3.0	57-pet/sok[1)]
293.15	895.5 ± 3.0	51-smi/cre[1)]	293.15	859.5 ± 1.0	Recommended

[1)] Not included in calculation of recommended value.

5,5-Dimethyl-4-(1,1-dimethylethyl)-1-hexanol **[900002-67-7]** **$C_{12}H_{26}O$** **MW = 186.34** **361**

Table 1. Experimental value with uncertainty.

$\frac{T}{K}$	$\frac{\rho_{exp} \pm 2\sigma_{est}}{kg \cdot m^{-3}}$	Ref.
293.15	865.2 ± 2.0	60-pet/sok

2,2-Dimethyl-4-ethyl-3-octanol **[124154-63-8]** **$C_{12}H_{26}O$** **MW = 186.34** **362**

Table 1. Experimental value with uncertainty.

$\frac{T}{K}$	$\frac{\rho_{exp} \pm 2\sigma_{est}}{kg \cdot m^{-3}}$	Ref.
293.15	840.9 ± 1.0	41-whi/whi

2,2-Dimethyl-3-(1-methylethyl)-3-heptanol **[500001-78-5]** **$C_{12}H_{26}O$** **MW = 186.34** **363**

Table 1. Experimental value with uncertainty.

$\frac{T}{K}$	$\frac{\rho_{exp} \pm 2\sigma_{est}}{kg \cdot m^{-3}}$	Ref.
293.15	856.5 ± 1.0	50-naz/kot-1

2,6-Dimethyl-3-(1-methylethyl)-3-heptanol **[5340-82-9]** **$C_{12}H_{26}O$** **MW = 186.34** **364**

Table 1. Experimental values with uncertainties.

$\frac{T}{K}$	$\frac{\rho_{exp} \pm 2\sigma_{est}}{kg \cdot m^{-3}}$	Ref.
289.15	860.6 ± 2.0	14-mur/amo
273.15	871.7 ± 2.0	14-mur/amo

2,2-Dimethyl-4-propyl-4-heptanol **[500001-79-6]** **$C_{12}H_{26}O$** **MW = 186.34** **365**

Table 1. Experimental and recommended values with uncertainties.

$\frac{T}{K}$	$\frac{\rho_{exp} \pm 2\sigma_{est}}{kg \cdot m^{-3}}$	Ref.
293.15	837.6 ± 1.0	8 -whi/pop -1
293.15	838.6 ± 1.0	2 -whi/for -0
293.15	838.1 ± 1.1	Recommended

1-Dodecanol **[112-53-8]** **$C_{12}H_{26}O$** **MW = 186.34** **366**

Table 1. Coefficients of the polynomial expansion equation.
Standard deviations (see introduction):
$\sigma_{c,w} = 2.7829 \cdot 10^{-1}$ (combined temperature ranges, weighted),
$\sigma_{c,uw} = 9.5525 \cdot 10^{-2}$ (combined temperature ranges, unweighted).

Coefficient	T = 296.90 to 533.15 K $\rho = A + BT + CT^2 + DT^3 + \ldots$
A	$9.42675 \cdot 10^{2}$
B	$-4.51517 \cdot 10^{-2}$
C	$-1.31349 \cdot 10^{-3}$
D	$6.54515 \cdot 10^{-7}$

cont.

1-Dodecanol (cont.)

Table 2. Experimental values with uncertainties and deviation from calculated values.

$\frac{T}{K}$	$\frac{\rho_{exp} \pm 2\sigma_{est}}{kg \cdot m^{-3}}$	$\frac{\rho_{exp} - \rho_{calc}}{kg \cdot m^{-3}}$	Ref. (Symbol in Fig. 1)	$\frac{T}{K}$	$\frac{\rho_{exp} \pm 2\sigma_{est}}{kg \cdot m^{-3}}$	$\frac{\rho_{exp} - \rho_{calc}}{kg \cdot m^{-3}}$	Ref. (Symbol in Fig. 1)
	crystal			303.14	826.39 ± 0.20	−0.13	73-fin(×)
78.15	1017.0 ± 3.0		30-bil/fis-1	313.13	819.60 ± 0.20	−0.24	73-fin(×)
194.15	981.0 ± 3.0		30-bil/fis-1	323.07	812.78 ± 0.20	−0.28	73-fin(×)
	liquid			333.20	805.77 ± 0.20	−0.25	73-fin(×)
297.15	830.90 ± 1.00	0.45	1883-kra[1]	298.15	829.65 ± 0.25	−0.15	76-hal/ell(○)
313.15	820.10 ± 1.00	0.27	1883-kra[1]	303.15	826.20 ± 0.25	−0.31	76-hal/ell(○)
372.15	778.10 ± 1.00	0.41	1883-kra(×)	320.00	814.81 ± 0.25	−0.36	76-hal/ell(○)
313.15	819.60 ± 0.60	−0.23	58-cos/bow[1]	340.00	800.96 ± 0.25	−0.25	76-hal/ell(○)
333.15	805.70 ± 0.60	−0.35	58-cos/bow[1]	360.00	786.54 ± 0.25	−0.19	76-hal/ell(○)
353.15	792.00 ± 0.60	0.26	58–cos/bow(×)	380.00	771.73 ± 0.30	−0.03	76-hal/ell(○)
373.15	777.40 ± 0.60	0.46	58-cos/bow(×)	400.00	756.33 ± 0.30	−0.01	76-hal/ell(○)
393.15	762.20 ± 0.60	0.53	58-cos/bow(×)	420.00	740.41 ± 0.30	−0.09	76-hal/ell(○)
413.15	746.80 ± 1.00	0.83	58-cos/bow(×)	440.00	724.02 ± 0.30	−0.25	76-hal/ell(○)
433.15	730.70 ± 1.00	0.83	58-cos/bow(×)	460.00	707.21 ± 0.30	−0.47	76-hal/ell(○)
453.15	714.40 ± 1.00	1.00	58-cos/bow(×)	470.00	698.99 ± 0.35	−0.27	76-hal/ell(○)
473.15	698.20 ± 1.00	1.61	58-cos/bow[1]	480.00	689.99 ± 0.35	−0.77	76-hal/ell(○)
493.15	681.00 ± 1.00	1.53	58-cos/bow(×)	490.00	681.24 ± 0.35	−0.94	76-hal/ell(○)
513.15	663.50 ± 1.50	1.43	58-cos/bow(×)	298.15	830.00 ± 0.20	0.20	78-jel/leo(□)
533.15	645.70 ± 1.50	1.27	58-cos/bow(×)	303.15	826.57 ± 0.20	0.06	78-jel/leo(□)
553.15	627.50 ± 1.50	0.92	58-cos/bow[1]	308.15	823.16 ± 0.20	−0.03	78-jel/leo(□)
573.15	609.70 ± 1.50	1.16	58-cos/bow[1]	298.15	829.95 ± 0.30	0.15	79-dia/tar(Δ)
293.15	834.60 ± 1.00	1.55	62-gei/qui[1]	308.15	823.20 ± 0.30	0.01	79-dia/tar(Δ)
298.15	831.20 ± 1.00	1.40	62-gei/qui[1]	318.15	816.33 ± 0.30	−0.11	79-dia/tar(Δ)
303.15	827.80 ± 1.00	1.29	62-gei/qui[1]	333.15	806.06 ± 0.30	0.01	79-dia/tar(Δ)
308.15	824.40 ± 1.00	1.21	62-gei/qui[1]	298.15	823.00 ± 0.40	−6.80	86-wag/hei[1]
313.15	821.10 ± 1.00	1.27	62-gei/qui[1]	333.15	805.81 ± 0.40	−0.24	86-wag/hei(∇)
318.15	817.80 ± 1.00	1.36	62-gei/qui[1]	298.15	828.50 ± 0.60	−1.30	92-lie/sen-1[1]
323.15	814.40 ± 1.00	1.39	62-gei/qui[1]	303.15	825.50 ± 0.60	−1.01	92-lie/sen-1[1]
328.15	811.00 ± 1.00	1.45	62-gei/qui[1]	308.15	822.50 ± 0.60	−0.69	92-lie/sen-1[1]
333.15	807.50 ± 1.00	1.45	62-gei/qui[1]	313.15	818.20 ± 0.60	−1.63	92-lie/sen-1[1]
338.15	804.00 ± 1.00	1.48	62-gei/qui[1]	318.15	815.20 ± 0.60	−1.24	92-lie/sen-1[1]
343.15	800.20 ± 1.00	1.24	62-gei/qui[1]	323.15	811.90 ± 0.60	−1.11	92-lie/sen-1(◆)
348.15	796.50 ± 1.00	1.13	62-gei/qui[1]	328.15	809.00 ± 0.60	−0.55	92-lie/sen-1(◆)
353.15	792.80 ± 1.00	1.06	62-gei/qui[1]	333.15	804.70 ± 0.60	−1.35	92-lie/sen-1[1]
358.15	789.00 ± 1.00	0.91	62-gei/qui(×)	338.15	801.50 ± 0.60	−1.02	92-lie/sen-1(◆)
363.15	785.20 ± 1.00	0.80	62-gei/qui(×)	343.15	798.40 ± 0.60	−0.56	92-lie/sen-1(◆)
296.90	830.62 ± 0.20	0.00	73-fin(×)	348.15	794.20 ± 0.60	−1.17	92-lie/sen-1(◆)
297.89	829.95 ± 0.20	−0.02	73-fin(×)	353.15	790.90 ± 0.60	−0.84	92-lie/sen-1(◆)
298.78	829.36 ± 0.20	−0.03	73-fin(×)				

[1] Not included in Fig. 1.

Further references: [27-ver/coo, 29-mah/das, 31-zaa, 36-pal/sab, 42-mul, 43-hsu, 44-sto/rou, 46-par/row, 48-wei, 49-tsv/mar, 50-sac/sau, 55-pet/sus, 56-rat/cur, 63-vil/gav, 67-seu/mor, 93-yan/mae].

cont.

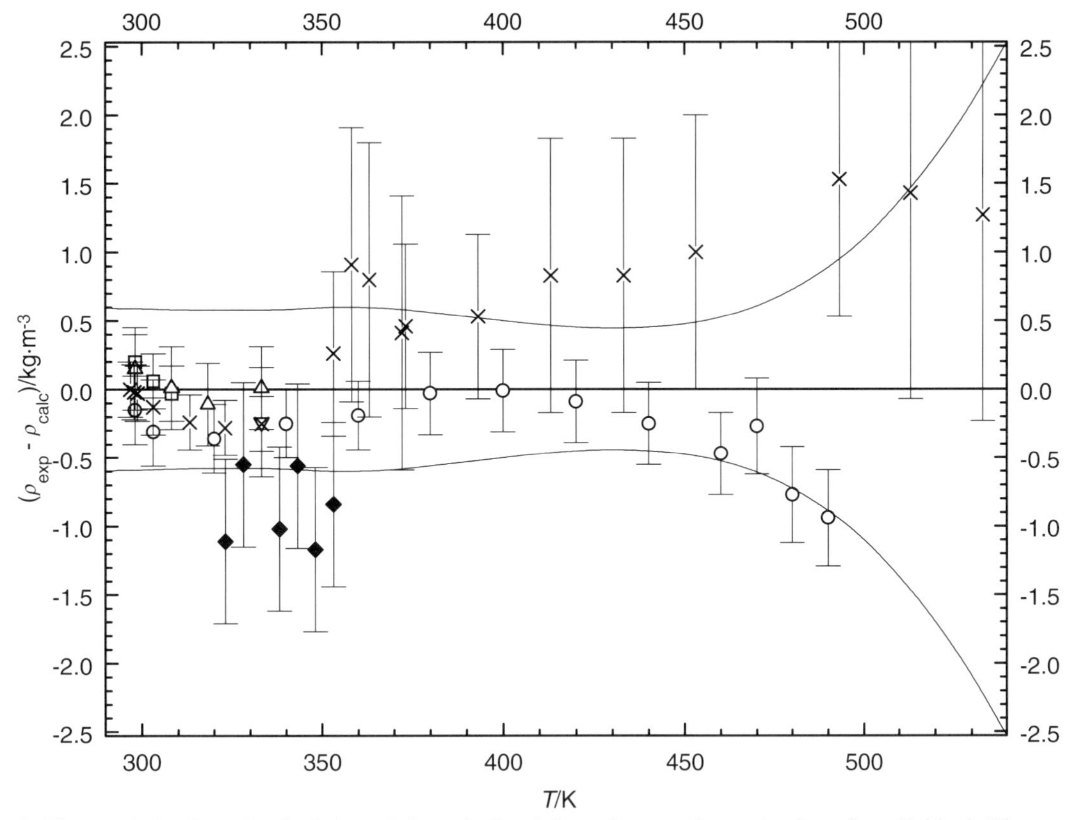

Fig. 1. The symbols show the deviation of the calculated from the experimental values from Table 2. The curves above and below the zero line indicate the calculated error region of the recommended values given in Table 3. The error bars represent the experimental errors. (Error bars smaller than the symbols are omitted for clarity of the figure.)

Table 3. Recommended values (fit to the reliable experimental values according to the equations $\rho = A + BT + CT^2 + DT^3 + \ldots$ or $\rho = [1 + 1.75(1 - T/T_c)^{1/3} + 0.75(1 - T/T_c)][\rho_c + A(T_c - T) + B(T_c - T)^2 + C(T_c - T)^3 + D(T_c - T)^4]$).

$\frac{T}{\mathrm{K}}$	$\frac{\rho \pm \sigma_{fit}}{\mathrm{kg \cdot m^{-3}}}$	$\frac{T}{\mathrm{K}}$	$\frac{\rho \pm \sigma_{fit}}{\mathrm{kg \cdot m^{-3}}}$	$\frac{T}{\mathrm{K}}$	$\frac{\rho \pm \sigma_{fit}}{\mathrm{kg \cdot m^{-3}}}$
290.00	835.08 ± 0.60	370.00	779.30 ± 0.59	470.00	699.26 ± 0.60
293.15	833.05 ± 0.59	380.00	771.76 ± 0.56	480.00	690.76 ± 0.72
298.15	829.80 ± 0.59	390.00	764.11 ± 0.53	490.00	682.18 ± 0.88
300.00	828.59 ± 0.59	400.00	756.34 ± 0.50	500.00	673.54 ± 1.09
310.00	821.95 ± 0.58	410.00	748.47 ± 0.47	510.00	664.83 ± 1.36
320.00	815.17 ± 0.58	420.00	740.50 ± 0.45	520.00	656.06 ± 1.69
330.00	808.26 ± 0.58	430.00	732.43 ± 0.44	530.00	647.23 ± 2.08
340.00	801.21 ± 0.58	440.00	724.27 ± 0.45	540.00	638.34 ± 2.53
350.00	794.03 ± 0.60	450.00	716.02 ± 0.47		
360.00	786.73 ± 0.60	460.00	707.68 ± 0.52		

2-Dodecanol **[10203-28-8]** **$C_{12}H_{26}O$** **MW = 186.34** **367**

Table 1. Coefficients of the polynomial expansion equation.
Standard deviations (see introduction):
$\sigma_{c,w} = 1.0204 \cdot 10^{-1}$ (combined temperature ranges, weighted),
$\sigma_{c,uw} = 2.8302 \cdot 10^{-2}$ (combined temperature ranges, unweighted).

Coefficient	T = 293.15 to 363.15 K $\rho = A + BT + CT^2 + DT^3 + \ldots$
A	$9.76605 \cdot 10^{2}$
B	$-3.16820 \cdot 10^{-1}$
C	$-6.34131 \cdot 10^{-4}$

Table 2. Experimental values with uncertainties and deviation from calculated values.

$\frac{T}{\mathrm{K}}$	$\frac{\rho_{exp} \pm 2\sigma_{est}}{\mathrm{kg \cdot m^{-3}}}$	$\frac{\rho_{exp} - \rho_{calc}}{\mathrm{kg \cdot m^{-3}}}$	Ref. (Symbol in Fig. 1)	$\frac{T}{\mathrm{K}}$	$\frac{\rho_{exp} \pm 2\sigma_{est}}{\mathrm{kg \cdot m^{-3}}}$	$\frac{\rho_{exp} - \rho_{calc}}{\mathrm{kg \cdot m^{-3}}}$	Ref. (Symbol in Fig. 1)
293.15	829.30 ± 1.00	0.07	62-gei/qui(□)	333.15	800.60 ± 1.00	–0.08	62-gei/qui(□)
298.15	825.80 ± 1.00	0.02	62-gei/qui(□)	338.15	796.90 ± 1.00	–0.06	62-gei/qui(□)
303.15	822.30 ± 1.00	0.02	62-gei/qui(□)	343.15	793.30 ± 1.00	0.08	62-gei/qui(□)
308.15	818.70 ± 1.00	–0.06	62-gei/qui(□)	348.15	789.40 ± 1.00	–0.04	62-gei/qui(□)
313.15	815.10 ± 1.00	–0.11	62-gei/qui(□)	353.15	785.80 ± 1.00	0.17	62-gei/qui(□)
318.15	811.50 ± 1.00	–0.12	62-gei/qui(□)	358.15	781.80 ± 1.00	0.00	62-gei/qui(□)
323.15	808.00 ± 1.00	–0.01	62-gei/qui(□)	363.15	777.80 ± 1.00	–0.12	62-gei/qui(□)
328.15	804.60 ± 1.00	0.24	62-gei/qui(□)				

Further references: [11-pic/ken].

Table 3. Recommended values (fit to the reliable experimental values according to the equations $\rho = A + BT + CT^2 + DT^3 + \ldots$ or $\rho = [1 + 1.75(1 - T/T_c)^{1/3} + 0.75(1 - T/T_c)][\rho_c + A(T_c - T) + B(T_c - T)^2 + C(T_c - T)^3 + D(T_c - T)^4]$).

$\frac{T}{\mathrm{K}}$	$\frac{\rho \pm \sigma_{fit}}{\mathrm{kg \cdot m^{-3}}}$	$\frac{T}{\mathrm{K}}$	$\frac{\rho \pm \sigma_{fit}}{\mathrm{kg \cdot m^{-3}}}$	$\frac{T}{\mathrm{K}}$	$\frac{\rho \pm \sigma_{fit}}{\mathrm{kg \cdot m^{-3}}}$
290.00	831.40 ± 1.45	310.00	817.45 ± 1.05	350.00	788.04 ± 1.05
293.15	829.23 ± 1.36	320.00	810.29 ± 0.97	360.00	780.37 ± 1.37
298.15	825.78 ± 1.24	330.00	803.00 ± 0.92	370.00	772.57 ± 1.94
300.00	824.49 ± 1.20	340.00	795.58 ± 0.92		

cont.

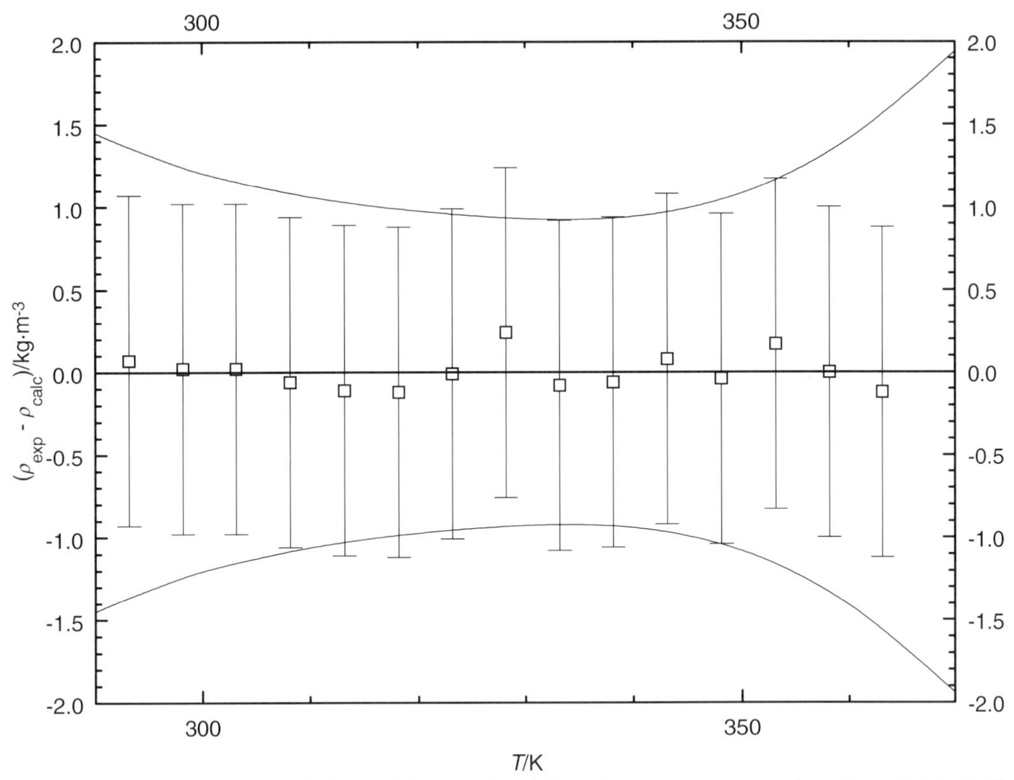

Fig. 1. The symbols show the deviation of the calculated from the experimental values from Table 2. The curves above and below the zero line indicate the calculated error region of the recommended values given in Table 3. The error bars represent the experimental errors. (Error bars smaller than the symbols are omitted for clarity of the figure.)

3-Dodecanol **[10203-30-2]** **$C_{12}H_{26}O$** **MW = 186.34** **368**

Table 1. Coefficients of the polynomial expansion equation.
Standard deviations (see introduction):
$\sigma_{c,w} = 3.8100 \cdot 10^{-2}$ (combined temperature ranges, weighted),
$\sigma_{c,uw} = 1.0998 \cdot 10^{-2}$ (combined temperature ranges, unweighted).

Coefficient	T = 293.15 to 363.15 K $\rho = A + BT + CT^2 + DT^3 + \ldots$
A	$1.00182 \cdot 10^{3}$
B	$-4.39256 \cdot 10^{-1}$
C	$-4.82504 \cdot 10^{-4}$

Table 2. Experimental values with uncertainties and deviation from calculated values.

$\frac{T}{K}$	$\frac{\rho_{exp} \pm 2\sigma_{est}}{kg \cdot m^{-3}}$	$\frac{\rho_{exp} - \rho_{calc}}{kg \cdot m^{-3}}$	Ref. (Symbol in Fig. 1)	$\frac{T}{K}$	$\frac{\rho_{exp} \pm 2\sigma_{est}}{kg \cdot m^{-3}}$	$\frac{\rho_{exp} - \rho_{calc}}{kg \cdot m^{-3}}$	Ref. (Symbol in Fig. 1)
293.15	831.60 ± 1.00	0.01	62-gei/qui(□)	308.15	820.60 ± 1.00	−0.05	62-gei/qui(□)
298.15	828.00 ± 1.00	0.03	62-gei/qui(□)	313.15	816.90 ± 1.00	−0.05	62-gei/qui(□)
303.15	824.30 ± 1.00	−0.02	62-gei/qui(□)	318.15	813.30 ± 1.00	0.06	62-gei/qui(□)

cont.

3-Dodecanol (cont.)

Table 2. (cont.)

$\frac{T}{K}$	$\frac{\rho_{exp} \pm 2\sigma_{est}}{kg \cdot m^{-3}}$	$\frac{\rho_{exp} - \rho_{calc}}{kg \cdot m^{-3}}$	Ref. (Symbol in Fig. 1)	$\frac{T}{K}$	$\frac{\rho_{exp} \pm 2\sigma_{est}}{kg \cdot m^{-3}}$	$\frac{\rho_{exp} - \rho_{calc}}{kg \cdot m^{-3}}$	Ref. (Symbol in Fig. 1)
323.15	809.70 ± 1.00	0.21	62-gei/qui[1)]	348.15	790.40 ± 1.00	−0.01	62-gei/qui(□)
328.15	805.80 ± 1.00	0.08	62-gei/qui(□)	353.15	786.50 ± 1.00	−0.03	62-gei/qui(□)
333.15	801.90 ± 1.00	−0.03	62-gei/qui(□)	358.15	782.60 ± 1.00	−0.01	62-gei/qui(□)
338.15	798.10 ± 1.00	−0.02	62-gei/qui(□)	363.15	778.70 ± 1.00	0.02	62-gei/qui(□)
343.15	794.30 ± 1.00	0.02	62-gei/qui(□)				

[1)] Not included in Fig. 1.

Further references: [13-pic/ken].

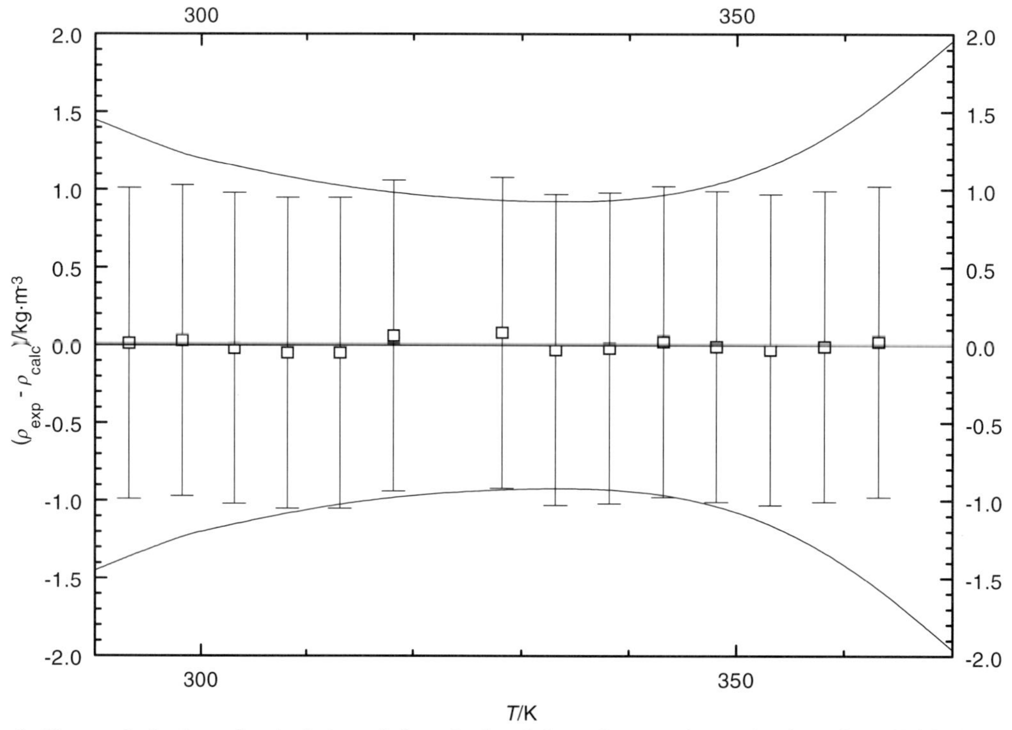

Fig. 1. The symbols show the deviation of the calculated from the experimental values from Table 2. The curves above and below the zero line indicate the calculated error region of the recommended values given in Table 3. The error bars represent the experimental errors. (Error bars smaller than the symbols are omitted for clarity of the figure.)

Table 3. Recommended values (fit to the reliable experimental values according to the equations
$\rho = A + BT + CT^2 + DT^3 + \ldots$ or $\rho = [1 + 1.75(1 - T/T_c)^{1/3} + 0.75(1 - T/T_c)][\rho_c + A(T_c - T) + B(T_c - T)^2 + C(T_c - T)^3 + D(T_c - T)^4]$).

$\frac{T}{K}$	$\frac{\rho \pm \sigma_{fit}}{kg \cdot m^{-3}}$	$\frac{T}{K}$	$\frac{\rho \pm \sigma_{fit}}{kg \cdot m^{-3}}$	$\frac{T}{K}$	$\frac{\rho \pm \sigma_{fit}}{kg \cdot m^{-3}}$
290.00	833.86 ± 1.45	310.00	819.29 ± 1.05	350.00	788.98 ± 1.04
293.15	831.59 ± 1.36	320.00	811.85 ± 0.96	360.00	781.16 ± 1.37
298.15	827.97 ± 1.23	330.00	804.32 ± 0.92	370.00	773.24 ± 1.96
300.00	826.62 ± 1.20	340.00	796.70 ± 0.92		

4-Dodecanol **[10203-32-4]** **$C_{12}H_{26}O$** **MW = 186.34** **369**

Table 1. Coefficients of the polynomial expansion equation.
Standard deviations (see introduction):
$\sigma_{c,w} = 4.1061 \cdot 10^{-2}$ (combined temperature ranges, weighted),
$\sigma_{c,uw} = 1.1853 \cdot 10^{-2}$ (combined temperature ranges, unweighted).

Coefficient	T = 293.15 to 363.15 K $\rho = A + BT + CT^2 + DT^3 + \dots$
A	$9.85212 \cdot 10^{2}$
B	$-3.51028 \cdot 10^{-1}$
C	$-6.18755 \cdot 10^{-4}$

Table 2. Experimental values with uncertainties and deviation from calculated values.

$\frac{T}{K}$	$\frac{\rho_{exp} \pm 2\sigma_{est}}{kg \cdot m^{-3}}$	$\frac{\rho_{exp} - \rho_{calc}}{kg \cdot m^{-3}}$	Ref. (Symbol in Fig. 1)	$\frac{T}{K}$	$\frac{\rho_{exp} \pm 2\sigma_{est}}{kg \cdot m^{-3}}$	$\frac{\rho_{exp} - \rho_{calc}}{kg \cdot m^{-3}}$	Ref. (Symbol in Fig. 1)
293.15	829.20 ± 1.00	0.07	62-gei/qui(□)	333.15	799.60 ± 1.00	0.01	62-gei/qui(□)
298.15	825.50 ± 1.00	−0.05	62-gei/qui(□)	338.15	795.70 ± 1.00	−0.06	62-gei/qui(□)
303.15	821.90 ± 1.00	−0.03	62-gei/qui(□)	343.15	791.90 ± 1.00	0.00	62-gei/qui(□)
308.15	818.30 ± 1.00	0.01	62-gei/qui(□)	348.15	788.10 ± 1.00	0.10	62-gei/qui(□)
313.15	814.60 ± 1.00	−0.01	62-gei/qui(□)	353.15	784.10 ± 1.00	0.02	62-gei/qui(□)
318.15	810.90 ± 1.00	−0.00	62-gei/qui(□)	358.15	780.10 ± 1.00	−0.02	62-gei/qui(□)
323.15	807.30 ± 1.00	0.14	62-gei/qui[1)]	363.15	776.10 ± 1.00	−0.04	62-gei/qui(□)
328.15	803.40 ± 1.00	0.01	62-gei/qui(□)				

[1)] Not included in Fig. 1.

Table 3. Recommended values (fit to the reliable experimental values according to the equations
$\rho = A + BT + CT^2 + DT^3 + \dots$ or $\rho = [1 + 1.75(1 - T/T_c)^{1/3} + 0.75(1 - T/T_c)][\rho_c + A(T_c - T) + B(T_c - T)^2 + C(T_c - T)^3 + D(T_c - T)^4]$).

$\frac{T}{K}$	$\frac{\rho \pm \sigma_{fit}}{kg \cdot m^{-3}}$	$\frac{T}{K}$	$\frac{\rho \pm \sigma_{fit}}{kg \cdot m^{-3}}$	$\frac{T}{K}$	$\frac{\rho \pm \sigma_{fit}}{kg \cdot m^{-3}}$
290.00	831.38 ± 1.45	310.00	816.93 ± 1.05	350.00	786.55 ± 1.04
293.15	829.13 ± 1.36	320.00	809.52 ± 0.96	360.00	778.65 ± 1.37
298.15	825.55 ± 1.23	330.00	801.99 ± 0.92	370.00	770.62 ± 1.96
300.00	824.22 ± 1.20	340.00	794.33 ± 0.92		

cont.

4-Dodecanol (cont.)

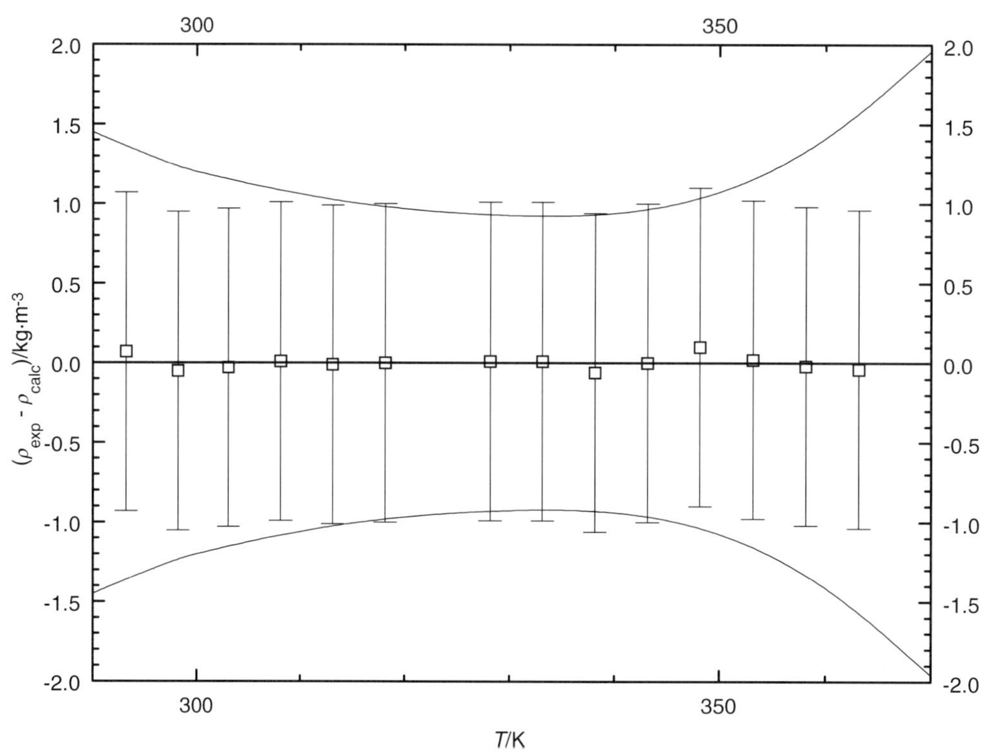

Fig. 1. The symbols show the deviation of the calculated from the experimental values from Table 2. The curves above and below the zero line indicate the calculated error region of the recommended values given in Table 3. The error bars represent the experimental errors. (Error bars smaller than the symbols are omitted for clarity of the figure.)

5-Dodecanol **[10203-33-5]** **$C_{12}H_{26}O$** **MW = 186.34** **370**

Table 1. Coefficients of the polynomial expansion equation.
Standard deviations (see introduction):
$\sigma_{c,w} = 7.5305 \cdot 10^{-2}$ (combined temperature ranges, weighted),
$\sigma_{c,uw} = 2.1739 \cdot 10^{-2}$ (combined temperature ranges, unweighted).

Coefficient	T = 293.15 to 363.15 K $\rho = A + BT + CT^2 + DT^3 + \ldots$
A	$9.59772 \cdot 10^{2}$
B	$-1.92462 \cdot 10^{-1}$
C	$-8.58906 \cdot 10^{-4}$

Table 2. Experimental values with uncertainties and deviation from calculated values.

$\frac{T}{K}$	$\frac{\rho_{exp} \pm 2\sigma_{est}}{kg \cdot m^{-3}}$	$\frac{\rho_{exp} - \rho_{calc}}{kg \cdot m^{-3}}$	Ref. (Symbol in Fig. 1)	$\frac{T}{K}$	$\frac{\rho_{exp} \pm 2\sigma_{est}}{kg \cdot m^{-3}}$	$\frac{\rho_{exp} - \rho_{calc}}{kg \cdot m^{-3}}$	Ref. (Symbol in Fig. 1)
293.15	829.50 ± 1.00	−0.04	62-gei/qui(□)	308.15	818.90 ± 1.00	−0.01	62-gei/qui(□)
298.15	826.00 ± 1.00	−0.04	62-gei/qui(□)	313.15	815.40 ± 1.00	0.12	62-gei/qui(□)
303.15	822.50 ± 1.00	0.01	62-gei/qui(□)	318.15	811.70 ± 1.00	0.10	62-gei/qui(□)

cont.

Table 2. (cont.)

$\frac{T}{K}$	$\frac{\rho_{exp} \pm 2\sigma_{est}}{kg \cdot m^{-3}}$	$\frac{\rho_{exp} - \rho_{calc}}{kg \cdot m^{-3}}$	Ref. (Symbol in Fig. 1)	$\frac{T}{K}$	$\frac{\rho_{exp} \pm 2\sigma_{est}}{kg \cdot m^{-3}}$	$\frac{\rho_{exp} - \rho_{calc}}{kg \cdot m^{-3}}$	Ref. (Symbol in Fig. 1)
323.15	807.90 ± 1.00	0.01	62-gei/qui(□)	348.15	788.60 ± 1.00	−0.06	62-gei/qui(□)
328.15	804.10 ± 1.00	−0.03	62-gei/qui(□)	353.15	784.80 ± 1.00	0.11	62-gei/qui(□)
333.15	800.30 ± 1.00	−0.02	62-gei/qui(□)	358.15	780.70 ± 1.00	0.03	62-gei/qui(□)
338.15	796.30 ± 1.00	−0.18	62-gei/qui(□)	363.15	776.60 ± 1.00	−0.01	62-gei/qui(□)
343.15	792.30 ± 1.00	−0.29	62-gei/qui[1)]				

[1)] Not included in Fig. 1.

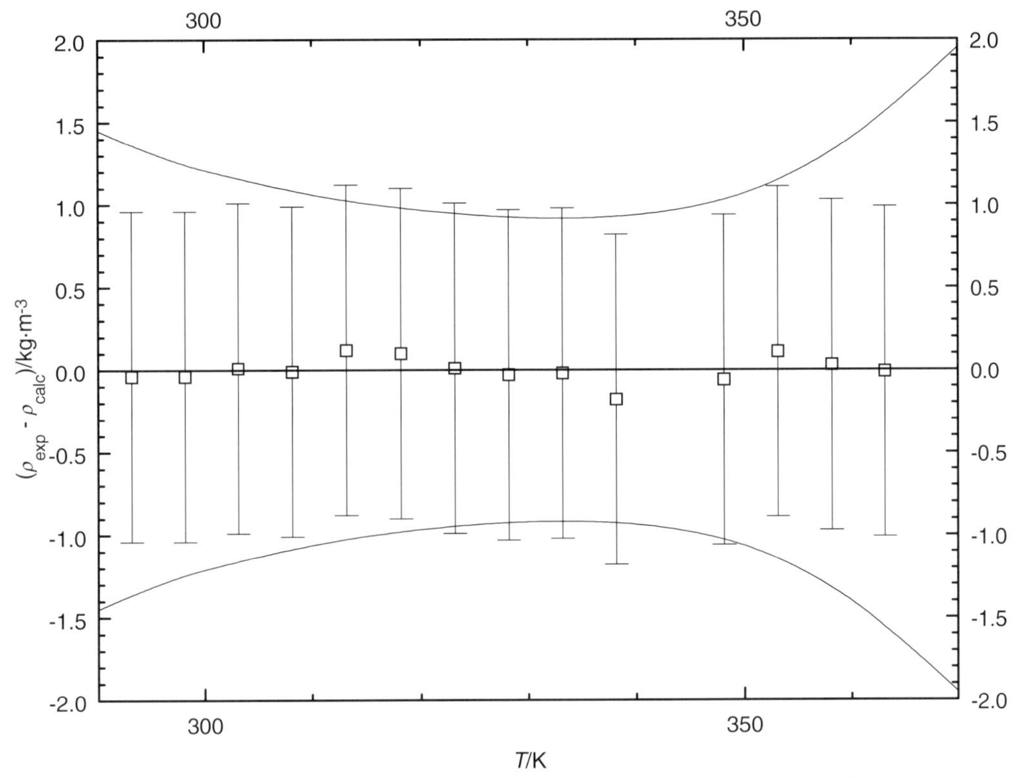

Fig. 1. The symbols show the deviation of the calculated from the experimental values from Table 2. The curves above and below the zero line indicate the calculated error region of the recommended values given in Table 3. The error bars represent the experimental errors. (Error bars smaller than the symbols are omitted for clarity of the figure.)

Table 3. Recommended values (fit to the reliable experimental values according to the equations
$\rho = A + BT + CT^2 + DT^3 + \ldots$ or $\rho = [1 + 1.75(1 - T/T_c)^{1/3} + 0.75(1 - T/T_c)][\rho_c + A(T_c - T) + B(T_c - T)^2 + C(T_c - T)^3 + D(T_c - T)^4]$).

$\frac{T}{K}$	$\frac{\rho \pm \sigma_{fit}}{kg \cdot m^{-3}}$	$\frac{T}{K}$	$\frac{\rho \pm \sigma_{fit}}{kg \cdot m^{-3}}$	$\frac{T}{K}$	$\frac{\rho \pm \sigma_{fit}}{kg \cdot m^{-3}}$
290.00	831.72 ± 1.45	310.00	817.57 ± 1.05	350.00	787.19 ± 1.03
293.15	829.54 ± 1.36	320.00	810.23 ± 0.96	360.00	779.17 ± 1.36
298.15	826.04 ± 1.24	330.00	802.73 ± 0.91	370.00	770.98 ± 1.95
300.00	824.73 ± 1.21	340.00	795.05 ± 0.92		

6-Dodecanol **[6836-38-0]** **$C_{12}H_{26}O$** **MW = 186.34** **371**

Table 1. Coefficients of the polynomial expansion equation.
Standard deviations (see introduction):
$\sigma_{c,w} = 8.7855 \cdot 10^{-2}$ (combined temperature ranges, weighted),
$\sigma_{c,uw} = 2.6489 \cdot 10^{-2}$ (combined temperature ranges, unweighted).

Coefficient	T = 303.15 to 363.15 K $\rho = A + BT + CT^2 + DT^3 + \ldots$
A	$9.92440 \cdot 10^{2}$
B	$-4.06419 \cdot 10^{-1}$
C	$-5.33467 \cdot 10^{-4}$

Table 2. Experimental values with uncertainties and deviation from calculated values.

$\frac{T}{\text{K}}$	$\frac{\rho_{exp} \pm 2\sigma_{est}}{\text{kg} \cdot \text{m}^{-3}}$	$\frac{\rho_{exp} - \rho_{calc}}{\text{kg} \cdot \text{m}^{-3}}$	Ref. (Symbol in Fig. 1)	$\frac{T}{\text{K}}$	$\frac{\rho_{exp} \pm 2\sigma_{est}}{\text{kg} \cdot \text{m}^{-3}}$	$\frac{\rho_{exp} - \rho_{calc}}{\text{kg} \cdot \text{m}^{-3}}$	Ref. (Symbol in Fig. 1)
303.15	820.10 ± 1.00	−0.11	62-gei/qui(□)	338.15	793.90 ± 1.00	−0.11	62-gei/qui(□)
308.15	816.50 ± 1.00	−0.05	62-gei/qui(□)	343.15	790.10 ± 1.00	−0.06	62-gei/qui(□)
313.15	812.90 ± 1.00	0.04	62-gei/qui(□)	348.15	786.20 ± 1.00	−0.08	62-gei/qui(□)
318.15	809.30 ± 1.00	0.16	62-gei/qui(□)	353.15	782.30 ± 1.00	−0.08	62-gei/qui(□)
323.15	805.50 ± 1.00	0.10	62-gei/qui(□)	358.15	778.50 ± 1.00	0.05	62-gei/qui(□)
328.15	801.70 ± 1.00	0.07	62-gei/qui(□)	363.15	774.60 ± 1.00	0.10	62-gei/qui(□)
333.15	797.80 ± 1.00	−0.03	62-gei/qui(□)				

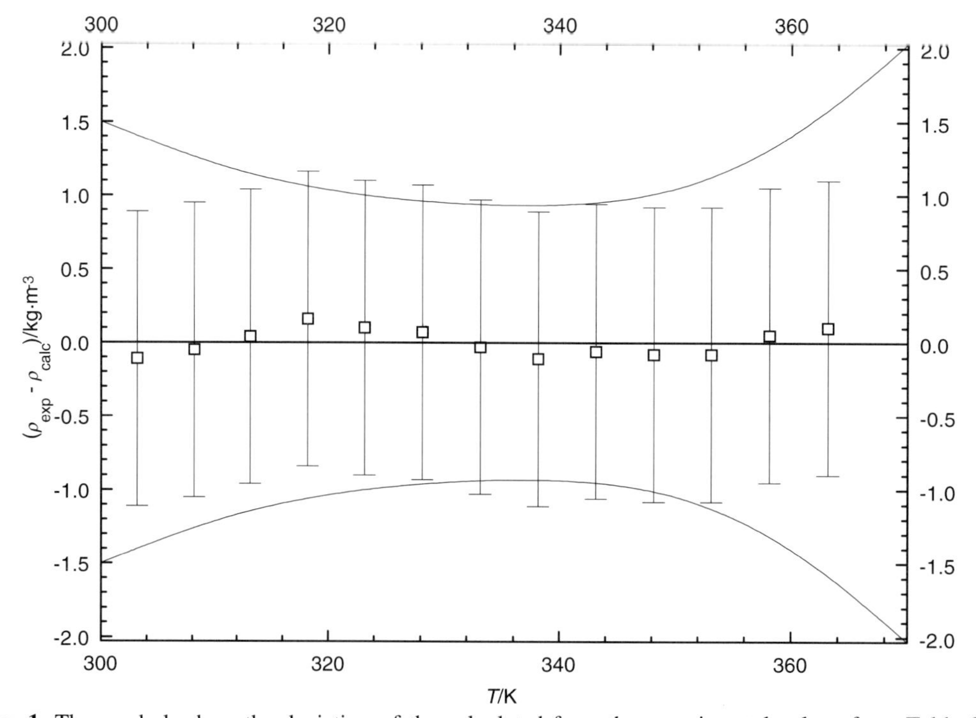

Fig. 1. The symbols show the deviation of the calculated from the experimental values from Table 2. The curves above and below the zero line indicate the calculated error region of the recommended values given in Table 3. The error bars represent the experimental errors. (Error bars smaller than the symbols are omitted for clarity of the figure.).

cont.

Table 3. Recommended values (fit to the reliable experimental values according to the equations $\rho = A + BT + CT^2 + DT^3 + \ldots$ or $\rho = [1 + 1.75(1 - T/T_c)^{1/3} + 0.75(1 - T/T_c)][\rho_c + A(T_c - T) + B(T_c - T)^2 + C(T_c - T)^3 + D(T_c - T)^4]$).

T/K	$\rho \pm \sigma_{fit}$ / kg·m^{-3}	T/K	$\rho \pm \sigma_{fit}$ / kg·m^{-3}	T/K	$\rho \pm \sigma_{fit}$ / kg·m^{-3}
300.00	822.50 ± 1.50	330.00	800.23 ± 0.94	360.00	776.99 ± 1.35
310.00	815.18 ± 1.19	340.00	792.59 ± 0.92	370.00	769.03 ± 2.03
320.00	807.76 ± 1.02	350.00	784.84 ± 1.00		

5-Ethyl-5-decanol **[91635-39-1]** $C_{12}H_{26}O$ **MW = 186.34** **372**

Table 1. Fit with estimated *B* coefficient for 3 accepted points. Deviation $\sigma_w = 1.122$.

Coefficient	$\rho = A + BT$
A	1054.83
B	–0.740

Table 2. Experimental values with uncertainties and deviation from calculated values.

T/K	$\rho_{exp} \pm 2\sigma_{est}$ / kg·m^{-3}	$\rho_{exp} - \rho_{calc}$ / kg·m^{-3}	Ref.
298.15	835.7 ± 1.0	1.50	44-qua/sma
318.15	819.1 ± 1.0	–0.30	44-qua/sma
328.15	810.8 ± 1.0	–1.20	44-qua/sma

Table 3. Recommended values.

T/K	$\rho_{exp} \pm 2\sigma_{est}$ / kg·m^{-3}	T/K	$\rho_{exp} \pm 2\sigma_{est}$ / kg·m^{-3}	T/K	$\rho_{exp} \pm 2\sigma_{est}$ / kg·m^{-3}
290.00	840.2 ± 1.9	298.15	834.2 ± 1.7	320.00	818.0 ± 1.5
293.15	837.9 ± 1.8	310.00	825.4 ± 1.5	330.00	810.6 ± 1.6

6-Ethyl-3-decanol **[19780-31-5]** $C_{12}H_{26}O$ **MW = 186.34** **373**

Table 1. Experimental value with uncertainty.

T/K	$\rho_{exp} \pm 2\sigma_{est}$ / kg·m^{-3}	Ref.
301.15	836.4 ± 1.0	36-pow/bal

4-Ethyl-2-methyl-3-(1-methylethyl)-3-hexanol **[500001-86-5]** $C_{12}H_{26}O$ **MW = 186.34** **374**

Table 1. Experimental value with uncertainty.

T/K	$\rho_{exp} \pm 2\sigma_{est}$ / kg·m^{-3}	Ref.
293.15	852.8 ± 1.0	43-geo

5-Ethyl-2-methyl-3-nonanol **[105902-95-2]** **$C_{12}H_{26}O$** **MW = 186.34** **375**

Table 1. Experimental value with uncertainty.

T / K	$\rho_{exp} \pm 2\sigma_{est}$ / $kg \cdot m^{-3}$	Ref.
293.15	847.1 ± 2.0	57-pet/nef-1

5-Ethyl-7-methyl-3-nonanol **[66634-87-5]** **$C_{12}H_{26}O$** **MW = 186.34** **376**

Table 1. Experimental value with uncertainty.

T / K	$\rho_{exp} \pm 2\sigma_{est}$ / $kg \cdot m^{-3}$	Ref.
273.15	921.0 ± 2.0	12-gue

4-Ethyl-2,2,3,4-tetramethyl-3-hexanol **[91635-45-9]** **$C_{12}H_{26}O$** **MW = 186.34** **377**

Table 1. Experimental value with uncertainty.

T / K	$\rho_{exp} \pm 2\sigma_{est}$ / $kg \cdot m^{-3}$	Ref.
291.15	892.0 ± 2.0	37-naz

3-Ethyl-3,4,5-trimethyl-4-heptanol **[500002-61-9]** **$C_{12}H_{26}O$** **MW = 186.34** **378**

Table 1. Experimental value with uncertainty.

T / K	$\rho_{exp} \pm 2\sigma_{est}$ / $kg \cdot m^{-3}$	Ref.
286.15	885.0 ± 2.0	37-naz

2,2,3,4,4,5-Hexamethyl-3-hexanol **[500002-63-1]** **$C_{12}H_{26}O$** **MW = 186.34** **379**

Table 1. Experimental value with uncertainty.

T / K	$\rho_{exp} \pm 2\sigma_{est}$ / $kg \cdot m^{-3}$	Ref.
286.15	893.0 ± 2.0	37-naz

2,2,4,4,5,5-Hexamethyl-3-hexanol **[500001-88-7]** **$C_{12}H_{26}O$** **MW = 186.34** **380**

Table 1. Fit with estimated B coefficient for 3 accepted points. Deviation $\sigma_w = 0.245$.

Coefficient	$\rho = A + BT$
A	1065.31
B	−0.700

cont.

Table 2. Experimental values with uncertainties and deviation from calculated values.

$\frac{T}{\mathrm{K}}$	$\frac{\rho_{exp} \pm 2\sigma_{est}}{\mathrm{kg \cdot m^{-3}}}$	$\frac{\rho_{exp} - \rho_{calc}}{\mathrm{kg \cdot m^{-3}}}$	Ref.
288.15	848.1 ± 4.0	−15.50	33-fav/naz[1)]
293.15	860.4 ± 0.6	0.30	53-per/wag
303.15	853.1 ± 0.6	−0.00	53-per/wag
313.15	845.8 ± 0.6	−0.30	53-per/wag

[1)] Not included in calculation of linear coefficients.

Table 3. Recommended values.

$\frac{T}{\mathrm{K}}$	$\frac{\rho_{exp} \pm 2\sigma_{est}}{\mathrm{kg \cdot m^{-3}}}$
290.00	862.3 ± 0.8
293.15	860.1 ± 0.7
298.15	856.6 ± 0.5
310.00	848.3 ± 0.6
320.00	841.3 ± 1.0

2,3,4,4,5,5-Hexamethyl-3-hexanol [100392-68-5] $C_{12}H_{26}O$ MW = 186.34 381

Table 1. Experimental value with uncertainty.

$\frac{T}{\mathrm{K}}$	$\frac{\rho_{exp} \pm 2\sigma_{est}}{\mathrm{kg \cdot m^{-3}}}$	Ref.
293.15	888.6 ± 2.0	60-pet/kao

5-Methyl-2-(3-methylbutyl)-1-hexanol [500001-82-1] $C_{12}H_{26}O$ MW = 186.34 382

Table 1. Experimental value with uncertainty.

$\frac{T}{\mathrm{K}}$	$\frac{\rho_{exp} \pm 2\sigma_{est}}{\mathrm{kg \cdot m^{-3}}}$	Ref.
293.15	834.0 ± 1.0	37-bra/kur

2-Methyl-3-(1-methylethyl)-3-octanol [19965-71-0] $C_{12}H_{26}O$ MW = 186.34 383

Table 1. Experimental value with uncertainty.

$\frac{T}{\mathrm{K}}$	$\frac{\rho_{exp} \pm 2\sigma_{est}}{\mathrm{kg \cdot m^{-3}}}$	Ref.
293.15	853.7 ± 1.0	43-geo

5-Methyl-5-propyl-4-octanol [500001-75-2] $C_{12}H_{26}O$ MW = 186.34 384

Table 1. Experimental value with uncertainty.

$\frac{T}{\mathrm{K}}$	$\frac{\rho_{exp} \pm 2\sigma_{est}}{\mathrm{kg \cdot m^{-3}}}$	Ref.
289.65	845.5 ± 1.0	21-ler

2-Methyl-2-undecanol **[32836-42-3]** $C_{12}H_{26}O$ **MW = 186.34** **385**

Table 1. Fit with estimated B coefficient for 2 accepted points. Deviation $\sigma_w = 0.410$.

Coefficient	$\rho = A + BT$
A	1046.24
B	–0.740

Table 2. Experimental values with uncertainties and deviation from calculated values.

$\frac{T}{K}$	$\frac{\rho_{exp} \pm 2\sigma_{est}}{kg \cdot m^{-3}}$	$\frac{\rho_{exp} - \rho_{calc}}{kg \cdot m^{-3}}$	Ref.
273.15	843.7 ± 1.0	–0.41	19-beh
286.15	834.9 ± 1.0	0.41	19-beh

Table 3. Recommended values.

$\frac{T}{K}$	$\frac{\rho_{exp} \pm 2\sigma_{est}}{kg \cdot m^{-3}}$
270.00	846.4 ± 1.1
280.00	839.0 ± 1.0
290.00	831.6 ± 1.1

2-Methyl-3-undecanol **[60671-36-5]** $C_{12}H_{26}O$ **MW = 186.34** **386**

Table 1. Experimental value with uncertainty.

$\frac{T}{K}$	$\frac{\rho_{exp} \pm 2\sigma_{est}}{kg \cdot m^{-3}}$	Ref.
293.15	832.7 ± 2.0	12-pic/ken

2-Methyl-5-undecanol **[33978-71-1]** $C_{12}H_{26}O$ **MW = 186.34** **387**

Table 1. Experimental values with uncertainties.

$\frac{T}{K}$	$\frac{\rho_{exp} \pm 2\sigma_{est}}{kg \cdot m^{-3}}$	Ref.
293.15	826.6 ± 1.0	44-pow/hag
293.15	825.1 ± 2.0	48-pet/old

3-Methyl-1-undecanol **[71526-27-7]** $C_{12}H_{26}O$ **MW = 186.34** **388**

Table 1. Fit with estimated B coefficient for 4 accepted points. Deviation $\sigma_w = 0.817$.

Coefficient	$\rho = A + BT$
A	1039.61
B	–0.700

cont.

Table 2. Experimental values with uncertainties and deviation from calculated values.

$\frac{T}{K}$	$\frac{\rho_{exp} \pm 2\sigma_{est}}{kg \cdot m^{-3}}$	$\frac{\rho_{exp} - \rho_{calc}}{kg \cdot m^{-3}}$	Ref.
293.15	834.1 ± 1.0	–0.30	48-pro/cas
298.15	830.9 ± 1.0	–0.00	48-pro/cas
303.15	827.5 ± 1.0	0.10	48-pro/cas
308.15	824.1 ± 1.0	0.20	48-pro/cas

Table 3. Recommended values.

$\frac{T}{K}$	$\frac{\rho_{exp} \pm 2\sigma_{est}}{kg \cdot m^{-3}}$
290.00	836.6 ± 1.1
293.15	834.4 ± 1.0
298.15	830.9 ± 0.9
310.00	822.6 ± 1.0

***[S-(R*,R*)]*-3-Methyl-5-undecanol** **[82749-56-2]** **$C_{12}H_{26}O$** **MW = 186.34** **389**

Table 1. Experimental value with uncertainty.

$\frac{T}{K}$	$\frac{\rho_{exp} \pm 2\sigma_{est}}{kg \cdot m^{-3}}$	Ref.
297.15	827.2 ± 2.0	35-lev/har

5-Methyl-5-undecanol **[21078-80-8]** **$C_{12}H_{26}O$** **MW = 186.34** **390**

Table 1. Coefficients of the polynomial expansion equation.
Standard deviations (see introduction):
$\sigma_{c,w} = 3.2352 \cdot 10^{-1}$ (combined temperature ranges, weighted),
$\sigma_{c,uw} = 1.4468 \cdot 10^{-1}$ (combined temperature ranges, unweighted).

Coefficient	T = 273.15 to 338.15 K $\rho = A + BT + CT^2 + DT^3 + \ldots$
A	$1.05408 \cdot 10^{3}$
B	$-7.68000 \cdot 10^{-1}$

Table 2. Experimental values with uncertainties and deviation from calculated values.

$\frac{T}{K}$	$\frac{\rho_{exp} \pm 2\sigma_{est}}{kg \cdot m^{-3}}$	$\frac{\rho_{exp} - \rho_{calc}}{kg \cdot m^{-3}}$	Ref. (Symbol in Fig. 1)	$\frac{T}{K}$	$\frac{\rho_{exp} \pm 2\sigma_{est}}{kg \cdot m^{-3}}$	$\frac{\rho_{exp} - \rho_{calc}}{kg \cdot m^{-3}}$	Ref. (Symbol in Fig. 1)
273.15	844.70 ± 1.00	0.40	39-owe/qua(□)	318.15	809.70 ± 1.00	–0.04	39-owe/qua(□)
298.15	824.70 ± 1.00	–0.40	39-owe/qua(□)	328.15	802.40 ± 1.00	0.34	39-owe/qua(□)
308.15	817.00 ± 1.00	–0.42	39-owe/qua(□)	338.15	794.50 ± 1.00	0.12	39-owe/qua(□)

cont.

5-Methyl-5-undecanol (cont.)

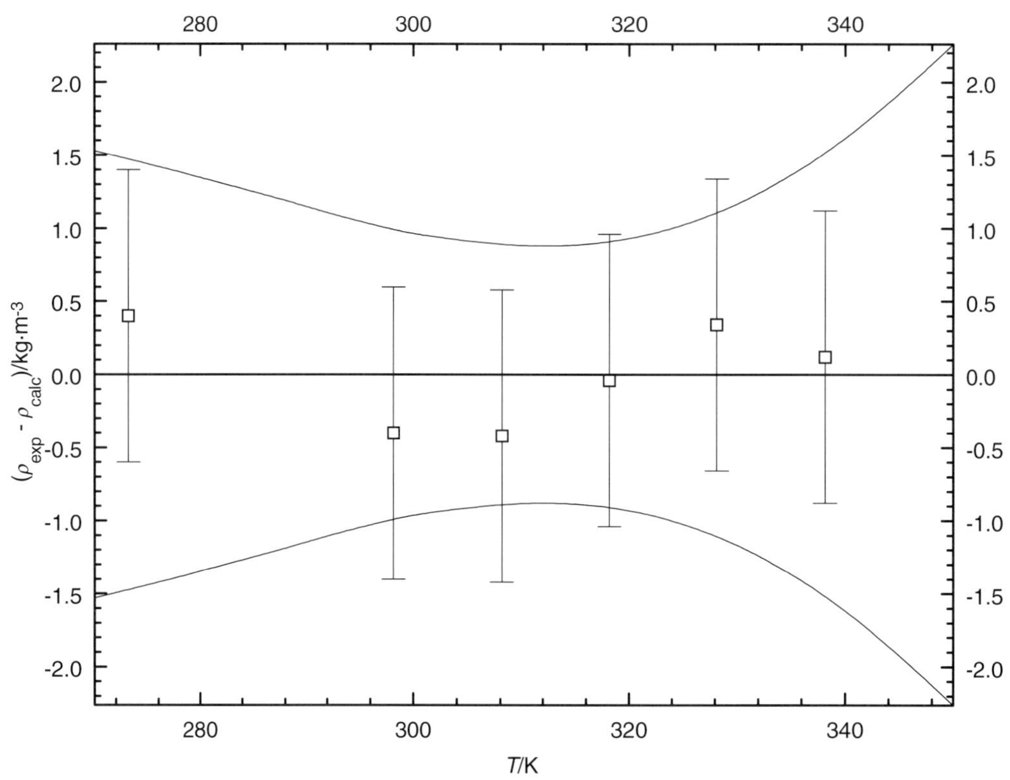

Fig. 1. The symbols show the deviation of the calculated from the experimental values from Table 2. The curves above and below the zero line indicate the calculated error region of the recommended values given in Table 3. The error bars represent the experimental errors. (Error bars smaller than the symbols are omitted for clarity of the figure.)

Table 3. Recommended values (fit to the reliable experimental values according to the equations $\rho = A + BT + CT^2 + DT^3 + \ldots$ or $\rho = [1 + 1.75(1 - T/T_c)^{1/3} + 0.75(1 - T/T_c)][\rho_c + A(T_c - T) + B(T_c - T)^2 + C(T_c - T)^3 + D(T_c - T)^4]$).

$\frac{T}{K}$	$\frac{\rho \pm \sigma_{fit}}{kg \cdot m^{-3}}$	$\frac{T}{K}$	$\frac{\rho \pm \sigma_{fit}}{kg \cdot m^{-3}}$	$\frac{T}{K}$	$\frac{\rho \pm \sigma_{fit}}{kg \cdot m^{-3}}$
270.00	846.72 ± 1.53	298.15	825.10 ± 0.99	330.00	800.64 ± 1.13
280.00	839.04 ± 1.35	300.00	823.68 ± 0.96	340.00	792.96 ± 1.58
290.00	831.36 ± 1.15	310.00	816.00 ± 0.86	350.00	785.28 ± 2.26
293.15	828.94 ± 1.08	320.00	808.32 ± 0.90		

6-Methyl-6-undecanol **[5340-31-8]** **$C_{12}H_{26}O$** **MW = 186.34** **391**

Table 1. Experimental value with uncertainty.

$\frac{T}{K}$	$\frac{\rho_{exp} \pm 2\sigma_{est}}{kg \cdot m^{-3}}$	Ref.
298.15	827.1 ± 1.0	33-whi/wil

9-Methyl-1-undecanol **[91635-46-0]** $C_{12}H_{26}O$ **MW = 186.34** **392**

Table 1. Experimental value with uncertainty.

$\frac{T}{K}$	$\frac{\rho_{exp} \pm 2\sigma_{est}}{kg \cdot m^{-3}}$	Ref.
298.15	831.6 ± 1.0	62-lar/sal

L(+)-**9-Methyl-1-undecanol** **[500006-92-8]** $C_{12}H_{26}O$ **MW = 186.34** **393**

Table 1. Experimental value with uncertainty.

$\frac{T}{K}$	$\frac{\rho_{exp} \pm 2\sigma_{est}}{kg \cdot m^{-3}}$	Ref.
298.15	831.6 ± 0.6	62-lar/sal

5-(1-Methylethyl)-5-nonanol **[76144-88-2]** $C_{12}H_{26}O$ **MW = 186.34** **394**

Table 1. Experimental value with uncertainty.

$\frac{T}{K}$	$\frac{\rho_{exp} \pm 2\sigma_{est}}{kg \cdot m^{-3}}$	Ref.
293.15	844.0 ± 2.0	33-whi/kru

3-(1-Methylethyl)-2,2,4,4-tetramethyl-3-pentanol **[5457-42-1]** $C_{12}H_{26}O$ **MW = 186.34** **395**

Table 1. Experimental and recommended values with uncertainties.

$\frac{T}{K}$	$\frac{\rho_{exp} \pm 2\sigma_{est}}{kg \cdot m^{-3}}$	Ref.
293.15	888.9± 1.0	51-smi/cre
293.15	885.3 ± 2.0	57-pet/sok
293.15	888.2 ± 1.6	Recommended

3-(1-Methylethyl)-2,2,5-trimethyl-3-hexanol **[500001-89-8]** $C_{12}H_{26}O$ **MW = 186.34** **396**

Table 1. Experimental value with uncertainty.

$\frac{T}{K}$	$\frac{\rho_{exp} \pm 2\sigma_{est}}{kg \cdot m^{-3}}$	Ref.
293.15	855.2 ± 2.0	57-pet/sus

2,2,3,3,4-Pentamethyl-4-heptanol **[500001-81-0]** $C_{12}H_{26}O$ **MW = 186.34** **397**

Table 1. Experimental value with uncertainty.

$\frac{T}{K}$	$\frac{\rho_{exp} \pm 2\sigma_{est}}{kg \cdot m^{-3}}$	Ref.
293.15	877.2 ± 2.0	60-pet/kao

2,2,4,6,6-Pentamethyl-3-heptanol **[105902-93-0]** **$C_{12}H_{26}O$** **MW = 186.34** **398**

Table 1. Experimental value with uncertainty.

T/K	$\rho_{exp} \pm 2\sigma_{est}$/kg·m^{-3}	Ref.
293.15	838.0 ± 1.0	41-whi/whi

4-Propyl-4-nonanol **[5340-77-2]** **$C_{12}H_{26}O$** **MW = 186.34** **399**

Table 1. Experimental value with uncertainty.

T/K	$\rho_{exp} \pm 2\sigma_{est}$/kg·m^{-3}	Ref.
297.65	832.4 ± 2.0	33-whi/wil

5-Propyl-5-nonanol **[5340-52-3]** **$C_{12}H_{26}O$** **MW = 186.34** **400**

Table 1. Fit with estimated B coefficient for 4 accepted points. Deviation σ_w = 1.052.

Coefficient	$\rho = A + BT$
A	1066.88
B	−0.780

Table 2. Experimental values with uncertainties and deviation from calculated values.

T/K	$\rho_{exp} \pm 2\sigma_{est}$/kg·m^{-3}	$\rho_{exp} - \rho_{calc}$/kg·m^{-3}	Ref.
293.15	837.0 ± 2.0	−1.23	33-whi/woo
298.15	834.0 ± 2.0	−0.33	33-whi/woo
288.15	842.0 ± 2.0	−0.13	37-pet/mal
293.15	839.9 ± 2.0	1.67	61-mes/erz

Table 3. Recommended values.

T/K	$\rho_{exp} \pm 2\sigma_{est}$/kg·m^{-3}
280.00	848.5 ± 2.5
290.00	840.7 ± 2.1
293.15	838.2 ± 2.1
298.15	834.3 ± 2.1

2,2,3,4,4-Pentamethyl-3-heptanol **[500002-60-8]** **$C_{12}H_{26}O$** **MW = 186.34** **401**

Table 1. Experimental value with uncertainty.

T/K	$\rho_{exp} \pm 2\sigma_{est}$/kg·m^{-3}	Ref.
286.15	885.0 ± 2.0	37-naz

2,2,5,6,6-Pentamethyl-3-heptanol **[500002-65-3]** **$C_{12}H_{26}O$** **MW = 186.34** **402**

Table 1. Experimental value with uncertainty.

$\frac{T}{K}$	$\frac{\rho_{exp} \pm 2\sigma_{est}}{kg \cdot m^{-3}}$	Ref.
289.15	845.0 ± 2.0	35-col

3,3,4,5,5-Pentamethyl-4-heptanol **[500002-62-0]** **$C_{12}H_{26}O$** **MW = 186.34** **403**

Table 1. Experimental value with uncertainty.

$\frac{T}{K}$	$\frac{\rho_{exp} \pm 2\sigma_{est}}{kg \cdot m^{-3}}$	Ref.
289.15	900.4 ± 1.0	37-naz

2,5,8-Trimethyl-5-nonanol **[64029-94-3]** **$C_{12}H_{26}O$** **MW = 186.34** **404**

Table 1. Experimental value with uncertainty.

$\frac{T}{K}$	$\frac{\rho_{exp} \pm 2\sigma_{est}}{kg \cdot m^{-3}}$	Ref.
293.15	828.9 ± 1.0	59-yur/bel

2,6,8-Trimethyl-4-nonanol **[123-17-1]** **$C_{12}H_{26}O$** **MW = 186.34** **405**

Table 1. Experimental value with uncertainty.

$\frac{T}{K}$	$\frac{\rho_{exp} \pm 2\sigma_{est}}{kg \cdot m^{-3}}$	Ref.
293.15	817.9 ± 1.0	53-ano-15
293.15	817.9 ± 1.0	58-ano-5
293.15	817.9 ± 1.0	68-ano

3,4,8-Trimethyl-1-nonanol **[18352-71-1]** **$C_{12}H_{26}O$** **MW = 186.34** **406**

Table 1. Experimental value with uncertainty.

$\frac{T}{K}$	$\frac{\rho_{exp} \pm 2\sigma_{est}}{kg \cdot m^{-3}}$	Ref.
293.15	846.3 ± 1.0	67-min/che

3,4,8-Trimethyl-3-nonanol **[18352-67-5]** **$C_{12}H_{26}O$** **MW = 186.34** **407**

Table 1. Experimental value with uncertainty.

$\frac{T}{K}$	$\frac{\rho_{exp} \pm 2\sigma_{est}}{kg \cdot m^{-3}}$	Ref.
293.15	836.6 ± 1.0	67-min/che

2.1.8 Alkanols, C_{13} - C_{22}

3-Butyl-2,2-dimethyl-3-heptanol **[900002-68-8]** **$C_{13}H_{28}O$** **MW = 200.36** **408**

Table 1. Experimental value with uncertainty.

$\frac{T}{K}$	$\frac{\rho_{exp} \pm 2\sigma_{est}}{kg \cdot m^{-3}}$	Ref.
293.15	849.8 ± 1.0	38-whi/pop

2-Butyl-1-nonanol **[51655-57-3]** **$C_{13}H_{28}O$** **MW = 200.36** **409**

Table 1. Experimental value with uncertainty.

$\frac{T}{K}$	$\frac{\rho_{exp} \pm 2\sigma_{est}}{kg \cdot m^{-3}}$	Ref.
293.15	835.9 ± 2.0	22-lev/tay-1

5-Butyl-5-nonanol **[597-93-3]** **$C_{13}H_{28}O$** **MW = 200.36** **410**

Table 1. Experimental and recommended values with uncertainties.

$\frac{T}{K}$	$\frac{\rho_{exp} \pm 2\sigma_{est}}{kg \cdot m^{-3}}$	Ref.	$\frac{T}{K}$	$\frac{\rho_{exp} \pm 2\sigma_{est}}{kg \cdot m^{-3}}$	Ref.
295.85	835.3 ± 2.0	19-eyk[1)]	293.15	844.2 ± 2.0	61-mes/erz[1)]
351.85	768.9 ± 3.0	19-eyk[1)]	298.15	835.0 ± 0.7	88-cac/cos
293.15	840.8 ± 2.0	33-whi/woo[1)]	298.15	835.0 ± 0.7	Recommended
298.15	836.8 ± 2.0	33-whi/woo[1)]			

[1)] Not included in calculation of recommended value.

4,4-Diethyl-2,2,3-trimethyl-3-hexanol **[500045-00-1]** **$C_{13}H_{28}O$** **MW = 200.36** **411**

Table 1. Experimental value with uncertainty.

$\frac{T}{K}$	$\frac{\rho_{exp} \pm 2\sigma_{est}}{kg \cdot m^{-3}}$	Ref.
283.15	906.0 ± 2.0	37-naz

2,2-Dimethyl-3-(1,1-dimethylethyl)-3-heptanol **[42930-67-6]** **$C_{13}H_{28}O$** **MW = 200.36** **412**

Table 1. Experimental and recommended values with uncertainties.

$\frac{T}{K}$	$\frac{\rho_{exp} \pm 2\sigma_{est}}{kg \cdot m^{-3}}$	Ref.
293.15	860.6 ± 1.0	54-mos/cox
293.15	859.6 ± 1.0	60-pet/sok
293.15	860.1 ± 1.1	Recommended

5,5-Dimethyl-2-neopentyl-1-hexanol **[109509-73-1]** **$C_{13}H_{28}O$** **MW = 200.36** **413**

Table 1. Experimental value with uncertainty.

T/K	$\rho_{exp} \pm 2\sigma_{est}$/$kg \cdot m^{-3}$	Ref.
293.15	847.0 ± 2.0	56-gol/kon

2,8-Dimethyl-5-ethyl-5-nonanol **[500001-93-4]** **$C_{13}H_{28}O$** **MW = 200.36** **414**

Table 1. Experimental value with uncertainty.

T/K	$\rho_{exp} \pm 2\sigma_{est}$/$kg \cdot m^{-3}$	Ref.
293.15	867.7 ± 2.0	14-hal

3,3-Dimethyl-5-ethyl-4-nonanol **[500000-49-7]** **$C_{13}H_{28}O$** **MW = 200.36** **415**

Table 1. Experimental value with uncertainty.

T/K	$\rho_{exp} \pm 2\sigma_{est}$/$kg \cdot m^{-3}$	Ref.
293.15	851.2 ± 1.0	41-whi/whi

2,4-Dimethyl-4-undecanol **[500045-01-2]** **$C_{13}H_{28}O$** **MW = 200.36** **416**

Table 1. Experimental value with uncertainty.

T/K	$\rho_{exp} \pm 2\sigma_{est}$/$kg \cdot m^{-3}$	Ref.
293.15	825.2 ± 1.5	59-pet/zak

3,5-Dimethyl-5-undecanol **[107618-96-2]** **$C_{13}H_{28}O$** **MW = 200.36** **417**

Table 1. Experimental value with uncertainty.

T/K	$\rho_{exp} \pm 2\sigma_{est}$/$kg \cdot m^{-3}$	Ref.
301.15	827.2 ± 1.0	60-tha/vas

3-(1,1-Dimethylethyl)-2,2,5-trimethyl-3-hexanol **[32579-70-7]** **$C_{13}H_{28}O$** **MW = 200.36** **418**

Table 1. Experimental and recommended values with uncertainties.

T/K	$\rho_{exp} \pm 2\sigma_{est}$/$kg \cdot m^{-3}$	Ref.
293.15	866.8 ± 1.0	54-mos/cox
293.15	867.6 ± 1.0	60-pet/sok
293.15	867.2 ± 1.0	Recommended

2-Ethyl-1-undecanol **[54381-03-2]** **$C_{13}H_{28}O$** **MW = 200.36** **419**

Table 1. Experimental value with uncertainty.

T/K	$\rho_{exp} \pm 2\sigma_{est}$ / $kg \cdot m^{-3}$	Ref.
293.15	840.4 ± 2.0	47-sto

6-Ethyl-6-undecanol **[5340-50-1]** **$C_{13}H_{28}O$** **MW = 200.36** **420**

Table 1. Experimental value with uncertainty.

T/K	$\rho_{exp} \pm 2\sigma_{est}$ / $kg \cdot m^{-3}$	Ref.
298.15	834.8 ± 2.0	33-whi/wil

4-Ethyl-3,3,5,5-tetramethyl-4-heptanol **[900002-19-9]** **$C_{13}H_{28}O$** **MW = 200.36** **421**

Table 1. Experimental value with uncertainty.

T/K	$\rho_{exp} \pm 2\sigma_{est}$ / $kg \cdot m^{-3}$	Ref.
293.15	892.8 ± 1.0	54-mos/cox

2-Methyl-1-dodecanol **[22663-61-2]** **$C_{13}H_{28}O$** **MW = 200.36** **422**

Table 1. Experimental value with uncertainty.

T/K	$\rho_{exp} \pm 2\sigma_{est}$ / $kg \cdot m^{-3}$	Ref.
298.15	844.0 ± 2.0	29-lev/mik

6-Methyl-6-dodecanol **[62958-40-1]** **$C_{13}H_{28}O$** **MW = 200.36** **423**

Table 1. Coefficients of the polynomial expansion equation.
Standard deviations (see introduction):
$\sigma_{c,w} = 3.0037 \cdot 10^{-1}$ (combined temperature ranges, weighted),
$\sigma_{c,uw} = 1.5019 \cdot 10^{-1}$ (combined temperature ranges, unweighted).

Coefficient	T = 273.15 to 338.15 K $\rho = A + BT + CT^2 + DT^3 + \ldots$
A	$9.86877 \cdot 10^{2}$
B	$-3.15853 \cdot 10^{-1}$
C	$-7.40820 \cdot 10^{-4}$

Table 2. Experimental values with uncertainties and deviation from calculated values.

T/K	$\rho_{exp} \pm 2\sigma_{est}$ / $kg \cdot m^{-3}$	$\rho_{exp} - \rho_{calc}$ / $kg \cdot m^{-3}$	Ref. (Symbol in Fig. 1)	T/K	$\rho_{exp} \pm 2\sigma_{est}$ / $kg \cdot m^{-3}$	$\rho_{exp} - \rho_{calc}$ / $kg \cdot m^{-3}$	Ref. (Symbol in Fig. 1)
273.15	845.20 ± 1.00	–0.13	39-owe/qua(□)	318.15	811.00 ± 1.00	–0.40	39-owe/qua(□)
298.15	827.40 ± 1.00	0.55	39-owe/qua(□)	328.15	803.50 ± 1.00	0.04	39-owe/qua(□)
308.15	819.00 ± 1.00	–0.20	39-owe/qua(□)	338.15	795.50 ± 1.00	0.14	39-owe/qua(□)

cont.

6-Methyl-6-dodecanol (cont.)

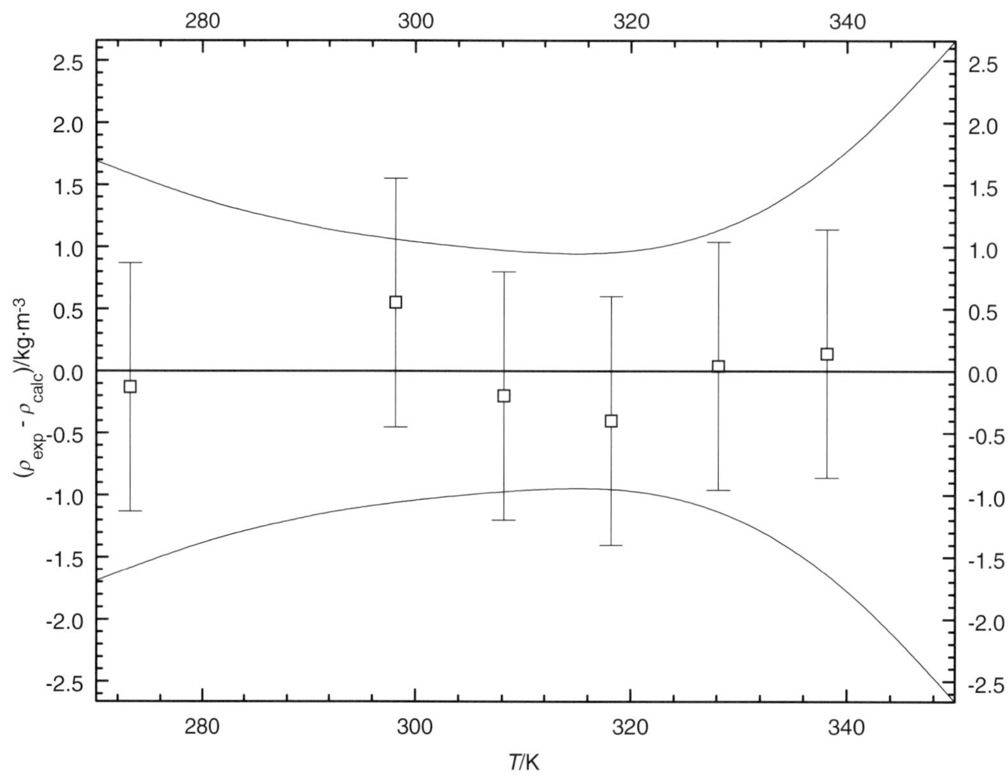

Fig. 1. The symbols show the deviation of the calculated from the experimental values from Table 2. The curves above and below the zero line indicate the calculated error region of the recommended values given in Table 3. The error bars represent the experimental errors. (Error bars smaller than the symbols are omitted for clarity of the figure.)

Table 3. Recommended values (fit to the reliable experimental values according to the equations
$\rho = A + BT + CT^2 + DT^3 + \ldots$ or $\rho = [1 + 1.75(1 - T/T_c)^{1/3} + 0.75(1 - T/T_c)][\rho_c + A(T_c - T) + B(T_c - T)^2 + C(T_c - T)^3 + D(T_c - T)^4]$).

$\frac{T}{\mathrm{K}}$	$\frac{\rho \pm \sigma_{fit}}{\mathrm{kg \cdot m^{-3}}}$	$\frac{T}{\mathrm{K}}$	$\frac{\rho \pm \sigma_{fit}}{\mathrm{kg \cdot m^{-3}}}$	$\frac{T}{\mathrm{K}}$	$\frac{\rho \pm \sigma_{fit}}{\mathrm{kg \cdot m^{-3}}}$
270.00	847.59 ± 1.69	298.15	826.85 ± 1.06	330.00	801.97 ± 1.14
280.00	840.36 ± 1.36	300.00	825.45 ± 1.04	340.00	793.85 ± 1.71
290.00	832.98 ± 1.17	310.00	817.77 ± 0.95	350.00	785.58 ± 2.66
293.15	830.62 ± 1.12	320.00	809.94 ± 0.93		

4-(1-Methylethyl)-3,3,6-trimethyl-4-heptanol **[900002-20-2]** **$C_{13}H_{28}O$** **MW = 200.36** **424**

Table 1. Experimental value with uncertainty.

$\frac{T}{\mathrm{K}}$	$\frac{\rho_{exp} \pm 2\sigma_{est}}{\mathrm{kg \cdot m^{-3}}}$	Ref.
293.15	864.8 ± 1.0	54-mos/cox

5-(2-Methylpropyl)-5-nonanol **[500013-26-3]** **$C_{13}H_{28}O$** **MW = 200.36** **425**

Table 1. Experimental value with uncertainty.

T / K	$\rho_{exp} \pm 2\sigma_{est}$ / $kg \cdot m^{-3}$	Ref.
288.15	843.8 ± 1.5	37-pet/mal

2,2,3,3,4-Pentamethyl-4-octanol **[100799-11-9]** **$C_{13}H_{28}O$** **MW = 200.36** **426**

Table 1. Experimental value with uncertainty.

T / K	$\rho_{exp} \pm 2\sigma_{est}$ / $kg \cdot m^{-3}$	Ref.
293.15	876.3 ± 2.0	60-pet/kao

5-Propyl-5-decanol **[62958-41-2]** **$C_{13}H_{28}O$** **MW = 200.36** **427**

Table 1. Experimental value with uncertainty.

T / K	$\rho_{exp} \pm 2\sigma_{est}$ / $kg \cdot m^{-3}$	Ref.
298.15	832.0 ± 2.0	44-qua/sma

4-Propyl-3,3,6-trimethyl-4-heptanol **[900002-21-3]** **$C_{13}H_{28}O$** **MW = 200.36** **428**

Table 1. Experimental value with uncertainty.

T / K	$\rho_{exp} \pm 2\sigma_{est}$ / $kg \cdot m^{-3}$	Ref.
293.15	859.8 ± 1.0	54-mos/cox

1-Tridecanol **[112-70-9]** **$C_{13}H_{28}O$** **MW = 200.36** **429**

Table 1. Fit with estimated B coefficient for 3 accepted points. Deviation $\sigma_w = 0.476$.

Coefficient	$\rho = A + BT$
A	1022.21
B	–0.640

Table 2. Experimental values with uncertainties and deviation from calculated values.

T / K	$\rho_{exp} \pm 2\sigma_{est}$ / $kg \cdot m^{-3}$	$\rho_{exp} - \rho_{calc}$ / $kg \cdot m^{-3}$	Ref.
323.15	816.5 ± 1.0	1.10	63-vil/gav
313.15	821.5 ± 0.6	–0.30	77-bel/bub
333.15	808.9 ± 0.6	–0.10	77-bel/bub

cont.

1-Tridecanol (cont.)

Table 3. Recommended values.

$\frac{T}{K}$	$\frac{\rho_{exp} \pm 2\sigma_{est}}{kg \cdot m^{-3}}$	$\frac{T}{K}$	$\frac{\rho_{exp} \pm 2\sigma_{est}}{kg \cdot m^{-3}}$
310.00	823.8 ± 1.0	330.00	811.0 ± 0.9
320.00	817.4 ± 0.8	340.00	804.6 ± 1.2

2-Tridecanol **[1653-31-2]** **$C_{13}H_{28}O$** **MW = 200.36** **430**

Table 1. Fit with estimated *B* coefficient for 4 accepted points. Deviation $\sigma_w = 0.944$.

Coefficient	$\rho = A + BT$
A	1035.14
B	−0.700

Table 2. Experimental values with uncertainties and deviation from calculated values.

$\frac{T}{K}$	$\frac{\rho_{exp} \pm 2\sigma_{est}}{kg \cdot m^{-3}}$	$\frac{\rho_{exp} - \rho_{calc}}{kg \cdot m^{-3}}$	Ref.
307.25	821.5 ± 2.0	1.43	11-pic/ken
320.45	810.9 ± 2.0	0.07	11-pic/ken
333.75	801.2 ± 2.0	−0.32	11-pic/ken
287.65	832.6 ± 2.0	−1.19	12-pic/ken

Table 3. Recommended values.

$\frac{T}{K}$	$\frac{\rho_{exp} \pm 2\sigma_{est}}{kg \cdot m^{-3}}$	$\frac{T}{K}$	$\frac{\rho_{exp} \pm 2\sigma_{est}}{kg \cdot m^{-3}}$	$\frac{T}{K}$	$\frac{\rho_{exp} \pm 2\sigma_{est}}{kg \cdot m^{-3}}$
280.00	839.1 ± 3.8	298.15	826.4 ± 2.5	330.00	804.1 ± 2.7
290.00	832.1 ± 3.0	310.00	818.1 ± 2.0	340.00	797.1 ± 3.4
293.15	829.9 ± 2.8	320.00	811.1 ± 2.2		

3-Tridecanol **[10289-68-6]** **$C_{13}H_{28}O$** **MW = 200.36** **431**

Table 1. Experimental value with uncertainty.

$\frac{T}{K}$	$\frac{\rho_{exp} \pm 2\sigma_{est}}{kg \cdot m^{-3}}$	Ref.
353.15	786.5 ± 3.0	13-pic/ken

4-Tridecanol **[26215-92-9]** **$C_{13}H_{28}O$** **MW = 200.36** **432**

Table 1. Experimental value with uncertainty.

$\frac{T}{K}$	$\frac{\rho_{exp} \pm 2\sigma_{est}}{kg \cdot m^{-3}}$	Ref.
293.15	823.4 ± 2.0	48-pet/old

2,2,3-Trimethyl-4-propyl-3-heptanol **[500002-64-2]** **$C_{13}H_{28}O$** **MW = 200.36** **433**

Table 1. Experimental value with uncertainty.

$\frac{T}{\text{K}}$	$\frac{\rho_{exp} \pm 2\sigma_{est}}{\text{kg}\cdot\text{m}^{-3}}$	Ref.
285.15	859.0 ± 2.0	37-naz

2,3,6-Trimethyl-1-decanol **[500002-12-0]** **$C_{13}H_{28}O$** **MW = 200.36** **434**

Table 1. Experimental value with uncertainty.

$\frac{T}{\text{K}}$	$\frac{\rho_{exp} \pm 2\sigma_{est}}{\text{kg}\cdot\text{m}^{-3}}$	Ref.
298.15	835.1 ± 2.0	38-wer/bog

2,5,9-Trimethyl-5-decanol **[500001-91-2]** **$C_{13}H_{28}O$** **MW = 200.36** **435**

Table 1. Experimental value with uncertainty.

$\frac{T}{\text{K}}$	$\frac{\rho_{exp} \pm 2\sigma_{est}}{\text{kg}\cdot\text{m}^{-3}}$	Ref.
284.15	844.4 ± 2.0	28-esc

5-Butyl-5-decanol **[5340-34-1]** **$C_{14}H_{30}O$** **MW = 214.39** **436**

Table 1. Fit with estimated *B* coefficient for 4 accepted points. Deviation $\sigma_w = 0.926$.

Coefficient	$\rho = A + BT$
A	1048.44
B	–0.720

Table 2. Experimental values with uncertainties and deviation from calculated values.

$\frac{T}{\text{K}}$	$\frac{\rho_{exp} \pm 2\sigma_{est}}{\text{kg}\cdot\text{m}^{-3}}$	$\frac{\rho_{exp} - \rho_{calc}}{\text{kg}\cdot\text{m}^{-3}}$	Ref.
298.15	834.5 ± 2.0	0.73	33-whi/wil
298.15	834.8 ± 2.0	1.03	44-qua/sma
318.15	818.9 ± 2.0	–0.47	44-qua/sma
328.15	810.9 ± 2.0	–1.27	44-qua/sma
293.15	845.4 ± 6.0	8.03	61-mes/erz [1)]

[1)] Not included in calculation of linear coefficients.

Table 3. Recommended values.

$\frac{T}{\text{K}}$	$\frac{\rho_{exp} \pm 2\sigma_{est}}{\text{kg}\cdot\text{m}^{-3}}$	$\frac{T}{\text{K}}$	$\frac{\rho_{exp} \pm 2\sigma_{est}}{\text{kg}\cdot\text{m}^{-3}}$	$\frac{T}{\text{K}}$	$\frac{\rho_{exp} \pm 2\sigma_{est}}{\text{kg}\cdot\text{m}^{-3}}$
290.00	839.6 ± 2.7	298.15	833.8 ± 2.1	320.00	818.0 ± 1.9
293.15	837.4 ± 2.4	310.00	825.2 ± 1.7	330.00	810.8 ± 2.6

2,2-Dimethyl-3-(1,1-dimethylethyl)-3-octanol **[500045-08-9]** **$C_{14}H_{30}O$** **MW = 214.39** **437**

Table 1. Experimental value with uncertainty.

$\frac{T}{K}$	$\frac{\rho_{exp} \pm 2\sigma_{est}}{kg \cdot m^{-3}}$	Ref.
293.15	861.4 ± 1.5	60-pet/sok

3,3-Dimethyl-4-(1,1-dimethylethyl)-4-octanol **[900002-34-8]** **$C_{14}H_{30}O$** **MW = 214.39** **438**

Table 1. Experimental value with uncertainty.

$\frac{T}{K}$	$\frac{\rho_{exp} \pm 2\sigma_{est}}{kg \cdot m^{-3}}$	Ref.
293.15	873.4 ± 1.0	54-mos/cox

2,2-Dimethyl-1-dodecanol **[92318-63-3]** **$C_{14}H_{30}O$** **MW = 214.39** **439**

Table 1. Experimental value with uncertainty.

$\frac{T}{K}$	$\frac{\rho_{exp} \pm 2\sigma_{est}}{kg \cdot m^{-3}}$	Ref.
293.15	836.3 ± 2.0	64-blo/hag

3,3-Dimethyl-4-(2-methylpropyl)-4-octanol **[900002-22-4]** **$C_{14}H_{30}O$** **MW = 214.39** **440**

Table 1. Experimental value with uncertainty.

$\frac{T}{K}$	$\frac{\rho_{exp} \pm 2\sigma_{est}}{kg \cdot m^{-3}}$	Ref.
293.15	856.8 ± 1.0	54-mos/cox

4-(1,1-Dimethylethyl)-3,3,6-trimethyl-4-heptanol **[900002-25-7]** **$C_{14}H_{30}O$** **MW = 214.39** **441**

Table 1. Experimental value with uncertainty.

$\frac{T}{K}$	$\frac{\rho_{exp} \pm 2\sigma_{est}}{kg \cdot m^{-3}}$	Ref.
293.15	879.4 ± 1.0	54-mos/cox

5-(2,2-Dimethylpropyl)-5-nonanol **[5340-38-5]** **$C_{14}H_{30}O$** **MW = 214.39** **442**

Table 1. Experimental and recommended values with uncertainties.

$\frac{T}{K}$	$\frac{\rho_{exp} \pm 2\sigma_{est}}{kg \cdot m^{-3}}$	Ref.
293.15	832.0 ± 3.0	38-whi/pop-1 [1)]
293.15	840.3 ± 2.0	42-whi/for
293.15	840.3 ± 2.0	Recommended

[1)] Not included in calculation of recommended value.

7-Ethyl-2-methyl-4-undecanol **[103-20-8]** **$C_{14}H_{30}O$** **MW = 214.39** **443**

Table 1. Experimental value with uncertainty.

$\frac{T}{K}$	$\frac{\rho_{exp} \pm 2\sigma_{est}}{kg \cdot m^{-3}}$	Ref.
20.00	834.1 ± 1.0	58-ano-5

2-Methyl-3-(1-methylethyl)-3-decanol **[57233-27-9]** **$C_{14}H_{30}O$** **MW = 214.39** **444**

Table 1. Experimental value with uncertainty.

$\frac{T}{K}$	$\frac{\rho_{exp} \pm 2\sigma_{est}}{kg \cdot m^{-3}}$	Ref.
293.15	856.2 ± 2.0	62-pet/kap

2-Methyl-2-tridecanol **[32836-44-5]** **$C_{14}H_{30}O$** **MW = 214.39** **445**

Table 1. Experimental values with uncertainties.

$\frac{T}{K}$	$\frac{\rho_{exp} \pm 2\sigma_{est}}{kg \cdot m^{-3}}$	Ref.
289.75	832.0 ± 2.0	19-eyk
354.95	784.1 ± 4.0	19-eyk

2-Methyl-3-tridecanol **[98930-89-3]** **$C_{14}H_{30}O$** **MW = 214.39** **446**

Table 1. Experimental value with uncertainty.

$\frac{T}{K}$	$\frac{\rho_{exp} \pm 2\sigma_{est}}{kg \cdot m^{-3}}$	Ref.
293.15	839.0 ± 2.0	12-pic/ken

4-Methyl-4-tridecanol **[116436-16-9]** **$C_{14}H_{30}O$** **MW = 214.39** **447**

Table 1. Experimental value with uncertainty.

$\frac{T}{K}$	$\frac{\rho_{exp} \pm 2\sigma_{est}}{kg \cdot m^{-3}}$	Ref.
297.15	829.2 ± 2.0	19-eyk

11-Methyl-1-tridecanol **[20194-46-1]** **$C_{14}H_{30}O$** **MW = 214.39** **448**

Table 1. Experimental value with uncertainty.

$\frac{T}{K}$	$\frac{\rho_{exp} \pm 2\sigma_{est}}{kg \cdot m^{-3}}$	Ref.
298.15	834.4 ± 1.0	62-lar/sal

***L*(+)-11-Methyl-1-tridecanol** **[500006-93-9]** **$C_{14}H_{30}O$** **MW = 214.39** **449**

Table 1. Experimental value with uncertainty.

$\frac{T}{K}$	$\frac{\rho_{exp} \pm 2\sigma_{est}}{kg \cdot m^{-3}}$	Ref.
298.15	834.4 ± 0.6	62-lar/sal

6-(1-Methylethyl)-6-undecanol **[500002-69-7]** **$C_{14}H_{30}O$** **MW = 214.39** **450**

Table 1. Experimental value with uncertainty.

$\frac{T}{K}$	$\frac{\rho_{exp} \pm 2\sigma_{est}}{kg \cdot m^{-3}}$	Ref.
293.15	842.5 ± 1.0	46-hus/bai

4-(2-Methylpropyl)-2,2,6-trimethyl-4-heptanol **[500002-08-4]** **$C_{14}H_{30}O$** **MW = 214.39** **451**

Table 1. Experimental value with uncertainty.

$\frac{T}{K}$	$\frac{\rho_{exp} \pm 2\sigma_{est}}{kg \cdot m^{-3}}$	Ref.
293.15	831.7 ± 2.0	42-whi/for

4-(2-Methylpropyl)-3,3,6-trimethyl-4-heptanol **[900002-24-6]** **$C_{14}H_{30}O$** **MW = 214.39** **452**

Table 1. Experimental value with uncertainty.

$\frac{T}{K}$	$\frac{\rho_{exp} \pm 2\sigma_{est}}{kg \cdot m^{-3}}$	Ref.
293.15	853.0 ± 1.0	54-mos/cox

2,2,3,3,4-Pentamethyl-4-nonanol **[108272-61-3]** **$C_{14}H_{30}O$** **MW = 214.39** **453**

Table 1. Experimental value with uncertainty.

$\frac{T}{K}$	$\frac{\rho_{exp} \pm 2\sigma_{est}}{kg \cdot m^{-3}}$	Ref.
293.15	875.1 ± 2.0	60-pet/kao

4,4,5,6,6-Pentamethyl-5-nonanol **[500002-06-2]** **$C_{14}H_{30}O$** **MW = 214.39** **454**

Table 1. Experimental value with uncertainty.

$\frac{T}{K}$	$\frac{\rho_{exp} \pm 2\sigma_{est}}{kg \cdot m^{-3}}$	Ref.
288.15	887.6 ± 2.0	37-naz

2-Pentyl-1-nonanol **[5333-48-2]** **$C_{14}H_{30}O$** **MW = 214.39** **455**

Table 1. Experimental values with uncertainties.

$\frac{T}{K}$	$\frac{\rho_{exp} \pm 2\sigma_{est}}{kg \cdot m^{-3}}$	Ref.
288.15	840.5 ± 2.0	01-gue
297.15	837.7 ± 2.0	38-mas

4-Propyl-3,3,5,5-tetramethyl-4-heptanol [900002-88-2] $C_{14}H_{30}O$ MW = 214.39 456

Table 1. Experimental value with uncertainty.

$\frac{T}{K}$	$\frac{\rho_{exp} \pm 2\sigma_{est}}{kg \cdot m^{-3}}$	Ref.
293.15	888.2 ± 1.0	54-mos/cox

6-Propyl-6-undecanol [500002-03-9] $C_{14}H_{30}O$ MW = 214.39 457

Table 1. Experimental value with uncertainty.

$\frac{T}{K}$	$\frac{\rho_{exp} \pm 2\sigma_{est}}{kg \cdot m^{-3}}$	Ref.
298.15	833.6 ± 2.0	33-whi/wil

1-Tetradecanol [112-72-1] $C_{14}H_{30}O$ MW = 214.39 458

Table 1. Coefficients of the polynomial expansion equation. Standard deviations (see introduction):
$\sigma_{c,w} = 8.9110 \cdot 10^{-1}$ (combined temperature ranges, weighted),
$\sigma_{c,uw} = 1.2677 \cdot 10^{-1}$ (combined temperature ranges, unweighted).

Coefficient	T = 293.15 to 573.15 K $\rho = A + BT + CT^2 + DT^3 + \ldots$
A	$1.01416 \cdot 10^{3}$
B	$-5.58592 \cdot 10^{-1}$
C	$-1.11840 \cdot 10^{-4}$
D	$-2.07120 \cdot 10^{-7}$

Table 2. Experimental values with uncertainties and deviation from calculated values.

$\frac{T}{K}$	$\frac{\rho_{exp} \pm 2\sigma_{est}}{kg \cdot m^{-3}}$	$\frac{\rho_{exp} - \rho_{calc}}{kg \cdot m^{-3}}$	Ref. (Symbol in Fig. 1)	$\frac{T}{K}$	$\frac{\rho_{exp} \pm 2\sigma_{est}}{kg \cdot m^{-3}}$	$\frac{\rho_{exp} - \rho_{calc}}{kg \cdot m^{-3}}$	Ref. (Symbol in Fig. 1)
311.15	823.60 ± 2.00	0.31	1883-kra(○)	433.15	734.00 ± 1.00	–0.39	58-cos/bow(□)
311.15	823.60 ± 1.00	0.31	1883-kra(○)	453.15	718.80 ± 1.00	0.00	58-cos/bow(□)
323.15	815.30 ± 1.00	0.31	1883-kra(○)	473.15	703.20 ± 1.00	0.31	58-cos/bow(□)
323.15	815.30 ± 2.00	0.31	1883-kra[1)]	493.15	686.50 ± 1.00	–0.15	58-cos/bow(□)
372.05	781.30 ± 1.00	1.11	1883-kra(○)	513.15	670.00 ± 1.00	–0.08	58-cos/bow(□)
372.05	781.30 ± 2.00	1.11	1883-kra(○)	533.15	653.40 ± 1.00	0.23	58-cos/bow(□)
311.55	824.00 ± 2.00	0.99	49-tsv/mar(◆)	553.15	636.20 ± 1.00	0.30	58-cos/bow(□)
313.15	822.00 ± 2.00	0.09	56-rat/cur(Δ)	573.15	617.80 ± 1.00	–0.47	58-cos/bow(□)
333.15	807.00 ± 2.00	–1.00	56-rat/cur[1)]	318.15	817.60 ± 0.60	–0.86	92-lie/sen-1(×)
353.15	794.00 ± 2.00	0.17	56-rat/cur[1)]	323.15	814.60 ± 0.60	–0.39	92-lie/sen-1(×)
293.15	834.00 ± 2.00	–1.58	58-ano-3(∇)	328.15	811.60 ± 0.60	0.10	92-lie/sen-1(×)
313.15	822.70 ± 1.00	0.79	58-cos/bow(□)	333.15	808.70 ± 0.60	0.70	92-lie/sen-1(×)
333.15	807.90 ± 1.00	–0.10	58-cos/bow(□)	338.15	804.00 ± 0.60	–0.48	92-lie/sen-1(×)
353.15	793.10 ± 1.00	–0.73	58-cos/bow(□)	343.15	801.10 ± 0.60	0.16	92-lie/sen-1(×)
373.15	778.40 ± 1.00	–0.99	58-cos/bow(□)	348.15	798.10 ± 0.60	0.71	92-lie/sen-1(×)
393.15	766.60 ± 1.00	1.92	58-cos/bow[1)]	353.15	793.40 ± 0.60	–0.43	92-lie/sen-1(×)
413.15	748.80 ± 1.00	–0.88	58-cos/bow(□)				

[1)] Not included in Fig. 1.

Further references: [60-pet/kao, 60-pet/sok, 63-vil/gav].

cont.

1-Tetradecanol (cont.)

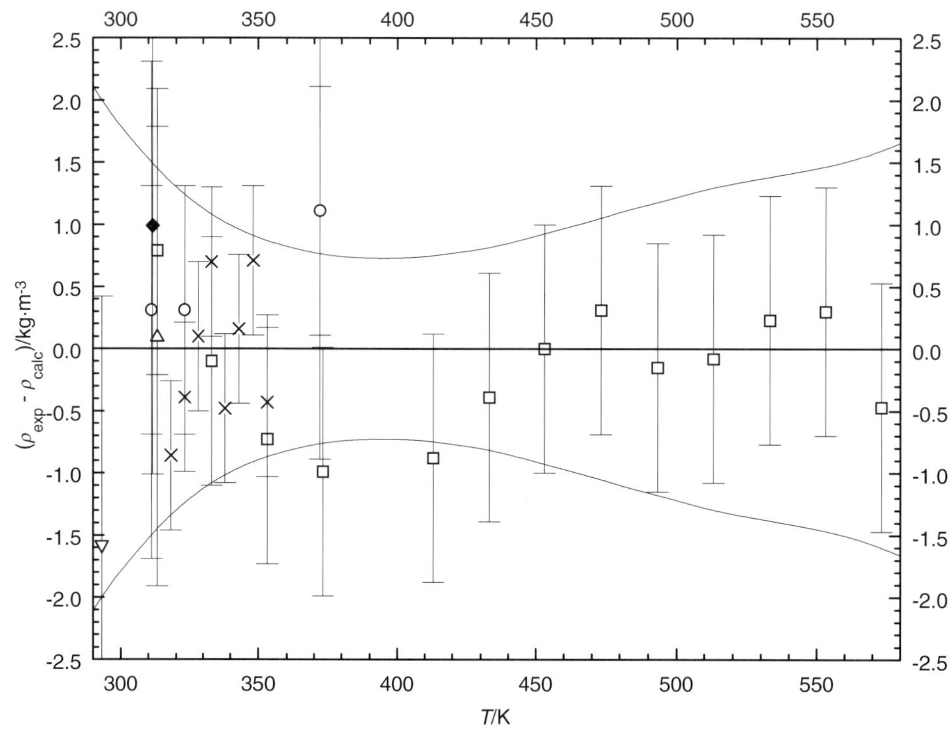

Fig. 1. The symbols show the deviation of the calculated from the experimental values from Table 2. The curves above and below the zero line indicate the calculated error region of the recommended values given in Table 3. The error bars represent the experimental errors. (Error bars smaller than the symbols are omitted for clarity of the figure.)

Table 3. Recommended values (fit to the reliable experimental values according to the equations $\rho = A + BT + CT^2 + DT^3 + \ldots$ or $\rho = [1 + 1.75(1 - T/T_c)^{1/3} + 0.75(1 - T/T_c)][\rho_c + A(T_c - T) + B(T_c - T)^2 + C(T_c - T)^3 + D(T_c - T)^4]$).

T/K	$\rho \pm \sigma_{fit}$/kg·m^{-3}	T/K	$\rho \pm \sigma_{fit}$/kg·m^{-3}	T/K	$\rho \pm \sigma_{fit}$/kg·m^{-3}
290.00	837.71 ± 2.11	380.00	774.38 ± 0.74	490.00	689.23 ± 1.16
293.15	835.58 ± 2.00	390.00	767.01 ± 0.73	500.00	681.02 ± 1.22
298.15	832.19 ± 1.84	400.00	759.58 ± 0.73	510.00	672.72 ± 1.28
300.00	830.93 ± 1.79	410.00	752.06 ± 0.74	520.00	664.33 ± 1.33
310.00	824.08 ± 1.52	420.00	744.48 ± 0.77	530.00	655.86 ± 1.37
320.00	817.17 ± 1.29	430.00	736.82 ± 0.80	540.00	647.30 ± 1.41
330.00	810.20 ± 1.12	440.00	729.09 ± 0.85	550.00	638.65 ± 1.45
340.00	803.17 ± 0.99	450.00	721.27 ± 0.91	560.00	629.90 ± 1.50
350.00	796.07 ± 0.89	460.00	713.38 ± 0.97	570.00	621.07 ± 1.57
360.00	788.91 ± 0.82	470.00	705.42 ± 1.03	580.00	612.14 ± 1.66
370.00	781.68 ± 0.77	480.00	697.36 ± 1.10		

3-Tetradecanol **[1653-32-3]** **$C_{14}H_{30}O$** **MW = 214.39** **459**

Table 1. Experimental value with uncertainty.

T / K	$\rho_{exp} \pm 2\sigma_{est}$ / $kg \cdot m^{-3}$	Ref.
353.15	788.5 ± 2.0	13-pic/ken

4-Tetradecanol **[1653-33-4]** **$C_{14}H_{30}O$** **MW = 214.39** **460**

Table 1. Experimental value with uncertainty.

T / K	$\rho_{exp} \pm 2\sigma_{est}$ / $kg \cdot m^{-3}$	Ref.
305.15	823.1 ± 3.0	48-pet/old

4-Butyl-3,3,5,5-tetramethyl-4-heptanol **[900002-89-3]** **$C_{15}H_{32}O$** **MW = 228.42** **461**

Table 1. Experimental value with uncertainty.

T / K	$\rho_{exp} \pm 2\sigma_{est}$ / $kg \cdot m^{-3}$	Ref.
293.15	884.9 ± 1.0	54-mos/cox

6-Butyl-6-undecanol **[5396-08-7]** **$C_{15}H_{32}O$** **MW = 228.42** **462**

Table 1. Experimental value with uncertainty.

T / K	$\rho_{exp} \pm 2\sigma_{est}$ / $kg \cdot m^{-3}$	Ref.
298.15	834.4 ± 2.0	33-whi/wil

2,2-Dimethyl-3-(1,1-dimethylethyl)-3-nonanol **[101082-11-5]** **$C_{15}H_{32}O$** **MW = 228.42** **463**

Table 1. Experimental values with uncertainties.

T / K	$\rho_{exp} \pm 2\sigma_{est}$ / $kg \cdot m^{-3}$	Ref.
293.15	860.3 ± 2.0	57-pet/sok
293.15	860.3 ± 2.0	60-pet/sok

2,8-Dimethyl-5-(2-methylpropyl)-5-nonanol **[500002-16-4]** **$C_{15}H_{32}O$** **MW = 228.42** **464**

Table 1. Fit with estimated B coefficient for 4 accepted points. Deviation $\sigma_w = 0.060$.

Coefficient	$\rho = A + BT$
A	1068.38
B	–0.800

cont.

2,8-Dimethyl-5-(2-methylpropyl)-5-nonanol (cont.)

Table 2. Experimental values with uncertainties and deviation from calculated values.

$\frac{T}{K}$	$\frac{\rho_{exp} \pm 2\sigma_{est}}{kg \cdot m^{-3}}$	$\frac{\rho_{exp} - \rho_{calc}}{kg \cdot m^{-3}}$	Ref.
273.15	849.8 ± 2.0	–0.06	04-gri
283.55	841.6 ± 2.0	0.06	04-gri
273.15	849.8 ± 2.0	–0.06	04-gri-1
283.55	841.6 ± 2.0	0.06	04-gri-1

Table 3. Recommended values.

$\frac{T}{K}$	$\frac{\rho_{exp} \pm 2\sigma_{est}}{kg \cdot m^{-3}}$
270.00	852.4 ± 1.8
280.00	844.4 ± 1.8
290.00	836.4 ± 1.9

6,10-Dimethyl-2-tridecanol **[101082-12-6]** **$C_{15}H_{32}O$** **MW = 228.42** **465**

Table 1. Experimental and recommended values with uncertainties.

$\frac{T}{K}$	$\frac{\rho_{exp} \pm 2\sigma_{est}}{kg \cdot m^{-3}}$	Ref.
293.15	830.3 ± 1.0	49-naz/zar-1
293.15	888.2 ± 2.0	61-shv/pet[1)]
293.15	830.3 ± 1.0	Recommended

[1)] Not included in calculation of recommended value.

2-Methyl-2-tetradecanol **[27570-83-8]** **$C_{15}H_{32}O$** **MW = 228.42** **466**

Table 1. Experimental and recommended values with uncertainties.

$\frac{T}{K}$	$\frac{\rho_{exp} \pm 2\sigma_{est}}{kg \cdot m^{-3}}$	Ref.
293.15	836.0 ± 2.0	42-pet/che
293.15	838.0 ± 2.0	57-for/lan
293.15	837.0 ± 2.1	Recommended

6-(2-Methylpropyl)-6-undecanol **[500002-70-0]** **$C_{15}H_{32}O$** **MW = 228.42** **467**

Table 1. Experimental value with uncertainty.

$\frac{T}{K}$	$\frac{\rho_{exp} \pm 2\sigma_{est}}{kg \cdot m^{-3}}$	Ref.
293.15	836.7 ± 2.0	46-hus/bai

1-Pentadecanol **[629-76-5]** **$C_{15}H_{32}O$** **MW = 228.42** **468**

Table 1. Fit with estimated B coefficient for 2 accepted points. Deviation $\sigma_w = 0.350$.

Coefficient	$\rho = A + BT$
A	1025.57
B	–0.640

Table 2. Experimental values with uncertainties and deviation from calculated values.

$\frac{T}{K}$	$\frac{\rho_{exp} \pm 2\sigma_{est}}{kg \cdot m^{-3}}$	$\frac{\rho_{exp} - \rho_{calc}}{kg \cdot m^{-3}}$	Ref.
295.55	836.4 ± 2.0	–0.01	47-sto[1)]
323.15	819.2 ± 2.0	0.45	63-vil/gav[1)]
313.15	824.8 ± 0.6	–0.35	77-bel/bub
333.15	812.7 ± 0.6	0.35	77-bel/bub

[1)] Not included in calculation of linear coefficients.

Table 3. Recommended values.

$\frac{T}{K}$	$\frac{\rho_{exp} \pm 2\sigma_{est}}{kg \cdot m^{-3}}$
310.00	827.2 ± 0.8
320.00	820.8 ± 0.5
330.00	814.4 ± 0.6
340.00	808.0 ± 1.0

3-Pentadecanol **[53346-71-7]** **$C_{15}H_{32}O$** **MW = 228.42** **469**

Table 1. Experimental value with uncertainty.

$\frac{T}{K}$	$\frac{\rho_{exp} \pm 2\sigma_{est}}{kg \cdot m^{-3}}$	Ref.
353.15	792.1 ± 3.0	13-pic/ken

3,7,11-Trimethyl-3-dodecanol **[7278-65-1]** **$C_{15}H_{32}O$** **MW = 228.42** **470**

Table 1. Experimental value with uncertainty.

$\frac{T}{K}$	$\frac{\rho_{exp} \pm 2\sigma_{est}}{kg \cdot m^{-3}}$	Ref.
293.15	838.7 ± 2.0	58-naz/gus

2-Butyl-1-dodecanol **[21078-85-3]** **$C_{16}H_{34}O$** **MW = 242.45** **471**

Table 1. Coefficients of the polynomial expansion equation.
Standard deviations (see introduction):
$\sigma_{c,w} = 7.9175 \cdot 10^{-1}$ (combined temperature ranges, weighted),
$\sigma_{c,uw} = 5.0211 \cdot 10^{-1}$ (combined temperature ranges, unweighted).

Coefficient	T = 293.15 to 373.15 K $\rho = A + BT + CT^2 + DT^3 + \ldots$
A	$1.03807 \cdot 10^3$
B	$-6.84381 \cdot 10^{-1}$

Table 2. Experimental values with uncertainties and deviation from calculated values.

$\frac{T}{\mathrm{K}}$	$\frac{\rho_{exp} \pm 2\sigma_{est}}{\mathrm{kg \cdot m^{-3}}}$	$\frac{\rho_{exp} - \rho_{calc}}{\mathrm{kg \cdot m^{-3}}}$	Ref. (Symbol in Fig. 1)	$\frac{T}{\mathrm{K}}$	$\frac{\rho_{exp} \pm 2\sigma_{est}}{\mathrm{kg \cdot m^{-3}}}$	$\frac{\rho_{exp} - \rho_{calc}}{\mathrm{kg \cdot m^{-3}}}$	Ref. (Symbol in Fig. 1)
293.15	837.94 ± 1.00	0.49	33-bin/ste(×)	353.15	796.81 ± 1.00	0.43	33-bin/ste(×)
303.15	831.46 ± 1.00	0.86	33-bin/ste(×)	373.15	782.72 ± 1.00	0.02	33-bin/ste(×)
313.15	824.54 ± 1.00	0.78	33-bin/ste(×)	293.15	834.50 ± 2.00	−2.95	35-cox/rei(□)
333.15	810.44 ± 1.00	0.37	33-bin/ste(×)				

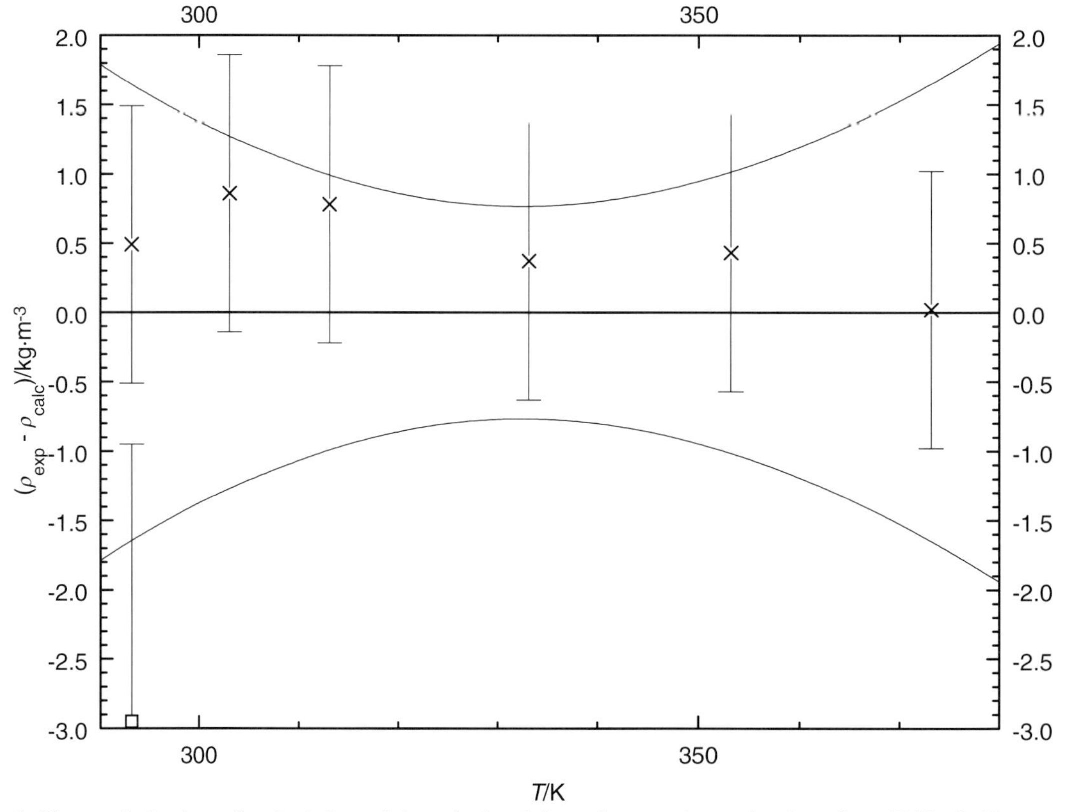

Fig. 1. The symbols show the deviation of the calculated from the experimental values from Table 2. The curves above and below the zero line indicate the calculated error region of the recommended values given in Table 3. The error bars represent the experimental errors. (Error bars smaller than the symbols are omitted for clarity of the figure.)

cont.

Table 3. Recommended values (fit to the reliable experimental values according to the equations
$\rho = A + BT + CT^2 + DT^3 + \dots$ or $\rho = [1 + 1.75(1 - T/T_c)^{1/3} + 0.75(1 - T/T_c)][\rho_c + A(T_c - T) + B(T_c - T)^2 + C(T_c - T)^3 + D(T_c - T)^4]$).

$\frac{T}{K}$	$\frac{\rho \pm \sigma_{fit}}{kg \cdot m^{-3}}$	$\frac{T}{K}$	$\frac{\rho \pm \sigma_{fit}}{kg \cdot m^{-3}}$	$\frac{T}{K}$	$\frac{\rho \pm \sigma_{fit}}{kg \cdot m^{-3}}$
290.00	839.60 ± 1.79	310.00	825.92 ± 1.05	350.00	798.54 ± 0.93
293.15	837.45 ± 1.64	320.00	819.07 ± 0.84	360.00	791.70 ± 1.18
298.15	834.03 ± 1.44	330.00	812.23 ± 0.75	370.00	784.85 ± 1.51
300.00	832.76 ± 1.37	340.00	805.38 ± 0.78	380.00	778.01 ± 1.94

5-Butyl-5-dodecanol **[93314-38-6]** **$C_{16}H_{34}O$** **MW = 242.45** **472**

Table 1. Experimental value with uncertainty.

$\frac{T}{K}$	$\frac{\rho_{exp} \pm 2\sigma_{est}}{kg \cdot m^{-3}}$	Ref.
293.15	841.5 ± 1.0	37-pet/and

2,2-Dimethyl-3-(1,1-dimethylethyl)-3-decanol **[93314-37-5]** **$C_{16}H_{34}O$** **MW = 242.45** **473**

Table 1. Experimental value with uncertainty.

$\frac{T}{K}$	$\frac{\rho_{exp} \pm 2\sigma_{est}}{kg \cdot m^{-3}}$	Ref.
293.15	865.6 ± 2.0	62-pet/kap

2,2-Dimethyl-1-tetradecanol **[5286-18-0]** **$C_{16}H_{34}O$** **MW = 242.45** **474**

Table 1. Experimental value with uncertainty.

$\frac{T}{K}$	$\frac{\rho_{exp} \pm 2\sigma_{est}}{kg \cdot m^{-3}}$	Ref.
293.15	840.0 ± 2.0	64-blo/hag

2,2-Dimethyl-3-tetradecanol **[500000-42-0]** **$C_{16}H_{34}O$** **MW = 242.45** **475**

Table 1. Experimental value with uncertainty.

$\frac{T}{K}$	$\frac{\rho_{exp} \pm 2\sigma_{est}}{kg \cdot m^{-3}}$	Ref.
293.15	835.6 ± 1.0	41-whi/whi

2-Ethyl-1-tetradecanol **[25354-99-8]** **$C_{16}H_{34}O$** **MW = 242.45** **476**

Table 1. Coefficients of the polynomial expansion equation. Standard deviations (see introduction):
$\sigma_{c,w} = 8.6253 \cdot 10^{-1}$ (combined temperature ranges, weighted),
$\sigma_{c,uw} = 5.4877 \cdot 10^{-1}$ (combined temperature ranges, unweighted).

Coefficient	T = 293.15 to 373.15 K $\rho = A + BT + CT^2 + DT^3 + \dots$
A	$1.04120 \cdot 10^{3}$
B	$-6.86917 \cdot 10^{-1}$

cont.

2-Ethyl-1-tetradecanol (cont.)

Table 2. Experimental values with uncertainties and deviation from calculated values.

$\frac{T}{K}$	$\frac{\rho_{exp} \pm 2\sigma_{est}}{kg \cdot m^{-3}}$	$\frac{\rho_{exp} - \rho_{calc}}{kg \cdot m^{-3}}$	Ref. (Symbol in Fig. 1)	$\frac{T}{K}$	$\frac{\rho_{exp} \pm 2\sigma_{est}}{kg \cdot m^{-3}}$	$\frac{\rho_{exp} - \rho_{calc}}{kg \cdot m^{-3}}$	Ref. (Symbol in Fig. 1)
293.15	840.48 ± 1.00	0.65	33-bin/ste(×)	353.15	798.98 ± 1.00	0.37	33-bin/ste(×)
303.15	833.61 ± 1.00	0.65	33-bin/ste(×)	373.15	784.87 ± 1.00	−0.00	33-bin/ste(×)
313.15	826.99 ± 1.00	0.90	33-bin/ste(×)	293.15	836.60 ± 2.00	−3.23	35-cox/rei(□)
333.15	813.01 ± 1.00	0.66	33-bin/ste(×)				

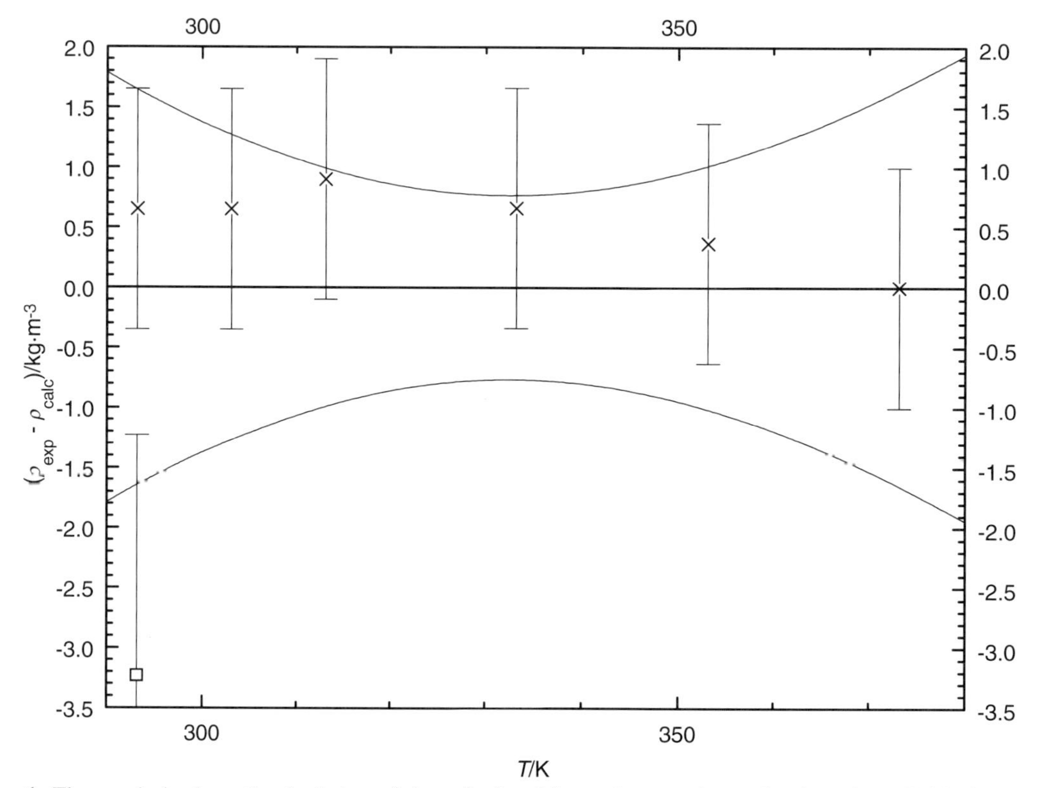

Fig. 1. The symbols show the deviation of the calculated from the experimental values from Table 2. The curves above and below the zero line indicate the calculated error region of the recommended values given in Table 3. The error bars represent the experimental errors. (Error bars smaller than the symbols are omitted for clarity of the figure.)

Table 3. Recommended values (fit to the reliable experimental values according to the equations $\rho = A + BT + CT^2 + DT^3 + \ldots$ or $\rho = [1 + 1.75(1 - T/T_c)^{1/3} + 0.75(1 - T/T_c)][\rho_c + A(T_c - T) + B(T_c - T)^2 + C(T_c - T)^3 + D(T_c - T)^4]$).

$\frac{T}{K}$	$\frac{\rho \pm \sigma_{fit}}{kg \cdot m^{-3}}$	$\frac{T}{K}$	$\frac{\rho \pm \sigma_{fit}}{kg \cdot m^{-3}}$	$\frac{T}{K}$	$\frac{\rho \pm \sigma_{fit}}{kg \cdot m^{-3}}$
290.00	841.99 ± 1.79	310.00	828.25 ± 1.05	350.00	800.78 ± 0.93
293.15	839.83 ± 1.64	320.00	821.38 ± 0.84	360.00	793.91 ± 1.18
298.15	836.39 ± 1.44	330.00	814.51 ± 0.75	370.00	787.04 ± 1.51
300.00	835.12 ± 1.37	340.00	807.65 ± 0.78	380.00	780.17 ± 1.94

3-Ethyl-3-tetradecanol [101433-18-5] $C_{16}H_{34}O$ MW = 242.45 477

Table 1. Experimental value with uncertainty.

$\frac{T}{K}$	$\frac{\rho_{exp} \pm 2\sigma_{est}}{kg \cdot m^{-3}}$	Ref.
288.45	846.0 ± 3.0	19-eyk

2-Heptyl-1-nonanol [25355-03-7] $C_{16}H_{34}O$ MW = 242.45 478

Table 1. Coefficients of the polynomial expansion equation.
Standard deviations (see introduction):
$\sigma_{c,w} = 1.3014 \cdot 10^{-1}$ (combined temperature ranges, weighted),
$\sigma_{c,uw} = 6.5072 \cdot 10^{-2}$ (combined temperature ranges, unweighted).

Coefficient	T = 293.15 to 373.15 K $\rho = A + BT + CT^2 + DT^3 + \ldots$
A	$1.00360 \cdot 10^{3}$
B	$-4.62418 \cdot 10^{-1}$
C	$-3.53389 \cdot 10^{-4}$

Table 2. Experimental values with uncertainties and deviation from calculated values.

$\frac{T}{K}$	$\frac{\rho_{exp} \pm 2\sigma_{est}}{kg \cdot m^{-3}}$	$\frac{\rho_{exp} - \rho_{calc}}{kg \cdot m^{-3}}$	Ref. (Symbol in Fig. 1)	$\frac{T}{K}$	$\frac{\rho_{exp} \pm 2\sigma_{est}}{kg \cdot m^{-3}}$	$\frac{\rho_{exp} - \rho_{calc}}{kg \cdot m^{-3}}$	Ref. (Symbol in Fig. 1)
293.15	837.59 ± 1.00	−0.09	33-bin/ste(□)	333.15	810.24 ± 1.00	−0.09	33-bin/ste(□)
303.15	830.91 ± 1.00	−0.04	33-bin/ste(□)	353.15	796.11 ± 1.00	−0.12	33-bin/ste(□)
313.15	824.40 ± 1.00	0.26	33-bin/ste(□)	373.15	781.92 ± 1.00	0.07	33-bin/ste(□)

Further references: [35-cox/rei].

Table 3. Recommended values (fit to the reliable experimental values according to the equations
$\rho = A + BT + CT^2 + DT^3 + \ldots$ or $\rho = [1 + 1.75(1 - T/T_c)^{1/3} + 0.75(1 - T/T_c)][\rho_c + A(T_c - T) + B(T_c - T)^2 + C(T_c - T)^3 + D(T_c - T)^4]$).

$\frac{T}{K}$	$\frac{\rho \pm \sigma_{fit}}{kg \cdot m^{-3}}$	$\frac{T}{K}$	$\frac{\rho \pm \sigma_{fit}}{kg \cdot m^{-3}}$	$\frac{T}{K}$	$\frac{\rho \pm \sigma_{fit}}{kg \cdot m^{-3}}$
290.00	839.78 ± 1.54	310.00	826.29 ± 1.06	350.00	798.47 ± 0.98
293.15	837.68 ± 1.42	320.00	819.44 ± 1.00	360.00	791.33 ± 1.13
298.15	834.32 ± 1.27	330.00	812.52 ± 0.98	370.00	784.13 ± 1.50
300.00	833.07 ± 1.22	340.00	805.53 ± 0.96	380.00	776.86 ± 2.13

cont.

2-Heptyl-1-nonanol (cont.)

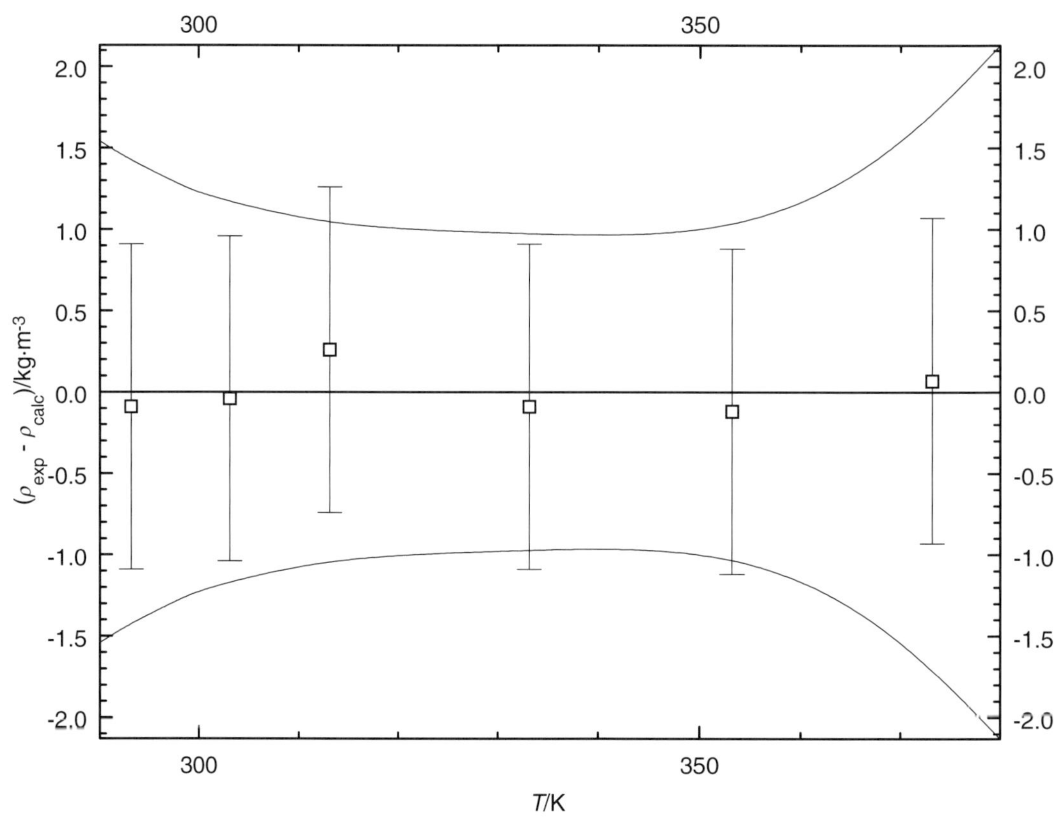

Fig. 1. The symbols show the deviation of the calculated from the experimental values from Table 2. The curves above and below the zero line indicate the calculated error region of the recommended values given in Table 3. The error bars represent the experimental errors. (Error bars smaller than the symbols are omitted for clarity of the figure.)

1-Hexadecanol **[36653-82-4]** **$C_{16}H_{34}O$** **MW = 242.45** **479**

Table 1. Coefficients of the polynomial expansion equation.
Standard deviations (see introduction):
$\sigma_{c,w} = 2.9130 \cdot 10^{-1}$ (combined temperature ranges, weighted),
$\sigma_{c,uw} = 5.8696 \cdot 10^{-2}$ (combined temperature ranges, unweighted).

Coefficient	T = 321.15 to 573.15 K $\rho = A + BT + CT^2 + DT^3 + \ldots$
A	$9.25311 \cdot 10^{2}$
B	$1.01083 \cdot 10^{-1}$
C	$-1.68710 \cdot 10^{-3}$
D	$1.06172 \cdot 10^{-6}$

cont.

Table 2. Experimental values with uncertainties and deviation from calculated values.

$\frac{T}{K}$	$\frac{\rho_{exp} \pm 2\sigma_{est}}{kg \cdot m^{-3}}$	$\frac{\rho_{exp} - \rho_{calc}}{kg \cdot m^{-3}}$	Ref. (Symbol in Fig. 1)	$\frac{T}{K}$	$\frac{\rho_{exp} \pm 2\sigma_{est}}{kg \cdot m^{-3}}$	$\frac{\rho_{exp} - \rho_{calc}}{kg \cdot m^{-3}}$	Ref. (Symbol in Fig. 1)
	crystal			513.15	677.00 ± 1.00	0.61	58-cos/bow(∇)
78.15	1000.0 ± 3.0		30-bil/fis-1	533.15	660.70 ± 1.00	0.15	58-cos/bow(∇)
194.15	974.0 ± 3.0		30-bil/fis-1	553.15	645.40 ± 1.00	0.69	58-cos/bow(∇)
	liquid			573.15	628.80 ± 1.00	−0.13	58-cos/bow(∇)
322.65	817.60 ± 1.00	−0.36	1883-kra[1)]	323.08	817.80 ± 0.20	0.13	73-fin(×)
322.65	817.60 ± 1.00	−0.36	1883-kra[1)]	333.09	811.04 ± 0.20	0.00	73-fin(×)
333.15	810.50 ± 1.00	−0.50	1883-kra[1)]	333.14	810.98 ± 0.20	−0.02	73-fin(×)
333.15	810.50 ± 1.00	−0.50	1883-kra[1)]	333.15	810.92 ± 0.20	−0.08	73-fin(×)
371.85	783.70 ± 1.00	−0.51	1883-kra(◆)	342.95	804.23 ± 0.20	−0.15	73-fin(×)
371.85	783.70 ± 1.00	−0.51	1883-kra(◆)	343.26	804.30 ± 0.20	0.14	73-fin(×)
298.15	888.60 ± 0.60	54.98	38-bak/smy[1)]	352.77	797.69 ± 0.20	0.06	73-fin(×)
323.15	817.60 ± 0.60	−0.03	38-bak/smy(○)	353.19	797.42 ± 0.20	0.08	73-fin(×)
321.15	819.00 ± 1.00	0.06	53-kre-1(Δ)	355.38	795.85 ± 0.20	0.04	73-fin(×)
323.15	817.60 ± 1.00	−0.03	53-kre-1[1)]	361.46	791.54 ± 0.20	−0.02	73-fin(×)
333.15	810.50 ± 1.00	−0.50	58-cos/bow[1)]	361.61	791.53 ± 0.20	0.07	73-fin(×)
353.15	797.10 ± 1.00	−0.26	58-cos/bow[1)]	367.39	787.42 ± 0.20	0.04	73-fin(×)
373.15	783.00 ± 1.00	−0.28	58-cos/bow(∇)	370.40	785.34 ± 0.20	0.10	73-fin(×)
393.15	768.40 ± 1.00	−0.40	58-cos/bow(∇)	333.15	811.30 ± 0.60	0.30	92-lie/sen-1(□)
413.15	753.70 ± 1.00	−0.27	58-cos/bow(∇)	338.15	806.80 ± 0.60	−0.83	92-lie/sen-1(□)
433.15	738.70 ± 1.00	−0.15	58-cos/bow(∇)	343.15	803.90 ± 0.60	−0.34	92-lie/sen-1(□)
453.15	723.30 ± 1.00	−0.17	58-cos/bow(∇)	348.15	800.90 ± 0.60	0.08	92-lie/sen-1(□)
473.15	708.20 ± 1.00	0.29	58-cos/bow(∇)	353.15	797.80 ± 0.60	0.44	92-lie/sen-1(□)
493.15	692.80 ± 1.00	0.60	58-cos/bow(∇)				

[1)] Not included in Fig. 1.

Further references: [1893-eyk, 22-tro, 31-del, 49-hos/ste, 50-boe/ned, 56-rat/cur, 63-vil/gav, 69-kat/pat].

Table 3. Recommended values (fit to the reliable experimental values according to the equations
$\rho = A + BT + CT^2 + DT^3 + \ldots$ or $\rho = [1 + 1.75(1 - T/T_c)^{1/3} + 0.75(1 - T/T_c)][\rho_c + A(T_c - T) + B(T_c - T)^2 + C(T_c - T)^3 + D(T_c - T)^4]$).

$\frac{T}{K}$	$\frac{\rho \pm \sigma_{fit}}{kg \cdot m^{-3}}$	$\frac{T}{K}$	$\frac{\rho \pm \sigma_{fit}}{kg \cdot m^{-3}}$	$\frac{T}{K}$	$\frac{\rho \pm \sigma_{fit}}{kg \cdot m^{-3}}$
320.00	819.69 ± 0.41	410.00	756.33 ± 0.75	500.00	686.79 ± 1.11
330.00	813.10 ± 0.43	420.00	748.82 ± 0.79	510.00	678.89 ± 1.15
340.00	806.38 ± 0.46	430.00	741.25 ± 0.84	520.00	670.97 ± 1.18
350.00	799.54 ± 0.50	440.00	733.61 ± 0.88	530.00	663.04 ± 1.22
360.00	792.59 ± 0.54	450.00	725.91 ± 0.93	540.00	655.12 ± 1.27
370.00	785.53 ± 0.58	460.00	718.16 ± 0.97	550.00	647.20 ± 1.33
380.00	778.36 ± 0.62	470.00	710.37 ± 1.01	560.00	639.30 ± 1.42
390.00	771.11 ± 0.67	480.00	702.54 ± 1.05	570.00	631.41 ± 1.53
400.00	763.76 ± 0.71	490.00	694.68 ± 1.08	580.00	623.55 ± 1.68

cont.

1-Hexadecanol (cont.)

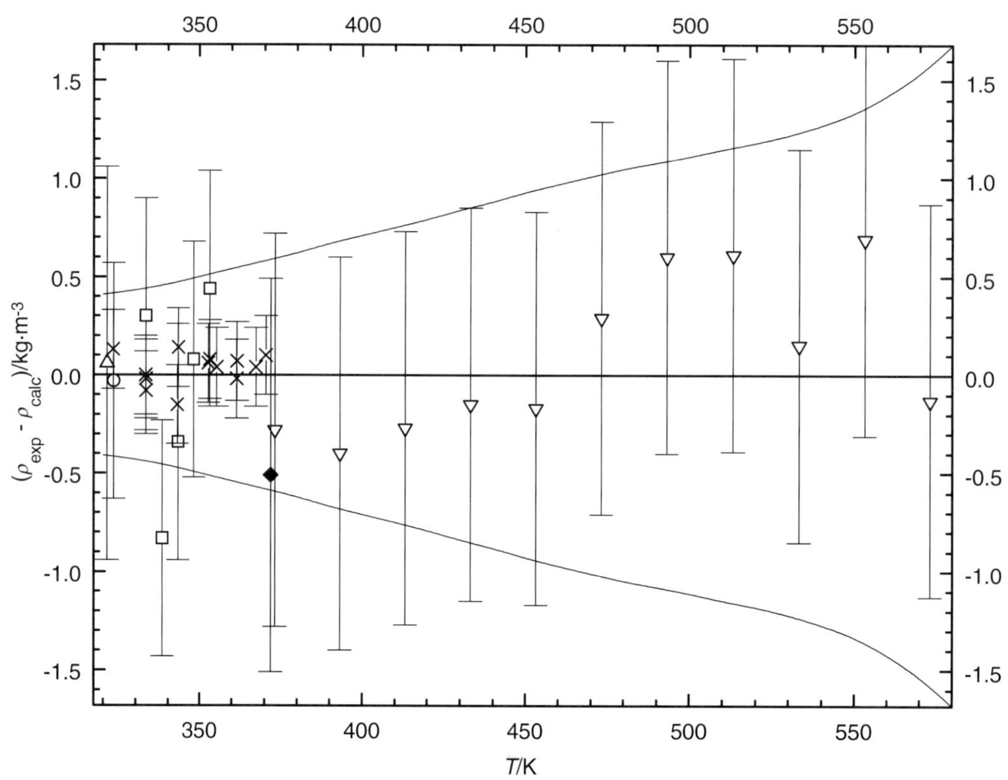

Fig. 1. The symbols show the deviation of the calculated from the experimental values from Table 2. The curves above and below the zero line indicate the calculated error region of the recommended values given in Table 3. The error bars represent the experimental errors. (Error bars smaller than the symbols are omitted for clarity of the figure.)

3-Hexadecanol **[593-03-3]** **$C_{16}H_{34}O$** **MW = 242.45** **480**

Table 1. Experimental value with uncertainty.

$\frac{T}{K}$	$\frac{\rho_{exp} \pm 2\sigma_{est}}{kg \cdot m^{-3}}$	Ref.
353.15	790.7 ± 2.0	13-pic/ken

2-Hexyl-1-decanol **[2425-77-6]** **$C_{16}H_{34}O$** **MW = 242.45** **481**

Table 1. Coefficients of the polynomial expansion equation.
Standard deviations (see introduction):
$\sigma_{c,w}$ = 1.1013 (combined temperature ranges, weighted),
$\sigma_{c,uw}$ = 5.0818 · 10^{-1} (combined temperature ranges, unweighted).

Coefficient	T = 289.65 to 373.15 K $\rho = A + BT + CT^2 + DT^3 + \ldots$
A	1.03893 · 10^3
B	−6.88973 · 10^{-1}

cont.

Table 2. Experimental values with uncertainties and deviation from calculated values.

$\frac{T}{K}$	$\frac{\rho_{exp} \pm 2\sigma_{est}}{kg \cdot m^{-3}}$	$\frac{\rho_{exp} - \rho_{calc}}{kg \cdot m^{-3}}$	Ref. (Symbol in Fig. 1)	$\frac{T}{K}$	$\frac{\rho_{exp} \pm 2\sigma_{est}}{kg \cdot m^{-3}}$	$\frac{\rho_{exp} - \rho_{calc}}{kg \cdot m^{-3}}$	Ref. (Symbol in Fig. 1)
293.15	837.24 ± 1.00	0.28	33-bin/ste(X)	353.15	795.67 ± 1.00	0.05	33-bin/ste(X)
303.15	830.56 ± 1.00	0.49	33-bin/ste(X)	373.15	781.86 ± 1.00	0.02	33-bin/ste(X)
313.15	823.79 ± 1.00	0.61	33-bin/ste(X)	293.15	833.60 ± 2.00	−3.36	35-cox/rei(O)
333.15	809.78 ± 1.00	0.38	33-bin/ste(X)	289.65	840.90 ± 2.00	1.53	38-mas(□)

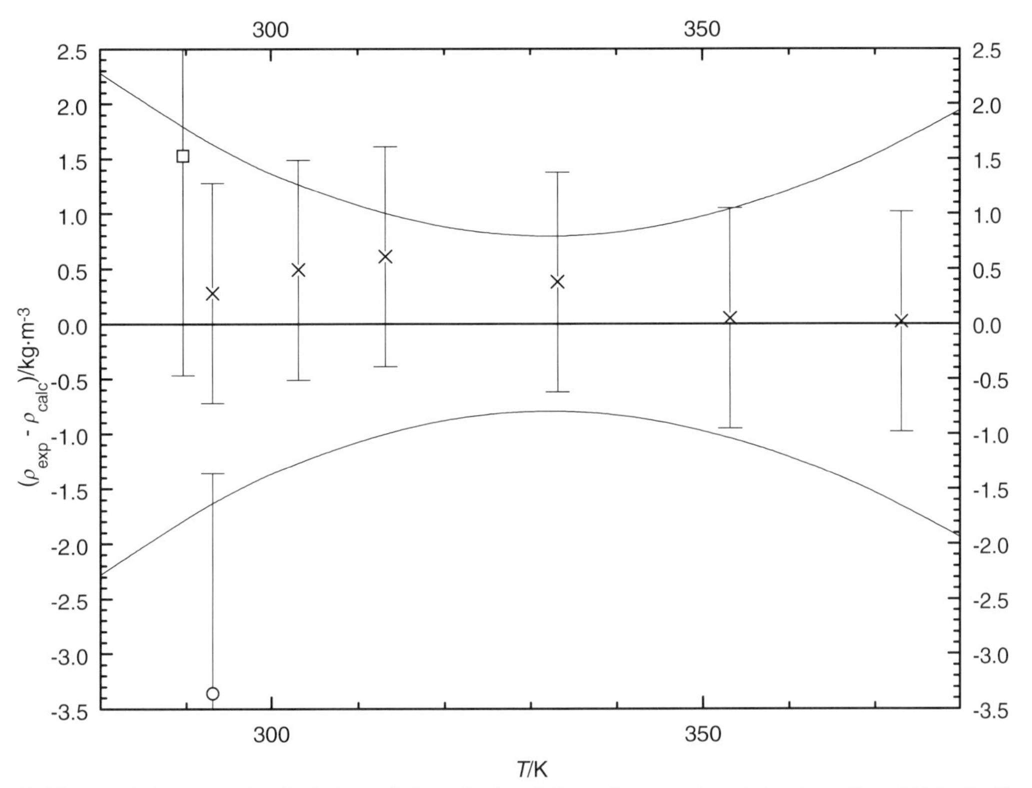

Fig. 1. The symbols show the deviation of the calculated from the experimental values from Table 2. The curves above and below the zero line indicate the calculated error region of the recommended values given in Table 3. The error bars represent the experimental errors. (Error bars smaller than the symbols are omitted for clarity of the figure.)

Table 3. Recommended values (fit to the reliable experimental values according to the equations
$\rho = A + BT + CT^2 + DT^3 + \ldots$ or $\rho = [1 + 1.75(1 - T/T_c)^{1/3} + 0.75(1 - T/T_c)][\rho_c + A(T_c - T) + B(T_c - T)^2 + C(T_c - T)^3 + D(T_c - T)^4]$).

$\frac{T}{K}$	$\frac{\rho \pm \sigma_{fit}}{kg \cdot m^{-3}}$	$\frac{T}{K}$	$\frac{\rho \pm \sigma_{fit}}{kg \cdot m^{-3}}$	$\frac{T}{K}$	$\frac{\rho \pm \sigma_{fit}}{kg \cdot m^{-3}}$
280.00	846.02 ± 2.28	310.00	825.35 ± 1.06	360.00	790.90 ± 1.20
290.00	839.13 ± 1.77	320.00	818.46 ± 0.86	370.00	784.01 ± 1.52
293.15	836.96 ± 1.63	330.00	811.57 ± 0.78	380.00	777.12 ± 1.94
298.15	833.51 ± 1.43	340.00	804.68 ± 0.81		
300.00	832.24 ± 1.36	350.00	797.79 ± 0.96		

2-Methyl-1-pentadecanol **[25354-98-7]** **$C_{16}H_{34}O$** **MW = 242.45** **482**

Table 1. Experimental value with uncertainty.

$\frac{T}{K}$	$\frac{\rho_{exp} \pm 2\sigma_{est}}{kg \cdot m^{-3}}$	Ref.
293.15	832.0 ± 1.0	35-cox/rei

2-Methyl-2-pentadecanol **[60129-23-9]** **$C_{16}H_{34}O$** **MW = 242.45** **483**

Table 1. Experimental values with uncertainties.

$\frac{T}{K}$	$\frac{\rho_{exp} \pm 2\sigma_{est}}{kg \cdot m^{-3}}$	Ref.
298.35	826.5 ± 2.0	19-eyk
353.95	764.0 ± 2.0	19-eyk

6-Methyl-6-pentadecanol **[108836-86-8]** **$C_{16}H_{34}O$** **MW = 242.45** **484**

Table 1. Experimental value with uncertainty.

$\frac{T}{K}$	$\frac{\rho_{exp} \pm 2\sigma_{est}}{kg \cdot m^{-3}}$	Ref.
298.15	831.6 ± 2.0	30-dav/dix

9-Methyl-7-pentadecanol **[500002-20-0]** **$C_{16}H_{34}O$** **MW = 242.45** **485**

Table 1. Experimental value with uncertainty.

$\frac{T}{K}$	$\frac{\rho_{exp} \pm 2\sigma_{est}}{kg \cdot m^{-3}}$	Ref.
288.15	835.1 ± 2.0	12-gue

2-Pentyl-1-undecanol **[25355-02-6]** **$C_{16}H_{34}O$** **MW = 242.45** **486**

Table 1. Coefficients of the polynomial expansion equation.
Standard deviations (see introduction):
$\sigma_{c,w} = 7.9132 \cdot 10^{-1}$ (combined temperature ranges, weighted),
$\sigma_{c,uw} = 4.9799 \cdot 10^{-1}$ (combined temperature ranges, unweighted).

Coefficient	T = 293.15 to 373.15 K $\rho = A + BT + CT^2 + DT^3 + \ldots$
A	$1.03716 \cdot 10^{3}$
B	$-6.82767 \cdot 10^{-1}$

Table 2. Experimental values with uncertainties and deviation from calculated values.

$\frac{T}{K}$	$\frac{\rho_{exp} \pm 2\sigma_{est}}{kg \cdot m^{-3}}$	$\frac{\rho_{exp} - \rho_{calc}}{kg \cdot m^{-3}}$	Ref. (Symbol in Fig. 1)	$\frac{T}{K}$	$\frac{\rho_{exp} \pm 2\sigma_{est}}{kg \cdot m^{-3}}$	$\frac{\rho_{exp} - \rho_{calc}}{kg \cdot m^{-3}}$	Ref. (Symbol in Fig. 1)
293.15	838.08 ± 1.00	1.07	33-bin/ste(×)	353.15	796.56 ± 1.00	0.52	33-bin/ste(×)
303.15	830.70 ± 1.00	0.52	33-bin/ste(×)	373.15	782.47 ± 1.00	0.08	33-bin/ste(×)
313.15	823.66 ± 1.00	0.31	33-bin/ste(×)	293.15	834.10 ± 2.00	−2.91	35-cox/rei(□)
333.15	810.11 ± 1.00	0.41	33-bin/ste(×)				

cont.

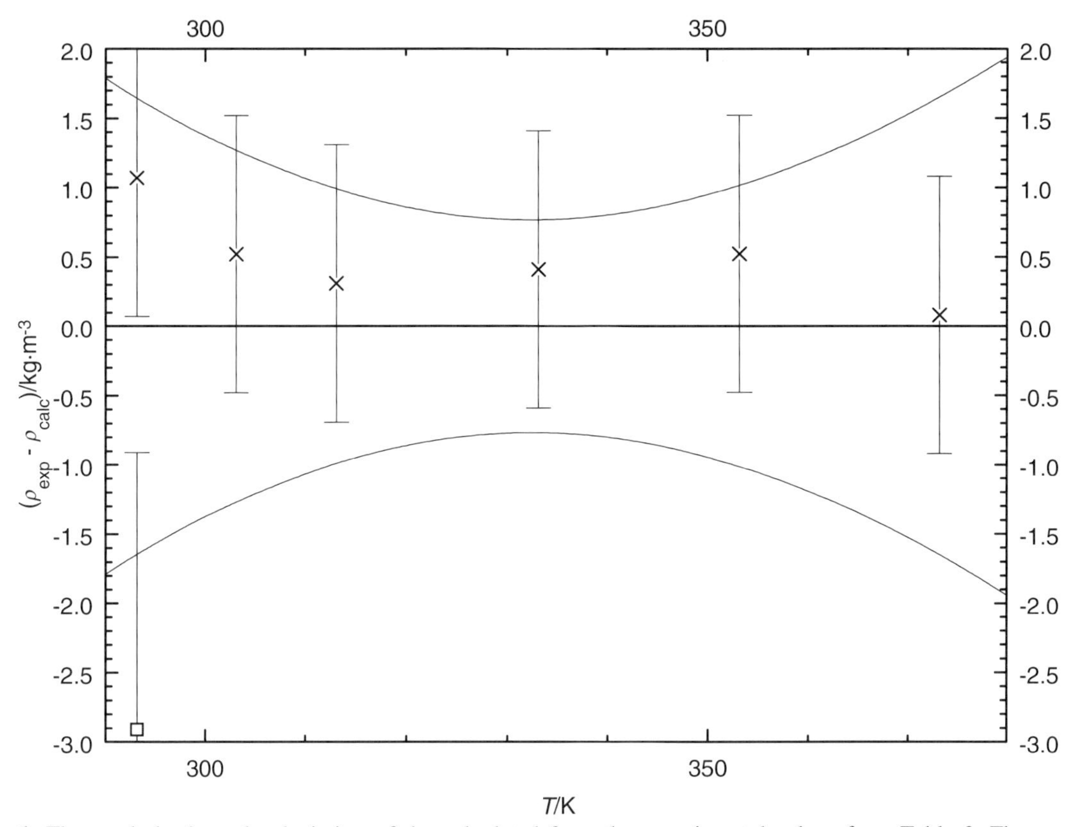

Fig. 1. The symbols show the deviation of the calculated from the experimental values from Table 2. The curves above and below the zero line indicate the calculated error region of the recommended values given in Table 3. The error bars represent the experimental errors. (Error bars smaller than the symbols are omitted for clarity of the figure.)

Table 3. Recommended values (fit to the reliable experimental values according to the equations

$$\rho = A + BT + CT^2 + DT^3 + \ldots \text{ or } \rho = [1 + 1.75(1 - T/T_c)^{1/3} + 0.75(1 - T/T_c)][\rho_c + A(T_c - T) + B(T_c - T)^2 + C(T_c - T)^3 + D(T_c - T)^4]\,).$$

$\frac{T}{\text{K}}$	$\frac{\rho \pm \sigma_{fit}}{\text{kg}\cdot\text{m}^{-3}}$	$\frac{T}{\text{K}}$	$\frac{\rho \pm \sigma_{fit}}{\text{kg}\cdot\text{m}^{-3}}$	$\frac{T}{\text{K}}$	$\frac{\rho \pm \sigma_{fit}}{\text{kg}\cdot\text{m}^{-3}}$
290.00	839.16 ± 1.79	310.00	825.50 ± 1.05	350.00	798.19 ± 0.93
293.15	837.01 ± 1.64	320.00	818.68 ± 0.84	360.00	791.37 ± 1.18
298.15	833.59 ± 1.44	330.00	811.85 ± 0.75	370.00	784.54 ± 1.51
300.00	832.33 ± 1.37	340.00	805.02 ± 0.78	380.00	777.71 ± 1.94

6-Pentyl-6-undecanol **[5331-63-5]** **$C_{16}H_{34}O$** **MW = 242.45** **487**

Table 1. Experimental value with uncertainty.

$\frac{T}{\text{K}}$	$\frac{\rho_{exp} \pm 2\sigma_{est}}{\text{kg}\cdot\text{m}^{-3}}$	Ref.
298.15	829.3 ± 2.0	33 -whi/wil -0

2-Propyl-1-tridecanol **[25355-00-4]** **$C_{16}H_{34}O$** **MW = 242.45** **488**

Table 1. Coefficients of the polynomial expansion equation.
Standard deviations (see introduction):
$\sigma_{c,w} = 2.6674 \cdot 10^{-1}$ (combined temperature ranges, weighted),
$\sigma_{c,uw} = 1.1929 \cdot 10^{-1}$ (combined temperature ranges, unweighted).

Coefficient	T = 293.15 to 373.15 K $\rho = A + BT + CT^2 + DT^3 + \ldots$
A	$1.04215 \cdot 10^{3}$
B	$-6.93516 \cdot 10^{-1}$

Table 2. Experimental values with uncertainties and deviation from calculated values.

$\frac{T}{\text{K}}$	$\frac{\rho_{exp} \pm 2\sigma_{est}}{\text{kg} \cdot \text{m}^{-3}}$	$\frac{\rho_{exp} - \rho_{calc}}{\text{kg} \cdot \text{m}^{-3}}$	Ref. (Symbol in Fig. 1)	$\frac{T}{\text{K}}$	$\frac{\rho_{exp} \pm 2\sigma_{est}}{\text{kg} \cdot \text{m}^{-3}}$	$\frac{\rho_{exp} - \rho_{calc}}{\text{kg} \cdot \text{m}^{-3}}$	Ref. (Symbol in Fig. 1)
293.15	838.43 ± 1.00	−0.41	33-bin/ste(□)	333.15	810.96 ± 1.00	−0.14	33-bin/ste(□)
303.15	832.09 ± 1.00	0.18	33-bin/ste(□)	353.15	797.19 ± 1.00	−0.04	33-bin/ste(□)
313.15	825.42 ± 1.00	0.45	33-bin/ste(□)	373.15	783.33 ± 1.00	−0.03	33-bin/ste(□)

Further references: [35-cox/rei].

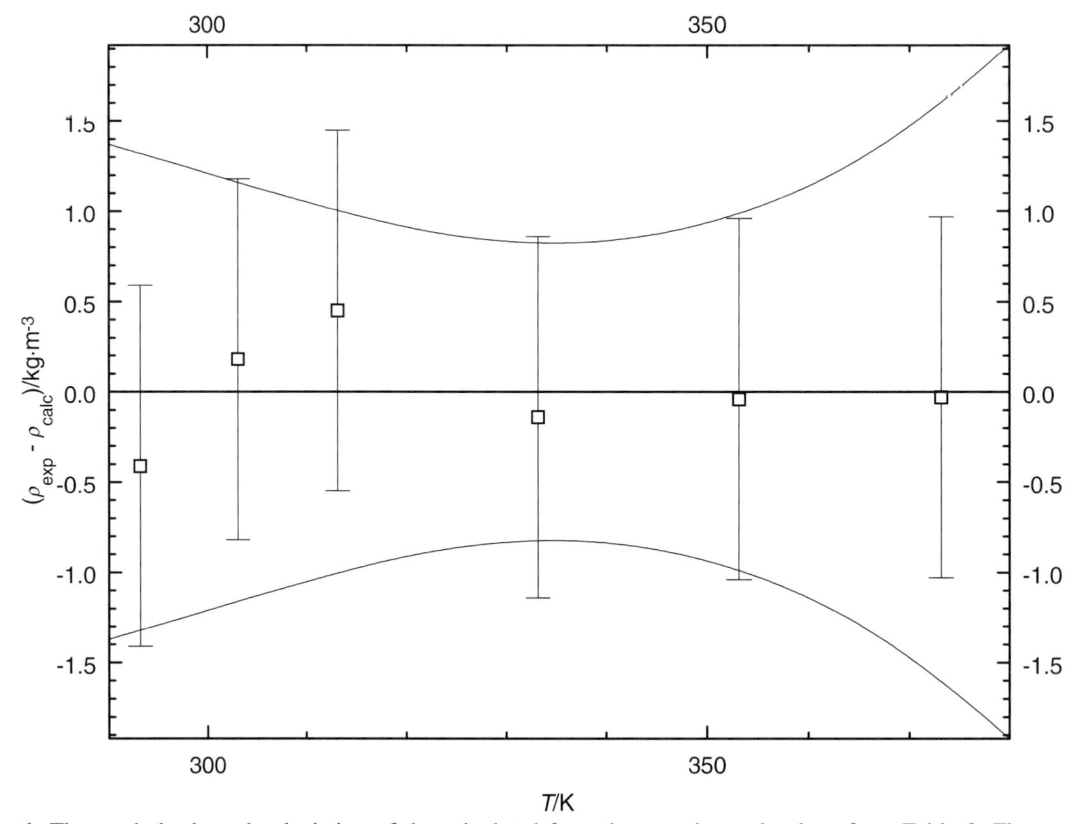

Fig. 1. The symbols show the deviation of the calculated from the experimental values from Table 2. The curves above and below the zero line indicate the calculated error region of the recommended values given in Table 3. The error bars represent the experimental errors. (Error bars smaller than the symbols are omitted for clarity of the figure.)

cont.

Table 3. Recommended values (fit to the reliable experimental values according to the equations $\rho = A + BT + CT^2 + DT^3 + \dots$ or $\rho = [1 + 1.75(1 - T/T_c)^{1/3} + 0.75(1 - T/T_c)][\rho_c + A(T_c - T) + B(T_c - T)^2 + C(T_c - T)^3 + D(T_c - T)^4]$).

T/K	$\rho \pm \sigma_{fit}$ / kg·m^{-3}	T/K	$\rho \pm \sigma_{fit}$ / kg·m^{-3}	T/K	$\rho \pm \sigma_{fit}$ / kg·m^{-3}
290.00	841.03 ± 1.37	310.00	827.16 ± 1.05	350.00	799.42 ± 0.92
293.15	838.84 ± 1.32	320.00	820.22 ± 0.90	360.00	792.48 ± 1.12
298.15	835.38 ± 1.24	330.00	813.29 ± 0.82	370.00	785.55 ± 1.45
300.00	834.09 ± 1.21	340.00	806.35 ± 0.82	380.00	778.61 ± 1.92

5-Propyl-4-tridecanol **[500002-21-1]** $C_{16}H_{34}O$ **MW = 242.45** **489**

Table 1. Experimental value with uncertainty.

T/K	$\rho_{exp} \pm 2\sigma_{est}$ / kg·m^{-3}	Ref.
288.15	831.0 ± 2.0	48-pet/old

3,9-Diethyl-6-tridecanol **[123-24-0]** $C_{17}H_{36}O$ **MW = 256.47** **490**

Table 1. Experimental values with uncertainties.

T/K	$\rho_{exp} \pm 2\sigma_{est}$ / kg·m^{-3}	Ref.
293.15	846.0 ± 1.0	53-ano-15
293.15	846.1 ± 1.0	58-ano-5

1-Heptadecanol **[1454-85-9]** $C_{17}H_{36}O$ **MW = 256.47** **491**

Table 1. Experimental value with uncertainty.

T/K	$\rho_{exp} \pm 2\sigma_{est}$ / kg·m^{-3}	Ref.
323.15	815.3 ± 2.0	63-vil/gav

2-Heptadecanol **[16813-18-6]** $C_{17}H_{36}O$ **MW = 256.47** **492**

Table 1. Experimental values with uncertainties.

T/K	$\rho_{exp} \pm 2\sigma_{est}$ / kg·m^{-3}	Ref.
293.15	837.0 ± 2.0	19-wil/sch
273.15	847.0 ± 2.0	19-wil/sch

8-Methyl-8-(2-methylpropyl)-6-dodecanol **[500002-24-4]** $C_{17}H_{36}O$ **MW = 256.47** **493**

Table 1. Experimental value with uncertainty.

T/K	$\rho_{exp} \pm 2\sigma_{est}$ / kg·m^{-3}	Ref.
293.15	846.1 ± 2.0	61-des/del

3-Ethyl-3-hexadecanol **[900002-70-2]** $C_{18}H_{38}O$ **MW = 270.5** **494**

Table 1. Experimental value with uncertainty.

T / K	$\rho_{exp} \pm 2\sigma_{est}$ / $kg \cdot m^{-3}$	Ref.
295.95	840.5 ± 2.0	19-eyk

2-Heptyl-1-undecanol **[5333-44-8]** $C_{18}H_{38}O$ **MW = 270.5** **495**

Table 1. Experimental value with uncertainty.

T / K	$\rho_{exp} \pm 2\sigma_{est}$ / $kg \cdot m^{-3}$	Ref.
288.15	844.6 ± 2.0	38-mas

1-Octadecanol **[112-92-5]** $C_{18}H_{38}O$ **MW = 270.5** **496**

Table 1. Coefficients of the polynomial expansion equation.
Standard deviations (see introduction):
$\sigma_{c,w} = 7.8077 \cdot 10^{-1}$ (combined temperature ranges, weighted),
$\sigma_{c,uw} = 1.5346 \cdot 10^{-1}$ (combined temperature ranges, unweighted).

Coefficient	T = 332.15 to 573.15 K $\rho = A + BT + CT^2 + DT^3 + \ldots$
A	$1.10078 \cdot 10^{3}$
B	-1.09517
C	$9.79822 \cdot 10^{-4}$
D	$-8.62705 \cdot 10^{-7}$

Table 2. Experimental values with uncertainties and deviation from calculated values.

T / K	$\rho_{exp} \pm 2\sigma_{est}$ / $kg \cdot m^{-3}$	$\rho_{exp} - \rho_{calc}$ / $kg \cdot m^{-3}$	Ref. (Symbol in Fig. 1)	T / K	$\rho_{exp} \pm 2\sigma_{est}$ / $kg \cdot m^{-3}$	$\rho_{exp} - \rho_{calc}$ / $kg \cdot m^{-3}$	Ref. (Symbol in Fig. 1)
332.15	812.40 ± 1.00	−1.10	1883-kra(○)	413.15	754.50 ± 1.00	−0.22	58-cos/bow(□)
332.15	812.40 ± 2.00	−1.10	1883-kra(○)	433.15	739.80 ± 1.00	−0.33	58-cos/bow(□)
343.15	804.80 ± 2.00	−0.69	1883-kra(○)	453.15	725.10 ± 1.00	−0.33	58-cos/bow(□)
343.15	804.80 ± 1.00	−0.69	1883-kra(○)	473.15	710.30 ± 1.00	−0.27	58-cos/bow(□)
372.25	784.90 ± 2.00	0.52	1883-kra(○)	493.15	695.60 ± 1.00	0.08	58-cos/bow(□)
372.25	784.90 ± 1.00	0.52	1883-kra(○)	513.15	680.80 ± 1.00	0.57	58-cos/bow(□)
332.15	813.30 ± 2.00	−0.20	1884-kra(▽)	533.15	666.10 ± 1.00	1.44	58-cos/bow(□)
332.25	812.00 ± 3.00	−1.43	49-tsv/mar(◆)	553.15	649.80 ± 1.00	1.02	58-cos/bow(□)
333.15	813.00 ± 2.00	0.23	56-rat/cur(△)	573.15	631.60 ± 1.00	−0.93	58-cos/bow(□)
342.15	806.00 ± 2.00	−0.22	56-rat/cur(△)	338.15	809.70 ± 0.60	0.57	92-lie/sen-1(×)
363.15	792.00 ± 2.00	1.03	56-rat/cur(△)	343.15	805.60 ± 0.60	0.11	92-lie/sen-1(×)
353.15	797.30 ± 1.00	−0.92	58-cos/bow(□)	348.15	802.60 ± 0.60	0.75	92-lie/sen-1(×)
373.15	783.90 ± 1.00	0.17	58-cos/bow(□)	353.15	799.50 ± 0.60	1.28	92-lie/sen-1(×)
393.15	769.40 ± 1.00	0.16	58-cos/bow(□)				

cont.

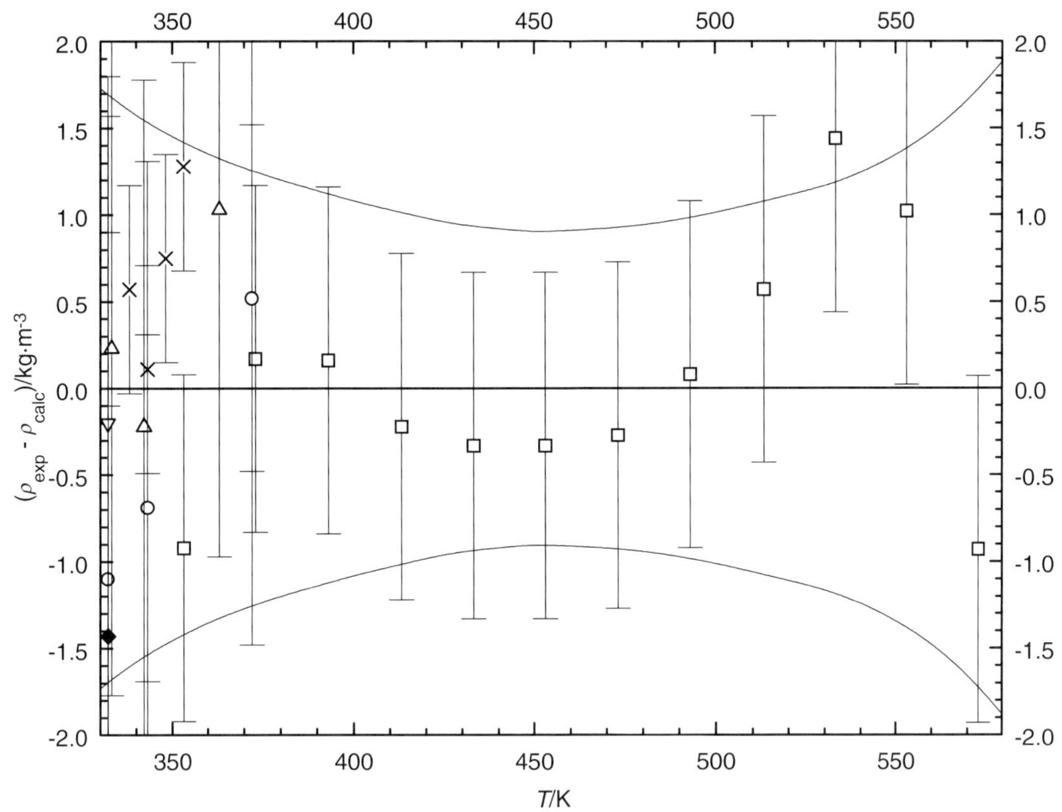

Fig. 1. The symbols show the deviation of the calculated from the experimental values from Table 2. The curves above and below the zero line indicate the calculated error region of the recommended values given in Table 3. The error bars represent the experimental errors. (Error bars smaller than the symbols are omitted for clarity of the figure.)

Table 3. Recommended values (fit to the reliable experimental values according to the equations

$$\rho = A + BT + CT^2 + DT^3 + \ldots \text{ or } \rho = [1 + 1.75(1 - T/T_c)^{1/3} + 0.75(1 - T/T_c)][\rho_c + A(T_c - T) + B(T_c - T)^2 + C(T_c - T)^3 + D(T_c - T)^4]).$$

$\frac{T}{\text{K}}$	$\frac{\rho \pm \sigma_{\text{fit}}}{\text{kg} \cdot \text{m}^{-3}}$	$\frac{T}{\text{K}}$	$\frac{\rho \pm \sigma_{\text{fit}}}{\text{kg} \cdot \text{m}^{-3}}$	$\frac{T}{\text{K}}$	$\frac{\rho \pm \sigma_{\text{fit}}}{\text{kg} \cdot \text{m}^{-3}}$
330.00	815.07 ± 1.73	420.00	749.73 ± 0.98	510.00	682.66 ± 1.06
340.00	807.78 ± 1.57	430.00	742.44 ± 0.94	520.00	674.93 ± 1.11
350.00	800.51 ± 1.45	440.00	735.11 ± 0.92	530.00	667.14 ± 1.16
360.00	793.25 ± 1.35	450.00	727.75 ± 0.90	540.00	659.26 ± 1.24
370.00	786.01 ± 1.27	460.00	720.36 ± 0.91	550.00	651.30 ± 1.34
380.00	778.76 ± 1.20	470.00	712.92 ± 0.92	560.00	643.25 ± 1.47
390.00	771.52 ± 1.14	480.00	705.44 ± 0.94	570.00	635.11 ± 1.65
400.00	764.27 ± 1.08	490.00	697.91 ± 0.97	580.00	626.87 ± 1.88
410.00	757.01 ± 1.03	500.00	690.31 ± 1.01		

3-Octadecanol **[26762-44-7]** **$C_{18}H_{38}O$** **MW = 270.5** **497**

Table 1. Experimental value with uncertainty.

$\frac{T}{K}$	$\frac{\rho_{exp} \pm 2\sigma_{est}}{kg \cdot m^{-3}}$	Ref.
353.15	785.8 ± 2.0	13-pic/ken

2,2-Dimethyl-3-(1,1-dimethylethyl)-3-tridecanol **[500045-09-0]** **$C_{19}H_{40}O$** **MW = 284.53** **498**

Table 1. Experimental value with uncertainty.

$\frac{T}{K}$	$\frac{\rho_{exp} \pm 2\sigma_{est}}{kg \cdot m^{-3}}$	Ref.
293.15	869.8 ± 1.5	60-pet/sok

7-Hexyl-7-tridecanol **[5340-59-0]** **$C_{19}H_{40}O$** **MW = 284.53** **499**

Table 1. Experimental value with uncertainty.

$\frac{T}{K}$	$\frac{\rho_{exp} \pm 2\sigma_{est}}{kg \cdot m^{-3}}$	Ref.
293.15	840.8 ± 2.0	61-mes/erz

4-Methyl-4-octadecanol **[54892-13-6]** **$C_{19}H_{40}O$** **MW = 284.53** **500**

Table 1. Experimental value with uncertainty.

$\frac{T}{K}$	$\frac{\rho_{exp} \pm 2\sigma_{est}}{kg \cdot m^{-3}}$	Ref.
293.15	839.4 ± 1.0	48-sor/sor

5-Methyl-5-octadecanol **[500020-99-5]** **$C_{19}H_{40}O$** **MW = 284.53** **501**

Table 1. Experimental value with uncertainty.

$\frac{T}{K}$	$\frac{\rho_{exp} \pm 2\sigma_{est}}{kg \cdot m^{-3}}$	Ref.
293.15	837.4 ± 0.6	48-sor/sor

6-Methyl-6-octadecanol **[500021-00-1]** **$C_{19}H_{40}O$** **MW = 284.53** **502**

Table 1. Experimental value with uncertainty.

$\frac{T}{K}$	$\frac{\rho_{exp} \pm 2\sigma_{est}}{kg \cdot m^{-3}}$	Ref.
293.15	838.5 ± 1.0	48-sor/sor

9-Methyl-9-octadecanol **[500021-01-2]** $C_{19}H_{40}O$ **MW = 284.53** **503**

Table 1. Experimental value with uncertainty.

$\frac{T}{K}$	$\frac{\rho_{exp} \pm 2\sigma_{est}}{kg \cdot m^{-3}}$	Ref.
293.15	835.5 ± 1.0	48-sor/sor

1-Eicosanol **[629-96-9]** $C_{20}H_{42}O$ **MW = 298.55** **504**

Table 1. Fit with estimated *B* coefficient for 3 accepted points. Deviation σ_w = 0.189.

Coefficient	$\rho = A + BT$
A	1056.70
B	−0.720

Table 2. Experimental values with uncertainties and deviation from calculated values.

$\frac{T}{K}$	$\frac{\rho_{exp} \pm 2\sigma_{est}}{kg \cdot m^{-3}}$	$\frac{\rho_{exp} - \rho_{calc}}{kg \cdot m^{-3}}$	Ref.
348.15	805.9 ± 0.50	−0.13	69-pat/kat
358.15	798.7 ± 0.50	−0.13	69-pat/kat
368.15	791.9 ± 0.50	0.27	69-pat/kat

Table 3. Recommended values.

$\frac{T}{K}$	$\frac{\rho_{exp} \pm 2\sigma_{est}}{kg \cdot m^{-3}}$
340.00	811.9 ± 1.0
350.00	804.7 ± 0.6
360.00	797.5 ± 0.5
370.00	790.3 ± 0.8

3-Ethyl-3-octadecanol **[35185-53-6]** $C_{20}H_{42}O$ **MW = 298.55** **505**

Table 1. Experimental value with uncertainty.

$\frac{T}{K}$	$\frac{\rho_{exp} \pm 2\sigma_{est}}{kg \cdot m^{-3}}$	Ref.
353.25	778.8 ± 3.0	19-eyk

2-Methyl-2-nonadecanol **[76695-48-2]** $C_{20}H_{42}O$ **MW = 298.55** **506**

Table 1. Experimental value with uncertainty.

$\frac{T}{K}$	$\frac{\rho_{exp} \pm 2\sigma_{est}}{kg \cdot m^{-3}}$	Ref.
352.95	771.7 ± 3.0	19-eyk

2-Octyl-1-dodecanol **[5333-42-6]** $C_{20}H_{42}O$ **MW = 298.55** **507**

Table 1. Experimental value with uncertainty.

$\frac{T}{K}$	$\frac{\rho_{exp} \pm 2\sigma_{est}}{kg \cdot m^{-3}}$	Ref.
292.15	846.3 ± 2.0	38-mas

2,6,11,15-Tetramethyl-8-hexadecanol **[500002-33-5]** $C_{20}H_{42}O$ **MW = 298.55** **508**

Table 1. Experimental value with uncertainty.

$\frac{T}{K}$	$\frac{\rho_{exp} \pm 2\sigma_{est}}{kg \cdot m^{-3}}$	Ref.
292.15	892.0 ± 2.0	23-von/kai

1-Docosanol **[661-19-8]** $C_{22}H_{46}O$ **MW = 326.61** **509**

Table 1. Fit with estimated B coefficient for 3 accepted points. Deviation $\sigma_w = 0.094$.

Coefficient	$\rho = A + BT$
A	1074.44
B	−0.770

Table 2. Experimental values with uncertainties and deviation from calculated values.

$\frac{T}{K}$	$\frac{\rho_{exp} \pm 2\sigma_{est}}{kg \cdot m^{-3}}$	$\frac{\rho_{exp} - \rho_{calc}}{kg \cdot m^{-3}}$	Ref.
348.15	806.3 ± 0.5	−0.07	69-pat/kat
358.15	798.6 ± 0.5	−0.07	69-pat/kat
368.15	791.1 ± 0.5	0.13	69-pat/kat

Table 3. Recommended values.

$\frac{T}{K}$	$\frac{\rho_{exp} \pm 2\sigma_{est}}{kg \cdot m^{-3}}$	$\frac{T}{K}$	$\frac{\rho_{exp} \pm 2\sigma_{est}}{kg \cdot m^{-3}}$
340.00	812.6 ± 1.0	360.00	797.2 ± 0.5
350.00	804.9 ± 0.6	370.00	789.5 ± 0.7

3-Ethyl-3-eicosanol **[95287-47-1]** $C_{22}H_{46}O$ **MW = 326.61** **510**

Table 1. Experimental value with uncertainty.

$\frac{T}{K}$	$\frac{\rho_{exp} \pm 2\sigma_{est}}{kg \cdot m^{-3}}$	Ref.
353.20	780.0 ± 3.0	19-eyk

2-Nonyl-1-tridecanol **[54439-52-0]** $C_{22}H_{46}O$ **MW = 326.61** **511**

Table 1. Experimental value with uncertainty.

$\frac{T}{K}$	$\frac{\rho_{exp} \pm 2\sigma_{est}}{kg \cdot m^{-3}}$	Ref.
290.65	847.6 ± 2.0	38-mas

2.2 Unsaturated Monoalcohols

2.2.1 Alcohols of General Formula $C_nH_{2n}O$

2-Propenol **[107-18-6]** **C_3H_6O** **MW = 58.08** **512**

Table 1. Coefficients of the polynomial expansion equation.
Standard deviations (see introduction):
$\sigma_{c,w} = 8.0998 \cdot 10^{-1}$ (combined temperature ranges, weighted),
$\sigma_{c,uw} = 1.6309 \cdot 10^{-1}$ (combined temperature ranges, unweighted).

Coefficient	T = 273.15 to 440.00 K $\rho = A + BT + CT^2 + DT^3 + \ldots$
A	$1.05862 \cdot 10^{3}$
B	-1.61233
C	$4.16485 \cdot 10^{-3}$
D	$-6.30190 \cdot 10^{-6}$

Table 2. Experimental values with uncertainties and deviation from calculated values.

$\frac{T}{\text{K}}$	$\frac{\rho_{exp} \pm 2\sigma_{est}}{\text{kg}\cdot\text{m}^{-3}}$	$\frac{\rho_{exp} - \rho_{calc}}{\text{kg}\cdot\text{m}^{-3}}$	Ref. (Symbol in Fig. 1)	$\frac{T}{\text{K}}$	$\frac{\rho_{exp} \pm 2\sigma_{est}}{\text{kg}\cdot\text{m}^{-3}}$	$\frac{\rho_{exp} - \rho_{calc}}{\text{kg}\cdot\text{m}^{-3}}$	Ref. (Symbol in Fig. 1)
273.15	799.50 ± 0.60	−1.02	1882-zan(✕)	293.15	786.00 ± 0.40	0.88	52-cap/mug[1)]
290.15	785.10 ± 0.60	−2.39	1882-zan[1)]	298.15	781.50 ± 0.40	0.39	52-cap/mug[1)]
283.15	794.00 ± 1.00	1.06	1890-gar[1)]	298.15	780.89 ± 0.20	−0.22	52-wil/sim(□)
293.15	785.40 ± 1.00	0.28	1890-gar[1)]	298.15	780.80 ± 0.30	−0.31	54-bre(◆)
303.15	776.50 ± 1.00	−0.52	1890-gar[1)]	298.15	780.81 ± 0.20	−0.30	54-kre/wie(Δ)
313.15	767.50 ± 1.00	−1.12	1890-gar[1)]	298.15	780.86 ± 0.20	−0.25	54-pur/bow(O)
323.15	758.20 ± 1.00	−1.65	1890-gar(✕)	293.15	785.50 ± 0.50	0.38	56-tor-1[1)]
286.15	790.90 ± 0.60	0.28	19-eyk(✕)	313.15	768.40 ± 0.50	−0.22	56-tor-1(✕)
273.15	799.90 ± 0.60	−0.62	25-pal/con(✕)	333.15	749.70 ± 0.50	−1.01	56-tor-1(✕)
355.35	728.30 ± 0.70	−0.51	27-arb-2(✕)	298.15	780.90 ± 0.30	−0.21	58-arn/was(✕)
355.35	728.40 ± 0.70	-0.41	27-arb-2(✕)	293.15	785.40 ± 0.30	0.28	63-amb/tow(∇)
356.45	727.70 ± 0.70	0.03	27-arb-2(✕)	300.00	779.00 ± 1.00	−0.61	64-dan/bah[1)]
356.45	727.70 ± 0.70	0.03	27-arb-2(✕)	320.00	763.00 ± 1.00	0.35	64-dan/bah(✕)
273.15	801.34 ± 0.60	0.82	34-tim/del(✕)	340.00	745.00 ± 1.00	0.81	64-dan/bah(✕)
288.15	789.14 ± 0.60	0.08	34-tim/del[1)]	360.00	724.00 ± 1.00	0.08	64-dan/bah(✕)
303.15	776.88 ± 0.60	−0.14	34-tim/del[1)]	380.00	702.00 ± 1.00	0.46	64-dan/bah(✕)
308.15	772.21 ± 0.60	−0.65	43-lan/key(✕)	400.00	676.00 ± 1.50	−0.74	64-dan/bah(✕)
293.15	786.40 ± 0.60	1.28	48-vog-2[1)]	420.00	648.00 ± 1.50	−1.23	64-dan/bah(✕)
314.35	769.70 ± 0.60	2.12	48-vog-2(✕)	440.00	619.00 ± 1.50	0.31	64-dan/bah(✕)
332.35	753.20 ± 0.60	1.75	48-vog-2(✕)	460.00	585.00 ± 1.50	0.17	64-dan/bah[1)]
296.15	783.84 ± 0.30	1.12	51-dim/lan(✕)	480.00	541.00 ± 2.00	−6.34	64-dan/bah[1)]
288.15	790.00 ± 0.40	0.94	52-cap/mug(✕)	500.00	470.00 ± 2.00	−35.93	64-dan/bah[1)]

[1)] Not included in Fig. 1.

cont.

2-Propenol (cont.)

Further references: [1872-lin/von, 1880-bru-1, 1880-bru-3, 1880-tho, 1881-pri/han, 1882-sch-1, 1883-sch-3, 1884-gla, 1884-per, 1884-sch-6, 1897-tho, 02-you/for, 08-dor/dvo, 13-atk/wal, 14-vav, 21-leb, 23-clo/joh, 25-nor/ash, 25-par/kel, 25-par/kel-1, 25-rak, 27-par/cha, 28-par/bar, 28-par/nel, 29-ber, 29-ber/reu, 29-swa, 31-ton/ueh, 33-tre/wat, 35-hen, 35-mah-1, 35-ols/was, 37-zep, 38-dol/gre, 40-mil/bli, 40-was/beg, 40-was/gra, 42-sch/hun, 42-was/bro, 43-bru/bog, 43-dur/roe, 48-wei, 49-hat, 50-jac, 50-par/gol, 50-pic/zie, 52-coo, 53-ani, 53-ani-1, 57-rom-1, 58-hil/hay, 61-ogi/cor, 62-bro/smi, 63-lab].

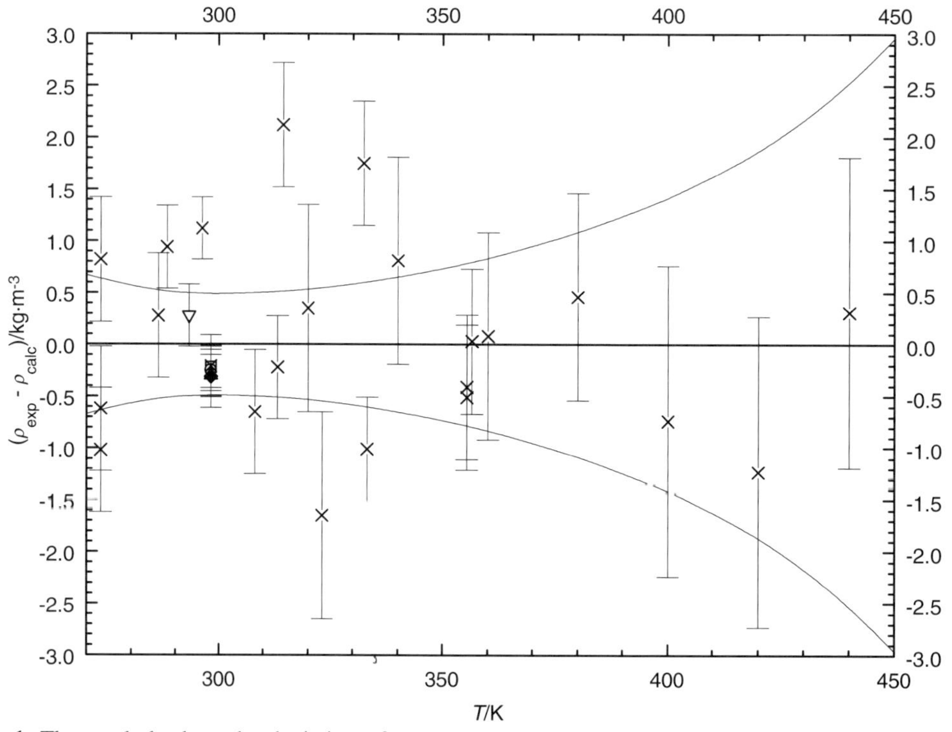

Fig. 1. The symbols show the deviation of the calculated from the experimental values from Table 2. The curves above and below the zero line indicate the calculated error region of the recommended values given in Table 3. The error bars represent the experimental errors. (Error bars smaller than the symbols are omitted for clarity of the figure.)

Table 3. Recommended values (fit to the reliable experimental values according to the equations $\rho = A + BT + CT^2 + DT^3 + \ldots$ or $\rho = [1 + 1.75(1 - T/T_c)^{1/3} + 0.75(1 - T/T_c)][\rho_c + A(T_c - T) + B(T_c - T)^2 + C(T_c - T)^3 + D(T_c - T)^4]$).

$\frac{T}{K}$	$\frac{\rho \pm \sigma_{fit}}{kg \cdot m^{-3}}$	$\frac{T}{K}$	$\frac{\rho \pm \sigma_{fit}}{kg \cdot m^{-3}}$	$\frac{T}{K}$	$\frac{\rho \pm \sigma_{fit}}{kg \cdot m^{-3}}$
270.00	802.87 ± 0.67	320.00	762.65 ± 0.53	390.00	689.46 ± 1.24
280.00	795.35 ± 0.56	330.00	753.63 ± 0.58	400.00	676.74 ± 1.41
290.00	787.61 ± 0.50	340.00	744.19 ± 0.65	410.00	663.34 ± 1.62
293.15	785.12 ± 0.50	350.00	734.30 ± 0.73	420.00	649.23 ± 1.86
298.15	781.11 ± 0.49	360.00	723.92 ± 0.83	430.00	634.35 ± 2.16
300.00	779.61 ± 0.49	370.00	713.02 ± 0.95	440.00	618.69 ± 2.52
310.00	771.30 ± 0.50	380.00	701.54 ± 1.08	450.00	602.19 ± 2.96

1-Buten-3-ol **[598-32-3]** **C_4H_8O** **MW = 72.11** **513**

Table 1. Experimental value with uncertainty.

$\frac{T}{K}$	$\frac{\rho_{exp} \pm 2\sigma_{est}}{kg \cdot m^{-3}}$	Ref.
298.15	830.0 ± 2.0	52-wib

(Z)-2-Buten-1-ol **[4088-60-2]** **C_4H_8O** **MW = 72.11** **514**

Table 1. Experimental value with uncertainty.

$\frac{T}{K}$	$\frac{\rho_{exp} \pm 2\sigma_{est}}{kg \cdot m^{-3}}$	Ref.
293.15	895.0 ± 3.0	58-hil/hay

2-Methyl-3-buten-2-ol **[115-18-4]** **$C_5H_{10}O$** **MW = 86.13** **515**

Table 1. Coefficients of the polynomial expansion equation.
Standard deviations (see introduction):
$\sigma_{c,w} = 3.3193 \cdot 10^{-1}$ (combined temperature ranges, weighted),
$\sigma_{c,uw} = 1.9173 \cdot 10^{-1}$ (combined temperature ranges, unweighted).

Coefficient	T = 293.15 to 343.15 K $\rho = A + BT + CT^2 + DT^3 + \ldots$
A	$8.68886 \cdot 10^{2}$
B	$5.71276 \cdot 10^{-1}$
C	$-2.47190 \cdot 10^{-3}$

Table 2. Experimental values with uncertainties and deviation from calculated values.

$\frac{T}{K}$	$\frac{\rho_{exp} \pm 2\sigma_{est}}{kg \cdot m^{-3}}$	$\frac{\rho_{exp} - \rho_{calc}}{kg \cdot m^{-3}}$	Ref. (Symbol in Fig. 1)	$\frac{T}{K}$	$\frac{\rho_{exp} \pm 2\sigma_{est}}{kg \cdot m^{-3}}$	$\frac{\rho_{exp} - \rho_{calc}}{kg \cdot m^{-3}}$	Ref. (Symbol in Fig. 1)
293.15	824.80 ± 1.00	0.87	41-cam/eby-1(□)	323.15	795.00 ± 0.50	–0.36	88-bag/gur(×)
293.15	823.40 ± 0.50	–0.53	88-bag/gur(×)	333.15	784.60 ± 0.50	–0.25	88-bag/gur(×)
303.15	814.90 ± 0.50	–0.00	88-bag/gur(×)	343.15	773.90 ± 0.50	0.05	88-bag/gur(×)
313.15	805.60 ± 0.50	0.22	88-bag/gur(×)				

Table 3. Recommended values (fit to the reliable experimental values according to the equations
$\rho = A + BT + CT^2 + DT^3 + \ldots$ or $\rho = [1 + 1.75(1 - T/T_c)^{1/3} + 0.75(1 - T/T_c)][\rho_c + A(T_c - T) + B(T_c - T)^2 + C(T_c - T)^3 + D(T_c - T)^4]$).

$\frac{T}{K}$	$\frac{\rho \pm \sigma_{fit}}{kg \cdot m^{-3}}$	$\frac{T}{K}$	$\frac{\rho \pm \sigma_{fit}}{kg \cdot m^{-3}}$	$\frac{T}{K}$	$\frac{\rho \pm \sigma_{fit}}{kg \cdot m^{-3}}$
290.00	826.67 ± 0.97	300.00	817.80 ± 0.63	330.00	788.22 ± 0.51
293.15	823.93 ± 0.84	310.00	808.43 ± 0.51	340.00	777.37 ± 0.70
298.15	819.48 ± 0.67	320.00	798.57 ± 0.48	350.00	766.02 ± 1.13

cont.

2-Methyl-3-buten-2-ol (cont.)

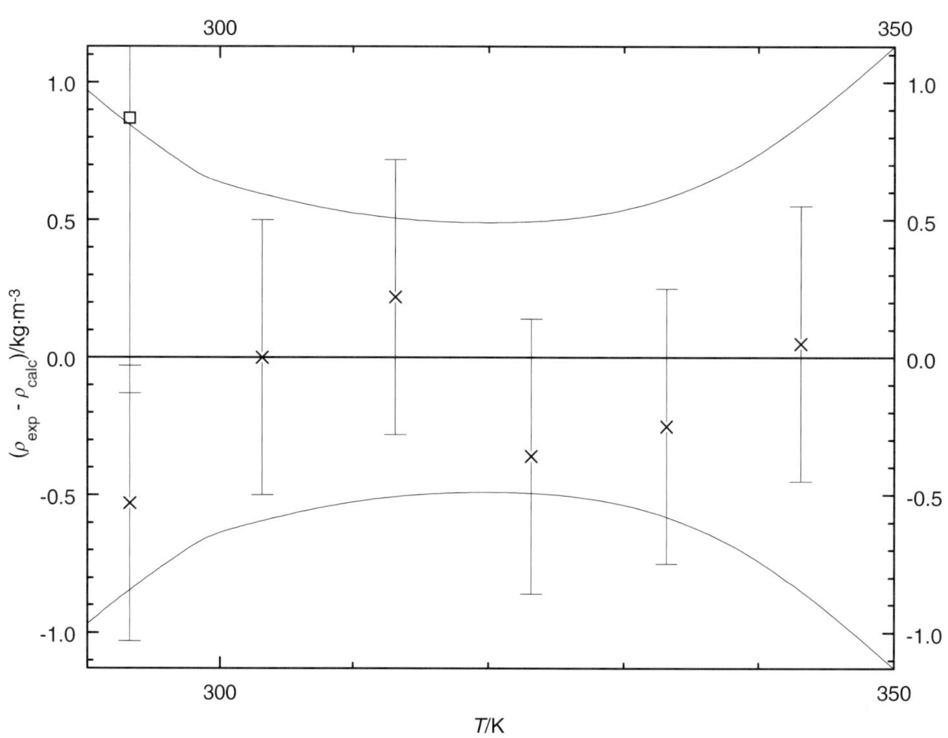

Fig. 1. The symbols show the deviation of the calculated from the experimental values from Table 2. The curves above and below the zero line indicate the calculated error region of the recommended values given in Table 3. The error bars represent the experimental errors. (Error bars smaller than the symbols are omitted for clarity of the figure.)

1-Penten-3-ol **[616-25-1]** **$C_5H_{10}O$** **MW = 86.13** **516**

Table 1. Experimental value with uncertainty.

$\frac{T}{K}$	$\frac{\rho_{exp} \pm 2\sigma_{est}}{kg \cdot m^{-3}}$	Ref.
293.15	833.0 ± 0.6	63-lab

3-Penten-1-ol **[39161-19-8]** **$C_5H_{10}O$** **MW = 86.13** **517**

Table 1. Experimental value with uncertainty.

$\frac{T}{K}$	$\frac{\rho_{exp} \pm 2\sigma_{est}}{kg \cdot m^{-3}}$	Ref.
293.15	849.2 ± 1.0	48-goe/cri

4-Penten-1-ol **[821-09-0]** **$C_5H_{10}O$** **MW = 86.13** **518**

Table 1. Experimental value with uncertainty.

$\frac{T}{K}$	$\frac{\rho_{exp} \pm 2\sigma_{est}}{kg \cdot m^{-3}}$	Ref.
298.15	839.5 ± 1.0	45-sch/gel

(*E*)-3,3-Dimethyl-2-buten-1-ol **[500029-65-2]** **$C_6H_{12}O$** **MW = 100.16** **519**

Table 1. Fit with estimated *B* coefficient for 3 accepted points. Deviation $\sigma_w = 0.047$.

Coefficient	$\rho = A + BT$
A	1037.77
B	–0.700

Table 2. Experimental values with uncertainties and deviation from calculated values.

$\frac{T}{K}$	$\frac{\rho_{exp} \pm 2\sigma_{est}}{kg \cdot m^{-3}}$	$\frac{\rho_{exp} - \rho_{calc}}{kg \cdot m^{-3}}$	Ref.
293.15	832.5 ± 0.7	–0.07	61-hat/wei
298.15	829.1 ± 0.7	0.03	61-hat/wei
303.15	825.6 ± 0.7	0.03	61-hat/wei

Table 3. Recommended values.

$\frac{T}{K}$	$\frac{\rho_{exp} \pm 2\sigma_{est}}{kg \cdot m^{-3}}$
290.00	834.8 ± 0.7
293.15	832.6 ± 0.7
298.15	829.1 ± 0.6
310.00	820.8 ± 0.9

(*Z*)-3,3-Dimethyl-2-buten-1-ol **[500029-64-1]** **$C_6H_{12}O$** **MW = 100.16** **520**

Table 1. Fit with estimated *B* coefficient for 3 accepted points. Deviation $\sigma_w = 0.094$.

Coefficient	$\rho = A + BT$
A	1041.21
B	–0.680

Table 2. Experimental values with uncertainties and deviation from calculated values.

$\frac{T}{K}$	$\frac{\rho_{exp} \pm 2\sigma_{est}}{kg \cdot m^{-3}}$	$\frac{\rho_{exp} - \rho_{calc}}{kg \cdot m^{-3}}$	Ref.
293.15	841.8 ± 0.7	–0.07	61-hat/wei
298.15	838.4 ± 0.7	–0.07	61-hat/wei
303.15	835.2 ± 0.7	0.13	61-hat/wei

Table 3. Recommended values.

$\frac{T}{K}$	$\frac{\rho_{exp} \pm 2\sigma_{est}}{kg \cdot m^{-3}}$
290.00	844.0 ± 0.8
293.15	841.9 ± 0.7
298.15	838.5 ± 0.6
310.00	830.4 ± 0.9

1-Hexen-3-ol **[4798-44-1]** $C_6H_{12}O$ **MW = 100.16** **521**

Table 1. Experimental value with uncertainty.

T/K	$\rho_{exp} \pm 2\sigma_{est}$ / kg·m^{-3}	Ref.
293.15	835.7 ± 0.6	63-lab

2-Hexen-1-ol **[500026-97-1]** $C_6H_{12}O$ **MW = 100.16** **522**

Table 1. Experimental value with uncertainty.

T/K	$\rho_{exp} \pm 2\sigma_{est}$ / kg·m^{-3}	Ref.
288.15	850.8 ± 1.0	30-wal/ros

(*E*)-2-Hexen-1-ol **[928-95-0]** $C_6H_{12}O$ **MW = 100.16** **523**

Table 1. Fit with estimated *B* coefficient for 3 accepted points. Deviation $\sigma_w = 0.094$.

Coefficient	$\rho = A + BT$
A	1039.35
B	−0.640

Table 2. Experimental values with uncertainties and deviation from calculated values.

T/K	$\rho_{exp} \pm 2\sigma_{est}$ / kg·m^{-3}	$\rho_{exp} - \rho_{calc}$ / kg·m^{-3}	Ref.
293.15	851.6 ± 0.7	−0.13	61-hat/wei
298.15	848.6 ± 0.7	0.07	61-hat/wei
303.15	845.4 ± 0.7	0.07	61-hat/wei

Table 3. Recommended values.

T/K	$\rho_{exp} \pm 2\sigma_{est}$ / kg·m^{-3}
290.00	853.7 ± 0.8
293.15	851.7 ± 0.7
298.15	848.5 ± 0.6
310.00	840.9 ± 0.9

2-Hexen-4-ol **[4798-58-7]** $C_6H_{12}O$ **MW = 100.16** **524**

Table 1. Experimental and recommended values with uncertainties.

T/K	$\rho_{exp} \pm 2\sigma_{est}$ / kg·m^{-3}	Ref.
298.15	836.7 ± 1.0	54-gay/cro[1)]
293.15	841.6 ± 0.6	63-lab
293.15	841.6 ± 0.6	Recommended

[1)] Not included in calculation of recommended value.

5-Hexen-2-ol **[626-94-8]** **$C_6H_{12}O$** **MW = 100.16** **525**

Table 1. Experimental value with uncertainty.

$\frac{T}{K}$	$\frac{\rho_{exp} \pm 2\sigma_{est}}{kg \cdot m^{-3}}$	Ref.
297.15	838.0 ± 2.0	49-col/lag

2-Methyl-1-penten-3-ol **[2088-07-5]** **$C_6H_{12}O$** **MW = 100.16** **526**

Table 1. Experimental value with uncertainty.

$\frac{T}{K}$	$\frac{\rho_{exp} \pm 2\sigma_{est}}{kg \cdot m^{-3}}$	Ref.
298.15	840.0 ± 2.0	65-col/des

2-Methyl-2-penten-1-ol **[1610-29-3]** **$C_6H_{12}O$** **MW = 100.16** **527**

Table 1. Experimental value with uncertainty.

$\frac{T}{K}$	$\frac{\rho_{exp} \pm 2\sigma_{est}}{kg \cdot m^{-3}}$	Ref.
298.15	851.0 ± 2.0	65-col/des

3-Methyl-1-penten-3-ol **[918-85-4]** **$C_6H_{12}O$** **MW = 100.16** **528**

Table 1. Experimental value with uncertainty.

$\frac{T}{K}$	$\frac{\rho_{exp} \pm 2\sigma_{est}}{kg \cdot m^{-3}}$	Ref.
298.15	833.4 ± 0.5	52-mye/col

3-Methyl-2-penten-1-ol **[500060-37-7]** **$C_6H_{12}O$** **MW = 100.16** **529**

Table 1. Fit with estimated B coefficient for 2 accepted points. Deviation $\sigma_w = 0.380$.

Coefficient	$\rho = A + BT$
A	1078.78
B	–0.780

Table 2. Experimental values with uncertainties and deviation from calculated values.

$\frac{T}{K}$	$\frac{\rho_{exp} \pm 2\sigma_{est}}{kg \cdot m^{-3}}$	$\frac{\rho_{exp} - \rho_{calc}}{kg \cdot m^{-3}}$	Ref.
273.15	866.1 ± 1.5	0.38	50-fav/fri
291.15	851.3 ± 1.5	–0.38	50-fav/fri

Table 3. Recommended values.

$\frac{T}{K}$	$\frac{\rho_{exp} \pm 2\sigma_{est}}{kg \cdot m^{-3}}$	$\frac{T}{K}$	$\frac{\rho_{exp} \pm 2\sigma_{est}}{kg \cdot m^{-3}}$
270.00	868.2 ± 1.5	293.15	850.1 ± 1.5
280.00	860.4 ± 1.4	298.15	846.2 ± 1.6
290.00	852.6 ± 1.4		

3-Methyl-4-penten-3-ol **[500060-36-6]** **$C_6H_{12}O$** **MW = 100.16** **530**

Table 1. Fit with estimated *B* coefficient for 2 accepted points. Deviation $\sigma_w = 2.090$.

Coefficient	$\rho = A + BT$
A	1096.92
B	–0.860

Table 2. Experimental values with uncertainties and deviation from calculated values.

$\frac{T}{\text{K}}$	$\frac{\rho_{exp} \pm 2\sigma_{est}}{\text{kg} \cdot \text{m}^{-3}}$	$\frac{\rho_{exp} - \rho_{calc}}{\text{kg} \cdot \text{m}^{-3}}$	Ref.
273.15	864.1 ± 1.5	2.09	50-fav/fri
295.15	841.0 ± 1.5	–2.09	50-fav/fri

Table 3. Recommended values.

$\frac{T}{\text{K}}$	$\frac{\rho_{exp} \pm 2\sigma_{est}}{\text{kg} \cdot \text{m}^{-3}}$
270.00	864.7 ± 2.9
280.00	856.1 ± 2.5
290.00	847.5 ± 2.6
293.15	844.8 ± 2.6
298.15	840.5 ± 2.9

2-Hepten-4-ol **[4798-59-8]** **$C_7H_{14}O$** **MW = 114.19** **531**

Table 1. Experimental value with uncertainty.

$\frac{T}{\text{K}}$	$\frac{\rho_{exp} \pm 2\sigma_{est}}{\text{kg} \cdot \text{m}^{-3}}$	Ref.
293.15	844.5 ± 0.6	63-lab

(*Z*)-3-Hepten-1-ol **[1708-81-2]** **$C_7H_{14}O$** **MW = 114.19** **532**

Table 1. Fit with estimated *B* coefficient for 3 accepted points. Deviation $\sigma_w = 0.249$.

Coefficient	$\rho = A + BT$
A	1064.56
B	–0.740

Table 2. Experimental values with uncertainties and deviation from calculated values.

$\frac{T}{\text{K}}$	$\frac{\rho_{exp} \pm 2\sigma_{est}}{\text{kg} \cdot \text{m}^{-3}}$	$\frac{\rho_{exp} - \rho_{calc}}{\text{kg} \cdot \text{m}^{-3}}$	Ref.
293.15	847.9 ± 0.7	0.27	61-hat/wei
298.15	843.6 ± 0.7	–0.33	61-hat/wei
303.15	840.3 ± 0.7	0.07	61-hat/wei

cont.

Table 3. Recommended values.

$\frac{T}{K}$	$\frac{\rho_{exp} \pm 2\sigma_{est}}{kg \cdot m^{-3}}$
290.00	850.0 ± 0.8
293.15	847.6 ± 0.7
298.15	843.9 ± 0.7
310.00	835.2 ± 0.9

4-Hepten-1-ol **[20851-55-2]** **$C_7H_{14}O$** **MW = 114.19** **533**

Table 1. Experimental value with uncertainty.

$\frac{T}{K}$	$\frac{\rho_{exp} \pm 2\sigma_{est}}{kg \cdot m^{-3}}$	Ref.
293.15	844.3 ± 2.0	64-ber/vav

5-Hepten-1-ol **[89794-36-5]** **$C_7H_{14}O$** **MW = 114.19** **534**

Table 1. Experimental value with uncertainty.

$\frac{T}{K}$	$\frac{\rho_{exp} \pm 2\sigma_{est}}{kg \cdot m^{-3}}$	Ref.
291.15	846.1 ± 1.0	56-gla/gau

2-Methyl-1-hexen-3-ol **[500029-09-4]** **$C_7H_{14}O$** **MW = 114.19** **535**

Table 1. Experimental value with uncertainty.

$\frac{T}{K}$	$\frac{\rho_{exp} \pm 2\sigma_{est}}{kg \cdot m^{-3}}$	Ref.
298.15	842.0 ± 2.0	65-col/des

2-Methyl-2-hexen-1-ol **[500029-07-2]** **$C_7H_{14}O$** **MW = 114.19** **536**

Table 1. Experimental value with uncertainty.

$\frac{T}{K}$	$\frac{\rho_{exp} \pm 2\sigma_{est}}{kg \cdot m^{-3}}$	Ref.
298.15	846.0 ± 2.0	65-col/des

3-Methyl-1-hexen-3-ol **[55145-28-3]** **$C_7H_{14}O$** **MW = 114.19** **537**

Table 1. Experimental value with uncertainty.

$\frac{T}{K}$	$\frac{\rho_{exp} \pm 2\sigma_{est}}{kg \cdot m^{-3}}$	Ref.
290.15	836.7 ± 1.0	46-shi

4-Methyl-1-hexen-4-ol **[500005-96-9]** **$C_7H_{14}O$** **MW = 114.19** **538**

Table 1. Experimental value with uncertainty.

T / K	$\rho_{exp} \pm 2\sigma_{est}$ / $kg \cdot m^{-3}$	Ref.
293.15	827.0 ± 0.6	63-lab

4-Methyl-4-hexen-3-ol **[101084-24-6]** **$C_7H_{14}O$** **MW = 114.19** **539**

Table 1. Experimental value with uncertainty.

T / K	$\rho_{exp} \pm 2\sigma_{est}$ / $kg \cdot m^{-3}$	Ref.
298.15	856.0 ± 2.0	50-pau/tch

5-Methyl-2-hexen-4-ol **[500025-43-4]** **$C_7H_{14}O$** **MW = 114.19** **540**

Table 1. Experimental value with uncertainty.

T / K	$\rho_{exp} \pm 2\sigma_{est}$ / $kg \cdot m^{-3}$	Ref.
293.15	842.0 ± 0.6	63-lab

1,1-Dimethyl-5-hexen-1-ol **[77437-98-0]** **$C_8H_{16}O$** **MW = 128.21** **541**

Table 1. Experimental value with uncertainty.

T / K	$\rho_{exp} \pm 2\sigma_{est}$ / $kg \cdot m^{-3}$	Ref.
293.15	842.3 ± 1.0	55-che/che

3,5-Dimethyl-4-hexen-3-ol **[1569-43-3]** **$C_8H_{16}O$** **MW = 128.21** **542**

Table 1. Fit with estimated B coefficient for 2 accepted points. Deviation $\sigma_w = 0.298$.

Coefficient	$\rho = A + BT$
A	1098.38
B	−0.820

Table 2. Experimental values with uncertainties and deviation from calculated values.

T / K	$\rho_{exp} \pm 2\sigma_{est}$ / $kg \cdot m^{-3}$	$\rho_{exp} - \rho_{calc}$ / $kg \cdot m^{-3}$	Ref.
273.15	874.7 ± 2.0	0.30	34-jac-6
290.35	860.0 ± 2.0	−0.30	34-jac-6

cont.

Table 3. Recommended values.

$\frac{T}{K}$	$\frac{\rho_{exp} \pm 2\sigma_{est}}{kg \cdot m^{-3}}$
270.00	877.0 ± 1.9
280.00	868.8 ± 1.8
290.00	860.6 ± 1.9
293.15	858.0 ± 1.9
298.15	853.9 ± 2.0

3-Ethyl-5-hexen-2-ol **[60091-37-4]** **$C_8H_{16}O$** **MW = 128.21** **543**

Table 1. Experimental values with uncertainties.

$\frac{T}{K}$	$\frac{\rho_{exp} \pm 2\sigma_{est}}{kg \cdot m^{-3}}$	Ref.
289.45	854.0 ± 2.0	47-col/lag
289.15	854.0 ± 2.0	49-col/lag

6-Methyl-3-hepten-2-ol **[51500-48-2]** **$C_8H_{16}O$** **MW = 128.21** **544**

Table 1. Experimental value with uncertainty.

$\frac{T}{K}$	$\frac{\rho_{exp} \pm 2\sigma_{est}}{kg \cdot m^{-3}}$	Ref.
293.15	830.1 ± 2.0	62-mir/fed

6-Methyl-5-hepten-2-ol **[1569-60-4]** **$C_8H_{16}O$** **MW = 128.21** **545**

Table 1. Experimental value with uncertainty.

$\frac{T}{K}$	$\frac{\rho_{exp} \pm 2\sigma_{est}}{kg \cdot m^{-3}}$	Ref.
293.15	858.8 ± 2.0	62-mir/fed

2-Octen-4-ol **[4798-61-2]** **$C_8H_{16}O$** **MW = 128.21** **546**

Table 1. Experimental value with uncertainty.

$\frac{T}{K}$	$\frac{\rho_{exp} \pm 2\sigma_{est}}{kg \cdot m^{-3}}$	Ref.
293.15	847.5 ± 0.6	63-lab

2,2-Dimethyl-6-hepten-3-ol **[54525-85-8]** **$C_9H_{18}O$** **MW = 142.24** **547**

Table 1. Experimental values with uncertainties.

$\frac{T}{K}$	$\frac{\rho_{exp} \pm 2\sigma_{est}}{kg \cdot m^{-3}}$	Ref.
294.15	843.0 ± 2.0	47-col/lag
297.15	843.0 ± 2.0	49-col/lag

2,4-Dimethyl-2-hepten-4-ol **[59673-20-0]** $C_9H_{18}O$ **MW = 142.24** **548**

Table 1. Fit with estimated B coefficient for 2 accepted points. Deviation $\sigma_w = 0.176$.

Coefficient	$\rho = A + BT$
A	1076.98
B	−0.780

Table 2. Experimental values with uncertainties and deviation from calculated values.

$\frac{T}{K}$	$\frac{\rho_{exp} \pm 2\sigma_{est}}{kg \cdot m^{-3}}$	$\frac{\rho_{exp} - \rho_{calc}}{kg \cdot m^{-3}}$	Ref.
273.15	864.1 ± 2.0	0.18	34-jac-6
289.75	850.8 ± 2.0	−0.18	34-jac-6

Table 3. Recommended values.

$\frac{T}{K}$	$\frac{\rho_{exp} \pm 2\sigma_{est}}{kg \cdot m^{-3}}$
270.00	866.4 ± 1.9
280.00	858.6 ± 1.8
290.00	850.8 ± 1.8

(*E*)-2-Methyl-3-octen-2-ol **[18521-06-7]** $C_9H_{18}O$ **MW = 142.24** **549**

Table 1. Experimental value with uncertainty.

$\frac{T}{K}$	$\frac{\rho_{exp} \pm 2\sigma_{est}}{kg \cdot m^{-3}}$	Ref.
293.15	830.1 ± 0.5	41-cam/eby-1

(*Z*)-2-Methyl-3-octen-2-ol **[18521-07-8]** $C_9H_{18}O$ **MW = 142.24** **550**

Table 1. Experimental value with uncertainty.

$\frac{T}{K}$	$\frac{\rho_{exp} \pm 2\sigma_{est}}{kg \cdot m^{-3}}$	Ref.
293.15	837.8 ± 0.5	41-cam/eby-1

4-Methyl-3-octen-5-ol **[500025-37-6]** $C_9H_{18}O$ **MW = 142.24** **551**

Table 1. Experimental value with uncertainty.

$\frac{T}{K}$	$\frac{\rho_{exp} \pm 2\sigma_{est}}{kg \cdot m^{-3}}$	Ref.
298.15	846.8 ± 0.8	12-bje

4-Methyl-4-octen-1-ol **[500000-06-6]** $C_9H_{18}O$ **MW = 142.24** **552**

Table 1. Experimental value with uncertainty.

$\frac{T}{K}$	$\frac{\rho_{exp} \pm 2\sigma_{est}}{kg \cdot m^{-3}}$	Ref.
298.15	852.0 ± 2.0	50-pau/tch

2,4-Dimethyl-2-octen-4-ol [76008-28-1] $C_{10}H_{20}O$ MW = 156.27 **553**

Table 1. Fit with estimated *B* coefficient for 2 accepted points. Deviation $\sigma_w = 0.120$.

Coefficient	$\rho = A + BT$
A	1078.70
B	−0.800

Table 2. Experimental values with uncertainties and deviation from calculated values.

$\frac{T}{K}$	$\frac{\rho_{exp} \pm 2\sigma_{est}}{kg \cdot m^{-3}}$	$\frac{\rho_{exp} - \rho_{calc}}{kg \cdot m^{-3}}$	Ref.
273.15	860.3 ± 2.0	0.12	34-jac-6
292.35	844.7 ± 2.0	−0.12	34-jac-6

Table 3. Recommended values.

$\frac{T}{K}$	$\frac{\rho_{exp} \pm 2\sigma_{est}}{kg \cdot m^{-3}}$
270.00	862.7 ± 1.9
280.00	854.7 ± 1.8
290.00	846.7 ± 1.8
293.15	844.2 ± 1.9
298.15	840.2 ± 2.0

3,5-Dimethyl-6-octen-2-ol [57785-04-3] $C_{10}H_{20}O$ MW = 156.27 **554**

Table 1. Experimental value with uncertainty.

$\frac{T}{K}$	$\frac{\rho_{exp} \pm 2\sigma_{est}}{kg \cdot m^{-3}}$	Ref.
293.15	864.00 ± 1.00	75-lee/che

3,7-Dimethyl-6-octen-1-ol [106-22-9] $C_{10}H_{20}O$ MW = 156.27 **555**

Table 1. Experimental and recommended values with uncertainties.

$\frac{T}{K}$	$\frac{\rho_{exp} \pm 2\sigma_{est}}{kg \cdot m^{-3}}$	Ref.
293.15	890.0 ± 6.0	1884-gla[1)]
293.15	865.3 ± 1.0	54-ser/voi-1
293.15	865.3 ± 1.0	Recommended

[1)] Not included in calculation of recommended value.

2,2,3-Trimethyl-6-hepten-3-ol [85924-69-2] $C_{10}H_{20}O$ MW = 156.27 **556**

Table 1. Experimental value with uncertainty.

$\frac{T}{K}$	$\frac{\rho_{exp} \pm 2\sigma_{est}}{kg \cdot m^{-3}}$	Ref.
294.15	862.0 ± 2.0	49-col/lag

2,3,4-Trimethyl-5-hepten-2-ol **[57785-05-4]** **$C_{10}H_{20}O$** **MW = 156.27** **557**

Table 1. Experimental value with uncertainty.

$\frac{T}{K}$	$\frac{\rho_{exp} \pm 2\sigma_{est}}{kg \cdot m^{-3}}$	Ref.
293.15	875.0 ± 1.0	75-lee/che

3-Ethyl-1-nonen-4-ol **[10544-97-5]** **$C_{11}H_{22}O$** **MW = 170.30** **558**

Table 1. Experimental value with uncertainty.

$\frac{T}{K}$	$\frac{\rho_{exp} \pm 2\sigma_{est}}{kg \cdot m^{-3}}$	Ref.
293.15	843.0 ± 2.0	65-mig/mig

3,4,8-Trimethyl-1-nonen-3-ol **[18352-66-4]** **$C_{12}H_{24}O$** **MW = 184.32** **559**

Table 1. Experimental value with uncertainty.

$\frac{T}{K}$	$\frac{\rho_{exp} \pm 2\sigma_{est}}{kg \cdot m^{-3}}$	Ref.
293.15	852.1 ± 1.0	67-min/che

(*E*)-9-Octadecen-1-ol **[506-42-3]** **$C_{18}H_{36}O$** **MW = 268.48** **560**

Table 1. Experimental value with uncertainty.

$\frac{T}{K}$	$\frac{\rho_{exp} \pm 2\sigma_{est}}{kg \cdot m^{-3}}$	Ref.
313.15	833.8 ± 1.0	26-boe/bel

(*Z*)-9-Octadecen-1-ol **[143-28-2]** **$C_{18}H_{36}O$** **MW = 268.48** **561**

Table 1. Experimental value with uncertainty.

$\frac{T}{K}$	$\frac{\rho_{exp} \pm 2\sigma_{est}}{kg \cdot m^{-3}}$	Ref.
313.15	836.7 ± 1.0	26-boe/bel

3,7,11-Trimethyl-2-hexadecen-1-ol **[102013-46-7]** **$C_{19}H_{38}O$** **MW = 282.51** **562**

Table 1. Experimental value with uncertainty.

$\frac{T}{K}$	$\frac{\rho_{exp} \pm 2\sigma_{est}}{kg \cdot m^{-3}}$	Ref.
293.15	856.1 ± 2.0	60-naz/mak

3,7,11,15-Tetramethyl-1-hexadecen-3-ol [60046-87-9] $C_{20}H_{40}O$ MW = 296.54 563

Table 1. Coefficients of the polynomial expansion equation.
Standard deviations (see introduction):
$\sigma_{c,w} = 2.9464 \cdot 10^{-1}$ (combined temperature ranges, weighted),
$\sigma_{c,uw} = 8.4869 \cdot 10^{-2}$ (combined temperature ranges, unweighted).

Coefficient	T = 293.15 to 343.15 K $\rho = A + BT + CT^2 + DT^3 + \ldots$
A	$1.06651 \cdot 10^3$
B	$-7.53664 \cdot 10^{-1}$

Table 2. Experimental values with uncertainties and deviation from calculated values.

$\frac{T}{\text{K}}$	$\frac{\rho_{exp} \pm 2\sigma_{est}}{\text{kg} \cdot \text{m}^{-3}}$	$\frac{\rho_{exp} - \rho_{calc}}{\text{kg} \cdot \text{m}^{-3}}$	Ref. (Symbol in Fig. 1)	$\frac{T}{\text{K}}$	$\frac{\rho_{exp} \pm 2\sigma_{est}}{\text{kg} \cdot \text{m}^{-3}}$	$\frac{\rho_{exp} - \rho_{calc}}{\text{kg} \cdot \text{m}^{-3}}$	Ref. (Symbol in Fig. 1)
293.15	845.40 ± 0.60	−0.18	62-mau/smi(○)	293.15	845.80 ± 0.50	0.22	88-bag/gur(□)
293.15	845.80 ± 0.50	0.22	85-bel/ber(Δ)	303.15	837.80 ± 0.50	−0.24	88-bag/gur(□)
303.15	837.80 ± 0.50	−0.24	85-bel/ber(Δ)	313.15	830.50 ± 0.50	−0.00	88-bag/gur(□)
313.15	830.50 ± 0.50	−0.00	85-bel/ber(Δ)	323.15	822.80 ± 0.50	−0.17	88-bag/gur(□)
323.15	822.80 ± 0.50	−0.17	85-bel/ber(Δ)	333.15	816.00 ± 0.50	0.57	88-bag/gur(□)
333.15	816.00 ± 0.50	0.57	85-bel/ber(Δ)	343.15	807.60 ± 0.50	−0.29	88-bag/gur(□)
343.15	807.60 ± 0.50	−0.29	85-bel/ber(Δ)				

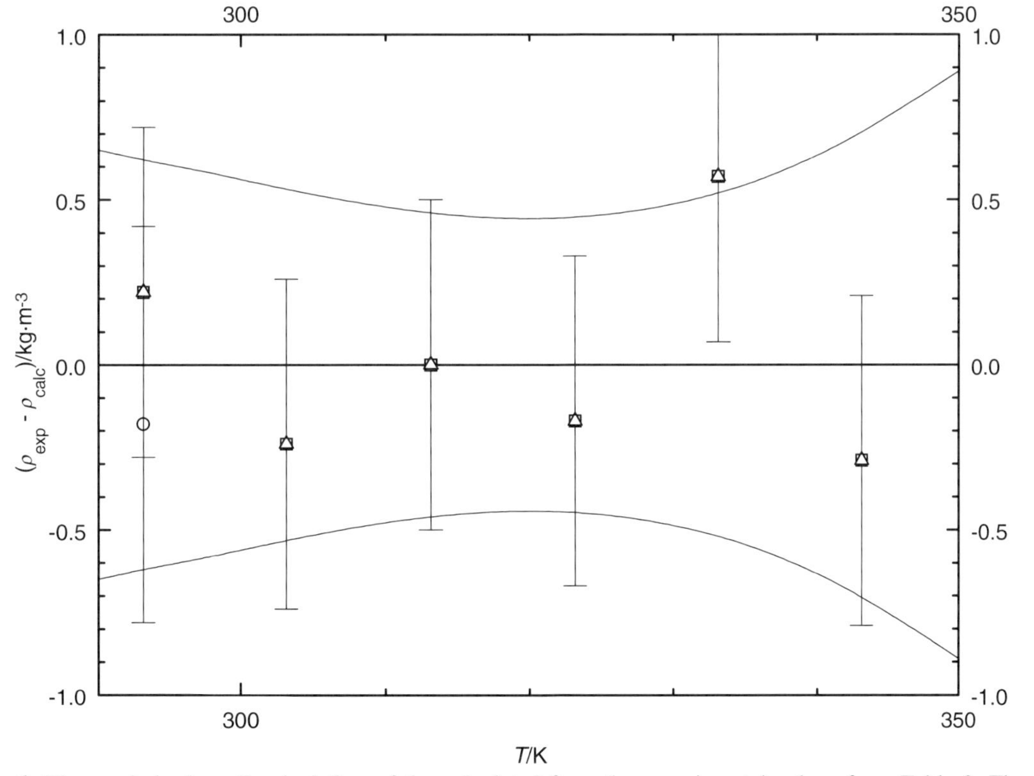

Fig. 1. The symbols show the deviation of the calculated from the experimental values from Table 2. The curves above and below the zero line indicate the calculated error region of the recommended values given in Table 3. The error bars represent the experimental errors. (Error bars smaller than the symbols are omitted for clarity of the figure.)

cont.

3,7,11,15-Tetramethyl-1-hexadecen-3-ol (cont.)

Table 3. Recommended values (fit to the reliable experimental values according to the equations $\rho = A + BT + CT^2 + DT^3 + \dots$ or $\rho = [1 + 1.75(1 - T/T_c)^{1/3} + 0.75(1 - T/T_c)][\rho_c + A(T_c - T) + B(T_c - T)^2 + C(T_c - T)^3 + D(T_c - T)^4]$).

$\frac{T}{\text{K}}$	$\frac{\rho \pm \sigma_{fit}}{\text{kg} \cdot \text{m}^{-3}}$	$\frac{T}{\text{K}}$	$\frac{\rho \pm \sigma_{fit}}{\text{kg} \cdot \text{m}^{-3}}$	$\frac{T}{\text{K}}$	$\frac{\rho \pm \sigma_{fit}}{\text{kg} \cdot \text{m}^{-3}}$
290.00	847.95 ± 0.65	300.00	840.41 ± 0.56	330.00	817.80 ± 0.47
293.15	845.58 ± 0.62	310.00	832.88 ± 0.47	340.00	810.27 ± 0.61
298.15	841.81 ± 0.58	320.00	825.34 ± 0.43	350.00	802.73 ± 0.89

3,7,11,15-Tetramethyl-2-heptadecen-1-ol, **[114161-31-8]** **$C_{21}H_{42}O$** **MW = 310.56** **564**

Table 1. Experimental value with uncertainty.

$\frac{T}{\text{K}}$	$\frac{\rho_{exp} \pm 2\sigma_{est}}{\text{kg} \cdot \text{m}^{-3}}$	Ref.
293.15	858.6 ± 2.0	60-naz/mak

3,7,11,15,16,16-Hexamethyl-1-heptadecen-3-ol **[113058-51-8]** **$C_{23}H_{46}O$** **MW = 338.62** **565**

Table 1. Experimental value with uncertainty.

$\frac{T}{\text{K}}$	$\frac{\rho_{exp} \pm 2\sigma_{est}}{\text{kg} \cdot \text{m}^{-3}}$	Ref.
293.15	851.9 ± 2.0	60-naz/mak

3,7,11,15,16,16-Hexamethyl-2-heptadecen-1-ol **[113057-97-9]** **$C_{23}H_{46}O$** **MW = 338.62** **566**

Table 1. Experimental value with uncertainty.

$\frac{T}{\text{K}}$	$\frac{\rho_{exp} \pm 2\sigma_{est}}{\text{kg} \cdot \text{m}^{-3}}$	Ref.
293.15	862.8 ± 2.0	60-naz/mak

2.2.2 Alcohols of General Formula $C_nH_{2n-2}O$

2-Propyn-1-ol **[107-19-7]** **C_3H_4O** **MW = 56.06** **567**

Table 1. Experimental and recommended values with uncertainties.

T/K	$\rho_{exp} \pm 2\sigma_{est}$/kg·m^{-3}	Ref.	T/K	$\rho_{exp} \pm 2\sigma_{est}$/kg·m^{-3}	Ref.
293.15	971.5 ± 5.00	1880-bru-3[1)]	293.15	983.5 ± 5.00	1880-bru-3[1)]
291.95	971.0 ± 5.00	1880-bru-3[1)]	293.15	947.6 ± 1.00	45-hen/gre[1)]
294.15	972.8 ± 5.00	1880-bru-3[1)]	298.15	945.2 ± 0.60	95-kum/rao-1
293.15	983.5 ± 5.00	1880-bru-3[1)]	298.15	945.2 ± 0.60	Recommended

[1)] Not included in calculation of recommended value.

3-Butyn-1-ol **[927-74-2]** **C_4H_6O** **MW = 70.09** **568**

Table 1. Experimental value with uncertainty.

T/K	$\rho_{exp} \pm 2\sigma_{est}$/kg·m^{-3}	Ref.
293.15	925.5 ± 0.5	45-hen/gre

2-Methyl-3-butyn-2-ol **[115-19-5]** **C_5H_8O** **MW = 84.12** **569**

Table 1. Experimental and recommended values with uncertainties.

T/K	$\rho_{exp} \pm 2\sigma_{est}$/kg·m^{-3}	Ref.
293.15	861.3 ± 0.6	41-cam/eby-1
293.15	861.2 ± 0.6	45-hen/gre
293.15	865.8 ± 2.0	62-mau/smi[1)]
293.15	861.3 ± 0.6	Recommended

[1)] Not included in calculation of recommended value.

3,4-Pentadien-2-ol **[17615-19-9]** **C_5H_8O** **MW = 84.12** **570**

Table 1. Experimental value with uncertainty.

T/K	$\rho_{exp} \pm 2\sigma_{est}$/kg·m^{-3}	Ref.
298.15	901.4 ± 2.0	62-ber/leg-1

2-Pentyn-1-ol **[6261-22-9]** **C_5H_8O** **MW = 84.12** **571**

Table 1. Experimental value with uncertainty.

T/K	$\rho_{exp} \pm 2\sigma_{est}$/kg·m^{-3}	Ref.
293.15	907.7 ± 1.0	64-zak/sta

4-Pentyn-1-ol **[5390-04-5]** **C_5H_8O** **MW = 84.12** **572**

Table 1. Experimental value with uncertainty.

T / K	$\rho_{exp} \pm 2\sigma_{est}$ / $kg \cdot m^{-3}$	Ref.
293.15	913.2 ± 2.0	55-mos

(*RS*)-4-Pentyn-2-ol **[2117-11-5]** **C_5H_8O** **MW = 84.12** **573**

Table 1. Experimental and recommended values with uncertainties.

T / K	$\rho_{exp} \pm 2\sigma_{est}$ / $kg \cdot m^{-3}$	Ref.
293.15	895.7 ± 0.6	54-pom/foo
293.15	905.0 ± 2.0	62-ber/leg-1 [1)]
293.15	895.7 ± 0.6	Recommended

[1)] Not included in calculation of recommended value.

3,3-Dimethyl-2-butyn-1-ol **[500029-63-0]** **$C_6H_{10}O$** **MW = 98.14** **574**

Table 1. Fit with estimated *B* coefficient for 3 accepted points. Deviation $\sigma_w = 0.047$.

Coefficient	$\rho = A + BT$
A	1074.80
B	−0.740

Table 2. Experimental values with uncertainties and deviation from calculated values.

T / K	$\rho_{exp} \pm 2\sigma_{est}$ / $kg \cdot m^{-3}$	$\rho_{exp} - \rho_{calc}$ / $kg \cdot m^{-3}$	Ref.
293.15	857.8 ± 0.7	−0.07	61-hat/wei
298.15	854.2 ± 0.7	0.03	61-hat/wei
303.15	850.5 ± 0.7	0.03	61-hat/wei

Table 3. Recommended values.

T / K	$\rho_{exp} \pm 2\sigma_{est}$ / $kg \cdot m^{-3}$	T / K	$\rho_{exp} \pm 2\sigma_{est}$ / $kg \cdot m^{-3}$
290.00	860.2 ± 0.7	298.15	854.2 ± 0.6
293.15	857.9 ± 0.7	310.00	845.4 ± 0.9

4,5-Hexadien-3-ol **[4376-43-6]** **$C_6H_{10}O$** **MW = 98.14** **575**

Table 1. Experimental value with uncertainty.

T / K	$\rho_{exp} \pm 2\sigma_{est}$ / $kg \cdot m^{-3}$	Ref.
293.15	880.7 ± 2.0	62-ber/leg-1

5-Hexyn-1-ol **[928-90-5]** **$C_6H_{10}O$** **MW = 98.14** **576**

Table 1. Experimental value with uncertainty.

$\frac{T}{K}$	$\frac{\rho_{exp} \pm 2\sigma_{est}}{kg \cdot m^{-3}}$	Ref.
293.15	902.0 ± 1.0	64-zak/sta

5-Hexyn-3-ol **[19780-84-8]** **$C_6H_{10}O$** **MW = 98.14** **577**

Table 1. Experimental value with uncertainty.

$\frac{T}{K}$	$\frac{\rho_{exp} \pm 2\sigma_{est}}{kg \cdot m^{-3}}$	Ref.
293.15	893.1 ± 2.0	62-ber/leg-1

3-Ethyl-1-pentyn-3-ol **[6285-06-9]** **$C_7H_{12}O$** **MW = 112.17** **578**

Table 1. Experimental value with uncertainty.

$\frac{T}{K}$	$\frac{\rho_{exp} \pm 2\sigma_{est}}{kg \cdot m^{-3}}$	Ref.
293.15	871.7 ± 0.6	45-hen/gre

1,2-Heptadien-4-ol **[4376-46-9]** **$C_7H_{12}O$** **MW = 112.17** **579**

Table 1. Experimental value with uncertainty.

$\frac{T}{K}$	$\frac{\rho_{exp} \pm 2\sigma_{est}}{kg \cdot m^{-3}}$	Ref.
294.15	868.8 ± 2.0	62-ber/leg-1

1-Heptyn-4-ol **[22127-83-9]** **$C_7H_{12}O$** **MW = 112.17** **580**

Table 1. Experimental value with uncertainty.

$\frac{T}{K}$	$\frac{\rho_{exp} \pm 2\sigma_{est}}{kg \cdot m^{-3}}$	Ref.
291.15	886.7 ± 2.0	62-ber/leg-1

2-Heptyn-1-ol **[1002-36-4]** **$C_7H_{12}O$** **MW = 112.17** **581**

Table 1. Fit with estimated *B* coefficient for 3 accepted points. Deviation σ_w = 0.062.

Coefficient	$\rho = A + BT$
A	1106.75
B	–0.750

cont.

2-Heptyn-1-ol (cont.)

Table 2. Experimental values with uncertainties and deviation from calculated values.

$\frac{T}{K}$	$\frac{\rho_{exp} \pm 2\sigma_{est}}{kg \cdot m^{-3}}$	$\frac{\rho_{exp} - \rho_{calc}}{kg \cdot m^{-3}}$	Ref.
293.15	886.8 ± 0.7	−0.08	61-hat/wei
298.15	883.2 ± 0.7	0.07	61-hat/wei
303.15	879.4 ± 0.7	0.02	61-hat/wei
293.15	885.8 ± 1.0	−1.08	64-zak/sta[1)]

[1)] Not included in calculation of linear coefficients.

Table 3. Recommended values.

$\frac{T}{K}$	$\frac{\rho_{exp} \pm 2\sigma_{est}}{kg \cdot m^{-3}}$
290.00	889.2 ± 0.6
293.15	886.9 ± 0.5
298.15	883.1 ± 0.5
310.00	874.2 ± 0.8

2-Methyl-3,5-hexadien-2-ol **[926-38-5]** **$C_7H_{12}O$** **MW = 112.17** **582**

Table 1. Experimental value with uncertainty.

$\frac{T}{K}$	$\frac{\rho_{exp} \pm 2\sigma_{est}}{kg \cdot m^{-3}}$	Ref.
293.15	864.9 ± 2.0	64-bog/kug

2-Methyl-4,5-hexadien-3-ol **[4376-49-2]** **$C_7H_{12}O$** **MW = 112.17** **583**

Table 1. Experimental value with uncertainty.

$\frac{T}{K}$	$\frac{\rho_{exp} \pm 2\sigma_{est}}{kg \cdot m^{-3}}$	Ref.
294.15	872.2 ± 2.0	62-ber/leg-1

5-Methyl-3,5-hexadien-1-ol **[19764-79-5]** **$C_7H_{12}O$** **MW = 112.17** **584**

Table 1. Experimental value with uncertainty.

$\frac{T}{K}$	$\frac{\rho_{exp} \pm 2\sigma_{est}}{kg \cdot m^{-3}}$	Ref.
293.15	889.9 ± 1.0	67-min/che

2-Methyl-5-hexyn-3-ol **[54838-77-6]** **$C_7H_{12}O$** **MW = 112.17** **585**

Table 1. Experimental value with uncertainty.

$\frac{T}{K}$	$\frac{\rho_{exp} \pm 2\sigma_{est}}{kg \cdot m^{-3}}$	Ref.
293.15	874.9 ± 2.0	62-ber/leg-1

3-Methyl-1-hexyn-3-ol **[4339-05-3]** **$C_7H_{12}O$** **MW = 112.17** **586**

Table 1. Experimental value with uncertainty.

T / K	$\rho_{exp} \pm 2\sigma_{est}$ / $kg \cdot m^{-3}$	Ref.
290.15	867.1 ± 1.0	46-shi

3,5-Dimethyl-1-hexyn-3-ol **[107-54-0]** **$C_8H_{14}O$** **MW = 126.2** **587**

Table 1. Experimental value with uncertainty.

T / K	$\rho_{exp} \pm 2\sigma_{est}$ / $kg \cdot m^{-3}$	Ref.
293.15	858.2 ± 1.0	57-hic/ken

2-Methyl-2,3-heptadien-1-ol **[14270-80-5]** **$C_8H_{14}O$** **MW = 126.20** **588**

Table 1. Experimental value with uncertainty.

T / K	$\rho_{exp} \pm 2\sigma_{est}$ / $kg \cdot m^{-3}$	Ref.
293.15	875.1 ± 1.0	66-per/bal

3-Methyl-1,6-heptadien-4-ol **[1838-74-0]** **$C_8H_{14}O$** **MW = 126.20** **589**

Table 1. Experimental value with uncertainty.

T / K	$\rho_{exp} \pm 2\sigma_{est}$ / $kg \cdot m^{-3}$	Ref.
298.15	868.0 ± 1.0	65-col/des

5-Methyl-3,5-heptadien-1-ol **[19756-78-6]** **$C_8H_{14}O$** **MW = 126.20** **590**

Table 1. Experimental value with uncertainty.

T / K	$\rho_{exp} \pm 2\sigma_{est}$ / $kg \cdot m^{-3}$	Ref.
293.15	898.2 ± 1.0	67-min/che

2,6-Dimethyl-3-heptyn-5-ol **[5923-00-2]** **$C_9H_{16}O$** **MW = 140.23** **591**

Table 1. Experimental value with uncertainty.

T / K	$\rho_{exp} \pm 2\sigma_{est}$ / $kg \cdot m^{-3}$	Ref.
293.15	848.2 ± 2.0	65-fav/nik

3-Methyl-3,4-octadien-2-ol **[14270-81-6]** **$C_9H_{16}O$** **MW = 140.23** **592**

Table 1. Experimental value with uncertainty.

T / K	$\rho_{exp} \pm 2\sigma_{est}$ / $kg \cdot m^{-3}$	Ref.
293.15	860.2 ± 1.0	66-per/bal

5-Methyl-3,5-octadien-1-ol **[19756-79-7]** **$C_9H_{16}O$** **MW = 140.23** **593**

Table 1. Experimental value with uncertainty.

$\frac{T}{K}$	$\frac{\rho_{exp} \pm 2\sigma_{est}}{kg \cdot m^{-3}}$	Ref.
293.15	890.2 ± 1.0	67-min/che

6-Methyl-2,6-octadien-8-ol **[900002-84-8]** **$C_9H_{16}O$** **MW = 140.23** **594**

Table 1. Experimental value with uncertainty.

$\frac{T}{K}$	$\frac{\rho_{exp} \pm 2\sigma_{est}}{kg \cdot m^{-3}}$	Ref.
293.15	877.0 ± 2.0	60-naz/mak

2-Methyl-3-octyn-2-ol **[20599-16-0]** **$C_9H_{16}O$** **MW = 140.23** **595**

Table 1. Experimental value with uncertainty.

$\frac{T}{K}$	$\frac{\rho_{exp} \pm 2\sigma_{est}}{kg \cdot m^{-3}}$	Ref.
293.15	850.6 ± 0.5	41-cam/eby-1

2-Methyl-3-octyn-5-ol **[5922-99-6]** **$C_9H_{16}O$** **MW = 140.23** **596**

Table 1. Experimental value with uncertainty.

$\frac{T}{K}$	$\frac{\rho_{exp} \pm 2\sigma_{est}}{kg \cdot m^{-3}}$	Ref.
293.15	851.8 ± 2.0	65-fav/nik

5,8-Nonadien-2-ol **[13175-61-6]** **$C_9H_{16}O$** **MW = 140.23** **597**

Table 1. Experimental value with uncertainty.

$\frac{T}{K}$	$\frac{\rho_{exp} \pm 2\sigma_{est}}{kg \cdot m^{-3}}$	Ref.
298.15	1086.5 ± 2.0	63-col/bue

4-Nonyn-3-ol **[999-70-2]** **$C_9H_{16}O$** **MW = 140.23** **598**

Table 1. Experimental value with uncertainty.

$\frac{T}{K}$	$\frac{\rho_{exp} \pm 2\sigma_{est}}{kg \cdot m^{-3}}$	Ref.
293.15	866.1 ± 0.7	63-lab

2,3-Dimethyl-3,4-octadien-2-ol **[14129-51-2]** **$C_{10}H_{18}O$** **MW = 154.25** **599**

Table 1. Experimental value with uncertainty.

$\frac{T}{K}$	$\frac{\rho_{exp} \pm 2\sigma_{est}}{kg \cdot m^{-3}}$	Ref.
293.15	867.8 ± 1.0	66-per/bal

(*E*)-3,7-Dimethyl-2,6-octadien-1-ol [106-24-1] $C_{10}H_{18}O$ MW = 154.25 **600**

Table 1. Experimental value with uncertainty.

$\frac{T}{K}$	$\frac{\rho_{exp} \pm 2\sigma_{est}}{kg \cdot m^{-3}}$	Ref.
298.15	879.2 ± 0.7	65-rum

(*Z*)-3,7-Dimethyl-2,6-octadien-1-ol [106-25-2] $C_{10}H_{18}O$ MW = 154.25 **601**

Table 1. Experimental value with uncertainty.

$\frac{T}{K}$	$\frac{\rho_{exp} \pm 2\sigma_{est}}{kg \cdot m^{-3}}$	Ref.
298.15	873.3 ± 0.7	65-rum

3,7-Dimethyl-1-octyn-3-ol [1604-26-8] $C_{10}H_{18}O$ MW = 154.25 **602**

Table 1. Experimental value with uncertainty.

$\frac{T}{K}$	$\frac{\rho_{exp} \pm 2\sigma_{est}}{kg \cdot m^{-3}}$	Ref.
293.15	852.1 ± 2.0	62-mau/smi

8-Methyl-5,8-nonadien-2-ol [13175-62-7] $C_{10}H_{18}O$ MW = 154.25 **603**

Table 1. Experimental value with uncertainty.

$\frac{T}{K}$	$\frac{\rho_{exp} \pm 2\sigma_{est}}{kg \cdot m^{-3}}$	Ref.
298.15	882.0 ± 2.0	63-col/bue

3-Propyl-1,2-heptadien-4-ol [900002-85-9] $C_{10}H_{18}O$ MW = 154.25 **604**

Table 1. Experimental value with uncertainty.

$\frac{T}{K}$	$\frac{\rho_{exp} \pm 2\sigma_{est}}{kg \cdot m^{-3}}$	Ref.
297.15	863.2 ± 2.0	62-ber/leg-1

3,5-Diethyl-1,6-heptadine-4-ol [10545-05-8] $C_{11}H_{20}O$ MW = 168.28 **605**

Table 1. Experimental value with uncertainty.

$\frac{T}{K}$	$\frac{\rho_{exp} \pm 2\sigma_{est}}{kg \cdot m^{-3}}$	Ref.
293.15	892.0 ± 2.0	65-mig/mig

3,7-Dimethyl-2,6-nonadien-1-ol [41865-30-9] $C_{11}H_{20}O$ MW = 168.28 **606**

Table 1. Experimental value with uncertainty.

$\frac{T}{K}$	$\frac{\rho_{exp} \pm 2\sigma_{est}}{kg \cdot m^{-3}}$	Ref.
293.15	881.0 ± 2.0	60-naz/mak

3-Ethenyl-1-nonen-4-ol **[13014-73-8]** **$C_{11}H_{20}O$** **MW = 168.28** **607**

Table 1. Experimental value with uncertainty.

$\frac{T}{K}$	$\frac{\rho_{exp} \pm 2\sigma_{est}}{kg \cdot m^{-3}}$	Ref.
293.15	854.0 ± 2.0	65-mig/mig

10-Undecyn-1-ol **[2774-84-7]** **$C_{11}H_{20}O$** **MW = 168.28** **608**

Table 1. Experimental value with uncertainty.

$\frac{T}{K}$	$\frac{\rho_{exp} \pm 2\sigma_{est}}{kg \cdot m^{-3}}$	Ref.
293.15	873.3 ± 1.0	62-ber/mol

3,4,8-Trimethyl-1-nonyn-3-ol **[18352-64-2]** **$C_{12}H_{22}O$** **MW = 182.31** **609**

Table 1. Experimental value with uncertainty.

$\frac{T}{K}$	$\frac{\rho_{exp} \pm 2\sigma_{est}}{kg \cdot m^{-3}}$	Ref.
293.15	858.6 ± 1.0	67-min/che

3,7,8-Trimethyl-2,6-nonadien-1-ol **[105906-02-3]** **$C_{12}H_{22}O$** **MW = 182.31** **610**

Table 1. Experimental value with uncertainty.

$\frac{T}{K}$	$\frac{\rho_{exp} \pm 2\sigma_{est}}{kg \cdot m^{-3}}$	Ref.
293.15	879.5 ± 2.0	60-naz/mak

2,8-Dimethyl-3-ethyl-3,4-nonadien-2-ol **[14270-84-9]** **$C_{13}H_{24}O$** **MW = 196.33** **611**

Table 1. Experimental value with uncertainty.

$\frac{T}{K}$	$\frac{\rho_{exp} \pm 2\sigma_{est}}{kg \cdot m^{-3}}$	Ref.
293.15	830.0 ± 1.0	66-per/bal

3-(1,1-Dimethylethyl)-2-methyl-3,4-octadien-2-ol **[14129-52-3]** **$C_{13}H_{24}O$** **MW = 196.33** **612**

Table 1. Experimental value with uncertainty.

$\frac{T}{K}$	$\frac{\rho_{exp} \pm 2\sigma_{est}}{kg \cdot m^{-3}}$	Ref.
293.15	833.9 ± 1.0	66-per/bal

3-Ethyl-2,7,7-trimethyl-3,4-octadien-2-ol [14270-83-8] $C_{13}H_{24}O$ MW = 196.33 613

Table 1. Experimental value with uncertainty.

T / K	$\rho_{exp} \pm 2\sigma_{est}$ / $kg \cdot m^{-3}$	Ref.
293.15	849.3 ± 1.0	66-per/bal

5-Tridecyn-7-ol [1846-65-7] $C_{13}H_{24}O$ MW = 196.33 614

Table 1. Experimental value with uncertainty.

T / K	$\rho_{exp} \pm 2\sigma_{est}$ / $kg \cdot m^{-3}$	Ref.
293.15	854.6 ± 0.6	63-lab

5-Pentadecyn-7-ol [92857-08-4] $C_{15}H_{28}O$ MW = 224.39 615

Table 1. Experimental value with uncertainty.

T / K	$\rho_{exp} \pm 2\sigma_{est}$ / $kg \cdot m^{-3}$	Ref.
293.15	853.6 ± 0.7	63-lab

3,7,11-Trimethyl-1-dodecyn-3-ol [1604-35-9] $C_{15}H_{28}O$ MW = 224.39 616

Table 1. Coefficients of the polynomial expansion equation. Standard deviations (see introduction):
$\sigma_{c,w} = 3.7211 \cdot 10^{-1}$ (combined temperature ranges, weighted),
$\sigma_{c,uw} = 1.5548 \cdot 10^{-1}$ (combined temperature ranges, unweighted).

Coefficient	T = 293.00 to 343.15 K $\rho = A + BT + CT^2 + DT^3 + \dots$
A	$1.07542 \cdot 10^{3}$
B	$-7.63342 \cdot 10^{-1}$

Table 2. Experimental values with uncertainties and deviation from calculated values.

T / K	$\rho_{exp} \pm 2\sigma_{est}$ / $kg \cdot m^{-3}$	$\rho_{exp} - \rho_{calc}$ / $kg \cdot m^{-3}$	Ref. (Symbol in Fig. 1)	T / K	$\rho_{exp} \pm 2\sigma_{est}$ / $kg \cdot m^{-3}$	$\rho_{exp} - \rho_{calc}$ / $kg \cdot m^{-3}$	Ref. (Symbol in Fig. 1)
293.15	851.00 ± 1.00	−0.65	62-mau/smi(○)	313.15	836.10 ± 0.50	−0.28	88-bag/gur(□)
293.00	851.90 ± 0.50	0.14	86-bae(×)	323.15	828.80 ± 0.50	0.05	88-bag/gur(□)
293.15	851.90 ± 0.50	0.25	88-bag/gur(□)	333.15	821.90 ± 0.50	0.79	88-bag/gur(□)
303.15	844.10 ± 0.50	0.09	88-bag/gur(□)	343.15	813.10 ± 0.50	−0.38	88-bag/gur(□)

Table 3. Recommended values (fit to the reliable experimental values according to the equations $\rho = A + BT + CT^2 + DT^3 + \dots$ or $\rho = [1 + 1.75(1 - T/T_c)^{1/3} + 0.75(1 - T/T_c)][\rho_c + A(T_c - T) + B(T_c - T)^2 + C(T_c - T)^3 + D(T_c - T)^4]$).

T / K	$\rho \pm \sigma_{fit}$ / $kg \cdot m^{-3}$	T / K	$\rho \pm \sigma_{fit}$ / $kg \cdot m^{-3}$	T / K	$\rho \pm \sigma_{fit}$ / $kg \cdot m^{-3}$
290.00	854.05 ± 0.73	300.00	846.42 ± 0.62	330.00	823.52 ± 0.47
293.15	851.65 ± 0.70	310.00	838.79 ± 0.50	340.00	815.89 ± 0.64
298.15	847.83 ± 0.64	320.00	831.15 ± 0.43	350.00	808.25 ± 0.97

cont.

3,7,11-Trimethyl-1-dodecyn-3-ol (cont.)

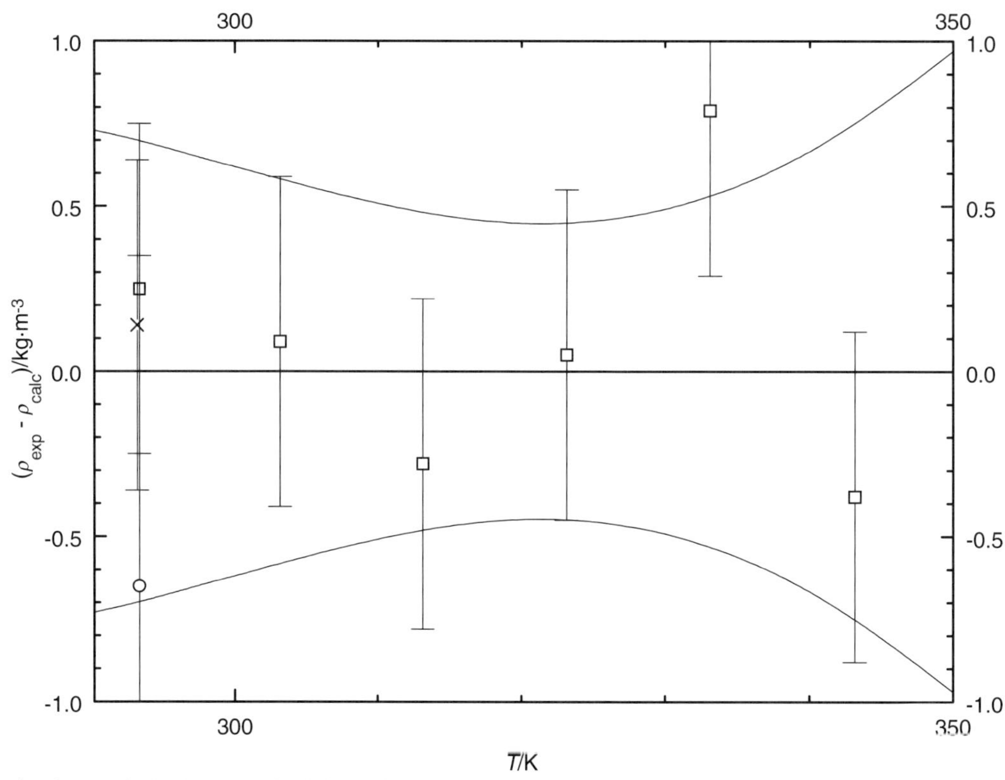

Fig. 1. The symbols show the deviation of the calculated from the experimental values from Table 2. The curves above and below the zero line indicate the calculated error region of the recommended values given in Table 3. The error bars represent the experimental errors. (Error bars smaller than the symbols are omitted for clarity of the figure.)

9,12-Octadecadien-1-ol **[1577-52-2]** **$C_{18}H_{34}O$** **MW = 266.47** **617**

Table 1. Experimental value with uncertainty.

$\frac{T}{K}$	$\frac{\rho_{exp} \pm 2\sigma_{est}}{kg \cdot m^{-3}}$	Ref.
293.15	861.2 ± 1.0	38-tur

3,7,11,15-Tetramethyl-1-hexadecyn-3-ol **[29171-23-1]** **$C_{20}H_{38}O$** **MW = 294.52** **618**

Table 1. Coefficients of the polynomial expansion equation.
Standard deviations (see introduction):
$\sigma_{c,w} = 4.5296 \cdot 10^{-1}$ (combined temperature ranges, weighted),
$\sigma_{c,uw} = 2.9444 \cdot 10^{-1}$ (combined temperature ranges, unweighted).

Coefficient	T = 293.15 to 343.15 K $\rho = A + BT + CT^2 + DT^3 + \ldots$
A	$1.06610 \cdot 10^{3}$
B	$-7.31308 \cdot 10^{-1}$

cont.

Table 2. Experimental values with uncertainties and deviation from calculated values.

$\frac{T}{K}$	$\frac{\rho_{exp} \pm 2\sigma_{est}}{kg \cdot m^{-3}}$	$\frac{\rho_{exp} - \rho_{calc}}{kg \cdot m^{-3}}$	Ref. (Symbol in Fig. 1)	$\frac{T}{K}$	$\frac{\rho_{exp} \pm 2\sigma_{est}}{kg \cdot m^{-3}}$	$\frac{\rho_{exp} - \rho_{calc}}{kg \cdot m^{-3}}$	Ref. (Symbol in Fig. 1)
293.15	853.30 ± 2.00	1.59	62-mau/smi(○)	323.15	829.10 ± 0.50	–0.67	88-bag/gur(□)
293.15	851.70 ± 0.50	–0.01	88-bag/gur(□)	333.15	822.00 ± 0.50	–0.46	88-bag/gur(□)
303.15	843.90 ± 0.50	–0.50	88-bag/gur(□)	343.15	815.50 ± 0.50	0.35	88-bag/gur(□)
313.15	836.80 ± 0.50	–0.29	88-bag/gur(□)				

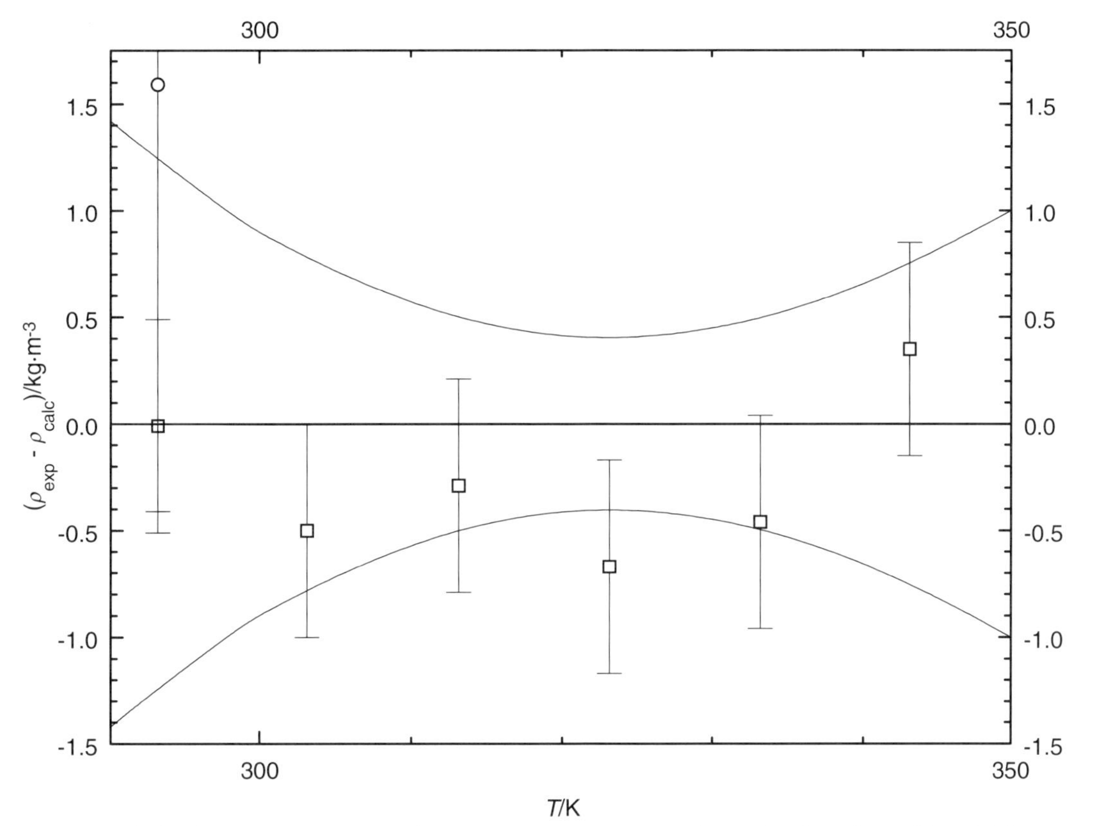

Fig. 1. The symbols show the deviation of the calculated from the experimental values from Table 2. The curves above and below the zero line indicate the calculated error region of the recommended values given in Table 3. The error bars represent the experimental errors. (Error bars smaller than the symbols are omitted for clarity of the figure.)

Table 3. Recommended values (fit to the reliable experimental values according to the equations $\rho = A + BT + CT^2 + DT^3 + \ldots$ or $\rho = [1 + 1.75(1 - T/T_c)^{1/3} + 0.75(1 - T/T_c)][\rho_c + A(T_c - T) + B(T_c - T)^2 + C(T_c - T)^3 + D(T_c - T)^4]$).

$\frac{T}{K}$	$\frac{\rho \pm \sigma_{fit}}{kg \cdot m^{-3}}$	$\frac{T}{K}$	$\frac{\rho \pm \sigma_{fit}}{kg \cdot m^{-3}}$	$\frac{T}{K}$	$\frac{\rho \pm \sigma_{fit}}{kg \cdot m^{-3}}$
290.00	854.02 ± 1.42	300.00	846.70 ± 0.89	330.00	824.77 ± 0.42
293.15	851.71 ± 1.24	310.00	839.39 ± 0.54	340.00	817.45 ± 0.63
298.15	848.06 ± 0.98	320.00	832.08 ± 0.38	350.00	810.14 ± 1.00

2.2.3 Alcohols of General Formula $C_nH_{2n-4}O$

1-Hexen-5-yn-3-ol **[1573-66-6]** **C_6H_8O** **MW = 96.13** **619**

Table 1. Experimental value with uncertainty.

T/K	$\rho_{exp} \pm 2\sigma_{est}$/kg·m^{-3}	Ref.
293.15	919.0 ± 0.7	63-lab

(*RS*)-1-Hexen-5-yn-3-ol **[126110-03-0]** **C_6H_8O** **MW = 96.13** **620**

Table 1. Experimental value with uncertainty.

T/K	$\rho_{exp} \pm 2\sigma_{est}$/kg·m^{-3}	Ref.
293.15	915.5 ± 1.0	65-alb/bry

2-Hepten-6-yn-4-ol **[500025-44-5]** **$C_7H_{10}O$** **MW = 110.16** **621**

Table 1. Experimental value with uncertainty.

T/K	$\rho_{exp} \pm 2\sigma_{est}$/kg·m^{-3}	Ref.
293.15	925.6 ± 0.7	63-lab

3-Methyl-1-hexen-5-yn-3-ol **[1573-67-7]** **$C_7H_{10}O$** **MW = 110.16** **622**

Table 1. Experimental value with uncertainty.

T/K	$\rho_{exp} \pm 2\sigma_{est}$/kg·m^{-3}	Ref.
293.15	906.6 ± 1.0	65-alb/bry

4-Methyl-1,2,6-heptatrien-4-ol **[90198-78-0]** **$C_8H_{12}O$** **MW = 124.18** **623**

Table 1. Experimental value with uncertainty.

T/K	$\rho_{exp} \pm 2\sigma_{est}$/kg·m^{-3}	Ref.
294.15	900.6 ± 2.0	62-ber/leg

6-Methyl-1,3,4-heptatrien-6-ol **[500025-52-5]** **$C_8H_{12}O$** **MW = 124.18** **624**

Table 1. Experimental value with uncertainty.

T/K	$\rho_{exp} \pm 2\sigma_{est}$/kg·m^{-3}	Ref.
293.15	882.5 ± 0.7	63-lab

4-Ethyl-1,2,6-heptatrien-4-ol [90611-15-7] $C_9H_{14}O$ MW = 138.21 **625**

Table 1. Experimental value with uncertainty.

T / K	$\rho_{exp} \pm 2\sigma_{est}$ / $kg \cdot m^{-3}$	Ref.
293.15	895.4 ± 2.0	62-ber/leg

6-Methyl-1,3,4-octatrien-6-ol [500025-51-4] $C_9H_{14}O$ MW = 138.21 **626**

Table 1. Experimental value with uncertainty.

T / K	$\rho_{exp} \pm 2\sigma_{est}$ / $kg \cdot m^{-3}$	Ref.
293.15	885.5 ± 0.7	63-lab

2,4,5-Nonatrien-7-ol [500025-50-3] $C_9H_{14}O$ MW = 138.21 **627**

Table 1. Experimental value with uncertainty.

T / K	$\rho_{exp} \pm 2\sigma_{est}$ / $kg \cdot m^{-3}$	Ref.
293.15	891.8 ± 0.7	63-lab

3,7-Dimethyl-6-octen-1-yn-3-ol [29171-20-8] $C_{10}H_{16}O$ MW = 152.2 **628**

Table 1. Coefficients of the polynomial expansion equation.
Standard deviations (see introduction):
$\sigma_{c,w} = 5.4842 \cdot 10^{-1}$ (combined temperature ranges, weighted),
$\sigma_{c,uw} = 1.6817 \cdot 10^{-1}$ (combined temperature ranges, unweighted).

Coefficient	T = 293.15 to 343.15 K $\rho = A + BT + CT^2 + DT^3 + \ldots$
A	$1.12321 \cdot 10^{3}$
B	$-8.36474 \cdot 10^{-1}$

Table 2. Experimental values with uncertainties and deviation from calculated values.

T / K	$\rho_{exp} \pm 2\sigma_{est}$ / $kg \cdot m^{-3}$	$\rho_{exp} - \rho_{calc}$ / $kg \cdot m^{-3}$	Ref. (Symbol in Fig. 1)	T / K	$\rho_{exp} \pm 2\sigma_{est}$ / $kg \cdot m^{-3}$	$\rho_{exp} - \rho_{calc}$ / $kg \cdot m^{-3}$	Ref. (Symbol in Fig. 1)
293.15	878.60 ± 1.00	0.60	85-bel/ber(□)	293.15	878.60 ± 1.00	0.60	88-bag/gur(○)
303.15	869.50 ± 1.00	−0.13	85-bel/ber(□)	303.15	869.50 ± 1.00	−0.13	88-bag/gur(○)
313.15	861.20 ± 1.00	−0.07	85-bel/ber(□)	313.15	861.20 ± 1.00	−0.07	88-bag/gur(○)
323.15	852.40 ± 1.00	−0.50	85-bel/ber(□)	323.15	852.40 ± 1.00	−0.50	88-bag/gur(○)
333.15	843.80 ± 1.00	−0.74	85-bel/ber(□)	333.15	843.80 ± 1.00	−0.74	88-bag/gur(○)
343.15	837.00 ± 1.00	0.83	85-bel/ber(□)	343.15	837.00 ± 1.00	0.83	88-bag/gur(○)

cont.

3,7-Dimethyl-6-octen-1-yn-3-ol (cont.)

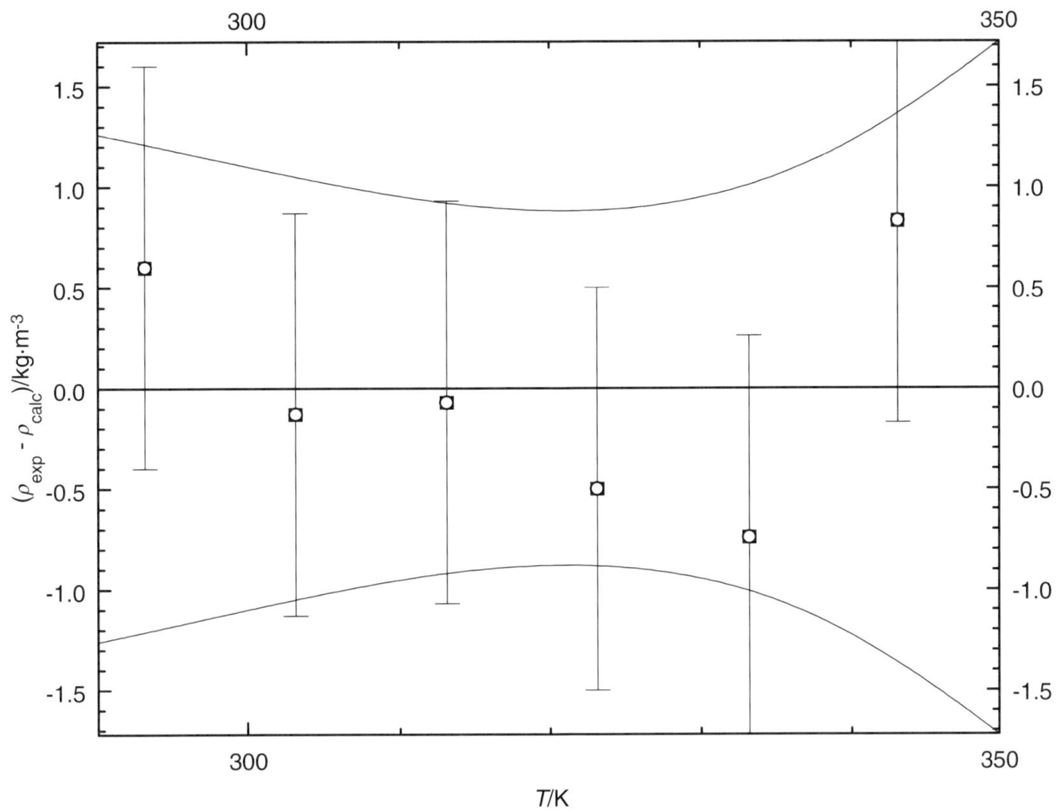

Fig. 1. The symbols show the deviation of the calculated from the experimental values from Table 2. The curves above and below the zero line indicate the calculated error region of the recommended values given in Table 3. The error bars represent the experimental errors. (Error bars smaller than the symbols are omitted for clarity of the figure.)

Table 3. Recommended values (fit to the reliable experimental values according to the equations

$$\rho = A + BT + CT^2 + DT^3 + \ldots \text{ or } \rho = [1 + 1.75(1 - T/T_c)^{1/3} + 0.75(1 - T/T_c)][\rho_c + A(T_c - T) + B(T_c - T)^2 + C(T_c - T)^3 + D(T_c - T)^4]\,).$$

$\frac{T}{K}$	$\frac{\rho \pm \sigma_{fit}}{kg \cdot m^{-3}}$	$\frac{T}{K}$	$\frac{\rho \pm \sigma_{fit}}{kg \cdot m^{-3}}$	$\frac{T}{K}$	$\frac{\rho \pm \sigma_{fit}}{kg \cdot m^{-3}}$
290.00	880.63 ± 1.26	300.00	872.27 ± 1.10	330.00	847.17 ± 0.91
293.15	878.00 ± 1.21	310.00	863.90 ± 0.94	340.00	838.81 ± 1.18
298.15	873.81 ± 1.13	320.00	855.54 ± 0.86	350.00	830.44 ± 1.72

3-Methyl-2-nonen-4-yn-6-ol **[500025-47-8]** **$C_{10}H_{16}O$** **MW = 152.24** **629**

Table 1. Experimental value with uncertainty.

$\frac{T}{K}$	$\frac{\rho_{exp} \pm 2\sigma_{est}}{kg \cdot m^{-3}}$	Ref.
293.15	881.3 ± 0.6	63-lab

4-(1-Methylethyl)-1,2,6-heptatrien-4-ol [91055-94-6] $C_{10}H_{16}O$ MW = 152.24 630

Table 1. Experimental value with uncertainty.

$\frac{T}{K}$	$\frac{\rho_{exp} \pm 2\sigma_{est}}{kg \cdot m^{-3}}$	Ref.
295.15	885.1 ± 2.0	62-ber/leg

4-Propyl-1,2,6-heptatrien-4-ol [91055-95-7] $C_{10}H_{16}O$ MW = 152.24 631

Table 1. Experimental value with uncertainty.

$\frac{T}{K}$	$\frac{\rho_{exp} \pm 2\sigma_{est}}{kg \cdot m^{-3}}$	Ref.
294.15	882.1 ± 2.0	62-ber/leg

3-Ethenyl-1-nonyn-4-ol [900002-83-7] $C_{11}H_{18}O$ MW = 166.26 632

Table 1. Experimental value with uncertainty.

$\frac{T}{K}$	$\frac{\rho_{exp} \pm 2\sigma_{est}}{kg \cdot m^{-3}}$	Ref.
293.15	879.0 ± 2.0	65-mig/mig

5-Methyl-2,5,6-decatrien-4-ol [14129-53-4] $C_{11}H_{18}O$ MW = 166.26 633

Table 1. Experimental value with uncertainty.

$\frac{T}{K}$	$\frac{\rho_{exp} \pm 2\sigma_{est}}{kg \cdot m^{-3}}$	Ref.
293.15	871.5 ± 1.0	66-per/bal

3-(Methylethenyl)-3,4-octadien-2-ol [14129-70-5] $C_{11}H_{18}O$ MW = 166.26 634

Table 1. Experimental value with uncertainty.

$\frac{T}{K}$	$\frac{\rho_{exp} \pm 2\sigma_{est}}{kg \cdot m^{-3}}$	Ref.
293.15	874.3 ± 1.0	66-per/bal

2-Methyl-3-(1-methylethenyl)-3,4-octadien-2-ol [14270-82-7] $C_{12}H_{20}O$ MW = 180.29 635

Table 1. Experimental value with uncertainty.

$\frac{T}{K}$	$\frac{\rho_{exp} \pm 2\sigma_{est}}{kg \cdot m^{-3}}$	Ref.
293.15	855.5 ± 1.0	66-per/bal

2,7-Dimethyl-3-(1-methylethenyl)-3,4-octadien-2-ol [14270-85-0] $C_{13}H_{22}O$ MW = 194.32 636

Table 1. Experimental value with uncertainty.

T / K	$\rho_{exp} \pm 2\sigma_{est}$ / $kg \cdot m^{-3}$	Ref.
293.15	855.8 ± 1.0	66-per/bal

5-Ethyl-9-methyl-2,5,6-decatrien-4-ol [14129-71-6] $C_{13}H_{22}O$ MW = 194.32 637

Table 1. Experimental value with uncertainty.

T / K	$\rho_{exp} \pm 2\sigma_{est}$ / $kg \cdot m^{-3}$	Ref.
293.15	867.1 ± 1.0	66-per/bal

2.2.4 Alcohols of General Formula $C_nH_{2n-6}O$

1,6-Heptadiyn-4-ol **[21972-06-5]** **C_7H_8O** **MW = 108.14** **638**

Table 1. Experimental value with uncertainty.

$\frac{T}{K}$	$\frac{\rho_{exp} \pm 2\sigma_{est}}{kg \cdot m^{-3}}$	Ref.
293.15	967.2 ± 1.0	63-lab

4-Methyl-1,6-heptadiyn-4-ol **[41005-07-6]** **$C_8H_{10}O$** **MW = 122.17** **639**

Table 1. Experimental value with uncertainty.

$\frac{T}{K}$	$\frac{\rho_{exp} \pm 2\sigma_{est}}{kg \cdot m^{-3}}$	Ref.
293.15	959.5 ± 1.0	63-lab

1,7-Octadien-5-yn-3-ol **[1573-68-8]** **$C_8H_{10}O$** **MW = 122.17** **640**

Table 1. Experimental value with uncertainty.

$\frac{T}{K}$	$\frac{\rho_{exp} \pm 2\sigma_{est}}{kg \cdot m^{-3}}$	Ref.
293.15	922.5 ± 1.0	65-alb/bry

3-Methyl-1,7-octadien-5-yn-3-ol **[1573-69-9]** **$C_9H_{12}O$** **MW = 136.19** **641**

Table 1. Experimental value with uncertainty.

$\frac{T}{K}$	$\frac{\rho_{exp} \pm 2\sigma_{est}}{kg \cdot m^{-3}}$	Ref.
293.15	910.5 ± 1.0	65-alb/bry

2-Methyl-3,5-octadiyn-2-ol **[500025-48-9]** **$C_9H_{12}O$** **MW = 136.19** **642**

Table 1. Experimental value with uncertainty.

$\frac{T}{K}$	$\frac{\rho_{exp} \pm 2\sigma_{est}}{kg \cdot m^{-3}}$	Ref.
293.15	870.0 ± 1.0	63-lab

2-Methyl-3,5-decadiyn-2-ol **[500025-49-0]** **$C_{11}H_{16}O$** **MW = 164.25** **643**

Table 1. Experimental value with uncertainty.

$\frac{T}{K}$	$\frac{\rho_{exp} \pm 2\sigma_{est}}{kg \cdot m^{-3}}$	Ref.
293.15	897.3 ± 1.0	63-lab

3 Tabulated Data on Density - Diols

3.1 Alkanediols

3.1.1 Alkanediols C_1 - C_6

1,2-Ethanediol **[107-21-1]** **$C_2H_6O_2$** **MW = 62.07** **644**

Table 1. Coefficients of the polynomial expansion equation.
Standard deviations (see introduction):
$\sigma_{c,w} = 6.0901 \cdot 10^{-1}$ (combined temperature ranges, weighted),
$\sigma_{c,uw} = 1.3696 \cdot 10^{-1}$ (combined temperature ranges, unweighted).

Coefficient	T = 250.25 to 513.15 K $\rho = A + BT + CT^2 + DT^3 + \ldots$
A	$1.31971 \cdot 10^{3}$
B	$-8.26843 \cdot 10^{-1}$
C	$8.03242 \cdot 10^{-4}$
D	$-1.30719 \cdot 10^{-6}$

Table 2. Experimental values with uncertainties and deviation from calculated values.

$\frac{T}{K}$	$\frac{\rho_{exp} \pm 2\sigma_{est}}{kg \cdot m^{-3}}$	$\frac{\rho_{exp} - \rho_{calc}}{kg \cdot m^{-3}}$	Ref. (Symbol in Fig. 1)	$\frac{T}{K}$	$\frac{\rho_{exp} \pm 2\sigma_{est}}{kg \cdot m^{-3}}$	$\frac{\rho_{exp} - \rho_{calc}}{kg \cdot m^{-3}}$	Ref. (Symbol in Fig. 1)
289.95	1118.00 ± 2.00	2.37	1893-ram/shi-3[1]	343.15	1077.40 ± 0.60	−0.34	57-ket/van(×)
319.25	1091.20 ± 2.00	−3.87	1893-ram/shi-3[1]	273.15	1126.00 ± 1.00	−1.15	58-cos/bow[1]
351.35	1068.00 ± 2.00	−3.66	1893-ram/shi-3[1]	293.15	1113.00 ± 1.00	−0.42	58-cos/bow[1]
405.05	1029.70 ± 2.00	−0.01	1893-ram/shi-3(×)	313.15	1100.00 ± 1.00	0.59	58-cos/bow[1]
292.45	1113.40 ± 1.00	−0.50	1895-eyk[1]	333.15	1086.00 ± 1.00	0.94	58-cos/bow[1]
411.95	1023.00 ± 2.00	−1.02	1895-eyk(×)	353.15	1071.00 ± 1.00	0.69	58-cos/bow(×)
293.15	1112.72 ± 1.00	−0.70	30-bin/for[1]	373.15	1056.00 ± 1.00	0.90	58-cos/bow(×)
303.15	1105.58 ± 1.00	−0.87	30-bin/for[1]	393.15	1041.00 ± 1.00	1.64	58-cos/bow(×)
313.15	1098.78 ± 1.00	−0.63	30-bin/for[1]	413.15	1024.00 ± 1.00	0.98	58-cos/bow(×)
333.15	1084.13 ± 1.00	−0.93	30-bin/for[1]	433.15	1007.00 ± 1.00	0.97	58-cos/bow(×)
353.15	1069.63 ± 1.00	−0.68	30-bin/for(×)	453.15	989.20 ± 1.00	0.87	58-cos/bow(×)
371.15	1054.52 ± 1.00	−2.12	30-bin/for(×)	473.15	970.80 ± 1.50	0.95	58-cos/bow(×)
393.15	1037.45 ± 1.00	−1.91	30-bin/for(×)	493.15	950.30 ± 1.50	−0.22	58-cos/bow(×)
298.15	1109.90 ± 0.60	−0.04	31-smy/wal[1]	513.15	930.40 ± 1.50	0.11	58-cos/bow(×)
323.15	1092.30 ± 0.60	0.02	31-smy/wal(×)	533.15	910.30 ± 1.50	1.20	58-cos/bow[1]
273.15	1127.60 ± 0.20	0.45	35-tim/hen(Δ)	553.15	889.20 ± 1.50	2.33	58-cos/bow[1]
288.15	1117.07 ± 0.20	0.20	35-tim/hen(Δ)	250.25	1140.70 ± 0.60	−1.91	59-kom/ros(×)
303.15	1106.61 ± 0.20	0.16	35-tim/hen(Δ)	366.25	1067.80 ± 0.60	7.40	59-kom/ros[1]
293.15	1113.10 ± 0.60	−0.32	53-dan/fad[1]	293.10	1113.96 ± 0.50	0.51	75-khi/gri[1]
333.15	1085.10 ± 0.60	0.04	53-dan/fad(×)	298.10	1110.28 ± 0.50	0.30	75-khi/gri[1]
293.15	1113.10 ± 0.60	−0.32	57-ket/van[1]	303.10	1106.72 ± 0.50	0.23	75-khi/gri[1]
318.15	1095.50 ± 0.60	−0.36	57-ket/van(×)	308.10	1103.06 ± 0.50	0.08	75-khi/gri[1]

[1]) Not included in Fig. 1.

cont.

1,2-Ethanediol (cont.)

Table 2. (cont.)

$\frac{T}{K}$	$\frac{\rho_{exp} \pm 2\sigma_{est}}{kg \cdot m^{-3}}$	$\frac{\rho_{exp} - \rho_{calc}}{kg \cdot m^{-3}}$	Ref. (Symbol in Fig. 1)	$\frac{T}{K}$	$\frac{\rho_{exp} \pm 2\sigma_{est}}{kg \cdot m^{-3}}$	$\frac{\rho_{exp} - \rho_{calc}}{kg \cdot m^{-3}}$	Ref. (Symbol in Fig. 1)
313.10	1099.58 ± 0.50	0.13	75-khi/gri(×)	393.40	1040.80 ± 0.70	1.64	89-taw/tej(×)
318.10	1095.99 ± 0.50	0.10	75-khi/gri(×)	422.80	1015.80 ± 0.80	0.89	89-taw/tej(×)
323.10	1092.24 ± 0.50	−0.08	75-khi/gri(×)	283.15	1119.29 ± 0.50	−1.02	90-lee/hon(×)
293.15	1113.50 ± 0.25	0.08	81-joo/arl(▽)	293.15	1112.02 ± 0.50	−1.40	90-lee/hon[1)]
298.00	1094.00 ± 1.00	−16.05	84-idr/fre[1)]	303.15	1105.82 ± 0.50	−0.63	90-lee/hon[1)]
318.00	1109.00 ± 1.00	13.04	84-idr/fre[1)]	288.15	1116.96 ± 0.20	0.09	91-dou/pal(○)
338.00	1080.00 ± 1.00	−1.53	84-idr/fre(×)	298.15	1109.99 ± 0.20	0.05	91-dou/pal(○)
358.00	1065.00 ± 1.00	−1.67	84-idr/fre(×)	303.15	1106.49 ± 0.20	0.04	91-dou/pal(○)
298.15	1109.82 ± 0.20	−0.12	88-dou/pal(□)	308.15	1102.96 ± 0.20	0.02	91-dou/pal(○)
303.15	1105.90 ± 0.60	−0.55	89-taw/tej[1)]	293.15	1113.47 ± 0.30	0.05	93-chi/pro(◆)
313.30	1099.90 ± 0.60	0.60	89-taw/tej[1)]	273.15	1126.90 ± 0.40	−0.25	93-kum/moc(×)
336.00	1083.80 ± 0.60	0.81	89-taw/tej(×)	293.15	1113.10 ± 0.40	−0.32	93-kum/moc(×)
354.40	1069.80 ± 0.70	0.42	89-taw/tej(×)	313.15	1099.00 ± 0.40	−0.41	93-kum/moc(×)
363.15	1063.70 ± 0.70	0.93	89-taw/tej(×)	333.15	1084.70 ± 0.40	−0.36	93-kum/moc(×)
373.15	1055.90 ± 0.70	0.80	89-taw/tej(×)				

[1)] Not included in Fig. 1.

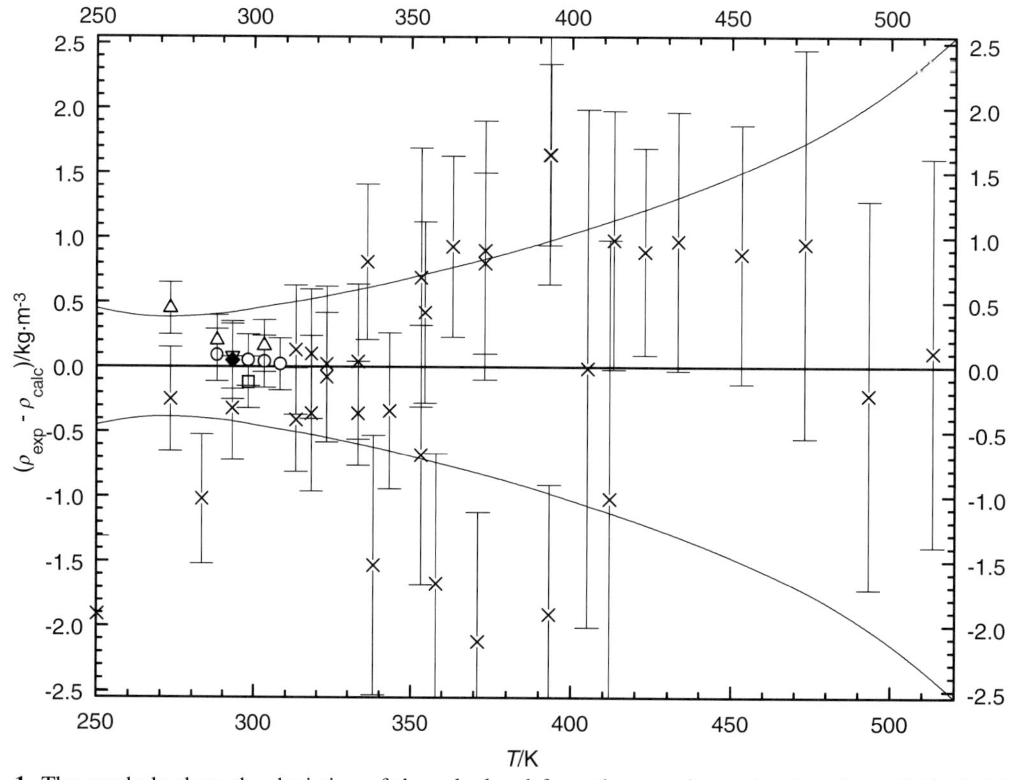

Fig. 1. The symbols show the deviation of the calculated from the experimental values from Table 2. The curves above and below the zero line indicate the calculated error region of the recommended values given in Table 3. The error bars represent the experimental errors. (Error bars smaller than the symbols are omitted for clarity of the figure.)

cont.

Further references: [1859-wur, 1864-lan, 1884-gla, 1884-per, 01-def-1, 05-dun, 06-wal-1, 07-wal-4, 09-sch-2, 14-low, 18-kar, 19-eyk, 25-rii/soe, 26-tay/rin, 27-kai/mel, 29-ber/reu, 32-bri-2, 35-sch/sta, 36-rom, 37-ewe, 37-gib/kin, 37-mou/dod, 37-puk, 41-gib/loe, 43-ish/kat, 44-ira, 48-lad/smi, 48-wei, 52-cur/joh, 53-ish/kat, 55-fog/hix, 56-koi/han, 57-cru/jos, 60-rab/mur, 61-lut/obu, 62-mel, 65-por/pli, 66-gar/kom, 66-koz/rab, 67-dav/fin, 67-sho/ata, 68-ano, 68-naz/tsy, 69-ada, 70-kun/cha, 70-paz/paz, 70-ras/gaz, 72-gla/gha, 74-jim/paz, 74-wol/ska, 75-esp/man, 77-zhu/zhu-2, 78-fro/ers-1, 78-mus/kan, 79-jim/paz, 79-sah/hay, 79-sin/siv, 82-diz/mar, 82-man/les, 85-zhu, 91-fen/wan, 98-pal/sha].

Table 3. Recommended values (fit to the reliable experimental values according to the equations $\rho = A + BT + CT^2 + DT^3 + \ldots$ or $\rho = [1 + 1.75(1 - T/T_c)^{1/3} + 0.75(1 - T/T_c)][\rho_c + A(T_c - T) + B(T_c - T)^2 + C(T_c - T)^3 + D(T_c - T)^4]$).

T / K	$\rho \pm \sigma_{fit}$ / $kg \cdot m^{-3}$	T / K	$\rho \pm \sigma_{fit}$ / $kg \cdot m^{-3}$	T / K	$\rho \pm \sigma_{fit}$ / $kg \cdot m^{-3}$
250.00	1142.78 ± 0.45	330.00	1087.35 ± 0.58	430.00	1008.76 ± 1.28
260.00	1136.05 ± 0.40	340.00	1080.06 ± 0.63	440.00	1000.05 ± 1.37
270.00	1129.29 ± 0.38	350.00	1072.67 ± 0.69	450.00	991.17 ± 1.47
280.00	1122.47 ± 0.39	360.00	1065.16 ± 0.75	460.00	982.09 ± 1.58
290.00	1115.60 ± 0.41	370.00	1057.53 ± 0.81	470.00	972.81 ± 1.69
293.15	1113.42 ± 0.42	380.00	1049.77 ± 0.88	480.00	963.33 ± 1.82
298.15	1109.94 ± 0.44	390.00	1041.87 ± 0.95	490.00	953.63 ± 1.97
300.00	1108.65 ± 0.45	400.00	1033.83 ± 1.03	500.00	943.70 ± 2.14
310.00	1101.64 ± 0.49	410.00	1025.64 ± 1.11	510.00	933.54 ± 2.33
320.00	1094.54 ± 0.53	420.00	1017.28 ± 1.19	520.00	923.15 ± 2.55

1,2-Propanediol [57-55-6] $C_3H_8O_2$ MW = 76.1 645

Table 1. Coefficients of the polynomial expansion equation.
Standard deviations (see introduction):
$\sigma_{c,w} = 3.0002 \cdot 10^{-1}$ (combined temperature ranges, weighted),
$\sigma_{c,uw} = 4.9012 \cdot 10^{-2}$ (combined temperature ranges, unweighted).

Coefficient	T = 273.15 to 323.15 K $\rho = A + BT + CT^2 + DT^3 + \ldots$
A	$1.25088 \cdot 10^3$
B	$-7.32395 \cdot 10^{-1}$

Table 2. Experimental values with uncertainties and deviation from calculated values.

T / K	$\rho_{exp} \pm 2\sigma_{est}$ / $kg \cdot m^{-3}$	$\rho_{exp} - \rho_{calc}$ / $kg \cdot m^{-3}$	Ref. (Symbol in Fig. 1)	T / K	$\rho_{exp} \pm 2\sigma_{est}$ / $kg \cdot m^{-3}$	$\rho_{exp} - \rho_{calc}$ / $kg \cdot m^{-3}$	Ref. (Symbol in Fig. 1)
298.15	1032.80 ± 0.80	0.28	31-smy/wal[1)]	296.15	1035.40 ± 0.50	1.42	62-mel[1)]
323.15	1013.80 ± 0.80	−0.41	31-smy/wal(×)	278.15	1047.20 ± 0.60	0.03	70-kun/cha(◆)
293.15	1038.10 ± 1.00	1.92	51-mac/tho[1)]	283.15	1043.50 ± 0.60	−0.00	70-kun/cha(◆)
308.15	1025.10 ± 1.00	−0.09	51-mac/tho(×)	288.15	1039.90 ± 0.60	0.06	70-kun/cha(◆)
273.15	1050.60 ± 1.00	−0.23	51-van-1(×)	293.15	1036.30 ± 0.60	0.12	70-kun/cha[1)]
293.15	1037.60 ± 1.00	1.42	51-van-1[1)]	298.15	1032.60 ± 0.60	0.08	70-kun/cha(◆)
273.15	1050.81 ± 0.40	−0.02	55-tim/hen(Δ)	303.15	1028.90 ± 0.60	0.04	70-kun/cha(◆)
288.15	1040.01 ± 0.40	0.17	55-tim/hen(Δ)	308.15	1025.30 ± 0.60	0.11	70-kun/cha(◆)
303.15	1029.08 ± 0.40	0.22	55-tim/hen(Δ)	313.15	1021.70 ± 0.60	0.17	70-kun/cha(◆)
293.15	1036.20 ± 0.40	0.02	57-ano(□)	318.15	1018.10 ± 0.80	0.23	70-kun/cha(◆)
293.15	1036.10 ± 0.50	−0.08	62-mel(∇)	298.15	1032.00 ± 0.40	−0.52	85-les/eic(○)
293.15	1036.40 ± 0.50	0.22	62-mel(∇)				

[1)] Not included in Fig. 1.

cont.

1,2-Propanediol (cont.)

Further references: [1857-wur-1, 1859-wur, 1879-bel, 1882-zan, 1890-gar, 27-tro/luk, 32-bri-2, 35-sch/sta, 36-dup-1, 37-puk, 46-puc/wis, 50-cle/mac, 52-cur/joh, 53-ano-5, 55-ano-2, 61-sch-3, 62-deg/lad, 65-wei/lan, 66-mil-1, 68-ano, 70-gar/kom, 73-gar/paz, 74-wol/ska, 78-mus/kan, 82-man/les].

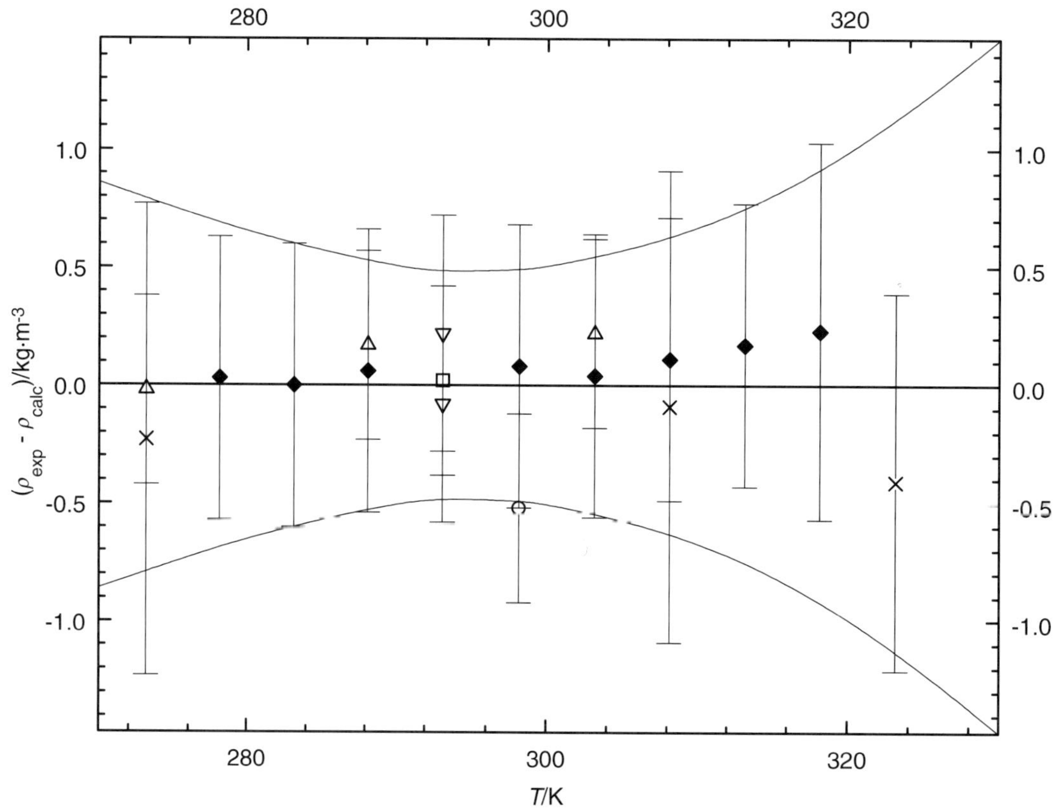

Fig. 1. The symbols show the deviation of the calculated from the experimental values from Table 2. The curves above and below the zero line indicate the calculated error region of the recommended values given in Table 3. The error bars represent the experimental errors. (Error bars smaller than the symbols are omitted for clarity of the figure.)

Table 3. Recommended values (fit to the reliable experimental values according to the equations $\rho = A + BT + CT^2 + DT^3 + \ldots$ or $\rho = [1 + 1.75(1 - T/T_c)^{1/3} + 0.75(1 - T/T_c)][\rho_c + A(T_c - T) + B(T_c - T)^2 + C(T_c - T)^3 + D(T_c - T)^4]$).

$\frac{T}{\text{K}}$	$\frac{\rho \pm \sigma_{\text{fit}}}{\text{kg}\cdot\text{m}^{-3}}$	$\frac{T}{\text{K}}$	$\frac{\rho \pm \sigma_{\text{fit}}}{\text{kg}\cdot\text{m}^{-3}}$	$\frac{T}{\text{K}}$	$\frac{\rho \pm \sigma_{\text{fit}}}{\text{kg}\cdot\text{m}^{-3}}$
270.00	1053.14 ± 0.86	293.15	1036.18 ± 0.48	310.00	1023.84 ± 0.64
280.00	1045.81 ± 0.64	298.15	1032.52 ± 0.49	320.00	1016.52 ± 0.96
290.00	1038.49 ± 0.50	300.00	1031.16 ± 0.50	330.00	1009.19 ± 1.47

1,3-Propanediol **[504-63-2]** **$C_3H_8O_2$** **MW = 76.1** **646**

Table 1. Coefficients of the polynomial expansion equation.
Standard deviations (see introduction):
$\sigma_{c,w} = 8.3438 \cdot 10^{-1}$ (combined temperature ranges, weighted),
$\sigma_{c,uw} = 1.6142 \cdot 10^{-1}$ (combined temperature ranges, unweighted).

Coefficient	T = 273.15 to 343.15 K $\rho = A + BT + CT^2 + DT^3 + \ldots$
A	$1.13576 \cdot 10^{3}$
B	$2.29954 \cdot 10^{-2}$
C	$-1.03995 \cdot 10^{-3}$

Table 2. Experimental values with uncertainties and deviation from calculated values.

$\frac{T}{\text{K}}$	$\frac{\rho_{exp} \pm 2\sigma_{est}}{\text{kg} \cdot \text{m}^{-3}}$	$\frac{\rho_{exp} - \rho_{calc}}{\text{kg} \cdot \text{m}^{-3}}$	Ref. (Symbol in Fig. 1)	$\frac{T}{\text{K}}$	$\frac{\rho_{exp} \pm 2\sigma_{est}}{\text{kg} \cdot \text{m}^{-3}}$	$\frac{\rho_{exp} - \rho_{calc}}{\text{kg} \cdot \text{m}^{-3}}$	Ref. (Symbol in Fig. 1)
273.15	1063.40 ± 2.00	−1.05	1874-reb-2(×)	343.15	1021.80 ± 0.60	0.60	57-ket/van(×)
292.15	1053.00 ± 2.00	−0.72	1874-reb-2[1)]	293.15	1053.80 ± 0.50	0.67	62-mel(∇)
298.15	1050.30 ± 0.80	0.13	31-smy/wal(×)	293.15	1052.90 ± 0.50	−0.23	62-mel(∇)
323.15	1034.40 ± 0.80	−0.20	31-smy/wal(×)	298.15	1049.00 ± 2.00	−1.17	80-bru[1)]
293.15	1052.80 ± 0.60	−0.33	32-bri-2(×)	323.15	1033.00 ± 2.00	−1.60	80-bru(×)
293.15	1053.80 ± 0.50	0.67	35-sch/sta(Δ)	373.15	1003.00 ± 2.00	3.46	80-bru[1)]
293.15	1053.60 ± 0.60	0.47	52-cur/joh(◆)	293.15	1053.70 ± 0.40	0.57	88-cze/zyw(○)
293.15	1052.90 ± 0.40	−0.23	53-ano-6(□)	303.15	1047.44 ± 0.40	0.28	88-cze/zyw(○)
293.15	1053.30 ± 0.60	0.17	57-ket/van(×)	313.15	1041.10 ± 0.40	0.12	88-cze/zyw(○)
318.15	1037.80 ± 0.60	−0.02	57-ket/van(×)				

[1)] Not included in Fig. 1.

Further references: [1874-reb-1, 1882-zan, 18-kar, 26-ray, 36-sto/rou, 50-cle/mac, 59-ale, 72-rak/isa, 74-paz/rom, 74-rak/mel, 74-rak/zlo, 75-nak/kom, 78-mus/kan].

Table 3. Recommended values (fit to the reliable experimental values according to the equations
$\rho = A + BT + CT^2 + DT^3 + \ldots$ or $\rho = [1 + 1.75(1 - T/T_c)^{1/3} + 0.75(1 - T/T_c)][\rho_c + A(T_c - T) + B(T_c - T)^2 + C(T_c - T)^3 + D(T_c - T)^4])$.

$\frac{T}{\text{K}}$	$\frac{\rho \pm \sigma_{fit}}{\text{kg} \cdot \text{m}^{-3}}$	$\frac{T}{\text{K}}$	$\frac{\rho \pm \sigma_{fit}}{\text{kg} \cdot \text{m}^{-3}}$	$\frac{T}{\text{K}}$	$\frac{\rho \pm \sigma_{fit}}{\text{kg} \cdot \text{m}^{-3}}$
270.00	1066.16 ± 1.82	298.15	1050.17 ± 0.56	330.00	1030.10 ± 1.04
280.00	1060.67 ± 1.16	300.00	1049.07 ± 0.55	340.00	1023.36 ± 1.56
290.00	1054.97 ± 0.73	310.00	1042.95 ± 0.56	350.00	1016.42 ± 2.29
293.15	1053.13 ± 0.64	320.00	1036.63 ± 0.72		

cont.

1,3-Propanediol (cont.)

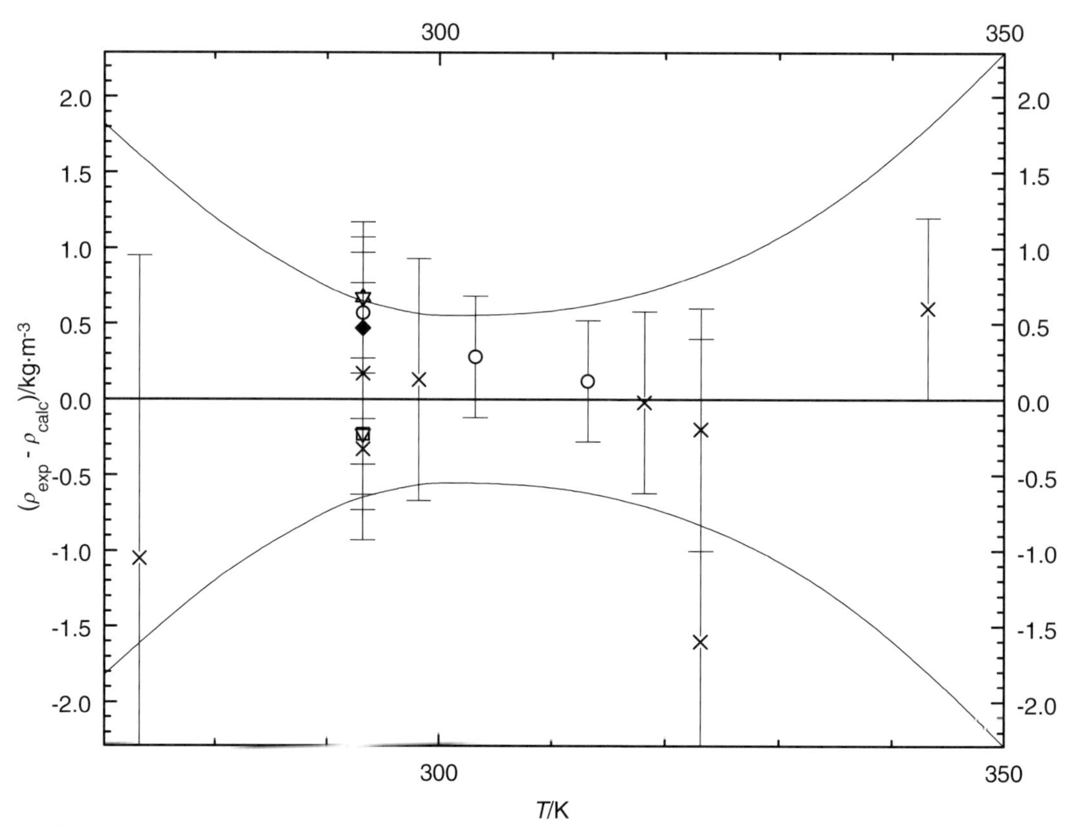

Fig. 1. The symbols show the deviation of the calculated from the experimental values from Table 2. The curves above and below the zero line indicate the calculated error region of the recommended values given in Table 3. The error bars represent the experimental errors. (Error bars smaller than the symbols are omitted for clarity of the figure.)

***R*-(-)-1,2-Propanediol** **[4254-14-2]** $C_3H_8O_2$ **MW = 76.1** **647**

Table 1. Experimental values with uncertainties.

$\frac{T}{\text{K}}$	$\frac{\rho_{exp} \pm 2\sigma_{est}}{\text{kg} \cdot \text{m}^{-3}}$	Ref.
293.15	1004.3 ± 3.0	18-abd/eic
299.15	1025.3 ± 3.0	56-pri/osg

(*S*)-(+)-1,2-Propanediol **[4254-15-3]** $C_3H_8O_2$ **MW = 76.1** **648**

Table 1. Experimental values with uncertainties.

$\frac{T}{\text{K}}$	$\frac{\rho_{exp} \pm 2\sigma_{est}}{\text{kg} \cdot \text{m}^{-3}}$	Ref.
293.15	1040.0 ± 3.0	48-bae/fis
295.15	1041.0 ± 3.0	62-kuh/kuh

1,2-Butanediol **[584-03-2]** **$C_4H_{10}O_2$** **MW = 90.12** **649**

Table 1. Coefficients of the polynomial expansion equation.
Standard deviations (see introduction):
$\sigma_{c,w}$ = 1.1759 (combined temperature ranges, weighted),
$\sigma_{c,uw}$ = 2.4556 · 10^{-1} (combined temperature ranges, unweighted).

Coefficient	T = 293.15 to 454.70 K $\rho = A + BT + CT^2 + DT^3 + \ldots$
A	1.09605 · 10^3
B	2.45483 · 10^{-2}
C	−1.17159 · 10^{-3}

Table 2. Experimental values with uncertainties and deviation from calculated values.

T/K	$\rho_{exp} \pm 2\sigma_{est}$ / kg·m^{-3}	$\rho_{exp} - \rho_{calc}$ / kg·m^{-3}	Ref. (Symbol in Fig. 1)	T/K	$\rho_{exp} \pm 2\sigma_{est}$ / kg·m^{-3}	$\rho_{exp} - \rho_{calc}$ / kg·m^{-3}	Ref. (Symbol in Fig. 1)
293.15	1004.60 ± 2.00	2.04	37-tis(✕)	318.80	984.80 ± 0.50	−0.00	92-sun/dig(○)
293.15	1001.00 ± 2.00	−1.56	37-tis/chu(✕)	337.70	970.00 ± 0.50	−0.73	92-sun/dig(○)
293.15	1002.40 ± 1.00	−0.16	50-cle/mac(◆)	357.70	954.70 ± 0.50	−0.23	92-sun/dig(○)
293.15	1002.40 ± 1.00	−0.16	62-mel(∇)	370.90	943.70 ± 0.50	−0.28	92-sun/dig(○)
293.15	1002.80 ± 1.00	0.24	65-wei/lan(Δ)	393.90	923.90 ± 0.50	−0.04	92-sun/dig(○)
293.15	1003.47 ± 0.40	0.91	88-cze/zyw(□)	413.50	907.40 ± 0.50	1.52	92-sun/dig(○)
303.15	995.44 ± 0.40	−0.38	88-cze/zyw(□)	434.60	886.50 ± 0.50	1.07	92-sun/dig(○)
313.15	987.87 ± 0.40	−0.98	88-cze/zyw(□)	454.70	863.40 ± 0.50	−1.58	92-sun/dig(○)
301.80	997.10 ± 0.50	0.35	92-sun/dig(○)				

Further references: [1886-hen, 73-sza/mys].

Table 3. Recommended values (fit to the reliable experimental values according to the equations
$\rho = A + BT + CT^2 + DT^3 + \ldots$ or $\rho = [1 + 1.75(1 - T/T_c)^{1/3} + 0.75(1 - T/T_c)][\rho_c + A(T_c - T) + B(T_c - T)^2 + C(T_c - T)^3 + D(T_c - T)^4]$).

T/K	$\rho \pm \sigma_{fit}$ / kg·m^{-3}	T/K	$\rho \pm \sigma_{fit}$ / kg·m^{-3}	T/K	$\rho \pm \sigma_{fit}$ / kg·m^{-3}
290.00	1004.64 ± 1.23	340.00	968.96 ± 0.38	410.00	909.17 ± 0.58
293.15	1002.56 ± 1.13	350.00	961.12 ± 0.36	420.00	899.69 ± 0.65
298.15	999.22 ± 0.98	360.00	953.05 ± 0.36	430.00	889.98 ± 0.72
300.00	997.97 ± 0.93	370.00	944.74 ± 0.38	440.00	880.03 ± 0.78
310.00	991.07 ± 0.71	380.00	936.20 ± 0.42	450.00	869.85 ± 0.84
320.00	983.93 ± 0.54	390.00	927.43 ± 0.46	460.00	859.43 ± 0.89
330.00	976.57 ± 0.44	400.00	918.42 ± 0.52		

cont.

1,2-Butanediol (cont.)

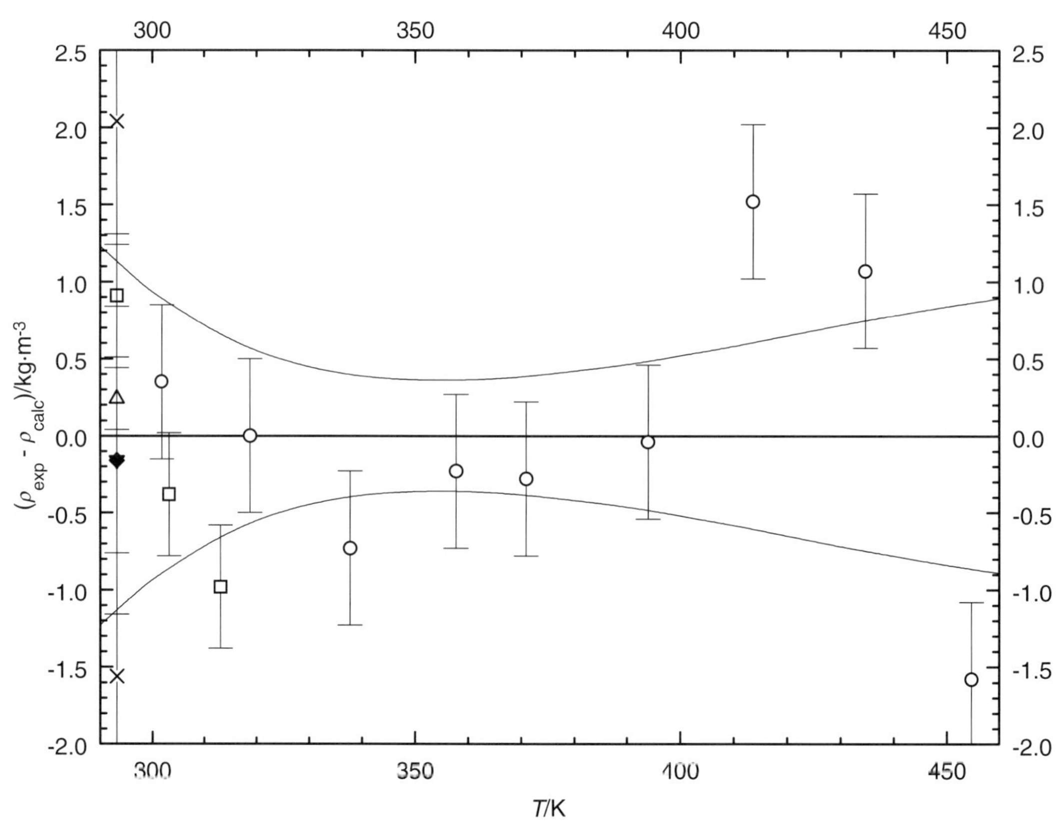

Fig. 1. The symbols show the deviation of the calculated from the experimental values from Table 2. The curves above and below the zero line indicate the calculated error region of the recommended values given in Table 3. The error bars represent the experimental errors. (Error bars smaller than the symbols are omitted for clarity of the figure.)

(*RS*)-1,2-Butanediol **[26171-83-5]** **$C_4H_{10}O_2$** **MW = 90.12** **650**

Table 1. Coefficients of the polynomial expansion equation.
Standard deviations (see introduction):
$\sigma_{c,w} = 4.4743 \cdot 10^{-1}$ (combined temperature ranges, weighted),
$\sigma_{c,uw} = 2.2372 \cdot 10^{-1}$ (combined temperature ranges, unweighted).

Coefficient	T = 323.14 to 448.11 K $\rho = A + BT + CT^2 + DT^3 + \ldots$
A	$1.11932 \cdot 10^3$
B	$-1.05834 \cdot 10^{-1}$
C	$-1.01373 \cdot 10^{-3}$

Table 2. Experimental values with uncertainties and deviation from calculated values.

$\frac{T}{K}$	$\frac{\rho_{exp} \pm 2\sigma_{est}}{kg \cdot m^{-3}}$	$\frac{\rho_{exp} - \rho_{calc}}{kg \cdot m^{-3}}$	Ref. (Symbol in Fig. 1)	$\frac{T}{K}$	$\frac{\rho_{exp} \pm 2\sigma_{est}}{kg \cdot m^{-3}}$	$\frac{\rho_{exp} - \rho_{calc}}{kg \cdot m^{-3}}$	Ref. (Symbol in Fig. 1)
323.14	979.00 ± 0.50	−0.27	96-ste/chi(□)	398.12	916.00 ± 0.50	−0.51	96-ste/chi(□)
348.14	959.80 ± 0.50	0.19	96-ste/chi(□)	423.11	892.60 ± 0.50	−0.46	96-ste/chi(□)
373.12	939.40 ± 0.50	0.70	96-ste/chi(□)	448.11	868.70 ± 0.50	0.36	96-ste/chi(□)

cont.

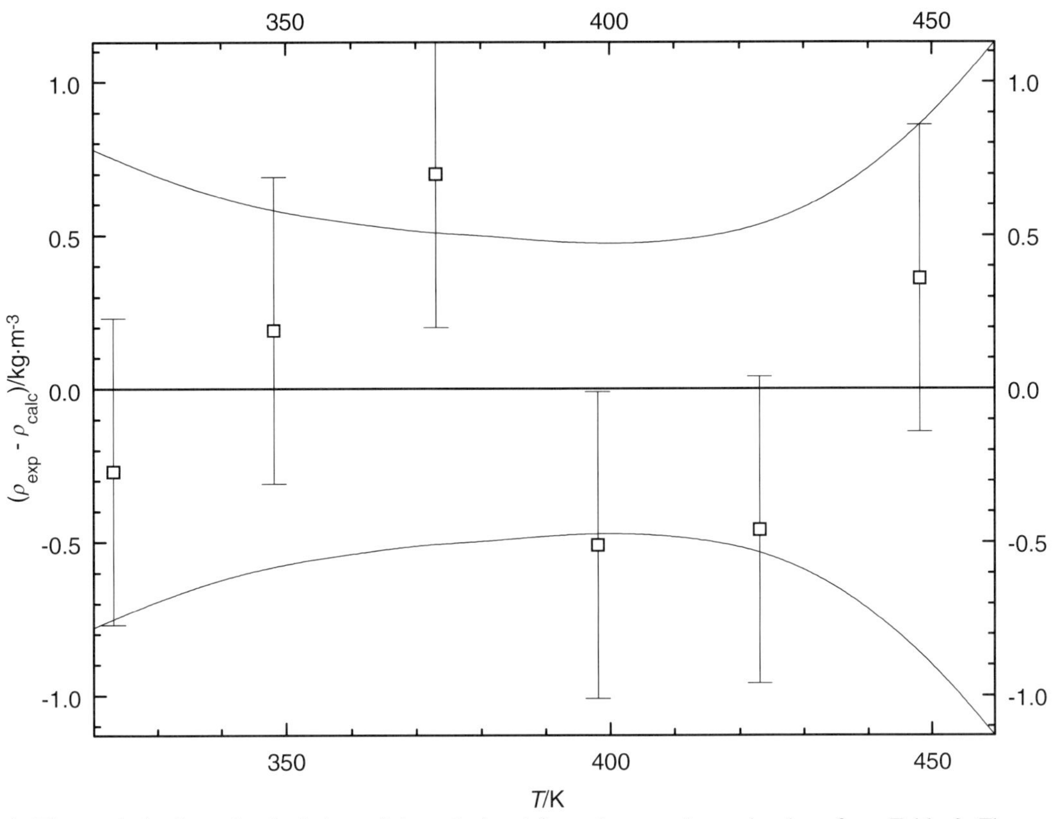

Fig. 1. The symbols show the deviation of the calculated from the experimental values from Table 2. The curves above and below the zero line indicate the calculated error region of the recommended values given in Table 3. The error bars represent the experimental errors. (Error bars smaller than the symbols are omitted for clarity of the figure.)

Table 3. Recommended values (fit to the reliable experimental values according to the equations $\rho = A + BT + CT^2 + DT^3 + \ldots$ or $\rho = [1 + 1.75(1 - T/T_c)^{1/3} + 0.75(1 - T/T_c)][\rho_c + A(T_c - T) + B(T_c - T)^2 + C(T_c - T)^3 + D(T_c - T)^4]$).

$\frac{T}{K}$	$\frac{\rho \pm \sigma_{fit}}{kg \cdot m^{-3}}$	$\frac{T}{K}$	$\frac{\rho \pm \sigma_{fit}}{kg \cdot m^{-3}}$	$\frac{T}{K}$	$\frac{\rho \pm \sigma_{fit}}{kg \cdot m^{-3}}$
320.00	981.65 ± 0.78	370.00	941.38 ± 0.51	420.00	896.05 ± 0.51
330.00	974.00 ± 0.69	380.00	932.72 ± 0.50	430.00	886.38 ± 0.58
340.00	966.15 ± 0.62	390.00	923.86 ± 0.48	440.00	876.50 ± 0.71
350.00	958.10 ± 0.57	400.00	914.79 ± 0.47	450.00	866.42 ± 0.89
360.00	949.84 ± 0.54	410.00	905.52 ± 0.48	460.00	856.13 ± 1.13

1,3-Butanediol **[107-88-0]** **$C_4H_{10}O_2$** **MW = 90.12** **651**

Table 1. Coefficients of the polynomial expansion equation.
Standard deviations (see introduction):
$\sigma_{c,w} = 8.4552 \cdot 10^{-1}$ (combined temperature ranges, weighted),
$\sigma_{c,uw} = 2.2006 \cdot 10^{-1}$ (combined temperature ranges, unweighted).

Coefficient	T = 293.15 to 458.90 K $\rho = A + BT + CT^2 + DT^3 + \ldots$
A	$1.09606 \cdot 10^{3}$
B	$-8.02845 \cdot 10^{-3}$
C	$-1.03339 \cdot 10^{-3}$

Table 2. Experimental values with uncertainties and deviation from calculated values.

$\frac{T}{\text{K}}$	$\frac{\rho_{exp} \pm 2\sigma_{est}}{\text{kg} \cdot \text{m}^{-3}}$	$\frac{\rho_{exp} - \rho_{calc}}{\text{kg} \cdot \text{m}^{-3}}$	Ref. (Symbol in Fig. 1)	$\frac{T}{\text{K}}$	$\frac{\rho_{exp} \pm 2\sigma_{est}}{\text{kg} \cdot \text{m}^{-3}}$	$\frac{\rho_{exp} - \rho_{calc}}{\text{kg} \cdot \text{m}^{-3}}$	Ref. (Symbol in Fig. 1)
298.15	1000.20 ± 1.00	−1.60	48-adk/bil(∇)	323.10	986.10 ± 0.60	0.52	92-sun/dig(O)
293.15	1003.50 ± 0.60	−1.40	62-mel(Δ)	342.90	970.80 ± 0.60	−1.00	92-sun/dig(O)
293.15	1002.00 ± 1.00	−2.90	62-mel[1)]	363.00	957.20 ± 0.60	0.23	92-sun/dig(O)
293.15	1003.70 ± 0.60	−1.20	62-mel(Δ)	383.20	941.50 ± 0.60	0.27	92-sun/dig(O)
293.15	1005.30 ± 0.70	0.40	62-mel(Δ)	400.90	927.00 ± 0.60	0.25	92-sun/dig(O)
293.15	1005.79 ± 0.40	0.89	88-cze/zyw(□)	420.30	910.40 ± 0.60	0.27	92-sun/dig(O)
303.15	999.04 ± 0.40	0.39	88-cze/zyw(□)	439.60	893.10 ± 0.60	0.27	92-sun/dig(O)
313.15	992.39 ± 0.40	0.19	88-cze/zyw(□)	458.90	874.80 ± 0.60	0.05	92-sun/dig(O)
302.40	1000.60 ± 0.60	1.47	92-sun/dig(O)				

[1)] Not included in Fig. 1.

Further references: [29-ber/mie, 35-sch/sta, 37-tis/chu, 43-gre/kel, 48-bou/nic, 48-mic/hop, 50-cle/mac, 52-cur/joh, 54-ros, 55-ser, 56-far/she, 67-pis/gas, 72-rak/isa, 74-rak/mak, 74-rak/mel, 74-wol/ska, 78-mus/kan, 82-man/les].

Table 3. Recommended values (fit to the reliable experimental values according to the equations
$\rho = A + BT + CT^2 + DT^3 + \ldots$ or $\rho = [1 + 1.75(1 - T/T_c)^{1/3} + 0.75(1 - T/T_c)][\rho_c + A(T_c - T) + B(T_c - T)^2 + C(T_c - T)^3 + D(T_c - T)^4]$).

$\frac{T}{\text{K}}$	$\frac{\rho \pm \sigma_{fit}}{\text{kg} \cdot \text{m}^{-3}}$	$\frac{T}{\text{K}}$	$\frac{\rho \pm \sigma_{fit}}{\text{kg} \cdot \text{m}^{-3}}$	$\frac{T}{\text{K}}$	$\frac{\rho \pm \sigma_{fit}}{\text{kg} \cdot \text{m}^{-3}}$
290.00	1006.82 ± 0.72	340.00	973.87 ± 0.54	410.00	919.05 ± 0.63
293.15	1004.90 ± 0.69	350.00	966.66 ± 0.54	420.00	910.39 ± 0.67
298.15	1001.80 ± 0.66	360.00	959.24 ± 0.54	430.00	901.53 ± 0.73
300.00	1000.64 ± 0.65	370.00	951.61 ± 0.55	440.00	892.46 ± 0.81
310.00	994.26 ± 0.60	380.00	943.78 ± 0.56	450.00	883.18 ± 0.90
320.00	987.67 ± 0.57	390.00	935.75 ± 0.58	460.00	873.70 ± 1.01
330.00	980.87 ± 0.55	400.00	927.50 ± 0.60	470.00	864.01 ± 1.14

cont.

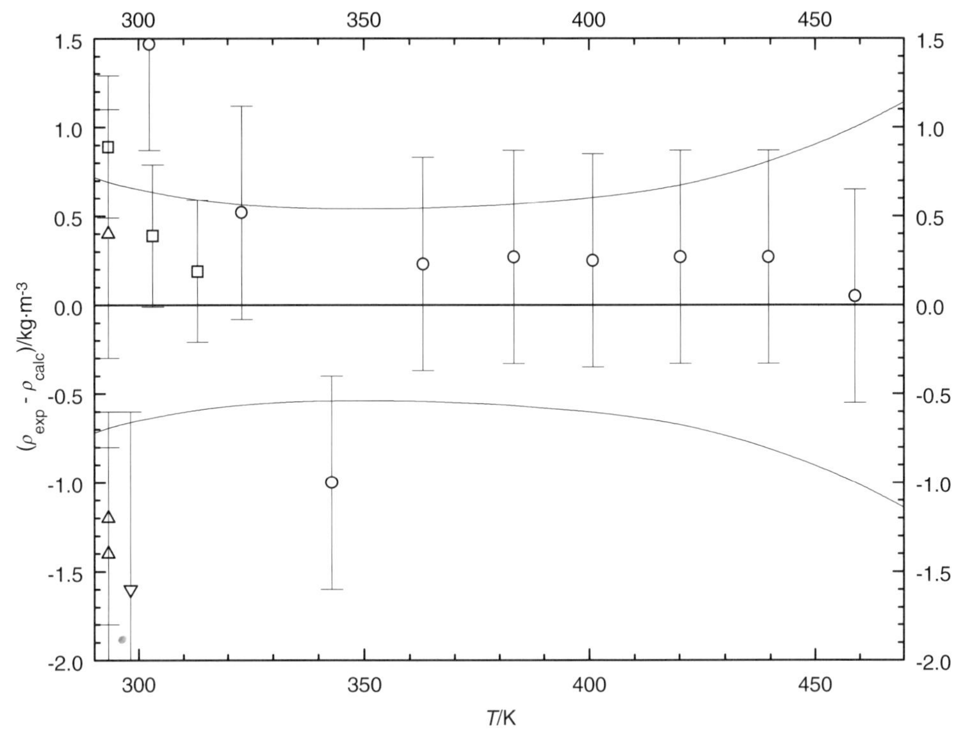

Fig. 1. The symbols show the deviation of the calculated from the experimental values from Table 2. The curves above and below the zero line indicate the calculated error region of the recommended values given in Table 3. The error bars represent the experimental errors. (Error bars smaller than the symbols are omitted for clarity of the figure.)

(*RS*)-1,3-Butanediol **[18826-95-4]** **$C_4H_{10}O_2$** **MW = 90.12** **652**

Table 1. Coefficients of the polynomial expansion equation.
Standard deviations (see introduction):
$\sigma_{c,w} = 4.2923 \cdot 10^{-1}$ (combined temperature ranges, weighted),
$\sigma_{c,uw} = 2.1462 \cdot 10^{-1}$ (combined temperature ranges, unweighted).

Coefficient	T = 323.14 to 448.11 K $\rho = A + BT + CT^2 + DT^3 + \ldots$
A	$1.07840 \cdot 10^3$
B	$8.73975 \cdot 10^{-2}$
C	$-1.18509 \cdot 10^{-3}$

Table 2. Experimental values with uncertainties and deviation from calculated values.

$\frac{T}{K}$	$\frac{\rho_{exp} \pm 2\sigma_{est}}{kg \cdot m^{-3}}$	$\frac{\rho_{exp} - \rho_{calc}}{kg \cdot m^{-3}}$	Ref. (Symbol in Fig. 1)	$\frac{T}{K}$	$\frac{\rho_{exp} \pm 2\sigma_{est}}{kg \cdot m^{-3}}$	$\frac{\rho_{exp} - \rho_{calc}}{kg \cdot m^{-3}}$	Ref. (Symbol in Fig. 1)
323.14	982.80 ± 0.50	−0.10	96-ste/chi(□)	448.11	880.00 ± 0.50	0.40	96-ste/chi(□)
348.13	965.10 ± 0.50	−0.10	96-ste/chi(□)	672.80	391.30 ± 0.00	−209.46	96-ste/chi[1)]
373.12	946.50 ± 0.50	0.48	96-ste/chi(□)	674.50	348.90 ± 0.00	−249.29	96-ste/chi[1)]
398.12	925.50 ± 0.50	0.14	96-ste/chi(□)	675.40	316.00 ± 0.00	−280.83	96-ste/chi[1)]
423.11	902.40 ± 0.50	−0.82	96-ste/chi(□)				

[1)] Not included in Fig. 1.

cont.

(*RS*)-1,3-Butanediol (cont.)

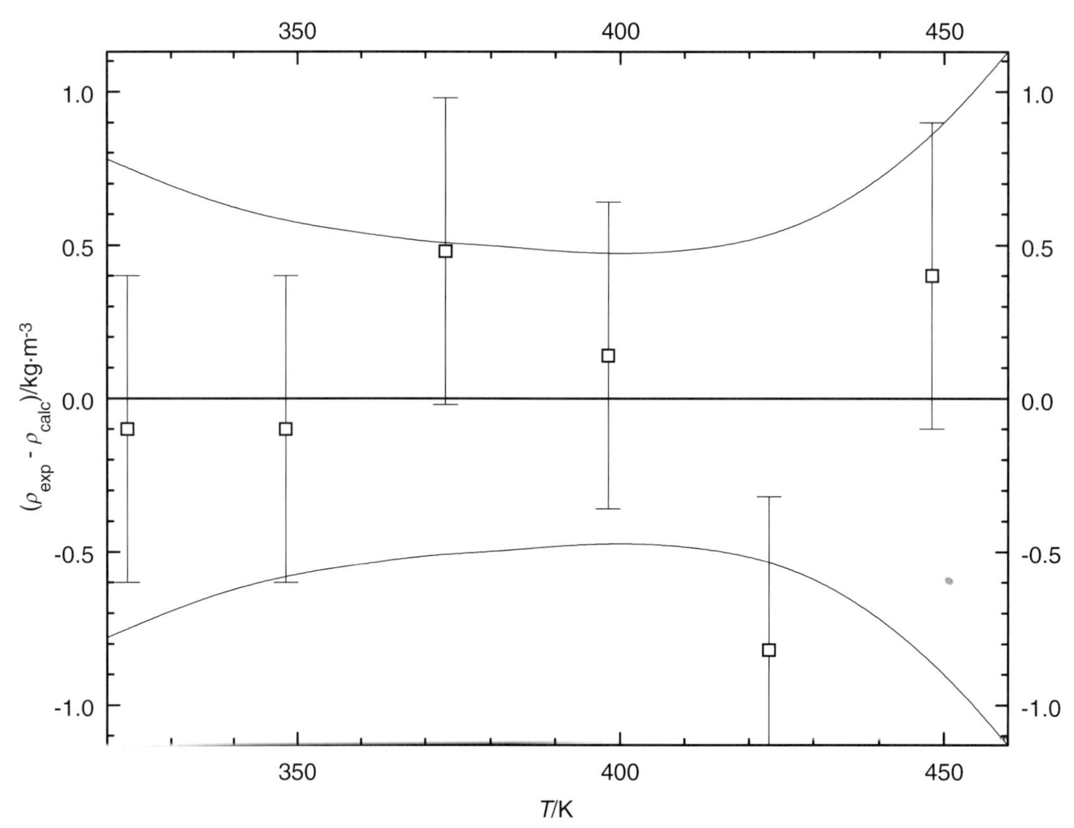

Fig. 1. The symbols show the deviation of the calculated from the experimental values from Table 2. The curves above and below the zero line indicate the calculated error region of the recommended values given in Table 3. The error bars represent the experimental errors. (Error bars smaller than the symbols are omitted for clarity of the figure.)

Table 3. Recommended values (fit to the reliable experimental values according to the equations $\rho = A + BT + CT^2 + DT^3 + \ldots$ or $\rho = [1 + 1.75(1 - T/T_c)^{1/3} + 0.75(1 - T/T_c)][\rho_c + A(T_c - T) + B(T_c - T)^2 + C(T_c - T)^3 + D(T_c - T)^4]$).

$\frac{T}{\mathrm{K}}$	$\frac{\rho \pm \sigma_{fit}}{\mathrm{kg \cdot m^{-3}}}$	$\frac{T}{\mathrm{K}}$	$\frac{\rho \pm \sigma_{fit}}{\mathrm{kg \cdot m^{-3}}}$	$\frac{T}{\mathrm{K}}$	$\frac{\rho \pm \sigma_{fit}}{\mathrm{kg \cdot m^{-3}}}$
320.00	985.02 ± 0.78	370.00	948.50 ± 0.51	420.00	906.06 ± 0.51
330.00	978.19 ± 0.69	380.00	940.49 ± 0.50	430.00	896.86 ± 0.58
340.00	971.12 ± 0.62	390.00	932.23 ± 0.48	440.00	887.42 ± 0.71
350.00	963.82 ± 0.57	400.00	923.75 ± 0.47	450.00	877.75 ± 0.89
360.00	956.28 ± 0.54	410.00	915.02 ± 0.48	460.00	867.84 ± 1.13

1,4-Butanediol **[110-63-4]** **$C_4H_{10}O_2$** **MW = 90.12** **653**

Table 1. Coefficients of the polynomial expansion equation.
Standard deviations (see introduction):
$\sigma_{c,w}$ = 1.0892 (combined temperature ranges, weighted),
$\sigma_{c,uw}$ = 2.6052 · 10^{-1} (combined temperature ranges, unweighted).

Coefficient	T = 293.15 to 493.15 K $\rho = A + BT + CT^2 + DT^3 + \ldots$
A	1.12256 · 10^3
B	−1.72321 · 10^{-1}
C	−5.89027 · 10^{-4}
D	−1.76600 · 10^{-7}

Table 2. Experimental values with uncertainties and deviation from calculated values.

$\frac{T}{\text{K}}$	$\frac{\rho_{exp} \pm 2\sigma_{est}}{\text{kg} \cdot \text{m}^{-3}}$	$\frac{\rho_{exp} - \rho_{calc}}{\text{kg} \cdot \text{m}^{-3}}$	Ref. (Symbol in Fig. 1)	$\frac{T}{\text{K}}$	$\frac{\rho_{exp} \pm 2\sigma_{est}}{\text{kg} \cdot \text{m}^{-3}}$	$\frac{\rho_{exp} - \rho_{calc}}{\text{kg} \cdot \text{m}^{-3}}$	Ref. (Symbol in Fig. 1)
293.15	1017.10 ± 0.80	0.13	29-kir/ric(X)	493.15	874.60 ± 1.00	1.45	78-apa/ker-2(X)
293.15	1016.00 ± 0.60	−0.97	48-bou/nic(∇)	293.15	1016.22 ± 0.40	−0.75	88-cze/zyw(□)
293.15	1017.30 ± 1.00	0.33	57-ket/van[1)]	303.15	1010.11 ± 0.40	−1.16	88-cze/zyw(□)
318.15	1001.80 ± 1.00	−0.63	57-ket/van(X)	313.15	1004.15 ± 0.40	−1.26	88-cze/zyw(□)
343.15	985.90 ± 1.00	−1.03	57-ket/van(X)	303.30	1011.40 ± 0.60	0.22	92-sun/dig(Δ)
293.15	1017.10 ± 0.70	0.13	62-mel(◆)	322.10	999.00 ± 0.60	−1.04	92-sun/dig(Δ)
293.15	1016.00 ± 0.60	−0.97	62-mel(◆)	343.00	986.40 ± 0.60	−0.63	92-sun/dig(Δ)
293.15	1018.50 ± 0.60	1.53	62-mel(◆)	363.60	973.30 ± 0.60	−0.24	92-sun/dig(Δ)
298.15	1015.40 ± 1.00	1.26	62-mel(◆)	384.80	959.30 ± 0.60	0.33	92-sun/dig(Δ)
298.15	1015.40 ± 0.60	1.26	66-fre/hor(O)	402.20	946.50 ± 0.60	0.02	92-sun/dig(Δ)
295.15	1017.80 ± 1.00	1.96	78-apa/ker-2(X)	423.30	930.50 ± 0.60	−0.18	92-sun/dig(Δ)
358.16	980.20 ± 1.00	3.03	78-apa/ker-2(X)	442.30	915.60 ± 0.60	−0.23	92-sun/dig(Δ)
451.11	913.00 ± 1.00	4.26	78-apa/ker-2[1)]	460.30	899.00 ± 0.60	−2.22	92-sun/dig(Δ)

[1)] Not included in Fig. 1.

Further references: [1890-dek, 01-ham, 32-hun, 50-cle/mac, 52-cur/joh, 54-ros, 60-kun/sak, 63-shu/bar, 66-mya/pya, 68-naz/tsy, 70-are/tav, 78-mus/kan, 82-man/les].

Table 3. Recommended values (fit to the reliable experimental values according to the equations
$\rho = A + BT + CT^2 + DT^3 + \ldots$ or $\rho = [1 + 1.75(1 - T/T_c)^{1/3} + 0.75(1 - T/T_c)][\rho_c + A(T_c - T) + B(T_c - T)^2 + C(T_c - T)^3 + D(T_c - T)^4]$).

$\frac{T}{\text{K}}$	$\frac{\rho \pm \sigma_{fit}}{\text{kg} \cdot \text{m}^{-3}}$	$\frac{T}{\text{K}}$	$\frac{\rho \pm \sigma_{fit}}{\text{kg} \cdot \text{m}^{-3}}$	$\frac{T}{\text{K}}$	$\frac{\rho \pm \sigma_{fit}}{\text{kg} \cdot \text{m}^{-3}}$
290.00	1018.74 ± 0.69	350.00	982.52 ± 0.76	430.00	925.51 ± 0.63
293.15	1016.97 ± 0.70	360.00	975.95 ± 0.73	440.00	917.66 ± 0.66
298.15	1014.14 ± 0.71	370.00	969.22 ± 0.69	450.00	909.64 ± 0.71
300.00	1013.08 ± 0.72	380.00	962.33 ± 0.66	460.00	901.46 ± 0.80
310.00	1007.27 ± 0.76	390.00	955.29 ± 0.63	470.00	893.12 ± 0.94
320.00	1001.31 ± 0.78	400.00	948.08 ± 0.61	480.00	884.60 ± 1.14
330.00	995.20 ± 0.79	410.00	940.72 ± 0.61	490.00	875.92 ± 1.40
340.00	988.94 ± 0.78	420.00	933.20 ± 0.61	500.00	867.07 ± 1.74

cont.

1,4-Butanediol (cont.)

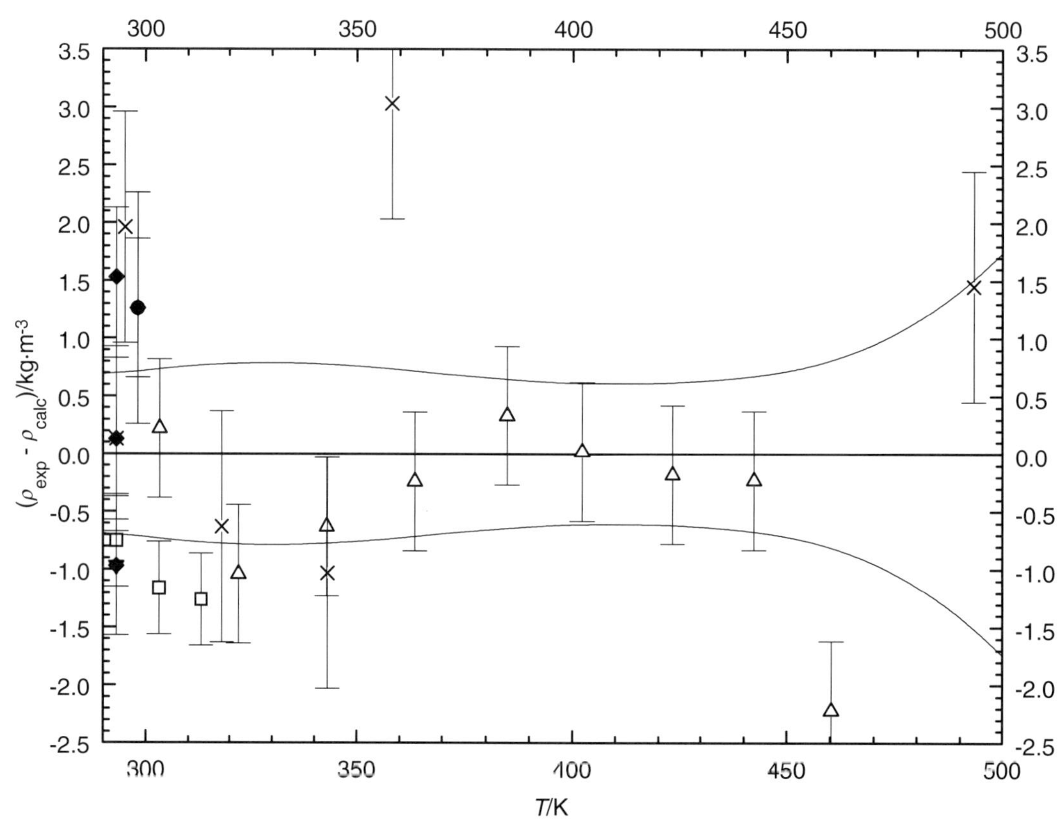

Fig. 1. The symbols show the deviation of the calculated from the experimental values from Table 2. The curves above and below the zero line indicate the calculated error region of the recommended values given in Table 3. The error bars represent the experimental errors. (Error bars smaller than the symbols are omitted for clarity of the figure.)

2,3-Butanediol (isomer not specified) [513-85-9] $C_4H_{10}O_2$ MW = 90.12 654

Table 1. Fit with estimated *B* coefficient for 5 accepted points. Deviation $\sigma_w = 0.509$.

Coefficient	$\rho = A + BT$
A	1238.27
B	−0.800

Table 2. Experimental values with uncertainties and deviation from calculated values.

$\frac{T}{K}$	$\frac{\rho_{exp} \pm 2\sigma_{est}}{kg \cdot m^{-3}}$	$\frac{\rho_{exp} - \rho_{calc}}{kg \cdot m^{-3}}$	Ref.	$\frac{T}{K}$	$\frac{\rho_{exp} \pm 2\sigma_{est}}{kg \cdot m^{-3}}$	$\frac{\rho_{exp} - \rho_{calc}}{kg \cdot m^{-3}}$	Ref.
293.15	1007.6 ± 3.0	3.85	62-mel[1)]	293.15	1005.0 ± 2.0	1.25	65-wei/lan
293.15	1003.3 ± 1.0	−0.45	64-mys/zie	293.15	1003.3 ± 1.0	−0.45	35-sch/sta
298.15	1000.3 ± 1.0	0.55	64-mys/zie	298.15	999.8 ± 1.0	0.05	46-cle-1

[1)] Not included in calculation of linear coefficients.

cont.

Table 3. Recommended values.

$\frac{T}{K}$	$\frac{\rho_{exp} \pm 2\sigma_{est}}{kg \cdot m^{-3}}$
290.00	1006.3 ± 1.1
293.15	1003.8 ± 1.0
298.15	999.8 ± 1.0

D-(-)-2,3-Butanediol [24347-58-8] $C_4H_{10}O_2$ MW = 90.12 655

Table 1. Experimental value with uncertainty.

$\frac{T}{K}$	$\frac{\rho_{exp} \pm 2\sigma_{est}}{kg \cdot m^{-3}}$	Ref.
298.15	987.2 ± 0.5	46-kno/sch

dl-2,3-Butanediol [6982-25-8] $C_4H_{10}O_2$ MW = 90.12 656

Table 1. Coefficients of the polynomial expansion equation.
Standard deviations (see introduction):
$\sigma_{c,w} = 7.3545 \cdot 10^{-1}$ (combined temperature ranges, weighted),
$\sigma_{c,uw} = 1.9800 \cdot 10^{-1}$ (combined temperature ranges, unweighted).

Coefficient	T = 293.15 to 452.20 K $\rho = A + BT + CT^2 + DT^3 + \ldots$
A	$1.16268 \cdot 10^{3}$
B	$-3.59946 \cdot 10^{-1}$
C	$-7.45938 \cdot 10^{-4}$

Table 2. Experimental values with uncertainties and deviation from calculated values.

$\frac{T}{K}$	$\frac{\rho_{exp} \pm 2\sigma_{est}}{kg \cdot m^{-3}}$	$\frac{\rho_{exp} - \rho_{calc}}{kg \cdot m^{-3}}$	Ref. (Symbol in Fig. 1)	$\frac{T}{K}$	$\frac{\rho_{exp} \pm 2\sigma_{est}}{kg \cdot m^{-3}}$	$\frac{\rho_{exp} - \rho_{calc}}{kg \cdot m^{-3}}$	Ref. (Symbol in Fig. 1)
298.15	990.00 ± 1.00	0.94	44-mor/aue(Δ)	381.20	917.70 ± 0.60	0.62	92-sun/dig(□)
293.15	993.00 ± 1.00	−0.06	75-nak/kom(○)	400.60	899.80 ± 0.60	1.02	92-sun/dig(□)
303.30	984.90 ± 0.60	0.01	92-sun/dig(□)	419.50	880.20 ± 0.60	−0.22	92-sun/dig(□)
322.90	968.50 ± 0.60	−0.18	92-sun/dig(□)	439.70	860.00 ± 0.60	−0.20	92-sun/dig(□)
343.60	950.20 ± 0.60	−0.74	92-sun/dig(□)	452.20	847.10 ± 0.60	−0.28	92-sun/dig(□)
361.70	934.00 ± 0.60	−0.90	92-sun/dig(□)				

Table 3. Recommended values (fit to the reliable experimental values according to the equations
$\rho = A + BT + CT^2 + DT^3 + \ldots$ or $\rho = [1 + 1.75(1 - T/T_c)^{1/3} + 0.75(1 - T/T_c)][\rho_c + A(T_c - T) + B(T_c - T)^2 + C(T_c - T)^3 + D(T_c - T)^4]$).

$\frac{T}{K}$	$\frac{\rho \pm \sigma_{fit}}{kg \cdot m^{-3}}$	$\frac{T}{K}$	$\frac{\rho \pm \sigma_{fit}}{kg \cdot m^{-3}}$	$\frac{T}{K}$	$\frac{\rho \pm \sigma_{fit}}{kg \cdot m^{-3}}$
290.00	995.57 ± 1.20	340.00	954.07 ± 0.58	410.00	889.71 ± 0.62
293.15	993.06 ± 1.14	350.00	945.33 ± 0.55	420.00	879.92 ± 0.68
298.15	989.06 ± 1.04	360.00	936.43 ± 0.54	430.00	869.98 ± 0.75
300.00	987.57 ± 1.01	370.00	927.39 ± 0.54	440.00	859.89 ± 0.84
310.00	979.42 ± 0.85	380.00	918.19 ± 0.54	450.00	849.66 ± 0.95
320.00	971.12 ± 0.73	390.00	908.85 ± 0.56	460.00	839.27 ± 1.09
330.00	962.67 ± 0.64	400.00	899.36 ± 0.59		

cont.

***dl*-2,3-Butanediol** (cont.)

Further references: [1859-wur, 51-wat/coo, 52-cur/joh].

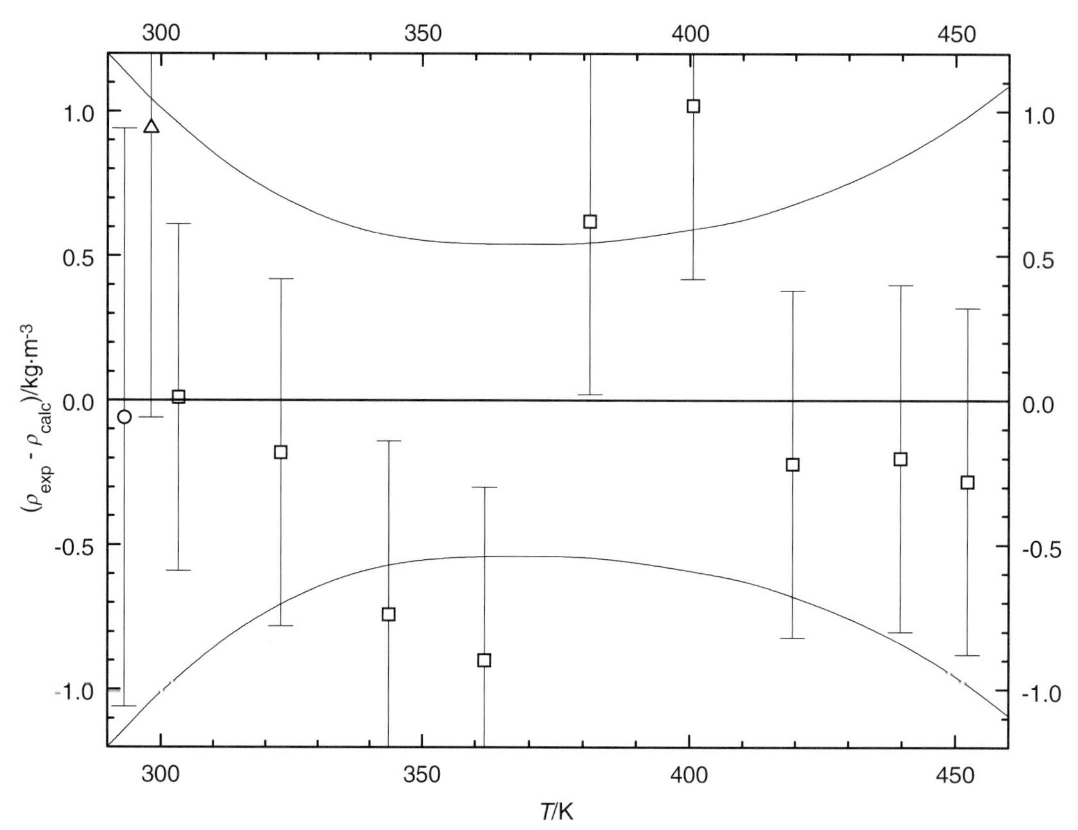

Fig. 1. The symbols show the deviation of the calculated from the experimental values from Table 2. The curves above and below the zero line indicate the calculated error region of the recommended values given in Table 3. The error bars represent the experimental errors. (Error bars smaller than the symbols are omitted for clarity of the figure.)

***L*-(+)-2,3-Butanediol** **[19132-06-0]** **$C_4H_{10}O_2$** **MW = 90.12** **657**

Table 1. Experimental value with uncertainty.

$\frac{T}{K}$	$\frac{\rho_{exp} \pm 2\sigma_{est}}{kg \cdot m^{-3}}$	Ref.
298.15	986.9 ± 0.5	46-kno/sch

(*R, *S**)-2,3-Butanediol** **[5341-95-7]** **$C_4H_{10}O_2$** **MW = 90.12** **658**

Table 1. Experimental value with uncertainty.

$\frac{T}{K}$	$\frac{\rho_{exp} \pm 2\sigma_{est}}{kg \cdot m^{-3}}$	Ref.
298.15	993.9 ± 2.0	46-kno/sch

2-Methyl-1,2-propanediol **[558-43-0]** **$C_4H_{10}O_2$** **MW = 90.12** **659**

Table 1. Experimental and recommended values with uncertainties.

T / K	$\rho_{exp} \pm 2\sigma_{est}$ / $kg \cdot m^{-3}$	Ref.	T / K	$\rho_{exp} \pm 2\sigma_{est}$ / $kg \cdot m^{-3}$	Ref.
293.15	1002.4 ± 3.0	37-dob/gut[1)]	293.15	996.7 ± 2.0	54-vor/tit
287.15	999.0 ± 3.0	37-mou/dod[1)]	293.15	997.0 ± 2.0	65-wei/lan
293.15	1003.0 ± 3.0	41-hea/tam[1)]	293.15	996.9 ± 2.0	Recommended
298.15	989.6 ± 3.0	48-adk/bil[1)]			

[1)] Not included in calculation of recommended value.

2-Methyl-1,3-propanediol **[2163-42-0]** **$C_4H_{10}O_2$** **MW = 90.12** **660**

Table 1. Experimental and recommended values with uncertainties.

T / K	$\rho_{exp} \pm 2\sigma_{est}$ / $kg \cdot m^{-3}$	Ref.	T / K	$\rho_{exp} \pm 2\sigma_{est}$ / $kg \cdot m^{-3}$	Ref.
273.15	1029.7 ± 3.0	07-fav/sok-3[1)]	293.15	1009.0 ± 3.0	65-bar/koz[1)]
273.15	1029.7 ± 3.0	07-sok[1)]	293.15	1020.0 ± 2.0	65-wei/lan
293.15	1029.0 ± 3.0	35-has/mcb[1)]	293.15	1020.0 ± 2.0	Recommended
293.15	1027.3 ± 3.0	42-pum/hah[1)]			

[1)] Not included in calculation of recommended value.

2,2-Dimethyl-1,3-propanediol **[126-30-7]** **$C_5H_{12}O_2$** **MW = 104.15** **661**

Table 1. Experimental values with uncertainties.

T / K	$\rho_{exp} \pm 2\sigma_{est}$ / $kg \cdot m^{-3}$	Ref.
	crystal	
298.15	1066.0 ± 5.0	57-ano-10
298.15	1066.0 ± 5.0	60-ano-13

2-Ethyl-1,3-propanediol **[2612-29-5]** **$C_5H_{12}O_2$** **MW = 104.15** **662**

Table 1. Experimental value with uncertainty.

T / K	$\rho_{exp} \pm 2\sigma_{est}$ / $kg \cdot m^{-3}$	Ref.
293.15	997.0 ± 1.0	48-adk/bil

2-Methyl-1,3-butanediol **[684-84-4]** **$C_5H_{12}O_2$** **MW = 104.15** **663**

Table 1. Experimental and recommended values with uncertainties.

T / K	$\rho_{exp} \pm 2\sigma_{est}$ / $kg \cdot m^{-3}$	Ref.	T / K	$\rho_{exp} \pm 2\sigma_{est}$ / $kg \cdot m^{-3}$	Ref.
293.15	994.0 ± 1.5	58-far/spe[1)]	293.15	991.5 ± 1.0	64-hel/dav
293.15	991.9 ± 1.0	56-far/she	293.15	991.7 ± 1.0	Recommended

[1)] Not included in calculation of recommended value.

2-Methyl-2,3-butanediol **[5396-58-7]** $C_5H_{12}O_2$ **MW = 104.15** **664**

Table 1. Experimental value with uncertainty.

T / K	$\rho_{exp} \pm 2\sigma_{est}$ / $kg \cdot m^{-3}$	Ref.
298.15	968.8 ± 1.0	58-hen/wat

3-Methyl-1,2-butanediol **[50468-22-9]** $C_5H_{12}O_2$ **MW = 104.15** **665**

Table 1. Experimental values with uncertainties.

T / K	$\rho_{exp} \pm 2\sigma_{est}$ / $kg \cdot m^{-3}$	Ref.
273.15	998.8 ± 2.0	1875-fla
294.65	984.2 ± 2.0	1875-fla

3-Methyl-1,3-butanediol **[2568-33-4]** $C_5H_{12}O_2$ **MW = 104.15** **666**

Table 1. Experimental and recommended values with uncertainties.

T / K	$\rho_{exp} \pm 2\sigma_{est}$ / $kg \cdot m^{-3}$	Ref.	T / K	$\rho_{exp} \pm 2\sigma_{est}$ / $kg \cdot m^{-3}$	Ref.
293.15	983.3 ± 3.0	07-kut[1)]	293.15	976.3 ± 2.0	56-far/she
273.15	995.4 ± 3.0	07-kut[1)]	293.15	977.7 ± 2.0	62-esa/shi
293.15	964.5 ± 6.0	55-sar/mor[1)]	293.15	977.0 ± 2.1	Recommended
293.15	986.7 ± 3.0	57-far/rot[1)]			

[1)] Not included in calculation of recommended value.

1,2-Pentanediol **[5343-92-0]** $C_5H_{12}O_2$ **MW = 104.15** **667**

Table 1. Fit with estimated B coefficient for 6 accepted points. Deviation $\sigma_w = 0.167$.

Coefficient	$\rho = A + BT$
A	1194.93
B	−0.760

Table 2. Experimental values with uncertainties and deviation from calculated values.

T / K	$\rho_{exp} \pm 2\sigma_{est}$ / $kg \cdot m^{-3}$	$\rho_{exp} - \rho_{calc}$ / $kg \cdot m^{-3}$	Ref.	T / K	$\rho_{exp} \pm 2\sigma_{est}$ / $kg \cdot m^{-3}$	$\rho_{exp} - \rho_{calc}$ / $kg \cdot m^{-3}$	Ref.
273.15	987.0 ± 1.0	−0.34	1859-wur	293.15	972.3 ± 1.0	0.16	50-cle/mac
293.15	978.5 ± 3.0	6.36	23-kau/ada[1)]	293.15	970.7 ± 2.0	−1.39	51-cop/fie[1)]
297.15	969.1 ± 1.0	0.00	45-len/dup	293.15	972.3 ± 1.0	0.16	62-mel
297.15	969.1 ± 1.0	0.00	45-sch/gel	297.15	969.1 ± 1.0	0.00	62-mel

[1)] Not included in calculation of linear coefficients.

Table 3. Recommended values.

T / K	$\rho_{exp} \pm 2\sigma_{est}$ / $kg \cdot m^{-3}$	T / K	$\rho_{exp} \pm 2\sigma_{est}$ / $kg \cdot m^{-3}$
270.00	989.7 ± 2.3	293.15	972.1 ± 0.7
280.00	982.1 ± 1.4	298.15	968.3 ± 0.9
290.00	974.5 ± 0.7		

1,3-Pentanediol **[3174-67-2]** $C_5H_{12}O_2$ **MW = 104.15** **668**

Table 1. Experimental value with uncertainty.

$\frac{T}{K}$	$\frac{\rho_{exp} \pm 2\sigma_{est}}{kg \cdot m^{-3}}$	Ref.
293.15	981.0 ± 1.0	64-hel/dav

1,4-Pentanediol **[626-95-9]** $C_5H_{12}O_2$ **MW = 104.15** **669**

Table 1. Experimental and recommended values with uncertainties.

$\frac{T}{K}$	$\frac{\rho_{exp} \pm 2\sigma_{est}}{kg \cdot m^{-3}}$	Ref.	$\frac{T}{K}$	$\frac{\rho_{exp} \pm 2\sigma_{est}}{kg \cdot m^{-3}}$	Ref.
273.00	1000.1 ± 2.0	1889-lip[1)]	293.15	988.6 ± 1.0	63-shu/bar
293.15	990.3 ± 1.0	57-fav/ser	293.15	989.5 ± 1.2	Recommended

[1)] Not included in calculation of recommended value.

1,5-Pentanediol **[111-29-5]** $C_5H_{12}O_2$ **MW = 104.15** **670**

Table 1. Fit with estimated B coefficient for 20 accepted points. Deviation $\sigma_w = 0.834$.

Coefficient	$\rho = A + BT$
A	1174.56
B	−0.620

Table 2. Experimental values with uncertainties and deviation from calculated values.

$\frac{T}{K}$	$\frac{\rho_{exp} \pm 2\sigma_{est}}{kg \cdot m^{-3}}$	$\frac{\rho_{exp} - \rho_{calc}}{kg \cdot m^{-3}}$	Ref.	$\frac{T}{K}$	$\frac{\rho_{exp} \pm 2\sigma_{est}}{kg \cdot m^{-3}}$	$\frac{\rho_{exp} - \rho_{calc}}{kg \cdot m^{-3}}$	Ref.
284.65	987.0 ± 6.0	−11.08	1893-eyk-1[1)]	318.15	978.5 ± 2.0	1.19	57-ket/van
351.15	944.5 ± 6.0	−12.35	1893-eyk-1[1)]	293.15	991.4 ± 1.0	−1.41	62-mel
291.15	994.0 ± 2.0	−0.05	04-ham	299.15	989.0 ± 1.0	−0.09	62-mel
293.15	992.1 ± 1.0	−0.71	23-kau/ada	293.15	990.4 ± 2.0	−2.41	67-ano-5
293.15	993.8 ± 2.0	0.99	27-tri	293.15	990.4 ± 2.0	−2.41	68-ano
284.15	998.0 ± 1.0	−0.39	34-pau	293.15	994.3 ± 2.0	1.49	68-naz/tsy
299.15	989.0 ± 2.0	−0.09	45-sch/gel	278.15	1005.9 ± 3.0	3.79	75-nak/kom[1)]
293.15	991.4 ± 1.0	−1.41	50-cle/mac	298.15	997.3 ± 3.0	7.59	75-nak/kom[1)]
293.15	992.3 ± 1.0	−0.51	52-cur/joh	318.15	986.0 ± 3.0	8.69	75-nak/kom[1)]
293.15	990.4 ± 2.0	−2.41	54-ano-12	293.15	992.8 ± 0.4	−0.02	88-cze/zyw
293.15	989.0 ± 2.0	−3.81	55-mos	303.15	986.7 ± 0.4	0.12	88-cze/zyw
293.15	994.0 ± 2.0	1.19	57-ket/van	313.15	981.2 ± 0.4	0.84	88-cze/zyw
343.15	962.8 ± 2.0	0.99	57-ket/van				

[1)] Not included in calculation of linear coefficients.

Table 3. Recommended values.

$\frac{T}{K}$	$\frac{\rho_{exp} \pm 2\sigma_{est}}{kg \cdot m^{-3}}$	$\frac{T}{K}$	$\frac{\rho_{exp} \pm 2\sigma_{est}}{kg \cdot m^{-3}}$	$\frac{T}{K}$	$\frac{\rho_{exp} \pm 2\sigma_{est}}{kg \cdot m^{-3}}$
280.00	1001.0 ± 1.7	298.15	989.7 ± 1.3	330.00	970.0 ± 2.0
290.00	994.8 ± 1.4	310.00	982.4 ± 1.4	340.00	963.8 ± 2.4
293.15	992.8 ± 1.4	320.00	976.2 ± 1.7	350.00	957.6 ± 2.8

2,3-Pentanediol **[42027-23-6]** $C_5H_{12}O_2$ **MW = 104.15** **671**

Table 1. Experimental and recommended values with uncertainties.

$\frac{T}{K}$	$\frac{\rho_{exp} \pm 2\sigma_{est}}{kg \cdot m^{-3}}$	Ref.
273.15	994.4 ± 4.0	1875-wag/say-1 [1)]
287.15	979.9 ± 4.0	1875-wag/say-1 [1)]
273.15	1005.0 ± 1.0	37-mil/sus
273.15	1005.0 ± 1.0	Recommended

[1)] Not included in calculation of recommended value.

2,4-Pentanediol **[625-69-4]** $C_5H_{12}O_2$ **MW = 104.15** **672**

Table 1. Fit with estimated B coefficient for 3 accepted points. Deviation $\sigma_w = 0.900$.

Coefficient	$\rho = A + BT$
A	1201.54
B	−0.820

Table 2. Experimental values with uncertainties and deviation from calculated values.

$\frac{T}{K}$	$\frac{\rho_{exp} \pm 2\sigma_{est}}{kg \cdot m^{-3}}$	$\frac{\rho_{exp} - \rho_{calc}}{kg \cdot m^{-3}}$	Ref.
293.15	963.5 ± 2.0	2.34	11-zel/arj [1)]
287.15	966.0 ± 1.0	−0.08	14-vav
293.15	962.3 ± 1.0	1.14	52-cur/joh
298.15	956.0 ± 1.0	−1.06	72-caz/mar

[1)] Not included in calculation of linear coefficients.

Table 3. Recommended values.

$\frac{T}{K}$	$\frac{\rho_{exp} \pm 2\sigma_{est}}{kg \cdot m^{-3}}$
280.00	971.9 ± 1.7
290.00	963.7 ± 1.2
293.15	961.2 ± 1.1
298.15	957.1 ± 1.2

2,2-Dimethyl-1,3-butanediol **[76-35-7]** $C_6H_{14}O_2$ **MW = 118.18** **673**

Table 1. Experimental value with uncertainty.

$\frac{T}{K}$	$\frac{\rho_{exp} \pm 2\sigma_{est}}{kg \cdot m^{-3}}$	Ref.
293.15	768.4 ± 1.0	61-ano-11

2,2-Dimethyl-1,4-butanediol **[32812-23-0]** $C_6H_{14}O_2$ **MW = 118.18** **674**

Table 1. Experimental value with uncertainty.

T/K	$\rho_{exp} \pm 2\sigma_{est}$ / $kg \cdot m^{-3}$	Ref.
277.15	996.0 ± 2.0	04-bou/bla

2,3-Dimethyl-1,3-butanediol **[24893-35-4]** $C_6H_{14}O_2$ **MW = 118.18** **675**

Table 1. Fit with estimated B coefficient for 3 accepted points. Deviation $\sigma_w = 0.082$.

Coefficient	$\rho = A + BT$
A	1179.17
B	−0.720

Table 2. Experimental values with uncertainties and deviation from calculated values.

T/K	$\rho_{exp} \pm 2\sigma_{est}$ / $kg \cdot m^{-3}$	$\rho_{exp} - \rho_{calc}$ / $kg \cdot m^{-3}$	Ref.
293.15	968.0 ± 0.6	−0.10	53-hat/jou
298.15	964.5 ± 0.6	−0.00	53-hat/jou
303.15	961.0 ± 0.6	0.10	53-hat/jou
293.15	998.0 ± 10.0	29.90	57-sar/vor[1)]

[1)] Not included in calculation of linear coefficients.

Table 3. Recommended values.

T/K	$\rho_{exp} \pm 2\sigma_{est}$ / $kg \cdot m^{-3}$
290.00	970.4 ± 0.6
293.15	968.1 ± 0.5
298.15	964.5 ± 0.4
310.00	956.0 ± 0.7

2,3-Dimethyl-1,4-butanediol **[57716-80-0]** $C_6H_{14}O_2$ **MW = 118.18** **676**

Table 1. Experimental value with uncertainty.

T/K	$\rho_{exp} \pm 2\sigma_{est}$ / $kg \cdot m^{-3}$	Ref.
293.15	977.1 ± 2.0	62-raz/bog

***dl*-2,3-Dimethyl-1,4-butanediol** **[66553-14-8]** $C_6H_{14}O_2$ **MW = 118.18** **677**

Table 1. Experimental value with uncertainty.

T/K	$\rho_{exp} \pm 2\sigma_{est}$ / $kg \cdot m^{-3}$	Ref.
293.15	974.0 ± 2.0	54-mcc/pro

meso-2,3-Dimethyl-1,4-butanediol **[500009-29-0]** $C_6H_{14}O_2$ **MW = 118.18** **678**

Table 1. Experimental value with uncertainty.

$\frac{T}{K}$	$\frac{\rho_{exp} \pm 2\sigma_{est}}{kg \cdot m^{-3}}$	Ref.
293.15	970.0 ± 2.0	54-mcc/pro

2,3-Dimethyl-2,3-butanediol **[76-09-5]** $C_6H_{14}O_2$ **MW = 118.18** **679**

Table 1. Fit with estimated *B* coefficient for 5 accepted points. Deviation $\sigma_w = 0.524$.

Coefficient	$\rho = A + BT$
A	1196.92
B	−0.800

Table 2. Experimental values with uncertainties and deviation from calculated values.

$\frac{T}{K}$	$\frac{\rho_{exp} \pm 2\sigma_{est}}{kg \cdot m^{-3}}$	$\frac{\rho_{exp} - \rho_{calc}}{kg \cdot m^{-3}}$	Ref.	$\frac{T}{K}$	$\frac{\rho_{exp} \pm 2\sigma_{est}}{kg \cdot m^{-3}}$	$\frac{\rho_{exp} - \rho_{calc}}{kg \cdot m^{-3}}$	Ref.
288.15	966.3 ± 2.0	−0.11	1884-per	290.15	964.1 ± 2.0	−0.70	46-lau/wie
298.15	958.0 ± 2.0	−0.39	1884-per	288.15	967.0 ± 2.0	0.60	52-cur/joh
288.15	967.0 ± 2.0	0.60	25-vor/wal	293.15	933.0 ± 3.0	−29.40	52-pet/she[1)]

[1)] Not included in calculation of linear coefficients.

Table 3. Recommended values.

$\frac{T}{K}$	$\frac{\rho_{exp} \pm 2\sigma_{est}}{kg \cdot m^{-3}}$
280.00	972.9 ± 1.7
290.00	964.9 ± 1.6
293.15	962.4 ± 1.6
298.15	958.4 ± 1.6

3,3-Dimethyl-1,2-butanediol **[59562-82-2]** $C_6H_{14}O_2$ **MW = 118.18** **680**

Table 1. Experimental value with uncertainty.

$\frac{T}{K}$	$\frac{\rho_{exp} \pm 2\sigma_{est}}{kg \cdot m^{-3}}$	Ref.
323.15	940.0 ± 2.0	09-cla

2-Ethyl-1,3-butanediol **[66553-17-1]** $C_6H_{14}O_2$ **MW = 118.18** **681**

Table 1. Experimental value with uncertainty.

$\frac{T}{K}$	$\frac{\rho_{exp} \pm 2\sigma_{est}}{kg \cdot m^{-3}}$	Ref.
298.15	967.7 ± 0.8	48-adk/bil

2-Ethyl-1,4-butanediol **[57716-79-7]** $C_6H_{14}O_2$ **MW = 118.18** **682**

Table 1. Experimental value with uncertainty.

$\frac{T}{K}$	$\frac{\rho_{exp} \pm 2\sigma_{est}}{kg \cdot m^{-3}}$	Ref.
293.15	982.5 ± 2.0	14-lon

2-Ethyl-2-methyl-1,3-propanediol **[77-84-9]** $C_6H_{14}O_2$ **MW = 118.18** **683**

Table 1. Experimental and recommended values with uncertainties.

$\frac{T}{K}$	$\frac{\rho_{exp} \pm 2\sigma_{est}}{kg \cdot m^{-3}}$	Ref.
323.15	958.0 ± 1.0	58-ano-13
323.15	958.0 ± 1.0	67-ano-5
323.15	958.0 ± 1.0	68-ano
323.15	958.2 ± 1.0	Recommended

1,2-Hexanediol **[6920-22-5]** $C_6H_{14}O_2$ **MW = 118.18** **684**

Table 1. Experimental value with uncertainty.

$\frac{T}{K}$	$\frac{\rho_{exp} \pm 2\sigma_{est}}{kg \cdot m^{-3}}$	Ref.
293.15	950.7 ± 2.0	51-cop/fie

1,3-Hexanediol **[21531-91-9]** $C_6H_{14}O_2$ **MW = 118.18** **685**

Table 1. Experimental value with uncertainty.

$\frac{T}{K}$	$\frac{\rho_{exp} \pm 2\sigma_{est}}{kg \cdot m^{-3}}$	Ref.
295.15	958.0 ± 2.0	44-gla

1,4-Hexanediol **[16432-53-4]** $C_6H_{14}O_2$ **MW = 118.18** **686**

Table 1. Experimental value with uncertainty.

$\frac{T}{K}$	$\frac{\rho_{exp} \pm 2\sigma_{est}}{kg \cdot m^{-3}}$	Ref.
289.45	982.0 ± 2.0	44-gla

1,5-Hexanediol **[928-40-5]** $C_6H_{14}O_2$ **MW = 118.18** **687**

Table 1. Fit with estimated B coefficient for 2 accepted points. Deviation $\sigma_w = 1.120$.

Coefficient	$\rho = A + BT$
A	1215.01
B	−0.840

cont.

1,5-Hexanediol (cont.)

Table 2. Experimental values with uncertainties and deviation from calculated values.

T / K	$\rho_{exp} \pm 2\sigma_{est}$ / $kg \cdot m^{-3}$	$\rho_{exp} - \rho_{calc}$ / $kg \cdot m^{-3}$	Ref.
293.15	971.0 ± 2.0	2.24	56-cri
298.15	964.0 ± 1.0	-0.56	64-ber/lon

Table 3. Recommended values.

T / K	$\rho_{exp} \pm 2\sigma_{est}$ / $kg \cdot m^{-3}$
290.00	971.4 ± 1.9
293.15	968.8 ± 1.8
298.15	964.6 ± 1.8

1,6-Hexanediol **[629-11-8]** $C_6H_{14}O_2$ **MW = 118.18** **688**

Table 1. Fit with estimated B coefficient for 3 accepted points. Deviation $\sigma_w = 1.678$.

Coefficient	$\rho = A + BT$
A	1183.31
B	−0.680

Table 2. Experimental values with uncertainties and deviation from calculated values.

T / K	$\rho_{exp} \pm 2\sigma_{est}$ / $kg \cdot m^{-3}$	$\rho_{exp} - \rho_{calc}$ / $kg \cdot m^{-3}$	Ref.
318.55	957.1 ± 4.0	−9.59	52-cur/joh [1)]
293.15	989.7 ± 2.0	5.73	63-tsy/sol [1)]
278.15	991.8 ± 1.0	−2.37	75-nak/kom
298.15	981.6 ± 1.0	1.03	75-nak/kom
318.15	968.3 ± 1.0	1.33	75-nak/kom

[1)] Not included in calculation of linear coefficients.

Table 3. Recommended values.

T / K	$\rho_{exp} \pm 2\sigma_{est}$ / $kg \cdot m^{-3}$	T / K	$\rho_{exp} \pm 2\sigma_{est}$ / $kg \cdot m^{-3}$	T / K	$\rho_{exp} \pm 2\sigma_{est}$ / $kg \cdot m^{-3}$
270.00	999.7 ± 2.3	293.15	984.0 ± 1.8	320.00	965.7 ± 2.1
280.00	992.9 ± 2.0	298.15	980.6 ± 1.8		
290.00	986.1 ± 1.8	310.00	972.5 ± 1.9		

2,3-Hexanediol **[617-30-1]** $C_6H_{14}O_2$ **MW = 118.18** **689**

Table 1. Experimental value with uncertainty.

T / K	$\rho_{exp} \pm 2\sigma_{est}$ / $kg \cdot m^{-3}$	Ref.
288.15	989.0 ± 2.0	37-tya

2,4-Hexanediol **[19780-90-6]** $C_6H_{14}O_2$ **MW = 118.18** **690**

Table 1. Experimental value with uncertainty.

T / K	$\rho_{exp} \pm 2\sigma_{est}$ / $kg \cdot m^{-3}$	Ref.
294.15	951.6 ± 1.0	32-les/wak

2,5-Hexanediol **[2935-44-6]** $C_6H_{14}O_2$ **MW = 118.18** **691**

Table 1. Coefficients of the polynomial expansion equation.
Standard deviations (see introduction):
$\sigma_{c,w} = 1.4746$ (combined temperature ranges, weighted),
$\sigma_{c,uw} = 5.9932 \cdot 10^{-1}$ (combined temperature ranges, unweighted).

Coefficient	$T = 273.00$ to 343.15 K $\rho = A + BT + CT^2 + DT^3 + \ldots$
A	$1.00102 \cdot 10^{3}$
B	$3.21158 \cdot 10^{-1}$
C	$-1.57871 \cdot 10^{-3}$

Table 2. Experimental values with uncertainties and deviation from calculated values.

T / K	$\rho_{exp} \pm 2\sigma_{est}$ / $kg \cdot m^{-3}$	$\rho_{exp} - \rho_{calc}$ / $kg \cdot m^{-3}$	Ref. (Symbol in Fig. 1)	T / K	$\rho_{exp} \pm 2\sigma_{est}$ / $kg \cdot m^{-3}$	$\rho_{exp} - \rho_{calc}$ / $kg \cdot m^{-3}$	Ref. (Symbol in Fig. 1)
273.00	966.89 ± 4.00	−4.14	1859-wur(×)	293.15	959.95 ± 0.40	0.46	96-gri/zhu(×)
273.00	966.90 ± 4.00	−4.13	1864-wur(×)	303.15	953.30 ± 0.40	0.01	96-gri/zhu(×)
293.15	961.00 ± 1.00	1.51	02-dud/lem(□)	313.15	946.60 ± 0.40	−0.17	96-gri/zhu(×)
293.15	960.10 ± 1.00	0.61	52-cur/joh(○)	323.15	939.80 ± 0.40	−0.14	96-gri/zhu(×)
293.15	960.00 ± 1.00	0.51	53-ano-5(Δ)	333.15	932.80 ± 0.40	0.01	96-gri/zhu(×)
293.15	961.70 ± 2.00	2.21	62-raz/bog(∇)	343.15	925.60 ± 0.40	0.28	96-gri/zhu(×)
293.15	962.50 ± 2.00	3.01	63-shu/bar(◆)				

Further references: [56-lev/sch, 62-mel].

Table 3. Recommended values (fit to the reliable experimental values according to the equations $\rho = A + BT + CT^2 + DT^3 + \ldots$ or $\rho = [1 + 1.75(1 - T/T_c)^{1/3} + 0.75(1 - T/T_c)][\rho_c + A(T_c - T) + B(T_c - T)^2 + C(T_c - T)^3 + D(T_c - T)^4]$).

T / K	$\rho \pm \sigma_{fit}$ / $kg \cdot m^{-3}$	T / K	$\rho \pm \sigma_{fit}$ / $kg \cdot m^{-3}$	T / K	$\rho \pm \sigma_{fit}$ / $kg \cdot m^{-3}$
270.00	972.64 ± 4.54	298.15	956.43 ± 0.89	330.00	935.08 ± 0.38
280.00	967.17 ± 2.76	300.00	955.28 ± 0.78	340.00	927.71 ± 0.58
290.00	961.38 ± 1.53	310.00	948.86 ± 0.43	350.00	920.03 ± 0.89
293.15	959.49 ± 1.25	320.00	942.13 ± 0.34		

cont.

2,5-Hexanediol (cont.)

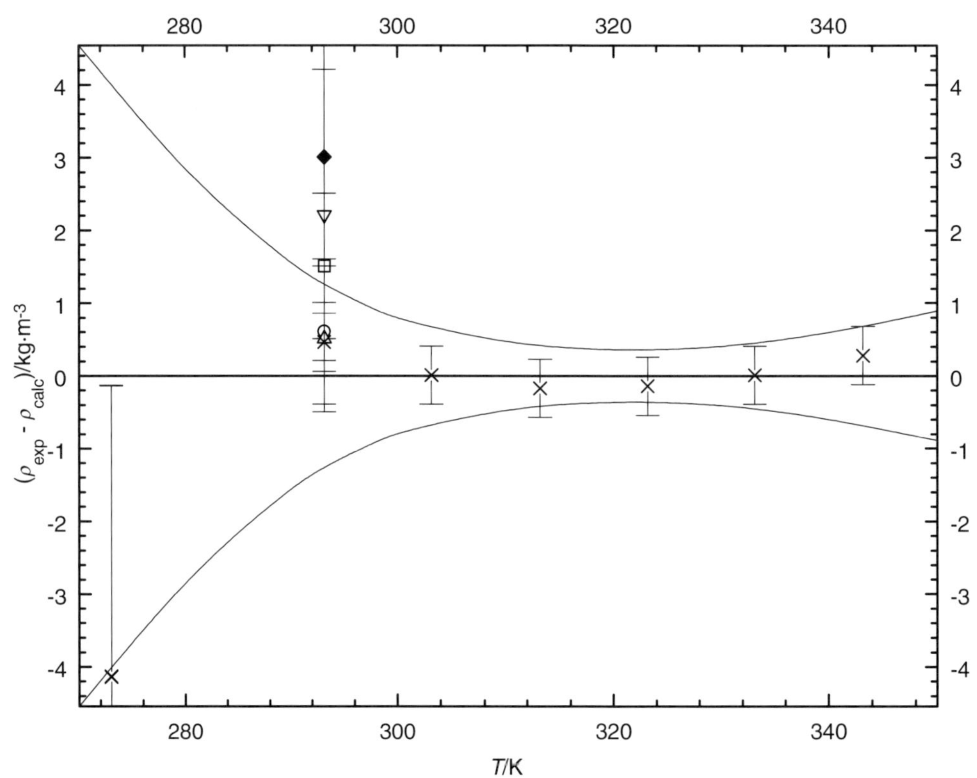

Fig. 1. The symbols show the deviation of the calculated from the experimental values from Table 2. The curves above and below the zero line indicate the calculated error region of the recommended values given in Table 3. The error bars represent the experimental errors. (Error bars smaller than the symbols are omitted for clarity of the figure.)

3,4-Hexanediol **[922-17-8]** **$C_6H_{14}O_2$** **MW = 118.18** **692**

Table 1. Experimental value with uncertainty.

$\frac{T}{K}$	$\frac{\rho_{exp} \pm 2\sigma_{est}}{kg \cdot m^{-3}}$	Ref.
273.15	799.3 ± 20.0	1859-wur

2-Methyl-1,3-pentanediol **[149-31-5]** **$C_6H_{14}O_2$** **MW = 118.18** **693**

Table 1. Experimental and recommended values with uncertainties.

$\frac{T}{K}$	$\frac{\rho_{exp} \pm 2\sigma_{est}}{kg \cdot m^{-3}}$	Ref.
295.15	973.7 ± 2.0	43-kul/nor[1)]
293.15	971.8 ± 1.0	52-cur/joh
293.15	972.9 ± 1.0	60-ano-13
293.15	972.3 ± 1.1	Recommended

[1)] Not included in calculation of recommended value.

2-Methyl-1,5-pentanediol **[42856-62-2]** $C_6H_{14}O_2$ **MW = 118.18** **694**

Table 1. Experimental and recommended values with uncertainties.

T/K	$\rho_{exp} \pm 2\sigma_{est}$ / $kg \cdot m^{-3}$	Ref.
293.15	971.9 ± 1.0	51-mca/cul
293.15	971.9 ± 1.0	51-whi/dea
293.15	975.0 ± 2.0	62-yur/rev[1)]
293.15	971.9 ± 1.0	Recommended

[1)] Not included in calculation of recommended value.

2-Methyl-2,3-pentanediol **[7795-80-4]** $C_6H_{14}O_2$ **MW = 118.18** **695**

Table 1. Fit with estimated B coefficient for 2 accepted points. Deviation $\sigma_w = 0.200$.

Coefficient	$\rho = A + BT$
A	1167.70
B	–0.700

Table 2. Experimental values with uncertainties and deviation from calculated values.

T/K	$\rho_{exp} \pm 2\sigma_{est}$ / $kg \cdot m^{-3}$	$\rho_{exp} - \rho_{calc}$ / $kg \cdot m^{-3}$	Ref.
293.15	962.7 ± 1.0	0.20	28-ven-1
273.15	976.3 ± 1.0	–0.20	28-ven-1

Table 3. Recommended values.

T/K	$\rho_{exp} \pm 2\sigma_{est}$ / $kg \cdot m^{-3}$
270.00	978.7 ± 1.1
280.00	971.7 ± 0.9
290.00	964.7 ± 1.0
293.15	962.5 ± 1.0
298.15	959.0 ± 1.2

2-Methyl-2,4-pentanediol **[107-41-5]** $C_6H_{14}O_2$ **MW = 118.18** **696**

Table 1. Coefficients of the polynomial expansion equation.
Standard deviations (see introduction):
$\sigma_{c,w} = 4.4579 \cdot 10^{-1}$ (combined temperature ranges, weighted),
$\sigma_{c,uw} = 1.7903 \cdot 10^{-1}$ (combined temperature ranges, unweighted).

Coefficient	T = 273.15 to 343.15 K $\rho = A + BT + CT^2 + DT^3 + \ldots$
A	$1.09017 \cdot 10^{3}$
B	$-4.33615 \cdot 10^{-1}$
C	$-4.76496 \cdot 10^{-4}$

cont.

2-Methyl-2,4-pentanediol (cont.)

Table 2. Experimental values with uncertainties and deviation from calculated values.

$\frac{T}{K}$	$\frac{\rho_{exp} \pm 2\sigma_{est}}{kg \cdot m^{-3}}$	$\frac{\rho_{exp} - \rho_{calc}}{kg \cdot m^{-3}}$	Ref. (Symbol in Fig. 1)	$\frac{T}{K}$	$\frac{\rho_{exp} \pm 2\sigma_{est}}{kg \cdot m^{-3}}$	$\frac{\rho_{exp} - \rho_{calc}}{kg \cdot m^{-3}}$	Ref. (Symbol in Fig. 1)
290.15	925.40 ± 2.00	1.16	01-zel/zel(X)	298.15	920.00 ± 1.50	1.47	62-ste/van(◆)
290.15	924.00 ± 1.00	−0.24	12-ost(Δ)	303.15	916.00 ± 1.50	1.07	62-ste/van(◆)
290.15	925.40 ± 2.00	1.16	39-dup/dar(X)	293.15	921.60 ± 0.60	−0.51	68-ano(O)
293.15	921.80 ± 1.00	−0.31	53-ano-5(∇)	293.15	921.80 ± 0.40	−0.31	96-gri/zhu(X)
293.15	924.00 ± 2.00	1.89	57-chi/tho[1)]	303.15	914.50 ± 0.40	−0.43	96-gri/zhu(X)
298.15	918.10 ± 1.00	−0.43	57-chi/tho(X)	313.15	907.20 ± 0.40	−0.46	96-gri/zhu(X)
273.15	935.50 ± 0.50	−0.68	62-mel(□)	323.15	899.90 ± 0.40	−0.39	96-gri/zhu(X)
293.15	921.60 ± 0.50	−0.51	62-mel(□)	333.15	892.60 ± 0.40	−0.22	96-gri/zhu(X)
303.15	914.50 ± 0.50	−0.43	62-mel(□)	343.15	885.30 ± 0.40	0.03	96-gri/zhu(X)

1) Not included in Fig. 1.

Further references: [52-cur/joh, 56-far/she].

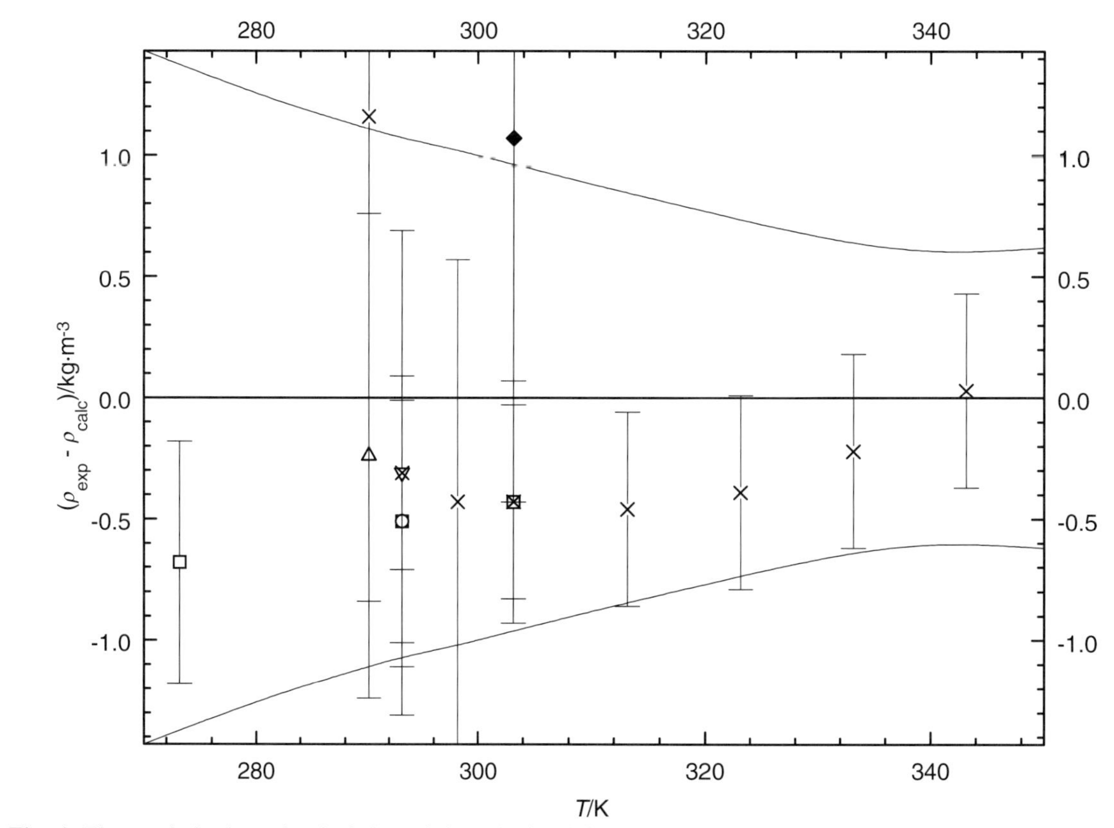

Fig. 1. The symbols show the deviation of the calculated from the experimental values from Table 2. The curves above and below the zero line indicate the calculated error region of the recommended values given in Table 3. The error bars represent the experimental errors. (Error bars smaller than the symbols are omitted for clarity of the figure.)

cont.

Table 3. Recommended values (fit to the reliable experimental values according to the equations $\rho = A + BT + CT^2 + DT^3 + \ldots$ or $\rho = [1 + 1.75(1 - T/T_c)^{1/3} + 0.75(1 - T/T_c)][\rho_c + A(T_c - T) + B(T_c - T)^2 + C(T_c - T)^3 + D(T_c - T)^4]$).

T/K	$\rho \pm \sigma_{fit}$/kg·m^{-3}	T/K	$\rho \pm \sigma_{fit}$/kg·m^{-3}	T/K	$\rho \pm \sigma_{fit}$/kg·m^{-3}
270.00	938.36 ± 1.43	298.15	918.53 ± 1.02	330.00	895.19 ± 0.66
280.00	931.40 ± 1.25	300.00	917.20 ± 1.00	340.00	887.66 ± 0.59
290.00	924.35 ± 1.11	310.00	909.96 ± 0.88	350.00	880.03 ± 0.62
293.15	922.11 ± 1.07	320.00	902.62 ± 0.77		

3-Methyl-1,3-pentanediol **[33879-72-0]** **$C_6H_{14}O_2$** **MW = 118.18** **697**

Table 1. Experimental and recommended values with uncertainties.

T/K	$\rho_{exp} \pm 2\sigma_{est}$/kg·m^{-3}	Ref.
293.15	972.6 ± 3.0	32-pfa/pla[1)]
273.15	975.5 ± 2.0	48-zal[1)]
290.75	965.5 ± 2.0	48-zal[1)]
293.15	969.0 ± 1.0	62-esa/shi
293.15	969.0 ± 1.0	Recommended

[1)] Not included in calculation of recommended value.

3-Methyl-1,5-pentanediol **[4457-71-0]** **$C_6H_{14}O_2$** **MW = 118.18** **698**

Table 1. Experimental and recommended values with uncertainties.

T/K	$\rho_{exp} \pm 2\sigma_{est}$/kg·m^{-3}	Ref.
293.15	973.8 ± 1.0	51-mca/cul
293.15	973.8 ± 1.0	51-whi/dea
293.15	972.6 ± 1.0	55-blo/ver-1
293.15	973.8 ± 1.0	56-ano-3
293.15	973.5 ± 1.0	Recommended

3-Methyl-2,3-pentanediol **[63521-37-9]** **$C_6H_{14}O_2$** **MW = 118.18** **699**

Table 1. Experimental and recommended values with uncertainties.

T/K	$\rho_{exp} \pm 2\sigma_{est}$/kg·m^{-3}	Ref.
293.15	970.1 ± 2.0	52-van[1)]
288.15	975.0 ± 2.0	52-van[1)]
298.15	963.8 ± 1.0	58-hen/wat
298.15	963.8 ± 1.0	Recommended

[1)] Not included in calculation of recommended value.

3-Methyl-2,4-pentanediol **[5683-44-3]** $C_6H_{14}O_2$ **MW = 118.18** **700**

Table 1. Experimental value with uncertainty.

$\frac{T}{K}$	$\frac{\rho_{exp} \pm 2\sigma_{est}}{kg \cdot m^{-3}}$	Ref.
287.15	990.6 ± 2.0	01-zel/zel

4-Methyl-1,4-pentanediol **[1462-10-8]** $C_6H_{14}O_2$ **MW = 118.18** **701**

Table 1. Experimental and recommended values with uncertainties.

$\frac{T}{K}$	$\frac{\rho_{exp} \pm 2\sigma_{est}}{kg \cdot m^{-3}}$	Ref.
290.15	965.1 ± 2.0	52-wil/sch[1)]
293.15	964.5 ± 1.0	55-sar/mor
293.15	970.0 ± 2.0	58-col/fal[1)]
293.15	964.5 ± 1.0	Recommended

[1)] Not included in calculation of recommended value.

2-(1-Methylethyl)-1,3-propanediol **[2612-27-3]** $C_6H_{14}O_2$ **MW = 118.18** **702**

Table 1. Experimental values with uncertainties.

$\frac{T}{K}$	$\frac{\rho_{exp} \pm 2\sigma_{est}}{kg \cdot m^{-3}}$	Ref.
293.15	976.2 ± 1.0	53-pin/hun
293.15	977.0 ± 0.5	62-bog/osi
293.15	976.8 ± 0.5	Recommended

2-Propyl-1,3-propanediol **[2612-28-4]** $C_6H_{14}O_2$ **MW = 118.18** **703**

Table 1. Experimental and recommended values with uncertainties.

$\frac{T}{K}$	$\frac{\rho_{exp} \pm 2\sigma_{est}}{kg \cdot m^{-3}}$	Ref.
298.15	963.6 ± 2.0	48-adk/bil[1)]
293.15	965.9 ± 0.5	62-bog/osi
293.15	965.9 ± 0.5	Recommended

[1)] Not included in calculation of recommended value.

3.1.2 Alkanediols, C_7 - C_{14}

2-Butyl-1,3-propanediol **[2612-26-2]** **$C_7H_{16}O_2$** **MW = 132.2** **704**

Table 1. Experimental value with uncertainty.

T/K	$\rho_{exp} \pm 2\sigma_{est}$ / $kg \cdot m^{-3}$	Ref.
298.15	946.1 ± 1.0	48-adk/bil

2,2-Diethyl-1,3-propanediol **[115-76-4]** **$C_7H_{16}O_2$** **MW = 132.2** **705**

Table 1. Experimental values with uncertainties.

T/K	$\rho_{exp} \pm 2\sigma_{est}$ / $kg \cdot m^{-3}$	Ref.
	crystal	
293.15	1052.0 ± 5.0	54-ano-12
293.15	1052.0 ± 5.0	60-ano-13
293.15	1052.0 ± 5.0	62-mel
	liquid	
334.45	949.0 ± 2.0	54-ano-12
334.45	949.0 ± 2.0	60-ano-13

2,3-Dimethyl-1,3-pentanediol **[66225-52-3]** **$C_7H_{16}O_2$** **MW = 132.2** **706**

Table 1. Experimental value with uncertainty.

T/K	$\rho_{exp} \pm 2\sigma_{est}$ / $kg \cdot m^{-3}$	Ref.
296.15	957.0 ± 1.0	53-col/dre

2,3-Dimethyl-2,3-pentanediol **[6931-70-0]** **$C_7H_{16}O_2$** **MW = 132.2** **707**

Table 1. Experimental value with uncertainty.

T/K	$\rho_{exp} \pm 2\sigma_{est}$ / $kg \cdot m^{-3}$	Ref.
293.15	961.3 ± 2.0	41-fav/oni

(R*,S*)-2,4-Dimethyl-1,5-Pentanediol **[3817-48-9]** **$C_7H_{16}O_2$** **MW = 132.2** **708**

Table 1. Experimental value with uncertainty.

T/K	$\rho_{exp} \pm 2\sigma_{est}$ / $kg \cdot m^{-3}$	Ref.
293.15	945.0 ± 2.0	55-nol/pan

2,4-Dimethyl-2,4-pentanediol **[24892-49-7]** $C_7H_{16}O_2$ **MW = 132.2** **709**

Table 1. Experimental value with uncertainty.

T/K	$\rho_{exp} \pm 2\sigma_{est}$ / $kg \cdot m^{-3}$	Ref.
293.15	920.6 ± 2.0	09-lem-1

3,4-Dimethyl-1,4-pentanediol **[63521-36-8]** $C_7H_{16}O_2$ **MW = 132.2** **710**

Table 1. Experimental value with uncertainty.

T/K	$\rho_{exp} \pm 2\sigma_{est}$ / $kg \cdot m^{-3}$	Ref.
293.15	952.8 ± 1.0	57-sar/vor

2-Ethyl-1,5-pentanediol **[14189-13-0]** $C_7H_{16}O_2$ **MW = 132.2** **711**

Table 1. Experimental value with uncertainty.

T/K	$\rho_{exp} \pm 2\sigma_{est}$ / $kg \cdot m^{-3}$	Ref.
293.15	967.9 ± 1.0	62-yur/rev

2-Ethyl-2,4-pentanediol **[38836-25-8]** $C_7H_{16}O_2$ **MW = 132.2** **712**

Table 1. Experimental and recommended values with uncertainties.

T/K	$\rho_{exp} \pm 2\sigma_{est}$ / $kg \cdot m^{-3}$	Ref.
293.15	920.6 ± 4.0	09-lem[1)]
293.15	929.6 ± 2.0	63-esa
293.15	929.6 ± 2.0	Recommended

[1)] Not included in calculation of recommended value.

3-Ethylpentane-2,3-diol **[66225-32-9]** $C_7H_{16}O_2$ **MW = 132.2** **713**

Table 1. Experimental value with uncertainty.

T/K	$\rho_{exp} \pm 2\sigma_{est}$ / $kg \cdot m^{-3}$	Ref.
298.15	961.2 ± 2.0	58-hen/wat

1,4-Heptanediol **[40646-07-9]** $C_7H_{16}O_2$ **MW = 132.2** **714**

Table 1. Fit with estimated B coefficient for 4 accepted points. Deviation $\sigma_w = 0.173$.

Coefficient	$\rho = A + BT$
A	1158.80
B	−0.700

cont.

Table 2. Experimental values with uncertainties and deviation from calculated values.

$\frac{T}{K}$	$\frac{\rho_{exp} \pm 2\sigma_{est}}{kg \cdot m^{-3}}$	$\frac{\rho_{exp} - \rho_{calc}}{kg \cdot m^{-3}}$	Ref.
293.15	954.3 ± 2.0	0.70	27-bra/ada[1)]
298.15	950.4 ± 1.0	0.30	34-bur/adk
288.15	957.0 ± 1.0	–0.10	51-gor
288.15	957.0 ± 1.0	–0.10	51-gor
288.15	957.0 ± 1.0	–0.10	52-roe/stu

[1)] Not included in calculation of linear coefficients.

Table 3. Recommended values.

$\frac{T}{K}$	$\frac{\rho_{exp} \pm 2\sigma_{est}}{kg \cdot m^{-3}}$	$\frac{T}{K}$	$\frac{\rho_{exp} \pm 2\sigma_{est}}{kg \cdot m^{-3}}$
280.00	962.8 ± 1.3	293.15	953.6 ± 0.8
290.00	955.8 ± 0.7	298.15	950.1 ± 1.1

1,5-Heptanediol **[60096-09-5]** **$C_7H_{16}O_2$** **MW = 132.2** **715**

Table 1. Experimental and recommended values with uncertainties.

$\frac{T}{K}$	$\frac{\rho_{exp} \pm 2\sigma_{est}}{kg \cdot m^{-3}}$	Ref.
293.15	970.5 ± 2.0	25-pie/ada
288.15	962.0 ± 6.0	35-pau-2[1)]
293.15	970.5 ± 2.0	Recommended

[1)] Not included in calculation of recommended value.

1,6-Heptanediol **[13175-27-4]** **$C_7H_{16}O_2$** **MW = 132.2** **716**

Table 1. Experimental value with uncertainty.

$\frac{T}{K}$	$\frac{\rho_{exp} \pm 2\sigma_{est}}{kg \cdot m^{-3}}$	Ref.
298.15	962.0 ± 2.0	66-bue

1,7-Heptanediol **[629-30-1]** **$C_7H_{16}O_2$** **MW = 132.2** **717**

Table 1. Coefficients of the polynomial expansion equation.
Standard deviations (see introduction):
$\sigma_{c,w} = 1.9518 \cdot 10^{-2}$ (combined temperature ranges, weighted),
$\sigma_{c,uw} = 9.7590 \cdot 10^{-3}$ (combined temperature ranges, unweighted).

Coefficient	T = 293.15 to 343.15 K $\rho = A + BT + CT^2 + DT^3 + \ldots$
A	$1.12164 \cdot 10^{3}$
B	$-5.46836 \cdot 10^{-1}$
C	$-7.14286 \cdot 10^{-5}$

cont.

1,7-Heptanediol (cont.)

Table 2. Experimental values with uncertainties and deviation from calculated values.

$\frac{T}{K}$	$\frac{\rho_{exp} \pm 2\sigma_{est}}{kg \cdot m^{-3}}$	$\frac{\rho_{exp} - \rho_{calc}}{kg \cdot m^{-3}}$	Ref. (Symbol in Fig. 1)	$\frac{T}{K}$	$\frac{\rho_{exp} \pm 2\sigma_{est}}{kg \cdot m^{-3}}$	$\frac{\rho_{exp} - \rho_{calc}}{kg \cdot m^{-3}}$	Ref. (Symbol in Fig. 1)
293.15	955.20 ± 0.50	0.00	96-gri/zhu(□)	323.15	937.50 ± 0.50	0.03	96-gri/zhu(□)
303.15	949.30 ± 0.50	–0.01	96-gri/zhu(□)	333.15	931.50 ± 0.50	–0.04	96-gri/zhu(□)
313.15	943.40 ± 0.50	0.00	96-gri/zhu(□)	343.15	925.60 ± 0.50	0.01	96-gri/zhu(□)

Further references: [34-bur/adk, 51-hub, 64-pol/bel].

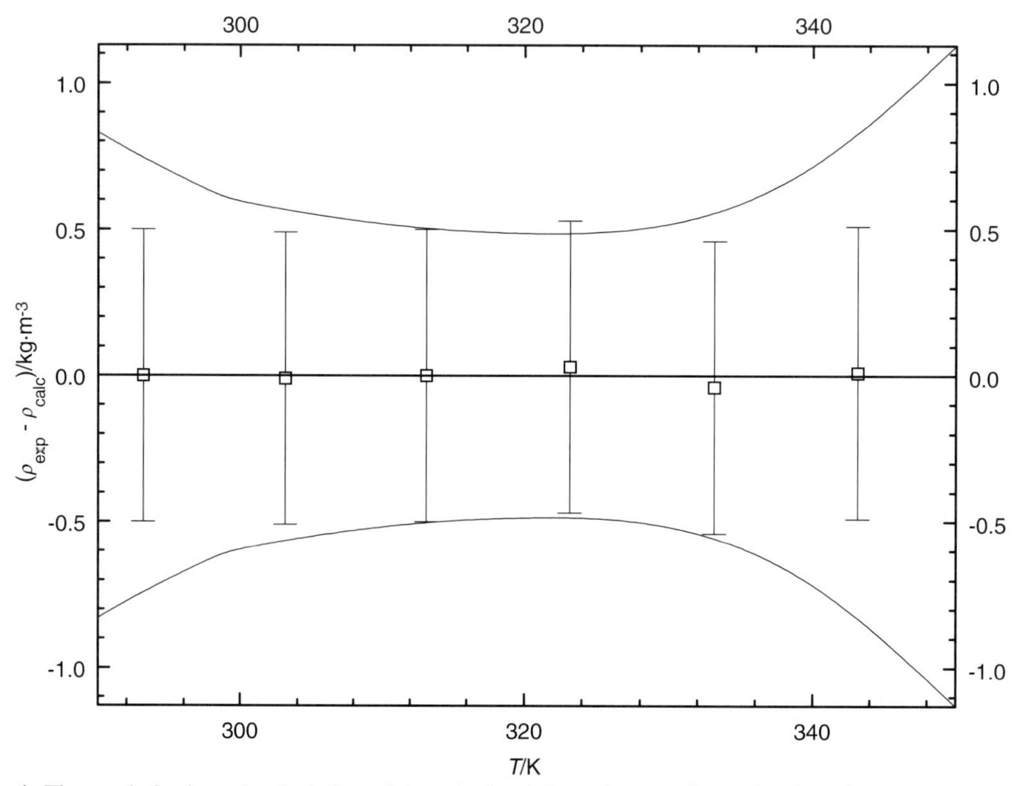

Fig. 1. The symbols show the deviation of the calculated from the experimental values from Table 2. The curves above and below the zero line indicate the calculated error region of the recommended values given in Table 3. The error bars represent the experimental errors. (Error bars smaller than the symbols are omitted for clarity of the figure.)

Table 3. Recommended values (fit to the reliable experimental values according to the equations

$$\rho = A + BT + CT^2 + DT^3 + \ldots \text{ or } \rho = [1 + 1.75(1 - T/T_c)^{1/3} + 0.75(1 - T/T_c)][\rho_c + A(T_c - T) + B(T_c - T)^2 + C(T_c - T)^3 + D(T_c - T)^4]\ ).$$

$\frac{T}{K}$	$\frac{\rho \pm \sigma_{fit}}{kg \cdot m^{-3}}$	$\frac{T}{K}$	$\frac{\rho \pm \sigma_{fit}}{kg \cdot m^{-3}}$	$\frac{T}{K}$	$\frac{\rho \pm \sigma_{fit}}{kg \cdot m^{-3}}$
290.00	957.05 ± 0.83	300.00	951.16 ± 0.59	330.00	933.41 ± 0.49
293.15	955.20 ± 0.74	310.00	945.26 ± 0.51	340.00	927.46 ± 0.67
298.15	952.25 ± 0.62	320.00	939.34 ± 0.48	350.00	921.50 ± 1.13

2,4-Heptanediol **[20748-86-1]** $C_7H_{16}O_2$ **MW = 132.2** **718**

Table 1. Fit with estimated B coefficient for 2 accepted points. Deviation σ_w = 0.100.

Coefficient	$\rho = A + BT$
A	1134.61
B	–0.700

Table 2. Experimental values with uncertainties and deviation from calculated values.

T / K	$\rho_{exp} \pm 2\sigma_{est}$ / kg·m^{-3}	$\rho_{exp} - \rho_{calc}$ / kg·m^{-3}	Ref.
298.15	926.0 ± 1.0	0.10	39-stu/adk
288.15	932.8 ± 1.0	–0.10	47-dev

Table 3. Recommended values.

T / K	$\rho_{exp} \pm 2\sigma_{est}$ / kg·m^{-3}	T / K	$\rho_{exp} \pm 2\sigma_{est}$ / kg·m^{-3}
280.00	938.6 ± 1.1	293.15	929.4 ± 0.9
290.00	931.6 ± 0.9	298.15	925.9 ± 0.9

3,4-Heptanediol **[62593-33-3]** $C_7H_{16}O_2$ **MW = 132.2** **719**

Table 1. Experimental and recommended values with uncertainties.

T / K	$\rho_{exp} \pm 2\sigma_{est}$ / kg·m^{-3}	Ref.
293.15	945.7 ± 2.0	25-pie/ada[1)]
295.15	943.0 ± 1.0	36-wie
295.15	943.0 ± 1.0	Recommended

[1)] Not included in calculation of recommended value.

2-Methyl-2,4-hexanediol **[66225-35-2]** $C_7H_{16}O_2$ **MW = 132.2** **720**

Table 1. Experimental value with uncertainty.

T / K	$\rho_{exp} \pm 2\sigma_{est}$ / kg·m^{-3}	Ref.
291.15	932.1 ± 1.0	26-pas/zam

2-Methyl-2,6-hexanediol **[1462-11-9]** $C_7H_{16}O_2$ **MW = 132.2** **721**

Table 1. Experimental value with uncertainty.

T / K	$\rho_{exp} \pm 2\sigma_{est}$ / kg·m^{-3}	Ref.
295.15	967.0 ± 2.0	56-cri

4-Methyl-1,5-hexanediol **[66225-37-4]** $C_7H_{16}O_2$ **MW = 132.2** **722**

Table 1. Experimental value with uncertainty.

T / K	$\rho_{exp} \pm 2\sigma_{est}$ / kg·m^{-3}	Ref.
293.15	966.3 ± 2.0	62-yur/pen

2-Methyl-2-propyl-1,3-propanediol **[78-26-2]** $C_7H_{16}O_2$ **MW = 132.2** **723**

Table 1. Experimental and recommended values with uncertainties.

$\frac{T}{K}$	$\frac{\rho_{exp} \pm 2\sigma_{est}}{kg \cdot m^{-3}}$	Ref.
293.15	857.0 ± 1.0	54-ano-12
293.15	857.1 ± 1.0	59-ano-7
293.15	858.5 ± 1.0	60-ano-13
293.15	857.5 ± 1.1	Recommended

2-(1-Methylethyl)-1,4-butanediol **[39497-66-0]** $C_7H_{16}O_2$ **MW = 132.2** **724**

Table 1. Experimental value with uncertainty.

$\frac{T}{K}$	$\frac{\rho_{exp} \pm 2\sigma_{est}}{kg \cdot m^{-3}}$	Ref.
293.15	967.2 ± 0.8	54-fre/lwo

2-Propyl-1,4-butanediol **[62946-68-3]** $C_7H_{16}O_2$ **MW = 132.2** **725**

Table 1. Experimental value with uncertainty.

$\frac{T}{K}$	$\frac{\rho_{exp} \pm 2\sigma_{est}}{kg \cdot m^{-3}}$	Ref.
293.15	962.5 ± 2.0	14-lon

2,4-Dimethyl-2,4-hexanediol **[29649-22-7]** $C_8H_{18}O_2$ **MW = 146.23** **726**

Table 1. Experimental value with uncertainty.

$\frac{T}{K}$	$\frac{\rho_{exp} \pm 2\sigma_{est}}{kg \cdot m^{-3}}$	Ref.
293.35	922.9 ± 2.0	34-jac-5

2,5-Dimethyl-2,4-hexanediol **[3899-89-6]** $C_8H_{18}O_2$ **MW = 146.23** **727**

Table 1. Experimental value with uncertainty.

$\frac{T}{K}$	$\frac{\rho_{exp} \pm 2\sigma_{est}}{kg \cdot m^{-3}}$	Ref.
293.15	917.2 ± 2.0	64-esa/das

3,4-Dimethyl-2,4-hexanediol **[900002-90-6]** $C_8H_{18}O_2$ **MW = 146.23** **728**

Table 1. Experimental value with uncertainty.

$\frac{T}{K}$	$\frac{\rho_{exp} \pm 2\sigma_{est}}{kg \cdot m^{-3}}$	Ref.
293.15	938.2 ± 2.0	62-fav/por

3,5-Dimethyl-2,3-hexanediol [99799-29-8] $C_8H_{18}O_2$ MW = 146.23 **729**

Table 1. Experimental and recommended values with uncertainties.

$\frac{T}{K}$	$\frac{\rho_{exp} \pm 2\sigma_{est}}{kg \cdot m^{-3}}$	Ref.
293.15	915.7 ± 2.0	51-ber-2[1)]
298.15	928.5 ± 1.0	58-hen/wat
298.15	928.5 ± 1.0	Recommended

[1)] Not included in calculation of recommended value.

2-Ethyl-1,3-hexanediol [94-96-2] $C_8H_{18}O_2$ MW = 146.23 **730**

Table 1. Fit with estimated *B* coefficient for 3 accepted points. Deviation $\sigma_w = 0.176$.

Coefficient	$\rho = A + BT$
A	1180.92
B	−0.820

Table 2. Experimental values with uncertainties and deviation from calculated values.

$\frac{T}{K}$	$\frac{\rho_{exp} \pm 2\sigma_{est}}{kg \cdot m^{-3}}$	$\frac{\rho_{exp} - \rho_{calc}}{kg \cdot m^{-3}}$	Ref.
295.15	938.4 ± 2.0	−0.50	64-bla/per-1
293.15	940.6 ± 1.0	0.06	53-ano-15
293.15	940.6 ± 1.0	0.06	62-mel

Table 3. Recommended values.

$\frac{T}{K}$	$\frac{\rho_{exp} \pm 2\sigma_{est}}{kg \cdot m^{-3}}$
290.00	943.1 ± 1.3
293.15	940.5 ± 1.2
298.15	936.4 ± 1.3

2-Ethyl-1-methyl-1,5-pentanediol [900002-72-4] $C_8H_{18}O_2$ MW = 146.23 **731**

Table 1. Experimental value with uncertainty.

$\frac{T}{K}$	$\frac{\rho_{exp} \pm 2\sigma_{est}}{kg \cdot m^{-3}}$	Ref.
293.15	960.0 ± 2.0	62-yur/rev

4-Ethyl-1,4-hexanediol [1113-00-4] $C_8H_{18}O_2$ MW = 146.23 **732**

Table 1. Experimental value with uncertainty.

$\frac{T}{K}$	$\frac{\rho_{exp} \pm 2\sigma_{est}}{kg \cdot m^{-3}}$	Ref.
292.15	970.4 ± 1.0	51-gor

4-Ethyl-1,5-hexanediol [90951-82-9] $C_8H_{18}O_2$ MW = 146.23 733

Table 1. Experimental value with uncertainty.

$\frac{T}{K}$	$\frac{\rho_{exp} \pm 2\sigma_{est}}{kg \cdot m^{-3}}$	Ref.
293.15	960.0 ± 2.0	62-yur/pen

2-Methyl-2,3-heptanediol [1068-81-1] $C_8H_{18}O_2$ MW = 146.23 734

Table 1. Experimental value with uncertainty.

$\frac{T}{K}$	$\frac{\rho_{exp} \pm 2\sigma_{est}}{kg \cdot m^{-3}}$	Ref.
298.15	929.0 ± 2.0	64-col/var

3-Methyl-3,5-heptanediol [99799-27-6] $C_8H_{18}O_2$ MW = 146.23 735

Table 1. Experimental values with uncertainties.

$\frac{T}{K}$	$\frac{\rho_{exp} \pm 2\sigma_{est}}{kg \cdot m^{-3}}$	Ref.
291.15	929.9 ± 2.0	26-pas/zam
294.15	928.0 ± 2.0	51-dub

5-Methyl-1,5-heptanediol [99799-26-5] $C_8H_{18}O_2$ MW = 146.23 736

Table 1. Experimental value with uncertainty.

$\frac{T}{K}$	$\frac{\rho_{exp} \pm 2\sigma_{est}}{kg \cdot m^{-3}}$	Ref.
292.15	961.0 ± 2.0	56-cri

5-Methyl-2,4-heptanediol [500014-48-2] $C_8H_{18}O_2$ MW = 146.23 737

Table 1. Experimental value with uncertainty.

$\frac{T}{K}$	$\frac{\rho_{exp} \pm 2\sigma_{est}}{kg \cdot m^{-3}}$	Ref.
298.15	928.0 ± 2.0	39-stu/adk

6-Methyl-1,6-heptanediol [5392-57-4] $C_8H_{18}O_2$ MW = 146.23 738

Table 1. Experimental value with uncertainty.

$\frac{T}{K}$	$\frac{\rho_{exp} \pm 2\sigma_{est}}{kg \cdot m^{-3}}$	Ref.
298.15	959.0 ± 2.0	66-bue

2-(1-Methylethyl)-1,5-pentanediol [90951-89-6] $C_8H_{18}O_2$ MW = 146.23 739

Table 1. Experimental value with uncertainty.

T / K	$\rho_{exp} \pm 2\sigma_{est}$ / $kg \cdot m^{-3}$	Ref.
293.15	956.1 ± 2.0	62-yur/rev

1,5-Octanediol [2736-67-6] $C_8H_{18}O_2$ MW = 146.23 740

Table 1. Experimental and recommended values with uncertainties.

T / K	$\rho_{exp} \pm 2\sigma_{est}$ / $kg \cdot m^{-3}$	Ref.	T / K	$\rho_{exp} \pm 2\sigma_{est}$ / $kg \cdot m^{-3}$	Ref.
290.15	949.0 ± 2.0	35-pau-2[1)]	298.15	949.0 ± 1.0	66-bue-1
289.15	946.0 ± 2.0	56-cri[1)]	298.15	949.0 ± 1.0	Recommended

[1)] Not included in calculation of recommended value.

1,6-Octanediol [4066-76-6] $C_8H_{18}O_2$ MW = 146.23 741

Table 1. Experimental values with uncertainties.

T / K	$\rho_{exp} \pm 2\sigma_{est}$ / $kg \cdot m^{-3}$	Ref.
288.15	943.0 ± 3.0	65-mor/lam
298.15	954.0 ± 3.0	66-bue

1,7-Octanediol [13175-32-1] $C_8H_{18}O_2$ MW = 146.23 742

Table 1. Experimental and recommended values with uncertainties.

T / K	$\rho_{exp} \pm 2\sigma_{est}$ / $kg \cdot m^{-3}$	Ref.
298.15	942.0 ± 1.0	62-col/gir
298.15	944.0 ± 1.0	66-bue
298.15	943.0 ± 1.2	Recommended

2,4-Octanediol [90162-24-6] $C_8H_{18}O_2$ MW = 146.23 743

Table 1. Experimental value with uncertainty.

T / K	$\rho_{exp} \pm 2\sigma_{est}$ / $kg \cdot m^{-3}$	Ref.
298.15	918.0 ± 2.0	39-stu/adk

2-Propyl-1,5-pentanediol [90951-90-9] $C_8H_{18}O_2$ MW = 146.23 744

Table 1. Experimental value with uncertainty.

T / K	$\rho_{exp} \pm 2\sigma_{est}$ / $kg \cdot m^{-3}$	Ref.
293.15	953.0 ± 2.0	62-yur/rev

2,2,4-Trimethyl-1,3-pentanediol **[144-19-4]** $C_8H_{18}O_2$ **MW = 146.23** **745**

Table 1. Experimental and recommended values with uncertainties.

$\frac{T}{K}$	$\frac{\rho_{exp} \pm 2\sigma_{est}}{kg \cdot m^{-3}}$	Ref.
296.15	961.0 ± 10.0	43-kul/nor[1)]
293.15	922.9 ± 1.0	59-ano-7
293.15	922.9 ± 1.0	Recommended

[1)] Not included in calculation of recommended value.

2-Butyl-2-ethyl-1,3-propanediol **[115-84-4]** $C_9H_{20}O_2$ **MW = 160.26** **746**

Table 1. Experimental and recommended values with uncertainties.

$\frac{T}{K}$	$\frac{\rho_{exp} \pm 2\sigma_{est}}{kg \cdot m^{-3}}$	Ref.	$\frac{T}{K}$	$\frac{\rho_{exp} \pm 2\sigma_{est}}{kg \cdot m^{-3}}$	Ref.
323.15	929.0 ± 1.0	53-ano-15	323.15	931.0 ± 2.0	62-mel[1)]
323.15	929.0 ± 1.0	60-ano-13	323.15	929.0 ± 1.0	Recommended

[1)] Not included in calculation of recommended value.

2-Butyl-1,5-pentanediol **[90724-91-7]** $C_9H_{20}O_2$ **MW = 160.26** **747**

Table 1. Experimental value with uncertainty.

$\frac{T}{K}$	$\frac{\rho_{exp} \pm 2\sigma_{est}}{kg \cdot m^{-3}}$	Ref.
293.15	941.4 ± 2.0	62-yur/rev

2,4-Dimethyl-2,4-heptanediol **[59194-83-1]** $C_9H_{20}O_2$ **MW = 160.26** **748**

Table 1. Experimental value with uncertainty.

$\frac{T}{K}$	$\frac{\rho_{exp} \pm 2\sigma_{est}}{kg \cdot m^{-3}}$	Ref.
290.35	913.8 ± 2.0	34-jac-5

2,4-Dimethyl-2,6-heptanediol **[73264-94-5]** $C_9H_{20}O_2$ **MW = 160.26** **749**

Table 1. Experimental value with uncertainty.

$\frac{T}{K}$	$\frac{\rho_{exp} \pm 2\sigma_{est}}{kg \cdot m^{-3}}$	Ref.
298.15	920.0 ± 1.5	64-hin/dre

2,5-Dimethyl-3,5-heptanediol **[3955-69-9]** $C_9H_{20}O_2$ **MW = 160.26** **750**

Table 1. Experimental value with uncertainty.

$\frac{T}{K}$	$\frac{\rho_{exp} \pm 2\sigma_{est}}{kg \cdot m^{-3}}$	Ref.
293.15	915.6 ± 2.0	64-esa/das

2,6-Dimethyl-2,4-heptanediol [73264-93-4] $C_9H_{20}O_2$ MW = 160.26 751

Table 1. Experimental value with uncertainty.

$\frac{T}{K}$	$\frac{\rho_{exp} \pm 2\sigma_{est}}{kg \cdot m^{-3}}$	Ref.
291.15	902.0 ± 2.0	26-pas/zam

4,4-Dimethyl-1,7-heptanediol [900002-74-6] $C_9H_{20}O_2$ MW = 160.26 752

Table 1. Experimental value with uncertainty.

$\frac{T}{K}$	$\frac{\rho_{exp} \pm 2\sigma_{est}}{kg \cdot m^{-3}}$	Ref.
293.15	939.0 ± 1.0	55-blo/whe

4-Ethyl-1,4-heptanediol [900002-73-5] $C_9H_{20}O_2$ MW = 160.26 753

Table 1. Experimental value with uncertainty.

$\frac{T}{K}$	$\frac{\rho_{exp} \pm 2\sigma_{est}}{kg \cdot m^{-3}}$	Ref.
298.15	970.0 ± 2.0	64-des/sou

5-Ethyl-1,5-heptanediol [57740-06-4] $C_9H_{20}O_2$ MW = 160.26 754

Table 1. Experimental value with uncertainty.

$\frac{T}{K}$	$\frac{\rho_{exp} \pm 2\sigma_{est}}{kg \cdot m^{-3}}$	Ref.
291.15	958.0 ± 2.0	56-cri

3-Methyl-3,5-octanediol [38836-28-1] $C_9H_{20}O_2$ MW = 160.26 755

Table 1. Experimental value with uncertainty.

$\frac{T}{K}$	$\frac{\rho_{exp} \pm 2\sigma_{est}}{kg \cdot m^{-3}}$	Ref.
293.15	916.0 ± 1.0	62-esa/zhu

4-Methyl-2,4-octanediol [38836-27-0] $C_9H_{20}O_2$ MW = 160.26 756

Table 1. Experimental value with uncertainty.

$\frac{T}{K}$	$\frac{\rho_{exp} \pm 2\sigma_{est}}{kg \cdot m^{-3}}$	Ref.
293.15	911.7 ± 1.0	63-esa

6-Methyl-1,6-octanediol [13175-25-2] $C_9H_{20}O_2$ MW = 160.26 757

Table 1. Experimental value with uncertainty.

$\frac{T}{K}$	$\frac{\rho_{exp} \pm 2\sigma_{est}}{kg \cdot m^{-3}}$	Ref.
298.15	952.0 ± 2.0	66-bue

6-Methyl-1,7-octanediol **[91391-44-5]** $C_9H_{20}O_2$ **MW = 160.26** **758**

Table 1. Experimental value with uncertainty.

$\frac{T}{K}$	$\frac{\rho_{exp} \pm 2\sigma_{est}}{kg \cdot m^{-3}}$	Ref.
298.15	953.0 ± 2.0	62-col/gir

7-Methyl-1,7-octanediol **[13175-30-9]** $C_9H_{20}O_2$ **MW = 160.26** **759**

Table 1. Experimental value with uncertainty.

$\frac{T}{K}$	$\frac{\rho_{exp} \pm 2\sigma_{est}}{kg \cdot m^{-3}}$	Ref.
298.15	938.0 ± 2.0	66-bue

1-Methyl-2-propyl-1,5-pentanediol **[900002-75-7]** $C_9H_{20}O_2$ **MW = 160.26** **760**

Table 1. Experimental value with uncertainty.

$\frac{T}{K}$	$\frac{\rho_{exp} \pm 2\sigma_{est}}{kg \cdot m^{-3}}$	Ref.
293.15	954.3 ± 2.0	62-yur/pen

2-(2-Methylpropyl)-1,5-pentanediol **[57740-10-0]** $C_9H_{20}O_2$ **MW = 160.26** **761**

Table 1. Experimental value with uncertainty.

$\frac{T}{K}$	$\frac{\rho_{exp} \pm 2\sigma_{est}}{kg \cdot m^{-3}}$	Ref.
293.15	941.6 ± 2.0	62-yur/rev

1,4-Nonanediol **[2430-73-1]** $C_9H_{20}O_2$ **MW = 160.26** **762**

Table 1. Experimental value with uncertainty.

$\frac{T}{K}$	$\frac{\rho_{exp} \pm 2\sigma_{est}}{kg \cdot m^{-3}}$	Ref.
293.15	929.5 ± 1.0	62-nik/vor

1,5-Nonanediol **[13686-96-9]** $C_9H_{20}O_2$ **MW = 160.26** **763**

Table 1. Experimental value with uncertainty.

$\frac{T}{K}$	$\frac{\rho_{exp} \pm 2\sigma_{est}}{kg \cdot m^{-3}}$	Ref.
293.15	937.0 ± 1.0	25-pie/ada

1,6-Nonanediol **[4066-78-8]** **$C_9H_{20}O_2$** **MW = 160.26** **764**

Table 1. Experimental values with uncertainties.

T / K	$\rho_{exp} \pm 2\sigma_{est}$ / $kg \cdot m^{-3}$	Ref.
294.15	935.0 ± 3.0	65-mor/lam
298.15	941.0 ± 3.0	66-bue

1,7-Nonanediol **[4469-84-5]** **$C_9H_{20}O_2$** **MW = 160.26** **765**

Table 1. Experimental value with uncertainty.

T / K	$\rho_{exp} \pm 2\sigma_{est}$ / $kg \cdot m^{-3}$	Ref.
298.15	944.0 ± 2.0	66-bue

4-Propyl-1,5-hexanediol **[13687-05-3]** **$C_9H_{20}O_2$** **MW = 160.26** **766**

Table 1. Experimental value with uncertainty.

T / K	$\rho_{exp} \pm 2\sigma_{est}$ / $kg \cdot m^{-3}$	Ref.
293.15	954.3 ± 2.0	62-yur/pen

2,4,5-Trimethyl-2,4-hexanediol **[36587-81-2]** **$C_9H_{20}O_2$** **MW = 160.26** **767**

Table 1. Experimental value with uncertainty.

T / K	$\rho_{exp} \pm 2\sigma_{est}$ / $kg \cdot m^{-3}$	Ref.
301.35	920.7 ± 2.0	34-jac-5

1,4-Decanediol **[37810-94-9]** **$C_{10}H_{22}O_2$** **MW = 174.28** **768**

Table 1. Experimental values with uncertainties.

T / K	$\rho_{exp} \pm 2\sigma_{est}$ / $kg \cdot m^{-3}$	Ref.
293.15	922.0 ± 2.0	62-nik/vor
293.15	917.6 ± 1.0	51-gor
293.15	917.6 ± 1.0	52-roe/stu

1,7-Decanediol **[13175-33-2]** **$C_{10}H_{22}O_2$** **MW = 174.28** **769**

Table 1. Experimental value with uncertainty.

T / K	$\rho_{exp} \pm 2\sigma_{est}$ / $kg \cdot m^{-3}$	Ref.
298.15	929.0 ± 2.0	66-bue

1,10-Decanediol **[112-47-0]** $C_{10}H_{22}O_2$ **MW = 174.28** **770**

Table 1. Fit with estimated B coefficient for 4 accepted points. Deviation σ_w = 2.320.

Coefficient	$\rho = A + BT$
A	1151.52
B	−0.755

Table 2. Experimental values with uncertainties and deviation from calculated values.

T / K	$\rho_{exp} \pm 2\sigma_{est}$ / $kg \cdot m^{-3}$	$\rho_{exp} - \rho_{calc}$ / $kg \cdot m^{-3}$	Ref.
	crystal		
298.15	1100.0 ± 5.0		62-par/mos
	liquid		
353.15	883.0 ± 1.5	−1.89	50-boe/ned
403.15	850.0 ± 2.0	2.86	50-boe/ned
453.15	812.0 ± 2.0	2.61	50-boe/ned
513.15	762.0 ± 2.0	−2.09	50-boe/ned

Table 3. Recommended values.

T / K	$\rho_{exp} \pm 2\sigma_{est}$ / $kg \cdot m^{-3}$	T / K	$\rho_{exp} \pm 2\sigma_{est}$ / $kg \cdot m^{-3}$	T / K	$\rho_{exp} \pm 2\sigma_{est}$ / $kg \cdot m^{-3}$
350.00	887.3 ± 4.4	410.00	842.0 ± 2.9	470.00	796.7 ± 3.9
360.00	879.7 ± 4.1	420.00	834.4 ± 2.9	480.00	789.1 ± 4.2
370.00	872.2 ± 3.7	430.00	826.9 ± 2.9	490.00	781.6 ± 4.6
380.00	864.6 ± 3.4	440.00	819.3 ± 3.1	500.00	774.0 ± 5.0
390.00	857.1 ± 3.2	450.00	811.8 ± 3.3	510.00	766.5 ± 5.4
400.00	849.5 ± 3.0	460.00	804.2 ± 3.5	520.00	758.9 ± 5.8

2,5-Diethyl-1,6-hexanediol **[91241-30-4]** $C_{10}H_{22}O_2$ **MW = 174.28** **771**

Table 1. Experimental value with uncertainty.

T / K	$\rho_{exp} \pm 2\sigma_{est}$ / $kg \cdot m^{-3}$	Ref.
298.15	930.7 ± 1.0	38-hil/adk

3,4-Diethyl-3,4-hexanediol **[6931-71-1]** $C_{10}H_{22}O_2$ **MW = 174.28** **772**

Table 1. Experimental value with uncertainty.

T / K	$\rho_{exp} \pm 2\sigma_{est}$ / $kg \cdot m^{-3}$	Ref.
298.15	943.5 ± 0.2	79-bal/fri

2,4-Dimethyl-2,4-octanediol **[7177-01-7]** $C_{10}H_{22}O_2$ **MW = 174.28** **773**

Table 1. Experimental value with uncertainty.

T / K	$\rho_{exp} \pm 2\sigma_{est}$ / $kg \cdot m^{-3}$	Ref.
294.65	902.0 ± 1.0	34-jac-5

2,5-Dimethyl-3,5-octanediol **[3899-88-5]** **$C_{10}H_{22}O_2$** **MW = 174.28** **774**

Table 1. Experimental value with uncertainty.

$\frac{T}{K}$	$\frac{\rho_{exp} \pm 2\sigma_{est}}{kg \cdot m^{-3}}$	Ref.
293.15	909.7 ± 2.0	64-esa/das

3,7-Dimethyl-1,6-octanediol **[53067-10-0]** **$C_{10}H_{22}O_2$** **MW = 174.28** **775**

Table 1. Experimental value with uncertainty.

$\frac{T}{K}$	$\frac{\rho_{exp} \pm 2\sigma_{est}}{kg \cdot m^{-3}}$	Ref.
293.15	948.1 ± 2.0	32-lon/kha

3,7-Dimethyl-1,7-octanediol **[107-74-4]** **$C_{10}H_{22}O_2$** **MW = 174.28** **776**

Table 1. Experimental value with uncertainty.

$\frac{T}{K}$	$\frac{\rho_{exp} \pm 2\sigma_{est}}{kg \cdot m^{-3}}$	Ref.
293.15	926.0 ± 2.0	42-mul

3,7-Dimethyl-3,5-octanediol **[56548-45-9]** **$C_{10}H_{22}O_2$** **MW = 174.28** **777**

Table 1. Experimental value with uncertainty.

$\frac{T}{K}$	$\frac{\rho_{exp} \pm 2\sigma_{est}}{kg \cdot m^{-3}}$	Ref.
285.15	911.8 ± 2.0	26-pas/zam

3,4-Dimethyl-3,4-octanediol **[91179-88-3]** **$C_{10}H_{22}O_2$** **MW = 174.28** **778**

Table 1. Experimental value with uncertainty.

$\frac{T}{K}$	$\frac{\rho_{exp} \pm 2\sigma_{est}}{kg \cdot m^{-3}}$	Ref.
293.15	932.6 ± 1.0	64-nog/rtv

3,7-Dimethyl-1,3-octanediol **[102880-60-4]** **$C_{10}H_{22}O_2$** **MW = 174.28** **779**

Table 1. Experimental value with uncertainty.

$\frac{T}{K}$	$\frac{\rho_{exp} \pm 2\sigma_{est}}{kg \cdot m^{-3}}$	Ref.
293.15	916.5 ± 2.0	32-pfa/pla

2,3-Dipropyl-1,4-butanediol **[74854-17-4]** **$C_{10}H_{22}O_2$** **MW = 174.28** **780**

Table 1. Experimental value with uncertainty.

$\frac{T}{K}$	$\frac{\rho_{exp} \pm 2\sigma_{est}}{kg \cdot m^{-3}}$	Ref.
293.15	936.2 ± 2.0	39-mar/wil

4-Ethyl-3,5-octanediol **[900002-77-9]** $C_{10}H_{22}O_2$ **MW = 174.28** **781**

Table 1. Experimental value with uncertainty.

$\frac{T}{K}$	$\frac{\rho_{exp} \pm 2\sigma_{est}}{kg \cdot m^{-3}}$	Ref.
293.15	926.3 ± 1.0	51-dra-1

6-Methyl-2-(1-methylethyl)-1,3-hexanediol **[900002-78-0]** $C_{10}H_{22}O_2$ **MW = 174.28** **782**

Table 1. Experimental value with uncertainty.

$\frac{T}{K}$	$\frac{\rho_{exp} \pm 2\sigma_{est}}{kg \cdot m^{-3}}$	Ref.
296.15	916.1 ± 2.0	43-kul/nor

2-Methyl-2,3-nonanediol **[900002-76-8]** $C_{10}H_{22}O_2$ **MW = 174.28** **783**

Table 1. Experimental value with uncertainty.

$\frac{T}{K}$	$\frac{\rho_{exp} \pm 2\sigma_{est}}{kg \cdot m^{-3}}$	Ref.
273.15	937.5 ± 2.0	26-nic

6-Methyl-1,6-nonanediol **[13175-26-3]** $C_{10}H_{22}O_2$ **MW = 174.28** **784**

Table 1. Experimental value with uncertainty.

$\frac{T}{K}$	$\frac{\rho_{exp} \pm 2\sigma_{est}}{kg \cdot m^{-3}}$	Ref.
298.15	942.0 ± 2.0	66-bue

7-Methyl-1,7-nonanediol **[13379-31-2]** $C_{10}H_{22}O_2$ **MW = 174.28** **785**

Table 1. Experimental value with uncertainty.

$\frac{T}{K}$	$\frac{\rho_{exp} \pm 2\sigma_{est}}{kg \cdot m^{-3}}$	Ref.
298.15	934.0 ± 2.0	66-bue

2-Propyl-1,3-heptanediol **[6628-65-5]** $C_{10}H_{22}O_2$ **MW = 174.28** **786**

Table 1. Experimental value with uncertainty.

$\frac{T}{K}$	$\frac{\rho_{exp} \pm 2\sigma_{est}}{kg \cdot m^{-3}}$	Ref.
301.15	915.5 ± 2.0	43-kul/nor

2,4,6-Trimethyl-2,4-heptanediol **[33070-42-7]** $C_{10}H_{22}O_2$ **MW = 174.28** **787**

Table 1. Fit with estimated *B* coefficient for 3 accepted points. Deviation $\sigma_w = 0.696$.

Coefficient	$\rho = A + BT$
A	1117.71
B	−0.720

cont.

Table 2. Experimental values with uncertainties and deviation from calculated values.

$\frac{T}{K}$	$\frac{\rho_{exp} \pm 2\sigma_{est}}{kg \cdot m^{-3}}$	$\frac{\rho_{exp} - \rho_{calc}}{kg \cdot m^{-3}}$	Ref.
293.15	906.1 ± 2.0	–0.54	31-deg
288.15	909.8 ± 2.0	–0.44	31-deg
292.35	908.2 ± 2.0	0.98	34-jac-5

Table 3. Recommended values.

$\frac{T}{K}$	$\frac{\rho_{exp} \pm 2\sigma_{est}}{kg \cdot m^{-3}}$	$\frac{T}{K}$	$\frac{\rho_{exp} \pm 2\sigma_{est}}{kg \cdot m^{-3}}$
280.00	916.1 ± 2.0	293.15	906.6 ± 1.9
290.00	908.9 ± 1.9	298.15	903.0 ± 2.0

4,4-Diethyl-1,7-heptanediol [72936-15-3] $C_{11}H_{24}O_2$ MW = 188.31 788

Table 1. Experimental value with uncertainty.

$\frac{T}{K}$	$\frac{\rho_{exp} \pm 2\sigma_{est}}{kg \cdot m^{-3}}$	Ref.
293.15	953.0 ± 2.0	79-zel/hub

2,4-Dimethyl-2,4-nonanediol [69201-96-3] $C_{11}H_{24}O_2$ MW = 188.31 789

Table 1. Experimental value with uncertainty.

$\frac{T}{K}$	$\frac{\rho_{exp} \pm 2\sigma_{est}}{kg \cdot m^{-3}}$	Ref.
299.75	896.3 ± 2.0	34-jac-5

3-Methyl-3,4-decanediol [900002-79-1] $C_{11}H_{24}O_2$ MW = 188.31 790

Table 1. Experimental value with uncertainty.

$\frac{T}{K}$	$\frac{\rho_{exp} \pm 2\sigma_{est}}{kg \cdot m^{-3}}$	Ref.
273.15	940.0 ± 2.0	26-nic

7-Methyl-1,7-decanediol [13175-31-0] $C_{11}H_{24}O_2$ MW = 188.31 791

Table 1. Experimental value with uncertainty.

$\frac{T}{K}$	$\frac{\rho_{exp} \pm 2\sigma_{est}}{kg \cdot m^{-3}}$	Ref.
298.15	929.0 ± 2.0	66-bue

2,4,7-Trimethyl-2,4-octanediol [900002-80-4] $C_{11}H_{24}O_2$ MW = 188.31 792

Table 1. Experimental value with uncertainty.

$\frac{T}{K}$	$\frac{\rho_{exp} \pm 2\sigma_{est}}{kg \cdot m^{-3}}$	Ref.
295.35	896.1 ± 2.0	34-jac-5

1,4-Undecanediol [4272-02-0] $C_{11}H_{24}O_2$ MW = 188.31 793

Table 1. Experimental value with uncertainty.

T / K	$\rho_{exp} \pm 2\sigma_{est}$ / $kg \cdot m^{-3}$	Ref.
293.15	915.7 ± 1.0	65-das/mae

1,5-Undecanediol [13686-98-1] $C_{11}H_{24}O_2$ MW = 188.31 794

Table 1. Experimental value with uncertainty.

T / K	$\rho_{exp} \pm 2\sigma_{est}$ / $kg \cdot m^{-3}$	Ref.
294.15	914.0 ± 2.0	56-cri

2-Butyl-1,3-octanediol [55109-62-1] $C_{12}H_{26}O_2$ MW = 202.34 795

Table 1. Experimental value with uncertainty.

T / K	$\rho_{exp} \pm 2\sigma_{est}$ / $kg \cdot m^{-3}$	Ref.
298.15	918.4 ± 2.0	43-kul/nor

4-Methyl-4,5-undecanediol [900002-81-5] $C_{12}H_{26}O_2$ MW = 202.34 796

Table 1. Experimental value with uncertainty.

T / K	$\rho_{exp} \pm 2\sigma_{est}$ / $kg \cdot m^{-3}$	Ref.
273.15	935.5 ± 2.0	26-nic

2-Methyl-1,4-dodecanediol [92153-96-3] $C_{13}H_{28}O_2$ MW = 216.36 797

Table 1. Experimental value with uncertainty.

T / K	$\rho_{exp} \pm 2\sigma_{est}$ / $kg \cdot m^{-3}$	Ref.
293.15	897.2 ± 2.0	62-nik/vor

3-Methyl-1,3-dodecanediol [900002-82-6] $C_{13}H_{28}O_2$ MW = 216.36 798

Table 1. Experimental value with uncertainty.

T / K	$\rho_{exp} \pm 2\sigma_{est}$ / $kg \cdot m^{-3}$	Ref.
293.15	902.0 ± 2.0	32-pfa/pla

2-Pentyl-1,3-nonanediol [55109-63-2] $C_{14}H_{30}O_2$ MW = 230.39 799

Table 1. Experimental value with uncertainty.

T / K	$\rho_{exp} \pm 2\sigma_{est}$ / $kg \cdot m^{-3}$	Ref.
296.15	898.4 ± 2.0	43-kul/nor

3.2 Unsaturated Diols

3.2.1 Unsaturated Diols of General Formula $C_nH_{2n}O_2$

2-Butene-1,4-diol **[110-64-5]** **$C_4H_8O_2$** **MW = 88.11** **800**

Table 1. Experimental values with uncertainties.

$\frac{T}{K}$	$\frac{\rho_{exp} \pm 2\sigma_{est}}{kg \cdot m^{-3}}$	Ref.
288.15	1082.0 ± 3.0	46-pre/val
293.15	1080.0 ± 3.0	48-pud
293.15	1080.0 ± 3.0	48-val
298.15	1080.0 ± 3.0	78-ovc/kry

(*E*)-2-Butene-1,4-diol **[821-11-4]** **$C_4H_8O_2$** **MW = 88.11** **801**

Table 1. Experimental and recommended values with uncertainties.

$\frac{T}{K}$	$\frac{\rho_{exp} \pm 2\sigma_{est}}{kg \cdot m^{-3}}$	Ref.
293.15	1068.7 ± 1.0	52-cur/joh
293.15	1068.5 ± 0.8	56-smi/ebe
293.15	1068.6 ± 0.8	Recommended

(*Z*)-2-Butene-1,4-diol **[6117-80-2]** **$C_4H_8O_2$** **MW = 88.11** **802**

Table 1. Experimental and recommended values with uncertainties.

$\frac{T}{K}$	$\frac{\rho_{exp} \pm 2\sigma_{est}}{kg \cdot m^{-3}}$	Ref.
293.15	1069.8 ± 2.0	28-pre-4[1)]
294.15	1069.9 ± 2.0	44-pre[1)]
293.15	1074.0 ± 0.8	56-smi/ebe
293.15	1074.0 ± 0.8	Recommended

[1)] Not included in calculation of recommended value.

(*Z*)-2-Pentene-1,4-diol **[500036-05-5]** **$C_5H_{10}O_2$** **MW = 102.13** **803**

Table 1. Experimental value with uncertainty.

$\frac{T}{K}$	$\frac{\rho_{exp} \pm 2\sigma_{est}}{kg \cdot m^{-3}}$	Ref.
291.15	1017.0 ± 2.0	51-gor

(*Z*)-4-Methyl-2-pentene-1,4-diol **[500036-09-9]** **$C_6H_{12}O_2$** **MW = 116.16** **804**

Table 1. Experimental value with uncertainty.

$\frac{T}{K}$	$\frac{\rho_{exp} \pm 2\sigma_{est}}{kg \cdot m^{-3}}$	Ref.
290.15	990.0 ± 2.0	51-gor

(*Z*)-2-Heptene-1,4-diol **[83726-19-6]** **$C_7H_{14}O_2$** **MW = 130.19** **805**

Table 1. Experimental value with uncertainty.

$\frac{T}{K}$	$\frac{\rho_{exp} \pm 2\sigma_{est}}{kg \cdot m^{-3}}$	Ref.
299.65	970.0 ± 2.0	51-gor

5-Methyl-3-hexene-1,5-diol **[19764-75-1]** **$C_7H_{14}O_2$** **MW = 130.19** **806**

Table 1. Experimental value with uncertainty.

$\frac{T}{K}$	$\frac{\rho_{exp} \pm 2\sigma_{est}}{kg \cdot m^{-3}}$	Ref.
293.15	980.2 ± 1.0	67-min/che

(*Z*)-4-Ethyl-2-hexene-1,4-diol **[500036-10-2]** **$C_8H_{16}O_2$** **MW = 144.21** **807**

Table 1. Experimental value with uncertainty.

$\frac{T}{K}$	$\frac{\rho_{exp} \pm 2\sigma_{est}}{kg \cdot m^{-3}}$	Ref.
288.15	974.0 ± 0.0	51-gor

5-Methyl-3-heptene-1,5-diol **[19764-76-2]** **$C_8H_{16}O_2$** **MW = 144.21** **808**

Table 1. Experimental value with uncertainty.

$\frac{T}{K}$	$\frac{\rho_{exp} \pm 2\sigma_{est}}{kg \cdot m^{-3}}$	Ref.
293.15	972.5 ± 1.0	67-min/che

5-Methyl-3-octene-1,5-diol **[19764-77-3]** **$C_9H_{18}O_2$** **MW = 158.24** **809**

Table 1. Experimental value with uncertainty.

$\frac{T}{K}$	$\frac{\rho_{exp} \pm 2\sigma_{est}}{kg \cdot m^{-3}}$	Ref.
293.15	955.5 ± 1.0	67-min/che

(*Z*)-2-Decene-1,4-diol **[500036-07-7]** **$C_{10}H_{20}O_2$** **MW = 172.27** **810**

Table 1. Experimental value with uncertainty.

$\frac{T}{K}$	$\frac{\rho_{exp} \pm 2\sigma_{est}}{kg \cdot m^{-3}}$	Ref.
295.15	930.0 ± 2.0	51-gor

3,4-Dimethyl-5-octene-3,4-diol **[91008-98-9]** **$C_{10}H_{20}O_2$** **MW = 172.27** **811**

Table 1. Experimental value with uncertainty.

$\frac{T}{K}$	$\frac{\rho_{exp} \pm 2\sigma_{est}}{kg \cdot m^{-3}}$	Ref.
293.15	970.4 ± 2.0	64-nog/rtv

2-Dodecene-1,4-diol **[97029-80-6]** **$C_{12}H_{24}O_2$** **MW = 200.32** **812**

Table 1. Experimental value with uncertainty.

$\frac{T}{K}$	$\frac{\rho_{exp} \pm 2\sigma_{est}}{kg \cdot m^{-3}}$	Ref.
293.15	934.3 ± 2.0	62-nik/vor

6,7-Diethyl-3-methyl-6-nonene-3,4-diol **[20368-03-0]** **$C_{14}H_{28}O_2$** **MW = 228.38** **813**

Table 1. Experimental value with uncertainty.

$\frac{T}{K}$	$\frac{\rho_{exp} \pm 2\sigma_{est}}{kg \cdot m^{-3}}$	Ref.
293.15	921.2 ± 1.0	67-zal/aru

3.2.2 Unsaturated Diols of General Formula $C_nH_{2n-2}O_2$

2-Butyne-1,4-diol **[110-65-6]** **$C_4H_6O_2$** **MW = 86.09** **814**

Table 1. Experimental value with uncertainty.

$\frac{T}{K}$	$\frac{\rho_{exp} \pm 2\sigma_{est}}{kg \cdot m^{-3}}$	Ref.
293.15	1100.1 ± 0.0	66 -koz/rab

2-Pentyne-1,4-diol **[927-57-1]** **$C_5H_8O_2$** **MW = 100.12** **815**

Table 1. Experimental value with uncertainty.

$\frac{T}{K}$	$\frac{\rho_{exp} \pm 2\sigma_{est}}{kg \cdot m^{-3}}$	Ref.
290.15	1072.0 ± 1.5	51-gor

2-Methyl-3-pentyne-2,5-diol **[900002-87-1]** **$C_6H_{10}O_2$** **MW = 114.14** **816**

Table 1. Experimental value with uncertainty.

$\frac{T}{K}$	$\frac{\rho_{exp} \pm 2\sigma_{est}}{kg \cdot m^{-3}}$	Ref.
293.15	1029.1 ± 1.0	66-vla/vas

4-Methyl-2-pentyne-1,4-diol **[10605-66-0]** **$C_6H_{10}O_2$** **MW = 114.14** **817**

Table 1. Experimental value with uncertainty.

$\frac{T}{K}$	$\frac{\rho_{exp} \pm 2\sigma_{est}}{kg \cdot m^{-3}}$	Ref.
290.15	1029.0 ± 2.5	51-gor

2-Heptyne-1,4-diol **[18864-39-6]** **$C_7H_{12}O_2$** **MW = 128.17** **818**

Table 1. Experimental value with uncertainty.

$\frac{T}{K}$	$\frac{\rho_{exp} \pm 2\sigma_{est}}{kg \cdot m^{-3}}$	Ref.
288.15	1012.4 ± 1.5	51-gor

4-Methyl-2-hexyne-1,4-diol **[920-09-2]** **$C_7H_{12}O_2$** **MW = 128.17** **819**

Table 1. Experimental value with uncertainty.

$\frac{T}{K}$	$\frac{\rho_{exp} \pm 2\sigma_{est}}{kg \cdot m^{-3}}$	Ref.
293.15	1022.40 ± 1.0	66-vla/vas

4-Ethyl-2-hexyne-1,4-diol **[163005-62-7]** $C_8H_{14}O_2$ **MW = 142.2** **820**

Table 1. Experimental value with uncertainty.

T / K	$\rho_{exp} \pm 2\sigma_{est}$ / $kg \cdot m^{-3}$	Ref.
291.15	1000.5 ± 2.0	51-gor

4-Methyl-2-heptyne-1,4-diol **[10605-67-1]** $C_8H_{14}O_2$ **MW = 142.2** **821**

Table 1. Experimental value with uncertainty.

T / K	$\rho_{exp} \pm 2\sigma_{est}$ / $kg \cdot m^{-3}$	Ref.
293.15	996.5 ± 1.0	66-vla/vas

2-Decyne-1,4-diol **[71393-78-7]** $C_{10}H_{18}O_2$ **MW = 170.25** **822**

Table 1. Fit with estimated B coefficient for 2 accepted points. Deviation σ_w = 0.200.

Coefficient	$\rho = A + BT$
A	1148.34
B	−0.680

Table 2. Experimental values with uncertainties and deviation from calculated values.

T / K	$\rho_{exp} \pm 2\sigma_{est}$ / $kg \cdot m^{-3}$	$\rho_{exp} - \rho_{calc}$ / $kg \cdot m^{-3}$	Ref.
273.15	962.4 ± 1.0	−0.20	51-gor
293.15	949.2 ± 1.0	0.20	51-gor

Table 3. Recommended values.

T / K	$\rho_{exp} \pm 2\sigma_{est}$ / $kg \cdot m^{-3}$
270.00	964.7 ± 1.1
280.00	957.9 ± 0.9
290.00	951.1 ± 1.0
293.15	949.0 ± 1.0
298.15	945.6 ± 1.2

3,4-Dimethyl-5-octyne-3,4-diol **[92490-84-1]** $C_{10}H_{18}O_2$ **MW = 170.25** **823**

Table 1. Experimental value with uncertainty.

T / K	$\rho_{exp} \pm 2\sigma_{est}$ / $kg \cdot m^{-3}$	Ref.
293.15	964.2 ± 2.0	64-nog/rtv

4 Tabulated Data on Density - Triols

1,2,3-Propanetriol **[56-81-5]** **$C_3H_8O_3$** **MW = 92.09** **824**

Table 1. Coefficients of the polynomial expansion equation.
Standard deviations (see introduction):
$\sigma_{c,w} = 5.4871 \cdot 10^{-1}$ (combined temperature ranges, weighted),
$\sigma_{c,uw} = 1.5204 \cdot 10^{-1}$ (combined temperature ranges, unweighted).

Coefficient	T = 288.15 to 533.15 K $\rho = A + BT + CT^2 + DT^3 + \ldots$
A	$1.43632 \cdot 10^{3}$
B	$-5.28889 \cdot 10^{-1}$
C	$-2.29556 \cdot 10^{-4}$

Table 2. Experimental values with uncertainties and deviation from calculated values.

$\frac{T}{K}$	$\frac{\rho_{exp} \pm 2\sigma_{est}}{kg \cdot m^{-3}}$	$\frac{\rho_{exp} - \rho_{calc}}{kg \cdot m^{-3}}$	Ref. (Symbol in Fig. 1)	$\frac{T}{K}$	$\frac{\rho_{exp} \pm 2\sigma_{est}}{kg \cdot m^{-3}}$	$\frac{\rho_{exp} - \rho_{calc}}{kg \cdot m^{-3}}$	Ref. (Symbol in Fig. 1)
293.15	1259.40 ± 1.00	–2.15	1864-lan(×)	333.15	1234.00 ± 1.00	–0.64	58-cos/bow(×)
291.15	1264.60 ± 1.00	1.72	07-che-1(×)	353.15	1220.00 ± 1.00	–0.91	58-cos/bow(×)
295.65	1257.20 ± 1.00	–2.69	07-che-1 [1)]	373.15	1206.00 ± 1.00	–1.00	58-cos/bow(×)
288.15	1264.40 ± 0.50	–0.46	35-tim/hen(○)	393.15	1192.00 ± 1.02	–0.91	58-cos/bow(×)
293.15	1261.31 ± 0.50	–0.24	35-tim/hen(○)	413.15	1178.00 ± 1.04	–0.63	58-cos/bow(×)
303.15	1255.09 ± 0.50	0.20	35-tim/hen(○)	433.15	1164.00 ± 1.06	–0.16	58-cos/bow(×)
298.15	1258.50 ± 0.50	0.27	37-alb(□)	453.15	1150.00 ± 1.08	0.48	58-cos/bow(×)
298.15	1258.30 ± 0.60	0.07	42-bri/rin(Δ)	473.15	1135.00 ± 1.10	0.31	58-cos/bow(×)
303.15	1254.90 ± 0.60	0.01	42-bri/rin(Δ)	493.15	1120.00 ± 1.12	0.33	58-cos/bow(×)
313.15	1248.60 ± 0.60	0.41	42-bri/rin(Δ)	513.15	1105.00 ± 1.14	0.53	58-cos/bow(×)
323.15	1242.20 ± 0.60	0.76	42-bri/rin(Δ)	533.15	1089.00 ± 1.16	–0.09	58-cos/bow(×)
333.15	1235.80 ± 0.60	1.16	42-bri/rin(Δ)	553.15	1068.00 ± 1.18	–5.53	58-cos/bow [1)]
296.75	1258.60 ± 0.50	–0.56	46-par/wes(×)	293.15	1261.90 ± 0.60	0.35	68-ano(∇)
293.15	1262.00 ± 1.00	0.45	58-cos/bow(×)	298.15	1258.10 ± 0.80	–0.13	69-ver/lau(◆)
313.15	1249.00 ± 1.00	0.81	58-cos/bow(×)				

[1)] Not included in Fig. 1.

Further references: [1885-sto, 1891-gla, 17-jae-1, 32-bri-2, 35-but/ram-1, 36-ern/wat, 70-che/tho, 77-cam/sch].

cont.

1,2,3-Propanetriol (cont.)

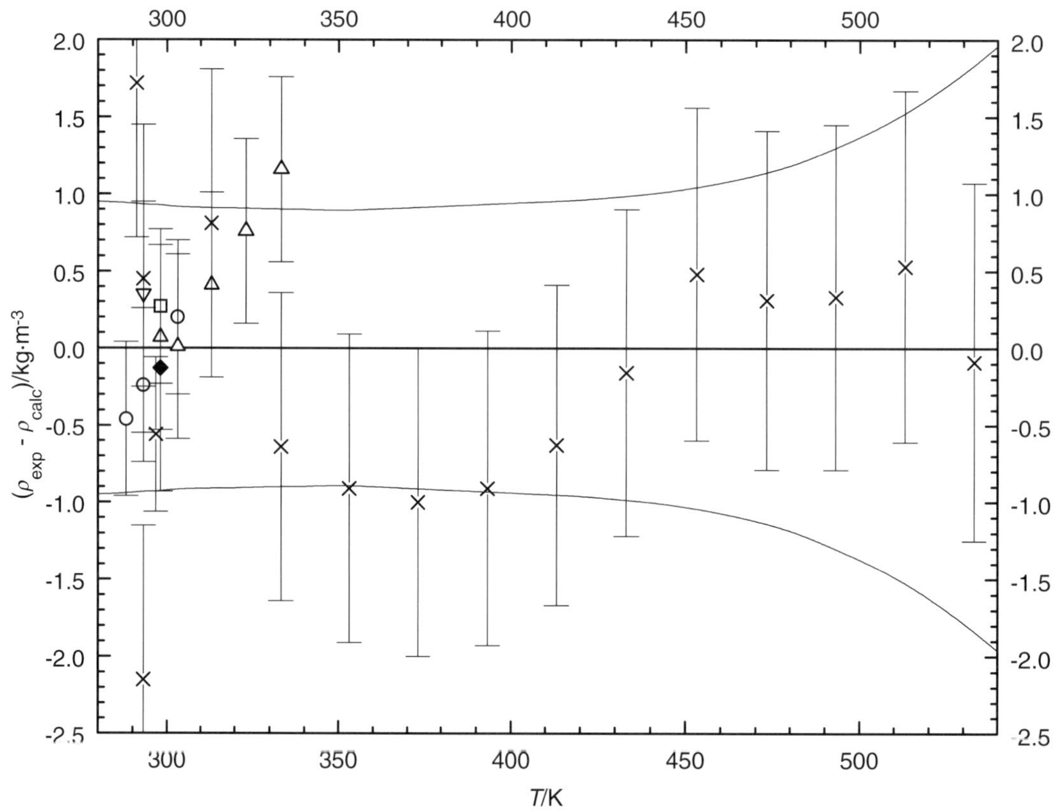

Fig. 1. The symbols show the deviation of the calculated from the experimental values from Table 2. The curves above and below the zero line indicate the calculated error region of the recommended values given in Table 3. The error bars represent the experimental errors. (Error bars smaller than the symbols are omitted for clarity of the figure.)

Table 3. Recommended values (fit to the reliable experimental values according to the equations $\rho = A + BT + CT^2 + DT^3 + \ldots$ or $\rho = [1 + 1.75(1 - T/T_c)^{1/3} + 0.75(1 - T/T_c)][\rho_c + A(T_c - T) + B(T_c - T)^2 + C(T_c - T)^3 + D(T_c - T)^4]$).

$\frac{T}{\text{K}}$	$\frac{\rho \pm \sigma_{\text{fit}}}{\text{kg}\cdot\text{m}^{-3}}$	$\frac{T}{\text{K}}$	$\frac{\rho \pm \sigma_{\text{fit}}}{\text{kg}\cdot\text{m}^{-3}}$	$\frac{T}{\text{K}}$	$\frac{\rho \pm \sigma_{\text{fit}}}{\text{kg}\cdot\text{m}^{-3}}$
280.00	1270.23 ± 0.95	360.00	1216.17 ± 0.90	460.00	1144.46 ± 1.07
290.00	1263.64 ± 0.94	370.00	1209.21 ± 0.91	470.00	1137.03 ± 1.12
293.15	1261.55 ± 0.93	380.00	1202.19 ± 0.92	480.00	1129.56 ± 1.18
298.15	1258.23 ± 0.93	390.00	1195.14 ± 0.93	490.00	1122.05 ± 1.27
300.00	1256.99 ± 0.92	400.00	1188.04 ± 0.94	500.00	1114.49 ± 1.37
310.00	1250.30 ± 0.91	410.00	1180.89 ± 0.95	510.00	1106.88 ± 1.48
320.00	1243.57 ± 0.91	420.00	1173.69 ± 0.96	520.00	1099.23 ± 1.62
330.00	1236.79 ± 0.90	430.00	1166.45 ± 0.98	530.00	1091.53 ± 1.78
340.00	1229.96 ± 0.90	440.00	1159.17 ± 1.00	540.00	1083.78 ± 1.96
350.00	1223.09 ± 0.89	450.00	1151.84 ± 1.03		

1,2,4-Butanetriol **[3068-00-6]** $C_4H_{10}O_3$ **MW = 106.12** **825**

Table 1. Experimental value with uncertainty.

T / K	$\rho_{exp} \pm 2\sigma_{est}$ / $kg \cdot m^{-3}$	Ref.
298.15	1184.0 ± 2.0	62-mel

1,3,5-Pentanetriol **[4328-94-3]** $C_5H_{12}O_3$ **MW = 120.15** **826**

Table 1. Experimental value with uncertainty.

T / K	$\rho_{exp} \pm 2\sigma_{est}$ / $kg \cdot m^{-3}$	Ref.
298.15	1103.6 ± 1.0	48-adk/bil

1,2,6-Hexanetriol **[106-69-4]** $C_6H_{14}O_3$ **MW = 134.18** **827**

Table 1. Coefficients of the polynomial expansion equation.
Standard deviations (see introduction):
$\sigma_{c,w} = 5.4613 \cdot 10^{-2}$ (combined temperature ranges, weighted),
$\sigma_{c,uw} = 1.4472 \cdot 10^{-2}$ (combined temperature ranges, unweighted).

Coefficient	T = 293.15 to 343.15 K $\rho = A + BT + CT^2 + DT^3 + \ldots$
A	$1.26157 \cdot 10^{3}$
B	$-4.84437 \cdot 10^{-1}$
C	$-1.92931 \cdot 10^{-4}$

Table 2. Experimental values with uncertainties and deviation from calculated values.

T / K	$\rho_{exp} \pm 2\sigma_{est}$ / $kg \cdot m^{-3}$	$\rho_{exp} - \rho_{calc}$ / $kg \cdot m^{-3}$	Ref. (Symbol in Fig. 1)	T / K	$\rho_{exp} \pm 2\sigma_{est}$ / $kg \cdot m^{-3}$	$\rho_{exp} - \rho_{calc}$ / $kg \cdot m^{-3}$	Ref. (Symbol in Fig. 1)
293.15	1103.00 ± 1.00	0.02	96-gri/zhu(□)	323.15	1084.90 ± 1.00	0.02	96-gri/zhu(□)
303.15	1097.00 ± 1.00	0.01	96-gri/zhu(□)	333.15	1078.80 ± 1.00	0.03	96-gri/zhu(□)
313.15	1090.90 ± 1.00	−0.05	96-gri/zhu(□)	343.15	1072.60 ± 1.00	−0.02	96-gri/zhu(□)

Table 3. Recommended values (fit to the reliable experimental values according to the equations
$\rho = A + BT + CT^2 + DT^3 + \ldots$ or $\rho = [1 + 1.75(1 - T/T_c)^{1/3} + 0.75(1 - T/T_c)][\rho_c + A(T_c - T) + B(T_c - T)^2 + C(T_c - T)^3 + D(T_c - T)^4]$).

T / K	$\rho \pm \sigma_{fit}$ / $kg \cdot m^{-3}$	T / K	$\rho \pm \sigma_{fit}$ / $kg \cdot m^{-3}$	T / K	$\rho \pm \sigma_{fit}$ / $kg \cdot m^{-3}$
290.00	1104.86 ± 1.38	300.00	1098.88 ± 1.12	330.00	1080.70 ± 0.91
293.15	1102.98 ± 1.29	310.00	1092.86 ± 0.94	340.00	1074.56 ± 1.20
298.15	1099.99 ± 1.16	320.00	1086.80 ± 0.85	350.00	1068.39 ± 1.83

cont.

1,2,6-Hexanetriol (cont.)

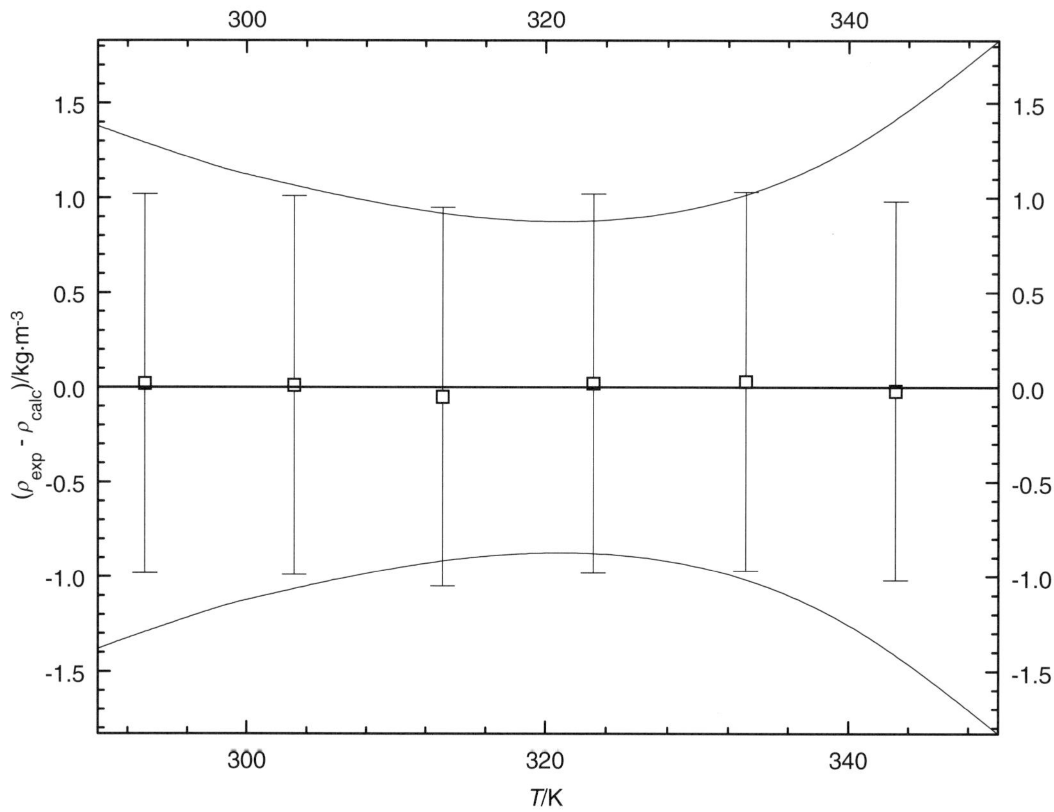

Fig. 1. The symbols show the deviation of the calculated from the experimental values from Table 2. The curves above and below the zero line indicate the calculated error region of the recommended values given in Table 3. The error bars represent the experimental errors. (Error bars smaller than the symbols are omitted for clarity of the figure.)

1,3,6-Hexanetriol **[18990-98-2]** **$C_6H_{14}O_3$** **MW = 134.18** **828**

Table 1. Experimental value with uncertainty.

$\frac{T}{K}$	$\frac{\rho_{exp} \pm 2\sigma_{est}}{kg \cdot m^{-3}}$	Ref.
293.15	1104.1 ± 10.0	62-nik/vor

1,4,7-Heptanetriol **[3920-53-4]** **$C_7H_{16}O_3$** **MW = 148.2** **829**

Table 1. Fit with estimated *B* coefficient for 2 accepted points. Deviation σ_w = 1.260.

Coefficient	$\rho = A + BT$
A	1259.98
B	−0.640

cont.

Table 2. Experimental values with uncertainties and deviation from calculated values.

$\frac{T}{K}$	$\frac{\rho_{exp} \pm 2\sigma_{est}}{kg \cdot m^{-3}}$	$\frac{\rho_{exp} - \rho_{calc}}{kg \cdot m^{-3}}$	Ref.
291.00	1075.0 ± 3.0	1.26	18-ham
273.00	1084.0 ± 3.0	−1.26	18-ham

Table 3. Recommended values.

$\frac{T}{K}$	$\frac{\rho_{exp} \pm 2\sigma_{est}}{kg \cdot m^{-3}}$
270.00	1087.2 ± 3.2
280.00	1080.8 ± 3.0
290.00	1074.4 ± 3.1
293.15	1072.4 ± 3.2
298.15	1069.2 ± 3.4

References

Reference codes are those used in the TRC SOURCE database. A reference code consists of the year prior to 1900, or the last two digits of the year after 1899, the first three letters of the first author , the first three letters of the second author. An additional sequence number is used when more than one reference in the database has an identical code.

1811-bra Brande, W. T.; Philos. Trans. R. Soc. London **101** (1811) 337.
1848-kop Kopp, H.; Justus Liebigs Ann. Chem. **64** (1848) 212.
1854-kop Kopp, H.; Justus Liebigs Ann. Chem. **92** (1854) 1.
1857-wur-1 Wurtz, A.; C. R. Hebd. Seances Acad. Sci. **45** (1857) 306.
1859-wur Wurtz, A.; Ann. Chim. Phys. **55** (1859) 400.
1863-gla/dal Gladstone, J. H.; Dale, T. P.; Philos. Trans. R. Soc. London **153** (1863) 317.
1864-lan Landolt, H.; Ann. Phys. (Leipzig) **122** (1864) 545.
1864-wur Wurtz, A.; Ann. Chim. Phys. **3** (1864) 129.
1865-men Mendeleev, D.; Z. Chem. **8** (1865) 257.
1869-dup/pag Dupre; Page; Philos. Trans. R. Soc. London **159** (1869) 591.
1869-lie Lieben, A.; Justus Liebigs Ann. Chem. **150** (1869) 87.
1869-zin Zincke, T.; Justus Liebigs Ann. Chem. **152** (1869) 1.
1870-gie Giesecke, A.; Z. Chem. **13** (1870) 430.
1871-lie/ros Lieben, A.; Rossi, A.; Justus Liebigs Ann. Chem. **158** (1871) 137.
1871-lie/ros-3 Lieben, A.; Rossi, A.; Justus Liebigs Ann. Chem. **159** (1871) 70.
1871-ros Rossi, A.; Justus Liebigs Ann. Chem. **159** (1871) 79.
1872-but Butlerov, A.; Justus Liebigs Ann. Chem. **162** (1872) 228.
1872-lin/von Linnemann, E.; von Zotta, V.; Justus Liebigs Ann. Chem. **161** (1872) 15.
1872-lin/von-2 Linnemann, E.; von Zotta, V.; Justus Liebigs Ann. Chem. **162** (1872) 3.
1872-pie/puc Pierre, J. -I.; Puchot, E.; Justus Liebigs Ann. Chem. **163** (1872) 253.
1873-fri/sil Friedel, C.; Silva, R. D.; C. R. Hebd. Seances Acad. Sci. **76** (1873) 226.
1874-reb-1 Reboul, C. E.; C. R. Hebd. Seances Acad. Sci. **79** (1874) 169.
1874-reb-2 Reboul, C. E.; Jahresbericht fur Chem. (1874) 336.
1875-fla Flavitsky, F.; Justus Liebigs Ann. Chem. **179** (1875) 340.
1875-lie Lieben, A.; Chem. Ber. **8** (1875) 1017.
1875-lie/vol Lieben, A.; Volker, O.; Chem. Ber. **8** (1875) 1019.
1875-wag/say Wagner, G.; Saytzeff, A.; Justus Liebigs Ann. Chem. **175** (1875) 351.
1875-wag/say-1 Wagner, G.; Saytzeff, A.; Justus Liebigs Ann. Chem. **179** (1875) 302.
1876-bal Balbiano, L.; Gazz. Chim. Ital. **6** (1876) 229.
1876-dec De Cininck, M. W. O.; Bull. Soc. Chim. Fr. **25** (1876) 7.
1876-lin Linnemann, E.; Chem. Ber. **9** (1876) 924.
1876-mun Munch, R.; Justus Liebigs Ann. Chem. **180** (1876) 327.
1877-cro Cross, C. J.; Justus Liebigs Ann. Chem. **189** (1877) 1.
1877-cro-1 Cross, C. J.; Chem. Ber. **10** (1877) 1601.
1877-lie/jan Lieben, A.; Janecek, G.; Justus Liebigs Ann. Chem. **187** (1877) 126.
1878-roh Rohn, W.; Justus Liebigs Ann. Chem. **190** (1878) 305.
1878-win Winogradow, W.; Justus Liebigs Ann. Chem. **191** (1878) 125.
1878-wis Wischnegradsky, A.; Justus Liebigs Ann. Chem. **190** (1878) 328.
1879-bel Belohoubek, A.; Chem. Ber. **12** (1879) 1872.
1879-bru Bruhl, J. W.; Justus Liebigs Ann. Chem. **200** (1879) 139.
1879-car Carleton-Williams, W.; J. Chem. Soc. **35** (1879) 125.
1879-pav Pavlov, D.; Justus Liebigs Ann. Chem. **196** (1879) 122.
1880-bru-1 Bruhl, J. W.; Justus Liebigs Ann. Chem. **203** (1880) 1.
1880-bru-3 Bruhl, J. W.; Justus Liebigs Ann. Chem. **200** (1880) 139.
1880-pry Prytz, K.; Ann. Phys. (Leipzig) **11** (1880) 104.
1880-tho Thorpe, T. E.; J. Chem. Soc. **37** (1880) 141.

1881-ber Berthelot, M.; Ann. Chim. Phys. **23** (1881) 176.
1881-nac/pag Naccari, A.; Pagliani, S.; Atti R. Accad. Sci. Torino, Cl. Sci. Fis. Mat. Nat. **16** (1881) 407.
1881-pri/han Pribram, A.; Handl, B.; Monatsh. Chem. **2** (1881) 643.
1882-sch-1 Schiff, R.; Chem. Ber. **15** (1882) 2965.
1882-zan Zander, A.; Justus Liebigs Ann. Chem. **214** (1882) 138.
1883-fre Frentzel, J.; Chem. Ber. **16** (1883) 743.
1883-kra Krafft, F.; Ber. Dtsch. Chem. Ges. **16** (1883) 1714.
1883-lie/zei Lieben, A.; Zeisel, B.; Monatsh. Chem. **4** (1883) 10.
1883-sch-3 Schiff, R.; Justus Liebigs Ann. Chem. **220** (1883) 71.
1883-wis Wislicenus, J.; Justus Liebigs Ann. Chem. **219** (1883) 307.
1884-gla Gladstone, J. H.; J. Chem. Soc., Trans. **45** (1884) 241.
1884-kra Krafft, F.; Chem. Ber. **17** (1884) 1627.
1884-per Perkin, W. H. ; J. Chem. Soc. **45** (1884) 421.
1884-sch-6 Schiff, R.; Justus Liebigs Ann. Chem. **223** (1884) 47.
1884-zan Zander, A.; Justus Liebigs Ann. Chem. **224** (1884) 56.
1885-sto Stohmann, F.; J. Prakt. Chem. **31** (1885) 273.
1886-gar Gartenmeister, R.; Justus Liebigs Ann. Chem. **233** (1886) 249.
1886-hen Henninger, A.; Ann. Chim. Phys. **7** (1886) 209.
1886-kra Krafft, F.; Chem. Ber. **19** (1886) 2218.
1886-tra Traube, J.; Chem. Ber. **19** (1886) 871.
1887-ram/you Ramsay, W.; Young, S.; Philos. Trans. R. Soc. London, A **178** (1887) 313.
1887-ref Reformatsky, A.; J. Prakt. Chem. **36** (1887) 340.
1888-ket Ketteler, E.; Ann. Phys. (Leipzig) **33** (1888) 506.
1889-lip Lipp, A.; Chem. Ber. **22** (1889) 2567.
1889-ram/you Ramsay, W.; Young, S.; Philos. Trans. R. Soc. London, A **180** (1889) 137.
1890-dek Dekkers, P. J.; Recl. Trav. Chim. Pays-Bas **9** (1890) 92.
1890-gar Gartenmeister, R.; Z. Phys. Chem., Stoechiom. Verwandtschaftsl. **6** (1890) 524.
1891-gla Gladstone, J. H.; J. Chem. Soc. **59** (1891) 290.
1891-gri/paw Grigorwitsch, A.; Pawlow, B.; Zh. Russ. Fiz.-Khim. O-va. **23** (1891) 172.
1891-jah Jahn, H.; Ann. Phys. (Leipzig) **43** (1891) 280.
1891-pol Poletaeff, G.; Chem. Ber. **24** (1891) 1308.
1891-sch/kos Schall, C.; Kossakowsky, L.; Z. Phys. Chem., Stoechiom. Verwandtschaftsl. **8** (1891) 241.
1892-lan/jah Landolt, H.; Jahn, H.; Z. Phys. Chem., Stoechiom. Verwandtschaftsl. **10** (1892) 289.
1892-sch-1 Schuett, F.; Z. Phys. Chem., Stoechiom. Verwandtschaftsl. **9** (1892) 349.
1893-eyk Eykman, J. F.; Recl. Trav. Chim. Pays-Bas **12** (1893) 157.
1893-eyk-1 Eykman, J. F.; Recl. Trav. Chim. Pays-Bas **12** (1893) 268.
1893-ram/shi-3 Ramsay, W.; Shields, J.; Z. Phys. Chem., Stoechiom. Verwandtschaftsl. **12** (1893) 433.
1893-tho/jon Thorpe, T. E.; Jones, L. M.; J. Chem. Soc., Trans. **63** (1893) 273.
1894-jah/mol Jahn, H.; Moller, G.; Z. Phys. Chem. (Leipzig) **13** (1894) 385.
1894-sch Schlamp, A.; Z. Phys. Chem. (Leipzig) **14** (1894) 272.
1895-eyk Eykman, J. F.; Recl. Trav. Chim. Pays-Bas **14** (1895) 185.
1896-zel/kra Zelinskii, N. D.; Krapiwin, S.; Z. Phys. Chem. (Leipzig) **21** (1896) 35.
1897-tho Thorpe, T. E.; J. Chem. Soc., Trans. **71** (1897) 920.
1897-zec Zecchini, F.; Gazz. Chim. Ital. **27** (1897) 358.
1898-kah Kahlbaum, G. W. A.; Z. Phys. Chem., Stoechiom. Verwandtschaftsl. **26** (1898) 577.
1898-lou Louguinine, W.; Ann. Chim. Phys. **13** (1898) 289.
1898-roh Rohland, P.; Z. Anorg. Chem. **18** (1898) 327.
00-car-1 Carleton-Williams, W.; J. Chem. Soc. **77** (1900) 125.
00-loo Loomis, E. H.; Z. Phys. Chem., Stoechiom. Verwandtschaftsl. **32** (1900) 578.
00-ste Stephan, K.; J. Prakt. Chem. **62** (1900) 523.
01-def-1 De Forcrand, M.; C. R. Hebd. Seances Acad. Sci. **132** (1901) 569.
01-gue Guerbet, M.; C. R. Hebd. Seances Acad. Sci. **132** (1901) 207.

01-gue-1 Guerbet, M.; C. R. Hebd. Seances Acad. Sci. **133** (1901) 1220.
01-ham Hamonet, J.; C. R. Hebd. Seances Acad. Sci. **132** (1901) 631.
01-mar/mck Marckwald, W.; McKenzie, A.; Chem. Ber. **34** (1901) 485.
01-rud Rudolphi, M.; Z. Phys. Chem. (Leipzig) **37** (1901) 426.
01-zel/zel Zelinskii, N. D.; Zelikov, J.; Ber. Dtsch. Chem. Ges. **34** (1901) 2856.
02-dud/lem Duden, P.; Lemme, R.; Ber. Dtsch. Chem. Ges. **35** (1902) 1335.
02-gri/tis Grignard, V.; Tissier, L.; C. R. Hebd. Seances Acad. Sci. **134** (1902) 107.
02-gue Guerbet, M.; C. R. Hebd. Seances Acad. Sci. **135** (1902) 172.
02-kon Konovalov, M. I.; Zh. Russ. Fiz.-Khim. O-va. **34** (1902) 26.
02-you Young, S.; J. Chem. Soc. **81** (1902) 707.
02-you/for Young, S.; Fortey, E. C.; J. Chem. Soc. **81** (1902) 717.
03-bla/gue Blaise, E. E.; Guerin; Bull. Soc. Chim. Fr. **29** (1903) 1202.
03-car/cop Carrara, G.; Coppadoro, A.; Gazz. Chim. Ital. **33** (1903) 329.
03-tho/man Thoms, H.; Mannich, C.; Chem. Ber. **36** (1903) 2544.
04-bou/bla Bouveault, L.; Blanc, G.; Bull. Soc. Chim. Fr. **31** (1904) 1203.
04-bou/bla-2 Bouveault, L.; Blanc, G.; Bull. Soc. Chim. Fr. **31** (1904) 748.
04-bru/sch Bruhl, J. W.; Schroder, H. ; Z. Phys. Chem. (Leipzig) **50** (1904) 1.
04-cri Crismer, L.; Bull. Soc. Chim. Belg. **18** (1904) 18.
04-dun Dunstan, A. E.; Z. Phys. Chem., Stoechiom. Verwandtschaftsl. **49** (1904) 590.
04-gri Grignard, V.; Bull. Soc. Chim. Fr. **31** (1904) 751.
04-gri-1 Grignard, V.; C. R. Hebd. Seances Acad. Sci. **138** (1904) 152.
04-ham Hamonet, J. L.; C. R. Hebd. Seances Acad. Sci. **139** (1904) 59.
04-kon Konovalov, M. I.; Zh. Russ. Fiz.-Khim. O-va. **36** (1904) 228.
05-dun Dunstan, A. E.; Z. Phys. Chem. (Leipzig) **51** (1905) 732.
05-win Winkler, L. W.; Ber. Dtsch. Chem. Ges. **38** (1905) 3612.
06-bou/loc Bouveault, L.; Locquin, R.; Bull. Soc. Chim. Fr. **35** (1906) 646.
06-car/fer Carrara, G.; Ferrari, G.; Gazz. Chim. Ital. **36** (1906) 1419.
06-kla/nor Klason, P.; Norlin, E.; Ark. Kemi, Mineral. Geol. **2**(24) (1906) 1.
06-kon/mil Konovalov, D.; Miller, E. F.; Timchenko; Zh. Russ. Fiz.-Khim. O-va. **38** (1906) 447.
06-mal-1 Malengraeu, F.; Bull. Cl. Sci. Acad. R. Belg. (1906) 796.
06-mus Muset, J.; Bull. Cl. Sci. Acad. R. Belg. (1906) 775.
06-van Van Gysegem, J.; Bull. Cl. Sci. Acad. R. Belg. (1906) 692.
06-wal-1 Walden, P.; Z. Phys. Chem., Stoechiom. Verwandtschaftsl. **55** (1906) 207.
07-che Cheneveau, C.; Ann. Chim. Phys. **12** (1907) 145.
07-che-1 Cheneveau, C.; Ann. Chim. Phys. **12** (1907) 289.
07-dun/tho Dunstan, A. E.; Thole, B. T.; Hunt, J. S.; J. Chem. Soc. **91** (1907) 1728.
07-fav/sok-3 Favorskii, A.; Sokovnin, N.; Justus Liebigs Ann. Chem. **354** (1907) 364.
07-fou/tif Fourneau, A. C.; Tiffeneau, D.; C. R. Hebd. Seances Acad. Sci. **145** (1907) 437.
07-kut Kutscherov, L.; Ann. **354** (1907) 376.
07-pic/ken Pickard, R. H.; Kenyon, J.; J. Chem. Soc. **91** (1907) 2058.
07-sok Sokovnin, N.; Justus Liebigs Ann. Chem. **354** (1907) 364.
07-tim Timmermans, J.; Bull. Soc. Chim. Belg. **21** (1907) 395.
07-wal-4 Walden, P.; Z. Phys. Chem., Stoechiom. Verwandtschaftsl. **59** (1907) 304.
08-and Andrew, L. W.; J. Am. Chem. Soc. **30** (1908) 353.
08-bue Buelens, A.; Bull. Cl. Sci. Acad. R. Belg. (1908) 921.
08-bue-1 Buelens, A.; Recl. Trav. Chim. Pays-Bas **28** (1908) 113.
08-dor/dvo Doroshevskii, A. G.; Dvorzhanchik, Z.; Zh. Russ. Fiz.-Khim. O-va., Chast Khim. **40** (1908) 908.
08-dor/rak Doroshevskii, A. G.; Rakovskii, A. V.; Zh. Russ. Fiz.-Khim. O-va., Chast Khim. **40** (1908) 860.
08-dun/stu Dunstan, A. E.; Stubbs, J. A.; J. Chem. Soc., Trans. **93** (1908) 1919.
08-get Getman, F. H.; J. Am. Chem. Soc. **30** (1908) 1077.
08-gyr Gyr, J.; Ber. Dtsch. Chem. Ges. **41** (1908) 4332.
08-har Hardin, D.; J. Chim. Phys. Phys.-Chim. Biol. **6** (1908) 584.

08-ric/mat	Richards, T. W.; Mathews, J. H.; J. Am. Chem. Soc. **30** (1908) 8.
08-zel/prz	Zelinskii, N. D.; Przheval'skii, E. C.; Zh. Russ. Fiz.-Khim. O-va., Chast Khim. **40** (1908) 1105.
09-bod/tab	Bodroux, F.; Taboury, F.; Bull. Soc. Chim. Fr. **5** (1909) 812.
09-cla	Claessens, F.; Bull. Soc. Chim. Fr. **5** (1909) 113.
09-dor	Doroshevskii, A. G.; Zh. Russ. Fiz.-Khim. O-va., Chast Khim. **41** (1909) 958.
09-dor/roz	Doroshevskii, A. G.; Rozhdestvenskii, M. S.; Zh. Russ. Fiz.-Khim. O-va., Chast Khim. **41** (1909) 1428.
09-fal-1	Falk, K. G.; J. Am. Chem. Soc. **31** (1909) 86.
09-gue	Guerbet, M.; C. R. Hebd. Seances Acad. Sci. **149** (1909) 129.
09-hen	Henry, L; Recl. Trav. Chim. Pays-Bas **28** (1909) 444.
09-hol/sag	Holmes, J.; Sageman, P. J.; J. Chem. Soc. **95** (1909) 1919.
09-kho	Khonin, G. V.; Zh. Russ. Fiz.-Khim. O-va., Chast Khim. **41** (1909) 327.
09-lem	Lemaire, J.; Bull. Cl. Sci. Acad. R. Belg. (1909) 83.
09-lem-1	Lemaire, J.; Recl. Trav. Chim. Pays-Bas **29** (1909) 22.
09-mas	Masson, H.; C. R. Hebd. Seances Acad. Sci. **149** (1909) 630.
09-sch-1	Schultz, G.; Ber. Dtsch. Chem. Ges. **42** (1909) 3609.
09-sch-2	Schwers, M. F.; Recl. Trav. Chim. Pays-Bas Belg. **28** (1909) 42.
10-daw	Dawson, H. M.; J. Chem. Soc. **97** (1910) 1041.
10-dor/pol	Doroshevskii, A.; Polansky, E.; Z. Phys. Chem., Stoechiom. Verwandtschaftsl. **73** (1910) 192.
10-dor/pol-1	Doroshevskii, A. G.; Polyanskii, E. V.; Zh. Russ. Fiz.-Khim. O-va., Chast Khim. **42** (1910) 109.
10-fon	Fontein, F.; Z. Phys. Chem., Stoechiom. Verwandtschaftsl. **73** (1910) 212.
10-fre	Freylon, G.; Ann. Chim. (Paris) **20** (1910) 58.
10-hal/las	Haller, A.; Lassieur, A.; C. R. Hebd. Seances Acad. Sci. **151** (1910) 697.
10-pol	Polowzow, V.; Z. Phys. Chem. (Leipzig) **75** (1910) 513.
10-ric	Richard, A.; Ann. Chim. Phys. **21** (1910) 323.
10-tho	Thole, F. B.; J. Chem. Soc. **97** (1910) 2596.
10 tim	Timmermans, J.; Bull. Soc. Chim. Belg. **24** (1910) 244.
11-del	De Leeuw, H. L.; Z. Phys. Chem., Stoechiom. Verwandtschaftsl. **77** (1911) 284.
11-dor	Doroshevskii, A. G.; Zh. Russ. Fiz.-Khim. O-va., Chast Khim. **43** (1911) 46.
11-dor-1	Doroshevskii, A. G.; Zh. Russ. Fiz.-Khim. O-va., Chast Khim. **43** (1911) 66.
11-kai	Kailan, A.; Ber. Dtsch. Chem. Ges. **44** (1911) 2881.
11-pic/ken	Pickard, R. H.; Kenyon, J.; J. Chem. Soc. **99** (1911) 45.
11-sch	Schwers, F.; J. Chim. Phys. Phys.-Chim. Biol. **9** (1911) 15.
11-wal	Wallach, O.; Justus Liebigs Ann. Chem. **381** (1911) 51.
11-zel/arj	Zelinskii, N. D.; Arjedinoff; J. Prakt. Chem. **84** (1911) 543.
12-bje	Bjelouss, E.; Ber. Dtsch. Chem. Ges. **45** (1912) 625.
12-gue	Guerbet, M.; Bull. Soc. Chim. Fr. **11** (1912) 276.
12-gue-1	Guerbet, M.; Bull. Soc. Chim. Fr. **11** (1912) 533.
12-gue-2	Guerbet, M.; C. R. Hebd. Seances Acad. Sci. **154** (1912) 1357.
12-ipa	Ipatieff, W.; Chem. Ber. **45** (1912) 3218.
12-kor	Korber, F.; Ann. Phys. (Leipzig) **37** (1912) 1014.
12-mal	Malosse, H.; C. R. Hebd. Seances Acad. Sci. **154** (1912) 1697.
12-mic	Michiels, L.; Bull. Cl. Sci. Acad. R. Belg. (1912) 10.
12-osb/mck	Osborne, N. S.; McKelvy, E. C.; Bearce, H. W.; J. Wash. Acad. Sci. **2** (1912) 95.
12-ost	Ostling, G. J.; J. Chem. Soc. **101** (1912) 457.
12-pic/ken	Pickard, R. H.; Kenyon, J.; J. Chem. Soc. **101** (1912) 620.
12-sch-1	Schwers, F.; Bull. Cl. Sci. Acad. R. Belg. (1912) 610.
12-sch-2	Schwers, F.; Bull. Cl. Sci. Acad. R. Belg. (1912) 719.
12-sch-3	Schwers, F.; J. Chem. Soc., Trans. **101** (1912) 1889.
12-tim	Timmermans, J.; Bull. Soc. Chim. Belg. **26** (1912) 205.
12-tim-1	Timmermans, J.; Sci. Proc. R. Dublin Soc. **13** (1912) 310.
12-tyr	Tyrer, D.; J. Chem. Soc., Trans. **101** (1912) 1104.

12-wad/mer Wade, J. W.; Merriman, R. W.; J. Chem. Soc. **101** (1912) 2429.
12-wre Wrewski, M. S.; Z. Phys. Chem. (Leipzig) **81** (1912) 20.
13-atk/wal Atkins, W. R. G.; Wallace, T. A.; J. Chem. Soc., Trans. **103** (1913) 1461.
13-bri-1 Bridgman, P. W.; Proc. Am. Acad. Arts Sci. **49** (1913) 3.
13-dup Dupont, G.; Ann. Chim. Phys. **30** (1913) 485.
13-fav Favorskii, A.; J. Prakt. Chem. **88** (1913) 641.
13-gor Gorskii, A.; Zh. Russ. Fiz.-Khim. O-va., Chast Khim. **45** (1913) 167.
13-har Hardy, W. B.; Proc. R. Soc. London, A **88** (1913) 303.
13-kis-3 Kishner, N.; Zh. Russ. Fiz.-Khim. O-va., Chast Khim. **45** (1913) 973.
13-mer Merriman, R. W.; J. Chem. Soc. **103** (1913) 628.
13-muc Muchin, G.; Z. Electrochem. Angew. Phys. Chem. **19** (1913) 819.
13-osb/mck Osborne, N. S.; McKelvy, E. C.; Bearce, H. W.; Bur. Stand. (U. S.), Bull. **9** (1913) 327.
13-pic/ken Pickard, R. H.; Kenyon, J.; J. Chem. Soc. **103** (1913) 1923.
13-rob/acr Robertson, H. C.; Acree, S. F.; Am. Chem. J. **49** (1913) 474.
13-rom Romanov, W.; Ann. Phys. (Leipzig) **40** (1913) 281.
13-ste Stern, O.; Z. Phys. Chem., Stoechiom. Verwandtschaftsl. **81** (1913) 441.
13-tho Thole, F. B.; J. Chem. Soc. **103** (1913) 19.
14-eng/tur English, S.; Turner, W. E. S.; J. Chem. Soc., Trans. **105** (1914) 1656.
14-gas Gascard, A.; C. R. Hebd. Seances Acad. Sci. **159** (1914) 259.
14-hal Halse, M. O.; J. Prakt. Chem. **89** (1914) 451.
14-kre/mei Kremann, R.; Meingast, R.; Gugl, F.; Monatsh. Chem. **35** (1914) 1235.
14-lon Longinov, V.; Zh. Russ. Fiz.-Khim. O-va., Chast Khim. **46** (1914) 1084.
14-low Lowry, T. M.; J. Chem. Soc. **105** (1914) 81.
14-mer/tur Merry, E. W.; Turner, W. E. S.; J. Chem. Soc. **105** (1914) 748.
14-mur/amo Murat, M.; Amouroux, G.; Bull. Soc. Chim. Fr. **15** (1914) 159.
14-smi Smith, C.; J. Chem. Soc., Trans. **105** (1914) 1703.
14-tyr Tyrer, D.; J. Chem. Soc., Trans. **105** (1914) 2534.
14-vav Vavon, G.; Ann. Chim. (Paris) **1** (1914) 144.
14-wal-1 Wallach, O.; Justus Liebigs Ann. Chem. **408** (1914) 183.
14-wor Worley, R. P.; J. Chem. Soc. **105** (1914) 260
15-pea Peacock, D. H.; J. Chem. Soc. **107** (1915) 1547.
15-pri Price, T. W.; J. Chem. Soc. **107** (1915) 188.
15-ric/coo Richards, T. W.; Coombs, L. B.; J. Am. Chem. Soc. **37** (1915) 1656.
15-sch-1 Schwers, F.; Bull. Cl. Sci. Acad. R. Belg. (1915) 525.
15-wal-3 Wallach, O.; Justus Liebigs Ann. Chem. **408** (1915) 200.
16-har-2 Hartung, E. J.; Trans. Faraday Soc. **12** (1916) 66.
16-lev/all Levene, P. A.; Allen, C. H.; J. Biol. Chem. **27** (1916) 433.
16-ric/shi Richards, T. W.; Shipley, J. W.; J. Am. Chem. Soc. **38** (1916) 989.
16-sei/alt Seitz, W.; Alterthum, H.; Lechner, G.; Ann. Phys. (Leipzig) **49** (1916) 85.
16-wil/bru Willcox, M.; Brunel, R. F.; J. Am. Chem. Soc. **38** (1916) 1821.
16-wro/rei Wroth, B. B.; Reid, E. E.; J. Am. Chem. Soc. **38** (1916) 2316.
17-jae Jaeger, F. M.; Z. Anorg. Allg. Chem. **101** (1917) 130.
17-jae-1 Jaeger, F. M.; Z. Anorg. Allg. Chem. **101** (1917) 1.
18-abd/eic Abderhalden, E.; Eichwald, E.; Chem. Ber. **51** (1918) 1312.
18-bro/hum Brooks, B. T.; Humphrey, I.; J. Am. Chem. Soc. **40** (1918) 822.
18-ham Hamonet, J.; Ann. Chim. Phys. **10** (1918) 28.
18-her-2 Herz, W.; Z. Anorg. Chem. **104** (1918) 47.
18-kar Karvonen, A.; Ann. Acad. Sci. Fenn., Ser. A **10** (1918) 9.
18-lev/tay Levene, P. A.; Taylor, F. A.; J. Biol. Chem. **35** (1918) 281.
19-beh Behal, A.; Bull. Soc. Chim. Fr. **25** (1919) 473.
19-eyk Eykman, J. F.; Natuurkd. Verh. Hollandsche Maatschappij Wet. Haarlem **8** (1919) 438.
19-ort/jon Orton, K. J. P.; Jones, D. C.; J. Chem. Soc. **115** (1919) 1194.

19-rei/ral Reilly, J.; Ralph, E. W.; Sci. Proc. R. Dublin Soc. **15** (1919) 597.
19-wil/hat Willstatter, R.; Hatt, D.; Justus Liebigs Ann. Chem. **418** (1919) 148.
19-wil/sch Willstatter, R.; Schuppli, O.; Mayer, E. W.; Justus Liebigs Ann. Chem. **418** (1919) 121
20-ric/dav Richards, T. W.; Davis, H. S.; J. Am. Chem. Soc. **42** (1920) 1599.
21-bar/bir Barr, G.; Bircumshaw, L. L.; Aeron. Res. Comm. Rep. Mem. **No. 746** (1921).
21-bru/cre Brunel, R. F.; Crenshaw, J. L.; Tobin, E.; J. Am. Chem. Soc. **43** (1921) 561.
21-leb Lebo, R. B.; J. Am. Chem. Soc. **43** (1921) 1005.
21-ler Leroide, J.; Ann. Chim. (Paris) **16** (1921) 354.
21-par/sim Pariselle, M.; Simon, L. J.; C. R. Hebd. Seances Acad. Sci. **173** (1921) 86.
21-rei/hic Reilly, J.; Hickinbottom, W. J.; Sci. Proc. R. Dublin Soc. **16** (1921) 233.
22-bae Baerts, F.; Bull. Soc. Chim. Belg. **31** (1922) 421.
22-her/sch Herz, W.; Schuftan, P.; Z. Phys. Chem. (Leipzig) **101** (1922) 269.
22-lev/tay-1 Levene, P. A.; Taylor, F. A.; J. Biol. Chem. **54** (1922) 351.
22-mck/sim McKelvy, E. C.; Simpson, D. H.; J. Am. Chem. Soc. **44** (1922) 105.
22-mic Michels, A.; Arch. Neerl. Sci. Exactes Nat., Ser. 3A **6** (1922) 127.
22-sch/reg Schoorl, N.; Regenboden, A.; Recl. Trav. Chim. Pays-Bas **41** (1922) 1.
22-tim Timmermans, J.; Arch. Neerl. Sci. Exactes Nat., Ser. 3A **6** (1922) 147.
22-tro Tromp, S. T. J.; Recl. Trav. Chim. Pays-Bas **41** (1922) 278.
23-bru Brunel, R. F.; J. Am. Chem. Soc. **45** (1923) 1334.
23-clo/joh Clough, W. W.; Johns, C. O.; Ind. Eng. Chem. **15** (1923) 1030.
23-kau/ada Kaufmann, W. E.; Adams, R.; J. Am. Chem. Soc. **45** (1923) 3029.
23-moe-1 Moesveld, A. L. Th.; Z. Phys. Chem., Stoechiom. Verwandtschaftsl. **105** (1923) 450.
23-pop Popelier, J.; Bull. Soc. Chim. Belg. **32** (1923) 179.
23-rii Riiber, C. N.; Z. Elektrochem. Angew. Phys. Chem. **29** (1923) 334.
23-tim Timmermans, J.; Bull. Soc. Chim. Belg. **32** (1923) 299.
23-vav/iva Vavon, G.; Ivanoff, D.; C. R. Hebd. Seances Acad. Sci. **177** (1923) 453.
23-von/kai Von Braun, J.; Kaiser, W.; Chem. Ber. **56** (1923) 2268.
23-wil/smi Willard, H. H.; Smith, G. F.; J. Am. Chem. Soc. **45** (1923) 286.
24-bou Bourgou, A.; Bull. Soc. Chim. Belg. **33** (1924) 101.
24-bus-1 Busse, W.; Ann. Phys. (Leipzig) **75** (1924) 657.
24-cha/deg Chavanne, G.; De Graef, H.; Bull. Soc. Chim. Belg. **33** (1924) 366.
24-dan Dancaster, E.; J. Chem. Soc. **125** (1924) 2036.
24-dew/wec Dewael, A.; Weckering, A.; Bull. Soc. Chim. Belg. **33** (1924) 495.
24-lie Lievens, G.; Bull. Soc. Chim. Belg. **33** (1924) 122.
24-lon/mar Longinov, V.; Margoliss, E.; Bull. Soc. Chim. Fr. **35** (1924) 753.
24-mar-1 Mardles, E.; J. Chem. Soc. **125** (1924) 2244.
24-mil Miller, C. C.; Proc. R. Soc. London, A **106** (1924) 724.
24-par/sch Parks, G. S.; Schwenk, J. R.; J. Phys. Chem. **28** (1924) 720.
24-pow Powell, S. G.; J. Am. Chem. Soc. **46** (1924) 2514.
24-ter Terent'ev, A.; Bull. Soc. Chim. Fr. **35** (1924) 1145.
25-deg De Graef, H.; Bull. Soc. Chim. Belg. **34** (1925) 427.
25-fai Faillebin, M.; Ann. Chim. (Paris) **4** (1925) 156.
25-fai-1 Faillebin, M.; Ann. Chim. (Paris) **4** (1925) 410.
25-fav/zal Favorskii, A.; Zalesskii-Kibardine, T.; Bull. Soc. Chim. Fr. **37** (1925) 1227.
25-gro/kel Groh, Von J.; Kelp, I.; Z. Anorg. Allg. Chem. **147** (1925) 321.
25-har/rai Hartley, H.; Raikes, H. R.; J. Chem. Soc. **127** (1925) 524.
25-hes/bap Hess, K.; Bappert, R.; Justus Liebigs Ann. Chem. **441** (1925) 137.
25-lew Lewis, J. R.; J. Am. Chem. Soc. **47** (1925) 626.
25-nor/ash Norris, J. F.; Ashdown, A. A.; J. Am. Chem. Soc. **47** (1925) 837.
25-pal/con Palmer, W. G.; Constable, F. H.; Proc. R. Soc. London, A **107** (1925) 255.
25-par Parks, G. S.; J. Am. Chem. Soc. **47** (1925) 338.
25-par/kel Parks, G. S.; Kelley, K. K.; J. Am. Chem. Soc. **47** (1925) 2089.

25-par/kel-1 Parks, G. S.; Kelley, K. K.; J. Phys. Chem. **29** (1925) 727.
25-per Perrakis, N.; J. Chim. Phys. Phys.-Chim. Biol. **22** (1925) 280.
25-pie/ada Pierce, J. S.; Adams, R.; J. Am. Chem. Soc. **47** (1925) 1098.
25-rak Rakshit, J. N.; Z. Elektrochem. Angew. Phys. Chem. **31** (1925) 320.
25-ric/cha Richards, T. W.; Chadwell, H. M.; J. Am. Chem. Soc. **47** (1925) 2283.
25-rii/soe Riiber, C. N.; Soerensen, T.; Thorkelsen, K.; Z. Elektrochem. Angew. Phys. Chem. **58** (1925) 964.
25-ter Terent'ev, A.; Bull. Soc. Chim. Fr. **37** (1925) 1553.
25-tho/kah Thoms, H.; Kahre, H.; Arch. Pharm. Ber. Dtsch. Pharm. **263** (1925) 241.
25-vor/wal Vorlander, D.; Walter, R.; Z. Phys. Chem. (Leipzig) **118** (1925) 1.
26-bar Barbaudy, J.; Bull. Soc. Chim. Fr. **39** (1926) 371.
26-boe/bel Boeseken, J.; Belinfante, A. H.; Recl. Trav. Chim. Pays-Bas **45** (1926) 914.
26-ewa/rai Ewart, F. K.; Raikes, H. R.; J. Chem. Soc. (1926) 1907.
26-gol/aar Goldschmidt, H.; Aarflot, H.; Z. Phys. Chem., Stoechiom. Verwandtschaftsl. **122** (1926) 371.
26-han Hannotte, T.; Bull. Soc. Chim. Belg. **35** (1926) 86.
26-mat Mathews, J. H.; J. Am. Chem. Soc. **48** (1926) 562.
26-mun Munch, J. C.; J. Am. Chem. Soc. **48** (1926) 994.
26-nic Nicolle, P.; Bull. Soc. Chim. Belg. **39** (1926) 55.
26-par/huf Parks, G. S.; Huffman, H. M.; J. Am. Chem. Soc. **48** (1926) 2788.
26-pas/zam Pastureau, M.; Zamenhof, M. S.; Bull. Soc. Chim. Fr. **39** (1926) 1430.
26-ray Rayne, G. J.; J. Soc. Chem. Ind., London, Trans. Commun. **45** (1926) 265.
26-ris/hic Rising, M. M.; Hicks, J. S.; J. Am. Chem. Soc. **48** (1926) 1929.
26-sch Schmidt, G. C.; Z. Phys. Chem., Stoechiom. Verwandtschaftsl. **121** (1926) 221.
26-sta Stas, J.; Bull. Soc. Chim. Belg. **35** (1926) 379.
26-tay/rin Taylor, C. A.; Rinkenbach, W. H.; Ind. Eng. Chem. **18** (1926) 676.
26-van Van Risseghem, H.; Bull. Soc. Chim. Belg. **35** (1926) 328.
27-arb-2 Arbusov, A. E.; Z. Phys. Chem., Stoechiom. Verwandtschaftsl. **131** (1927) 49.
27-bra/ada Bray, R. H.; Adams, R.; J. Am. Chem. Soc. **49** (1927) 2101.
27-cla/rob Clarke, J. T.; Robinson, R.; Smith, J. C.; J. Chem. Soc. (1927) 2647.
27-del De Lattre, G.; J. Chim. Phys. Phys.-Chim. Biol. **24** (1927) 289.
27-kai/mel Kailan, A.; Melkus, K.; Monatsh. Chem. **48** (1927) 9.
27-krc/wil Krchma, I. J.; Williams, J. W.; J. Am. Chem. Soc. **49** (1927) 2408.
27-mou/duf Moureu, C.; Dufraisse, C.; Ann. Chim. (Paris) **7** (1927) 1.
27-nor/cor Norris, J. F.; Cortese, F.; J. Am. Chem. Soc. **49** (1927) 2640.
27-nor/reu Norris, J. F.; Reuter, R.; J. Am. Chem. Soc. **49** (1927) 2624.
27-par/cha Parks, G. S.; Chaffee, C. S.; J. Phys. Chem. **31** (1927) 439.
27-tri Trickey, J. P.; Ind. Eng. Chem. **19** (1927) 643.
27-tro/luk Tronov, B. V.; Lukanin, A. A.; Pavlinov, I. I.; Zh. Russ Fiz.-Khim. O-va., Chast Khim. **59** (1927) 1174.
27-ver/coo Verkade, P. E.; Coops, J.; Recl. Trav. Chim. Pays-Bas **46** (1927) 903.
28-dil/luc Dillon, R. T. ; Lucas, H. J. ; J. Am. Chem. Soc. **50** (1928) 1711.
28-esc Escourrou, R.; Bull. Soc. Chim. Fr. **43** (1928) 1101.
28-har-2 Harkins, W. D.; Z. Phys. Chem., Abt. A **139** (1928) 647.
28-kro/see Krollpfeiffer, F.; Seebaum, H.; J. Prakt. Chem. **119** (1928) 131.
28-llo/bro Lloyd, E.; Brown, C. B.; Bonnell, D. G. R.; Jones, W. J.; J. Chem. Soc. (1928) 658.
28-mon Mondain-Monval, P.; C. R. Hebd. Seances Acad. Sci. **187** (1928) 444.
28-par/bar Parks, G. S.; Barton, B.; J. Am. Chem. Soc. **50** (1928) 24.
28-par/kel Parks, G. S.; Kelley, K. K.; J. Phys. Chem. **32** (1928) 734.
28-par/nel Parks, G. S.; Nelson, W. K.; J. Phys. Chem. **32** (1928) 61.
28-pre-4 Prevost, C.; Ann. Chim. (Paris) **10** (1928) 356.
28-tim/mar Timmermans, J.; Martin, F.; J. Chim. Phys. Phys.-Chim. Biol. **25** (1928) 411.
28-ven-1 Venus-Danilova, E. D.; Bull. Soc. Chim. Fr. **43** (1928) 582.

29-ber Berner, E.; Z. Phys. Chem., Abt. A **141** (1929) 91.
29-ber/mie Bergmann, M.; Miekeley, A.; Lippmann; Ber. Dtsch. Chem. Ges. A **62** (1929) 1467.
29-ber/reu Berlin, E.; Reuter, R.; Kolloid-Z. **47** (1929) 251.
29-con/bla Conant, J. B.; Blatt, A. H.; J. Am. Chem. Soc. **51** (1929) 1227.
29-edg/cal Edgar, G.; Calingaert, G.; Marker, R. E.; J. Am. Chem. Soc. **51** (1929) 1483.
29-ham/and Hammick, D. L.; Andrew, L. W.; J. Chem. Soc. (1929) 754.
29-jon Jones, D. C.; J. Chem. Soc. (1929) 799.
29-kel Kelley, K. K.; J. Am. Chem. Soc. **51** (1929) 1145.
29-kel-1 Kelley, K. K.; J. Am. Chem. Soc. **51** (1929) 180.
29-kel-3 Kelley, K. K.; J. Am. Chem. Soc. **51** (1929) 779.
29-kir/ric Kirner, W. R.; Richter, G. H.; J. Am. Chem. Soc. **51** (1929) 2505.
29-lev/mik Levene, P. A.; Mikeska, L. A.; J. Biol. Chem. **84** (1929) 571.
29-luc Lucas, H. J. ; J. Am. Chem. Soc. **51** (1929) 248.
29-mah/das Mahanti, P. C.; Das-Gupta, R. N.; Indian J. Phys. **3** (1929) 467.
29-pre Prentiss, S. W.; J. Am. Chem. Soc. **51** (1929) 2825.
29-sim Simon, I.; Bull. Soc. Chim. Belg. **38** (1929) 47.
29-smy/sto Smyth, C. P.; Stoops, W. N.; J. Am. Chem. Soc. **51** (1929) 3312.
29-swa Swarts, F.; Bull. Soc. Chim. Belg. **38** (1929) 99.
29-tim/hen Timmermans, J.; Hennaut-Roland, M.; An. R. Soc. Esp. Fis. Quim. **27** (1929) 460
30-bil/fis-1 Biltz, W.; Fischer, W.; Wunnenberg, E.; Z. Phys. Chem., Abt. A **151** (1930) 13.
30-bin/dar Bingham, E. C.; Darrall, L. B.; J. Rheol. (Easton, Pa.) **1** (1930) 174.
30-bin/for Bingham, E. C.; Fornwalt, H. J.; J. Rheol. (N. Y.) **1** (1930) 372.
30-dav/dix Davies, W. C.; Dixon, R. S.; Jones, W. J.; J. Chem. Soc. (1930) 468.
30-err/she Errera, J.; Sherrill, M. L.; J. Am. Chem. Soc. **52** (1930) 1993.
30-fro Frost, A. V.; Tr. Inst. Chist. Khim. Reakt. **No. 9** (1930) 105.
30-mon Montagne, M.; Ann. Chim. (Paris) **13** (1930) 40.
30-rak/fro Rakovskii, A. V.; Frost, A. V.; Tr. Inst. Chist. Khim. Reakt. **9** (1930) 95.
30-she Sherrill, M. L.; J. Am. Chem. Soc. **52** (1930) 1982.
30-tim/hen Timmermans, J.; Hennaut-Roland, M.; J. Chim. Phys. Phys.-Chim. Biol. **27** (1930) 401.
30-van Van Risseghem, H.; Bull. Soc. Chim. Belg. **39** (1930) 369.
30-wal/ros Walbaum, H.; Rosenthal, A.; J. Prakt. Chem. **124** (1930) 63.
31-bea/mcv Beare, W. G.; McVicar, G. A.; Ferguson, J. B.; J. Phys. Chem. **35** (1931) 1068.
31-chu/mar Chu, T. T.; Marvel, C. S.; J. Am. Chem. Soc. **53** (1931) 4449.
31-def Deffet, L.; Bull. Soc. Chim. Belg. **40** (1931) 385.
31-deg De Graef, H.; Bull. Soc. Chim. Belg. **40** (1931) 315.
31-del Delcourt, Y.; Bull. Soc. Chim. Belg. **40** (1931) 284.
31-fio/gin Fiock, E. F.; Ginnings, D. C.; Holton, W. B.; J. Res. Natl. Bur. Stand. (U. S.) **6** (1931) 881.
31-hom Homeyer, A. H.; Thesis, Pennsylvania State University, University Park, PA (1931)
31-lev/mar-1 Levene, P. A.; Marker, R. E.; J. Biol. Chem. **91** (1931) 1931.
31-lev/mar-2 Levene, P. A.; Marker, R. E.; J. Biol. Chem. **91** (1931) 405.
31-lev/mar-3 Levene, P. A.; Marker, R. E.; J. Biol. Chem. **91** (1931) 687.
31-lev/mar-4 Levene, P. A.; Marker, R. E.; J. Biol. Chem. **91** (1931) 761.
31-lev/mar-5 Levene, P. A.; Marker, R. E.; J. Biol. Chem. **91** (1931) 77.
31-lun/bje Lund, H.; Bjerrum, J.; Ber. Dtsch. Chem. Ges. A **64** (1931) 210.
31-pow/sec Powell, S. G.; Secoy, C. H.; J. Am. Chem. Soc. **53** (1931) 765.
31-smy/dor-1 Smyth, C. P.; Dornte, R. W.; J. Am. Chem. Soc. **53** (1931) 545.
31-smy/wal Smyth, C. P.; Walls, W. S.; J. Am. Chem. Soc. **53** (1931) 2115.
31-sto/hul Stockhardt, J. S.; Hull, C. M.; Ind. Eng. Chem. **23** (1931) 1438.
31-ton/ueh Tonomura, T.; Uehara, K.; Bull. Chem. Soc. Jpn. **6** (1931) 118.
31-zaa Zaar, B.; J. Prakt. Chem. **132** (1931) 163.
32-boe/wil Boeseken, J.; Wildschut, A. J.; Recl. Trav. Chim. Pays-Bas **51** (1932) 168.
32-bri-2 Bridgman, P. W.; Proc. Am. Acad. Arts Sci. **67** (1932) 1.

32-byl Bylewski, T.; Rocz. Chem. **12** (1932) 311.
32-con/adk Connor, R.; Adkins, H.; J. Am. Chem. Soc. **54** (1932) 4678.
32-ell/rei Ellis, L. M.; Reid, E. E.; J. Am. Chem. Soc. **54** (1932) 1674.
32-ern/lit Ernst, R. C.; Litkenhous, E. E.; Spanyer, J. W.; J. Phys. Chem. **36** (1932) 842.
32-ess/cla Essex, H.; Clark, J. D.; J. Am. Chem. Soc. **54** (1932) 1290.
32-hun Hunt, R.; Ind. Eng. Chem. **24** (1932) 361.
32-kom/tal Komppa, G.; Talvitie, Y.; J. Prakt. Chem. **135** (1932) 193.
32-les/wak Lespieau, R.; Wakeman, R. L.; Bull. Soc. Chim. Fr. **51** (1932) 384.
32-lev/mar Levene, P. A.; Marker, R. E.; J. Biol. Chem. **95** (1932) 1.
32-lon/kha Longinov, V.; Khasanova, R.; Bull. Soc. Chim. Fr. **51** (1932) 636.
32-mor/har Morgan, G. T.; Hardy, D. V. N.; Proctor, R. A.; J. Soc. Chem. Ind., London, Trans. Commun. **51** (1932) 1.
32-pfa/pla Pfau, A. S.; Plattaner, Pl.; Helv. Chim. Acta **15** (1932) 1250.
32-ros Rossini, F. D.; J. Res. Natl. Bur. Stand. (U. S.) **8** (1932) 119.
32-sol/mol Solana, L.; Moles, E.; An. R. Soc. Esp. Fis. Quim. **30** (1932) 886.
32-swi/zma Swietoslawski, W.; Zmaczynski, A. L.; Usakiewicz, J; C. R. Hebd. Seances Acad. Sci. **194** (1932) 357.
32-tim/hen Timmermans, J.; Hennaut-Roland, M.; J. Chim. Phys. Phys.-Chim. Biol. **29** (1932) 529.
33-ano Schimmel & Co., Ann. Rep. (1933) .
33-azi/bha Azim, M. A.; Bhatnagar, S. S.; Mathur, R. N.; Philos. Mag. **16** (1933) 580.
33-bin/ste Bingham, E. C.; Stephens, R. A.; Physics (N. Y.) **4** (1933) 206.
33-bri Bridgman, P. W.; Proc. Am. Acad. Arts Sci. **68** (1933) 1.
33-but/tho Butler, J. A. V.; Thomson, D. W.; Maclennan, W. H.; J. Chem. Soc. (1933) 674.
33-fav/naz Favorskii, A.; Nazarov, I. N.; C. R. Hebd. Seances Acad. Sci. **196** (1933) 1229.
33-gin/her Ginnings, P. M.; Herring, E.; Webb, B.; J. Am. Chem. Soc. **55** (1933) 875.
33-har Harris, L.; J. Am. Chem. Soc. **55** (1933) 1940.
33-hov/lan Hovorka, F.; Lankelma, H. P.; Naujoks, C. K.; J. Am. Chem. Soc. **55** (1933) 4820.
33-huc/ack Huckel, W.; Ackermann, P.; J. Prakt. Chem. **136** (1933) 15.
33-kil Kilpi, S.; Z. Phys. Chem. (Leipzig) **166** (1933) 285.
33-koz/koz Kozakevich, P. P.; Kozakevich, N. S.; Z. Phys. Chem., Abt. A **166** (1933) 113.
33-lev/mar-1 Levene, P. A.; Marker, R. E.; J. Biol. Chem. **103** (1933) 299.
33-mey/tuo Meyer, A.; Tuot, M.; C. R. Hebd. Seances Acad. Sci. **196** (1933) 1231.
33-nat/bac Natta, G.; Baccaredda, M.; G. Chim. Ind. Appl. **15** (1933) 273.
33-nev/jat Nevgi, G. V.; Jatkar, S. K. K.; Indian J. Phys. **8** (1933) 397.
33-pow/mur Powell, S. G.; Murray, H. C.; Baldwin, M. M.; J. Am. Chem. Soc. **55** (1933) 1153.
33-ste Stevens, P. G.; J. Am. Chem. Soc. **55** (1933) 4237.
33-tre/wat Trew, V. C. G.; Watkins, G. M. C.; Trans. Faraday Soc. **29** (1933) 1310.
33-van Van Risseghem, H.; Bull. Soc. Chim. Belg. **42** (1933) 219.
33-vos/con Vosburgh, W. C.; Connell, L. C.; Butler, J. A. V.; J. Chem. Soc. (1933) 933.
33-whi/eve Whitmore, F. C.; Evers, W. L.; J. Am. Chem. Soc. **55** (1933) 812.
33-whi/her Whitmore, F. C.; Herndon, J. M.; J. Am. Chem. Soc. **55** (1933) 3428.
33-whi/hom Whitmore, F. C.; Homeyer, A. H.; J. Am. Chem. Soc. **55** (1933) 4194.
33-whi/hom-1 Whitmore, F. C.; Homeyer, A. H.; J. Am. Chem. Soc. **55** (1933) 4555.
33-whi/kru Whitmore, F. C.; Krueger, P. A.; J. Am. Chem. Soc. **55** (1933) 1528.
33-whi/lau Whitmore, F. C.; Laughlin, K. C.; J. Am. Chem. Soc. **55** (1933) 3732.
33-whi/wil Whitmore, F. C.; Williams, F. E.; J. Am. Chem. Soc. **55** (1933) 406.
33-whi/woo Whitmore, F. C.; Woodburn, H. M.; J. Am. Chem. Soc. **55** (1933) 361.
34-bur/adk Burdick, H. E.; Adkins, H.; J. Am. Chem. Soc. **56** (1934) 438.
34-car/jon Carter, E. G.; Jones, D. C.; Trans. Faraday Soc. **30** (1934) 1027.
34-cor/arc Cornish, R. E.; Archibald, R. C.; Murphy, E. A.; Evans, H. M.; Ind. Eng. Chem. **26** (1934) 397.
34-gre Green, J. H.; J. Am. Chem. Soc. **56** (1934) 1167.
34-jac-5 Jacquemain, R.; Compt. rend. **199** (1934) 1315.

34-jac-6 Jacquemain, R.; Compt. rend. **198** (1934) 482.
34-lal Lalande, A.; Bull. Soc. Chim. Fr. **1** (1934) 236.
34-les/lom Lespieau, R.; Lombard, R.; C. R. Hebd. Seances Acad. Sci. **198** (1934) 2179.
34-pau Paul, R.; Bull. Soc. Chim. Fr. **1** (1934) 971.
34-smi Smith, D. M.; Ind. Eng. Chem. **26** (1934) 392.
34-sta Stahly, E. E.; PhD Dissertation, Pennsylvania State Univeristy (1934) .
34-tim/del Timmermans, J.; Delcourt, Y.; J. Chim. Phys. Phys.-Chim. Biol. **31** (1934) 85.
34-von/man Von Braun, J.; Manz, G.; Chem. Ber. **67** (1934) 1696.
34-was/spe Washburn, E. R.; Spencer, H. C.; J. Am. Chem. Soc. **56** (1934) 361.
35-bil/gis Bilterys, R.; Gisseleire, J.; Bull. Soc. Chim. Belg. **44** (1935) 567.
35-bra/fel Bratton, A. C.; Felsing, W. A.; Bailey, J. R.; Ind. Eng. Chem. **28** (1935) 424.
35-bru/fur Brunjes, A. S.; Furnas, C. C.; Ind. Eng. Chem. **27** (1935) 396.
35-but/ram Butler, J. A. V.; Ramchandani, C. N.; Thomson, D. W.; J. Chem. Soc. (1935) 280.
35-but/ram-1 Butler, J. A. V.; Ramchandani, C. N.; J. Chem. Soc. (1935) 952.
35-col Colonge, J.; Bull. Soc. Chim. Fr. **2** (1935) 57.
35-col-1 Colonge, J.; Bull. Soc. Chim. Fr. **2** (1935) 754.
35-cou/hop Coull, J.; Hope, H. B.; J. Phys. Chem. **39** (1935) 967.
35-cox/rei Cox, W. M.; Reid, E. E.; J. Am. Chem. Soc. **57** (1935) 1801.
35-gib Gibson, R. E.; J. Am. Chem. Soc. **57** (1935) 1551.
35-gre-2 Gredy, B. ; Bull. Soc. Chim. Fr. **2** (1935) 1038.
35-gui Guillemonat, A.; C. R. Hebd. Seances Acad. Sci. **200** (1935) 1416.
35-has/mcb Hass, H. B.; McBee, E. T.; Weber, P.; Ind. Eng. Chem. **27** (1935) 1190.
35-hen Hennings, C.; Z. Phys. Chem., Abt. B **28** (1935) 267.
35-kef/mcl Keffler, L.; MacLean, J. H.; J. Soc. Chem. Ind., London, Trans. Commun. **54** (1935) 178.
35-les/lom Lespieau, R.; Lombard, R.; Bull. Soc. Chim. Fr. **2** (1935) 369.
35-lev/har Levene, P. A.; Harris, S. A.; J. Biol. Chem. **112** (1935) 195.
35 lev/mar Levene, P. A.; Marker, R. E.; J. Biol. Chem. **110** (1935) 299.
35-mah-1 Mahanti, P. C.; Z. Phys. **94** (1935) 220.
35-ols/was Olsen, A. L.; Washburn, E. R.; J. Am. Chem. Soc. **57** (1935) 303.
35-pau-2 Paul, R.; Bull. Soc. Chim. Fr. **2** (1935) 311.
35-sav Savard, J.; Bull. Soc. Chim. Fr. **2** (1935) 633.
35-saw Sawai, I.; Trans. Faraday Soc. **31** (1935) 765.
35-sch/sta Schierholtz, O. J.; Staples, M. L.; J. Am. Chem. Soc. **57** (1935) 2709.
35-tim/hen Timmermans, J.; Hennaut-Roland, M.; J. Chim. Phys. Phys.-Chim. Biol. **32** (1935) 501
36-dup-1 Dupire, A.; C. R. Hebd. Seances Acad. Sci. **202** (1936) 2086.
36-ern/wat Ernst, R. C.; Watkins, C. H.; Ruwe, H. H.; J. Phys. Chem. **40** (1936) 627.
36-ipa/cor Ipatieff, V. N.; Corson, B. B.; Pines, H.; J. Am. Chem. Soc. **58** (1936) 919.
36-lev/rot-1 Levene, P. A.; Rothen, A.; Meyer, G. M.; Kuna, M.; J. Biol. Chem. **115** (1936) 401.
36-naz Nazarov, I. N.; Chem. Ber. **69** (1936) 21.
36-nor/has Norton, F. H.; Hass, H. B.; J. Am. Chem. Soc. **58** (1936) 2147.
36-oli Olivier, S. C. J.; Recl. Trav. Chim. Pays-Bas **55** (1936) 1027.
36-pal/sab Palfray, L.; Sabetay, S.; Bull. Soc. Chim. Fr. **3** (1936) 682.
36-par Parthasarathy, P. S.; Proc. Indian Acad. Sci., Sect. A **4** (1936) 59.
36-pow/bal Powell, S. G.; Baldwin, M. M.; J. Am. Chem. Soc. **58** (1936) 1871.
36-rom Romstatt, G.; Ind. Chim. (Paris) **23** (1936) 567.
36-spe Spells, K. E.; Trans. Faraday Soc. **32** (1936) 530.
36-sto/rou Stoll, M.; Rouve, A.; Helv. Chim. Acta **19** (1936) 253.
36-tom Tomonari, T.; Z. Phys. Chem., Abt. B **32** (1936) 202.
36-tuo Tout, M.; C. R. Hebd. Seances Acad. Sci. **202** (1936) 1339.
36-wie Wiemann, J.; Ann. Chim. (Paris) **5** (1936) 267.
37-alb Albright, P. S.; J. Am. Chem. Soc. **59** (1937) 2098.
37-bra Brauns, D. H.; J. Res. Natl. Bur. Stand. (U. S.) **18** (1937) 315.

37-bra/kur	Braun, J. V.; Kurtz, P.; Chem. Ber. **70** (1937) 1224.
37-dob/gut	Dobrjanski, A. F.; Gutner, R. A.; Shchigel'skaya, M. K.; Zh. Obshch. Khim. **7** (1937) 132.
37-dol/bri	Dolian, F. E.; Briscoe, H. T.; J. Phys. Chem. **41** (1937) 1129.
37-ewe	Ewert, M.; Bull. Soc. Chim. Belg. **46** (1937) 90.
37-gib/kin	Gibson, R. E.; Kincaid, J. F.; J. Am. Chem. Soc. **59** (1937) 579.
37-gin/bau	Ginnings, P. M.; Baum, R.; J. Am. Chem. Soc. **59** (1937) 1111.
37-mas	Mastagli, P.; C. R. Hebd. Seances Acad. Sci. **204** (1937) 1656.
37-mil/sus	Milas, N. A.; Sussman, S.; J. Am. Chem. Soc. **59** (1937) 2345.
37-mou/dod	Moureu, H.; Dode, M.; Bull. Soc. Chim. Fr. **4** (1937) 637.
37-naz	Nazarov, I. N.; Chem. Ber. **70** (1937) 599.
37-oli	Olivier, S. C. J.; Recl. Trav. Chim. Pays-Bas **56** (1937) 247.
37-pat/hol	Patterson, T. S.; Holmes, G. M.; J. Chem. Soc. (1937) 1007.
37-pet/and	Petrov, A. D.; Andreev, D. N.; J. Gen. Chem. USSR (Engl. Transl.) **7** (1937) 570.
37-pet/mal	Petrov, A. D.; Malinovskii, M. S.; J. Gen. Chem. USSR (Engl. Transl.) **7** (1937) 565.
37-puk	Pukirev, A. B.; Tr. Inst. Chist. Khim. Reakt. **15** (1937) 45.
37-she/mat	Sherrill, M. L.; Matlack, L. S.; J. Am. Chem. Soc. **59** (1937) 2134.
37-tis	Tishchenko, D. V.; Zh. Obshch. Khim. **7** (1937) 658.
37-tis/chu	Tishchenko, D. V.; Churbakov, A. N.; Zh. Obshch. Khim. **7** (1937) 893.
37-tya	Tyazhelova, A. A.; Tr. Voronezh. Gos. Univ. **9**(3) (1937) 139.
37-zep	Zepalova-Mikhailova, L. A.; Tr. Inst. Chist. Khim. Reakt. **No. 15** (1937) 3.
38-bak/smy	Baker, W. O.; Smyth, C. P.; J. Am. Chem. Soc. **60** (1938) 1229.
38-dol/gre	Dolliver, M. A.; Gresham, T. L.; Kistiakowsky, G. B.; Smith, E. A.; Vaughan, W. E.; J.; Am. Chem. Soc. **60** (1938) 440.
38-gin/hau	Ginnings, P. M.; Hauser, M.; J. Am. Chem. Soc. **60** (1938) 2581.
38-gin/web	Ginnings, P. M.; Webb, R.; J. Am. Chem. Soc. **60** (1938) 1388.
38-hil/adk	Hill, R. M.; Adkins, H.; J. Am. Chem. Soc. **60** (1938) 1033.
38-hov/lan	Hovorka, F.; Lankelma, H. P.; Stanford, S. C.; J. Am. Chem. Soc. **60** (1938) 820.
38-mas	Mastagli, P.; Ann. Chim. (Paris) **10** (1938) 281.
38-ols/was	Olsen, A. L.; Washburn, E. R.; J. Phys. Chem. **42** (1938) 275.
38-sca/ray	Scatchard, G.; Raymond, C. L.; J. Am. Chem. Soc. **60** (1938) 1278.
38-tur	Turpeinen, O.; J. Am. Chem. Soc. **60** (1938) 56.
38-wer/bog	Werner, J.; Bogert, M. T.; J. Org. Chem. **3** (1938) 578.
38-whi/joh	Whitmore, F. C.; Johnston, F.; J. Am. Chem. Soc. **60** (1938) 2265.
38-whi/kar	Whitmore, F. C.; Karnatz, F. A.; J. Am. Chem. Soc. **60** (1938) 2533.
38-whi/mey	Whitmore, F. C.; Meyer, R. E.; Pedlow, G. W.; Popkin, A. H.; J. Am. Chem. Soc. **60** (1938) 2788.
38-whi/ole	Whitmore, F. C.; Olewine, J. H.; J. Am. Chem. Soc. **60** (1938) 2569.
38-whi/ole-1	Whitmore, F. C.; Olewine, J. H.; J. Am. Chem. Soc. **60** (1938) 2570.
38-whi/ore	Whitmore, F. C.; Orem, H. P.; J. Am. Chem. Soc. **60** (1938) 2573.
38-whi/pop	Whitmore, F. C.; Popkin, A. H.; Whitaker, J. S.; Mattil, K. F.; Zech, J. D.; J. Am. Chem. Soc. **60** (1938) 2458.
38-whi/pop-1	Whitmore, F. C.; Popkin, A. H.; Whitaker, J. S.; Mattil, K. F.; Zech, J. D.; J. Am. Chem. Soc. **60** (1938) 2462.
39-all/lin	Allen, B. B.; Lingo, S. P.; Felsing, W. A.; J. Phys. Chem. **43** (1939) 425.
39-bar/atk	Barrow, F.; Atkinson, R. G.; J. Chem. Soc. (1939) 638.
39-cop/gos	Coppock, J. B. M.; Goss, F. R.; J. Chem. Soc. (1939) 1789.
39-dup/dar	Dupont, G.; Darmon, M.; Bull. Soc. Chim. Fr. **6** (1939) 1208.
39-gin/col	Ginnings, P. M.; Coltrane, D.; J. Am. Chem. Soc. **61** (1939) 525.
39-gol/tay	Goldwasser, S.; Taylor, H. S.; J. Am. Chem. Soc. **61** (1939) 1751.
39-hus/gui	Huston, R. C.; Guile, R. L.; J. Am. Chem. Soc. **61** (1939) 69.
39-jon/chr	Jones, G.; Christian, S. M.; J. Am. Chem. Soc. **61** (1939) 82.
39-ken/pla	Kenyon, J.; Platt, B. C.; J. Chem. Soc. (1939) 633.

39-lar/hun Larson, R. G.; Hunt, H. ; J. Phys. Chem. **43** (1939) 417.
39-mar/wil Marvel, C. S.; Williams, W. W.; J. Am. Chem. Soc. **61** (1939) 2714.
39-owe/qua Owen, K.; Quayle, O. R.; Beavers, E. M.; J. Am. Chem. Soc. **61** (1939) 900.
39-par/moo Parks, G. S.; Moore, G. E.; J. Chem. Phys. **7** (1939) 1066.
39-pet/sum Petrov, A. D.; Sumin, I. G.; Meerovich, Z. A.; Kudrina, K. N.; Tikhonova, G. N.; Zh. Obshch. Khim. **9** (1939) 2144.
39-spi/tin-1 Spiegler, L.; Tinker, J. M.; J. Am. Chem. Soc. **61** (1939) 940.
39-ste/mcn Stevens, P. G.; McNiven, N. L.; J. Am. Chem. Soc. **61** (1939) 1295
39 stu/adk Stutsman, P. S.; Adkins, H.; J. Am. Chem. Soc. **61** (1939) 3303
40-gos-1 Goss, F. R.; J. Chem. Soc. (1940) 883.
40-hov/lan Hovorka, F.; Lankelma, H. P.; Schneider, I.; J. Am. Chem. Soc. **62** (1940) 1096.
40-hov/lan-1 Hovorka, F.; Lankelma, H. P.; Axelrod, A. E.; J. Am. Chem. Soc. **62** (1940) 187.
40-hov/lan-2 Hovorka, F.; Lankelma, H. P.; Smith, W. R.; J. Am. Chem. Soc. **62** (1940) 2372.
40-mil/bli Miller, H. C.; Bliss, H.; Ind. Eng. Chem. **32** (1940) 123.
40-mon/qui Mondain-Monval, P.; Quiquerez, J.; Bull. Soc. Chim. Fr. **7** (1940) 240.
40-mos Mosher, W. A.; Ph.D. Dissertation, Penn. State Univ., Univ. Park, PA (1940) .
40-pal Palfray, L.; Bull. Soc. Chim. Fr. **7** (1940) 401.
40-sut Sutherland, L. H.; Ph.D. Thesis, Penn. State Univ., Univ. Park, PA, (1940) .
40-tri/ric Trimble, H. M.; Richardson, E. L.; J. Am. Chem. Soc. **62** (1940) 1018.
40-was/beg Washburn, E. R.; Beguin, A. E.; J. Am. Chem. Soc. **62** (1940) 579.
40-was/gra Washburn, E. R.; Graham, C. L.; Arnold, G. B.; Transue, L. F.; J. Am. Chem. Soc. **62** (1940) 1454.
40-whi/sur Whitmore, F. C.; Surmatis, J. D.; J. Am. Chem. Soc. **62** (1940) 995.
41-boh Bohnsack, H.; Chem. Ber. **74** (1941) 1575.
41-cam/eby-1 Campbell, K. N.; Eby, L. T.; J. Am. Chem. Soc. **63** (1941) 2683.
41-dor/gla Dorough, G. L.; Glass, H. B.; Gresham, T. L.; Malone, G. B.; Reid, E. E.; J. Am. Chem. Soc. **63** (1941) 3100.
41-fav/oni Favorskii, A. E.; Onishchenko, A. S.; Zh. Obshch. Khim. **11** (1941) 1111.
41-gib/loe Gibson, R. E.; Loeffler, O. H.; J. Am. Chem. Soc. **63** (1941) 898.
41-gle Gleim, C. E.; Ph.D. Thesis, Penn. State Univ., Univ. Park, PA (1941) .
41-hea/tam Hearne. G.; Tamele, M.; Converse, W.; Ind. Eng. Chem. **33** (1941) 805.
41-hov/lan Hovorka, F.; Lankelma, H. P.; Bishop, J. W.; J. Am. Chem. Soc. **63** (1941) 1097.
41-hus/age Huston, R. C.; Agett, A. H.; J. Org. Chem. **6** (1941) 123.
41-hus/gui Huston, R. C.; Guile, R. L.; Sculati, J. J.; Wasson, W. N.; J. Org. Chem. **6** (1941) 252.
41-whe Wheeler, W. R.; Thesis, Penn. State University, (1941) .
41-whi/mos Whitmore, F. C.; Mosher, W. A.; J. Am. Chem. Soc. **63** (1941) 1120.
41-whi/whi Whitmore, F. C.; Whitaker, J. S.; Mosher, W. A.; Breivik, O. N.; Wheeler, W. R.; Miner, C. S.; Sutherland, L. H.; Wagner, R. B.; Clapper, T. W.; Lewis, C. E.; Lux, A. R.; Popkin, A. H.; J. Am. Chem. Soc. **63** (1941) 643.
41-whi/wil Whitmore, F. C.; Wilson, C. D.; Capinjola, J. V.; Tongberg, C. O.; Fleming, G. H.; McGrew, R. V.; Cosby, J. N.; J. Am. Chem. Soc. **63** (1941) 2035.
42-air/bal Airs, R. S.; Balfe, M. P.; Kenyon, J.; J. Chem. Soc. (1942) 18.
42-boe/han Boeseken, J.; Hanegraaff, J. A.; Recl. Trav. Chim. Pays-Bas **61** (1942) 69.
42-boe/han-1 Boeke, J.; Hanewald, K. H.; Recl. Trav. Chim. Pays-Bas **61** (1942) 881.
42-boh Bohnsack, H.; Chem. Ber. **75** (1942) 502.
42-bri/rin Briscoe, H. T.; Rinehart, W. T.; J. Phys. Chem. **46** (1942) 387.
42-hen/all Henze, H. R.; Allen, B. B.; Leslie, W. B.; J. Org. Chem. **7** (1942) 326.
42-moe Moersch, G. W.; Ph.D. Thesis, Penn. State Univ., Univ. Park, PA, (1942).
42-mul Muller, A.; Fette Seifen **49** (1942) 572.
42-pet/che Petrov, A. D.; Chel'tsova, M. A.; J. Gen. Chem. USSR (Engl. Transl.) **12** (1942) 87.
42-pum/hah Pummerer, R.; Hahn, H.; Johne, F.; Kehlen, H.; Ber. Dtsch. Chem. Ges. **75** (1942) 867
42-sch/hun Schumacher, J. E.; Hunt, H. ; Ind. Eng. Chem. **34** (1942) 701.

42-sny/gil Snyder, H. B.; Gilbert, E. C.; Ind. Eng. Chem. **34** (1942) 1519.
42-was/bro Washburn, E. R.; Brockway, C. E.; Graham, C. L.; Deming, P.; J. Am. Chem. Soc. **64** (1942) 1886.
42-whi/for Whitmore, F. C.; Forster, W. S.; J. Am. Chem. Soc. **64** (1942) 2966.
43-boh Bohnsack, H.; Chem. Ber. **76** (1943) 564.
43-bra Brauns, D. H.; J. Res. Natl. Bur. Stand. (U. S.) **31** (1943) 83.
43-bru/bog Brunjes, A. S.; Bogart, M. J. P.; Ind. Eng. Chem. **35** (1943) 255.
43-col/jol Colonge, J.; Joly, D.; Ann. Chim. (Paris) **18** (1943) 286.
43-dur/roe Durodie, J.; Roelens, E.; Bull. Soc. Chim. Fr. **10** (1943) 169.
43-geo George, R. S.; Ph.D. Thesis, Penn. State Univ., Univ. Park, PA (1943) .
43-gre/kel Green, M. W.; Kelly, K. L.; Steinmetz, C. A.; Bull. Nat. Formul. Comm. **11** (1943) 91.
43-hsu Hsu, P. T.; Ann. Chim. (Paris) **18** (1943) 185.
43-ish/kat Ishiguro, T.; Kato, S.; Akazawa, Y.; Yakugaku Zasshi **63** (1943) 282.
43-jam James, W. H.; Ph.D. Thesis, Penn. State Univ., Univ. Park, PA, (1943) .
43-kul/nor Kulpinski, M. S.; Nord, F. F.; J. Org. Chem. **8** (1943) 256.
43-lan/key Langdon, W. M.; Keyes, D. B.; Ind. Eng. Chem. **35** (1943) 459.
43-ste/gre Stevens, P. G.; Greenwood, S. H. J.; J. Am. Chem. Soc. **65** (1943) 2153.
44-app/dob Appleby, W. G.; Dobratz, C. J.; Kapranos, S. W.; J. Am. Chem. Soc. **66** (1944) 1938.
44-gla Glacet, C.; C. R. Hebd. Seances Acad. Sci. **218** (1944) 283.
44-hen/mat Henne, A. L.; Matuszak, A. H.; J. Am. Chem. Soc. **66** (1944) 1649.
44-hus/awu Huston, R. C.; Awuapara, J.; J. Org. Chem. **9** (1944) 401.
44-ira Irany, E. P.; J. Am. Chem. Soc. **65** (1944) 1392.
44-mor/aue Morell, S. A.; Auernheimer, A. H.; J. Am. Chem. Soc. **66** (1944) 792.
44-pow/hag Powell, S. G.; Hagemann, F.; J. Am. Chem. Soc. **66** (1944) 372.
44-pre Prevost, C.; Bull. Soc. Chim. Fr. **11** (1944) 218.
44-pre/zal Prelog, V.; Zalan, E.; Helv. Chim. Acta **27** (1944) 545.
44-qua/sma Quayle, O. R.; Smart, K. O.; J. Am. Chem. Soc. **66** (1944) 935.
44-sto/rou Stoll, M.; Rouve, A.; Helv. Chim. Acta **27** (1944) 950.
44-was/str Washburn, E. R.; Strandskov, C. V.; J. Phys. Chem. **48** (1944) 241.
44-you/rob Young, W. G.; Roberts, J. D.; J. Am. Chem. Soc. **66** (1944) 1444.
45-add Addison, C. C.; J. Chem. Soc. (1945) 98.
45-alb/was Alberty, R. A.; Washburn, E. R.; J. Phys. Chem. **49** (1945) 4.
45-dul Dulitskaya, K. A.; Zh. Obshch. Khim. **15** (1945) 9.
45-hen/gre Henne, A. L.; Greenlee, K. W.; J. Am. Chem. Soc. **67** (1945) 484.
45-kol/bur Kolb, H. J.; Burwell, R. L.; J. Am. Chem. Soc. **67** (1945) 1084.
45-len/dup Lenth, C. W.; Dupuis, R. N.; Ind. Eng. Chem. **37** (1945) 152.
45-sch Schmerling, L.; J. Am. Chem. Soc. **67** (1945) 1152.
45-sch/gel Schniepp, L. E.; Geller, H. H.; J. Am. Chem. Soc. **67** (1945) 54
46-bra Brauns, H.; Recl. Trav. Chim. Pays-Bas **65** (1946) 799.
46-cle-1 Clendenning, K. A.; Can. J. Res., Sect. B **24** (1946) 269.
46-dou-1 Doumani, T.; Patent, U.S. 2,394,848, Feb. 12, (1946) .
46-fri/sto Fritzsche, R. H.; Stockton, D. L.; Ind. Eng. Chem. **38** (1946) 737.
46-haf/lov Hafslund, E. R.; Lovell, C. L.; Ind. Eng. Chem. **38** (1946) 556.
46-hus/bai Huston, R. C.; Bailey, D. L.; J. Am. Chem. Soc. **68** (1946) 1382.
46-kno/sch Knowlton, J. W.; Schieltz, N. C.; Macmillan, D.; J. Am. Chem. Soc. **68** (1946) 208.
46-kre/now Kretschmer, C. B.; Nowakowska, J.; Wiebe, R.; Ind. Eng. Chem. **38** (1946) 506.
46-lau/wie Laude, G.; Wiemann, J.; Bull. Soc. Chim. Fr. (1946) 256.
46-par/row Parks, G. S.; Rowe, R. D.; J. Chem. Phys. **14** (1946) 507.
46-par/wes Parks, G. S.; West, T. J.; Naylor, B. F.; Fujii, P. S.; McClaine, L. A.; J. Am. Chem. Soc. **68** (1946) 2524.
46-pre/val Prevost, C.; Valette, A.; C. R. Hebd. Seances Acad. Sci. **222** (1946) 326.
46-puc/wis Puck, T. T.; Wise, H.; J. Phys. Chem. **50** (1946) 329.

46-shi	Shikheev, I. A.; Zh. Obshch. Khim. **16** (1946) 657.
46-sim/was	Simonsen, D. R.; Washburn, E. R.; J. Am. Chem. Soc. **68** (1946) 235.
46-vav/col	Vavon, G.; Colin, H.; C. R. Hebd. Seances Acad. Sci. **222** (1946) 801.
47-col/lag	Colonge, J.; Lagier, A.; C. R. Hebd. Seances Acad. Sci. **224** (1947) 572.
47-det/cra	Detling, K. D.; Crawford, C. C.; Yabroff, D. L.; Peterson, W. H.; Patent, Brit. 591,632, Aug. 25, (1947).
47-dev	De Vrieze, J. J.; Recl. Trav. Chim. Pays-Bas **66** (1947) 486.
47-how/mea	Howard, F. L.; Mears, T. W.; Fookson, A.; Pomerantz, P.; Brooks, D. B.; J. Res. Natl. Bur. Stand. (U. S.) **38** (1947) 365.
47-kor/lic	Kornblum, N.; Lichtin, N. N.; Patton, J. T.; Iffland, D. C.; J. Am. Chem. Soc. **69** (1947) 307.
47-rei/dem	Reinders, W.; De Minjer, C. H.; Recl. Trav. Chim. Pays-Bas **66** (1947) 552.
47-sch/sch	Schinz, H.; Schappi, G.; Helv. Chim. Acta **30** (1947) 1483.
47-sho/pri	Shostakovskii, M. F.; Prilezhaeva, E. N.; Zh. Obshch. Khim. **17** (1947) 1129.
47-sto	Stoll, M.; Helv. Chim. Acta **30** (1947) 991.
47-str/gab	Stross, F. H.; Gable, C. M.; Rounds. G. C.; J. Am. Chem. Soc. **69** (1947) 1629.
47-tuo/guy	Tuot, M.; Guyard, M.; Bull. Soc. Chim. Fr. (1947) 1086.
48-adk/bil	Adkins, H.; Billica, H. R.; J. Am. Chem. Soc. **70** (1948) 3121.
48-bae/fis	Baer, E. B.; Fischer, H. O. L.; J. Am. Chem. Soc. **70** (1948) 609.
48-bou/nic	Bourns, A. N.; Nicholls, R. V. V.; Can. J. Res., Sect. B **26** (1948) 81.
48-bro/bro	Brokaw, G. Y.; Brode, W. R.; J. Org. Chem. **13** (1948) 194.
48-cad/foo	Cadwallader, E. A.; Fookson, A.; Mears, T. W.; Howard, F. L.; J. Res. Natl. Bur. Stand. (U. S.) **41** (1948) 111.
48-con/elv	Conner, A. Z.; Elving, P. J.; Steingiser, S.; Ind. Eng. Chem. **40** (1948) 497.
48-goe/cri	Goering, H. L.; Cristol, S. J.; Dittmer, K.; J. Am. Chem. Soc. **70** (1948) 3314.
48-hus/goe	Huston, R. C.; Goerner, G. L.; Breining, E. R.; Bostwick, C. O.; Cline, K. D.; Snyder, L. J.; J. Am. Chem. Soc. **70** (1948) 1090.
48-hus/kra	Huston, R. C.; Krantz, R. J.; J. Org. Chem. **13** (1948) 63.
48-jon/bow	Jones, W. J.; Bowden, S. T.; Yarnold, W. W.; Jones, W. H.; J. Phys. Colloid Chem. **52** (1948) 753.
48-kor/pat	Kornblum, N.; Patton, J. T.; Nordmann, J. B.; J. Am. Chem. Soc. **70** (1948) 746.
48-kre/now	Kretschmer, C. B.; Nowakowska, J.; Wiebe, R.; J. Am. Chem. Soc. **70** (1948) 1785.
48-lad/smi	Laddha, G. S.; Smith, J. M.; Ind. Eng. Chem. **40** (1948) 494.
48-laz	Lazzari, G.; Ann. Chim. Appl. **38** (1948) 287.
48-mal/kon	Malinovskii, M. S.; Konevichev, E. N.; Zh. Obshch. Khim. **18** (1948) 1833.
48-mcm/rop	Mcmahon, E. M.; Roper, J. N.; Utermohlen, W. P.; Hasek, R. H.; Harris, R. C.; Brant, J. H.; J. Am. Chem. Soc. **70** (1948) 2971.
48-mic/hop	Michael, T. H. G.; Hopkins, C. Y.; Can. Chem. Process Ind. **32** (1948) 341.
48-naz/tor	Nazarov, I. N.; Torgov, I. V.; Zh. Obshch. Khim. **18** (1948) 1480.
48-pet/old	Petrov, A. D.; Ol'dekop, Y. A.; Zh. Obshch. Khim. **18** (1948) 859.
48-pow/nie	Powell, S.; Nielsen, A.; J. Am. Chem. Soc. **70** (1948) 3627.
48-pro/cas	Prout, F. S.; Cason, J.; Ingersoll, A. W.; J. Am. Chem. Soc. **70** (1948) 298.
48-pud	Pudovik, A. N.; Zh. Obshch. Khim. **19** (1948) 1187.
48-rit	Ritter, J. J.; J. Am. Chem. Soc. **70** (1948) 4253.
48-ruo	Ruof, C. H.; Ph.D. Thesis, Penn. State Univ., Univ. Park, PA, (1948)
48-sor/sor	Sorensen, J. S.; Sorensen, N. A.; Acta Chem. Scand. **2** (1948) 166.
48-val	Valette, P. A.; Ann. Chim. (Paris) **3** (1948) 644.
48-vog-2	Vogel, A. I.; J. Chem. Soc. (1948) 1814.
48-wei	Weissler, A.; J. Am. Chem. Soc. **70** (1948) 1634.
48-zal	Zal'manovich, M. Z.; Zh. Obshch. Khim. **18** (1948) 2103.
48-zas	Zaslavskii, I. I.; Zh. Prikl. Khim. (Leningrad) **21** (1948) 732.
49-ber/ped	Bernstein, H. J.; Pedersen, E. E.; J. Chem. Phys. **17** (1949) 885.

49-boo/gre	Boord, C. E.; Greenlee, K. W.; Perilstein, W. L.; Derfer, J. M.; Am. Pet. Inst. Res. Proj. 45, Eleventh Annu. Rep., Ohio State Univ., June (1949) .
49-bru	Bruner, W. M.; Ind. Eng. Chem. **41** (1949) 2860.
49-col/lag	Colonge, J.; Lagier, A.; Bull. Soc. Chim. Fr. (1949) 15.
49-dre/mar	Dreisbach, R. R.; Martin, R. A.; Ind. Eng. Chem. **41** (1949) 2875.
49-gri/chu	Griswold, J.; Chu, P. L.; Winsauer, W. O.; Ind. Eng. Chem. **41** (1949) 2352.
49-hat	Hatem, S.; Bull. Soc. Chim. Fr. (1949) 337.
49-hat-1	Hatem, S.; Bull. Soc. Chim. Fr. (1949) 483.
49-hat-2	Hatem, S.; Bull. Soc. Chim. Fr. (1949) 599.
49-hos/ste	Hosman, B. B. A.; Van Steenis, J.; Waterman, H. I.; Recl. Trav. Chim. Pays-Bas **68** (1949) 939.
49-ken/str	Kenyon, J.; Strauss, H. E.; J. Chem. Soc. (1949) 2153.
49-kre/wie-1	Kretschmer, C. B.; Wiebe, R.; J. Am. Chem. Soc. **71** (1949) 3176.
49-mal/vol	Malinovskii, M. S.; Volkova, E. E.; Morozova, N. M.; Zh. Obshch. Khim. **19** (1949) 114.
49-moe/whi	Moersch, G. W.; Whitmore, F. C.; J. Am. Chem. Soc. **71** (1949) 819.
49-naz/pin	Nazarov, I. N.; Pinkina, L. N.; Zh. Obshch. Khim. **19** (1949) 1870.
49-naz/zar-1	Nazarov, I. N.; Zaretskaya, I. I.; Izv. Akad. Nauk SSSR Ser. Khim. (1949) 184.
49-pra/dra	Pratt, E. F.; Draper, J. D.; J. Am. Chem. Soc. **71** (1949) 2846.
49-tsc/ric	Tschamler, H.; Richter, E.; Wettig, F.; Monatsh. Chem. **80** (1949) 749.
49-tsv/mar	Tsvetkov, V. N.; Marinin, V. A.; Dokl. Akad. Nauk SSSR **68** (1949) 49
49-udo/kal	Udovenko, V. V.; Kalabanovskaya, E. I.; Prokop'eva, M. F.; Zh. Obshch. Khim. **19** (1949) 165.
49-vve/iva	Vvedenskii, A. A.; Ivannikov, P. Ya.; Nekrasova, V. A.; Zh. Obshch. Khim. **19** (1949) 1094
50-ada/van	Adams, R. M.; Vanderwerf, C. A.; J. Am. Chem. Soc. **72** (1950) 4368.
50-boe/ned	Boelhouwer, J. W. M.; Nederbragt, G. W.; Verberg, G. W.; Appl. Sci. Res., Sect. A **2** (1950) 249.
50-cle/mac	Clendenning, K. A.; Macdonald, F. J.; Wright, D. E.; Can. J. Res., Sect. B **28** (1950) 608.
50-cro/van	Croxall, W. J.; Van Hook, J. O.; J. Am. Chem. Soc. **72** (1950) 803.
50-doe/zei	Doering, W. von E.; Zeiss, H. H.; J. Am. Chem. Soc. **72** (1950) 147.
50-fav/fri	Favorskaya, T. A.; Fridman, Sh. A.; Zh. Obshch.Khim. **20** (1950) 413.
50-gor	Gorin, Y.; Zh. Obshch. Khim. **20** (1950) 1596.
50-hou/mas-1	Hough, E. W.; Mason, D. M.; Sage, B. H.; J. Am. Chem. Soc. **72** (1950) 5775.
50-jac	Jacobson, B.; Ark. Kemi **2** (1950) 177.
50-lad/smi	Laddha, G. S.; Smith, J. M.; Chem. Eng. Prog. **46** (1950) 195.
50-let/tra	Letsinger, R. L.; Traynham, J. G.; J. Am. Chem. Soc. **72** (1950) 849.
50-mea/foo	Mears, T. W.; Fookson, A.; Pomerantz, P.; Rich, E. H.; Dussinger, C. S.; Howard, F. L.; J. Res. Natl. Bur. Stand. (U. S.) **44** (1950) 299.
50-mos/lac	Mosher, H. S.; La Combe, E.; J. Am. Chem. Soc. **72** (1950) 3994.
50-mum/phi	Mumford, S. A.; Phillips, J. W. C.; J. Chem. Soc. (1950) 75.
50-naz/bak	Nazarov, I. N.; Bakhmutskaya, S. S.; Zh. Obshch. Khim. **20** (1950) 1837.
50-naz/fis	Nazarov, I. N.; Fisher, L. B.; Zh. Obshch. Khim. **20** (1950) 1107.
50-naz/kot-1	Nazarov, I. N.; Kotlyarevskii, P. A.; Zh. Obshch. Khim. **20** (1950) 1449.
50-par/gol	Parker, E. D.; Goldblatt, L. A.; J. Am. Chem. Soc. **72** (1950) 2151.
50-pau/tch	Paul, R.; Tchelitcheff, S.; Bull. Soc. Chim. Fr. (1950) 520.
50-pic/zie	Pichler, H.; Ziesecke, K. H.; Traeger, B.; Brennst.-Chem. **31** (1950) 361.
50-sac/sau	Sackmann, H.; Sauerwald, F.; Z. Phys. Chem. (Leipzig) **195** (1950) 295.
50-ste/coo	Stehman, C. J.; Cook, N. C.; Percival, W. C.; Whitmore, F. C.; J. Am. Chem. Soc. **72** (1950) 4163.
50-tei/gor	Teitel'baum, B. Ya.; Gortalova, T. A.; Ganelina, S. G.; Zh. Obshch. Khim. **20** (1950) 1422.
50-vie	Vierk, A.-L ; Z. Anorg. Chem. **261** (1950) 283.
50-wol/sau	Wolf, F.; Sauerwald, F.; Kolloid-Z. **118** (1950) 1.
51-ami/wei	Amick, E. H.; Weiss, M. A. ; Kirshenbaum, M. S.; Ind. Eng. Chem. **43** (1951) 969.
51-ber-2	Bergmann, E. D.; Patent, U.S. 2,539,806, Jan. 30, (1951) .
51-bot	Bothner-By, A. A.; J. Am. Chem. Soc. **73** (1951) 846.

51-cop/fie Copet, A.; Fierens-Snoeck, M. P.; Van Risseghem, H.; Bull. Soc. Chim. Fr. (1951) 902.
51-dim/lan Dimmling, W.; Lange, E.; Z. Elektrochem. **55** (1951) 322.
51-dra-1 Drahowzal, F.; Monatsh. Chem. **82** (1951) 794.
51-dub Dubois, J. E. ; Ann. Chim. (Paris) **6** (1951) 406.
51-gor Gorge, M.; Ann. Chim. (Paris) **6** (1951) 649.
51-hau Hausermann, M.; Helv. Chim. Acta **34** (1951) 1211.
51-hub Huber, W. F.; J. Am. Chem. Soc. **73** (1951) 2730.
51-kre Kretschmer, C. B.; J. Phys. Colloid Chem. **55** (1951) 1351.
51-ler/luc Leroux, P. J.; Lucas, H. J. ; J. Am. Chem. Soc. **73** (1951) 41.
51-lev/fai Levina, R. Ya.; Fainzil'berg, A. A.; Tantsyreva, T. I.; Treshchova, E. G.; Izv. Akad. Nauk SSSR Ser. Khim. (1951) 321.
51-lyu/ter Lyubomilov, V. I.; Terent'ev, A. P.; Zh. Obshch. Khim. **21** (1951) 1479.
51-mac/tho Macbeth, G.; Thompson, A. R.; Anal. Chem. **23** (1951) 618.
51-mca/cul McAllan, D. T.; Cullum, T. V.; Dean, R. A.; Fidler, F. A.; J. Am. Chem. Soc. **73** (1951) 3627.
51-sew Seward, R. P.; J. Am. Chem. Soc. **73** (1951) 515.
51-smi/cre Smith, F. A.; Creitz, E. C.; J. Res. Natl. Bur. Stand. (U. S.) **46** (1951) 145.
51-sor/suc Sorm, F.; Suchy, M.; Herout, V.; Chem. Listy **45** (1951) 218.
51-tei/gor Teitel'baum, B. Ya.; Gortalova, T. A.; Sidorova, E. E.; Zh. Fiz. Khim. **25** (1951) 911.
51-van-1 Van Risseghem, H.; Bull. Soc. Chim. Fr. **18** (1951) 908.
51-wat/coo Watson, R. W.; Coope, J. A. R.; Barnwell, J. L.; Can. J. Chem. **29** (1951) 885.
51-whi/dea Whitehead, E. V.; Dean, R. A.; Fidler, F. A.; J. Am. Chem. Soc. **73** (1951) 3632.
52-cap/mug Capitani, C.; Mugnaini, E.; Chim. Ind. (Milan) **34** (1952) 193.
52-coo Cook, N. C.; Unpublished, Final Rep. Stand. Proj. on Oxygenated Compounds, Penn. State Univ., College Park, PA, (1952) .
52-cur/joh Curme, G. O.; Johnston, F.; Glycols, Reinhold Publishing Corp., New York (1952) .
52-doe/far Doering, W. von E.; Farber, M.; Sprecher, M.; Wiberg, K. B.; J. Am. Chem. Soc. **74** (1952) 3000.
52-dun/was Dunning, H. N.; Washburn, E. W.; J. Phys. Chem. **56** (1952) 235.
52-eri-1 Erichsen, L.; Brennst.-Chem. **33** (1952) 166.
52-gri Griffiths, V. S.; J. Chem. Soc. (1952) 1326.
52-hel Heller, H. E.; J. Am. Chem. Soc. **74** (1952) 4858.
52-her/zao Herout, V.; Zaoral, M.; Sorm, F.; Chem. Listy **46** (1952) 156.
52-hug/mal Hughes, H. E.; Maloney, J. O.; Chem. Eng. Prog. **48** (1952) 192.
52-ino Inoue, H.; Yakugaku Zasshi **72** (1952) 731.
52-kip/tes Kipling, J. J.; Tester, D. A.; J. Chem. Soc. (1952) 4123.
52 lev/shu Levina, R. Ya.; Shusherina, N. P.; Treshchova, E. G.; Vestn. Mosk. Univ., Ser. Fiz.-Mat. Estestv. Nauk **7** (1952) 105
52-lev/tan Levina, R. Ya.; Tantsyreva, T. I.; Fainzil'berg, A. A.; Zh. Obshch. Khim. **22** (1952) 571.
52-mye/col Myers, R. T.; Collett, A. R.; Lazzell, C. L.; J. Phys. Chem. **56** (1952) 461.
52-ove/ber Overberger, C. G.; Berenbaum, M. B.; J. Am. Chem. Soc. **74** (1952) 3293.
52-pet/she Petrov, A. D.; Shebanova, M. P.; Dokl. Akad. Nauk SSSR **84** (1952) 721.
52-pom Pomerantz, P.; J. Res. Natl. Bur. Stand. (U. S.) **48** (1952) 76.
52-roe/stu Roe, E. T.; Stutzman, J. M.; Scanlan, J. T.; Swern, D.; J. Am. Oil Chem. Soc. **29** (1952) 18.
52-sta/spi Staveley, L. A. K.; Spice, B.; J. Chem. Soc. (1952) 406.
52-van Van Risseghem, H.; Bull. Soc. Chim. Fr. (1952) 177.
52-von Von Erichsen, L.; Brennst.-Chem. **33** (1952) 166.
52-wib Wiberg, K. B.; J. Am. Chem. Soc. **74** (1952) 3891.
52-wil/sch Willimann, L.; Schinz, H.; Helv. Chim. Acta **35** (1952) 2401.
52-wil/sim Wilson, A.; Simons, E. L.; Ind. Eng. Chem. **44** (1952) 2214.
53-ame/pax Amer, H. H.; Paxton, R. R.; Van Winkle, M.; Anal. Chem. **25** (1953) 1204.
53-ani Anisimov, V. I.; Zh. Fiz. Khim. **27** (1953) 1797.
53-ani-1 Anisimov, V. I.; Zh. Fiz. Khim. **27** (1953) 674.

53-ano-1 Am. Pet. Inst. Res. Proj. 45, Fifteenth Annu. Rep., Ohio State Univ., June (1953) .
53-ano-5 Am. Pet. Inst. Res. Proj. 6, Carnegie-Mellon Univ., May (1953) .
53-ano-6 Coal Tar Data Book, 1st ed., Coal Tar Research Assoc.: Gomersal, Leeds, (1953) .
53-ano-15 Union Carbide & Carbon Corporation, unpublished (1953) .
53-bar-5 Barkan, A. S.; Proc. Byelorussian State Univ. Issue 14 (1953) 108.
53-bar/bro Barker, J. A.; Brown, I.; Smith, F.; Discuss. Faraday Soc. **No. 15** (1953) 142.
53-cha/mou Chang, Y.; Moulton, R. W.; Ind. Eng. Chem. **45** (1953) 2350.
53-col/dre Colonge, J.; Dreux, J.; C. R. Hebd. Seances Acad. Sci. **236** (1953) 1791.
53-dan/fad Danusso, F.; Fadigati, E.; Atti Accad. Naz. Lincei, Cl. Sci. Fis., Mat. Nat., Rend. **9** (1953) 81.
53-gru/ost Grubb, W. T.; Osthoff, R. C.; J. Am. Chem. Soc. **75** (1953) 2230.
53-hag/dec Hagemeyer, H. J.; DeCroes, G. C.; The Chemistry of Isobutylaldehyde and Its Derivatives, Eastman Co.: Tenn., (1953) .
53-hat/jou Hatch, L. F.; Journeay, G. E. ; J. Am. Chem. Soc. **75** (1953) 3712.
53-her/zao Herout, V.; Zaoral, M.; Sorm, F.; Collect. Czech. Chem. Commun. **18** (1953) 122.
53-ish/kat Ishiguro, T.; Kato, S.; Sakata, Y.; Akazawa, Y.; Yakugaku Zasshi **73** (1953) 1167.
53-kha/kud Kharasch, M. S.; Kuderna, J.; Nudenberg, W.; J. Org. Chem. **18** (1953) 1225.
53-kre-1 Kremmling, G.; Z. Naturforsch., A: Astrophys., Phys., Phys. Chem. **8** (1953) 708.
53-mcc/jon McCants, J. F.; Jones, J. H.; Hopson, W. H.; Ind. Eng. Chem. **45** (1953) 454.
53-mck/tar McKenna, F. E.; Tartar, H. V.; Lingafelter, E. C.; J. Am. Chem. Soc. **75** (1953) 604.
53-ner/hen Nerdel, F.; Henkel, E.; Chem. Ber. **86** (1953) 1002.
53-ots/wil Otsuki, H.; Williams, F. C.; Chem. Eng. Prog., Symp. Ser. **49** (1953) 55.
53-par/cha Parthasarathy, P. S.; Chari, S. S.; Srinivasan, D.; J. Phys. Radium **14** (1953) 541.
53-per/wag Percival, W. C.; Wagner, R. B.; Cook, N. C.; J. Am. Chem. Soc. **75** (1953) 3731.
53-pin/hun Pines, H.; Huntsman, W. D. ; Ipatieff, V. N.; J. Am. Chem. Soc. **75** (1953) 2311.
53-raz/old Razuvaev, G. A.; Ol'dekop, Yu. A.; Zh. Obshch. Khim. **23** (1953) 1173.
53-sok Sokolova, E. B.; Zh. Obshch. Khim. **23** (1953) 2002.
53-sut Sutherland, M. D.; J. Am. Chem. Soc. **75** (1953) 5944
54-ano-12 Laboratory Data Sheets, Union Carbide Corp, South Charleston WV (1954) .
54-bre Brey, W. S.; Anal. Chem. **26** (1954) 838.
54-dub/luf Dubois, J. E. ; Luft, R. ; Bull. Soc. Chim. Fr. (1954) 1148.
54-fre/lwo Freudenberg, K.; Lwowski, W.; Justus Liebigs Ann. Chem. **587** (1954) 213.
54-gay/cro Gaylord, N. G.; Crowdle, J. H.; Himmler, W. A.; Pepe, H. J.; J. Am. Chem. Soc. **76** (1954) 59.
54-gri Griffiths, V. S.; J. Chem. Soc. (1954) 860.
54-jon/mcc Jones, J. H.; McCants, J. F.; Ind. Eng. Chem. **46** (1954) 1956.
54-kre/wie Kretschmer, C. B.; Wiebe, R.; J. Am. Chem. Soc. **76** (1954) 2579.
54-mcc/pro McCasland, G. E.; Proskow, S.; J. Am. Chem. Soc. **76** (1954) 3486.
54-mos/cox Mosher, W. A.; Cox, J. C.; Grant-in-Aid Application to A.P.I., Univ. of Delaware, April 28, (1954) .
54-naz/kak-1 Nazarov, I. N.; Kakhniashvili, A. I.; Sb. Statei Obshch. Khim. **2** (1954) 896.
54-naz/kak-2 Nazarov, I. N.; Kakhniashvili, A. I.; Ryabchenko, V. F.; Sb. Statei Obshch. Khim. **2** (1954) 913.
54-naz/kak-3 Nazarov, I. N.; Kakhniashvili, A. I.; Sb. Statei Obshch. Khim. **2** (1954) 919.
54-naz/kak-4 Nazarov, I. N.; Kakhniashvili, A. I.; Sb. Statei Obshch. Khim. **2** (1954) 905.
54-pom/foo Pomerantz, P.; Fookson, A.; Mears, T. W.; Rothberg, S.; Howard, F. L.; J. Res. Natl. Bur. Stand. (U. S.) **52** (1954) 51.
54-pom/foo-1 Pomerantz, P.; Fookson, A.; Mears, T. W.; Rothberg, S.; Howard, F. L.; J. Res. Natl. Bur. Stand. (U. S.) **52** (1954) 59
54-pur/bow Purnell, J. H.; Bowden, S. T.; J. Chem. Soc. (1954) 539.
54-rin/ari Riniker, B.; Arigoni, D.; Jeger, O.; Helv. Chim. Acta **37** (1954) 546.
54-ros Ross, H. K.; Ind. Eng. Chem. **46** (1954) 601.
54-ser/voi-1 Serpinskii, V. V.; Voitkevich, S. A.; Lyuboshich, N. Yu.; Zh. Fiz. Khim. **28** (1954) 1969.
54-ski/flo Skinner, G. S.; Florentine, F. P.; J. Am. Chem. Soc. **76** (1954) 3200.
54-skr/mur Skrzec, A. E.; Murphy, N. F.; Ind. Eng. Chem. **46** (1954) 2245.

54-smi/otv Smith, V. N.; Otvos, J. W.; Anal. Chem. **26** (1954) 359.
54-tha/row Thacker, R.; Rowlinson, J. S.; Trans. Faraday Soc. **50** (1954) 1036.
54-vor/tit Voronkov, M. G.; Titlinova, E. S.; Zh. Obshch. Khim. **24** (1954) 613.
54-wes/aud Westwater, J. W.; Audrieth, L. F.; Ind. Eng. Chem. **46** (1954) 1281.
55-ano Am. Pet. Inst. Res. Proj. 45, Tech. Rep. 14, No. 1, Ohio State Univ., (1955) .
55-ano-2 Am. Pet. Inst. Res. Proj. 6, Carnegie-Mellon Univ., Aug. (1955).
55-bak Baker, E. M.; J. Phys. Chem. **59** (1955) 1182.
55-ber/sch Beringer, F. M.; Schultz, H. S.; J. Am. Chem. Soc. **77** (1955) 5533.
55-blo/ver-1 Blomquist, A. T.; Verdol, J. A.; J. Am. Chem. Soc. **77** (1955) 78.
55-blo/whe Blomquist, A. T.; Wheeler, E. S.; Chu, Y.; J. Am. Chem. Soc. **77** (1955) 6307.
55-bou/cle Boud, A. H.; Cleverdon, D.; Collins, G. B.; Smith, J. O.; J. Chem. Soc. (1955) 3793.
55-che/che Chel'tsova, M. A.; Chernyshev, E. A.; Petrov, A. D.; Izv. Akad. Nauk SSSR Ser. Khim. (1955) 522.
55-dan/col Dannhauser, W.; Cole, R. H.; J. Chem. Phys. **23** (1955) 1762.
55-fle/sau Fleming, R.; Saunders, L.; J. Chem. Soc. (1955) 4147.
55-fog/hix Fogg, E. T. ; Hixson, A. N.; Thompson, A. R.; Anal. Chem. **27** (1955) 1609.
55-gay/cau Gaylord, N. G.; Caul, L. D.; J. Am. Chem. Soc. **77** (1955) 3132.
55-ham/sto Hammond, B. R.; Stokes, R. H.; Trans. Faraday Soc. **51** (1955) 1641.
55-kay/don Kay, W. B.; Donham, W. E.; Chem. Eng. Sci. **4** (1955) 1.
55-kus Kuss, E.; Z. Angew. Phys. **7** (1955) 372.
55-mes/pet Meshcheryakov, A. P.; Petrova, L. V.; Izv. Akad. Nauk SSSR Ser. Khim. (1955) 1057.
55-mos Moshkin, P. A.; Vpr. Isolz. Pentozansoderzh. Syrya Tr. Vses. Soveshch. Riga (1955) 225
55-nol/pan Noller, C. R.; Pannell, C. E.; J. Am. Chem. Soc. **77** (1955) 1862.
55-pet Petrov, A. D.; Izv. Akad. Nauk SSSR Ser. Khim. (1955) 639.
55-pet/sus Petrov, A. D.; Sushchinskii, V. L.; Konoval'chikov, L. D.; Zh. Obshch. Khim. **25** (1955) 1566.
55-sar/mor Sarycheva, I. K.; Morozova, N. G.; Abramovich, V. A.; Breitburt, S. A.; Sergienko, L. F.; Preobrazhenskii, N. A.; Zh. Obshch. Khim. **25** (1955) 2001.
55-ser Serck-Hanssen, K.; Ark. Kemi **8** (1955) 401.
55-sin/she Singh, R.; Shemilt, L. W.; J. Chem. Phys. **23** (1955) 1370.
55-soe/fre Soehring, K.; Frey, H. H.; Endres, G.; Arzneim.-Forsch. **5** (1955) 161.
55-tim/hen Timmermans, J.; Hennaut-Roland, M.; J. Chim. Phys. Phys.-Chim. Biol. **52** (1955) 223.
55-wes Westwater, J. H.; Ind. Eng. Chem. **47** (1955) 451.
56-ame/pax Amer, H. H.; Paxton, R. R.; Van Winkle, M.; Ind. Eng. Chem. **48** (1956) 142.
56-ano-1 Br. Stand. Vol. 555, Brit. Standards Inst., London (1956) .
56-ano-3 Physical Properties: Synthetic Organic Chemicals, Carbide and Carbon Chemical Company, (1956) .
56-ano-4 Am. Pet. Inst. Res. Proj. 6, Carnegie-Mellon Univ., April (1956) .
56-cri Crisan, C.; Ann. Chim. (Paris) **1** (1956) 436.
56-fai/win Fainberg, A. H.; Winstein, S.; J. Am. Chem. Soc. **78** (1956) 2770.
56-far/she Farberov, M. I.; Shemyakina, N. K.; Zh. Obshch. Khim. **26** (1956) 2749.
56-gla/gau Glacet, C.; Gaumenton, A.; Bull. Soc. Chim. Fr. (1956) 1425.
56-goe/mcc Goering, H. L.; McCarron, F. H.; J. Am. Chem. Soc. **78** (1956) 2270.
56-gol/kon Gol'dfarb, Ya. L.; Konstantinov, P. A.; Izv. Akad. Nauk SSSR Ser. Khim. (1956) 992.
56-ike/kep Ikeda, R. M.; Kepner, R. E.; Webb, A. D.; Anal. Chem. **28** (1956) 1335.
56-kat/new Katz, K.; Newman, M.; Ind. Eng. Chem. **48** (1956) 137.
56-koi/han Koizumi, N.; Hanai, T.; J. Phys. Chem. **60** (1956) 1496.
56-kru/cho Krug, R. C.; Choate, W. L.; Walbert, J. T.; Leedy, R. C.; J. Am. Chem. Soc. **78** (1956) 1698.
56-lev/sch Levina, R. Ya.; Schabarov, Yu. S.; Vestn. Mosk. Univ., Ser. 2 Khim. **11** (1956) 61.
56-lib/lap-1 Liberman, A. L.; Lapshina, T. V.; Kazanskii, B. A.; Zh. Obshch. Khim. **26** (1956) 46
56-lic/dur Lichtenberger, J.; Durr, L.; Bull. Soc. Chim. Fr. (1956) 664.
56-nav/des Naves, Y. -R.; Desalbres, L.; Ardizio, P.; Bull. Soc. Chim. Fr. (1956) 1713.
56-pri/osg Price, C. C.; Osgan, M.; J. Am. Chem. Soc. **78** (1956) 4787.

56-rat/cur Rathmann, G. B.; Curtis, A. J.; McGeer, P. L.; Smyth, C. P.; J. Am. Chem. Soc. **78** (1956) 2035.
56-rus/ame Rush, R. I.; Ames, D. C.; Horst, R. W.; Mackay, J. R.; J. Phys. Chem. **60** (1956) 1591.
56-sar/new Sarel, S.; Newman, M. S.; J. Am. Chem. Soc. **78** (1956) 5416.
56-shu/bel Shuikin, N. I.; Bel'skii, I. F.; J. Gen. Chem. USSR (Engl. Transl.) **26** (1956) 3025.
56-smi/ebe Smith, W. M.; Eberly, K. C.; Hanson, E. E.; Binder, J. L.; J. Am. Chem. Soc. **78** (1956) 626.
56-sok/fed Sokolova, E. B.; Fedotov, N. S.; Tr. Inst. - Mosk. Khim. -Tekhnol. Inst. im. D. I. Mendeleeva **No. 23** (1956) 31.
56-tar/tai Tarasova, G. A.; Taits, G. S.; Plate, A. F.; Izv. Akad. Nauk SSSR Ser. Khim. (1956)1267.
56-tor-1 Toropov, A. P.; Zh. Obshch. Khim. **26** (1956) 1453.
56-woo/vio Woods, G. F.; Viola, A.; J. Am. Chem. Soc. **78** (1956) 4380.
57-ano Patent, Brit. 776,073, Esso Res. Eng. Co., June 5, (1957) .
57-ano-1 Patent, Ger. 1,007,772, Farbenfabriken Bayer Akt.-Ges., May 9, (1957) .
57-ano-11 Laboratory Data Sheets, Union Carbide Corp, South Charleston, WV, (1957) .
57-bol/ego Bol'shukhin, A. I.; Egorov, A. G.; Zh. Obshch. Khim. **27** (1957) 647.
57-bol/ego-1 Bol'shukhin, A. I.; Egorov, A. G.; Zh. Obshch. Khim. **27** (1957) 933.
57-chi/tho Chiao, T. T.; Thomson, A. R.; Anal. Chem. **29** (1957) 1678.
57-col/fal Colonge, J.; Falcotet, R.; Bull. Soc. Chim. Fr. (1957) 1166.
57-cru/jos Cruetzen, J. L.; Jost, W.; Sieg, L.; Z. Elektrochem. **61** (1957) 230.
57-far/rot Farberov, M. I.; Rotshtein, Y. I.; Kut'in, A. M.; Shemyakina, N. K.; Zh. Obshch. Khim. **27** (1957) 2806.
57-fav/ser Favorskaya, T. I.; Sergievskaya, O. V.; Ryzhova, N. P.; Zh. Obshch. Khim. **27** (1957) 937.
57-for/lan Foreman, R. W.; Lankelma, H. P.; J. Am. Chem. Soc. **79** (1957) 409.
57-gol/kon Gol'dfarb, Ya. L.; Konstantinov, P. A.; Izv. Akad. Nauk SSSR Ser. Khim. (1957) 217.
57-hic/ken Hickman, J. R.; Kenyon, J.; J. Chem. Soc. (1957) 4677.
57-ket/van Ketelaar, J. A. A.; Van Meurs, N.; Recl. Trav. Chim. Pays-Bas **76** (1957) 437.
57-kit Kitoaka, R.; Nippon Kagaku Zasshi **78** (1957) 1594.
57-luk/lan Lukes, R.; Langthaler, J.; Chem. Listy **51** (1957) 1869.
57-mal/mal Malyusov, V. A.; Malafeev, N. A.; Zhavoronkov, N. M.; Zh. Fiz. Khim. **31** (1957) 699.
57-mur/las Murphy, N. F.; Lastovica, J. E.; Fallis, J. G.; Ind. Eng. Chem. **49** (1957) 1035.
57-naz/kra Nazarov, I. N.; Krasnaya, Zh. A.; Makin, S. M.; Dokl. Akad. Nauk SSSR **114** (1957) 553.
57-pet/nef-1 Petrov, A. D.; Nefedov, O. M.; Grigor'ev, F. I.; Zh. Obshch. Khim. **27** (1957) 1876.
57-pet/sok Petrov, A. D.; Sokolova, E. B.; Kao, C. L.; Izv. Akad. Nauk SSSR Ser. Khim. (1957) 871.
57-pet/sus Petrov, A. D.; Sushchinskii, V. L.; Zakharov, E. P.; Rogozhnikova, T. I.; Zh. Obshch. Khim. **27** (1957) 467.
57-rao/rao Rao, M. R.; Rao, C. V.; J. Appl. Chem. **7** (1957) 659.
57-rom Romadane, I.; Zh. Obshch. Khim. **27** (1957) 1833.
57-rom-1 Romadane, I.; Zh. Obshch. Khim. **27** (1957) 1939.
57-sar/vor Sarycheva, I. K.; Vorob'eva, G. A.; Kucheryavenko, L. G.; Preobrazhenskii, N. A.; Zh. Obshch. Khim. **27** (1957) 2994.
57-shu/bel Shuikin, N. I.; Bel'skii, I. F.; Zh. Obshch. Khim. **27** (1957) 402.
57-tak/nak Takemoto, T.; Nakajima, T.; Yakugaku Zasshi **77** (1957) 1307.
57-tra/bat Traynham, J. G.; Battiste, M. A.; J. Org. Chem. **22** (1957) 1551
58-ano-3 Patent, Brit. 804,057, Monsanto Chemical Co., Nov. 5, (1958) .
58-ano-5 Alcohols, Union Carbide Corp., (1958) .
58-ano-13 Laboratory Data Sheets, Union Carbide Corp, South Charlston, WV, (1958) .
58-arn/was Arnold, V. W.; Washburn, E. R.; J. Phys. Chem. **62** (1958) 1088.
58-col/fal Colonge, J.; Falcotet, R.; Gaumont, R.; Bull. Soc. Chim. Fr. (1958) 211.
58-cos/bow Costello, J. M.; Bowden, S. T.; Recl. Trav. Chim. Pays-Bas **77** (1958) 36.
58-ego Egorov, A. G.; Uch. Zap., Leningr. Gos. Pedagog. Inst. im. A. I. Gertsena **179** (1958) 211.
58-far/spe Farberov, M. I.; Speranskaya, V. A.; Zh. Obshch. Khim. **28** (1958) 2151.
58-hag/hud Hagemeyer, H. J.; Hudson, G. V.; Patent, U.S. 2,852,563, Sept. 16, (1958) .
58-hen/wat Hennion, G.; Watson, E.; J. Org. Chem. **23** (1958) 656.

58-hil/hay	Hill, C. M.; Haynes, L.; Simmons, D. E.; Hill, M. E.; J. Am. Chem. Soc. **80** (1958) 3623.
58-hol/len	Hollo, J.; Lengyel, T.; Uzonyi, G.; Magy. Kem. Lapja **13** (1958) 440.
58-kut/lyu	Kutsenko, A. I.; Lyubomilov, V. I.; Zh. Prikl. Khim. (Leningrad) **31** (1958) 1419.
58-leb/kuk	Lebedeva, A. I.; Kukhareva, L. V.; Zh. Obshch. Khim. **28** (1958) 2782.
58-leg/ulr	LeGoff, E.; Ulrich, S. E.; Denney, D. B.; J. Am. Chem. Soc. **80** (1958) 622.
58-lin/tua	Lin, W.-C.; Tuan, F.-T.; J. Chin. Chem. Soc. (Taipei) **5** (1958) 33.
58-lin/van	Ling, T. D.; Van Winkle, M.; Chem. Eng. Data Ser. **3** (1958) 88.
58-lyu	Lyubomilov, V. I.; Zh. Obshch. Khim. **28** (1958) 328.
58-lyu/bel	Lyubomilov, V. I.; Belyanina, E. T.; Zh. Obshch. Khim. **28** (1958) 326.
58-muk/gru	Mukherjee, L. M.; Grunwald, E.; J. Phys. Chem. **62** (1958) 1311.
58-mur/van	Murti, P. S.; Van Winkle, M.; Chem. Eng. Data Ser. **3** (1958) 72.
58-naz/gus	Nazarov, I. N.; Gussev, B. P.; Gunar, V. I.; Zh. Obshch. Khim. **28** (1958) 1444.
58-oga/tak	Ogata, Y.; Takagi, Y.; J. Am. Chem. Soc. **80** (1958) 3591.
58-pan/osi	Pansevich-Kolyada, V. I.; Osipenko, I. F.; Zh. Obshch. Khim. **28** (1958) 641.
58-per/can	Perry, M. A.; Canter, F. C.; DeBusk, R. E.; Robinson, A. G.; J. Am. Chem. Soc. **80** (1958) 3618.
58-rao/ram	Rao, M. R.; Ramamurty, M.; Rao, C. V.; Chem. Eng. Sci. **8** (1958) 265.
58-sok/kra	Sokolova, E. B.; Krasnova, G. V.; Zhurleva, T. A.; Nauchn. Dokl. Vyssh. Shk. Khim. Khim. Tekhnol. (1958) 330.
58-wag/web	Wagner, I. F.; Weber, J. H.; Chem. Eng. Data **3** (1958) 220.
58-wie/thu	Wiemann, J.; Thuan, L. T.; Bull. Soc. Chim. Fr. (1958) 199.
58-yam/kun	Yamamoto, T.; Kunimoto, S.; Yuki Gosei Kagaku Kyokaishi **16** (1958) 258.
59-ale	Alexander, D. M.; J. Chem. Eng. Data **4** (1959) 252.
59-ano-7	Laboratory Data Sheets, Union Carbide Corp, South Charleston WV (1959) .
59-asi/gei	Asinger, F.; Geiseler, G.; Schmiedel, K.; Chem. Ber. **92** (1959) 3085.
59-bar/dod	Barr-David, F.; Dodge, B. F.; J. Chem. Eng. Data **4** (1959) 107.
59-bur/can	Bures, E.; Cano, C.; de Wirth, A.; J. Chem. Eng. Data **4** (1959) 199.
59-col/gau	Colonge, J.; Gaumont, R.; Bull. Soc. Chim. Fr. (1959) 939.
59-ell/raz	Ellis, S. R. M.; Razavipour, M.; Chem. Eng. Sci. **11** (1959) 99.
59-fol/wel	Foley, W. M.; Welch, F. J.; La Combe, E.; Mosher, H. S.; J. Am. Chem. Soc. **81** (1959) 2779.
59-her	Herbertz, T.; Chem. Ber. **92** (1959) 541.
59-hos/nis	Hoshiai, K.; Nishio, A.; Patent, Jpn. 2211-2, April 9, (1959) .
59-hou/lev	Houlihan, W. J.; Levy, J.; Mayer, J.; J. Am. Chem. Soc. **81** (1959) 4692.
59-kom/ros	Komandin, A. V.; Rosolovskii, V. Ya.; Zh. Fiz. Khim. **33** (1959) 1280.
59-lis/kor	Lishanskii, I. S.; Korotkov, A. A.; Andreev, G. A.; Zak, A. G.; Zh. Prikl. Khim. (Leningrad) **32** (1959) 2344.
59-mac/bar	Machinskaya, I. V.; Barkhash, V. A.; Zh. Obshch. Khim. **29** (1959) 2786.
59-mck/ski	McKinney, W. P.; Skinner, G. F.; Staveley, L. A. K.; J. Chem. Soc. (1959) 2415.
59-nie/web	Nielsen, R. L.; Weber, J. H.; J. Chem. Eng. Data **4** (1959) 145.
59-nog/dza	Nogaideli, A. I.; Dzagnidze, K. Ya.; Tr. Tbilis. Univ. **74** (1959) 159.
59-pet/zak	Petrov, A. D.; Zakharov, E. P.; Izv. Vyssh. Uchebn. Zaved., Khim. Khim. Tekhnol. **2** (1959) 384.
59-pet/zak-1	Petrov, A. D.; Zakharov, E. P.; Krasnova, T. L.; Zh. Obshch. Khim. **29** (1959) 49.
59-pin/lar	Pino, P.; Lardicci, L.; Centoni, L.; J. Org. Chem. **24** (1959) 1399.
59-tsu/hay	Tsuda, K. ; Hayatsu, R.; Kishida, Y.; Chem. Ind. (London) (1959) 1411.
59-yur/bel	Yur'ev, Yu. K.; Belyakova, Z. V.; Zh. Obshch. Khim. **29** (1959) 2960.
59-yur/bel-1	Yur'ev, Yu. K.; Belyakova, Z. V.; Volkov, V. P.; Zh. Obshch. Khim. **29** (1959) 3652
60-ano-13	Laboratory Data Sheets, Union Carbide Corp, South Charleston WV (1960) .
60-cop/fin	Copp, J. L.; Findlay, T. J. V.; Trans. Faraday Soc. **56** (1960) 13.
60-fro/shr	Frontas'ev, V. P.; Shraiber, L. S.; Uch. Zap. Sarat. Gos. Univ. **69** (1960) 225.
60-kor/pet	Kormer, V. A.; Petrov, A. A.; Zh. Obshch. Khim. **30** (1960) 3890.
60-kun/sak	Kunitika, S.; Sakakibara, Y.; Nippon Kagaku Zasshi **81** (1960) 140.
60-naz/mak	Nazarov, I. N.; Makin, S. M.; Nazarova, D. V.; Mochalin, V. B.; Shavrygina, O. A.; Zh. Obshch. Khim. **30** (1960) 1168.

60-oak/web	Oakeson, G. O.; Weber, J. H.; J. Chem. Eng. Data **5** (1960) 279.
60-pet/kao	Petrov, A. D.; Kao, C. L.; Semenkin, V. M.; Zh. Obshch. Khim. **30** (1960) 363.
60-pet/sok	Petrov, A. D.; Sokolova, E. B.; Kao, C. L.; Zh. Obshch. Khim. **30** (1960) 1107.
60-por/far	Porsch, F.; Farnow, H.; Dragoco Rep. (Engl. Ed.) **7** (1960) 215.
60-rab/mur	Rabinovich, I. B.; Murzin, V. I.; Zhilkin, L. S.; Russ. J. Phys. Chem. (Engl. Transl.) **34** (1960) 937.
60-ter/kep	Terry, T. D.; Kepner, R. E.; Dinsmore, W.; J. Chem. Eng. Data **5** (1960) 403.
60-tha/vas	Thaker, K. A.; Vasi, I. G.; J. Sci. Ind. Res., Sect. B **19** (1960) 322.
60-tje	Tjebbes, J.; Acta Chem. Scand. **14** (1960) 180.
60-tsu/kis	Tsuda, K. ; Kishida, Y.; Hayatsu, R.; J. Am. Chem. Soc. **82** (1960) 3396.
61-ano-11	Laboratory Data Sheets, Union Carbide Corp, South Charleston WV (1961) .
61-bel/shu-1	Bel'chev, F. V.; Shuikin, N. I.; Novikov, S. S.; Bull. Acad. Sci. USSR Div. Chem. Sci. (1961) 599.
61-bel/web	Belknap, R. C.; Weber, J. H.; J. Chem. Eng. Data **6** (1961) 485.
61-des/del	Desgrandchamps, G.; Deluzarche, A.; Maillard, A.; Bull. Soc. Chim. Fr. (1961) 264.
61-dyk/sep	Dykyj, J.; Seprakova, M.; Paulech, J.; Chem. Zvesti **15** (1961) 465.
61-hat/wei	Hatch, L. F.; Weiss, H. D.; Li, T. P.; J. Org. Chem. **26** (1961) 61.
61-lut/obu	Lutskii, A. E.; Obukhova, E. M.; J. Gen. Chem. USSR (Engl. Transl.) **31** (1961) 2522.
61-mar/pet	Maretina, I. A.; Petrov, A. A.; Zh. Obshch. Khim. **31** (1961) 419.
61-mes/erz	Meshcheryakov, A. P.; Erzyutova, E. I.; Kuo, Chun-I; Izv. Akad. Nauk SSSR Ser. Khim. (1961) 2198.
61-mil/ben	Miller, R. E.; Bennett, G. E.; Ind. Eng. Chem. **53** (1961) 33.
61-ogi/cor	Ogimachi, N. N.; Corcoran, J. M.; Kruse. H. W.; J. Chem. Eng. Data **6** (1961) 238.
61-sch-3	Schwenker, G.; Arch. Pharm. Ber. Dtsch. Pharm. **294** (1961) 661.
61-shv/pet	Shvarts, E. Yu.; Petrov, A. A.; Zh. Obshch. Khim. **31** (1961) 433.
61-sok/she	Sokolova, E. B.; Shebanova, M. P.; Shchepinov; Izv. Vyssh. Uchebn. Zaved., Khim. Khim. Tekhnol. **4** (1961) 617.
61-tis/sta	Tishchenko, I. G.; Stanishevskii, L. S.; Zhidkofazn. Okislenie Nepredel'nykh Org. Soedin. **No. 1** (1961) 133.
61-wib/fos	Wiberg, K. B.; Foster, G.; J. Am. Chem. Soc. **83** (1961) 423.
61-yeh	Yeh, P. H.; Riechst. Aromen **11** (1961) 285.
62-and/kuk-1	Andreevskii, D. N.; Kukharskaya, E. V.; Zh. Obshch. Khim. **32** (1962) 1353.
62-ber/leg	Bertrand, M.; Le Gras, J.; Bull. Soc. Chim. Fr. (1962) 2136.
62-ber/leg-1	Bertrand, M.; Le Gras, J.; C. R. Hebd. Seances Acad. Sci. **255** (1962) 1305.
62-ber/mol	Bergel'son, L. D.; Molotkovskii, Yu. G.; Shemyakin, M. M.; Zh. Obshch. Khim. **32** (1962) 58.
62-bog/osi	Bogatskaya, Z. D.; Osipchuk, V. P.; Ti, F.-P.; Gol'mov, V. P.; Zh. Obshch. Khim. **32** (1962) 2282.
62-bro/smi	Brown, I.; Smith, F.; Aust. J. Chem. **15** (1962) 3.
62-chu/tho	Chu, K.-Y.; Thompson, A. R.; J. Chem. Eng. Data **7** (1962) 358.
62-col/des	Colonge, J.; Descotes, G.; Soula, J. C.; C. R. Hebd. Seances Acad. Sci. **254** (1962) 887.
62-col/gir	Colonge, J.; Girantet, A.; Bull. Soc. Chim. Fr. (1962) 1002.
62-cuv/nor	Cuvigny, T.; Normant, H.; C. R. Hebd. Seances Acad. Sci. **254** (1962) 316.
62-deg/lad	Degaleesan, T. E.; Laddha, G. S.; J. Appl. Chem. **12** (1962) 111.
62-esa/shi	Esafov, V. I.; Shitoz, G. P.; Zh. Obshch. Khim. **32** (1962) 2819.
62-esa/zhu	Esafov, V. I.; Zhukova, L. P.; J. Gen. Chem. USSR (Engl. Transl.) **32** (1962) 2775.
62-fav/por	Favorskaya, T. A.; Portnyagin, Yu. M.; Zh. Obshch. Khim. **32** (1962) 2122.
62-gei/fru	Geiseler, G.; Fruwert, J.; Stockel, E.; Z. Phys. Chem. (Munich) **32** (1962) 330.
62-gei/qui	Geiseler, G.; Quitzsch, K.; Hesselbach, J.; Huettig, R.; Z. Phys. Chem. (Leipzig) **220** (1962) 79.
62-kae/web	Kaes, G. L.; Weber, J. H.; J. Chem. Eng. Data **7** (1962) 344.
62-kuh/kuh	Kuhn, R.; Kum, K.; Chem. Ber. **95** (1962) 2009.
62-lar/sal	Lardicci, L.; Salvadori, R.; Pino, P.; Ann. Chim. (Rome) **52** (1962) 652.

62-mau/smi Maurit, M. E.; Smirnova, G. V.; Parfenov, E. A.; Vinkovskaya, T. M.; Preobrazhenskii, N. A.; Zh. Obshch. Khim. **32** (1962) 2483.
62-mel Mellan, I.; Polyhydric Alcohols, Spartan Books: Washington, DC, (1962) .
62-mir/fed Miropol'skaya, M. A.; Fedotova, N. I.; Veinberg, A. Ya.; Yanotovskii, M. T.; Samokhvalov, G. I.; Zh. Obshch. Khim. **32** (1962) 2214.
62-nik/vor Nikishin, G. I.; Vorobev, V. D.; Izv. Akad. Nauk SSSR Ser. Khim. (1962) 892.
62-par/mis Paraskevopoulos, G. C.; Missen, R. W.; Trans. Faraday Soc. **58** (1962) 869.
62-par/mos Parks, G. S.; Mosher, H. P.; J. Chem. Phys. **37** (1962) 919.
62-pet/kap Petrov, A. D.; Kaplan, E. P.; Zh. Obshch. Khim. **32** (1962) 693.
62-raz/bog Razuvayev, G.; Boguslavskaya, L. S.; Zh. Obshch. Khim. **32** (1962) 2320.
62-ste/van Stephenson, R. W.; Van Winkle, M.; J. Chem. Eng. Data **7** (1962) 510.
62-thi Thi, T. S. L.; C. R. Hebd. Seances Acad. Sci. **254** (1962) 3873.
62-yur/pen Yur'ev, Yu. K.; Pentin, Y. A.; Revenko, O. M.; Lebedeva, E. I.; Neftekhimiya **2** (1962) 137.
62-yur/rev Yur'ev, Y.; Revenko, O. M.; Vestn. Mosk. Univ., Ser. 2: Khim. **17** (1962) 68.
63-aga/men Agarwal, M. M.; Mene, P. S.; Indian Chem. Eng. **5** (1963) 71.
63-amb/tow Ambrose, D.; Townsend, R.; J. Chem. Soc. **54** (1963) 3614.
63-brz/har Brzostowski, W.; Hardman, T. M.; Bull. Acad. Pol. Sci., Ser. Sci. Chim. **11** (1963) 447.
63-col/bue Colonge, J.; Buendia, J.; C. R. Hebd. Seances Acad. Sci. **257** (1963) 2663.
63-esa Esafov, V. I.; Zh. Obshch. Khim. **33** (1963) 3755.
63-gol/bag Golubev, I. F.; Bagina, E. N.; Tr. GIAP (1963) 39.
63-hov/sea Hovermale, R. A.; Sears, P. G.; Plucknett, W. K.; J. Chem. Eng. Data **8** (1963) 490.
63-kud/sus-1 Kudryavtseva, L. S.; Susarev, M. P.; Zh. Prikl. Khim. (Leningrad) **36** (1963) 1471.
63-lab Labarre, J. F.; Ann. Chim. **8** (1963) 45.
63-lyu/mer Lyubomilov, V. I.; Merkula, N. N.; Zh. Obshch. Khim. **33** (1963) 22.
63-man/she Mann, R. S.; Shemilt, L. W.; J. Chem. Eng. Data **8** (1963) 189.
63-man/she-1 Mann, R. S.; Shemilt, L. W.; Waldichuck, M.; J. Chem. Eng. Data **8** (1963) 502.
63-mcc/lai McCurdy, K. G.; Laidler, K. J.; Can. J. Chem. **41** (1963) 1867.
63-mik/kim Mikhail, S. Z.; Kimel, W. R.; J. Chem. Eng. Data **8** (1963) 323.
63-nik Niklas, H.; Chem. Ber. **96** (1963) 818.
63-pra/van Prabhu, P. S.; Van Winkle, M.; J. Chem. Eng. Data **8** (1963) 210.
63-pra/van-1 Prabhu, P. S.; Van Winkle, M.; J. Chem. Eng. Data **8** (1963) 14.
63-raj/ran Raju, B. N.; Ranganathan, R.; Narasinga Rao, M.; Indian Chem. Eng. **5** (1963) 82.
63-shu/bar Shuikin, N. I.; Bartok, M.; Karakhanov, R. A.; Shostakovskii, V. M.; Acta Phys. Chem. **9** (1963) 124.
63-sub/rao-1 Subbarao, B. V.; Rao, C. V.; J. Chem. Eng. Data **8** (1963) 368.
63-tho/mea Thomas, L. H.; Meatyard, R.; J. Chem. Soc. (1963) 1986.
63-tho/pal Thomas, A. F.; Palluy, E.; Willhalm, B.; Stoll, M.; Helv. Chim. Acta **46** (1963) 2089.
63-tsy/sol Tsyrkel, T. M.; Solov'eva, N. P.; Voltkevich, S. A.; Dushestykh Veshchestv **No. 6** (1963) 15.
63-vil/gav Vil'shau, K. V.; Gavrilova, V. A.; Tr. IREA **No. 25** (1963) 347.
63-yeh Yeh, P. H.; Am. Perfum. Cosmet. **78** (1963) 32.
64-ber/lon Bertocchio, R.; Longeray, R.; Dreux, J.; Bull. Soc. Chim. Fr. (1964) 60.
64-ber/vav Bergel'son, L. D.; Vaver, V. A.; Bezzubov, A. A.; Shemyakin, M. M.; Izv. Akad. Nauk SSSR Ser. Khim. (1964) 1453.
64-bla/per-1 Blanc, P. Y.; Perret, A.; Teppa, F.; Helv. Chim. Acta **47** (1964) 567.
64-blo/hag Blood, A. E.; Hagemeyer, H. H.; Patent, Fr. 1,359,089, April 24, (1964) .
64-bog/kug Bogdanova, A. V.; Kugatova, G. P.; Volkov, A. N.; Arakelyan, V. G.; Izv. Akad. Nauk. SSSR Ser. Khim. (1964) 174.
64-col/var Colonge, J.; Varagnat, A.; Bull. Soc. Chim. Fr. (1964) 2499.
64-dan/bah Dannhauser, W.; Bahe, L. W.; J. Chem. Phys. **40** (1964) 3058.
64-des/sou Descotes, G.; Soula, J. C.; Bull. Soc. Chim. Fr. (1964) 2636.
64-esa/das Esafov, V. I.; Dashko, V. N.; Narek, E. M.; Zh. Obshch. Khim. **34** (1964) 4094.

64-hel/dav	Hellin, M.; Davidson, M.; Lumbroso, D.; Giuliani, P.; Coussemant, F.; Bull. Soc. Chim. Fr. (1964) 2974.
64-hin/dre	Hinnen, A.; Dreux, J.; Bull. Soc. Chim. Fr. (1964) 1492.
64-mys/zie	Myszkowski, J.; Zielinski, A. Z.; Mikolajewicz, M.; Przem. Chem. **43** (1964) 383.
64-nog/rtv	Nogaideli, A. I.; Rtveliashvilli, N. A.; Zh. Obshch. Khim. **34** (1964) 1737.
64-ohl/sei	Ohloff, G.; Seibl, J.; Kovats, E.; Justus Liebigs Ann. Chem. **675** (1964) 83.
64-pol/bel	Polyakova, K. S.; Belov, V. N.; Zh. Obshch. Khim. **34** (1964) 565.
64-sca/sat	Scatchard, G.; Satkiewicz, F. G.; J. Am. Chem. Soc. **86** (1964) 130.
64-ska/kay	Skaates, J. M.; Kay, W. B.; Chem. Eng. Sci. **19** (1964) 431.
64-sta-1	Stanley, H. M.; Patent, Fr. 1,361,772, May 22, (1964) .
64-sta/kor	Stadnichuk, T. V.; Kormer, V. A.; Petrov, A. A.; Zh. Obshch. Khim. **34** (1964) 3279.
64-tis/sta	Tishchenko, I. G.; Stanishevskii, L. S.; Geterog. Reakts. Reakts. Sposobn. (1964) 254.
64-zak/sta	Zakharkin, L. I.; Stanko, V. I.; Brattsev, V. A.; Izv. Akad. Nauk SSSR Ser. Khim. (1964) 931.
65-alb/bry	Al'bitskaya, V. M.; Bryskovskaya, A. V.; Zh. Org. Khim. **1** (1965) 429.
65-bar/koz	Bartok, M.; Kozma, B.; Gilde, A. S.; Acta Phys. Chem. **11** (1965) 35.
65-col/des	Colonge, J.; Descotes, G.; Bahurel, Y.; Bull. Soc. Chim. Fr. (1965) 619.
65-das/mae	Dashunin, V. M.; Maeva, R. V.; Zh. Org. Khim. **1** (1965) 996.
65-fav/nik	Favorskaya, I. A.; Nikitina, A. A.; Zh. Organ. Khim. **1** (1965) 2094.
65-fin/kid	Findlay, T. J. V.; Kidman, A. D.; Aust. J. Chem. **18** (1965) 521.
65-mig/mig	Miginiac-Groizeleau, L.; Miginiac, P.; Prevost, C.; Bull. Soc. Chim. Fr. (1965) 3560.
65-mor/lam	Mornet, R.; Lamant, M.; Riobe, O.; Bull. Soc. Chim. Fr. (1965) 3043.
65-por/pli	Pormale, M. Ya.; Plisko, E. A.; Danilov, S. N.; J. Org. Chem. USSR **1** (1965) 1788.
65-red/rao	Reddy, M. S.; Rao, C. V.; J. Chem. Eng. Data **10** (1965) 309.
65-rum	Rummens, F. H. A.; Recl. Trav. Chim. Pays-Bas **84** (1965) 1003.
65-shu/puz	Shutikova, L. A.; Puzitskii, K. V.; Cherkaev, V. G.; Eidus, Y. T.; Tr. Vses. Nauchno-Issled. Inst. Sint. Nat. Dushistykh Veshchestv. **No. 7** (1965) 16.
65-vij/des	Vijayaraghavan, S. V.; Deshpande, P. K.; Kuloor, N. R.; Indian J. Technol. **3** (1965) 267.
65-wei/lan	Weiss, F.; Lantz, A.; Lakodey, A.; France Patent 1,409,376 (1965)
66-are/tav	Areshidze, Kh. I.; Tavartkiladze, E. K.; Soobshch. Akad. Nauk Gruz. SSR **41** (1966) 315
66-bue	Buendia, J.; Bull. Soc. Chim. Fr. (1966) 2781.
66-bue-1	Buendia, J.; Bull. Soc. Chim. Fr. (1966) 3600.
66-efr	Efremov, Yu. V.; Zh. Fiz. Khim. **40** (1966) 1240.
66-far/per	Farina, M.; Peronaci, E. M.; Chim. Ind. (Milan) **48** (1966) 602.
66-fre/hor	Freifeld, M.; Hort, E. V.; Kirk-Othmer Encyc. Chem. Technol., 2nd Ed. **10** (1966) 667.
66-gar/kom	Garber, Yu. N.; Komarova, L. F.; Zh. Prikl. Khim. (Leningrad) **39** (1966) 1366.
66-gur/raj	Gurukul, S. M. K. A.; Raju, B. N.; J. Chem. Eng. Data **11** (1966) 501.
66-kat/pra-1	Katti, P. K.; Prakash, O.; J. Chem. Eng. Data **11** (1966) 46.
66-kat/shi	Katti, P. K.; Shil, S. K.; J. Chem. Eng. Data **11** (1966) 601.
66-kil/che	Killgore, C. A.; Chew, W. W.; Orr, V.; J. Chem. Eng. Data **11** (1966) 535.
66-koz/rab	Kozlov, N. A.; Rabenovich, I. B.; Melkii, A. V.; J. Appl. Chem. USSR (Engl. Transl.) **39** (1966) 1290.
66-mil-1	Miller, P. H.; Kirk-Othmer Encyc. Chem. Technol., 2nd Ed. **10** (1966) 638.
66-mya/pya	Myagkova, G. I.; Pyatnova, Yu. B.; Sarycheva, I. K.; Preobrazhenskii, N. A.; J. Org. Chem. USSR (Engl. Transl.) **2** (1966) 1961.
66-per/bal	Perepelkin, O. V.; Bal'yan, K. V.; Zh. Org. Khim. **2** (1966) 1928.
66-rob/edm	Robinson, R. L.; Edmister, W. C.; Dullien, F. A. L.; Ind. Eng. Chem. Fundam. **5** (1966) 74.
66-vij/des	Vijayaraghavan, S. V.; Deshpande, P. K.; Kuloor, N. R.; J. Chem. Eng. Data **11** (1966) 147.
66 vla/vas	Vlasov, V. M.; Vasil'eva, A. A.; Semenova, E. F.; Zh. Org. Khim. **2** (1966) 595.
67-ano-5	Laboratory Data Sheets, Union Carbide Corp, South Charleston WV (1967) .
67-dav/fin	Davids, E. L.; Findlay, T. J. V.; Aust. J. Chem. **20** (1967) 1343.
67-dei	Deich, A. Y.; Latv. PSR Zinat. Akad. Vestis Kim. Ser. **No. 1** (1967) 9.
67-fre	Freshwater, D. C.; J. Chem. Eng. Data **12** (1967) 179.

67-gol/per Gold, P. I.; Perrine, R. L.; J. Chem. Eng. Data **12** (1967) 4.
67-kom/man Komarenko, V. G.; Manzhelii, V. G.; Radtsig, A. V.; Ukr. Fiz. Zh. (Ukr. Ed.) **12** (1967) 676.
67-min/che Min'kovskii, M. M.; Cherkaev, V. G.; Maslozhir. Prom. **33** (1967) 31.
67-mis/sub Mishchenko, K. P.; Subbotina, V. V.; Zh. Prikl. Khim. (Leningrad) **40** (1967) 1156.
67-mur/rao Muralimohan, S.; Rao, P. B.; J. Chem. Eng. Data **12** (1967) 494.
67-nat/rao Nataraj, V.; Rao, M. R.; Indian J. Technol. **5** (1967) 212.
67-pis/gas Pischnamassade, B. F.; Gasanova, S. D.; Azerb. Khim. Zh. **2** (1967) 60.
67-seu/mor Seucan, S.; Moraru, E.; Avramescu, E.; Metrol. Apl. **14** (1967) 313.
67-sho/ata Shostakovskii, M. F.; Atavin, A. S.; Vasil'ev, N. P.; Mikhaleva, A. I.; Dmitrieva, L. P.; J. Org. Chem. USSR (Engl. Transl.) **3** (1967) 2072.
67-vij/des-1 Vijayaraghavan, S. V.; Deshpande, P. K.; Kuloor, N. R.; J. Chem. Eng. Data **12** (1967) 13.
67-zal/aru Zalinyan, M. G.; Arutyunyan, V. S.; Dangyan, M. T.; Arm. Khim. Zh. **20** (1967) 620.
68-ana/rao Ananthanarayanan, P.; Rao, P. B.; J. Chem. Eng. Data **13** (1968) 194.
68-ano Chemicals and Plastics Physical Properties, Union Carbide Corp. (product bulletin), (1968) .
68-eva/lin Evans, L. R.; Lin, J. S.; J. Chem. Eng. Data **13** (1968) 14.
68-joh Johari, G. P.; J. Chem. Eng. Data **13** (1968) 541.
68-kaw/min Kawasaki, T.; Minowa, Z.; Inamatsu, T.; Bull. Nat. Res. Lab. Metrol., Tokyo **No. 17** (1968) 114.
68-naz/tsy Nazarova, S. S.; Tsyskovskii, V. K.; Ogurtsova, N. A.; Zh. Prikl. Khim. (Leningrad) **41** (1968) 144.
68-pfl/pop Pflug, H. D.; Pope, A. E.; Benson, G. C.; J. Chem. Eng. Data **13** (1968) 408.
68-rao/chi Rao, P. R.; Chiranjivi, C.; Dasarao, C. J.; J. Appl. Chem. **18** (1968) 166.
68-sin/ben Singh, J.; Benson, G. C.; Can. J. Chem. **46** (1968) 1249.
68-ver Verhoeye, L. A. J.; J. Chem. Eng. Data **13** (1968) 462.
69-ada Adamcova, Z.; Coll. Czech. Chem. Comm. **34** (1969) 3149.
69-bro/foc Brown, I.; Fock, W.; Smith, F.; J. Chem. Thermodyn. **1** (1969) 273.
69-fin/cop Findlay, T. J. V.; Copp, J. L.; Trans. Faraday Soc. **65** (1969) 1463.
69-kat/pat Katti, P. K.; Pathak, C. M.; J. Chem. Eng. Data **14** (1969) 73.
69-kom/kri Komarov, V. M.; Krichevtsov, B. K.; Zh. Prikl. Khim. (Leningrad) **42** (1969) 2772.
69-nav/tul Naves, Y. -R.; Tullen, P.; Bull. Soc. Chim. Fr. (1969) 586.
69-pat/kat Pathak, S.; Katti, S. S.; J. Chem. Eng. Data **14** (1969) 359.
69-smi/kur Smirnova, N. A.; Kurtynina, L. M.; Zh. Fiz. Khim. **43** (1969) 1883.
69-sub/nag Subramanian, D.; Nageshwar, G. D.; Mene, P. S.; J. Chem. Eng. Data **14** (1969) 421.
69-ver/lau Verhoeye, L. A. J.; Lauwers, E.; J. Chem. Eng. Data **14** (1969) 306.
69-zub/bag Zubarev, V. N.; Bagdonas, A.; Teploenergetika (Moscow) **16** (1969) 88
70-are/tav Areshidze, Kh. I.; Tavartkiladze, E. K.; Chivadze, G. O.; J. Appl. Chem. USSR (Engl. Transl.) **43** (1970) 606.
70-che/tho Chen, D. H. T.; Thompson, A. R.; J. Chem. Eng. Data **15** (1970) 471.
70-gal Galska-Krajewska, A.; Rocz. Chem. **44** (1970) 1255.
70-gar/kom Garber, Yu. N.; Komarova, L. F.; Aleinikova, L. I.; Fomina, L. S.; Zh. Prikl. Khim. (Leningrad) **43** (1970) 2658.
70-gur/raj Gurukul, S. M. K. A.; Raju, B. N.; J. Chem. Eng. Data **15** (1970) 361.
70-kat/kon Kato, M.; Konishi, H.; Hirata, M.; J. Chem. Eng. Data **15** (1970) 435.
70-kon/lya Konobeev, B. I.; Lyapin, V. V.; Zh. Prikl. Khim. (Leningrad) **43** (1970) 803.
70-kri/kom Krichevtsov, B. K.; Komarov, V. M.; Zh. Prikl. Khim. (Leningrad) **43** (1970) 112.
70-kri/kom-1 Krichevtsov, B. K.; Komarov, V. M.; Zh. Prikl. Khim. (Leningrad) **43** (1970) 703.
70-kun/cha Kundu, K. K.; Chattopadhyay, P. K.; Jana, D. ; Das, M. N.; J. Chem. Eng. Data **15** (1970) 209.
70-min/kaw Minowa, Z.; Kawasaki, T.; Inamatsu, T.; Bull. Nat. Res. Lab. Metrol. (Tokyo) **No. 21** (1970) 1.
70-mye/cle Myers, R. S.; Clever, H. L.; J. Chem. Thermodyn. **2** (1970) 53.
70-nak/shi Nakanishi, K.; Shirai, H.; Bull. Chem. Soc. Jpn. **43** (1970) 1634.
70-paz/paz Paz Andrade, M. I.; Paz Fernandez, J. M.; Recacho, E.; An. Quim. **66** (1970) 961.

70-puz/bul Puzitskii, K. V.; Bulanova, T. F.; Bin, Y. Y.; Eidus, Y. T.; Izv. Akad. Nauk SSSR Ser. Khim. (1970) 1872.
70-ras/gaz Rastorguev, Yu. L.; Gazdiev, M. A.; Russ. J. Phys. Chem. (Engl. Transl.) **44** (1970) 1758.
70-str/svo Strubl, K.; Svoboda, V.; Holub, R.; Pick, J.; Collect. Czech. Chem. Commun. **35** (1970) 3004.
70-sus/hol Suska, J.; Holub, R.; Vonka, P.; Pick, J.; Collect. Czech. Chem. Commun. **35** (1970) 385.
70-ver Verhoeye, L. A. J.; J. Chem. Eng. Data **15** (1970) 222.
71-abr/ber Abraham, T.; Bery, V.; Kudchadker, A. P.; J. Chem. Eng. Data **16** (1971) 355.
71-bra/joh Brandreth, D. A.; Johnson, R. E.; J. Chem. Eng. Data **16** (1971) 325.
71-des/bha-1 Deshpande, D. D.; Bhalgadde, L. G.; Oswal, S.; Prabhu, C. S.; J. Chem. Eng. Data **16** (1971) 469.
71-gol/dob Golubev, I. F.; Dobrovolskii, O. A.; Demin, G. P.; Tr. GIAP (1971) 5.
71-kat/lob Katz, M.; Lobo, P. W.; Minano, A. S.; Solimo, H.; Can. J. Chem. **49** (1971) 2605.
71-nag/oht Nagata, I.; Ohta, T.; J. Chem. Eng. Data **16** (1971) 164.
71-tha/rao Thayumanasundaram, R.; Rao, P. B.; J. Chem. Eng. Data **16** (1971) 323.
71-yer/swi Yergovich, T. W.; Swift, G. W.; Kurata, F.; J. Chem. Eng. Data **16** (1971) 222.
72-bon/pik Bonnet, J. C.; Pike, F. P.; J. Chem. Eng. Data **17** (1972) 145.
72-caz/mar Cazaux, L.; Maroni, P.; Bull. Soc. Chim. Fr. (1972) 773.
72-gla/gha Gladden, J. K.; Ghaffari, F.; J. Chem. Eng. Data **17** (1972) 468.
72-nie/nov Niepel, W.; Novak, J. P.; Matous, J.; Sobr, J.; Chem. Zvesti **26** (1972) 44.
72-pol/lu Polak, J.; Lu, B. C. Y.; J. Chem. Eng. Data **17** (1972) 456.
72-pol/lu-1 Polak, J.; Lu, B. C. Y.; J. Chem. Thermodyn. **4** (1972) 469.
72-rak/isa Rakhmankulov, D. L.; Isagulyants, V. I.; Zlotskii, S. S.; J. Appl. Chem. USSR (Engl. Transl.) **45** (1972) 2907.
72-rei/eis Reisler, E.; Eisenberg, H.; J. Chem. Soc., Faraday Trans. 2, **68** (1972) 1001.
72-udo/maz Udovenko, V. V.; Mazanko, T. F.; Plyngeu, V. Ya.; Zh. Fiz. Khim. **46** (1972) 218.
73-dak/rao Dakshinamurty, P.; Rao, K. V.; Venkateswara Rao, P.; Chiranjivi, C.; J. Chem. Eng. Data **18** (1973) 39.
73-dak/vee Dakshinamurty, P.; Veerabhadra Rao, K.; Venkateswara Rao, P.; Chiranjivi, C.; J. Chem. Eng. Data **18** (1973) 39.
73-daw/new Dawe, R. A.; Newsham, D. M. T.; Ng, S. B.; J. Chem. Eng. Data **18** (1973) 44.
73-fin Findenegg, G. H.; Monatsh. Chem. **104** (1973) 998.
73-gar/paz Garcia, M.; Paz Andrade, M. I.; An. Quim. **69** (1973) 708.
73-khi/ale Khimenko, M. T.; Aleksandrov, V. V.; Gritsenko, N. N.; Zh. Fiz. Khim. **47** (1973) 2914.
73-min/rue Minh, D. C.; Ruel, M.; J. Chem. Eng. Data **18** (1973) 41.
73-nag/oht Nagata, I.; Ohta, T.; Uchiyama, Y.; J. Chem. Eng. Data **18** (1973) 54.
73-nay/kud Nayar, S.; Kudchadker, A. P.; J. Chem. Eng. Data **18** (1973) 356.
73-svo/ves Svoboda, V.; Vesely, F.; Holub, R.; Pick, J.; Collect. Czech. Chem. Commun. **38** (1973) 3539.
73-sza/mys Szafraniak, K.; Myszkowski, J.; Zielinski, A. Z.; Pyc, W.; Przem. Chem. **52** (1973) 744.
74-dut/mat Dutta-Choudhury, M. K.; Mathur, H. B.; J. Chem. Eng. Data **19** (1974) 145.
74-jim/paz Jimenez Cuesta, E.; Paz Andrade, M. I.; An. Quim. **70** (1974) 103.
74-moo/wel Moore, J. W.; Wellek, R. M.; J. Chem. Eng. Data **19** (1974) 136.
74-mye/cle Myers, R. S.; Clever, H. L.; J. Chem. Thermodyn. **6** (1974) 949.
74-paz/rom Paz Andrade, M. I.; Romani, L.; Perez, R.; An. Quim. **70** (1974) 419.
74-pur/pol Puri, P. S.; Polak, J.; Ruether, J. A.; J. Chem. Eng. Data **19** (1974) 87.
74-rak/mak Rakhmankulov, D. L.; Maksimova, N. E.; Melikyan, V. R.; Isagulyants, V. I.; J. Appl. Chem. USSR (Engl. Transl.) **47** (1974) 471.
74-rak/mel Rakhmankulov, D. L.; Melikyan, V. R.; Maksimova, N. E.; Isagulyants, V. I.; J. Appl. Chem. USSR (Engl. Transl.) **47** (1974) 2333.
74-rak/zlo Rakhmankulov, D. L.; Zlotskii, S. S.; Agisheva, S. A.; Maksimova, N. E.; Isagulyants, V. I.; J. Appl. Chem. USSR (Engl. Transl.) **47** (1974) 1472.
74-rao/nai Rao, M. V. P.; Naidu, P. R.; Can. J. Chem. **52** (1974) 788.
74-rao/nai-1 Rao, M. V. P.; Naidu, P. R.; J. Chem. Thermodyn. **6** (1974) 1195.

74-wol/ska Wollmann, H.; Skaletki, B.; Schaaf, A.; Pharmazie **29** (1974) 708.
75-esp/man Espanol, M.; Manisse, A.; Beaudoin, M. T.; C. R. Hebd. Seances Acad. Sci. C **281** (1975) 445.
75-hsu/cle Hsu, K.-Y.; Clever, H. L.; J. Chem. Eng. Data **20** (1975) 268.
75-khi/gri Khimenko, M. T.; Gritsenko, N. N.; Tsybizova, L. P.; Bezpalyi, B. N.; Russ. J. Phys. Chem. **49** (1975) 146.
75-kub/tan Kubota, H.; Tanaka, Y.; Makita, T.; Kagaku Kogaku Ronbunshu **1** (1975) 176.
75-lee/che Leets, K. V.; Chernyshev, V. O.; Rang, K. A.; Erm, A. Y.; Zh. Org. Khim. **11** (1975) 1811.
75-mat/fer Mato, F.; Fernandez-Polonco, F.; An. Quim. **71** (1975) 815.
75-mus/ver Mussche, M. J.; Verhoeye, L. A. J.; J. Chem. Eng. Data **20** (1975) 46.
75-nak/kom Nakajima, T.; Komatsu, T.; Nakagawa, T.; Bull. Chem. Soc. Jpn. **48** (1975) 783.
75-tok Tokunaga, J.; J. Chem. Eng. Data **20** (1975) 41.
76-for/ben Fortier, J.-L.; Benson, G. C.; Picker, P.; J. Chem. Thermodyn. **8** (1976) 289.
76-hal/ell Hales, J. L.; Ellender, J. H.; J. Chem. Thermodyn. **8** (1976) 1177.
76-hon/sin Hon, H. C.; Singh, R. P.; Kudchadker, A. P.; J. Chem. Eng. Data **21** (1976) 430.
76-kat/nit Katayama, T.; Nitta, T.; J. Chem. Eng. Data **21** (1976) 194.
76-kow/kas Kowalski, B.; Kasprzycka-Guttman, T.; Orszagh, A.; Rocz. Chem. **50** (1976) 1445.
76-nag/oht Nagata, I.; Ohta, T.; Ogura, M.; Yasuda, S.; J. Chem. Eng. Data **21** (1976) 310.
76-sri/kul Srinivasan, D.; Kulkarni, P. L.; Chem. Eng. World **11**(12) (1976) 63.
76-sub/nai Subramanyam Reddy, K.; Naidu, P. R.; J. Chem. Thermodyn. **8** (1976) 1208.
76-tri/kri Tripathi, R. P.; Krishna, S.; Gulati, I. B.; J. Chem. Eng. Data **21** (1976) 44.
76-wes Westmeier, S.; Chem. Tech. (Leipzig) **28** (1976) 350.
76-wes-1 Westmeier, S.; Chem. Tech. (Leipzig) **28** (1976) 480.
77-bel/bub Belousov, V. P.; Bubnov, V. I.; Pfestorf, R.; Schumichen, A.; Z. Chem. **17** (1977) 382.
77-cam/sch Cammenga, H. K.; Schulze, F. W.; Theuerl, W.; J. Chem. Eng. Data **22** (1977) 131.
77-gov/and-1 Govindaswamy, S.; Andiappan, A. N.; Lakshmanan, S. M.; J. Chem. Eng. Data **22** (1977) 264.
77-gup/han Gupta, A. C.; Hanks, R. W.; Thermochim. Acta **21** (1977) 143.
77-hwa/rob Hwang, S.-C.; Robinson, R. L.; J. Chem. Eng. Data **22** (1977) 319.
77-tan/yam Tanaka, Y.; Yamamoto, T.; Satomi, Y.; Kubota, H.; Makita, T.; Rev. Phys. Chem. Jpn. **47** (1977) 12.
77-tre/ben Treszczanowicz, A. J.; Benson, G. C.; J. Chem. Thermodyn. **9** (1977) 1189.
77-zhu/zhu-2 Zhuravleva, I. K.; Zhuravlev, E. F.; J. Gen. Chem. USSR (Engl. Transl.) **47** (1977) 1774.
78-amb/cou-1 Ambrose, D.; Counsell, J. F.; Lawrenson, I. J.; Lewis, G. B.; J. Chem. Thermodyn. **10** (1978) 1033.
78-apa/ker-2 Apaev, T. A.; Kerimov, A. M.; Imanova, I. G.; High Temp. **16** (1978) 575.
78-ast Astrup, E. E.; Acta Chem. Scand., Ser. A **32** (1978) 115.
78-dap/don D'Aprano, A.; Donato, I. D.; Caponetti, E.; Agrigento, V.; Gazz. Chim. Ital. **108** (1978) 601.
78-fro/ers-1 Frolova, E. A.; Ershova, T. P.; Ustavshchikov, B. F.; Kiselev, V. Ya.; J. Appl. Chem. USSR (Engl. Transl.) **51** (1978) 2251.
78-jel/leo Jelinek, R. M.; Leopold, H.; Monatsh. Chem. **109** (1978) 387.
78-mus/kan Musavirov, R. S.; Kantor, E. A.; Rakhmankulov, D. L.; J. Appl. Chem. USSR (Engl. Transl.) **51** (1978) 2184.
78-nit/fuj Nitta, T.; Fujio, J.; Katayama, T.; J. Chem. Eng. Data **23** (1978) 157.
78-ovc/kry Ovcschinnikova, T. F.; Kryukov, S. I.; Simanov, N. A.; Farberov, M. I.; Neftekhimiya **18** (1978) 80.
78-paz/gar Paz Andrade, M. I.; Garcia, M.; Baluja, M. C.; Sanchez Bana, M. E.; Acta Cient. Compostelana **15** (1978) 33.
78-red/nai-1 Subramanyam Reddy, K.; Naidu, P. R.; J. Chem. Thermodyn. **10** (1978) 201.
78-sac/pes Sachek, A. I.; Peshchenko, A. D.; Andreevskii, D. N.; Vestsi Akad. Navuk BSSR Ser. Khim. Navuk (1978) 124.
78-tre/ben Treszczanowicz, A. J.; Benson, G. C.; J. Chem. Thermodyn. **10** (1978) 967.
79-bal/fri Balish, M.; Fried, V.; J. Chem. Eng. Data **24** (1979) 91.
79-cha/ses-1 Chandrashekara, M. N.; Seshadri, D. N.; Indian J. Technol. **17** (1979) 243.

79-dia/tar Diaz Pena, M.; Tardajos, G.; J. Chem. Thermodyn. **11** (1979) 441.
79-ern/gli Ernst, S.; Glinski, J.; Jezowska-Trzebiatowska, B.; Acta Phys. Pol., A **55** (1979) 501.
79-gyl/apa Gylmanov, A. A.; Apaev, T. A.; Akhmedov, L. A.; Lipovetskii, S. I.; Izv. Vyssh. Uchebn. Zaved., Neft Gaz **22** (1979) 55.
79-jim/paz Jimenez Cuesta, E.; Paz Andrade, M. I.; Casanova, C.; J. Chim. Phys. Phys.-Chim. Biol. **76** (1979) 46.
79-kiy/ben Kiyohara, O.; Benson, G. C.; J. Chem. Thermodyn. **11** (1979) 861.
79-sah/hay Sahgal, A.; Hayduk, W.; J. Chem. Eng. Data **24** (1979) 222.
79-sin/siv Singh, S.; Sivanarayana, K.; Kushwaha, R.; Prakash, S.; J. Chem. Eng. Data **24** (1979) 279.
79-sub/rao Subba Rao, D.; Rao, K. V.; Ravi Prasad, A.; Chiranjivi, C.; J. Chem. Eng. Data **24** (1979) 241.
79-tho/nag Thorat, R. T.; Nageshwar, G. D.; Mene, P. S.; J. Chem. Eng. Data **24** (1979) 270.
79-zel/hub Zeltner, P.; Huber, G. A.; Peters, R.; Tatrai, F.; Boksanyi, L.; Kovats, E. S.; Helv. Chim. Acta **62** (1979) 2495
80-arc/bla Arce, A.; Blanco, A.; Antorrena, G.; Quintela, M. D.; An. Quim., Ser. A **76** (1980) 405.
80-ben/kiy Benson, G. C.; Kiyohara, O.; J. Solution Chem. **9** (1980) 791.
80-bru Brunner, E.; J. Chem. Thermodyn. **12** (1980) 993.
80-cha/ses Chandrashekara, M. N.; Seshadri, D. N.; J. Chem. Eng. Data **25** (1980) 124.
80-edu/boy Eduljee, G. H.; Boyes, A. P.; J. Chem. Eng. Data **25** (1980) 249.
80-fuk/ogi Fukuchi, K.; Ogiwara, K.; Yonezawa, S.; Arai, Y.; Kogaku Shuho - Kyushu Daigaku **53** (1980) 187.
80-kas/izy Kasprzycka-Guttman, T.; Izycka, K.; Czelej, M.; Pol. J. Chem. **54** (1980) 1775.
80-mar/ric Marsh, K. N.; Richards, A. E.; Aust. J. Chem. **33** (1980) 2121.
80-oza/ooy Ozawa, S.; Ooyatsu, N.; Yamabe, M.; Honmo, S.; Ogino, Y.; J. Chem. Thermodyn. **12** (1980) 229.
80-pik Pikkarainen, L.; Finn. Chem. Lett. **No. 7-8** (1980) 185.
80-rig/ube Riggio, R.; Ubeda, M. H.; Ramos, J. F.; Martinez, H. E.; J. Chem. Eng. Data **25** (1980) 318.
80-sue/mul Suehnel, K.; Muller, S.; Z. Phys. Chem. (Leipzig) **261** (1980) 60.
80-tre/ben Treszczanowicz, A. J.; Benson, G. C.; J. Chem. Thermodyn. **12** (1980) 173.
80-yos/tak Yoshikawa, Y.; Takagi, A.; Kato, M.; J. Chem. Eng. Data **25** (1980) 344.
80-yu/ish Yu, J.-M.; Ishikawa, T.; Lu, B. C. Y.; J. Chem. Thermodyn. **12** (1980) 57.
81-ben/han Benson, G. C.; Handa, Y. P.; J. Chem. Thermodyn. **13** (1981) 887.
81-han/hal Handa, Y. P.; Halpin, C. J.; Benson, G. C.; J. Chem. Thermodyn. **13** (1981) 875.
81-joo/arl Joo, H.-J.; Arlt, W.; J. Chem. Eng. Data **26** (1981) 138.
81-kim/ben Kimura, F.; Benson, G. C.; J. Chem. Eng. Data **26** (1981) 317.
81-kiy/ben Kiyohara, O.; Benson, G. C.; J. Chem. Eng. Data **26** (1981) 263.
81-kor/kov Korosi, G.; Kovats, E.; J. Chem. Eng. Data **26** (1981) 323.
81-kum/pra Kumar, A.; Prakash, O.; Prakash, S.; J. Chem. Eng. Data **26** (1981) 64.
81-nai/nai Naidu, G. R.; Naidu, P. R.; J. Chem. Eng. Data **26** (1981) 197.
81-nar/dha Narayanaswamy, G.; Dharmaraju, G.; Raman, G. K.; J. Chem. Thermodyn. **13** (1981) 327.
81-oht/koy Ohta, T.; Koyabu, J.; Nagata, I.; Fluid Phase Equilib. **7** (1981) 65.
81-sjo/dyh Sjoblom, J.; Dyhr, H.; Hansen O.; Finn. Chem. Lett. (1981) 110
81-tas/ara Tashima, Y.; Arai, Y.; Mem. Fac. Eng., Kyushu Univ. **41** (1981) 217.
81-tre/kiy Treszczanowicz, A. J.; Kiyohara, O.; Benson, G. C.; J. Chem. Thermodyn. **13** (1981) 253.
81-won/chu Won, Y. S.; Chung, D. K.; Mills, A. F.; J. Chem. Eng. Data **26** (1981) 140.
82-aww/pet Awwad, A. M.; Pethrick, R. A.; J. Chem. Soc., Faraday Trans. 1 **78** (1982) 3203.
82-ber/rog-1 Berro, C.; Rogalski, M.; Peneloux, A.; J. Chem. Eng. Data **27** (1982) 352.
82-dap/don D'Aprano, A.; Donato, I. D.; Agrigento, V.; J. Solution Chem. **11** (1982) 259.
82-diz/mar Dizechi, M.; Marschell, E.; J. Chem. Eng. Data **27** (1982) 358.
82-dom/rat Domonkos, L.; Ratkovics, F.; Monatsh. Chem. **113** (1982) 1119.
82-kar/red Karunakar, J.; Reddy, K. D.; Rao, M. V. P.; J. Chem. Eng. Data **27** (1982) 346.
82-kat/wat Katayama, H.; Watanabe, I.; J. Chem. Eng. Data **27** (1982) 91.
82-man/les Mandik, L.; Lesek, F.; Coll. Czech. Chem. Comm. **47** (1982) 1686.

82-nai/nai Naidu, G. R.; Naidu, P. R.; Indian J. Pure Appl. Phys. **20** (1982) 313.
82-ort Ortega, J.; J. Chem. Eng. Data **27** (1982) 312.
82-sch/pol Schroeder, M. R.; Poling, B. E.; Manley, D. B.; J. Chem. Eng. Data **27** (1982) 256.
82-sin/sin Singh, R. P.; Sinha, C. P.; J. Chem. Eng. Data **27** (1982) 283.
82-tre/han Treszczanowicz, A. J.; Handa, Y. P.; Benson, G. C.; J. Chem. Thermodyn. **14** (1982) 871.
82-ven/dha Venkateswarlu, P.; Dharmaraju, G.; Raman, G. K.; Acoust. Lett. **6** (1982) 1.
82-zak Zakharyaev, Z. R.; Inzh.-Fiz. Zh. **43** (1982) 796.
83-alb/edg Albright, J. G.; Edge, A. V. J.; Mills, R.; J. Chem. Soc., Faraday Trans. 1 **79** (1983) 1327.
83-dap/del D'Aprano, A.; De Lisi, R.; Donato, I. D.; J. Solution Chem. **12** (1983) 383.
83-fuk/ogi Fukuchi, K.; Ogiwara, K.; Tashima, Y.; Yonezawa, S.; Ube Kogyo Koto Senmon Gakko Kenkyu Hokoku **29** (1983) 93.
83-gop/rao Gopal, K.; Rao, N. P.; Acustica **54** (1983) 115.
83-hal/gun Hales, J. L.; Gundry, H. A.; Ellender, J. H.; J. Chem. Thermodyn. **15** (1983) 211.
83-kim/ben Kimura, F.; Benson, G. C.; J. Chem. Eng. Data **28** (1983) 157.
83-lin Linek, J.; Collect. Czech. Chem. Commun. **48** (1983) 2446.
83-pik-1 Pikkarainen, L.; Finn. Chem. Lett. **No. 3-4** (1983) 63.
83-pik-2 Pikkarainen, L.; J. Chem. Eng. Data **28** (1983) 344.
83-rau/ste Rauf, M. A.; Stewart, G. H.; Farhat-Aziz; J. Chem. Eng. Data **28** (1983) 324.
83-tri Triday, J. O.; J. Chem. Eng. Data **28** (1983) 307.
83-wec Weclawski, J.; Fluid Phase Equilib. **12** (1983) 155.
83-wec/byl Weclawski, J.; Bylicki, A.; Fluid Phase Equilib. **12** (1983) 143.
84-ber/pen Berro, C.; Peneloux, A.; J. Chem. Eng. Data **29** (1984) 206.
84-bra/pin Bravo, R.; Pintos, M.; Baluja, M. C.; Paz Andrade, M. I.; Roux-Desgranges, G.; Grolier, J.-P. E.; J. Chem. Thermodyn. **16** (1984) 73.
84-cer/bou Cervenkova, I.; Boublik, T.; J. Chem. Eng. Data **29** (1984) 425.
84-eas/woo Easteal, A. J.; Woolf, L. A.; J. Chem. Thermodyn. **16** (1984) 391.
84-idr/fre Idriss-Ali, K. M.; Freeman, G. R.; Can. J. Chem. **62** (1984) 2217.
84-kim/ben Kimura, F.; Benson, G. C.; Fluid Phase Equilib. **16** (1984) 77.
84-kum/ben Kumaran, M. K.; Benson, G. C.; J. Chem. Thermodyn. **16** (1984) 175.
84-ort/ang Ortega, J.; Angulo, M. C.; J. Chem. Eng. Data **29** (1984) 340.
84-ped/sal Pedrosa, G. C.; Salas, J. A.; Davolio, F.; Katz, M.; An. Asoc. Quim. Argent. **72** (1984) 541.
84-sak/nak Sakurai, M.; Nakagawa, T.; J. Chem. Thermodyn. **16** (1984) 171.
85-bel/ber Belousov, V. P.; Beregovykh, V. V.; Gurarii, L. L.; Vses. Konf. po Termodinamike Organicheskikh Soyedinenii, 4th, Kuibyshev (1985) 100.
85-chi/lin Chiu, R. M.-H.; Lin, H. B.; Kan, P. Y.; J. Chin. Inst. Chem. Eng. **16** (1985) 307.
85-dap/don D'Aprano, A.; Donato, I. D.; D'Arrigo, G.; Bertolini, D.; Cassettari, M.; Salvetti, G.; Mol. Phys. **55** (1985) 475.
85-dri/ras Dribika, M. M.; Rashed, I. G.; Biddulph, M. W.; J. Chem. Eng. Data **30** (1985) 146.
85-fer/ber Fernandez, J.; Berro, C.; Paz Andrade, M. I.; Fluid Phase Equilib. **20** (1985) 145.
85-fer/pin Fernandez, J.; Pintos, M.; Baluja, M. C.; Jimenez Cuesta, E.; Paz Andrade, M. I.; J. Chem. Eng. Data **30** (1985) 318.
85-kov/svo Kovac, A.; Svoboda, J.; Ondrus, I.; Chem. Zvesti **39** (1985) 729.
85-kum/ben Kumaran, M. K.; Benson, G. C.; J. Chem. Thermodyn. **17** (1985) 699.
85-les/eic Lesek, F.; Eichler, J.; Balcar, M.; Collect. Czech. Chem. Commun. **50** (1985) 153.
85-mat/ben Mato, F.; Benito, G. G.; An. Quim., Ser. A **81** (1985) 116.
85-mat/ben-1 Mato, F.; Benito, G. G.; Sobron, F.; An. Quim., Ser. A **81** (1985) 400.
85-nag Nagata, I.; J. Chem. Eng. Data **30** (1985) 201.
85-nag-1 Nagata, I.; J. Chem. Eng. Data **30** (1985) 363.
85-nag-2 Nagata, I.; Fluid Phase Equilib. **24** (1985) 279.
85-nag-3 Nagata, I.; Fluid Phase Equilib. **19** (1985) 13.
85-ogi/ara Ogiwara, K.; Arai, Y.; Saito, S.; J. Chem. Eng. Jpn. **18** (1985) 273.
85-ort Ortega, J.; J. Chem. Eng. Data **30** (1985) 5.

85-ort-1 Ortega, J.; J. Chem. Eng. Data **30** (1985) 462.
85-ort/paz-1 Ortega, J.; Paz Andrade, M. I.; Rodriguez-Nunez, E.; Romani, L.; Aust. J. Chem. **38** (1985) 1435.
85-ort/paz-2 Ortega, J.; Paz Andrade, M. I.; Bravo, R.; J. Chem. Thermodyn. **17** (1985) 1199.
85-ped/dav Pedrosa, G. C.; Davolio, F.; Schaefer, C. O.; Katz, M.; Rev. Latinoam. Ing. Quim. Quim. Apl. **15** (1985) 189.
85-rao/red Rao, K. P. C.; Reddy, K. S.; Thermochim. Acta **91** (1985) 321.
85-sar/paz Sarmiento, F.; Paz Andrade, M. I.; Fernandez, J.; Bravo, R.; Pintos, M.; J. Chem. Eng. Data **30** (1985) 321.
85-sin/sin Singh, R. P.; Sinha, C. P.; J. Chem. Eng. Data **30** (1985) 38.
85-sin/sin-1 Singh, R. P.; Sinha, C. P.; J. Chem. Eng. Data **30** (1985) 470.
85-tre/ben Treszczanowicz, A. J.; Benson, G. C.; J. Chem. Thermodyn. **17** (1985) 123.
85-wie/sip Wieczorek, S. A.; Sipowska, J. T.; J. Chem. Thermodyn. **17** (1985) 255.
85-zhu Zhuravleva, I. K.; Zh. Obshch. Khim. **55** (1985) 1685.
85-zhu/dur Zhuravlev, V. I.; Durov, V. A.; Usacheva, T. M.; Shakhparonov, M. I.; Zh. Fiz. Khim. **59** (1985) 1677
86-ash/sri Ashraf, S. M.; Srivastava, R.; Hussain, A.; J. Chem. Eng. Data **31** (1986) 100.
86-bae Baev, A. A.; Termodin. Org. Soedin. (1986) 74.
86-ber/wec Berro, C.; Weclawski, J.; Int. DATA Ser., Sel. Data Mixtures Ser. A **No. 3** (1986) 224.
86-ber/wec-1 Berro, C.; Weclawski, J.; Int. DATA Ser., Sel. Data Mixtures Ser. A **No. 3** (1986) 221.
86-cha/lam Chauhdry, M. S.; Lamb, J. A.; J. Chem. Thermodyn. **18** (1986) 665.
86-dew/meh Dewan, R. K.; Mehta, S. K.; J. Chem. Thermodyn. **18** (1986) 697.
86-hei/sch Heintz, A.; Schmittecker, B.; Wagner, D.; Lichtenthaler, R. N.; J. Chem. Eng. Data **31** (1986) 487.
86-hne/cib Hnedkovsky, L.; Cibulka, I.; J. Chem. Thermodyn. **18** (1986) 331.
86-kar/cam Kartzmark, E. M.; Campbell, A. N.; J. Chem. Eng. Data **31** (1986) 241.
86-lep/mat Lepori, L.; Matteoli, E.; J. Chem. Thermodyn. **18** (1986) 13.
86-mah/daw Mahers, E. G.; Dawe, R. A.; J. Chem. Eng. Data **31** (1986) 28.
86-miy/hay Miyano, Y.; Hayduk, W.; J. Chem. Eng. Data **31** (1986) 81.
86-mou/nai Mouli, J. C.; Naidu, P. R.; Choudary, N. V.; J. Chem. Eng. Data **31** (1986) 493.
86-oga/mur Ogawa, H.; Murakami, S.; Thermochim. Acta **109** (1986) 145.
86-ort/paz Ortega, J.; Paz Andrade, M. I.; Rodriguez-Nunez, E.; J. Chem. Eng. Data **31** (1986) 336.
86-ort/paz-1 Ortega, J.; Paz Andrade, M. I.; J. Chem. Eng. Data **31** (1986) 231.
86-ort/pen Ortega, J.; Pena, J. A.; de Alfonso, C.; J. Chem. Eng. Data **31** (1986) 339.
86-rig/mar Riggio, R.; Martinez, H. E.; Solimo, H. N.; J. Chem. Eng. Data **31** (1986) 235.
86-san/sha Sandhu, J. S.; Sharma, A. K.; Wadi, R. K.; J. Chem. Eng. Data **31** (1986) 152.
86-sin/sin Singh, R. P.; Sinha, C. P.; Singh, B. N.; J. Chem. Eng. Data **31** (1986) 107.
86-tan/toy Tanaka, R.; Toyama, S.; Murakami, S.; J. Chem. Thermodyn. **18** (1986) 63.
86-wag/hei Wagner, D.; Heintz, A.; J. Chem. Eng. Data **31** (1986) 483.
86-zha/ben-1 Zhang, D.; Benson, G. C.; Kumaran, M. K.; Lu, B. C. Y.; J. Chem. Thermodyn. **18** (1986) 149.
87-ber-1 Berro, C.; Int. DATA Ser., Sel. Data Mixtures Ser. A **No. 2** (1987) 96.
87-ber-2 Berro, C.; Int. DATA Ser., Sel. Data Mixtures Ser. A **No. 2** (1987) 85.
87-ber-3 Berro, C.; Int. DATA Ser., Sel. Data Mixtures Ser. A **No. 1** (1987) 64.
87-ber-4 Berro, C.; Int. DATA Ser., Sel. Data Mixtures Ser. A **No. 2** (1987) 87.
87-ber-8 Berro, C.; Int. DATA Ser., Sel. Data Mixtures Ser. A **No. 1** (1987) 62.
87-ber-9 Berro, C.; Int. DATA Ser., Sel. Data Mixtures Ser. A **No. 1** (1987) 63.
87-dew/meh Dewan, R. K.; Mehta, S. K.; J. Chem. Thermodyn. **19** (1987) 819.
87-fer/ber Fernandez, J.; Berro, C.; Peneloux, A.; J. Chem. Eng. Data **32** (1987) 17.
87-isl/qua Islam, M. R.; Quadri, S. K.; Thermochim. Acta **115** (1987) 335.
87-kri/cho Krishnaiah, A.; Choudary, N. V.; J. Chem. Eng. Data **32** (1987) 196.
87-kub/tan Kubota, H.; Tanaka, Y.; Makita, T.; Int. J. Thermophys. **8** (1987) 47.
87-mou Mousa, A. H. N.; J. Chem. Eng. Jpn. **20** (1987) 635.

87-oga/mur Ogawa, H.; Murakami, S.; J. Solution Chem. **16** (1987) 315.
87-ogi/ara Ogiwara, K.; Arai, Y.; Netsu Bussei **1** (1987) 52.
87-pap/pap Papanastasiou, G. E.; Papoutsis, A. D.; Kokkinidis, I.; J. Chem. Eng. Data **32** (1987) 377.
87-pik Pikkarainen, L.; J. Chem. Eng. Data **32** (1987) 429.
87-rat/sin-3 Rattan, V. K.; Singh, S.; Sethi, B. P. S.; Raju, K. S. N.; J. Chem. Thermodyn. **19** (1987) 535.
88-bag/gur Baglai, A. K.; Gurarii, L. L.; Kuleshov, G. G.; J. Chem. Eng. Data **33** (1988) 512.
88-cab/bar Cabezas, J. L.; Barcena, L. A.; Coca, J.; Cockrem, M.; J. Chem. Eng. Data **33** (1988) 435.
88-cac/cos Caceres Alonso, M.; Costas, M.; Andreoli-Ball, L.; Patterson, D.; Can. J. Chem. **66** (1988) 989.
88-cze/zyw Czechowski, G.; Zywucki, B.; Jadzyn, J.; J. Chem. Eng. Data **33** (1988) 55.
88-dou/pal Douheret, G.; Pal, A.; J. Chem. Eng. Data **33** (1988) 40.
88-jad/fra Jadot, R.; Fraiha, M.; J. Chem. Eng. Data **33** (1988) 237.
88-kim/mar Kim, E. S.; Marsh, K. N.; J. Chem. Eng. Data **33** (1988) 288.
88-nag Nagata, I.; Thermochim. Acta **126** (1988) 107.
88-nag-1 Nagata, I.; Thermochim. Acta **127** (1988) 109.
88-nag-2 Nagata, I.; Thermochim. Acta **127** (1988) 337.
88-nag-4 Nagata, I.; J. Chem. Eng. Data **33** (1988) 286.
88-oka/oga Okano, T.; Ogawa, H.; Murakami, S.; Can. J. Chem. **66** (1988) 713.
88-ort/gar Ortega, J.; Garcia, J. D.; Can. J. Chem. **66** (1988) 1520.
88-ort/mat Ortega, J.; Matos, J. S.; Pena, J. A.; Paz Andrade, M. I.; Pias, L.; Fernandez, J.; Thermochim. Acta **131** (1988) 57.
88-sak Sakurai, M.; J. Solution Chem. **17** (1988) 267.
88-sip/wie Sipowska, J. T.; Wieczorek, S. A.; J. Chem. Thermodyn. **20** (1988) 333.
88-sun/bis Sun, T.; Biswas, S. N.; Trappeniers, N. J.; Ten Seldam, C. A.; J. Chem. Eng. Data **33** (1988) 395.
88-sun/sch-1 Sun, T. F.; Schouten, J. A.; Trappeniers, N. J.; Biswas, S. N.; J. Chem. Thermodyn. **20** (1988) 1089.
88-sun/ten Sun, T. F.; Ten Seldam, C. A.; Kortbeek, P. J.; Trappeniers, N. J.; Biswas, S. N.; Phys. Chem. Liq. **18** (1988) 107.
88-tan/luo Tanaka, R.; Luo, B.; Benson, G. C.; Lu, B. C. Y.; Thermochim. Acta **127** (1988) 15.
89-ala/sal Al-Azzawi, S. F.; Salman, M. A.; Fluid Phase Equilib. **45** (1989) 95.
89-dew/gup Dewan, R. K.; Gupta, S. P.; Mehta, S. K.; J. Solution Chem. **18** (1989) 13.
89-kac/rad Kaczmarek, B.; Radecki, A.; J. Chem. Eng. Data **34** (1989) 195.
89-kat/tan Kato, M.; Tanaka, H.; J. Chem. Eng. Data **34** (1989) 203.
89-mat/mak-1 Matsuo, S.; Makita, T.; Int. J. Thermophys. **10** (1989) 885.
89-nao/sur Naorem, H.; Suri, S. K.; J. Chem. Eng. Data **34** (1989) 395.
89-ort/sus Ortega, J.; Susial, P.; Can. J. Chem. **67** (1989) 1120.
89-pae/con Paez, S.; Contreras-Slotosch, M.; J. Chem. Eng. Data **34** (1989) 455.
89-sin/sin Singh, R. P.; Sinha, C. P.; Das, J. C.; Ghosh, P.; J. Chem. Eng. Data **34** (1989) 335.
89-sus/ort Susial, P.; Ortega, J.; de Alfonso, C.; Alonso, C.; J. Chem. Eng. Data **34** (1989) 247.
89-taw/tej Tawfik, W. Y.; Teja, A. S.; Chem. Eng. Sci. **44** (1989) 921.
89-vij/nai Vijayalakshmi, T. S.; Naidu, P. R.; J. Chem. Eng. Data **34** (1989) 413
90-apa/gyl Apaev, T. A.; Gylmanov, A. A.; Izv. Vyssh. Uchebn. Zaved., Neft Gaz **33** (1990) 22.
90-bar/paz Bartolome, A. L.; Pazos, C.; Coca, J.; J. Chem. Eng. Data **35** (1990) 285.
90-cha/kat Chaudhari, S. K.; Katti, S. S.; Thermochim. Acta **158** (1990) 99.
90-klo/pal Klofutar, C.; Paljk, S.; Domanska, U.; Thermochim. Acta **158** (1990) 301.
90-lee/hon Lee, H.; Hong, W. H.; Kim, H.; J. Chem. Eng. Data **35** (1990) 371.
90-mal/rao Mallu, B. V.; Rao, Y. V. C.; J. Chem. Eng. Data **35** (1990) 444.
90-sri/nai Srinivasulu, U.; Naidu, P. R.; J. Chem. Eng. Data **35** (1990) 33.
90-sun/sch-1 Sun, T. F.; Schouten, J. A.; Biswas, S. N.; Ber. Bunsen-Ges. Phys. Chem. **94** (1990) 528.
90-vij/nai Vijayalakshmi, T. S.; Naidu, P. R.; J. Chem. Eng. Data **35** (1990) 338.
91-ace/ped-1 Acevedo, I. L.; Pedrosa, G. C.; Arancibia, E. L.; Katz, M.; J. Chem. Eng. Data **36** (1991) 137.
91-cab/bel Cabezas, J. L.; Beltran, S.; Coca, J.; J. Chem. Eng. Data **36** (1991) 184.

91-dou/pal	Douheret, G.; Pal, A.; Hoiland, H.; Anowi, O.; Davis, M. I.; J. Chem. Thermodyn. **23** (1991) 569.
91-fen/wan	Feng, H.; Wang, Y.; Shi, J.; Benson, G. C.; Lu, B. C. Y.; J. Chem. Thermodyn. **23** (1991) 169.
91-gar/her	Garcia, B.; Herrera, C.; Leal, J. M.; J. Chem. Eng. Data **36** (1991) 269.
91-kat/tan	Kato, M.; Tanaka, H.; Yoshikawa, H.; J. Chem. Eng. Data **36** (1991) 387.
91-ram/muk	Ramprasad, G.; Mukherjee, A. K.; Das, T. R.; J. Chem. Eng. Data **36** (1991) 124.
91-sun/sch	Sun, T. F.; Schouten, J. A.; Biswas, S. N.; Int. J. Thermophys. **12** (1991) 381.
91-yos/kat	Yoshikawa, H.; Kato, M.; J. Chem. Eng. Data **36** (1991) 57.
92-ard/say-1	Arda, N.; Sayar, A. A.; Fluid Phase Equilib. **73** (1992) 129.
92-kum/sre	Kumar, V. C.; Sreenivasulu, B.; Naidu, P. R.; J. Chem. Eng. Data **37** (1992) 71.
92-lie/sen-1	Liew, K. Y.; Seng, C. E.; Ng, B. H.; J. Solution Chem. **21** (1992) 1177.
92-sun/dig	Sun, T.; DiGuilio, R. M.; Teja, A. S.; J. Chem. Eng. Data **37** (1992) 246.
92-tan/mur	Tanaka, H.; Muramatsu, T.; Kato, M.; J. Chem. Eng. Data **37** (1992) 164.
93-ami/ara	Aminabhavi, T. M.; Aralaguppi, M. I.; Harogoppad, S. B.; Balundgi, R. H.; J. Chem. Eng. Data **38** (1993) 31.
93-ami/rai	Aminabhavi, T. M.; Raikar, S. K.; J. Chem. Eng. Data **38** (1993) 310.
93-chi/pro	Chiavone-Filho, O.; Proust, P.; Rasmussen, P.; J. Chem. Eng. Data **38** (1993) 128.
93-gar/ban-1	Garg, S. K.; Banipal, T. S.; Ahluwalia. J. C.; J. Chem. Eng. Data **38** (1993) 227.
93-kum/moc	Kumagai, A.; Mochida, H.; Takahashi, S.; Int. J. Thermophys. **14** (1993) 45.
93-yan/mae	Yanes, C.; Maestre, A.; Perez-Tejeda, P.; Calvente, J. J.; J. Chem. Eng. Data **38** (1993) 512.
94-ben/car	Benito, G. G.; Carton, A.; Uruena, M. A.; J. Chem. Eng. Data **39** (1994) 249.
94-gil	Gillis, K. A.; Int. J. Thermophys. **15** (1994) 821.
94-hia/tak	Hiaki, T.; Takahashi, K.; Tsuji, T.; Hongo, M.; Kojima, K.; J. Chem. Eng. Data **39** (1994) 602.
94-hia/tak-1	Hiaki, T.; Takahashi, K.; Tsuji, T.; Hongo, M.; Kojima, K.; J. Chem. Eng. Data **39** (1994) 605.
94-kim/lee	Kim, K.-J.; Lee, C.-H.; Ryu, S.-K.; J. Chem. Eng. Data **39** (1994) 228.
94-kum/nai	Kumar, K. S.; Naidu, P. R.; J. Chem. Eng. Data **39** (1994) 5.
94-kum/nai-1	Kumar, K. S.; Naidu, P. R.; Acree, W. E.; J. Chem. Eng. Data **39** (1994) 2.
94-pap/pan	Papaioannou, D.; Panayiotou, C. G.; J. Chem. Eng. Data **39** (1994) 463.
94-pap/pan-1	Papaioannou, D.; Panayiotou, C. G.; J. Chem. Eng. Data **39** (1994) 457.
94-rom/pel	Romani, L.; Peleteiro, J.; Iglesias, T. P.; Carballo, E.; Escudero, R.; Legido, J. L.; J. Chem. Eng. Data **39** (1994) 19.
94-sin/kal	Singh, K. C.; Kalra, K. C.; Maken, S.; Yadav, B. L.; J. Chem. Eng. Data **39** (1994) 241.
94-tar/jun	Tardajos, G.; Junquera, E.; Aicart, E.; J. Chem. Eng. Data **39** (1994) 349.
94-ven/ven	Venkatesu, P.; Venkatesulu, D.; Rao, M. V. P.; J. Chem. Eng. Data **39** (1994) 140.
94-yu/tsa	Yu, C. H.; Tsai, F. N.; J. Chem. Eng. Data **39** (1994) 125.
94-yu/tsa-1	Yu, C. H.; Tsai, F. N.; J. Chem. Eng. Data **39** (1994) 441.
95-arc/bla	Arce, A.; Blanco, A.; Souza, P.; Vidal, I.; J. Chem. Eng. Data **40** (1995) 225.
95-cas/cal	Castro, I.; Calvo, E.; Bravo, R.; Pintos, M.; Amigo, A.; J. Chem. Eng. Data **40** (1995) 230.
95-fra/jim	Franjo, C.; Jimenez Cuesta, E.; Iglesias, T. P.; Legido, J. L.; Paz Andrade, M. I.; J. Chem. Eng. Data **40** (1995) 68.
95-fra/men	Franjo, C.; Menaut, C. P.; Jimenez Cuesta, E.; Legido, J. L.; Paz Andrade, M. I.; J. Chem. Eng. Data **40** (1995) 992.
95-hia/tak	Hiaki, T.; Takahashi, K.; Tsuji, T.; Hongo, M.; Kojima, K.; J. Chem. Eng. Data **40** (1995) 274.
95-hia/tak-1	Hiaki, T.; Takahashi, K.; Tsuji, T.; Hongo, M.; Kojima, K.; J. Chem. Eng. Data **40** (1995) 271.
95-kum/rao-1	Kumar, R. V.; Rao, M. A.; Prasad, D. H. L.; Rao, M. V.; J. Chem. Eng. Data **40** (1995) 1056
95-org/igl	Orge, B.; Iglesias, M.; Tojo, J.; J. Chem. Eng. Data **40** (1995) 260.
95-red/ram-1	Reddy, V. K.; Rambabu, K.; Devarajulu, T.; Krishnaiah, A.; J. Chem. Eng. Data **40** (1995) 124.
95-sen/say	Senol, A.; Sayar, A. A.; Fluid Phase Equilib. **106** (1995) 169.
96-bha/mak	Bhardwaj, U.; Maken, S.; Singh, K. C.; J. Chem. Eng. Data **41** (1996) 1043.
96-dej/gon-1	Dejoz, A.; Gonzalez-Alfaro, V.; Miguel, P. J.; Vazquez, M. I.; J. Chem. Eng. Data **41** (1996) 89.
96-dom/rod	Dominguez, M.; Rodriguez, S.; Lopez, M. C.; Royo, F. M.; Urieta, J. S.; J. Chem. Eng. Data **41** (1996) 37.

96-elb	El-Banna, M. M.; J. Chem. Eng. Data **42** (1996) 31.
96-gri/zhu	Grineva, O. V.; Zhuravlev, V. I.; Lifanova, N. V.; J. Chem. Eng. Data **41** (1996) 155.
96-nik/jad	Nikam, P. S.; Jadhay, M. C.; Hasan, M.; J. Chem. Eng. Data **41** (1996) 1028.
96-nik/mah	Nikam, P. S.; Mahale, T. R.; Hasan, M.; J. Chem. Eng. Data **41** (1996) 1055.
96-ste/chi	Steele, W. V.; Chirico, R. D.; Knipmeyer, S. E.; Nguyen, A.; J. Chem. Eng. Data **41** (1996) 1255.
96-tak/uem	Takiguchi, Y.; Uematsu, M.; J. Chem. Thermodyn. **28** (1996) 7.
97-com/fra	Comelli, F.; Francesconi, R.; J. Chem. Eng. Data **42** (1997) 705.
98-ami/ban	Aminabhavi, T. M.; Banerjee, K.; J. Chem. Eng. Data **43** (1998) 509.
98-ami/pat-1	Aminabhavi, T. M.; Patil, V. B.; J. Chem. Eng. Data **43** (1998) 504.
98-art/dom	Artigas, H.; Dominguez, M.; Mainar, A. M.; Lopez, M. C.; Royo, F. M.; J. Chem. Eng. Data **43** (1998) 580.
98-fen/cho	Feng, L.-C.; Chou, C.-H.; Tang, M.; Chen, Y. P.; J. Chem. Eng. Data **43** (1998) 658.
98-nik/shi	Nikam, P. S.; Shirsat, L. N.; Hasan, M.; J. Chem. Eng. Data **43** (1998) 732.
98-pai/che	Pai, Y.-H.; Chen, L.-J.; J. Chem. Eng. Data **43** (1998) 665.
98-pal/sha	Pal, A.; Sharma, S.; J. Chem. Eng. Data **43** (1998) 532.
98-sen	Senol, A.; J. Chem. Eng. Data **43** (1998) 763

Chemical Name Index

Primary chemical names used in the tables together with other (alternative and trade) names are ordered alphabetically in this index. Chemical Abstracts Service Registry Numbers (CASRN) are provided for each name used.

Chemical Abstracts Service Registry Number Index

Chemical Abstracts Service Registry Numbers (CASRN) for each compound used in the tables are ordered numerically in this index. Registry numbers greater than 500000-00-0 are not assigned by Chemical Abstracts. They are assigned by TRC for indexing purposes.

Landolt-Börnstein License Agreement for

LB CD-ROM

Springer-Verlag Berlin Heidelberg New York Tokyo and the end-user of LB CD-ROM approve and enter into the following agreement for the use and registration of LB CD-ROM

§ I. User´s Rights and Copyrights
1. The data on the CD-ROM, the retrieval system (the software) and the handbook are protected by copyright; all rights to them or their use accrue to the user exclusively through Springer-Verlag. Independently of this, both sides agree to apply the rules of copyright to LB CD-ROM.
2. The user has the non-exclusive authorization to use LB CD-ROM as described in the handbook. Any alternation, reworking, decompiling or reverse engineering of the software is not permitted.
3. The CD-ROM and the software may be used only on one computer and at one workstation at any one time. Any person using LB CD-ROM must be a member of the user's institution, for example an employee of the user's university or a library user.
4. The user may save or print items from the LB CD-ROM databank only for temporary purposes and only as described in the handbook. The CD-ROM, the software and the handbook may not be reproduced in any other respects.

§ II. Transfer
1. Any transfer (e.g., sale) of LB CD-ROM, or any of its parts, and the use thereof requires the written approval of Springer-Verlag.
2. Springer-Verlag will give this permission if the previous user makes a written application and the subsequent User makes a declaration that he or she will be bound by the terms of this agreement. Receipt of this permission ends any rights of the previous user to LB CD-ROM and the transfer may take place.

§ III. Registration
By returning the accompanying card, the user may register with Springer-Verlag. He or she will then receive regular information about the latest up-dates of LB CD-ROM and may make use of the Help Service, as described in § IV.

§ IV. Help Service
1. Springer-Verlag has created the possibility for registered users, as described in § III, to pose questions about the installation and use of LB CD-ROM. There is, however, no legal claim to this service.
2. Questions may be posed through either regular mail, electronic mail, FAX or telephone. Springer-Verlag will respond appropriately.

§ V. Warranty
1. Springer-Verlag is not the author of the data or of the software but is making them available for use. A warranty of the completeness of the data cannot be guaranteed. The user is aware that databanks and software of this type cannot be made completely error-free; the user will check the accuracy of any results of his or her searches in a suitable manner.
2. In cases of faulty materials, production defects or of damage in delivery, Springer-Verlag will replace the product. The user has rights to further claims only when he or she has purchased LB CD-ROM directly from Springer-Verlag. The warranty assumes that the user gives an exact description of the fault immediately and in writing.

§ VI. Liability of Springer-Verlag
Springer-Verlag is only liable for wilful intent, gross negligence and wilful misrepresentation. Representations are only valid when made in writing. There is no warranty for any information obtained according to § IV. Liability according to product liability law is not affected. Springer-Verlag reserves the right to assert that the user was acting at his or her own risk and the user must assume full or partial responsibility.

§ VII. Responsibilities of the user
1. The user is obliged to abide by the terms and conditions (§§ I and II) governing use and transfer of the product. Any violation is punishable and may be pursued under civil and criminal law. The user may then be made to pay damages to Springer-Verlag or to one of the grantors of its licenses.
2. In cases of severe violation, Springer-Verlag reserves the right to immediately withdraw the user's authorization to use LB CD-ROM and to demand the return of the product.

§ VIII. Data Protection
The user agrees that his or her name and address may be stored electronically.

§ IX. Conclusion of the Agreement
The user waives any claim to a written acknowledgement from Springer-Verlag of its agreement to the terms of this agreement.

§ X. Conclusion
1. This agreement is valid for the product that has already been delivered and for any versions of it delivered in the future.
2. Should any provision of this agreement be or become ineffective or should the agreement be incomplete, then the remainder of the agreement shall remain in effect. The ineffective provisions will be assumed to have been replaced by an effective provision that in meaning and purpose comes closest to the ineffective provision in financial terms. The same holds for any omission in the agreement.
3. The place of jurisdiction is Heidelberg, if the user is a registered trader or equivalent, is a legal entity according to public law, is a public separate property, or has no legal residence or place of business in Germany.
4. This agreement is governed exclusively by the laws of the Federal Republic of Germany, with the exception of the UNCITRAL laws of trade and commerce.

Please ask your librarian for the LB CD-ROM

For installation and further use, please see readme.txt on the CD.